# Studies in Computational Intelligence

Volume 806

The series "Studies in Computational Intelligence" (SCI) publishes new developments and advances in the various areas of computational intelligence—quickly and with a high quality. The intent is to cover the theory, applications, and design methods of computational intelligence, as embedded in the fields of engineering, computer science, physics, and life sciences, as well as the methodologies behind them. The series contains monographs, lecture notes, and edited volumes in computational intelligence spanning the areas of neural networks, connectionist systems, genetic algorithms, evolutionary computation, artificial intelligence, cellular automata, self-organizing systems, soft computing, fuzzy systems, and hybrid intelligent systems. Of particular value to both the contributors and the readership are the short publication timeframe and the world-wide distribution, which enable both wide and rapid dissemination of research output.

Indexed by SCOPUS, DBLP, WTI AG (Switzerland), zbMATH, SCImago.

All books published in the series are submitted for consideration in Web of Science.

Diego Oliva · Arturo Valdivia ·
Seyed Jalaleddin Mousavirad · Kanak Kalita
Editors

# Advances in Optimization Algorithms for Multidisciplinary Engineering Applications: From Classical Methods to AI-Enhanced Solutions

Springer

*Editors*
Diego Oliva
Departamento De
Ingeniería Electro-Fotónica
Universidad de Guadalajara
Guadalajara, Jalisco, Mexico

Arturo Valdivia
Departamento Innovación Basada en la
Información y el
Universidad de Guadalajara
Conocimiento, Mexico

Seyed Jalaleddin Mousavirad
Department of Computer and Electrical
Engineering (DET)
Mid Sweden University
Holmgatan, Sweden

Kanak Kalita
Vel Tech Rangarajan Dr. Sagunthala R&D
Institute of Science and Technology
Chennai, Tamil Nadu, India

ISSN 1860-949X ISSN 1860-9503 (electronic)
Studies in Computational Intelligence
ISBN 978-3-031-78439-2 ISBN 978-3-031-78440-8 (eBook)
https://doi.org/10.1007/978-3-031-78440-8



This Springer imprint is published by the registered company Springer Nature Switzerland AG
The registered company address is: Gewerbestrasse 11, 6330 Cham, Switzerland

# Preface

Optimization is crucial in solving engineering problems. Whether we are designing advanced aircraft or managing energy systems, finding the best solutions is essential for efficiency and performance. This book explores the importance of optimization in engineering and how both traditional and modern optimization techniques can significantly improve problem-solving.

Optimization techniques can be broadly categorized into mathematical optimization and metaheuristic optimization. Mathematical optimization relies on precise mathematical formulations and algorithms to find optimal solutions. This approach involves defining an objective function and constraints, then applying methods such as linear programming, nonlinear programming, or integer programming to solve the problem. These techniques are powerful and can provide exact solutions when the problem structure is well-understood and the solution space is manageable. They are especially useful in scenarios where the relationships between variables are clearly defined and can be expressed mathematically.

On the other hand, metaheuristic optimization uses more flexible, problem-independent methods that can efficiently search large solution spaces. Unlike mathematical optimization, metaheuristic algorithms do not require a precise mathematical model of the problem. Instead, they use strategies inspired by natural processes, such as evolution, swarm behavior, and annealing, to explore the solution space and find near-optimal solutions. Techniques such as genetic algorithms, particle swarm optimization, and simulated annealing fall into this category. These methods are particularly useful for solving complex, high-dimensional problems where traditional methods may struggle due to the size or complexity of the solution space.

In this book, we focus primarily on metaheuristic algorithms due to their versatility and ability to tackle complex, real-world problems where traditional methods may fall short. They can handle nonlinearities, discontinuities, and large-scale problems, which are common in practical engineering scenarios. However, we also cover mathematical optimization techniques in certain cases to provide a comprehensive understanding of the field. By including both approaches, we aim to give readers a complete toolkit for tackling optimization problems.

The book is focused on the theory and application of optimization algorithms for various applications. Topics are selected based on their importance and complexity in this field. For instance, we provide a separate section on optimization for machine learning.

This book is designed so that each chapter can be read independently. It is organized into three sections, each grouping chapters with similar themes.

The first section of this book is dedicated to metaheuristic optimization. Different techniques, such as partitioning, and single and multi-objective optimization, are discussed in this section. We also provide examples of modified metaheuristic optimization applied to simple applications.

The second section explores the exciting intersection of optimization and machine learning. This section covers various domains in machine learning, such as neural networks, deep learning, feature selection, and clustering, showing how optimization approaches can solve even more complex problems in this domain.

The third section focuses on engineering optimization. Here, we consider practical applications of optimization in various engineering fields. From improving smart home energy systems to failure detection in quadcopter unmanned aerial vehicles, this section illustrates how the discussed methods can be applied to real-world engineering challenges.

The material in this book is primarily aimed at undergraduate and postgraduate students studying Science, Engineering, or Computational Mathematics. It is suitable for courses like artificial intelligence, evolutionary computation, and computational intelligence. Additionally, researchers in the fields of optimization, evolutionary computation, and artificial intelligence will find this book valuable.

Finally, it is important to note that this book is just a small part of the vast field of optimization in engineering. We encourage readers to further explore and expand their knowledge to develop implementations tailored to their specific needs.

Guadalajara, México — Diego Oliva
Conocimiento, México — Arturo Valdivia
Sundsvall, Sweeden — Seyed Jalaleddin Mousavirad
Chennai, India — Kanak Kalita
August 2024

# Contents

# Metaheuristics, Theory and Applications

# Multilevel Thresholding Color Image Segmentation Solved with Metaheuristics

**Jorge Ramos-Frutos, Israel Miguel-Andrés, Teresa Alonso-Rasgado, Noé Ortega-Sánchez, Gonzalo Pájares, and Juan Carlos Barragán-Barajas**

**Abstract** The chapter of this book exposes the methodology to segment color images using metaheuristics. There are several methods for segmentation and the one used in this case is multilevel thresholding color image segmentation. In this method, the frequency histogram of the pixel intensities is used as a criterion to locate the thresholds that allow the images to be divided into completely different regions. In this case, the objective function established by Otsu is used for the segmentation of the color images and the pelican optimization algorithm for its optimization. This algorithm shows to be superior over the performance of the objective function in more than 70% of the cases and in the cases in which it does not show that superiority it is very close to the optimal results. With this algorithm, there is a balance in the performance of the objective function and the metrics (PSNR, SSIM, and

J. Ramos-Frutos (✉)
Centro de Innovación Aplicada en Tecnologías Competitivas, Pograndos, Guanajuato, Mexico
e-mail: jramos.estudiantepicyt@ciatec.mx

I. Miguel-Andrés
Laboratorio Nacional CONAHCYT en Biomecánica del Cuerpo Humano, CIATEC, Guanajuato, Mexico
e-mail: imiguel@ciatec.mx

T. Alonso-Rasgado
School of Engineering and Materials Science, Queen Mary University of London, London, England
e-mail: t.alonso@qmul.ac.uk

N. Ortega-Sánchez
Depto. de Ciencias Computacionales, Universidad de Guadalajara, CUCEI, Guadalajara, Mexico
e-mail: noe.ortega@academicos.udg.mx

G. Pájares
Depto. de Ingeniería de Software e Inteligencia Artificial, Universidad Complutense de Madrid, Madrid, Spain
e-mail: pajares@ucm.es

J. C. Barragán-Barajas
Depto. de Ingeniería Industrial, Tecnológico Nacional de México/Instituto Tecnológico de Jiquilpan, Michoacán, Mexico
e-mail: juancarlos.bb@jiquilpan.tecnm.mx



D. Oliva et al. (eds.), *Advances in Optimization Algorithms for Multidisciplinary Engineering Applications: From Classical Methods to AI-Enhanced Solutions*, Studies in Computational Intelligence 806, https://doi.org/10.1007/978-3-031-78440-8_1

FSIM). Furthermore, it can be successfully applied in medical images to facilitate the diagnosis of health professionals, as shown in the case of application in polyps in the intestine which can lead to cancer.

## 1 Introduction

Image segmentation is an operation used in digital image processing to divide an image into different regions [1, 2]. In recent years, a large number of techniques and methods have been proposed for segmentation on gray-scale and color images. One of these techniques is multilevel thresholding segmentation which works using the histogram of the image pixels and a set of thresholds placed on it. The thresholding-based segmentation approach has efficient processing speed, less storage space, and ease of handling [3]. Applying multi-threshold segmentation to color image segmentation can obtain a more accurate segmentation image, this is important because you can get the details of the regions of interest in an image [4]. In these regions, it is necessary to consider some criterion that indicates how different the two regions are when placing the thresholds in controlled positions. Some of the criteria used to separate the image into regions are Otsu's between-class variance [5], Kapur's entropy [6], minimum cross entropy [7], among many others. For this case, the Otsu criterion will be used because it is one of the most used in the literature [8–12].

In solving the problem of color image segmentation by multilevel thresholding, metaheuristics are used because they are simple to implement and generate effective results on the images. Maximizing the Otsu objective function allows a classification of pixels within the different classes that form the regions of interest, and this is where these algorithms come into action. Recently, many algorithms have been developed to solve the problem. Guo and Peng [11] work on the color image segmentation problem using an improved firefly algorithm and energy curve. This manuscript takes the Tsallis entropy, Otsu's between-class variance, and Kapur's entropy as objective functions. Rangu et al. [13] carry out a study on satellite images using multilevel thresholding on color images with the Otsu and Kapur objective functions. They use the electromagnetism optimization algorithm considering spatial contextual information obtaining satisfactory results. Wu et al. [14] propose an enhanced Giza pyramid construction algorithm. The proposed method on different categories of image datasets outperforms a set of state-of-the-art algorithms. These are very recent works on multilevel thresholding color image segmentation, but there are a large number of metaheuristics applied to the problem. Some of them are particle swarm optimization (PSO) [15], differential evolution (DE) [16], grey wolf optimizer (GWO) [17], cuckoo search (CS) [18], genetic algorithm (GA) [19], artificial hummingbird algorithm (AHA) [20], ant colony optimizer (ACO) [21], artificial bee colony (ABC) [22], sine cosine algorithm (SCA) [23], firefly algorithm (FA) [24], whale optimization algorithm (WOA) [25], moth-flame optimization (MFO) [26], bacterial foraging optimizer (BFO) [27], among others. The continuous development of these algorithms occurs because it has not been demonstrated that one of them can solve all problems by outperforming all the others; some algorithms perform well in some scientific fields but poorly in others. other fields. This is due to the No Free Lunch theorem [28]. Therefore, the development of new algorithms that improve the results of existing ones will occur until the dominance of an algorithm in a specific area is demonstrated.

An important field for the application of digital image processing is medicine. In this field, processing serves to assist health professionals in making decisions. In this way, findings about people are achieved with greater precision and can positively affect the diagnoses issued to patients.

This Chapter addresses the optimization problem applied in the segmentation of color images with the following structure. In Sect. 2 the multilevel thresholding problem and the separability criterion between regions are discussed. The pelican optimization algorithm that gives very competitive results concerning some state-of-the-art algorithms is described in Sect. 3. Section 4 illustrates how metaheuristics are applied to the segmentation problem. Section 5 shows the experimental conditions and metrics used to evaluate the compared algorithms. Section 6 contains the numerical results of nine algorithms and are discussed. Section 7 shows a medical case applying color image segmentation by multilevel thresholding in polyps. Finally, the conclusion of the work is in Sect. 8.

## 2 Multilevel Color Image Segmentation

Thresholding is a method for image segmentation to classify the pixels of a gray image or a color image using criteria such as Otsu. In 1975, Otsu proposed a non-parametric technique to segment an image into different regions, the criteria are based on the maximization of between-class-variance [5]. Equation 1 shows the probability distribution of the histogram pixels for each channel:

$$P_i^c = \frac{h_i^c}{N} \tag{1}$$

suppose, $i$ represents the intensity of a given pixel such that $0 \leq i \leq L$, $N$ represents the total number of pixels, $h$ indicates the image histogram; $h_i^c$ indicates the number of pixels having $i$ intensity in each channel, and $P_i^c$ is the probability distribution of the $i$th level of the channel $c$. Otsu method can incorporate multi-level functionality using a given number of threshold values ($nt$), which divide the image $I$ into $K$ classes. Then the objective function is shown in Eq. 2.

$$f(\mathbf{th}^c) = \max\left(\sigma^2(\mathbf{th}^c)\right) \tag{2}$$

where $\mathbf{th}^c$ are the thresholds vector values equal to $[th_1^c, th_2^c, \ldots, th_{nt}^c]$ in each channel of the image. The between-class-variance is defined by Eq. 3.

$$\sigma^2(\mathbf{th}^c) = \sum_{i=1}^{K} w_i(th_i^c)\left(\mu_i(th_i^c) - \mu_T(th_i^c)\right)^2 \tag{3}$$

here $w_i(th_i^c)$ value is obtained with:

$$\begin{aligned} w_1(th_1^c) &= \sum_{i=1}^{th_1^c} P_i^c \\ w_2(th_2^c) &= \sum_{i=th_1^c+1}^{th_2^c} P_i^c \\ &\vdots \\ w_K(th_{nt}^c) &= \sum_{i=th_{nt}^c+1}^{L} P_i^c \end{aligned} \tag{4}$$

and the means for each class are computed as:

$$\begin{aligned} \mu_1(th_1^c) &= \sum_{i=1}^{th_1^c} \frac{i \cdot P_i^c}{w_1(th_1^c)} \\ \mu_2(th_2^c) &= \sum_{i=th_1^c+1}^{th_2^c} \frac{i \cdot P_i^c}{w_1(th_2^c)} \\ &\vdots \\ \mu_K(th_{nt}^c) &= \sum_{i=th_{nt}^c+1}^{L} \frac{i \cdot P_i^c}{w_K(th_{nt}^c)} \end{aligned} \tag{5}$$

Procedure to maximise $\sigma^2(\mathbf{th}^c)$ is replicated in each channel of the image $I$ to obtain the segmented color image.

## 3 Pelican Optimization Algorithm

Pelican optimization algorithm (POA) is a new bio-inspired metaheuristic [29]. POA simulates the behavior and strategy of pelicans when attacking and hunting prey to update candidate solutions. This hunting strategy is simulated in two stages:

1. Moving towards prey (exploration phase).
2. Winging on the water surface (exploitation phase).

The first phase of the algorithms permits the exploration of the space. The important point in POA is that the location of the prey is generated randomly in the search space. The Eq. (6) simulates the pelican behavior.

$$th_{i,j}^{P_1} = \begin{cases} th_{i,j} + rand \cdot (p_j - I \cdot th_{i,j}) \; if \; F_p < F_i \\ th_{i,j} + rand \cdot (th_{i,j} - p_j) \quad if \; F_p \geq F_i \end{cases} \tag{6}$$

where $th_{i,j}^{P_1}$ is the new status of the $i$th pelican in the $j$th dimension based on phase 1, $p_j$ is the location of prey in the $j$th dimension, and $F_p$ and $F_i$ are evaluations of the objective function. The parameter $I$ is a number randomly equal to 1 or 2. This parameter is randomly selected for each iteration and for each member. When the value of this parameter is equal to two, it brings more displacement for a member, which can lead that member to newer areas of the search space. Therefore, parameter $I$ affects the POA exploration power to scan the search space accurately.

The new position for a pelican is accepted if the value of the objective function is improved in that position. In this type of updating, which is called effective updating, the algorithm is prevented from moving to non-optimal areas. This process is modeled using Eq. (7)

$$th_{i,j} = \begin{cases} th_{i,j}^{P_1} \; if \; F_i^{P_1} < F_i \\ th_{i,j} \; if \; F_i^{P_1} \geq F_i \end{cases} \tag{7}$$

$th_{i,j}^{P_1}$ is the new status of the $i$th pelican, and $F_i^{P_1}$ is its objective function value based on the operation that corresponds at the first phase.

In the second phase, the local search power and the exploitation ability of POA increases. The Eq. (8) simulated the behavior of pelicans hunting on the surface of the water to move the fish upwards.

$$th_{i,j}^{P_2} = th_{i,j} + R \cdot \left(1 - \frac{t}{T}\right) \cdot (2rand - 1) \cdot th_{i,j} \tag{8}$$

here $th_{i,j}^{P_2}$ is the new status of the $i$th pelican in the $j$th dimension based on phase 2, $R$ is a constant equal to 0.2, $1 - \frac{t}{T}$ is a part of the neighborhood radius of $th_{i,j}$, $t$ is the iteration counter, and $T$ is the maximum quantity of iterations.

At this phase, the algorithm uses a criterion (Eq. (9)) to accept or reject the new pelican position.

$$th_{i,j} = \begin{cases} th_{i,j}^{P_2} \; if \; F_i^{P_2} < F_i \\ C_{i,j} \; if \; F_i^{P_2} \geq F_i \end{cases} \tag{9}$$

where $th_{i,j}^{P_2}$ is the new status of the $i$th pelican and $F_i^{P_2}$ is its evaluated objective function on the second phase.

## 4 Multilevel Color Image Segmentation Metaheuristic Implementation

For multilevel color image segmentation, the implementation of POA is generated how as shown in Fig. 1. The first step is to read the image in the RGB color channels. Then the histogram of each of them is calculated. The following process is performed for each color channel: A population of vectors ($\mathbf{X} = [\mathbf{Th}_1, \mathbf{Th}_2, \ldots, \mathbf{Th}_N]$) containing the thresholds ($\mathbf{Th}_i = [th_1, th_2, \ldots, th_k]^T$) that meet the constraints; thresholds are subject to $th_1 < th_2 < \cdots < th_k < L$ where $L = 255$, is generated. The algorithm used to optimize the threshold positions within each histogram is the Pelican-inspired algorithm (POA). The result of the operations performed by the POA in the main loop gives the maximization of an objective function (Otsu) and the set of thresholds that maximize this function by separating the image regions in a way that is very different from each other. The process ends with the image segmented into

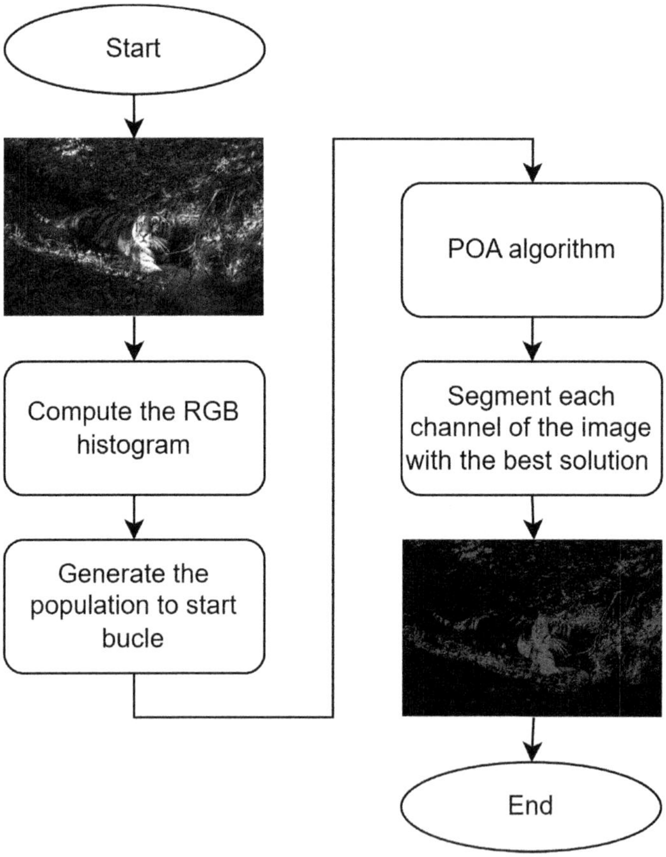


**Fig. 1** POA implementation flow chart

different regions, highlighting the available details and that can be analyzed together or by each channel, depending on the application problem.

## 5 Experimental Conditions

This section presents the experimental conditions with which the algorithms were run, the dataset used and the metrics used to see the quality of the segmentation of the algorithms. A given set of algorithms is compared: artificial hummingbird algorithm (AHA) [20], differential evolution (DE) [16], gradient-based optimization (GBO) [30], golden search algorithm (GSA) [31], gorilla troops optimization (GTO) [32], jellyfish (JF) [33], pelican optimization algorithm (POA) [29], and runge kutta (RUN) [34].

### *5.1 Experiments Setup and Benchmark Images for Testing*

The execution of the experiments was conceived in the Matlab 2019 test environment, with a computer equipped with a Core i7 processor, and 12 GB of RAM. All algorithms were implemented using the Otsu method as an objective function; 30 runs of each method were performed for 10 different images commonly used in the related literature [35] extracted from Berkeley database BSDS300. Images 12003, 37073, 56028, 108070, 160068, and 296059 have a size 481 × 321 pixels, and the rest of the images have a size 321 × 481 pixels. The images used in this paper are shown in Fig. 2. The images and their corresponding histograms are displayed in the three channels (RGB). This set of images is used in the literature [36–38].

### *5.2 Metrics for Segmentation*

To validate the results obtained by the optimization algorithms is necessary use different metrics. The average (Mean) and standard deviation (Std) are two metrics that should be used as a global overview of the performance and stability of the algorithms, respectively. Those are computed using Eq. 10.

$$\mu = \frac{1}{N}\sum_{i=1}^{N} X_i$$
$$\sigma = \sqrt{\sum_{i=1}^{N}\frac{(X_i - \mu)^2}{N}} \tag{10}$$

(a) Image 12003.

(b) Image 37073.

(c) Image 56028.

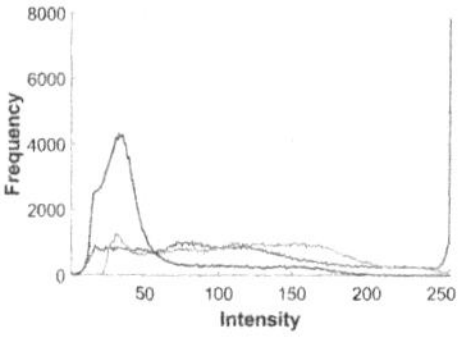


(d) RGB histogram 12003.

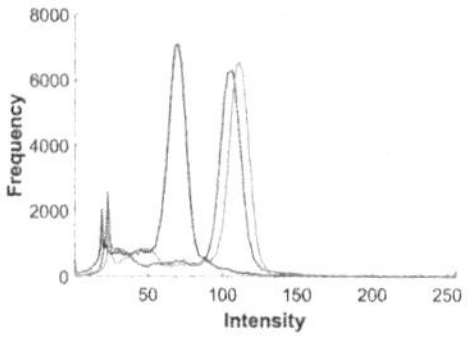


(e) RGB histogram 37073.

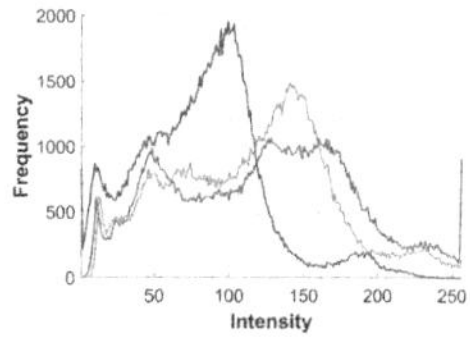


(f) RGB histogram 56028.

(g) Image 66075.

(h) Image 101087.

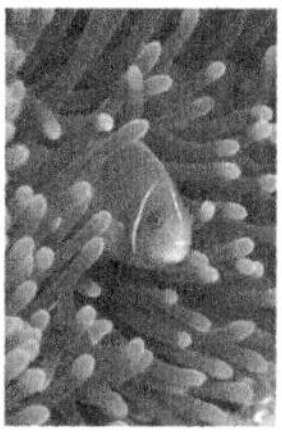

(i) Image 210088.

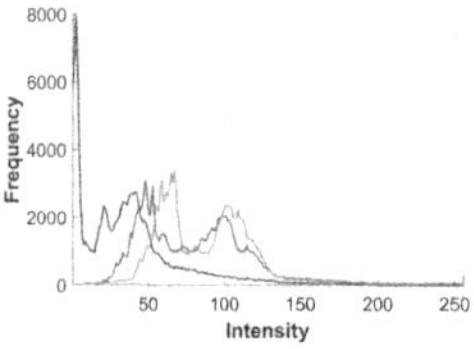


(j) RGB histogram 66075.

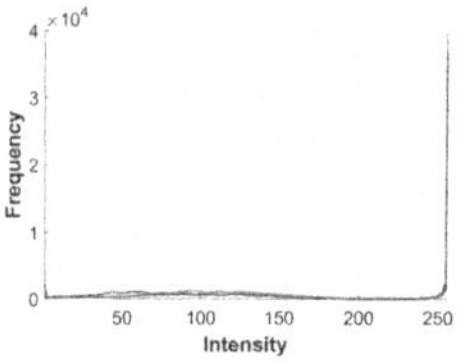


(k) RGB histogram 101087.

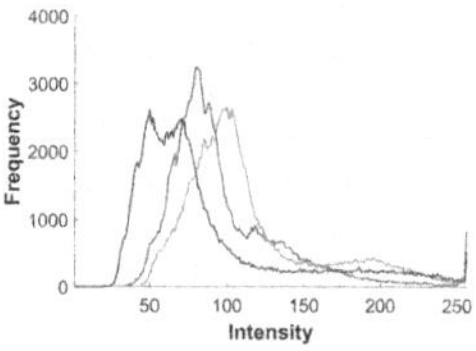


(l) RGB histogram 210088.

(m) Image 108070.

(n) Image 160068.

(o) Image 296059.

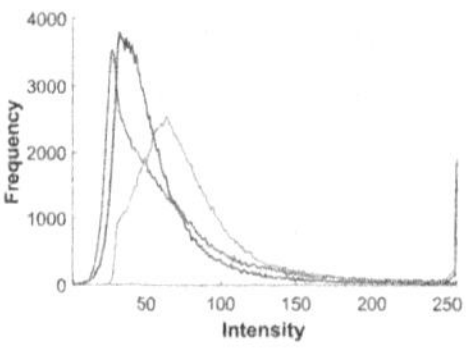


(p) RGB histogram 108070.

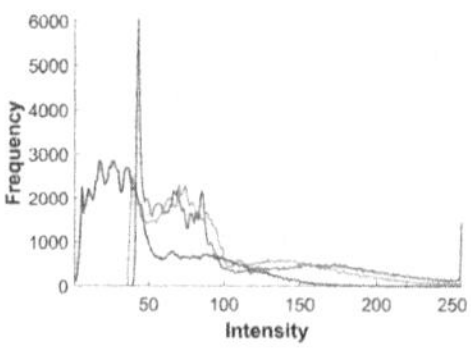


(q) RGB histogram 160068.

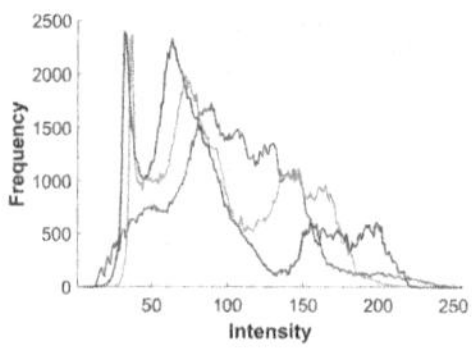


(r) RGB histogram 296059.

**Fig. 2** Images and histograms

$\mu$ and $\sigma$ are the mean and standard deviation of the objective function obtained by the algorithms.

Besides the evaluation, and the performance regarding the optimization process, it is necessary to verify the quality of the segmented images. One of the metrics used for this purpose is the peak signal-to-noise ratio (PSNR) [39], which is defined by Eq. 11. The PSNR uses the root mean square error (RMSE) to compare the original and the segmented image in terms of the pixel information.

$$PSNR = 20 \log \left( \frac{255}{RSME} \right) \tag{11}$$

and

$$RSME = \sqrt{\frac{1}{R \times C} \sum_{i=1}^{R} \sum_{j=1}^{C} (I(i,j) - I_S(i,j))^2} \tag{12}$$

$R, C$ are the dimensions of the image, $I(i,j)$ is the unsegmented image and $I_S(i,j)$ is the segmented image with a given thresholding level. PSNR must be high to obtain better results in the segmentation process.

Another metric mentioned is the SSIM, which measures the similarity between the two images. The SSIM is calculated by Eq. 13:

$$SSIM = \frac{(2\mu_I \mu_S + c_1)(2\sigma_{IS} + c_2)}{(\mu_I^2 + \mu_S^2 + c_1)(\sigma_I^2 + \sigma_S^2 + c_2)} \tag{13}$$

$\mu_I$ and $\mu_S$ are the mean values of original and segmented images, respectively. $\sigma_I^2$ and $\sigma_S^2$ are the variances, $\sigma_{IS}$ the covariance. The constants $c_1 = (0.01 \times 255)^2$ and $c_2 = (0.03 \times 255)^2$ are employed to stabilize the division with weak denominator [40]. A SSIM near one indicates better results in segmentation.

Another metric is the Featured Similarity Index (FSIM). FSIM evaluates the importance of the local structure between the non-segmented image and segmented image [41]. The maximum FSIM value that can be reached is 1. Equation 14 introduces the FSIM.

$$FSIM = \frac{\sum_{I \in \Omega} S_L(I) PC_m(I)}{\sum_{I \in \Omega} PC_m(I)} \tag{14}$$

where $\Omega$ represents the entire domain of the image, $S_L(I)$ is the product of the similarity measure of the Gradient Magnitude map and Phase Congruency map and is calculated with Eq. 15.

$$S_L(I) = S_{PC(I)} S_{G(I)} \tag{15}$$

In this Eq. $S_{PC(I)}$ and $S_{G(I)}$ are calculated with the next equations.

$$S_{PC(I)} = \frac{2PC_{1(I)}2PC_{2(I)} + T_1}{PC^2_{1(I)}PC^2_{2(I)} + T_1} \tag{16}$$

To calculate $PC$ is necessary the Eq. 17.

$$PC(I) = \frac{E(I)}{\varepsilon + \sum_n A_n(I)} \tag{17}$$

where $A_n(I)$ is the local amplitude on scale $n$, and $E(I)$ is the magnitude of the response vector in $I$ over $n$, $\varepsilon$ is a small positive number. To continue with the description of Eq. 15, we define $S_{G(I)}$ as:

$$S_{G(I)} = \frac{2G_{1(I)}G_{2(I)} + T_2}{G^2_{1(I)}G^2_{2(I)} + T_2} \tag{18}$$

Here $T_1$ and $T_2$ are constants with values 0.85 and 160, respectively. $G$ is the magnitude of the gradient of a digital image and is calculated by Eq. 19.

$$G = \sqrt{G_x^2 + G_y^2} \tag{19}$$

$G_x$ is the gradient in the $x$ direction and $Gy$ is the gradient in the $y$ direction. To complete the description of Eq. 14, $PC_m(I)$ is the maximum of the Phase Congruency map between segmented and non-segmented images.

$$PC_m(I) = \max\left(PC_1(I), PC_2(I)\right) \tag{20}$$

## 6 Results and Discussion

Table 1 shows the thresholds that POA found by maximizing the Otsu objective function. A column is displayed for each color in the RGB color space. Four thresholding levels (2, 3, 4, and 5 thresholds) were used for each image. At each thresholding level, the corresponding thresholds are shown on the image mentioned in one of the columns.

Table 2 shows ten images with 4 thresholding levels each and the scores obtained by each of the 9 algorithms proposed to be compared. The POA is the algorithm that in 72.5% of the cases obtains a higher score when optimizing the objective function. Three algorithms win in the same number of cases, these algorithms are GBO, GTO, and RUN. They coincide in successful cases. The JF algorithm is an algorithm that is successful in finding the maximum reported in these images at 40%. And the DE obtains better results on 14 occasions. The algorithms not mentioned are not among the best-positioned algorithms in this table. Algorithms that have a tie generate good

**Table 1** Thresholds obtained with POA for each channel

| Image | Thresholds | | | Otsu |
|---|---|---|---|---|
| | R | G | B | |
| 12003 | 137 230 | 119 136 | 83 225 | 6.94E+03 |
| | 88 176 223 | 89 156 166 | 58 117 184 | 8.43E+03 |
| | 5 83 136 224 | 20 62 88 116 | 43 72 96 218 | 1.11E+04 |
| | 6 52 93 197 224 | 12 94 115 129 227 | 37 67 92 109 253 | 1.23E+04 |
| 37073 | 72 84 | 77 189 | 51 158 | 1.81E+03 |
| | 71 135 238 | 77 148 181 | 49 96 210 | 2.07E+03 |
| | 15 50 65 145 | 15 78 86 170 | 2 37 61 108 | 3.09E+03 |
| | 4 41 71 162 170 | 21 76 87 156 186 | 12 33 43 57 161 | 3.32E+03 |
| 56028 | 115 137 | 108 197 | 77 104 | 5.42E+03 |
| | 90 161 207 | 94 163 197 | 67 134 138 | 6.73E+03 |
| | 14 91 125 212 | 1 79 119 188 | 78 84 121 225 | 8.62E+03 |
| | 1 78 100 149 187 | 11 34 83 113 192 | 62 113 133 172 243 | 9.94E+03 |
| 66075 | 86 109 | 91 142 | 62 108 | 2.46E+03 |
| | 77 145 182 | 84 142 175 | 28 80 96 | 3.34E+03 |
| | 82 98 123 242 | 93 95 129 242 | 61 79 210 216 | 4.19E+03 |
| | 73 108 134 214 234 | 86 135 142 158 249 | 7 74 82 221 244 | 4.84E+03 |
| 101087 | 173 221 | 179 191 | 162 186 | 1.53E+04 |
| | 99 190 223 | 114 198 213 | 88 190 237 | 1.74E+04 |
| | 2 109 109 170 | 23 120 184 201 | 2 120 160 167 | 2.18E+04 |
| | 6 31 108 108 215 | 5 104 127 183 206 | 6 77 78 170 196 | 2.37E+04 |
| 108070 | 109 153 | 117 145 | 93 174 | 3.73E+03 |
| | 67 150 171 | 80 151 176 | 61 129 179 | 4.73E+03 |
| | 110 122 175 249 | 107 124 150 252 | 111 119 164 251 | 6.08E+03 |
| | 85 111 190 208 252 | 101 109 143 242 255 | 42 110 127 163 249 | 6.91E+03 |
| 160068 | 124 145 | 114 183 | 64 143 | 5.13E+03 |
| | 103 176 205 | 80 137 181 | 49 97 103 | 5.81E+03 |
| | 112 150 188 245 | 112 114 239 251 | 10 57 74 202 | 8.55E+03 |
| | 11 113 124 145 250 | 14 124 124 178 232 | 45 74 78 140 200 | 9.69E+03 |
| 210088 | 118 232 | 117 142 | 126 178 | 4.53E+03 |
| | 103 164 189 | 98 156 205 | 82 153 161 | 5.24E+03 |
| | 117 155 205 254 | 24 127 153 245 | 119 134 161 244 | 7.70E+03 |
| | 42 109 111 177 253 | 54 146 146 208 253 | 116 120 192 228 251 | 8.46E+03 |
| 296059 | 105 252 | 108 153 | 119 126 | 4.52E+03 |
| | 76 125 164 | 71 120 200 | 70 133 145 | 5.26E+03 |
| | 22 115 131 241 | 78 101 125 227 | 114 123 134 231 | 7.66E+03 |
| | 1 62 87 119 144 | 18 115 118 171 206 | 9 36 127 139 236 | 8.73E+03 |
| 302008 | 92 188 | 78 210 | 86 144 | 1.08E+04 |
| | 65 149 174 | 53 132 144 | 51 135 165 | 1.21E+04 |
| | 66 105 127 251 | 60 98 103 237 | 11 85 115 239 | 1.87E+04 |
| | 73 113 118 191 243 | 63 73 92 121 247 | 7 77 87 104 232 | 1.96E+04 |

**Table 2** Comparison of the Average Otsu's between-class variance obtained with different meta-heuristics

| Image | Level | AHA | DE | EDA | GBO | GSA | GTO | JF | RUN | POA |
|---|---|---|---|---|---|---|---|---|---|---|
| 12003 | 2 | 6.94E+03 | 6.94E+03 | 6.94E+03 | **6.94E+03** | 6.92E+03 | **6.94E+03** | **6.94E+03** | **6.94E+03** | **6.94E+03** |
| | 3 | 8.36E+03 | 8.43E+03 | 8.40E+03 | **8.43E+03** | 8.35E+03 | **8.43E+03** | **8.43E+03** | **8.43E+03** | **8.43E+03** |
| | 4 | 8.85E+03 | 9.91E+03 | 8.89E+03 | 8.96E+03 | 8.83E+03 | 8.96E+03 | 8.96E+03 | 8.96E+03 | **1.00E+04** |
| | 5 | 9.10E+03 | **1.09E+04** | 9.12E+03 | 9.21E+03 | 9.07E+03 | 9.21E+03 | 9.21E+03 | 9.21E+03 | 1.08E+04 |
| 37073 | 2 | 1.81E+03 | **1.81E+03** | 1.81E+03 | **1.81E+03** | 1.81E+03 | **1.81E+03** | **1.81E+03** | **1.81E+03** | **1.81E+03** |
| | 3 | 2.05E+03 | 2.07E+03 | 2.06E+03 | **2.07E+03** | 2.04E+03 | **2.07E+03** | 2.07E+03 | **2.07E+03** | **2.07E+03** |
| | 4 | 2.18E+03 | **2.38E+03** | 2.19E+03 | 2.21E+03 | 2.17E+03 | 2.21E+03 | 2.21E+03 | 2.21E+03 | 2.35E+03 |
| | 5 | 2.24E+03 | 2.68E+03 | 2.26E+03 | 2.29E+03 | 2.24E+03 | 2.29E+03 | 2.29E+03 | 2.29E+03 | **2.69E+03** |
| 56028 | 2 | 5.42E+03 | **5.42E+03** | 5.42E+03 | **5.42E+03** | 5.41E+03 | **5.42E+03** | **5.42E+03** | **5.42E+03** | **5.42E+03** |
| | 3 | 6.67E+03 | 6.73E+03 | 6.70E+03 | **6.73E+03** | 6.65E+03 | **6.73E+03** | **6.73E+03** | **6.73E+03** | **6.73E+03** |
| | 4 | 7.18E+03 | **7.74E+03** | 7.21E+03 | 7.29E+03 | 7.15E+03 | 7.29E+03 | 7.29E+03 | 7.29E+03 | 7.64E+03 |
| | 5 | 7.43E+03 | 8.48E+03 | 7.45E+03 | 7.56E+03 | 7.38E+03 | 7.56E+03 | 7.56E+03 | 7.56E+03 | **8.50E+03** |
| 66075 | 2 | 2.45E+03 | 2.46E+03 | 2.45E+03 | **2.46E+03** | 2.45E+03 | **2.46E+03** | **2.46E+03** | **2.46E+03** | **2.46E+03** |
| | 3 | 3.30E+03 | 3.34E+03 | 3.32E+03 | **3.34E+03** | 3.28E+03 | **3.34E+03** | 3.34E+03 | **3.34E+03** | **3.34E+03** |
| | 4 | 3.52E+03 | **3.80E+03** | 3.53E+03 | 3.58E+03 | 3.50E+03 | 3.58E+03 | 3.58E+03 | 3.58E+03 | 3.78E+03 |
| | 5 | 3.63E+03 | 4.20E+03 | 3.65E+03 | 3.70E+03 | 3.63E+03 | 3.70E+03 | 3.70E+03 | 3.70E+03 | **4.24E+03** |
| 101087 | 2 | 1.53E+04 | 1.53E+04 | 1.53E+04 | **1.53E+04** | 1.53E+04 | **1.53E+04** | **1.53E+04** | **1.53E+04** | **1.53E+04** |
| | 3 | 1.73E+04 | 1.74E+04 | 1.74E+04 | **1.74E+04** | 1.73E+04 | **1.74E+04** | **1.74E+04** | **1.74E+04** | **1.74E+04** |
| | 4 | 1.78E+04 | **2.02E+04** | 1.79E+04 | 1.79E+04 | 1.78E+04 | 1.79E+04 | 1.79E+04 | 1.79E+04 | 2.01E+04 |
| | 5 | 1.81E+04 | 2.16E+04 | 1.81E+04 | 1.82E+04 | 1.80E+04 | 1.82E+04 | 1.82E+04 | 1.82E+04 | **2.16E+04** |
| 108070 | 2 | 3.73E+03 | **3.73E+03** | 3.73E+03 | **3.73E+03** | 3.72E+03 | **3.73E+03** | **3.73E+03** | **3.73E+03** | **3.73E+03** |
| | 3 | 4.69E+03 | 4.72E+03 | 4.71E+03 | **4.73E+03** | 4.67E+03 | **4.73E+03** | **4.73E+03** | **4.73E+03** | **4.73E+03** |
| | 4 | 5.03E+03 | **5.61E+03** | 5.06E+03 | 5.10E+03 | 5.00E+03 | 5.10E+03 | 5.10E+03 | 5.10E+03 | 5.46E+03 |
| | 5 | 5.19E+03 | **6.12E+03** | 5.22E+03 | 5.28E+03 | 5.17E+03 | 5.28E+03 | 5.28E+03 | 5.28E+03 | 6.10E+03 |
| 160068 | 2 | 5.12E+03 | 5.13E+03 | 5.13E+03 | **5.13E+03** | 5.12E+03 | **5.13E+03** | **5.13E+03** | **5.13E+03** | **5.13E+03** |
| | 3 | 5.77E+03 | 5.81E+03 | 5.79E+03 | **5.81E+03** | 5.75E+03 | **5.81E+03** | **5.81E+03** | **5.81E+03** | **5.81E+03** |
| | 4 | 6.10E+03 | 7.26E+03 | 6.13E+03 | 6.19E+03 | 6.07E+03 | 6.19E+03 | 6.19E+03 | 6.19E+03 | **7.49E+03** |
| | 5 | 6.26E+03 | 8.16E+03 | 6.28E+03 | 6.36E+03 | 6.23E+03 | 6.36E+03 | 6.36E+03 | 6.36E+03 | **8.28E+03** |
| 210088 | 2 | 4.52E+03 | 4.53E+03 | 4.53E+03 | **4.53E+03** | 4.52E+03 | **4.53E+03** | **4.53E+03** | **4.53E+03** | **4.53E+03** |
| | 3 | 5.19E+03 | 5.24E+03 | 5.22E+03 | **5.24E+03** | 5.18E+03 | **5.24E+03** | 5.24E+03 | **5.24E+03** | **5.24E+03** |
| | 4 | 5.49E+03 | **6.55E+03** | 5.52E+03 | 5.58E+03 | 5.47E+03 | 5.58E+03 | 5.58E+03 | 5.58E+03 | 6.44E+03 |
| | 5 | 5.65E+03 | 7.19E+03 | 5.67E+03 | 5.75E+03 | 5.62E+03 | 5.75E+03 | 5.75E+03 | 5.75E+03 | **7.32E+03** |
| 296059 | 2 | 4.51E+03 | 4.52E+03 | 4.52E+03 | **4.52E+03** | 4.51E+03 | **4.52E+03** | **4.52E+03** | **4.52E+03** | **4.52E+03** |
| | 3 | 5.22E+03 | 5.26E+03 | 5.24E+03 | **5.26E+03** | 5.20E+03 | **5.26E+03** | 5.26E+03 | **5.26E+03** | **5.26E+03** |
| | 4 | 5.51E+03 | **6.81E+03** | 5.52E+03 | 5.58E+03 | 5.48E+03 | 5.58E+03 | 5.58E+03 | 5.58E+03 | 6.66E+03 |
| | 5 | 5.65E+03 | 7.42E+03 | 5.67E+03 | 5.77E+03 | 5.63E+03 | 5.77E+03 | 5.77E+03 | 5.77E+03 | **7.53E+03** |
| 302008 | 2 | 1.08E+04 | 1.08E+04 | 1.08E+04 | **1.08E+04** | 1.08E+04 | **1.08E+04** | **1.08E+04** | **1.08E+04** | **1.08E+04** |
| | 3 | 1.20E+04 | 1.21E+04 | 1.20E+04 | **1.21E+04** | 1.20E+04 | **1.21E+04** | **1.21E+04** | **1.21E+04** | **1.21E+04** |
| | 4 | 1.24E+04 | **1.72E+04** | 1.24E+04 | 1.24E+04 | 1.23E+04 | 1.24E+04 | 1.24E+04 | 1.24E+04 | 1.71E+04 |
| | 5 | 1.25E+04 | **1.84E+04** | 1.25E+04 | 1.26E+04 | 1.25E+04 | 1.26E+04 | 1.26E+04 | 1.26E+04 | 1.83E+04 |

results when the number of thresholds is low, as does JF. While DE wins when the number of thresholds is larger. This is due to the interaction that exists in the operations performed within each algorithm. It should be noted that the POA has good results with few thresholds, but also with many thresholds. Therefore, it is a balanced algorithm.

The PSNR is shown in Table 3. In this metric, the algorithm with the best results is the JF. Although no algorithm conclusively dominates this matrix. The JF wins in 16 of the 40 cases. In second place is the POA with 9 cases won and third place is won by the GTO. The other algorithms have less than 3 accumulated wins in the PSNR. No algorithm wins in at least 50% of the cases.

Table 4 shows the SSIM indicator, in which the best algorithm is JF with 11 won cases. The GTO is in second place with 8 cases. The third place in this case is the POA. The other algorithms win less than 5 times, with DE being the best of the rest. On this occasion, the JF also wins, but it is not in a formidable way. The POA and the GTO follow closely.

The last metric shown in Table 5 is the FSIM. In this metric, once again the JF is the winner with 57.5% of cases won. Second place goes to the POA with 20% of the cases won and third place goes to the GTO with 15% of the victories. The rest is taken by the AHA, DE, and GSA algorithms.

It is observed that the POA is the dominant algorithm in the optimization of the objective function, which is what it was designed for, to maximize this parameter. But the results obtained in the metrics are not the same as with Otsu, in the metrics it oscillates between second and third place. This is not a bad thing, since the objective for which the POA was designed is met. Therefore, the regions obtained in multilevel thresholding segmentation using POA are separated more effectively compared to the other algorithms by obtaining better results in the fitness function (Table 2).

The Friedman test is performed to determine if there is a significant difference in the results between the algorithms. The Friedman test is a non-parametric test used to compare various treatments (algorithms) and determine if there is a significant difference [42]. Figure 3 shows the ranking obtained for each algorithm. The algorithm with the highest score is that it maximizes the Otsu function most effectively. Furthermore, we have a p-value equal to $3.123 \times 10^{-53}$. Therefore, the conclusion is reached that the POA is the best algorithm and that the difference between the algorithms; in terms of the objective function, is significant.

## 7 Medical Images Application

This part addresses medical images (CVC-ClinicDB) that are segmented by POA. Only this algorithm is used in a medical application because it was the winner in the general dataset. CVC-ClinicDB is a database of frames extracted from colonoscopy videos. These frames contain examples of polyps [43]. Colon polyps are small clumps of cells that form on the lining of the colon or rectum. They are usually harmless, but

**Table 3** Average PSNR results with different algorithms

| Image | Level | AHA | DE | EDA | GBO | GSA | GTO | JF | RUN | POA |
|---|---|---|---|---|---|---|---|---|---|---|
| 12003 | 2 | 1.31E+01 | **1.33E+01** | 1.32E+01 | 1.31E+01 | 1.31E+01 | 1.32E+01 | 1.31E+01 | 1.31E+01 | 1.32E+01 |
| | 3 | 1.57E+01 | 1.57E+01 | 1.58E+01 | 1.58E+01 | 1.59E+01 | 1.59E+01 | **1.60E+01** | 1.59E+01 | 1.59E+01 |
| | 4 | 1.78E+01 | 1.62E+01 | 1.80E+01 | 1.83E+01 | 1.78E+01 | 1.83E+01 | **1.84E+01** | 1.83E+01 | 1.63E+01 |
| | 5 | 1.93E+01 | 1.70E+01 | 1.95E+01 | 2.02E+01 | 1.92E+01 | 2.03E+01 | **2.03E+01** | 2.02E+01 | 1.68E+01 |
| 37073 | 2 | 1.79E+01 | 1.84E+01 | 1.84E+01 | 1.81E+01 | 1.81E+01 | 1.82E+01 | **1.88E+01** | 1.82E+01 | 1.85E+01 |
| | 3 | 1.80E+01 | 1.82E+01 | 1.81E+01 | 1.82E+01 | 1.81E+01 | 1.83E+01 | **1.83E+01** | 1.83E+01 | 1.83E+01 |
| | 4 | 2.03E+01 | 2.04E+01 | **2.11E+01** | 2.10E+01 | 2.05E+01 | 2.10E+01 | 2.10E+01 | 2.10E+01 | 2.03E+01 |
| | 5 | 2.19E+01 | 2.15E+01 | 2.25E+01 | 2.33E+01 | 2.21E+01 | 2.33E+01 | **2.33E+01** | 2.33E+01 | 2.16E+01 |
| 56028 | 2 | 1.38E+01 | 1.40E+01 | 1.40E+01 | 1.40E+01 | 1.40E+01 | 1.40E+01 | 1.40E+01 | 1.41E+01 | **1.42E+01** |
| | 3 | 1.54E+01 | 1.55E+01 | 1.55E+01 | 1.55E+01 | 1.55E+01 | 1.56E+01 | **1.56E+01** | 1.55E+01 | 1.56E+01 |
| | 4 | 1.73E+01 | 1.66E+01 | 1.78E+01 | 1.80E+01 | 1.74E+01 | **1.81E+01** | 1.81E+01 | 1.80E+01 | 1.68E+01 |
| | 5 | 1.90E+01 | 1.75E+01 | 1.95E+01 | 2.02E+01 | 1.89E+01 | **2.03E+01** | 2.03E+01 | 2.01E+01 | 1.73E+01 |
| 66075 | 2 | 1.52E+01 | 1.55E+01 | 1.55E+01 | 1.54E+01 | 1.54E+01 | 1.55E+01 | 1.54E+01 | 1.54E+01 | **1.56E+01** |
| | 3 | 1.72E+01 | 1.73E+01 | 1.73E+01 | 1.73E+01 | 1.71E+01 | 1.75E+01 | 1.75E+01 | 1.74E+01 | **1.75E+01** |
| | 4 | 1.82E+01 | 1.71E+01 | 1.84E+01 | 1.86E+01 | 1.84E+01 | **1.87E+01** | 1.87E+01 | 1.86E+01 | 1.74E+01 |
| | 5 | 1.94E+01 | 1.82E+01 | 2.00E+01 | 1.97E+01 | 1.98E+01 | 1.97E+01 | 1.97E+01 | **2.00E+01** | 1.78E+01 |
| 101087 | 2 | **9.42E+00** | 9.27E+00 | 9.21E+00 | 9.18E+00 | 9.34E+00 | 9.25E+00 | 9.04E+00 | 9.32E+00 | 9.20E+00 |
| | 3 | 1.39E+01 | 1.37E+01 | 1.39E+01 | 1.39E+01 | 1.39E+01 | 1.39E+01 | 1.40E+01 | **1.41E+01** | 1.38E+01 |
| | 4 | 1.66E+01 | 1.36E+01 | 1.66E+01 | 1.67E+01 | 1.62E+01 | 1.66E+01 | **1.70E+01** | 1.68E+01 | 1.40E+01 |
| | 5 | 1.83E+01 | 1.56E+01 | 1.89E+01 | 1.93E+01 | 1.79E+01 | 1.93E+01 | **2.00E+01** | 1.99E+01 | 1.52E+01 |
| 108070 | 2 | 1.31E+01 | 1.31E+01 | 1.31E+01 | 1.31E+01 | 1.31E+01 | 1.30E+01 | 1.30E+01 | 1.31E+01 | **1.31E+01** |
| | 3 | 1.56E+01 | 1.55E+01 | 1.56E+01 | 1.56E+01 | 1.55E+01 | 1.56E+01 | **1.57E+01** | 1.56E+01 | 1.56E+01 |
| | 4 | 1.74E+01 | 1.55E+01 | 1.75E+01 | 1.76E+01 | 1.76E+01 | 1.76E+01 | **1.76E+01** | 1.76E+01 | 1.61E+01 |
| | 5 | 1.87E+01 | 1.64E+01 | 1.89E+01 | 1.93E+01 | 1.85E+01 | 1.93E+01 | **1.93E+01** | 1.93E+01 | 1.64E+01 |
| 160068 | 2 | 1.38E+01 | **1.40E+01** | 1.38E+01 | 1.38E+01 | 1.39E+01 | 1.38E+01 | 1.39E+01 | 1.38E+01 | 1.40E+01 |
| | 3 | 1.52E+01 | 1.55E+01 | 1.55E+01 | 1.55E+01 | 1.55E+01 | 1.56E+01 | **1.56E+01** | 1.56E+01 | 1.56E+01 |
| | 4 | 1.77E+01 | 1.62E+01 | 1.81E+01 | 1.84E+01 | 1.76E+01 | 1.84E+01 | **1.85E+01** | 1.84E+01 | 1.57E+01 |
| | 5 | 1.89E+01 | 1.69E+01 | 1.90E+01 | 1.95E+01 | 1.92E+01 | **1.96E+01** | 1.96E+01 | 1.95E+01 | 1.62E+01 |
| 210088 | 2 | 1.05E+01 | 1.06E+01 | 1.06E+01 | 1.06E+01 | 1.06E+01 | 1.06E+01 | 1.06E+01 | 1.06E+01 | **1.07E+01** |
| | 3 | 1.25E+01 | 1.27E+01 | 1.28E+01 | 1.28E+01 | 1.28E+01 | 1.28E+01 | **1.28E+01** | 1.28E+01 | 1.28E+01 |
| | 4 | 1.48E+01 | 1.34E+01 | 1.48E+01 | 1.52E+01 | 1.51E+01 | 1.52E+01 | **1.52E+01** | 1.52E+01 | 1.35E+01 |
| | 5 | 1.66E+01 | 1.40E+01 | 1.65E+01 | 1.66E+01 | **1.67E+01** | 1.66E+01 | 1.66E+01 | 1.65E+01 | 1.42E+01 |
| 296059 | 2 | 1.23E+01 | 1.24E+01 | 1.23E+01 | 1.23E+01 | 1.23E+01 | 1.24E+01 | 1.24E+01 | 1.23E+01 | **1.25E+01** |
| | 3 | 1.59E+01 | 1.64E+01 | 1.64E+01 | 1.64E+01 | 1.63E+01 | 1.66E+01 | 1.66E+01 | 1.64E+01 | **1.66E+01** |
| | 4 | 1.75E+01 | 1.51E+01 | 1.78E+01 | 1.85E+01 | 1.79E+01 | **1.86E+01** | 1.85E+01 | 1.85E+01 | 1.59E+01 |
| | 5 | 1.90E+01 | 1.62E+01 | 1.94E+01 | 2.01E+01 | 1.90E+01 | **2.01E+01** | 2.01E+01 | 2.01E+01 | 1.62E+01 |
| 302008 | 2 | 1.62E+01 | 1.64E+01 | 1.67E+01 | 1.65E+01 | 1.66E+01 | 1.66E+01 | 1.61E+01 | 1.67E+01 | **1.69E+01** |
| | 3 | 1.87E+01 | 1.90E+01 | 1.89E+01 | 1.90E+01 | 1.89E+01 | 1.92E+01 | 1.92E+01 | 1.91E+01 | **1.94E+01** |
| | 4 | 2.02E+01 | 1.81E+01 | 2.09E+01 | 2.11E+01 | 2.07E+01 | **2.14E+01** | 2.13E+01 | 2.12E+01 | 1.74E+01 |
| | 5 | 2.18E+01 | 1.88E+01 | 2.22E+01 | 2.28E+01 | 2.18E+01 | **2.29E+01** | 2.27E+01 | 2.27E+01 | 1.90E+01 |

**Table 4** Average SSIM results with different algorithms

| Image | Level | AHA | DE | EDA | GBO | GSA | GTO | JF | RUN | POA |
|---|---|---|---|---|---|---|---|---|---|---|
| 12003 | 2 | 3.53E–01 | **3.60E–01** | 3.45E–01 | 3.46E–01 | 3.46E–01 | 3.48E–01 | 3.42E–01 | 3.46E–01 | 3.54E–01 |
| | 3 | 4.90E–01 | 4.90E–01 | 4.94E–01 | 4.93E–01 | **5.02E–01** | 4.96E–01 | 4.99E–01 | 4.96E–01 | 4.97E–01 |
| | 4 | 6.26E–01 | 5.50E–01 | 6.26E–01 | 6.36E–01 | 6.19E–01 | 6.38E–01 | **6.40E–01** | 6.37E–01 | 5.44E–01 |
| | 5 | 6.92E–01 | 6.10E–01 | 7.00E–01 | 7.23E–01 | 6.85E–01 | **7.25E–01** | 7.24E–01 | 7.23E–01 | 5.94E–01 |
| 37073 | 2 | 7.20E–01 | 7.35E–01 | 7.34E–01 | 7.30E–01 | 7.30E–01 | 7.32E–01 | **7.38E–01** | 7.31E–01 | 7.34E–01 |
| | 3 | 7.16E–01 | 7.40E–01 | 7.38E–01 | 7.41E–01 | 7.35E–01 | 7.41E–01 | **7.41E–01** | 7.41E–01 | 7.41E–01 |
| | 4 | 7.74E–01 | 7.92E–01 | 7.97E–01 | 8.01E–01 | 7.79E–01 | 8.01E–01 | **8.01E–01** | 8.01E–01 | 7.88E–01 |
| | 5 | 8.02E–01 | 8.18E–01 | 8.13E–01 | 8.14E–01 | 8.06E–01 | 8.14E–01 | 8.14E–01 | 8.14E–01 | **8.22E–01** |
| 56028 | 2 | 4.98E–01 | 5.16E–01 | 5.08E–01 | 5.09E–01 | 5.07E–01 | 5.09E–01 | 5.07E–01 | 5.12E–01 | **5.23E–01** |
| | 3 | 5.96E–01 | 5.97E–01 | 5.96E–01 | 5.96E–01 | 5.98E–01 | 5.97E–01 | **6.00E–01** | 5.97E–01 | 5.99E–01 |
| | 4 | 6.80E–01 | 6.58E–01 | 7.04E–01 | 7.08E–01 | 6.89E–01 | **7.09E–01** | 7.09E–01 | 7.07E–01 | 6.70E–01 |
| | 5 | 7.53E–01 | 7.07E–01 | 7.71E–01 | 7.92E–01 | 7.51E–01 | **7.93E–01** | 7.93E–01 | 7.92E–01 | 6.96E–01 |
| 66075 | 2 | 4.64E–01 | 4.76E–01 | 4.71E–01 | 4.71E–01 | 4.69E–01 | 4.73E–01 | 4.72E–01 | 4.71E–01 | **4.86E–01** |
| | 3 | 6.01E–01 | 6.08E–01 | 6.05E–01 | 6.05E–01 | 5.95E–01 | 6.08E–01 | **6.08E–01** | 6.07E–01 | 6.08E–01 |
| | 4 | 6.48E–01 | 5.70E–01 | 6.50E–01 | 6.52E–01 | 6.53E–01 | 6.53E–01 | **6.53E–01** | 6.52E–01 | 5.81E–01 |
| | 5 | 7.01E–01 | 6.43E–01 | **7.24E–01** | 6.96E–01 | 7.13E–01 | 6.97E–01 | 6.97E–01 | 7.13E–01 | 6.12E–01 |
| 101087 | 2 | 3.59E–01 | **3.67E–01** | 3.46E–01 | 3.45E–01 | 3.49E–01 | 3.47E–01 | 3.43E–01 | 3.48E–01 | 3.56E–01 |
| | 3 | **6.27E–01** | 6.20E–01 | 6.19E–01 | 6.20E–01 | 6.17E–01 | 6.23E–01 | 6.24E–01 | 6.23E–01 | 6.21E–01 |
| | 4 | **7.49E–01** | 6.99E–01 | 7.38E–01 | 7.28E–01 | 7.28E–01 | 7.31E–01 | 7.32E–01 | 7.28E–01 | 7.13E–01 |
| | 5 | 8.04E–01 | 7.59E–01 | 8.19E–01 | 8.42E–01 | 7.97E–01 | 8.44E–01 | **8.45E–01** | 8.41E–01 | 7.72E–01 |

(continued)

**Table 4** (continued)

| Image | Level | AHA | DE | EDA | GBO | GSA | GTO | JF | RUN | POA |
|---|---|---|---|---|---|---|---|---|---|---|
| 108070 | 2 | 2.39E–01 | **2.45E–01** | 2.35E–01 | 2.35E–01 | 2.35E–01 | 2.35E–01 | 2.32E–01 | 2.36E–01 | 2.42E–01 |
| | 3 | 4.48E–01 | 4.42E–01 | 4.46E–01 | 4.46E–01 | 4.37E–01 | 4.46E–01 | **4.49E–01** | 4.47E–01 | 4.46E–01 |
| | 4 | 5.74E–01 | 4.34E–01 | 5.78E–01 | 5.85E–01 | **5.87E–01** | 5.85E–01 | 5.86E–01 | 5.85E–01 | 4.76E–01 |
| | 5 | 6.55E–01 | 5.05E–01 | 6.57E–01 | 6.80E–01 | 6.30E–01 | 6.80E–01 | 6.80E–01 | **6.80E–01** | 4.98E–01 |
| 160068 | 2 | 3.48E–01 | **3.76E–01** | 3.42E–01 | 3.42E–01 | 3.50E–01 | 3.46E–01 | 3.44E–01 | 3.43E–01 | 3.58E–01 |
| | 3 | 4.48E–01 | 4.52E–01 | 4.50E–01 | 4.53E–01 | **4.60E–01** | 4.55E–01 | 4.58E–01 | 4.54E–01 | 4.57E–01 |
| | 4 | 6.06E–01 | 5.27E–01 | 6.33E–01 | 6.43E–01 | 6.00E–01 | 6.44E–01 | **6.45E–01** | 6.43E–01 | 4.91E–01 |
| | 5 | 6.81E–01 | 5.82E–01 | 6.71E–01 | 6.92E–01 | 6.89E–01 | **6.93E–01** | 6.93E–01 | 6.93E–01 | 5.37E–01 |
| 210088 | 2 | 2.02E–01 | 2.04E–01 | 1.97E–01 | 1.99E–01 | 1.98E–01 | 1.98E–01 | 1.98E–01 | 1.99E–01 | **2.11E–01** |
| | 3 | 3.47E–01 | 3.55E–01 | 3.58E–01 | 3.56E–01 | **3.60E–01** | 3.56E–01 | 3.58E–01 | 3.56E–01 | 3.57E–01 |
| | 4 | 4.98E–01 | 4.21E–01 | 5.01E–01 | 5.28E–01 | 5.23E–01 | 5.29E–01 | **5.30E–01** | 5.28E–01 | 4.27E–01 |
| | 5 | 6.07E–01 | 4.79E–01 | 6.00E–01 | 6.12E–01 | **6.14E–01** | 6.11E–01 | 6.11E–01 | 6.10E–01 | 4.92E–01 |
| 296059 | 2 | 3.39E–01 | 3.49E–01 | 3.41E–01 | 3.42E–01 | 3.42E–01 | 3.43E–01 | 3.45E–01 | 3.42E–01 | **3.62E–01** |
| | 3 | 5.73E–01 | 5.87E–01 | **5.93E–01** | 5.89E–01 | 5.90E–01 | 5.90E–01 | 5.89E–01 | 5.87E–01 | 5.90E–01 |
| | 4 | 6.42E–01 | 5.20E–01 | 6.57E–01 | 6.94E–01 | 6.66E–01 | **6.94E–01** | 6.93E–01 | 6.94E–01 | 5.73E–01 |
| | 5 | 7.14E–01 | 5.93E–01 | 7.23E–01 | 7.47E–01 | 7.22E–01 | **7.48E–01** | 7.46E–01 | 7.47E–01 | 5.93E–01 |
| 302008 | 2 | 6.50E–01 | 6.58E–01 | 6.57E–01 | 6.56E–01 | 6.55E–01 | 6.56E–01 | 6.53E–01 | 6.59E–01 | **6.61E–01** |
| | 3 | 7.14E–01 | 7.18E–01 | 7.16E–01 | 7.17E–01 | 7.15E–01 | 7.19E–01 | 7.18E–01 | 7.18E–01 | **7.20E–01** |
| | 4 | 7.42E–01 | 7.04E–01 | 7.51E–01 | 7.55E–01 | 7.50E–01 | **7.57E–01** | 7.56E–01 | 7.55E–01 | 6.94E–01 |
| | 5 | 7.71E–01 | 7.24E–01 | 7.76E–01 | 7.85E–01 | 7.76E–01 | **7.86E–01** | 7.85E–01 | 7.85E–01 | 7.25E–01 |

**Table 5** Average FSIM results with different algorithms

| Image | Level | AHA | DE | EDA | GBO | GSA | GTO | JF | RUN | POA |
|---|---|---|---|---|---|---|---|---|---|---|
| 12003 | 2 | 5.76E–01 | 5.76E–01 | 5.74E–01 | 5.74E–01 | 5.76E–01 | **5.78E–01** | 5.73E–01 | 5.76E–01 | 5.78E–01 |
| | 3 | 6.57E–01 | 6.60E–01 | 6.64E–01 | 6.63E–01 | 6.64E–01 | 6.69E–01 | **6.72E–01** | 6.68E–01 | 6.70E–01 |
| | 4 | 7.16E–01 | 6.64E–01 | 7.24E–01 | 7.31E–01 | 7.18E–01 | 7.35E–01 | **7.38E–01** | 7.33E–01 | 6.68E–01 |
| | 5 | 7.55E–01 | 6.86E–01 | 7.66E–01 | 7.85E–01 | 7.57E–01 | **7.89E–01** | 7.89E–01 | 7.84E–01 | 6.84E–01 |
| 37073 | 2 | 7.39E–01 | 7.45E–01 | 7.46E–01 | 7.43E–01 | 7.43E–01 | 7.44E–01 | **7.47E–01** | 7.44E–01 | 7.45E–01 |
| | 3 | 7.47E–01 | 7.56E–01 | 7.56E–01 | 7.57E–01 | 7.54E–01 | 7.58E–01 | **7.58E–01** | 7.58E–01 | 7.58E–01 |
| | 4 | 7.83E–01 | 7.82E–01 | 7.96E–01 | 7.97E–01 | 7.83E–01 | 7.98E–01 | **7.99E–01** | 7.98E–01 | 7.82E–01 |
| | 5 | 8.07E–01 | 7.97E–01 | 8.19E–01 | 8.33E–01 | 8.12E–01 | 8.34E–01 | **8.34E–01** | 8.33E–01 | 8.00E–01 |
| 56028 | 2 | 6.09E–01 | 6.12E–01 | 6.12E–01 | 6.13E–01 | 6.12E–01 | 6.12E–01 | 6.05E–01 | 6.18E–01 | **6.21E–01** |
| | 3 | 6.80E–01 | 6.83E–01 | 6.85E–01 | 6.86E–01 | 6.84E–01 | 6.89E–01 | **6.95E–01** | 6.89E–01 | 6.90E–01 |
| | 4 | 7.40E–01 | 7.08E–01 | 7.60E–01 | 7.64E–01 | 7.45E–01 | 7.69E–01 | **7.71E–01** | 7.63E–01 | 7.21E–01 |
| | 5 | 7.88E–01 | 7.35E–01 | 8.00E–01 | 8.23E–01 | 7.89E–01 | 8.27E–01 | **8.28E–01** | 8.19E–01 | 7.34E–01 |
| 66075 | 2 | 7.62E–01 | 7.65E–01 | 7.66E–01 | 7.63E–01 | 7.63E–01 | 7.66E–01 | 7.62E–01 | 7.64E–01 | **7.67E–01** |
| | 3 | 8.02E–01 | 8.07E–01 | 8.06E–01 | 8.06E–01 | 8.01E–01 | 8.10E–01 | **8.11E–01** | 8.08E–01 | 8.10E–01 |
| | 4 | 8.13E–01 | 7.98E–01 | 8.16E–01 | 8.21E–01 | 8.14E–01 | 8.23E–01 | **8.24E–01** | 8.21E–01 | 8.02E–01 |
| | 5 | 8.17E–01 | 8.01E–01 | 8.24E–01 | 8.31E–01 | 8.24E–01 | 8.33E–01 | **8.33E–01** | 8.29E–01 | 7.99E–01 |
| 101087 | 2 | 6.01E–01 | **6.06E–01** | 5.98E–01 | 5.97E–01 | 5.99E–01 | 5.98E–01 | 5.98E–01 | 5.98E–01 | 6.02E–01 |
| | 3 | 7.02E–01 | 7.04E–01 | 7.02E–01 | 7.03E–01 | 7.02E–01 | 7.06E–01 | **7.07E–01** | 7.05E–01 | 7.05E–01 |
| | 4 | **7.49E–01** | 7.31E–01 | 7.44E–01 | 7.32E–01 | 7.42E–01 | 7.36E–01 | 7.38E–01 | 7.33E–01 | 7.37E–01 |
| | 5 | 7.82E–01 | 7.61E–01 | 7.88E–01 | 8.01E–01 | 7.76E–01 | 8.04E–01 | **8.05E–01** | 8.00E–01 | 7.64E–01 |

(continued)

**Table 5** (continued)

| Image | Level | AHA | DE | EDA | GBO | GSA | GTO | JF | RUN | POA |
|---|---|---|---|---|---|---|---|---|---|---|
| 108070 | 2 | 5.70E–01 | 5.65E–01 | 5.67E–01 | 5.67E–01 | 5.66E–01 | 5.66E–01 | 5.58E–01 | 5.70E–01 | **5.71E–01** |
| | 3 | 7.48E–01 | 7.52E–01 | 7.56E–01 | 7.56E–01 | 7.43E–01 | 7.57E–01 | **7.63E–01** | 7.60E–01 | 7.56E–01 |
| | 4 | 7.99E–01 | 7.08E–01 | 8.15E–01 | 8.24E–01 | 8.07E–01 | 8.25E–01 | **8.28E–01** | 8.26E–01 | 7.27E–01 |
| | 5 | 8.29E–01 | 7.33E–01 | 8.45E–01 | 8.60E–01 | 8.33E–01 | 8.62E–01 | **8.64E–01** | 8.61E–01 | 7.32E–01 |
| 160068 | 2 | 7.45E–01 | 7.45E–01 | 7.48E–01 | 7.49E–01 | **7.57E–01** | 7.53E–01 | 7.47E–01 | 7.51E–01 | 7.55E–01 |
| | 3 | 7.72E–01 | 7.91E–01 | 7.89E–01 | 7.89E–01 | 7.83E–01 | 7.94E–01 | **8.01E–01** | 7.93E–01 | 7.96E–01 |
| | 4 | 7.93E–01 | 7.72E–01 | 7.96E–01 | 8.03E–01 | 8.02E–01 | 8.04E–01 | **8.12E–01** | 8.05E–01 | 7.79E–01 |
| | 5 | 8.11E–01 | 7.89E–01 | 8.22E–01 | 8.38E–01 | 8.15E–01 | 8.40E–01 | **8.41E–01** | 8.38E–01 | 7.83E–01 |
| 210088 | 2 | 6.57E–01 | 6.57E–01 | 6.59E–01 | 6.62E–01 | 6.60E–01 | 6.61E–01 | 6.59E–01 | 6.61E–01 | **6.63E–01** |
| | 3 | 6.82E–01 | 6.85E–01 | 6.87E–01 | 6.91E–01 | 6.88E–01 | 6.91E–01 | **6.95E–01** | 6.92E–01 | 6.92E–01 |
| | 4 | 7.13E–01 | 6.94E–01 | 7.14E–01 | 7.14E–01 | 7.11E–01 | 7.18E–01 | **7.21E–01** | 7.16E–01 | 6.96E–01 |
| | 5 | 7.37E–01 | 7.06E–01 | 7.38E–01 | 7.46E–01 | 7.38E–01 | 7.49E–01 | **7.51E–01** | 7.45E–01 | 7.06E–01 |
| 296059 | 2 | 6.27E–01 | 6.30E–01 | 6.28E–01 | 6.29E–01 | 6.29E–01 | 6.30E–01 | 6.32E–01 | 6.29E–01 | **6.34E–01** |
| | 3 | 6.99E–01 | 7.05E–01 | 7.07E–01 | 7.07E–01 | 7.07E–01 | 7.09E–01 | 7.09E–01 | 7.06E–01 | **7.10E–01** |
| | 4 | 7.37E–01 | 6.84E–01 | 7.47E–01 | 7.66E–01 | 7.48E–01 | **7.68E–01** | 7.66E–01 | 7.66E–01 | 7.06E–01 |
| | 5 | 7.71E–01 | 7.08E–01 | 7.82E–01 | 8.05E–01 | 7.76E–01 | **8.06E–01** | 8.05E–01 | 8.05E–01 | 7.04E–01 |
| 302008 | 2 | 7.45E–01 | 7.50E–01 | 7.52E–01 | 7.50E–01 | 7.51E–01 | 7.51E–01 | 7.45E–01 | 7.54E–01 | **7.55E–01** |
| | 3 | 7.91E–01 | 7.95E–01 | 7.94E–01 | 7.95E–01 | 7.94E–01 | 7.98E–01 | 7.98E–01 | 7.96E–01 | **7.99E–01** |
| | 4 | 8.11E–01 | 7.80E–01 | 8.21E–01 | 8.22E–01 | 8.19E–01 | **8.26E–01** | 8.26E–01 | 8.23E–01 | 7.71E–01 |
| | 5 | 8.33E–01 | 7.92E–01 | 8.40E–01 | 8.47E–01 | 8.35E–01 | **8.50E–01** | 8.47E–01 | 8.47E–01 | 7.94E–01 |

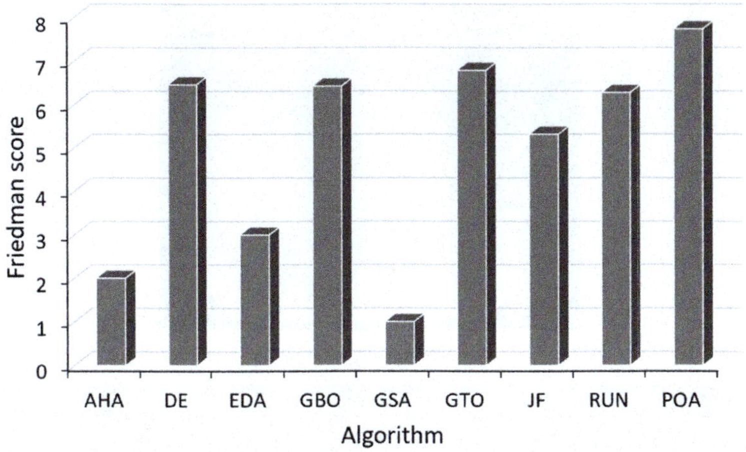


**Fig. 3** Comparison of the algorithms using the non-parametric Friedman test

some types can turn into colon cancer after many years [44]. An example of polyp images segmented in color using the color image thresholding technique is shown in Fig. 4.

The image in Fig. 4a has two polyps that are barely noticeable in the central part. Using two thresholding levels, regions with a green tone can be observed, but details of the surrounding regions are lost. With three thresholds, polyps are lost and the algorithm does not segment regions. Using four and five thresholds, the protuberances in the colon and the regions surrounding them are again observed in greater detail. This can help the specialist make a better decision about the type of polyp they are seeing in a patient. Figure 4f, compared with the previous one, shows a big polyp. This is to see if the algorithm behaves similarly with small regions or with large regions. In this case, using three thresholds the polyp region can be observed in greater detail. It can be seen that at the thresholding level where the small polyp region is not shown, the large polyps are shown and vice versa. One of the advantages of the multilevel thresholding segmentation method is that it can be adapted to any type of image, without the need for a large number of them.

Figure 5 shows 105 medical image with two and four thresholding levels and their respective histograms for each color of the RGB space. It is observed that the histograms load to the left with the largest number of pixels for the three colors, they have similar structures. However, the positions of the thresholds change in each color. The red histogram has two thresholds loaded to the right, and four thresholds are loaded further to the left. The green histogram for two and four thresholds remains in its positions. In the blue histogram, the positions of the thresholds are also similar for two and four. By locating the thresholds in the three histograms for each thresholding level in this way, it is possible to segment the affected regions.

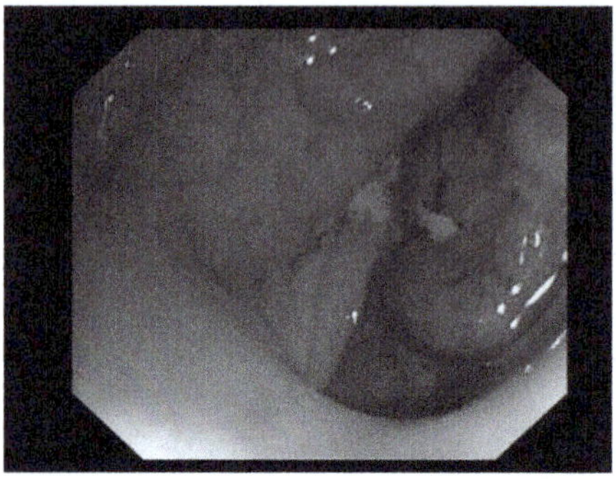

(a) Image 105 polyps.

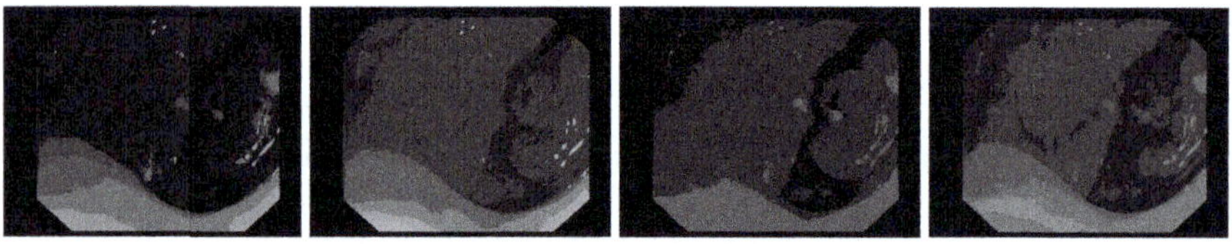

(b) 105 with 2 th. (c) 105 with 3 th. (d) 105 with 4 th. (e) 105 with 5 th.

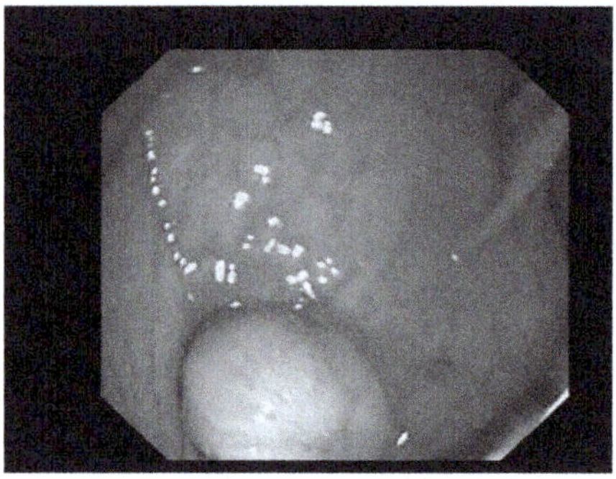

(f) Image 487 polyps.

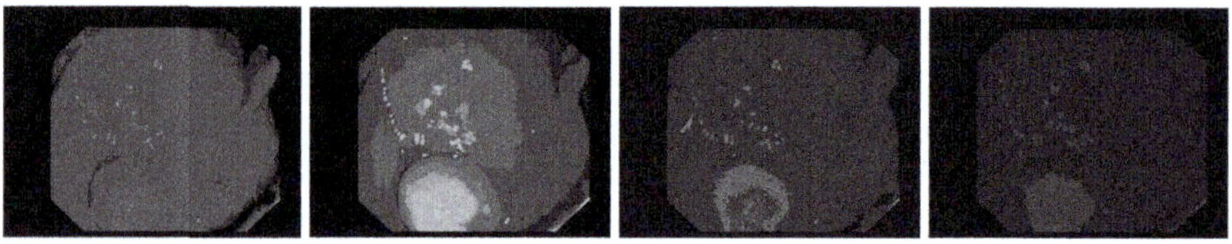

(g) 487 with 2 th. (h) 487 with 3 th. (i) 487 with 4 th. (j) 487 with 5 th.

**Fig. 4** Original and segmented polyp images with different number of thresholds

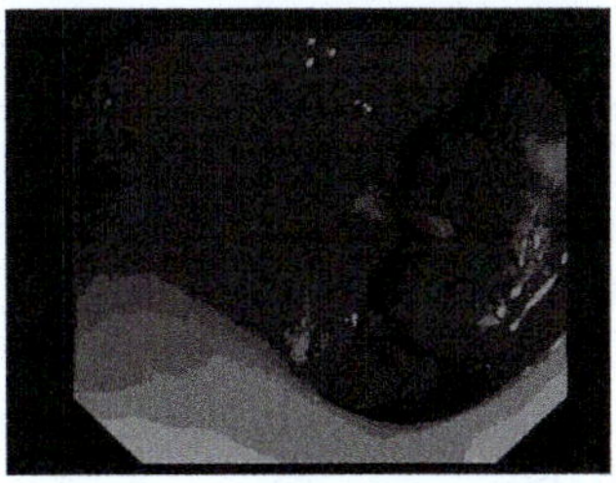

(a) Image 105 with 2 th.

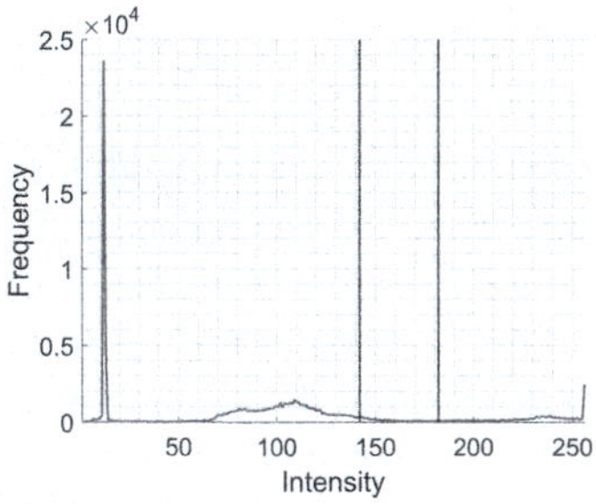


(b) Red histogram 105 image with 2 th.

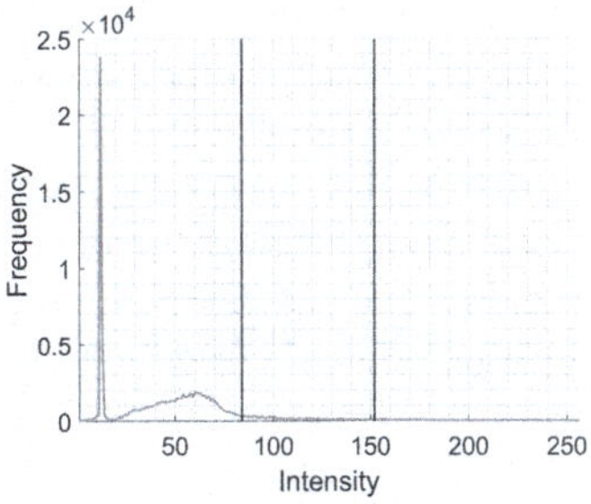


(c) Green histogram 105 image with 2 th.

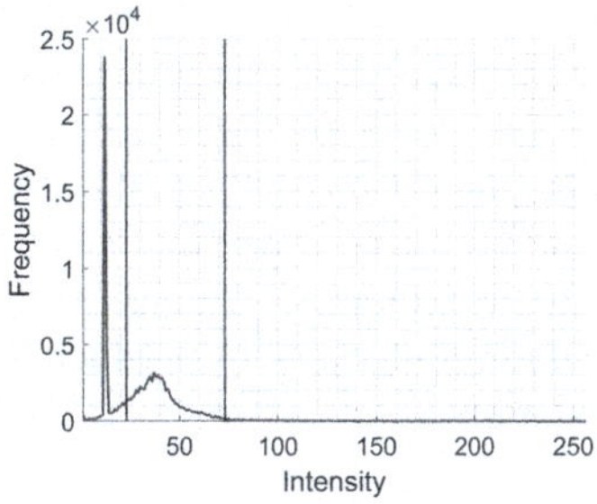


(d) Blue histogram 105 image with 2 th.

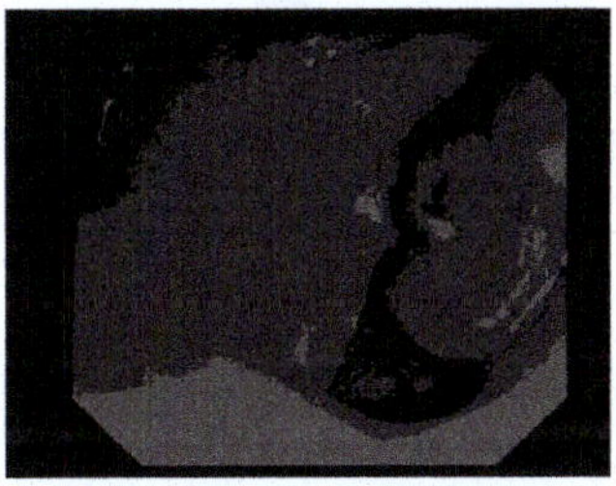

(e) Image 105 with 4 th.

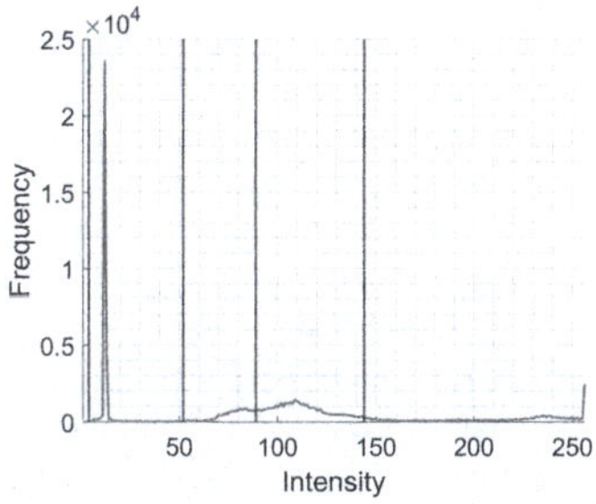


(f) Red histogram 105 image with 4 th.

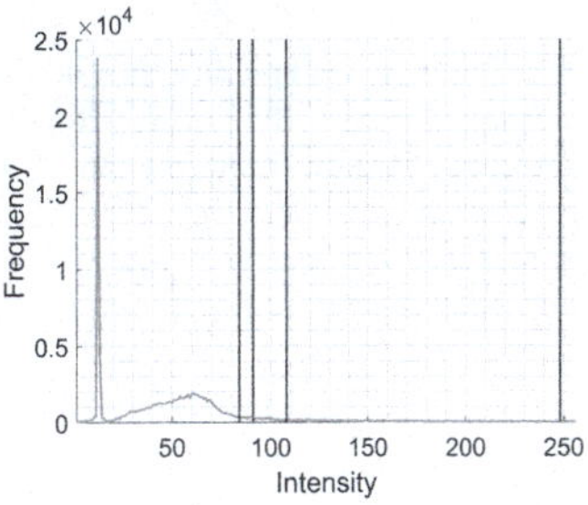


(g) Green histogram 105 image with 4 th.

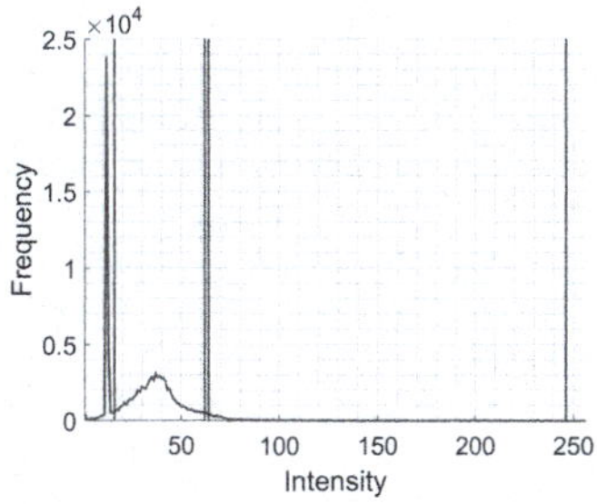


(h) Blue histogram 105 image with 4 th.

**Fig. 5** Thresholding image and their histograms

## 8 Conclusion

This chapter presents the methodology to address the segmentation problem by thresholding in the RGB color space. It is observed that POA is a competitive algorithm for maximizing the Otsu objective function compared to eight other algorithms. In 72.5% of the cases studied with different thresholding levels, this algorithm obtains the best score. Within the evaluation metrics, it maintains second and third place. But what it was designed for, it serves its purpose.

There are limitations of multilevel thresholding color image segmentation such as it is not possible to perform segmentation in real-time, increasing the thresholding level causes the execution time of the algorithm to increase non-linearly, there may be algorithms with better conditions that can be developed in the future as established by the No Free Lunch Theorem. But the advantages are: a simple method to apply for any color segmentation problem, it is possible to extrapolate it to medical images (as shown), and industrial images, among other areas of application, the image can be divided into the number of regions. As desired, there are many objective functions to be considered for segmentation, which gives the possibility of using multiple criteria in segmentation.

## References

1. Kwon, M.J., Han, Y., Shin, I., Park, H.W.: Hierarchical fuzzy segmentation of brain mr images. Int. J. Imaging Syst. Technol. **13**(2), 115–125 (2003)
2. Gharehchopogh, F.S., Ibrikci, T.: An improved african vultures optimization algorithm using different fitness functions for multi-level thresholding image segmentation. Multimed. Tools Appl. 1–47 (2023)
3. Bhandari, A.K.: A novel beta differential evolution algorithm-based fast multilevel thresholding for color image segmentation. Neural Comput. Appl. **32**(9), 4583–4613 (2020)
4. Wang, S., Jia, H., Peng, X.: Modified salp swarm algorithm based multilevel thresholding for color image segmentation. Math. Biosci. Eng. **17**(1), 700–724 (2020)
5. Otsu, N.: A threshold selection method from gray-level histograms. IEEE Trans. Syst. Man Cybernet. **9**(1), 62–66 (1979)
6. Kapur, J.N., Sahoo, P.K., Wong, A.K.: A new method for gray-level picture thresholding using the entropy of the histogram. Comput. Vis. Graph. Image Process. **29**(3), 273–285 (1985)
7. Li, C.H., Lee, C.: Minimum cross entropy thresholding. Pattern Recognit. **26**(4), 617–625 (1993)
8. Shahabi, F., Poorahangaryan, F., Edalatpanah, S., Beheshti, H.: A multilevel image thresholding approach based on crow search algorithm and otsu method. Int. J. Comput. Intell. Appl. **19**(02), 2050015 (2020)
9. Xu, L., Jia, H., Lang, C., Peng, X., Sun, K.: A novel method for multilevel color image segmentation based on dragonfly algorithm and differential evolution. IEEE Access **7**, 19502–19538 (2019)
10. Ma, G., Yue, X.: An improved whale optimization algorithm based on multilevel threshold image segmentation using the otsu method. Eng. Appl. Artif. Intell. **113**, 104960 (2022)
11. Guo, Q., Peng, H.: A novel multilevel color image segmentation technique based on an improved firefly algorithm and energy curve. Evolving Syst. **14**(4), 685–733 (2023)

12. Abualigah, L., Almotairi, K.H., Elaziz, M.A.: Multilevel thresholding image segmentation using meta-heuristic optimization algorithms: comparative analysis, open challenges and new trends. Appl. Intell. **53**(10), 11654–11704 (2023)
13. Rangu, S., Veramalla, R., Salkuti, S.R., Kalagadda, B.: Efficient approach to color image segmentation based on multilevel thresholding using emo algorithm by considering spatial contextual information. J. Imaging **9**(4), 74 (2023)
14. Wu, B., Zhu, L., Li, X.: Giza pyramids construction algorithm with gradient contour approach for multilevel thresholding color image segmentation. Appl. Intell. 1–20 (2023)
15. Kennedy, J., Eberhart, R.: Particle swarm optimization. In: Proceedings of ICNN'95-International Conference on Neural Networks, vol. 4, pp. 1942–1948. IEEE (1995)
16. Storn, R.: Differential evolution-a simple and efficient adaptive scheme for global optimization over continuous spaces, Technical report, International Computer Science Institute 11 (1995)
17. Mirjalili, S., Mirjalili, S.M., Lewis, A.: Grey wolf optimizer. Adv. Eng. Softw. **69**, 46–61 (2014)
18. Yang, X.-S., Deb, S.: Cuckoo search via lévy flights. In: World Congress on nature and Biologically Inspired Computing (NaBIC), pp. 210–214 . IEEE (2009)
19. Holland, J.H.: Genetic algorithms. Sci. Am. **267**(1), 66–73 (1992)
20. Zhao, W., Wang, L., Mirjalili, S.: Artificial hummingbird algorithm: a new bio-inspired optimizer with its engineering applications. Comput. Methods Appl. Mech. Eng. **388**, 114194 (2022)
21. Socha, K., Dorigo, M.: Ant colony optimization for continuous domains. Euro. J. Oper. Res. **185**(3), 1155–1173 (2008)
22. Karaboga, D., Basturk, B.: A powerful and efficient algorithm for numerical function optimization: artificial bee colony (abc) algorithm. J. Global Optim. **39**, 459–471 (2007)
23. Mirjalili, S.: Sca: a sine cosine algorithm for solving optimization problems. Knowl.-Based Syst. **96**, 120–133 (2016)
24. Yang, X.-S.: Multiobjective firefly algorithm for continuous optimization. Eng. Comput. **29**, 175–184 (2013)
25. Mirjalili, S., Lewis, A.: The whale optimization algorithm. Adv. Eng. Soft. **95**, 51–67 (2016)
26. Mirjalili, S.: Moth-flame optimization algorithm: a novel nature-inspired heuristic paradigm. Knowl.-Based Syst. **89**, 228–249 (2015)
27. Passino, K.M.: Biomimicry of bacterial foraging for distributed optimization and control. IEEE Control Syst. Mag. **22**(3), 52–67 (2002)
28. Wolpert, D.H., Macready, W.G.: No free lunch theorems for optimization. IEEE Trans. Evol. Comput. **1**(1), 67–82 (1997)
29. Trojovskỳ, P., Dehghani, M.: Pelican optimization algorithm: A novel nature-inspired algorithm for engineering applications. Sensors **22**(3), 855 (2022)
30. Ahmadianfar, I., Bozorg-Haddad, O., Chu, X.: Gradient-based optimizer: a new metaheuristic optimization algorithm. Inf. Sci. **540**, 131–159 (2020)
31. Noroozi, M., Mohammadi, H., Efatinasab, E., Lashgari, A., Eslami, M., Khan, B.: Golden search optimization algorithm. IEEE Access **10**, 37515–37532 (2022)
32. Abdollahzadeh, B., Soleimanian Gharehchopogh, F., Mirjalili, S.: Artificial gorilla troops optimizer: a new nature-inspired metaheuristic algorithm for global optimization problems. Int. J. Intell. Syst. **36**(10), 5887–5958 (2021)
33. Chou, J.-S., Truong, D.-N.: A novel metaheuristic optimizer inspired by behavior of jellyfish in ocean. Appl. Math. Comput. **389**, 125535 (2021)
34. Ahmadianfar, I., Heidari, A.A., Gandomi, A.H., Chu, X., Chen, H.: Run beyond the metaphor: an efficient optimization algorithm based on runge kutta method. Expert Syst. Appl. **181**, 115079 (2021)
35. Martin, D., Fowlkes, C., Tal, D., Malik, J.: A database of human segmented natural images and its application to evaluating segmentation algorithms and measuring ecological statistics. In: Proceedings of 8th International Conference Computer Vision, vol. 2, pp. 416–423 (2001)
36. Wang, S., Sun, K., Zhang, W., Jia, H.: Multilevel thresholding using a modified ant lion optimizer with opposition-based learning for color image segmentation. Math. Biosci. Eng. **18**(4), 3092–3143 (2021)

37. Zhao, X., Turk, M., Li, W., Lien, K.-C., Wang, G.: A multilevel image thresholding segmentation algorithm based on two-dimensional k-l divergence and modified particle swarm optimization. Appl. Soft Comput. **48**, 151–159 (2016)
38. Olmez, Y., Sengur, A., Koca, G.O., Rao, R.V.: An adaptive multilevel thresholding method with chaotically-enhanced rao algorithm. Multimed. Tools Appl. **82**(8), 12351–12377 (2023)
39. Huynh-Thu, Q., Ghanbari, M.: Scope of validity of psnr in image/video quality assessment. Electron. Lett. **44**(13), 800–801 (2008)
40. Wang, Z., Bovik, A.C., Sheikh, H.R., Simoncelli, E.P.: Image quality assessment: from error visibility to structural similarity. IEEE Trans. Image Process. **13**(4), 600–612 (2004)
41. Zhang, L., Zhang, L., Mou, X., Zhang, D.: Fsim: a feature similarity index for image quality assessment. IEEE Trans. Image Process. **20**(8), 2378–2386 (2011)
42. Derrac, J., García, S., Molina, D., Herrera, F.: A practical tutorial on the use of nonparametric statistical tests as a methodology for comparing evolutionary and swarm intelligence algorithms. Swarm Evol. Comput. **1**(1), 3–18 (2011)
43. Bernal, J., Sánchez, F.J., Fernández-Esparrach, G., Gil, D., Rodríguez, C., Vilariño, F.: Wm-dova maps for accurate polyp highlighting in colonoscopy: Validation versus saliency maps from physicians. Comput. Med. Imaging Graph. **43**, 99–111 (2015)
44. Allen, J.I.: Molecular biology of colon polyps and colon cancer. In: Seminars in Surgical Oncology, vol. 11, pp. 399–405. Wiley Online Library (1995)

# Optimization and Improve Image Contrast: A Comparative Study of Classical Metaheuristic Algorithms

**Beatriz A. Rivera-Aguilar, Diego Campos Peña, Noé Ortega-Sánchez, and Alma Nayeli Rodríguez Vázquez**

**Abstract** This chapter compares different optimization algorithms for contrast enhancement in grayscale images. The algorithms DE, HS, PSO, TLBO, ABC, FA and IWO which are classics in the literature for conducting this analysis, were addressed. All these methods were adapted with the same enhancement technique to observe the results they can offer. For the tests, a dataset commonly found in the literature was used, and 8 metrics were taken to measure the performance of the contrasted images: RMS, FSIM, ENTROPY, PSNR, MSE, VIF, IFC and Intensity levels. A deep analysis is carried out regarding metrics, fitness, and convergence, aiming to observe and contrast the results that each algorithm offers. Finally, conclusions and recommendations regarding the best method to improve the contrast in the images are presented.

**Keywords** Improve image contrast · Algorithm metaheuristics · DE · HS · PSO · TLBO · ABC · FA · IWO · Gray scale image

B. A. Rivera-Aguilar (✉) · D. C. Peña · N. Ortega-Sánchez · A. N. R. Vázquez
División de Tecnologías para la integración ciber-humana, CUCEI, Universidad de Guadalajara, Guadalajara, Mexico
e-mail: beatriz.rivera@academicos.udg.mx

D. C. Peña
e-mail: diego.campos7896@alumnos.udg.mx

N. Ortega-Sánchez
e-mail: noe.ortega@academicos.udg.mx

A. N. R. Vázquez
e-mail: alma.rvazquez@academicos.udg.mx

 
D. Oliva et al. (eds.), *Advances in Optimization Algorithms for Multidisciplinary Engineering Applications: From Classical Methods to AI-Enhanced Solutions*, Studies in Computational Intelligence 806, https://doi.org/10.1007/978-3-031-78440-8_2

# 1 Introduction

Improving contrast in images has become a very important area of study in science. In everyday life, we have encountered images of poor quality according to the criteria of the human eye; this can pose significant problems if important decisions need to be made based on them. Therefore, using existing tools, improving them, or developing new methods to help us perceive images with higher visual quality is necessary. Various disciplines, such as medicine, agriculture, astronomy, forestry, etc., require the support of these contrast enhancement techniques [1–5].

In a low-quality grayscale image, it can be observed that all intensity levels are accumulated in a certain area in its histogram within the range [0–255]. If they are very close to 0, the image will tend to be very dark; otherwise, if they are very close to 255, it will be very bright. In either case, the image will exhibit a visual deficiency Fig. 1.

There are different techniques to improve the contrast of an image; among them are histogram equalizers. These methods mainly aim to distribute those accumulations of pixel intensity throughout the entire range of the histogram [0–255] as evenly as possible. This way, more details can be visualized in the image. Some of the histogram equalizers found in the state of the art are: Averaging Histogram Equalization (AVHEQ) [6], Recursive Sub-Image Histogram Equalization (RSIHE) [7], Exposure-Based Sub Image Histogram Equalization (ESIHE) [8], Minimum Mean Brightness Error Bi-Histogram Equalization in contrast enhancement (MMBEBHE) [9], and Brightness Preserving Bi-histogram Equalization (BBHE) [10].

However, using this histogram equalization techniques often does not solve the problem, as it has certain disadvantages. Enhancing the contrast of images can often amplify certain visually unpleasant details, making the image appear even lower in quality. Another issue is that by redistributing pixels to enhance contrast, there is a risk of losing the relationship between pixels and the image, and in many cases, preserving the fidelity of the image is much more important. Additionally, using equalizers in real-time can often be detrimental in terms of execution time, especially when processing high-resolution images, which may take longer computational processing

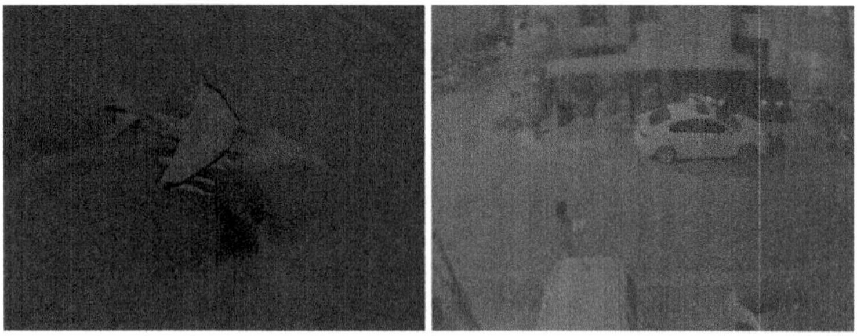

**Fig. 1** Example of low-quality images

time than expected. Furthermore, these equalizers do not always provide the best contrast in the image; there are occasions when the images have excessive visual artifacts after equalization, making them appear unnatural rather than natural. Finally, another disadvantage of using these equalizers is that they cannot be applied to any image, such as those with non-uniform lighting, resulting in poor visual quality, even worse than the original. Therefore, while it is a very useful tool in some cases, it is important to note and consider that it is not a generic method, leading us to search for other methods that can help improve contrast in images, such as metaheuristic algorithms.

Metaheuristic algorithms are computational techniques that help us find solutions close to the optimum within reasonable time frames in optimization problems that are difficult to solve. Therefore, they are also applicable in this type of image processing, where enhancing the contrast as effectively as possible is required. Some of the most recent works in the literature suggest two approaches. First, a mathematical transformation function is proposed, and a metaheuristic algorithm is used to optimize the parameters of this equation. The second approach involves using a transformation function from the literature and modifying a metaheuristic to optimize and provide better solutions.

Some recent works found in the state of the art that apply this type of process are [11], in this work utilizes the Artificial Electric Field Algorithm (AEFA) to optimize and find the best control parameters of the Incomplete Beta Function (IBF) to enhance image contrast. Another approach is the one proposed by [12]; this work first utilizes the Mean-shift algorithm, which removes any noise present in the images and subsequently employs the metaheuristic Moth Swarm Algorithm (MSA) to distribute pixel intensities and thereby enhance contrast evenly. In [13], a hybrid metaheuristic algorithm is proposed, combining the Bat Algorithm (BA) and the Artificial Bee Colony (ABC) algorithm, which merge their parameters to optimize the statistical transformation function more effectively, thereby enhancing the contrast of medical images specifically applied in skin cancer. Another recent work is the one proposed by [14], where they use a hybrid optimization algorithm combining Particle Swarm Optimization (PSO) and the Black Hole Algorithm (BHA) to find the parameters that enhance the parameterized mapping function they use for contrast enhancement, along with a new proposed objective function. In [15], a contrast enhancement work in images is presented using the Political Optimizer (PO) algorithm, which is combined with the β-hill climbing algorithm (AβHC), thus creating a potential hybrid to find the best parameters of the Incomplete Beta Function (IBF). Similarly, in [16], a contrast enhancement work in images is presented by modifying the Gray Wolf Optimization (GWO) metaheuristic algorithm, altering the initialization method with a low-discrepancy sequence, aiming to start with the referenced algorithm rather than randomly.

According to the previously presented works, which were found in state of the art, it is evident that metaheuristic algorithms have performed very well in contrast enhancement in grayscale images, unlike equalizers, as not only applied methods have been found in the literature, but also hybridizations of different algorithms have been proposed to enhance the solutions or parameters of the transformation functions used to improve image quality. For that, a contrast enhancement technique is addressed in this chapter. It is adapted to 7 classical optimization algorithms to show the behavior that each of these algorithms presented and be able to conclude and recommend the method that achieved the best evaluations according to the measured metrics, as well as to highlight the solution capabilities and vulnerabilities of each one.

For the remainder of this chapter, a brief description of the metaheuristic algorithms used in this study is provided in Sect. 2. Section 3 outlines the methodology used to generate image contrast and presents the objective function to optimize contrast in grayscale images. The experiments of the 7 tested algorithms are presented in Sect. 4, along with the results they obtained. Finally, in Sect. 5, the conclusions of this study are outlined, and future work is proposed.

## 2 Metaheuristic Algorithms

The literature contains many metaheuristic algorithms for solving problems. To understand the general overview, it could recognize classic metaheuristic algorithms such as DE, HSA, PSO, TLBO, ABC, FA and IWO. Each algorithm will be described briefly and shown in its capabilities.

### *2.1 Differential Evolution (DE)*

The algorithm, based on population evolution, was proposed in 1995 [17, 18]. The first idea was to solve problems involving global optimization over continuous space. DE has four stages. In the first stage, the algorithm initializes a population within the problem boundaries.The second stage involves mutating the initial population based on three individuals. These are selected randomly to form a new value, but noise is also applied to conform to the new value. In the third stage, the new values are evaluated, and the vector will be replaced by the newer one only if it is better. The new generation moves in the problem space. The last stage is the selection of the best candidates compared to the originals. In other words, the candidates with the highest fitness will survive to the next generation, and the cycle will begin again until the fitness is near the optimal value.

### *2.2 Harmony Search (HS)*

Geem [19, 20] proposed the Harmony Search Algorithm since the publication was used in areas of interest like engineering, processing images, and other kinds of problems. The inspiration for the algorithm was musician jazz, designed as an analogy of how players made adjustments in relationships with the public and the reaction of their moods. When musicians play their instruments, they reach the public's emotions and change the rhythm and tone to find a response to the public and make them feel better.

The process begins by making a vector with a blend of data sets and calculating the fitness reaction. The vector is sorted in descendent or ascendent depending on the problem, representing a maximization or minimization. In order to make new harmonies, a PAR (Pitch Adjustable Rate) creates a disturbance and a new set based on the memory. This parameter controls the exploration and exploitation of the problem area. The PAR is decreased in the timeline to be finer and near the optimal solution. Then, the new harmony is compared with the last element of the pool of data, and if the new harmony is better than the last element, this is inserted in the vector of the best solutions. All these operations are repeated if they reach the criteria of termination.

## 2.3 *Particle Swarm Optimization (PSO)*

The PSO analogy is based on swarm populations [21], where simple rules manage the behavior of a set of individuals.An extensive set of individuals is dispersed into an area to evaluate their fitness; this set will generate a new generation based on their fitness, best global value, and speed factor; in the beginning, it is high, but with each cycle, the speed of movements is reduced, and change the positions of population.

The algorithm's capabilities include fast convergence due to a large set, simple programming, and acceptable performance. On the other hand, a weakness is premature convergence, which makes for suboptimal fitness. It is necessary to tune their values to avoid problems, which is highly used in the literature.

## 2.4 *Teaching Learning Based Optimization (TLBO)*

Rao [22] presented the teaching-learning-based optimization algorithm based on the natural social phenomena of teaching and learning. One of their advantages is that they only need standard controlling parameters. In the case of the population and the number of generations, this means not necessarily setting controlling parameters like other algorithms based on populations or swarms but also is an algorithm where the use of random values is present. The students are vector solutions in the first stage, initiated by the problem's boundaries. The improvements in the class are based on the student subset influenced by lectures and the teacher's ideas.

In this way, the function objective that represents a student's performance greater than the function of the old student is not taken to replace in the class. The teaching phase is to improve the values or, in this case, the skills of students to level the teacher's grade.

## 2.5 Artificial Bee Colony (ABC)

There has been a place for swam algorithms in the literature. Indeed, a popular one is the ABC [23, 24], which is considered a classic based on the bee's analogy of finding food and communicating with the colony where the resource is. The model proposed by Karaboga is composed of a group of bees: employees, onlookers, and scouts.

It creates a subset of the half of the colony with employees; the other subset is the onlookers. Each employee has only one source of food. When a bee finishes food, it becomes a scout to escape from the area and improve the probability of resources. Creating a cycle: moving employers and onlookers bees to another food area and calculating the nectar. Second, the direction of the scouts should be determined, and possible foods that represent a solution to the problem should be identified. The colony's survival depends on the efficiency of discovering the best solutions; this means improving the fitness of each bee.

## 2.6 Firefly Algorithm (FA)

Yang was the developer of the algorithm [25, 26] since the launch was a disruptor and applied a long of problems in engineering and solving problems. The inspiration was an insect. To be more specific, a firefly is a kind of insect with a capacity of bioluminescence produced in the lower abdomen for reproduction, communication, and defense purposes; In essence, biosignals are created to communicate with the swarm population. Three kinds of behaviors are based on the emission light that represents the objective function and are modified over time. First, a pool of individuals is initialized in the boundaries of the problem and is calculated as the fitness value to establish the best candidates. It can be remembered that light is the fitness and represents the best candidates. The second stage creates a new candidate using the old values to move into an area and recalculate their light. With this in mind, the best fireflies survive to the next generation due to being ordered in a descending manner to stay with the best candidates. The movements are influenced by other best fireflies and attracted to other more attractive fireflies; this generates an interesting phenomenon, creating a subset of fireflies given suboptimal values concerning function fitness.

## 2.7 Invasive Weed Optimization (IWO)

Mehrabian and Lucas proposed the Invasive weed Algorithm (IWO) base in 2006 [27], using the analogy of how weeds grow and invade other territories to gain space and explore new areas. As the analogy means, nature propagates and evolves a plant organism.

IWO became popular because the algorithm has advantages: easy implementation, rapid convergence, and simple calculations. Nevertheless, there are some problems, like premature convergence, which means it is possible to get a local minimum without being able to escape. The algorithm is generally established in the following way: at the beginning, the weeds are spared in the soil; it is the random initialization in the problem. After some time, each weed produces seed based on fitness, which means that with higher fitness, the reproduction of seed is more intensive and locally generates expansive growth. The diffusion of the seed is based on taking a parent generation, applying a modification in random parameters, and developing an exploration in a local area near the parents. However, the uncontrollable growth creates exclusion and consequently discards the weak weed, but this is not bad since it ensures that the next generation of weed is stronger in fitness.

## 3 Transformation Functions and Objective Function as a Fundamental Basis to Improve Contrast in Images

Generally, a transformation function must be used to improve the contrast in an image; this mathematical function attempts to distribute the pixel intensities of the image in the best possible way across its histogram. As previously addressed, several state-of-the-art works use predefined functions [11, 13–16]. In contrast, others propose their own like [28, 29], and both approaches rely on some metaheuristic method to find the best parameters that adapt to enhancing the contrast of a given image. Therefore, these transformation functions are fundamental as they determine the new values that will be mapped to achieve a visually improved image quality.

This chapter uses a method to construct the transformation function in a simple and without any mathematical equation to use in each optimization algorithm. This method is shown graphically see Fig. 2, and each point is described below:

a. A vector is built with three points, where 0 and 255 are the first and third points, respectively, and are fixed to ensure that the function covers the entire grayscale histogram of the image. The second point is generated randomly between 0 and 255 for each $x$ and $y$ axis.
b. The points generated in (a) are located for both axes.
c. The previous points are interpolated to generate the transformation function.
d. The transformation function generated with this technique is shown.
e. Example of a histogram of an image without contrast.
f. Using the transformation function generated in (d) and the histogram in (e), the new pixel intensities are obtained by mapping each intensity level of the "x" axis with the new intensity level generated by the function on the "y" axis.
g. The result of the pixel mapping with the transformation function is shown. This obtains an image with more contrast since the intensity levels are now more distributed in the range 0–255.

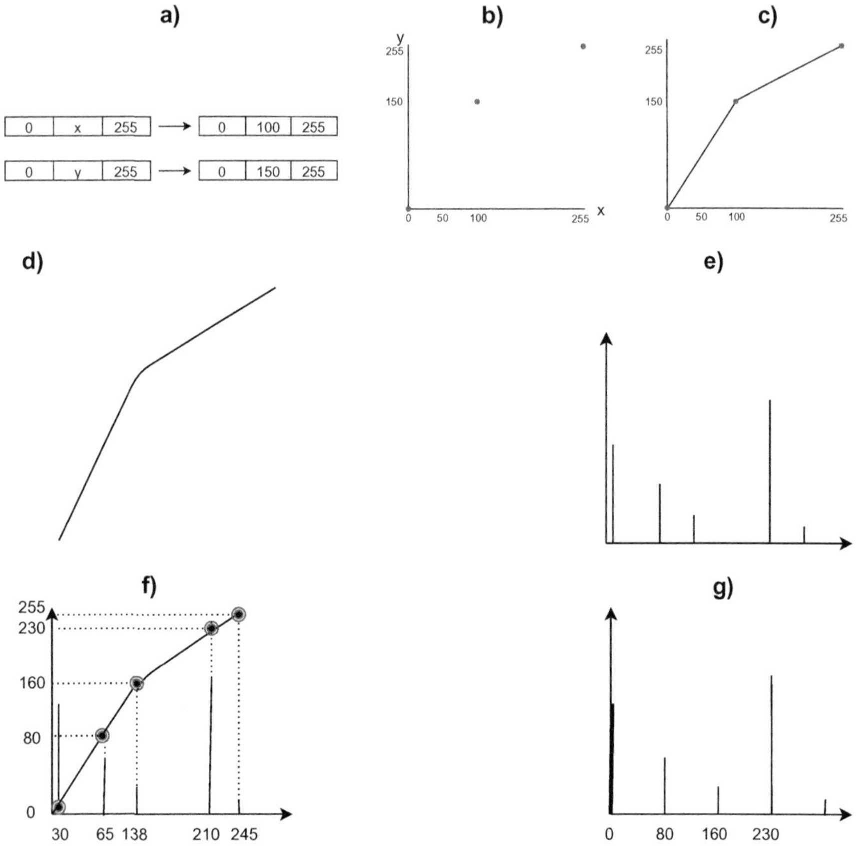


**Fig. 2** Example of contrast enhancement technique

As detailed in Sect. 2, the optimization algorithms work in iterations. In each iteration, they evaluate the quality of the solutions and store them so that at the end of the process, they will have the best of all.

For the contrast improvement presented in this chapter, all the optimization methods addressed use the same optimization function shown in Eq. 1 so that they evaluate the quality of the image by mapping the new values offered in each iteration by the generated transformation functions and, in the end, obtain the best of them.

$$F(V) = -\left(w_1\left(log(log(E(I(V)))) * \left(\frac{ne(I(V))}{PH*PV}\right) * H(I(V))\right) + w_2L\right) \quad (1)$$

where

- The solution vector V represents a mapping used to obtain the output image quality, which is denoted by $F(V)$. This output image is obtained by applying the mapping represented by $V$ on the input image $I$.
- The intensity of edges in the resulting image obtained by filtering the output image $I(V)$ using the Sobel filter is equal to the sum of $E(I(V))$ edge intensities.
- The resulting image has a certain number of edges, which $ne(I(V))$ represents.
- The entropy of the output image is represented by $H(I(V))$.
- The dimensions of an image can be described by $PH$ and $PV$, which represent the number of horizontal and vertical pixels, respectively.
- The value of $L$ is computed by subtracting the image's minimum contrast level from its maximum contrast level.
- The weights $w_1$ and $w_2$ can be adjusted to modify the weights within the objective function.

## 4 Experimental Results

In order to perform the comparative study, a set of experiments consisting of the following was created:

- A set of 22 reference images taken from TID2013 [30] was used to perform the tests.
- Seven different classical and popular metaheuristic algorithms such as Differential Evolution (DE) [17], Harmony Search(HS) [19], Particle Swarm Optimization(PSO) [21], Teaching Lerning Based Optimization(TLBO) [22], Artificial Bee Colony(ABC) [23], Firefly Algorithm(FA) [25], and Invasive Weed Optimization (IWO) [27] were programmed with the same objective function see Eq. 1, and with the same contrast enhancement technique.
- The experiments were conducted with the following parameters: 30 independent runs in each algorithm and for each of the test images and 500 iterations for each run.
- Each algorithm was configured with the best external parameters reported in their original publications, however, in the case of the number of population of each algorithm the parameter set was 30 search agents in all the algorithms under study, the latter in order that all of them compete under the same conditions without giving advantage to any of the algorithms.
- Parametric tests were performed with measures of central tendency of the data such as average, standard deviation, best solution and worst solution, extracted from the data set of the 30 executions of each of the algorithms.
- 8 metrics were used as indicators of contrast enhancement.
- The discussion is divided into four parts: in the first part, the performance of each algorithm is analyzed; in the second part, the results obtained by the metrics are evaluated; in the third part, the statistical results.

### 4.1 Experimental Configuration

Different experiments were designed to evaluate the performance of each algorithm used and to determine which one yielded the best results.

The image dataset used is the one found in [30], they are grayscale images and each of them has a different type of contrast. However, to display results, an extract of them was taken, showing only 12.

To evaluate the results of the images provided by each algorithm, universal contrast metrics found in the state of the art were used, such as Root Mean Square error (RMS). This measure assesses the difference between images, indicating that the lower the error, the better the quality. Equation 2 defines the equation representing this metric.

$$RMS = \sqrt{\frac{1}{MN}\sum_{i=0}^{M-1}\sum_{j=0}^{N-1}\left(Ic_{ij} - Io_{ij}\right)^2} \tag{2}$$

where the intensity values of the corresponding pixels in two images at position (i, j) are represented by $I_c$ as the contrasted image and $I_o$ as an original reference image. Additionally, each image contains M $\times$ N total pixels.

The metric Feature Similarity (FSIM) [31] is based in how we perceive images is influenced by their overall structure, local characteristics, and significant details. This metric aims to improve the accuracy of image evaluation by including high-frequency features and local details. It is particularly useful when image quality depends on preserving fine details and textures. The FSIM value scale ranges from 0 to 1, with 1 indicating perfect similarity between the features of the compared images. Thus, values closer to 1 indicate better image quality regarding feature similarity. The mathematical equation is represented by Eq. 3.

$$FSIM = \frac{\sum x \epsilon \Omega S_L(x) PC_m(x)}{\sum x \epsilon \Omega PC_m(x)} \tag{3}$$

where $\Omega$ represents the entire spatial domain of the image, and $PC_m(x)$ is computed within the brightness channel of the original image. It is worth noting that $PC$ is perceived by maximizing the Fourier component in phase, while $S_L(x)$ relies on the $PC$ value and reflects the overall similarity between images $Ic$ and $Io$.

Image Entropy (E) is calculated by examining the gray level distribution in an image and calculating the associated uncertainty. If the distribution of gray levels or color values is uniform, the entropy will be higher. Conversely, the entropy will be lower if the distribution is highly concentrated. Equation 4 is used to calculate entropy.

$$E = -\sum_{i=1}^{N} p(x_i) * log(p(x_i)) \tag{4}$$

where $E$ represents the entropy of the image, $p(x_i)$ denotes the probability of the occurrence of gray level $x_i$, and $N$ represents the total number of possible gray levels in the image.

Peak Signal to Noise Ratio (PSNR) [32] is a metric expressed in decibels (dB). It represents the ratio between the maximum power of the input signal and the signal noise. This ratio is used to measure the quality of a signal and is defined as Eq. 5:

$$PSNR = 10 * log_{10}\left(\frac{N^2}{MSE}\right) \tag{5}$$

where $N$ is the number of possible gray levels in the image, and $MSE$ is represented by the Eq. 6.

$$MSE = \frac{1}{mn}\sum_{i=0}^{m-1}\sum_{j=0}^{n-1}[Io(i, j)) - Ic(i, j)]^2 \tag{6}$$

where the pixel intensity values at position $(i, j)$ in the original and processed images are represented by $Io$ and $Ic$, respectively, and $m$ and $n$ represent the width and height of the images.

The quality of images can be measured using a metric called Visual Information Fidelity (VIF) [33], which is determined by an index factor. This index factor is based on statistics of natural scenes and visual image data. VIF is calculated using three models: (1) a model for image distortion, (2) a statistical model for natural scene images, and (3) an HVS model in an information-theoretic way. The equation that defines VIF is as follows:

$$VIF = \frac{\sum_{K\epsilon sub-band} Io\left(C^{N,k}; F^{N,k} \mid s^{N,k}\right)}{\sum_{K\epsilon sub-band} Ic\left(C^{N,k}; E^{N,k} \mid s^{N,k}\right)} \tag{7}$$

where the VIF value depends on $Io$ and $Ic$, which refer to the reference image information and enhanced image information in $k$th sub-band, respectively.

The IFC (Information Fidelity Criterion) [34] evaluates the preservation of relevant information while adjusting an image's contrast. It is an essential tool for ensuring that contrast adjustments enhance an image's visual appearance without damaging the quality of the information contained within it.

Finally, each contrasted image's intensity levels are calculated to compare the results.

### *4.2 Results Performance*

For the first experiment, all the algorithms used for this study implemented the technique described in Sect. 3 to generate transformation functions that map and enhance the contrast in images, as well as the objective function described in the same section, which was further detailed in Sect. 2. This experiment aims to visually analyze the results obtained by each algorithm, the best transformation function they could generate, and the final histogram when contrasting images.

In Tables 1, 2, 3, 4, 5 and 6, the results of the images generated by each algorithm used in images $I_1$ to $I_{12}$ are shown. In the first column of each table, the original image is displayed along with its corresponding histogram of low contrast, where we can observe that the histogram does not cover all intensity levels in the range 0–255. From the second column onwards, the best image obtained by the algorithms DE, HS, PSO, TLBO, ABC, FA, and IWO is presented, along with its histogram. Additionally, each one shows the transformation function used for mapping.

As can be observed visually, all the algorithms improve the original image, as characteristic features can be seen in the images that cannot be perceived in the original due to their darkness. However, although the resulting images are very similar, the difference between them is evident when we visualize the transformation function they used for their mapping. All the functions are different; therefore, the generated histograms are also different. If we observe each histogram of image $I_1$, we can see that the PSO algorithm takes more contrast levels; the same happens with the remaining images $I_2$ to $I_{12}$; all of them have a different transformation function and a different generated histogram; however, only with this observation is it impossible to determine that this algorithm was better than the others. Hence, it was necessary to carry out the following experiment.

The second experiment includes the 8 metrics described in the experimental configuration subsection. As mentioned earlier, it is not possible to conclude thus far that any of the 7 algorithms is superior to the others solely based on visual perception. The primary objective of this experiment is to mathematically demonstrate which algorithm achieved the best results in the images in terms of RMS, FSIM, ENTROPY, PSNR, MSE, VIF, IFC, and Intensity Levels. The results produced by the metrics of the images $I_1$ to $I_{12}$ generated by each of the 7 algorithms are shown in Tables 7 and 8.

As observed in Tables 7 and 8, the IWO algorithm achieved the best results in the PSNR and MSE metrics, surpassing all other algorithms for all calculated images. For the IFC and FSIM metrics, the IWO algorithm achieved the best results, followed by PSO and DE, which only obtained the best result for images $I_1$ and $I_5$, respectively. Regarding Entropy, the IWO algorithm obtained the best results again, with the DE and PSO algorithms following in second and third place, respectively. Regarding RMS, IWO again had the majority of good results, although other algorithms also achieved good results in this metric, notably ABC, TLBO, DE, and PSO. Despite the excellent results shown by IWO in all the metrics above, in terms of the VIF metric, IWO failed to obtain any good values for all test images. However, all other algorithms achieved good results in at least one or more images for this metric, with PSO and HS standing out. Finally, although intensity levels are not typically considered a metric in literature, we include them here as an observation measure that can lead to a more precise analysis. As we can see, in most images, the algorithm

**Table 1** Comparing contrast enhancement results between algorithms DE, HS, PSO, TLBO, ABC, FA, and IWO in images $I_1$ and $I_2$

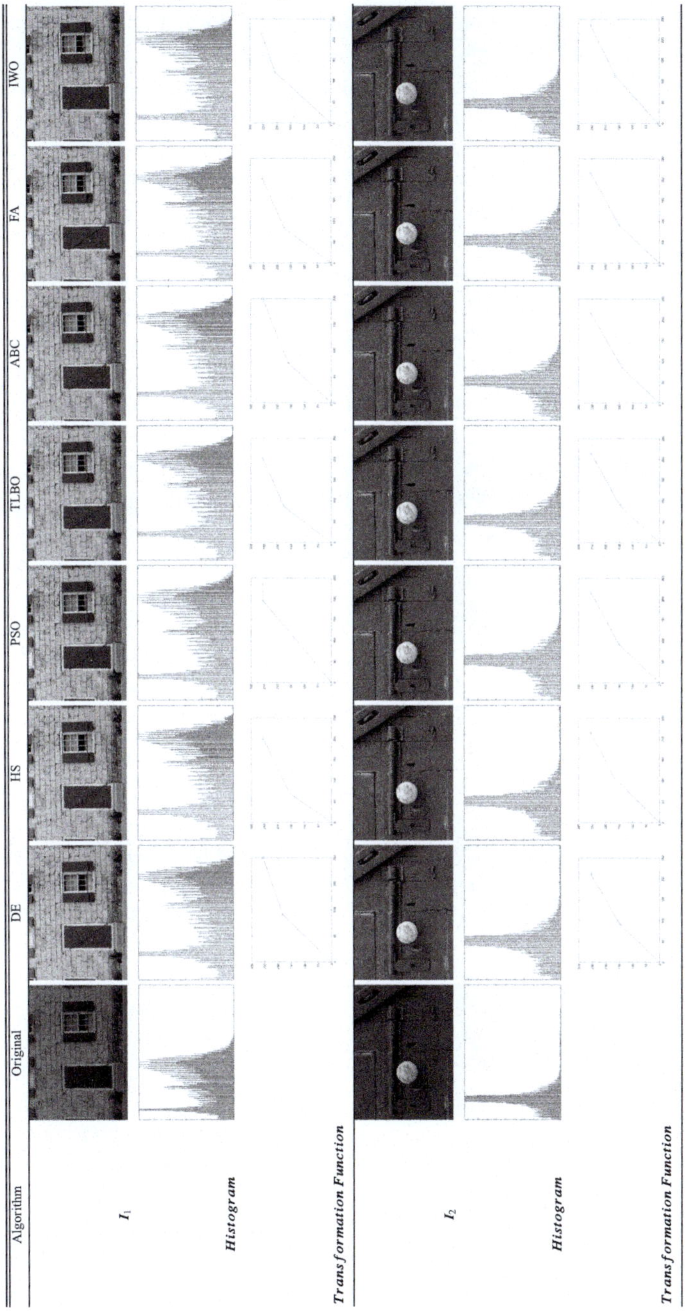

**Table 2** Comparing contrast enhancement results between algorithms DE, HS, PSO, TLBO, ABC, FA, and IWO in images $I_3$ and $I_4$

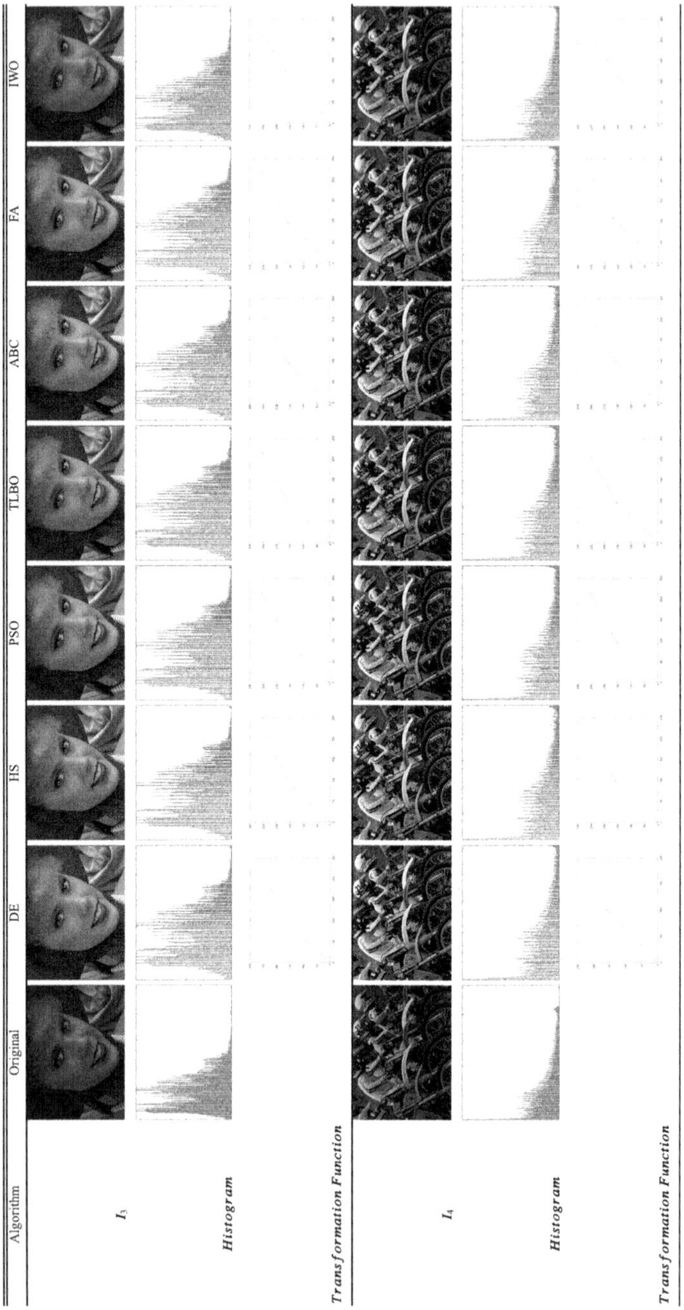

**Table 3** Comparing contrast enhancement results between algorithms DE, HS, PSO, TLBO, ABC, FA, and IWO in images $I_5$ and $I_6$

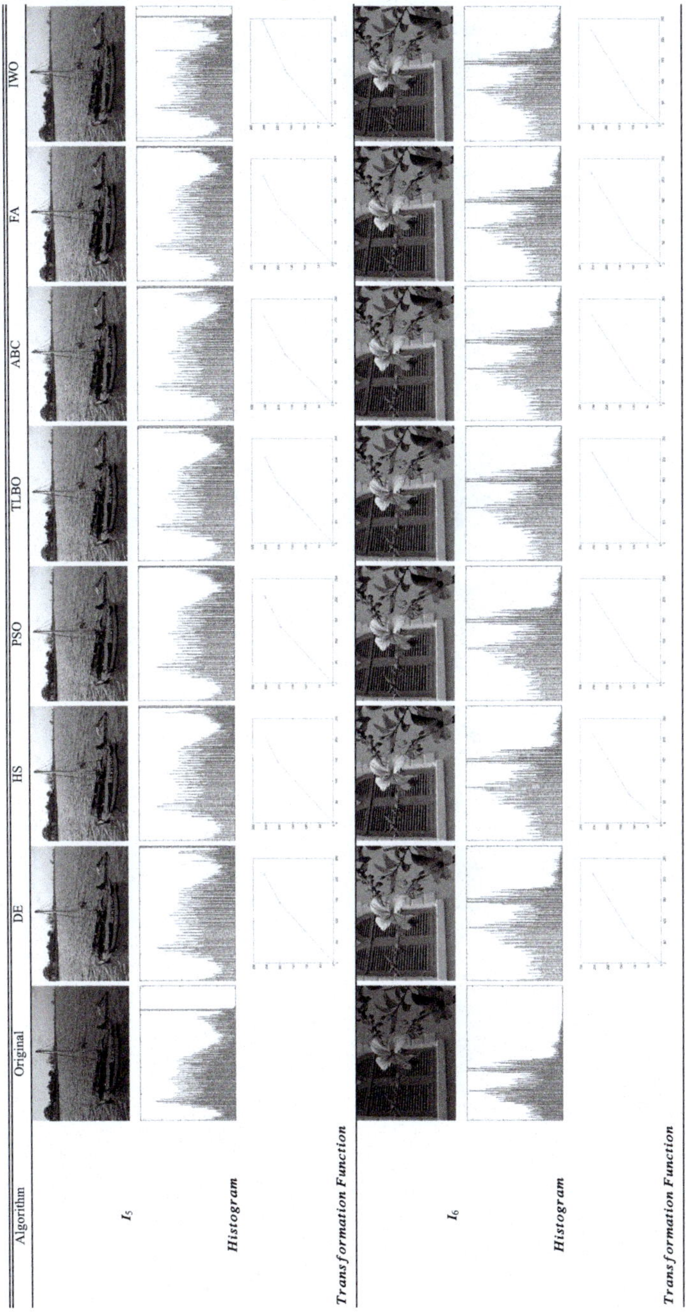

**Table 4** Comparing contrast enhancement results between algorithms DE, HS, PSO, TLBO, ABC, FA, and IWO in images $I_7$ and $I_8$

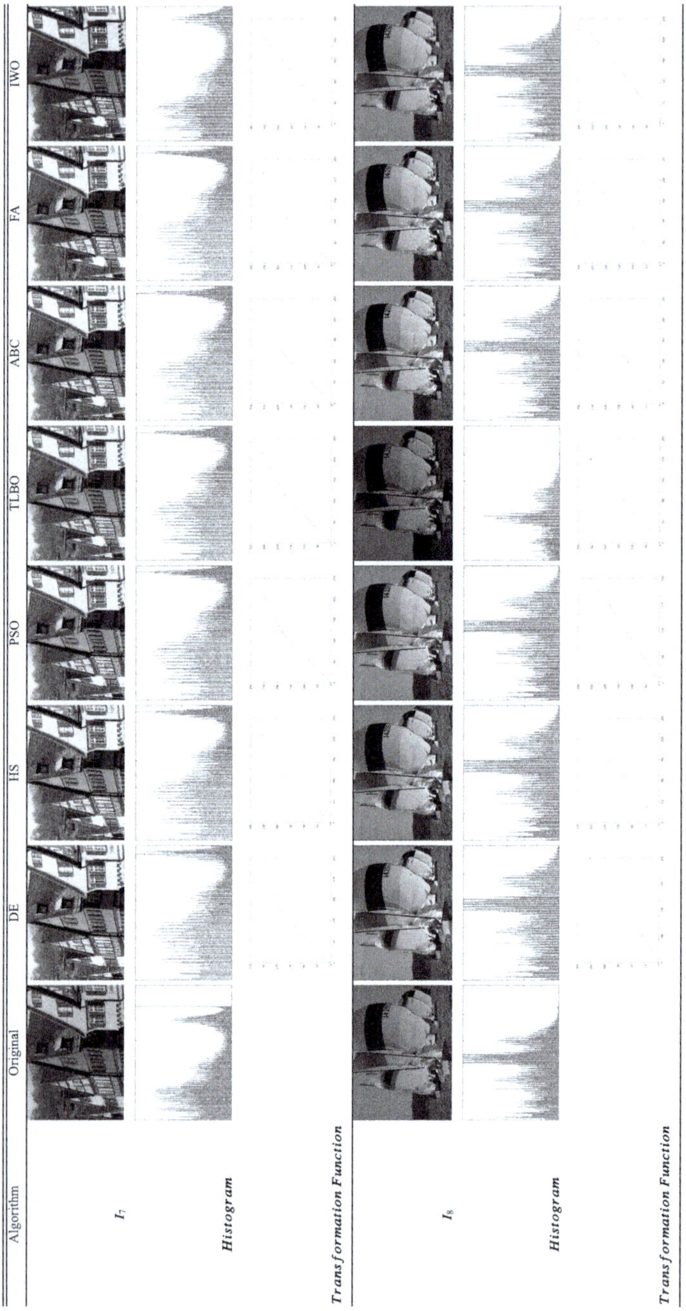

**Table 5** Comparing contrast enhancement results between algorithms DE, HS, PSO, TLBO, ABC, FA, and IWO in images $I_9$ and $I_{10}$

| Algorithm | Original | DE | HS | PSO | TLBO | ABC | FA | IWO |
|---|---|---|---|---|---|---|---|---|
| $I_9$ | 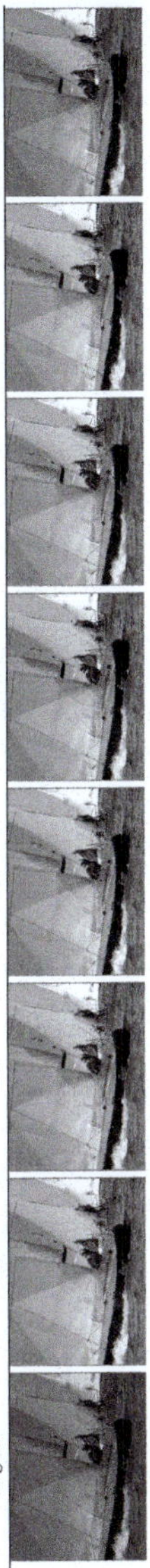 | | | | | | | |
| *Histogram* |  | | | | | | | |
| *Transformation Function* | 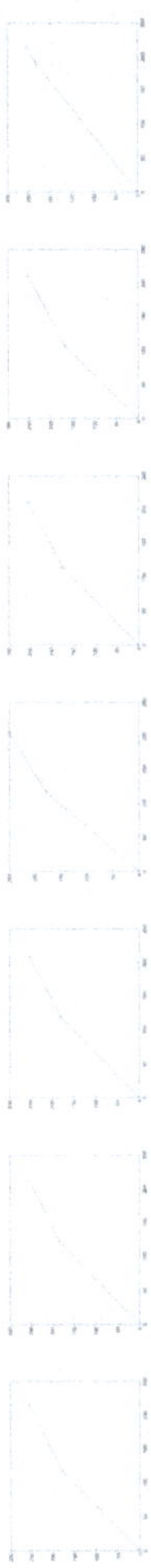 | | | | | | | |
| $I_{10}$ | 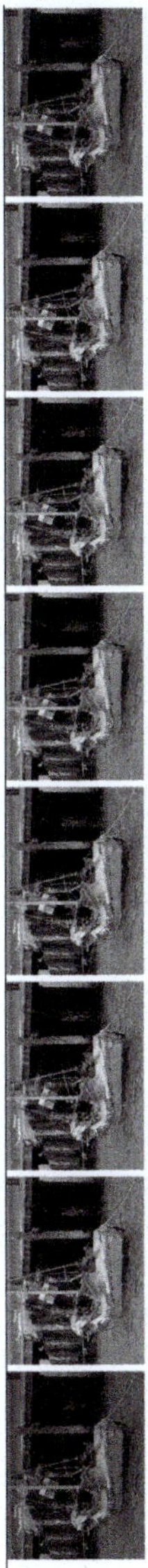 | | | | | | | |
| *Histogram* |  | | | | | | | |
| *Transformation Function* | 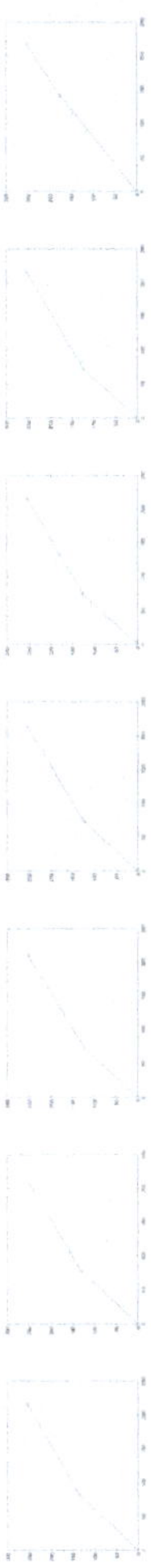 | | | | | | | |

**Table 6** Comparing contrast enhancement results between algorithms DE, HS, PSO, TLBO, ABC, FA, and IWO in images $I_{11}$ and $I_{12}$

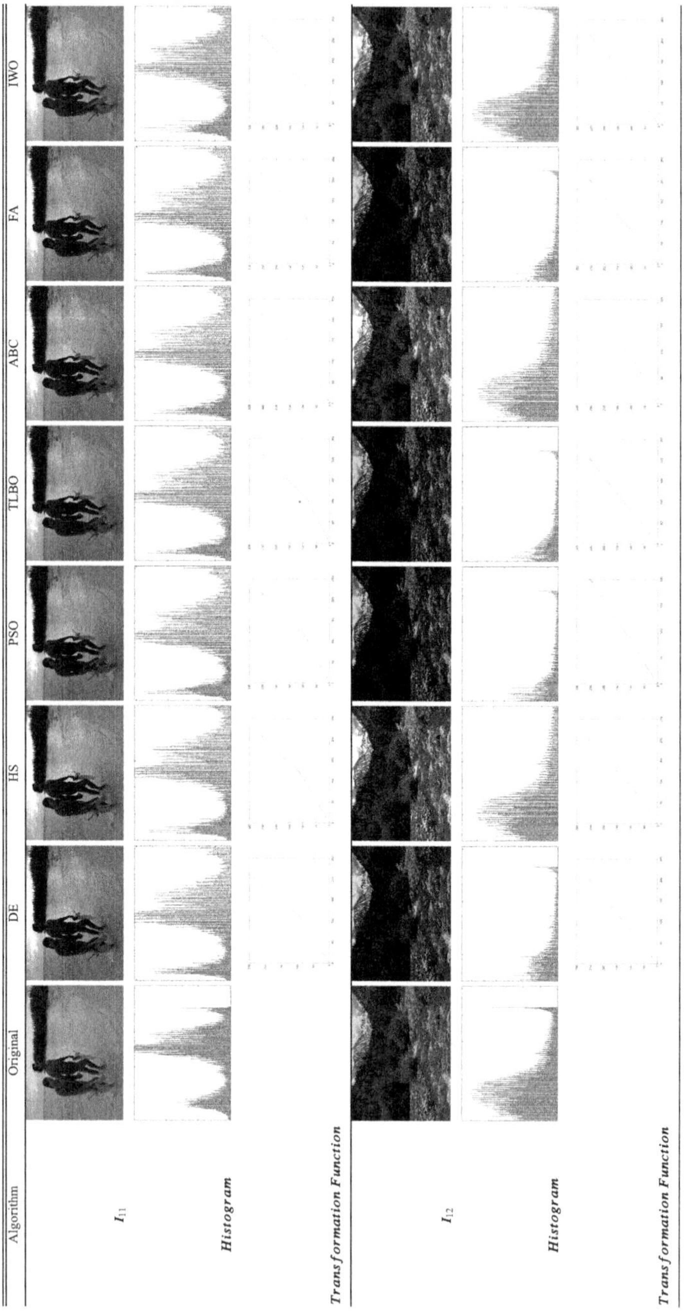

**Table 7** The RMS, FSIM, ENTROPY, PSNR, MSE, VIF and IFC metrics were used to evaluate the performance of DE, HS, PSO, TLBO, ABC, FA and IWO in image $I_1$ to $I_6$

| Image | Algorithm | RMS | FSIM | ENTROPY | PSNR | MSE | VIF | IFC | Intensity levels |
|---|---|---|---|---|---|---|---|---|---|
| $I_1$ | DE | 1.67545 | 0.99713 | 6.91084 | 10.59571 | 0.08718 | 2.44865 | 8.33022 | 153 |
| | HS | 1.63654 | 0.99649 | 6.92466 | 10.41412 | 0.09090 | 2.32997 | 7.95217 | 155 |
| | PSO | 1.71063 | **0.99889** | **6.93698** | 11.19215 | 0.07600 | **2.62353** | **10.57548** | **163** |
| | TLBO | 1.66043 | 0.99686 | 6.92283 | 10.60310 | 0.08703 | 2.39887 | 8.16257 | 155 |
| | ABC | **1.63526** | 0.99721 | 6.92867 | 10.88045 | 0.08165 | 2.34826 | 8.44588 | 157 |
| | FA | 1.66193 | 0.99662 | 6.92011 | 10.46525 | 0.08984 | 2.39533 | 8.02333 | 155 |
| | IWO | 1.65955 | 0.99832 | 6.90707 | **11.43764** | **0.07182** | 2.45452 | 9.48046 | 160 |
| $I_2$ | DE | 1.40731 | 0.99917 | 5.83171 | 17.57270 | 0.01749 | 2.04507 | 13.74280 | 195 |
| | HS | 1.41338 | 0.99915 | 5.83065 | 17.42253 | 0.01810 | 2.06899 | 13.92949 | 194 |
| | PSO | 1.41945 | 0.99899 | 5.83062 | 17.04063 | 0.01977 | **2.09542** | 13.34776 | 194 |
| | TLBO | 1.41845 | 0.99898 | 5.83062 | 17.04607 | 0.01974 | 2.09144 | 13.25428 | 194 |
| | ABC | 1.40551 | 0.99901 | 5.83089 | 17.32089 | 0.01853 | 2.04726 | 13.14773 | 195 |
| | FA | 1.41930 | 0.99899 | 5.83060 | 17.04097 | 0.01977 | 2.09516 | 13.34718 | 194 |
| | IWO | **1.28620** | **0.99940** | **5.83173** | **18.52657** | **0.01404** | 1.72425 | **14.55402** | **196** |
| $I_3$ | DE | 1.48711 | 0.99889 | 6.80768 | 15.34878 | 0.02918 | 1.96436 | 13.84648 | 193 |
| | HS | 1.49601 | 0.99901 | 6.80697 | 15.34665 | 0.02920 | 1.98497 | 13.40052 | 192 |
| | PSO | 1.51523 | 0.99904 | 6.80698 | 15.15578 | 0.03051 | 2.02292 | 13.55464 | 192 |
| | TLBO | 1.51077 | 0.99912 | 6.80757 | 15.31459 | 0.02941 | 2.01161 | 13.32613 | 192 |
| | ABC | 1.51259 | 0.99908 | 6.80718 | 15.21988 | 0.03006 | 2.01852 | 13.56643 | 192 |
| | FA | 1.51772 | 0.99904 | 6.80771 | 15.13860 | 0.03063 | **2.02882** | 13.67835 | 192 |
| | IWO | **1.32084** | **0.99985** | **6.81086** | **20.25398** | **0.00943** | 1.62466 | **15.33308** | **210** |
| $I_4$ | DE | 1.38596 | 0.99857 | 6.62483 | 17.16021 | 0.01923 | 1.80675 | 10.61991 | 195 |
| | HS | 1.38526 | 0.99817 | 6.63228 | 16.87352 | 0.02054 | 1.80546 | 9.97096 | 195 |
| | PSO | 1.38776 | 0.99811 | 6.63389 | 16.80723 | 0.02086 | **1.81096** | 9.93090 | 195 |
| | TLBO | 1.38590 | 0.99826 | 6.63394 | 16.93018 | 0.02028 | 1.80664 | 10.12149 | 195 |
| | ABC | 1.38605 | 0.99831 | 6.63350 | 16.95718 | 0.02015 | 1.80723 | 10.20332 | 195 |
| | FA | 1.38700 | 0.99820 | 6.63302 | 16.89195 | 0.02046 | 1.80939 | 10.14245 | 195 |
| | IWO | **1.28080** | **0.99960** | **6.65836** | **20.57288** | **0.00876** | 1.56800 | **13.30836** | **210** |
| $I_5$ | DE | 1.18729 | **0.99868** | **7.36846** | 15.44318 | 0.02855 | 1.62805 | **12.46057** | **200** |
| | HS | 1.18106 | 0.99828 | 7.35623 | 14.96809 | 0.03186 | 1.65012 | 11.86861 | 197 |
| | PSO | 1.18456 | 0.99834 | 7.36360 | 14.97780 | 0.03178 | 1.65159 | 11.90584 | 197 |
| | TLBO | 1.18129 | 0.99828 | 7.35626 | 14.96589 | 0.03187 | 1.65116 | 11.85915 | 197 |
| | ABC | 1.17672 | 0.99820 | 7.34694 | 14.97097 | 0.03183 | 1.63642 | 11.53537 | 198 |
| | FA | 1.18178 | 0.99807 | 7.35441 | 14.65096 | 0.03427 | **1.66475** | 11.46778 | 195 |
| | IWO | **1.06619** | 0.99851 | 7.34030 | **17.16816** | **0.01919** | 1.34340 | 11.79030 | 196 |
| $I_6$ | DE | 1.41865 | 0.99845 | **6.93439** | 15.00536 | 0.03158 | 1.93834 | 9.99646 | 195 |
| | HS | 1.42450 | 0.99836 | 6.93344 | 14.90496 | 0.03232 | **1.95503** | 10.03579 | 195 |
| | PSO | 1.41569 | 0.99850 | 6.93426 | 15.03880 | 0.03134 | 1.92793 | 10.05443 | 195 |
| | TLBO | 1.41749 | 0.99831 | 6.93331 | 14.90876 | 0.03229 | 1.93744 | 9.81991 | 195 |
| | ABC | 1.41236 | 0.99857 | 6.93378 | 15.17818 | 0.03035 | 1.91962 | 10.08167 | 195 |
| | FA | 1.41706 | 0.99830 | 6.93331 | 14.90249 | 0.03234 | 1.93695 | 9.79819 | 195 |
| | IWO | **1.38750** | **0.99858** | 6.93433 | **15.48597** | **0.02828** | 1.85532 | **10.17922** | 195 |

**Table 8** The RMS, FSIM, ENTROPY, PSNR, MSE, VIF and IFC metrics were used to evaluate the performance of DE, HS, PSO, TLBO, ABC, FA and IWO in image $I_7$ to $I_{12}$

| Image | Algorithm | RMS | FSIM | ENTROPY | PSNR | MSE | VIF | IFC | Intensity levels |
|---|---|---|---|---|---|---|---|---|---|
| $I_7$ | DE | 1.21927 | 0.99781 | 7.21826 | 15.49330 | 0.02823 | 1.44221 | 10.30456 | 195 |
| | HS | 1.22051 | 0.99778 | 7.23361 | 15.40914 | 0.02878 | 1.44326 | 10.22587 | 195 |
| | PSO | 1.22175 | 0.99785 | 7.23263 | 15.46069 | 0.02844 | **1.44583** | 10.31650 | 195 |
| | TLBO | 1.21969 | 0.99774 | 7.23054 | 15.38588 | 0.02893 | 1.44234 | 10.19998 | 195 |
| | ABC | 1.21972 | 0.99783 | 7.22703 | 15.46840 | 0.02839 | 1.44204 | 10.29185 | 195 |
| | FA | 1.21692 | 0.99765 | 7.23526 | 15.31324 | 0.02942 | 1.43671 | 10.05635 | 195 |
| | IWO | **1.21673** | **0.99997** | **7.30089** | **21.39436** | **0.00725** | 1.38812 | **21.21027** | **210** |
| $I_8$ | DE | 1.26163 | 0.99995 | 7.26230 | 20.06243 | 0.00986 | **1.50737** | 16.55905 | 201 |
| | HS | 1.24353 | 0.99994 | 7.26459 | 20.98585 | 0.00797 | 1.47609 | 16.05195 | 203 |
| | PSO | 1.24211 | 0.99995 | 7.26459 | 20.79008 | 0.00834 | 1.47145 | 16.42967 | 203 |
| | TLBO | **0.97344** | 0.99512 | 6.34411 | 17.51928 | 0.01770 | 0.97350 | 9.65425 | 171 |
| | ABC | 1.24389 | 0.99995 | 7.26459 | 20.77076 | 0.00837 | 1.47472 | 16.24298 | 203 |
| | FA | 1.25334 | 0.99995 | 7.26361 | 20.36937 | 0.00918 | 1.49265 | 16.32923 | 202 |
| | IWO | 1.16902 | **0.99997** | **7.27455** | **24.08563** | **0.00390** | 1.32545 | **16.82291** | **216** |
| $I_9$ | DE | 1.30601 | 0.99803 | 7.08433 | 13.76490 | 0.04203 | 1.63496 | 12.43842 | 195 |
| | HS | 1.30727 | 0.99805 | 7.08682 | 13.75961 | 0.04208 | 1.63727 | 12.90539 | 195 |
| | PSO | 1.30760 | 0.99808 | 7.08749 | 13.76251 | 0.04205 | 1.63815 | 13.04463 | 195 |
| | TLBO | 1.30801 | 0.99810 | 7.09219 | 13.75830 | 0.04209 | **1.63924** | 13.16578 | 195 |
| | ABC | 1.30359 | 0.99818 | 7.08901 | 13.89862 | 0.04075 | 1.62990 | 12.90714 | 196 |
| | FA | 1.30321 | 0.99792 | 7.09206 | 13.64181 | 0.04323 | 1.63602 | 13.48840 | 195 |
| | IWO | **1.22995** | **0.99968** | **7.12471** | **17.95338** | **0.01602** | 1.45449 | **14.48234** | **214** |
| $I_{10}$ | DE | 1.50583 | 0.99883 | 6.34596 | 16.41320 | 0.02284 | 1.85036 | 9.97311 | 196 |
| | HS | 1.51041 | 0.99880 | 6.34612 | 16.32165 | 0.02333 | **1.85918** | 9.95469 | 195 |
| | PSO | 1.50751 | 0.99876 | 6.34447 | 16.34817 | 0.02318 | 1.85294 | 9.96840 | 195 |
| | TLBO | 1.50899 | 0.99876 | 6.34544 | 16.32747 | 0.02329 | 1.85483 | 9.91502 | 195 |
| | ABC | 1.50872 | 0.99871 | 6.34520 | 16.29883 | 0.02345 | 1.85090 | 9.85092 | 195 |
| | FA | 1.51142 | 0.99865 | 6.34602 | 16.22832 | 0.02383 | 1.85300 | 9.84974 | 195 |
| | IWO | **1.27242** | **0.99987** | **6.35484** | **22.53132** | **0.00558** | 1.48320 | **14.61312** | **214** |
| $I_{11}$ | DE | **1.17334** | 0.99798 | 7.02898 | 24.34801 | 0.00367 | 1.32606 | 12.27295 | 195 |
| | HS | 1.22229 | 0.99837 | 7.00286 | 25.14764 | 0.00306 | **1.42163** | 12.37080 | 195 |
| | PSO | 1.17491 | 0.99799 | 7.03238 | 24.43815 | 0.00360 | 1.32931 | 12.38786 | 195 |
| | TLBO | 1.17477 | 0.99799 | 7.03378 | 24.41443 | 0.00362 | 1.32925 | 12.47661 | 195 |
| | ABC | 1.19579 | 0.99814 | 7.03809 | 24.96593 | 0.00319 | 1.36809 | 12.53900 | 195 |
| | FA | 1.17573 | 0.99798 | 7.03238 | 24.41890 | 0.00362 | 1.33058 | 12.33963 | 195 |
| | IWO | 1.20148 | **0.99903** | **7.07085** | **26.20683** | **0.00240** | 1.37376 | **13.95920** | **203** |
| $I_{12}$ | DE | 1.03831 | 0.99335 | 6.02867 | 19.93353 | 0.01015 | 1.10141 | 8.48214 | 177 |
| | HS | 1.21797 | 0.99930 | 7.13002 | 25.39646 | 0.00289 | 1.46681 | 15.94847 | 203 |
| | PSO | **0.98283** | 0.98850 | 5.65195 | 17.52170 | 0.01769 | 1.01424 | 6.55585 | 167 |
| | TLBO | 0.98463 | 0.98858 | 5.60312 | 17.54904 | 0.01758 | 1.01594 | 6.55784 | 169 |
| | ABC | 1.22801 | 0.99933 | 7.11859 | 24.89314 | 0.00324 | **1.48770** | 15.62918 | 201 |
| | FA | 1.00223 | 0.98980 | 5.64617 | 18.09432 | 0.01551 | 1.04444 | 6.99919 | 172 |
| | IWO | 1.16335 | **0.99969** | **7.25070** | **27.87581** | **0.00163** | 1.34210 | **17.78476** | **216** |

that achieves the best results in most metrics also has greater intensity levels. With this, we conclude that IWO is the best algorithm for producing the best-contrasted images in this experiment.

In the last experiment, the statistical calculations performed regarding the best solution, worst solution, mean, and standard deviation will be analyzed to verify which algorithm offers a suitable solution for contrast enhancement. For this experiment, 30 independent executions were conducted to obtain 30 solutions for each test image. The results are shown in Table 10.

As can be observed, similar to the previous experiment, the lower value represents the best solution. Thus, IWO achieves effective performance concerning the objective function. For the worst solution, some very poor statistical results can be found in images $I_4$, $I_5$, $I_8$, $I_{10}$, and $I_{12}$. This is due to the penalty applied to the objective function in the algorithms' programming to prevent the methods from diverging from the search space. This penalty is weighted with a very large value, which affects both the median and the standard deviation. However, it can be shown that the algorithms are consistent with minimal variation because the statistics of the other test images show low differences. Nonetheless, as observed in the best solution results, despite encountering a solution that is not as good, they manage to steer the algorithm back towards a very good solution.

### *4.3 Convergence Analysis*

In the third experiment, convergence graphs were obtained; the results are shown in Table 9. For practical purposes, only the results of images $I_1$ to $I_4$ and the behavior of the DE, HS, PSO, TLBO, ABC, FA, and IWO algorithms in each are shown.

The study and analysis of convergence in metaheuristic algorithms is a fundamental pillar, as it involves meticulously analyzing the algorithm's progress toward optimal or satisfactory solutions over time. This process encompasses a series of crucial aspects, from stopping criteria to the solution trajectory itself, encompassing diversification and intensification, revealing its ability to adapt and evolve in the face of the challenges posed by the problem.

Before analyzing the convergence of the algorithms, it is worth noting that although this is a contrast maximization problem, all algorithms were adapted as minimization problems for practical programming reasons. Therefore, in this case, lower results represent the best solution. Analyzing Table 9, we can observe that for image $I_1$, the PSO algorithm shows good convergence to an optimal result. This aligns with the analysis of the first and second experiments, where the PSO algorithm visually exhibits a more distributed histogram than the other algorithms. Additionally, in calculating metrics, this same image has the majority of the best results obtained, indicating that this algorithm has a good capacity for convergence to offer good solutions. However, reviewing the following images $I_2$ to $I_4$, we can observe that the IWO algorithm exhibits rapid convergence until reaching the final optimal value.

**Table 9** Results of the convergence of the DE, HS, PSO, TLBO, ABC, FA, and IWO algorithms on images $I_1$ and $I_4$

| Algorithm | DE | HS | PSO | TLBO | ABC | FA | IWO |
|---|---|---|---|---|---|---|---|
| $I_1$ |  |  |  |  |  |  |  |
| $I_2$ |  |  |  |  |  |  |  |
| $I_3$ |  |  |  |  |  |  |  |
| $I_4$ |  |  |  |  |  |  |  |

**Table 10** Statistical results, best solution, worst solution, average, and standard deviation of each algorithm in 30 executions each for images $I_1$ to $I_{12}$

| Image | Algorithm | Best | Worst | Mean | Std | Image | Algorithm | Best | Worst | Mean | Std |
|---|---|---|---|---|---|---|---|---|---|---|---|
| $I_1$ | DE | 1.00010 | 0.99584 | 0.99862 | 0.00162 | $I_7$ | DE | 1.01019 | 1.00819 | 1.00921 | 0.00086 |
| | HS | 0.99863 | 0.98792 | 0.99343 | 0.00404 | | HS | 1.01086 | 1.00710 | 1.00864 | 0.00173 |
| | PSO | 0.99941 | 0.98915 | 0.99380 | 0.00426 | | PSO | 1.01071 | 1.00759 | 1.00899 | 0.00136 |
| | TLBO | 1.00079 | 1.00019 | 1.00053 | 0.00023 | | TLBO | 1.01073 | 1.00920 | 1.01004 | 0.00066 |
| | ABC | 0.99772 | 0.99772 | 0.99772 | 0.00000 | | ABC | 1.01033 | 1.01033 | 1.01033 | 0.00000 |
| | FA | 1.00089 | 0.99583 | 0.99885 | 0.00273 | | FA | 1.01121 | 1.00767 | 1.01037 | 0.00151 |
| | IWO | 0.98545 | 0.94549 | 0.96344 | 0.01499 | | IWO | 1.00236 | 0.84779 | 0.93862 | 0.07428 |
| $I_2$ | DE | 0.91719 | 0.91564 | 0.91654 | 0.00056 | $I_8$ | DE | 0.94500 | 0.94488 | 0.94496 | 0.00006 |
| | HS | 0.91802 | 0.91618 | 0.91714 | 0.00070 | | HS | 0.94511 | 0.94483 | 0.94499 | 0.00012 |
| | PSO | 0.91840 | 0.91354 | 0.91559 | 0.00228 | | PSO | 0.94517 | 0.94508 | 0.94514 | 0.00004 |
| | TLBO | 0.91836 | 0.91708 | 0.91794 | 0.00052 | | TLBO | 0.94621 | 0.94497 | 0.94531 | 0.00051 |
| | ABC | 0.91751 | 0.91751 | 0.91751 | 0.00000 | | ABC | 0.94512 | 0.94512 | 0.94512 | 0.00000 |
| | FA | 0.91846 | 0.91832 | 0.91839 | 0.00005 | | FA | 0.94525 | 0.94517 | 0.94520 | 0.00003 |
| | IWO | 0.86901 | 0.80917 | 0.83708 | 0.02608 | | IWO | 0.94476 | 10000.0 | 1999.30 | 4472.53 |
| $I_3$ | DE | 0.90480 | 0.90439 | 0.90457 | 0.00018 | $I_9$ | DE | 0.93399 | 0.93284 | 0.93327 | 0.00045 |
| | HS | 0.90505 | 0.90071 | 0.90306 | 0.00159 | | HS | 0.93402 | 0.93246 | 0.93342 | 0.00063 |
| | PSO | 0.90544 | 0.90377 | 0.90502 | 0.00070 | | PSO | 0.93409 | 0.93384 | 0.93401 | 0.00010 |
| | TLBO | 0.90545 | 0.90434 | 0.90500 | 0.00046 | | TLBO | 0.93418 | 0.93291 | 0.93365 | 0.00057 |
| | ABC | 0.90514 | 0.90514 | 0.90514 | 0.00000 | | ABC | 0.93372 | 0.93372 | 0.93372 | 0.00000 |
| | FA | 0.90565 | 0.90547 | 0.90557 | 0.00007 | | FA | 0.93423 | 0.93418 | 0.93419 | 0.00002 |
| | IWO | 0.90084 | 0.81477 | 0.87904 | 0.03618 | | IWO | 0.92884 | 0.77858 | 0.86204 | 0.06182 |
| $I_4$ | DE | 0.97651 | 0.96783 | 0.97436 | 0.00368 | $I_{10}$ | DE | 0.96405 | 0.96110 | 0.96272 | 0.00110 |
| | HS | 0.97870 | 0.97713 | 0.97808 | 0.00066 | | HS | 0.96429 | 0.95971 | 0.96282 | 0.00187 |
| | PSO | 0.97867 | 0.97562 | 0.97744 | 0.00159 | | PSO | 0.96446 | 0.96268 | 0.96357 | 0.00064 |
| | TLBO | 0.97860 | 0.97563 | 0.97670 | 0.00121 | | TLBO | 0.96458 | 0.96319 | 0.96375 | 0.00056 |
| | ABC | 0.97836 | 0.97836 | 0.97836 | 0.00000 | | ABC | 0.96449 | 0.96449 | 0.96449 | 0.00000 |
| | FA | 0.97899 | 0.97863 | 0.97878 | 0.00018 | | FA | 0.96495 | 0.96490 | 0.96493 | 0.00002 |
| | IWO | 0.96650 | 13999.2 | 2799.15 | 6261.04 | | IWO | 0.94417 | 10000.0 | 1999.29 | 4472.53 |
| $I_5$ | DE | 0.99973 | 9999.01 | 1999.00 | 4472.14 | $I_{11}$ | DE | 0.89488 | 0.89252 | 0.89393 | 0.00091 |
| | HS | 1.00048 | 0.99904 | 0.99984 | 0.00058 | | HS | 0.89439 | 0.88980 | 0.89259 | 0.00198 |
| | PSO | 1.00047 | 1.00017 | 1.00033 | 0.00015 | | PSO | 0.89537 | 0.89511 | 0.89525 | 0.00010 |
| | TLBO | 1.00042 | 1.00001 | 1.00024 | 0.00019 | | TLBO | 0.89529 | 0.89507 | 0.89516 | 0.00011 |
| | ABC | 1.00001 | 1.00001 | 1.00001 | 0.00000 | | ABC | 0.89475 | 0.89475 | 0.89475 | 0.00000 |
| | FA | 1.00083 | 1.00066 | 1.00074 | 0.00007 | | FA | 0.89537 | 0.89522 | 0.89534 | 0.00006 |
| | IWO | 0.92966 | 0.85167 | 0.88344 | 0.03070 | | IWO | 0.89193 | 0.71380 | 0.84789 | 0.07545 |
| $I_6$ | DE | 0.99503 | 0.93064 | 0.98110 | 0.02825 | $I_{12}$ | DE | 0.99389 | 0.98885 | 0.99055 | 0.00209 |
| | HS | 0.99464 | 0.98940 | 0.99173 | 0.00246 | | HS | 0.98925 | 0.98880 | 0.98899 | 0.00019 |
| | PSO | 0.99584 | 0.99456 | 0.99524 | 0.00047 | | PSO | 1.00075 | 0.98902 | 0.99148 | 0.00518 |
| | TLBO | 0.99672 | 0.99431 | 0.99562 | 0.00114 | | TLBO | 0.99956 | 0.98850 | 0.99125 | 0.00466 |
| | ABC | 0.99479 | 0.99479 | 0.99479 | 0.00000 | | ABC | 0.98927 | 0.98927 | 0.98927 | 0.00000 |
| | FA | 0.99687 | 0.99664 | 0.99673 | 0.00012 | | FA | 1.00038 | 0.98949 | 0.99168 | 0.00486 |
| | IWO | 0.98641 | 0.77996 | 0.91611 | 0.08554 | | IWO | 0.98699 | 13999.2 | 2799.07 | 6261.07 |

From this, we can deduce that the IWO method is the best optimization algorithm that adapts very well to the contrast technique applied.

## 5 Conclusions and Future Work

This chapter investigated the effectiveness of several optimization algorithms for contrast enhancement in grayscale images. Differential Evolution (DE), Harmony Search (HS), Particle Swarm Optimization (PSO), Teaching-Learning Based Optimization (TLBO), Artificial Bee Colony (ABC), Firefly Algorithm (FA), and Invasive Weed Optimization (IWO) were evaluated.

The comparative study employed a common image enhancement technique and a standardized objective function to ensure a fair comparison between the algorithms. The performance of each algorithm was evaluated using eight different metrics to measure the quality of the contrasted images.

The comprehensive tests detailed the results by showing and identifying the best algorithms proposed to solve this problem based on the evaluation metrics and convergence analysis. The analysis also provided indications of the strengths and weaknesses of each optimization algorithm selected for image contrast enhancement.

As a reference for future researchers in this field, we present the ranking of the three best algorithms to solve the contrast enhancement problem according to the experimental results obtained: the IWO is in first place, the DE algorithm is in second place, and the PSO is in third place. The indicators used to measure the quality of optimization suggest that these three algorithms are suitable to solve this class of problems.

Future work is also envisioned to involve exploring alternative image enhancement techniques, incorporating new optimization algorithms, or investigating the impact of parameter tuning on the performance of these algorithms.

**Acknowledgements** If you want to include acknowledgments of assistance and the like at the end of an individual chapter please use the `acknowledgement` environment—it will automatically be rendered in line with the preferred layout.
**Competing Interests** Please declare any competing interests in the context of your chapter. The following sentences can be regarded as examples.
This study was funded by [X] [grant number X].
[Author A] has a received research grant from [Company W].
[Author B] has received a speaker honorarium from [Company X] and owns stock in [Company Y].
[Author C] is a member of [committee Z].
The authors have no conflicts of interest to declare that are relevant to the content of this chapter.
**Ethics Approval** If your chapter includes primary studies with humans please declare the adherence of ethical standards. Example text: This study was performed in line with the principles of the Declaration of Helsinki. Approval was granted by the Ethics Committee of University B (Date.../ No. ...).
In addition, for human participants, authors are required to include a statement that informed consent (to participate and/or to publish) was obtained from individual participants or parents/guardians if the participant is minor or incapable.

If animals are studied, authors should make sure that the legal requirements or guidelines in the country and/or state or province for the care and use of animals have been followed or specify that no ethics approval was required.

## References


1. Li, Q., Nie, J., Qu, S.: A small target detection algorithm in infrared image by combining multi-response fusion and local contrast enhancement. Optik **241**, 166919 (2021)
2. Mamatov, N., Niyozmatova, N., Jalelova, M., Samijonov, A., Tojiboyeva, S.X.: Methods for improving contrast of agricultural images. In: E3S Web of Conferences, vol. 401, p. 04020. EDP Sciences (2023)
3. Liu, Z., Zhuang, Y., Jia, P., Wu, C., Xu, H., Liu, Z.: A novel underwater image enhancement algorithm and an improved underwater biological detection pipeline. J. Mar. Sci. Eng. **10**(9), 1204 (2022)
4. Rahman, T., Khandakar, A., Qiblawey, Y., Tahir, A., Kiranyaz, S., Abul Kashem, S.B., Islam, M.T., Al Maadeed, S., Zughaier, S.M., Khan, M.S., Chowdhury, M.E.: Exploring the effect of image enhancement techniques on Covid-19 detection using chest x-ray images. Comput. Biol. Med. **132**, 104319 (2021)
5. Rao, G.S., Srikrishna, A.: Contrast enhancement of poor-quality satellite images through morphological operations. Traitement du Signal **38**(3) (2021)
6. Lin, S.C.-F., Wong, C.Y., Rahman, M.A., Jiang, G., Liu, S., Kwok, N., Shi, H., Yu, Y.-H., Wu, T.: Image enhancement using the averaging histogram equalization (avheq) approach for contrast improvement and brightness preservation. Comput. Electri. Eng. **46**, 356–370 (2015)
7. Sim, K.S., Tso, C.P., Tan, Y.Y.: Recursive sub-image histogram equalization applied to gray scale images. Pattern Recognit. Lett. **28**(10), 1209–1221 (2007)
8. Tan, S.F., Isa, N.A.M.: Exposure based multi-histogram equalization contrast enhancement for non-uniform illumination images. IEEE Access **7**, 70842–70861 (2019)
9. Chen, S.-D., Ramli, A.R.: Minimum mean brightness error bi-histogram equalization in contrast enhancement. IEEE Trans. Consum. Electron. **49**(4), 1310–1319 (2003)
10. Mousania, Y., Karimi, S., Farmani, A.: Optical remote sensing, brightness preserving and contrast enhancement of medical images using histogram equalization with minimum cross-entropy-otsu algorithm. Opt. Quantum Electron. **55**(2), 1–22 (2023)
11. Mukhopadhyay, S., Hossain, S., Malakar, S., Cuevas, E., Sarkar, R.: Image contrast improvement through a metaheuristic scheme. Soft Comput. **27**(18), 13657–13676 (2023)
12. Luque-Chang, A., Cuevas, E., Pérez-Cisneros, M., Fausto, F., Valdivia-González, A., Sarkar, R.: Moth swarm algorithm for image contrast enhancement. Knowl.-Based Syst. **212**, 106607 (2021)
13. Malik, S., Akram, T., Awais, M., Khan, M.A., Hadjouni, M., Elmannai, H., Alasiry, A., Marzougui, M., Tariq, U.: An improved skin lesion boundary estimation for enhanced-intensity images using hybrid metaheuristics. Diagnostics **13**(7), 1285 (2023)
14. Pashaei, E., Pashaei, E.: A fusion approach based on black hole algorithm and particle swarm optimization for image enhancement. Multimed. Tools Appl. **82**(1), 297–325 (2023)
15. Khan, A.H., Ahmed, S., Bera, S.K., Mirjalili, S., Oliva, D., Sarkar, R.: Enhancing the contrast of the grey-scale image based on meta-heuristic optimization algorithm. Soft Comput. **26**(13), 6293–6315 (2022)
16. Elewi, A., Kahveci, S., Avaroğlu, E.: Image contrast enhancement using a low-discrepancy population initialized gray wolf optimization algorithm. Multimed. Tools Appl. 1–22 (2023)
17. Storn, R., Price, K.: Minimizing the real functions of the icec'96 contest by differential evolution. In: Proceedings of IEEE International Conference on Evolutionary Computation, pp. 842–844. IEEE (1996)

18. Pant, M., Zaheer, H., Garcia-Hernandez, L., Abraham, A., et al.: Differential evolution: a review of more than two decades of research. Eng. Appl. Artif. Intell. **90**, 103479 (2020)
19. Geem, Z.W.: Harmony search algorithm for solving sudoku. In: Knowledge-Based Intelligent Information and Engineering Systems: 11th International Conference, KES 2007, XVII Italian Workshop on Neural Networks, Vietri sul Mare, Italy. Proceedings, Part I 11, pp. 371–378. Springer (2007)
20. Geem, Z.W.: Recent Advances in Harmony Search Algorithm (2010)
21. Kennedy, J., Eberhart, R.: Particle swarm optimization. In: Proceedings of ICNN'95-International Conference on Neural Networks, vol. 4, pp. 1942–1948. IEEE (19950
22. Rao, R.V., Savsani, V.J., Vakharia, D.P.: Teaching-learning-based optimization: a novel method for constrained mechanical design optimization problems. Comput.-Aided Des. **43**(3), 303–315 (2011)
23. Karaboga, D., et al.: An idea based on honey bee swarm for numerical optimization," technical report, Technical report-tr06, Erciyes university, engineering faculty, computer (2005)
24. Karaboga, D., Gorkemli, B., Ozturk, C., Karaboga, N.: A comprehensive survey: artificial bee colony (abc) algorithm and applications. Artif. Intell. Rev. **42**, 21–57 (2014)
25. Yang, X.-S.: Nature-Inspired Metaheuristic Algorithms. Luniver Press (2010)
26. Fister, I., Fister, I., Jr., Yang, X.-S., Brest, J.: A comprehensive review of firefly algorithms. Swarm Evol. Comput. **13**, 34–46 (2013)
27. Mehrabian, A.R., Lucas, C.: A novel numerical optimization algorithm inspired from weed colonization. Ecol. Inf. **1**(4), 355–366 (2006)
28. Gupta, R., Sehgal, S.: Gray-scale image enhancement using differential evolution optimization algorithm. In: 2016 International Conference on Information Technology (InCITe)-The Next Generation IT Summit on the Theme-Internet of Things: Connect your Worlds, pp. 103–108. IEEE (2016)
29. Sarangi, P., Mishra, B., Majhi, B., Dehuri, S.: Gray-level image enhancement using differential evolution optimization algorithm. In: 2014 International Conference on Signal Processing and Integrated Networks (SPIN), pp. 95–100. IEEE (2014)
30. Ponomarenko, N., Jin, L., Ieremeiev, O., Lukin, V., Egiazarian, K., Astola, J., Vozel, B., Chehdi, K., Carli, M., Battisti, F., Jay Kuo, C.-C.: Image database tid2013: peculiarities, results and perspectives. Signal Process. Image Commun. **30**, 57–77 (2015)
31. Zhang, L., Zhang, L., Mou, X., Zhang, D.: Fsim: a feature similarity index for image quality assessment. IEEE Trans. Image Process. **20**(8), 2378–2386 (2011)
32. Poobathy, D., Chezian, R.M.: Edge detection operators: peak signal to noise ratio based comparison. IJ Image, Graph. Signal Process. **10**, 55–61 (2014)
33. Sheikh, H.R., Bovik, A.C.: Image information and visual quality. IEEE Trans. Image Process. **15**(2), 430–444 (2006)
34. Sheikh, H.R., Bovik, A.C., De Veciana, G.: An information fidelity criterion for image quality assessment using natural scene statistics. IEEE Trans. Image Process. **14**(12), 2117–2128 (2005)

# Grouping and Partitioning Methods in Metaheuristic Algorithms

**Luis A. Beltran, Mario A. Navarro-Velázquez, Hector Escobar-Cuevas, Nahum Juda Aguirre, Mohamed Abd Elaziz, and Laith Abualigah**

**Abstract** Metaheuristic algorithms have gained popularity in addressing various optimization challenges in diverse fields such as digital image processing, energy, machine learning, robotics, and data analytics. Most metaheuristic methods are population-based, where a population of search agents (or individuals) explores different candidate solutions within a solution space. These optimization frameworks offer several advantages, including the interaction between individuals, which is crucial for effectively exploring the search space and overcoming local optima. Metaheuristic algorithms must balance exploring the search space for promising new solutions and exploiting regions containing high-quality solutions. Clustering

L. A. Beltran (✉) · M. A. Navarro-Velázquez · H. Escobar-Cuevas · N. J. Aguirre
Centro Universitario de Ciencias Exactas e Ingenierias CUCEI, Av. Revolucion 100, Guadalajara, Mexico
e-mail: luis.beltran7882@alumnos.udg.mx

M. A. Navarro-Velázquez
e-mail: mario.navarro@academicos.udg.mx

H. Escobar-Cuevas
e-mail: hector.escobar@academicos.udg.mx

N. J. Aguirre
e-mail: Nahum.aguirre7431@alumnos.udg.mx

M. A. Elaziz
Department of Mathematics, Faculty of Science, Zagazig University, Zagazig, Egypt

L. Abualigah
Hourani Center for Applied Scientific Research, Al-Ahliyya Amman University, Amman 19328, Jordan

Artificial Intelligence and Sensing Technologies (AIST) Research Center, University of Tabuk, Tabuk 71491, Saudi Arabia

MEU Research Unit, Middle East University, Amman 11831, Jordan

Department of Electrical and Computer Engineering, Lebanese American University, Byblos 13-5053, Lebanon

Applied Science Research Center, Applied Science Private University, Amman 11931, Jordan

 
D. Oliva et al. (eds.), *Advances in Optimization Algorithms for Multidisciplinary Engineering Applications: From Classical Methods to AI-Enhanced Solutions*, Studies in Computational Intelligence 806, https://doi.org/10.1007/978-3-031-78440-8_3

and partitioning help divide the search space into smaller, more manageable regions, allowing effective exploration and intensive exploitation of promising regions. These techniques can be employed in a variety of metaheuristic algorithms. This chapter provides a comprehensive review of clustering and partitioning techniques within metaheuristic algorithms. Its purpose is to provide a detailed analysis of these techniques, identify their applications, advantages, and limitations, and facilitate their understanding and use in optimization problems.

## 1 Introduction

Optimization uses techniques and methods to solve quantitative problems and find the best solutions. This concept is common in scientific and engineering challenges, where resource efficiency is critical to finding optimal results [1]. Traditional optimization techniques are based on gradient information and use derivatives to determine the optimal solution. However, this approach has limitations in complex problems with many locally optimal solutions since the problem must be unimodal and these problems must be doubly derivable, which is rarely the case in engineering problems, causing traditional gradient methods to stagnate without finding the global optimal solution [2]. In recent years, metaheuristic algorithms have gained popularity as tools for solving a wide range of optimization problems in many application areas such as digital image processing, energy, machine learning, robotics, data analysis and others applications. Various metaheuristic algorithms have been used to resolve global optimization problems. Some examples of algorithms may be covariance matrix adaptation evolution strategy (CMA-ES) [3], differential evolution (DE) [4], self-adaptive differential evolution (JADE) [5], linear population size reduction with adaptive differential evolution (L-SHADE) [6], particle swarm optimization (PSO) [7], biogeography-based optimization (BBO) [8], differential search algorithm (DSA) [9], genetic algorithm (GA) [10], artificial bee colony (ABC) [11], harmony search (HS) [12], among many others.

Most metaheuristic algorithms mentioned above are methods belonging to the class of Evolutionary Algorithms (EA) that use operators such as mutation, crossover and selection on a population of individuals containing candidate solutions in order to iterative evolve the population to converge the most promising areas of the search space but without neglecting the exploration of the search space. Therefore, a satisfactory balance between exploration and exploitation of the search space must be achieved to produce a competitive set of candidate solutions [13]. Clustering and partitioning technologies can divide the search space into smaller and more manageable regions. This procedure promotes an effective exploration and intensive exploitation in promising regions. These techniques can be employed in a variety of metaheuristic algorithms, such as genetic algorithms (GA) [10], simulated annealing (SA) [14], tabu search (TS) [15], particle swarm optimization (PSO) [7], among many others.

Clustering consists of dividing a data set into groups or clusters, where the elements within each group are more similar than elements in other groups. In the context of metaheuristic algorithms, clustering can help reduce the complexity of the search space by grouping similar solutions. This facilitates exploration and exploitation of the search space by focusing on smaller, homogeneous regions. Partitioning involves dividing elements into disjointed subsets so that each element belongs to exactly one subset. In the context of metaheuristic algorithms, partitioning divides the search space into smaller, more manageable regions. This can facilitate the exploration of the search space by allowing algorithms to focus on specific subsets of solutions.

This chapter presents a comprehensive review of clustering and partitioning techniques within metaheuristic algorithms. The remainder of the chapter is structured as follows: Sect. 2 presents a short description of optimization concepts and explains how optimization algorithms are divided, Sect. 3 discusses the main clustering techniques and shows optimization algorithms using the techniques described above and Sect. 4 discusses the main partitioning techniques.

## 2 Metaheuristic Algorithms for Optimization

Most scientists and engineers in all disciplines frequently confront classic search and optimization problems. Optimization consists of finding the most appropriate solution to a problem within the constraints and flexibilities that compose it. When we optimize the performance of a system, our goal is to find a set of system parameter values for which the overall system performance is simply the best concerning the constraints addressed by the problem. Typically, the parameters controlling the system performance are represented in a vector as shown in the Eq. 1:

$$\overrightarrow{X} = \{x_1, x_2, x_3, \ldots, x_D\} \tag{1}$$

An objective function (or fitness function) is designed to measure the "best" performance of the system. This function takes the parameter vector as an argument. It returns a value that allows us to evaluate the quality of the parameters (or solutions to the problem) to ensure that the system's response is optimal. The best parameter vector is the one for which the value of the objective function is either minimized or maximized, depending on whether it is a minimization or maximization problem. Therefore, the optimization techniques consist in the search for the vector of parameters $\overrightarrow{X}^*$, which minimizes (or maximizes) an objective function $f(\overrightarrow{X})$.

An optimization algorithm can be implemented to solve an optimization problem (minimization or maximization). Many optimization algorithms in the literature can be classified according to their approach and characteristics as shown in the Fig. 1.

Modern optimization algorithms can be classified as deterministic or probabilistic. If an algorithm works without any random nature, it is called deterministic. These algorithms will always arrive at the same final solution if we start with the same starting point. Tabu search and hill climbing are good examples of deterministic algorithms. On the other hand, if there is some randomness in the algorithm, it will

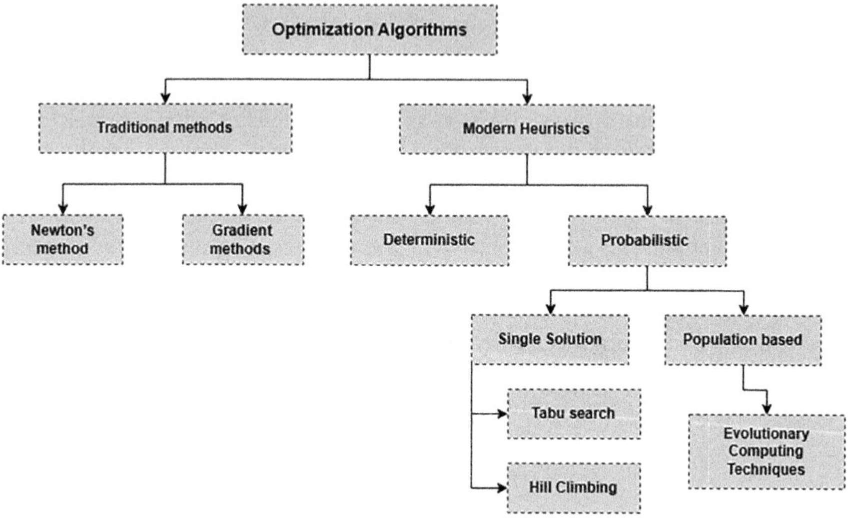


**Fig. 1** Taxonomy of different optimization methods

usually reach a different point each time it is run, even if the same starting point is used. Genetic algorithms and PSO are good examples of stochastic algorithms that belong to evolutionary computing techniques.

## *2.1 Evolutionary Computing Techniques*

For years, computer scientists have looked to nature for inspiration as both a model and a metaphor for tackling complex search problems, as well as many other challenging computational tasks. A new approach in computational intelligence has been developed by looking carefully at the relationship between optimization and biological evolution: evolutionary computation techniques, which allow for extremely complex searches and optimizations.

Evolutionary computation involves an iterative process similar to a population's natural growth or evolution, which undergoes a random search directed by parallel processing to reach a specific goal. The mechanisms of biological evolution inspire these methods. Evolutionary computing emerged in the 1950s when the idea of applying Darwinian principles to automated problem-solving was first proposed. Since the early 1990s, these techniques have been unified as different representations of a single technology, known as evolutionary computation. Also, in the 1990s, a fourth stream emerged that followed similar ideas: Genetic Programming. At present, evolutionary computation techniques are mainly based on metaheuristic optimization algorithms such as:

1. Evolutionary algorithms.
   Some examples include genetic algorithms, evolutionary programming, evolutionary strategy, genetic programming and differential evolution.
2. Swarm intelligence.
   Examples include ant colony optimization (ACO) and particle swarm optimization (PSO).

Metaheuristic algorithms (MA) are important tools for solving complex optimization problems. They are powerful tools that do not use gradient information from the objective function. MAs work independently from the problem, which makes them flexible and widely used in engineering and other applications [16]. In the related literature it is possible to find different MAs [17]; they have different operators that allow to search for the global optimum. Such operators have two roles: (1) to explore the search space and (2) to exploit the prominent regions. MA can be based on a single solution or population-based. The population-based MA have become more popular because they take advantage of a set of solutions to search for the optimal solution. In general, an MA has four main steps: Initialization, Evaluation, Update, and Selection.

## 3 Clustering Methods

### *3.1 Hierarchical Clustering Methods*

Several clustering problems can be solved with traditional clustering techniques due to their implementation of partitioning procedures. Nevertheless, complex problems require not only the partition of the objects but also understand the relationship among the partitions. This perspective can be reached by the hierarchical clustering methodology. Hierarchical clustering implements two different types of classification known as exclusive and inclusive classifications [18]. In an exclusive classification, the partitions are grouped by their linear ordering relation. This sequence information will be the only non-hierarchical classification storage. This procedure can be exemplified with the population's academic degrees in a university. Several university students of different areas could be working together in an academic project. However, this student group is subordinated to a group of Ph.D. researchers. Another example can be found in Nature, where several animals are grouped according to their characteristics. Nevertheless, according to the food chain, some groups prevail to others generating a subordinating scheme. On the other hand, Inclusive classifications generates small groups that compound bigger groups until all the elements are grouped. However, an element can be part of two or more groups according to their hierarchical value. This technique is widely applied in field such as biology in which a taxonomy rank needs to be created [19].

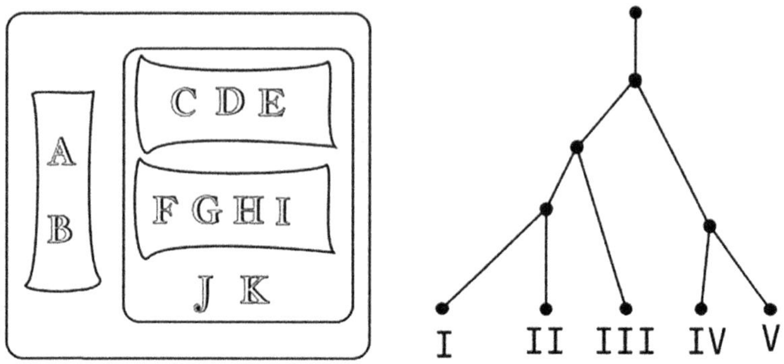


**Fig. 2** Graphical schemes to represent hierarchical clustering

Hierarchical clustering can be drawn in different forms such as a system of contour lines as can be observed in Fig. 2. However, the understanding and inclusion of the existent relationships between the partition or clusters can be difficult to show due to only the topological connections are expressed. To address this problem, a graphical representation know as dendrogram is implemented. A dendrogram is a diagram or graphical representation in which the characteristic distance between each pair of sequentially merged groups is plotted. This representation allows the correctly interpretation and understanding of the relationships between certain similar groups. Contrary to the traditional diagrams, dendrograms not only shown the relationship between the clusters but also express the relationship such as the distance or the similarity numerically. Commonly, the vertical axis of the dendrogram represent the hierarchical level while the horizontal axis represents the existing clusters. An example of dendrograms can be found in Fig. 3.

There are three different methods to create dendrograms: Single-Link, Complete-Link, and Average-Link. In Single-Link clustering, we consider the distance between

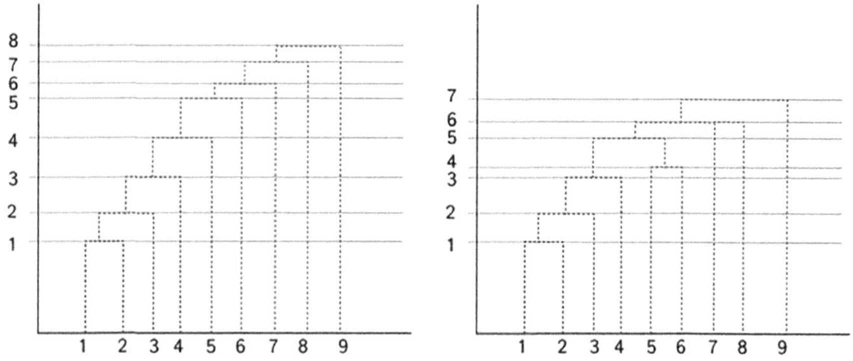

**Fig. 3** Examples of dendrograms

one cluster and another equal to the shortest distance from any member of one cluster to any member of the other. If the data consists of similarities, we consider the similarity between one cluster and another equal to the greatest similarity from any member of one cluster to any member of the other. In Complete-Link clustering, we consider the distance between one cluster and another equal to the greatest distance from any member of one cluster to any member of the other. On the other hand, in Average-Link clustering, the distance between two clusters is calculated as the average distance of any member from one cluster to any member from the other cluster. This method differs from the Single-Link and Complete-Link methods regarding cluster shape. The Single-Link method creates elongated clusters, while the Complete-Link method usually results in more compact clusters. The Average-Link method, on the other hand, creates clusters that are somewhere in between the two previous methods.

Hierarchical algorithms have two fundamental advantages: they do not require the number of classes to be specified beforehand and are independent of the initial conditions. However, there are some drawbacks to hierarchical methods. Firstly, they can be computationally expensive, making them unsuitable for large datasets. Secondly, they are static, meaning that once data points are assigned to a cluster, they cannot move to another cluster. Lastly, they may need more information about the global shape or size of the clusters to separate overlapping clusters. Nevertheless, hierarchical algorithms have proven to be effective in resolving different kinds of problems. In hierarchical clustering, the data is not immediately divided into distinct clusters. Instead, a series of partitions occur, ranging from a single cluster containing all objects to k clusters, each containing a single object. Hierarchical clustering is divided into two types of methods: agglomerative and divisive. Agglomerative methods fuse n objects into groups, while divisive methods separate n objects into finer groupings. Agglomerative techniques are more commonly used. These schemes are applied in several field due to their impressive performance [20].

Zhang et al. [21] have developed a widely-used hierarchical clustering algorithm known as CF-tree. This algorithm partitions incoming data points in an incremental and dynamic manner, utilizing a height-balanced tree structure that stores clustering features. CF-tree relies on two important parameters: branching factor $B$ and threshold $T$. The threshold $T$ defines the maximum diameter of a cluster, ensuring that the radius of each cluster is less than $T$. The developed methodology was capable of finding a high-quality clustering with just one scan of the data, and can further improve its accuracy with a few additional scans. Notably, the proposed scheme is also the first clustering algorithm that is able to effectively handle noise. Another interesting work was proposed by Guha et al. [20, 21] in this method, each cluster is represented by certain points generated by selecting well-dispersed points and then shrinking them towards the cluster centroid by a specified fraction. The algorithm employs a mix of random sampling and partition clustering to manage large databases effectively. This approach is a robust clustering algorithm for Boolean and categorical data. The study also introduces two novel concepts: neighbors of a point and links. The algorithm utilizes these concepts to measure the similarity or proximity between pairs of data points.

## 3.2 Hierarchical Clustering Implementations with Metaheuristic Approach

In response to the growing complexity of real-world optimization problems, a multitude of optimization methodologies have emerged. Classical approaches based on gradient-based techniques remain popular due to their ability to solve problems quickly. However, limitations arise when dealing with unimodal and twice-differentiable objective functions, necessitating the development of new alternatives. Metaheuristic approaches implements exploration and exploitation strategies to identify optimal solutions. Therefore, Metaheuristic approaches consists of a robust and effective methodologies for optimization problems and their application in various areas.

Metaheuristic Algorithms also can be categorized as physics-based or bio-inspired physics-based metaheuristic methods can be distinguished by the use of physics, mathematics, and chemistry principles and processes. The Covariance Matrix Adaptation Evolution Strategy (CMAES) or States of Matter Search (SMS) are examples of this category. On the other hand, Bio-inspired metaheuristic algorithms have been characterized to simulate biological principles and mechanisms. Most algorithms fall into this category. These techniques can be grouped into evolutionary, swarm intelligence, and immune system techniques. The Artificial Bee Colony (ABC) or Particle Swarm Optimization algorithm (PSO) are some examples of bio-inspired metaheuristic algorithms. Metaheuristic algorithms have proven to be more effective than classical techniques in various real-world applications. However, despite their efficiency and robustness, these algorithms still exhibit certain drawbacks that hinder the accurate detection of promising areas within the search space, as well as the refinement of the best candidate solutions.

From this perspective, Metaheuristic algorithms and Hierarchical clustering methods are often combined in various ways, particularly in optimization and pattern recognition. The integration of these two approaches can be found in literature in five distinct forms: optimization with metaheuristics, application of metaheuristics to clustering, hybrid approaches, algorithmic diversity, and objective function optimization, as can be observed in Fig. 4. Optimization with metaheuristics involves using the hierarchical clustering method to determine the best way to group data points based on a defined similarity or distance metric. Optimization with metaheuristics is particularly useful when dealing with large datasets where an exhaustive search is impractical. In such scenarios, metaheuristic approaches effectively navigate the solution space and discover approximations to the optimal clustering.

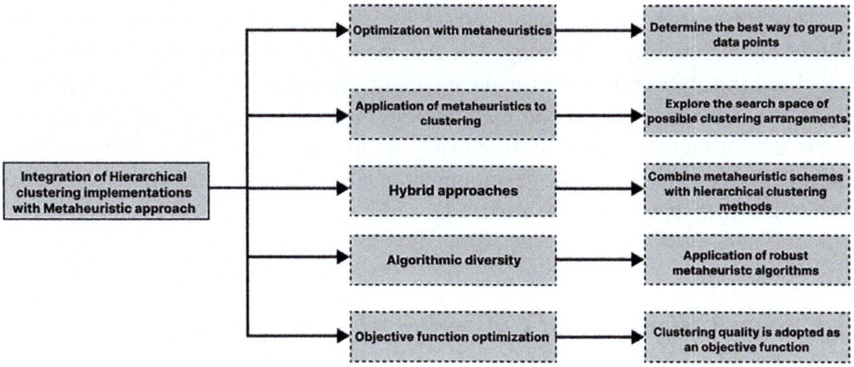


**Fig. 4** Taxonomy reviewed of integration of hierarchical clustering and metaheuristic approaches

On the other hand, applying metaheuristics to clustering involves using metaheuristic algorithms to explore the search space of possible clustering arrangements to find the ones that optimize a given criterion, such as minimizing or maximizing the inter-cluster distance. Hybrid approaches combine metaheuristic schemes with hierarchical clustering methods. This integration can determine the initial clustering, refine the results obtained from traditional hierarchical clustering, or optimize the linkage criteria. Algorithmic diversity introduces the idea of applying metaheuristic methodologies to problems where hierarchical clustering methods have limitations, such as being sensitive to initial conditions or getting stuck in local minima. In such cases, metaheuristic approaches can overcome these limitations by providing a robust search mechanism. Finally, in objective function optimization, the clustering quality is adopted as an objective function, and metaheuristic algorithms can search for clustering parameters that enhance these metrics.

In order to address drawbacks that hinder the accurate detection of promising areas within the search space, from the metaheuristic algorithms, it is beneficial to combine different metaheuristic algorithms to identify and refine solutions. These combinations are known as hybrid methodologies and are commonly used in literature. Hybrid algorithms can overcome the limitations of individual algorithms by implementing two or more methodologies. The algorithms used in hybrid schemes are chosen based on their ability to promote a global search strategy. However, the combination of these methodologies lacks a mechanism to properly regulate, combine, or integrate them to fully utilize their advantages. A powerful combination of metaheuristic algorithms and hierarchical clustering methods is utilized to tackle intricate optimization problems, particularly those involving the identification of optimal or near-optimal cluster structures within datasets. The metaheuristic algorithms are adept at searching and optimizing, leading to enhanced clustering outcomes.

This subsection reviews two methodologies that integrate metaheuristic algorithms and hierarchical clustering. The first methodology proposes a new hierarchical clustering algorithm based on metaheuristic algorithms for software modularization [22]. Similarly, the second methodology proposes an improved metaheuristic approach for hierarchical clustering using the quartet method [23]. These methods are analyzed to understand the development of novel ideas that incorporate important methodologies such as metaheuristic algorithms and hierarchical clustering.

## 3.3 Novel Hierarchical Clustering Algorithm Based on Metaheuristic Algorithms for Software Modularization

According to the structural refactoring literature, there is currently no search-based algorithm available that uses a hierarchical approach for modularization. To address this gap, a new approach called TDHC has been proposed. TDHC employs both global and local search strategies to modularize the system in a top-down hierarchical clustering approach [24]. The algorithm outputs a tree structure, wherein each node represents an artifact composed of all artifacts in its subtrees. These nodes are potential software modules or clusters. The tree structure helps software maintainers get a better understanding of the source code structure, allowing them to decide on appropriate compositions of artifacts to create modules such as files, packages, and components. Experimental results exhibit that TDHC produces modularization closer to the decomposition done by human experts than other existing algorithms.

A revolutionary genetic algorithm that employs a cutting-edge encoding and fitness function to achieve a highly effective modularization is proposed int his work. By constructing a tree from the source code's artifacts and using a branch and bound approach to determine appropriate levels in the tree, the encoding and fitness functions work in tandem to deliver unparalleled results. Furthermore, a hill-climbing algorithm [25] to optimize the quality of the modularization is applied. This scheme takes an artifact dependency graph (ADG) constructed from source code as input and provides a recommended modularization for software maintainers as output. The reviewed approach supports "big-bang" modularization, which considers all software system artifacts for modularization without considering the current structure. The proposed scheme is reviewed in three procedures such as encoding, evaluation, and neighboring search as can be analyzed in the following subsections.

### 3.3.1 Encoding

In genetic algorithms (GA) [26], a chromosome is a collection of parameters representing a potential solution to a given problem. GA aims to find the best possible chromosome, which will, in turn, represent an optimal or near-optimal solution. The parameters that make up a chromosome can take many forms, such as a binary string or other data structures. In this study, the Prufer sequence encoded a tree as a sequence of numbers, which then served as a chromosome. The Prufer sequence is a one-to-one mapping between a sequence of numbers and a labeled tree. Algorithm 1 outlines the steps in constructing Prufer numbers for a given tree. Likewise, the Prufer Sequence is constructed according to Algorithm 2.

**Algorithm 1** Encoding a labeled tree to Prufer sequence procedure

Imput: $T(V, E)$
$PS \leftarrow Empty\_List$
**while** $size(V) > 2$ **do**
    $V_i \leftarrow SmallNumber(V)$
    $V_J \leftarrow Conection(V_i)$
    $V \leftarrow Remove(V, V_i)$
    $E \leftarrow Remove(E, V_i, V_J)$
    $PS \leftarrow Add(PS, V_j)$
**end while**
Output: $PS$

### 3.3.2 Evaluation

Each chromosome needs to be evaluated to determine the quality of solutions in a GA population. Applying a new quality function to assess the chromosomes is suggested in this context. The developed quality function calculates the fitness of a chromosome by examining the dependencies between modules extracted from its corresponding tree. To analyze the structure of a tree and suggest a modularization approach, the breadth-first search (BFS) algorithm can be used starting from the root. During the traversal, if the sum of the proposed fitness function for two child nodes is greater than or equal to the proposed fitness function of their parent node, these child nodes will be added to the processing queue. However, if the sum of the proposed fitness function for the child nodes is less than the proposed fitness function of their parent node, then that parent node is the optimal location to create modules in the leaves of the subtree. In Algorithm 3 the pseudocode for evaluating the custom genetic algorithm is analyzed. To assess the tree and propose modularization based on its structure, the breadth-first search (BFS) algorithm is utilized to traverse the tree from the root node. As the traversal takes place, if the proposed fitness function sum for two child nodes is greater than or equal to the proposed fitness function of that node, they are included in the process queue. If not, that node serves as the optimal location to create artifacts in the leaves of the subtree, producing a module.

**Algorithm 2** Converting Prufer sequence to correspond tree procedure

Imput: $PS$
$D \leftarrow List$
**for** $i \in P$ **do**
$D_i \leftarrow D_i + 1$
**end for**
$D_i \leftarrow 1$
$T \leftarrow Empty\_List$
$ID \leftarrow Non\_Zero(D)$
**while** $ID > 3$ **do**
$V \leftarrow Small\_item(D)$
$D_i \leftarrow D_i - 1$
$T \leftarrow Add(V, A_i)$
$i \leftarrow i + 1$
**end while**
$A, B \leftarrow Non\_Zero(D)$
Output: $LT, T, PS$

### 3.3.3 Neighboring Search

A genetic algorithm is a specialized search technique that seeks to identify solutions to complex problems. To ensure optimal results, it is incorporate a local search approach called hill-climbing at the conclusion of the genetic algorithm process. This strategy involves analyzing adjacent solutions in an effort to find improved outcomes. The steepest ascent method is utilized to pinpoint the best possible neighbor solution. Each potential solution is assessed for quality, and the process continues until no further improvements can be made. The key to the success of this strategy is properly defining the relevant neighborhood, which varies based on the specific problem at hand.

**Algorithm 3** Chromosome evaluation procedure

Imput: $Ch$
$Ch\_fit \leftarrow 0$
$Tr \leftarrow Decode\_tree(Ch)$
$Tr\_root \leftarrow 1$
$Q \leftarrow Push(Tr\_root)$
**while** $Q \neq Empty$ **do**
$Pt \leftarrow Pop(Q)$
$C_1, C_2 \leftarrow Children(Tr)$
$C_1\_cf \leftarrow exCF(C_1, Pt)$
$C_2\_cf \leftarrow exCF(C_2, Pt)$
**if** $C_1\_cf + C_2\_cf \geq Pt\_cf$ **then**
$Q \leftarrow Push(C_1)$
$Q \leftarrow Push(C_2)$
**else**
$Ch\_fit \leftarrow Ch\_fit + Pt\_cf$
**end if**
**end while**
Output: $Ch\_fit$

The maintenance and evolution of software often require changes to the software's structure, necessitating the need for source code refactoring. This is a vital process for ensuring the software's efficiency. The reviewed scheme proposes a new clustering algorithm that takes into account the cohesion and coupling between artifacts. The algorithm uses a top-down hierarchical approach and a metaheuristic algorithm that combines genetic algorithm and hill-climbing. By analyzing the ADG input, which is independent of the source code's programming language, The algorithm suggests an optimal starting point for modularizing artifacts. It's worth noting that while the refactoring method is automatic and efficient, design decisions can often be more intricate than simply maximizing cohesion and minimizing coupling during the modularization process.

## *3.4 Improved Metaheuristic Approach for Hierarchical Clustering Implemented the Quartet Method*

The quartet method is a novel approach to hierarchical data grouping. It involves taking a set of n objects and their respective differences, and then constructing a tree comprising the best quartet combinations possible. The ultimate aim is to ensure that the quartet topologies embedded in the tree have the lowest possible sum of differences, which is a challenging task also known as the minimum quartet tree cost (MQTC) problem. To tackle this problem, a straightforward heuristic method that utilizes matrices to represent solution structures more efficiently is proposed in this work. Likewise, shortcuts for identifying all feasible solutions in a given configuration, as well as a means of merging highly connected objects to minimize the quartet function cost in a partial solution are generated. The reviewed scheme is analyzed generally in three interesting implementations such as greedy constructive heuristics, reduced variable neighborhood search, and Simulated annealing.

### 3.4.1 Greedy Constructive Heuristics

The proposed greedy constructive heuristic [27] aims to generate initial solutions of high quality, achieved in a timely manner, which can significantly enhance the overall performance of the MQTC heuristic that is later used to solve the problem. The purpose of this approach is to provide a fast and efficient method for producing starting solutions that are of good quality, thus increasing the chances of finding optimal solutions to complex problems. By utilizing this heuristic, businesses and academics alike can benefit from a more streamlined and effective process for problem-solving

### 3.4.2 Reduced Variable Neighborhood Search (RVNS)

The goal behind the implementation of reduced variable neighborhood search [28] is to modify the measurement used in each step to allow the search direction to better accommodate the current shape of the function. If random points are selected from the current area being explored and there is no descent, the RVNS method is used. Instead, the values of these new points are compared to the current best and updated if there is an improvement.

### 3.4.3 Simulated Annealing

Simulated annealing (SA) is a method that differs from classical descent methods in that it utilizes a gradual cooling process to decrease the likelihood of accepting suboptimal solutions as the search progresses [29]. At each step, the SA algorithm selects a neighboring solution, evaluates its quality, and determines whether to transition to it or remain with the current solution based on two probabilities. If the new solution is superior to the current one, the probability of transitioning to it remains at one. If it is inferior, the probability is gradually decreased towards 0 as the temperature parameter T decreases from an initial positive value to 0 (cooling schedule). At the same time, the probability of transitioning to a better solution remains at 1 or is increased to a positive value.

### 3.4.4 Applications Results

In the reviewed investigation is introduced an advanced metaheuristics for the quartet method of clustering, a novel hierarchical clustering technique based on the minimum quartet tree cost (MQTC) problem. This problem is known to be NP-hard, and its goal is to derive an optimal tree from the various possible quartet topology combinations of input objects n. Optimality refers to the minimum summation of dissimilarities of consistent quartet topologies. A new basic greedy heuristic that runs at a remarkable speed is analyzed. However, this method was inferior to some well-known solution schemes found in literature, such as a Reduced Variable Neighborhood Search (RVNS). Hence, simulated Annealing (SA) approach was implemented to significantly enhance the performance of the proposed method. This was accomplished by using it to construct initial high-quality solutions rather than random ones.

## 4 Partitional Clustering Algorithms

Algorithms for partitional clustering make an effort to split the dataset up into several distinct groups. By allocating clusters to peaks in the probability density function, for example, or by focusing on the global structure, they aim to optimize a specific criterion function that may highlight the local structure of the data. The global criteria often involve maximizing the dissimilarity across distinct clusters while reducing some measure of dissimilarity in the samples within each cluster. The partitional algorithms' drawbacks are the hierarchical algorithms' benefits, and vice versa [30].

Although multiple distinct clusters may be formed internally among the various algorithms, only one set is created. The user must provide the appropriate number "$k$" of clusters because only one set of clusters is output. Additionally, the quality of any suggested solution is assessed using a criterion function or metric. The average distance between clusters or other metrics could be used to calculate this quality indicator. Throughout the literature, there are a lot of partitional algorithms for diverse applications. Nevertheless, the most popular and representative methods, such as are reviewed.

## *4.1 Squared Error Clustering Algorithm*

The squared error clustering procedure minimizes the squared error. The total Euclidean distances between each cluster member and the cluster centroid, $C_k$, is the squared error for that cluster. Let $\{t_{i1}, t_{i2}, \ldots, t_{im}\}$ be the set of things mapped to a given cluster $K_i$. The definition of the squared error is as shown in the Eq. 2.

$$s.e._{ki} = \sum_{j=1}^{k} \|t_i j - C_k\|^2 \tag{2}$$

The squared error for a set of clusters $K = \{K_1, K_2, \ldots, K_k\}$ is defined as follows:

$$s.e._k = \sum_{j=1}^{k} s.e._{Kj} \tag{3}$$

The following steps give the implementation of this algorithm:

1. Define the inputs $D = \{t_1, t_2, \ldots, t_n\}$ as the set of elements and $k$ as the desired cluster number.
2. Define the output $K$ as the set of clusters.
3. Place every item $t_i$ to a cluster.
4. Put each item $t_i$ in the cluster whose center is closest to it.
5. Generate a new center for every cluster.
6. Calculate squared error.
7. Repeat steps 4 to 6 until the difference between consecutive squared errors falls below a predetermined level.

Every tuple in the squared error algorithm is allocated to the cluster whose center is closest to iterate. Considering that there are $n$ items and $k$ clusters, the operation is $O(kn)$. This algorithm becomes $O(tkn)$ given $t$ iterations. Since the distance between every item is not used, an adjacency matrix may not be necessary, resulting in a space need of merely $O(n)$. This algorithm is very simple to implement and computationally efficient, however this simplicity makes it sensitive to outliers, so that the groups formed can be affected by these values, and the effectiveness of the Euclidean distance is affected in higher dimensions.

## *4.2 K-Means Algorithm*

Proposed by Hugo Steinhaus in 1956 and published by Edward W. Forgy in 1965, the k-means clustering algorithm is a partition-based cluster analysis method that divides n elements into $k$ clusters. Each element in a cluster is a prototype of the cluster and belongs to the cluster with the closest mean, also known as the cluster centroid or cluster center. It can, therefore, be considered a particular kind of squared error algorithm, even though the convergence criteria don't have to be established using the squared error. It is possible to attain both a high degree of similarity and a high degree of dissimilarity between elements in separate clusters simultaneously [31].

K-means algorithm aims at minimizing the ICS (Intra cluster Spread), which for $k$ cluster centers can be defined as shown in the Eq. 4.

$$ICS\left(C_1, C_2, \ldots, C_K\right) = \sum_{j=1}^{K} \sum_{\overrightarrow{Z_i} \in C_j} d^2\left(\overrightarrow{Z_i}, \overrightarrow{m_j}\right) \tag{4}$$

where $\overrightarrow{m_j}$ represents the centroid of the $j$-th cluster $C_j$ and $\overrightarrow{Z_i}$ is a data point belonging to cluster $C_j$.

The k-means algorithm begins with $k$ centroids that are predetermined or randomly chosen. The nearest cluster center is then assigned to each pattern in the data set. Finally, the mean of the related patterns is used to update the centroids. Until a certain halting criterion is satisfied, the process is repeated. For k-means, the membership and weight functions are defined as shown in the Eq. 5.

$$u\left(\overrightarrow{m_j} | \overrightarrow{Z_i}\right) = \begin{cases} 1 & \text{if } d^2\left(\overrightarrow{Z_i}, \overrightarrow{m_j}\right) = argmin_j \left\{d^2\left(\overrightarrow{Z_i}, \overrightarrow{m_j}\right)\right\} \\ 0 & \text{otherwise} \end{cases}$$
$$\text{and} \tag{5}$$
$$W(\overrightarrow{z_i}) = 1$$

We have then that this algorithm is described according to the following steps to be followed [32, 33]:

1. Select $k$ objects as initial cluster centers $\left(\overrightarrow{m_1}, \ldots, \overrightarrow{m_j}\right)$
2. Calculate the distance between each object and each cluster center, then assign each object to the nearest cluster, with the formulas for calculating distance as shown in the Eq. 6.

$$d\left(\overrightarrow{Z_i}, \overrightarrow{m_i}\right) = \sqrt{\sum_{j=1}^{d}\left(\overrightarrow{Z_i}, \overrightarrow{m_j}\right)} \tag{6}$$

3. Calculate the mean of objects in each cluster as the new cluster centers as shown in the Eq. 7.

$$m_i = \frac{1}{N_i}\sum_{j=1}^{N_i}\overrightarrow{Z_{ij}} \tag{7}$$

where $N_i$ is the number of samples of the current cluster $i$.

This algorithm has two major advantages for which it has gained popularity in clustering [33]:

1. It is quite simple to use.
2. It is suitable for huge data sets due to low time complexity.

But, despite its simplicity, this method suffers from several problems:

1. The user needs to indicate how many classes they want ahead of time.
2. The algorithm's performance depends on the data.
3. The algorithm relies on the initial conditions and takes a greedy approach. K-means frequently converge to sub-optimal solutions as a result.

### *4.3 Fuzzy C-Means Algorithm*

The fuzzy c-means (FCM) algorithm, also referred to as the fuzzy k-means algorithm at times, was developed by James Bezdek as a fuzzy variant of the k-means algorithm [34]. In fuzzy clustering, it appears to be the most widely used algorithm. A fuzzy expansion of the least-square error criteria serves as the foundation for the FCM. To evolve the correct cluster centers in the standard FCM algorithm, a within-cluster sum function $J_q$ is minimized:

$$\overrightarrow{m_i} = \frac{\sum_{j=1}^{n}(u_{ij})^q\overrightarrow{Z_J}}{\sum_{j=1}^{n}(u_{ij})^q} \tag{8}$$

$\overrightarrow{Z_i}$ is the $i$th d-dimensional data vector, $u_{ij}$ is the membership of the $i$th data point to the $j$th cluster, and $\overrightarrow{m_i}$ is the $j$th cluster center. Any real number that affects the membership grade is represented here by $q(q > 1)$. Increasing $q's$ value further muddies the algorithm. The following expression can be used to get the cluster centers $\overrightarrow{m_j}$ of k classes for $i = 1$ to $k$:

$$\overrightarrow{m_i} = \frac{\sum_{j=1}^{n} (u_{ij})^q \overrightarrow{X_J}}{\sum_{j=1}^{n} (u_{ij})^q} \tag{9}$$

The following relationship may now be found by differentiating the performance criterion with regard to $\overrightarrow{m_i}$ and with respect to $u_{ij}$ and setting them to zero:

$$u_{ij} = \left[ \sum_{p=1}^{k} \left( \frac{\| \overrightarrow{Z_i} + \overrightarrow{m_j} \|^2}{\| \overrightarrow{Z_i} + \overrightarrow{m_p} \|^2} \right)^{1/(q-1)} \right]^{-1} \tag{10}$$

FCM generally outperforms the k-means algorithm, particularly in cases where there is natural cluster overlap [35]. But just like with k-means, we also have to state upfront how many classes are possible. Furthermore, FCM might reach a few local maxima within its fitness landscape [33].

## *4.4 PAM Algorithm*

The PAM (Partitioning Around Medoids) program's algorithm is predicated on finding $k$ representative objects within the data set's objects. As their names suggest, these items should symbolize different facets of the data structure. Such representative items are frequently referred to as centrotypes in the literature on cluster analysis. The so-called medoids of the clusters are the representative objects in the PAM algorithm. Each object in the data set is assigned to the nearest representative object once a set of $k$ representative objects has been identified. This process creates the $k$ clusters [36].

The two algorithms that make up "Program PAM" are BUILD, which selects an initial clustering, and SWAP, which enhances the clustering further in the direction of a local optimum (finding the global optimum of the k-medoids issue is, unfortunately, NP-hard) [37].

The algorithms require a dissimilarity matrix for many popular distance functions in d dimensional data, which can be computed using the routine DAISY of Kaufman and Rousseeuw (1990). This requires $O(n^2)$ memory and $O(n^2 d)$ time to compute. As a result, calculating the distance matrix will frequently account for a large portion of the computing expense.

### 4.4.1 BUILD Initialization Algorithm

BUILD selects $k$ times the point that provides the shortest distance total $TD$ to discover a decent starting clustering. In Algorithm 4, we provide a pseudocode in which the change in $TD$ is represented by $\Delta TD$, and the best values discovered thus far are represented by $*$. It is already a rather costly approach because BUILD initialization requires $O(n^2k)$ time if we cache the nearest medoid in line 13. The goal was to identify a solid beginning point so that the refining process would need fewer iterations.

**Algorithm 4** PAM BUILD Algorithm Pseudocode

1: $(TD, m_1) \leftarrow (\infty, null)$
2: **for each** $X_j$ **do**
3: $\quad TD_j \leftarrow 0$
4: $\quad$ **for each** $X_0 \neq X_j$ **do** $TD_j \leftarrow TD_j + d(X_0, X_j)$
5: $\quad$ **if** $TD_j < TD$ **then** $(TD, m_1) \leftarrow (TD_j, X_j)$
6: **for** $1...k-1$ **do**
7: $\quad (\Delta TD^*, X^*) \leftarrow (\infty, null)$
8: $\quad$ **for each** $X_j \notin \{m_1, ..., m_i\}$ **do**
9: $\quad\quad \Delta TD \leftarrow 0$
10: $\quad\quad$ **for each** $X_0 \notin \{m_1, ..., m_i\}$ **do**
11: $\quad\quad\quad \delta \leftarrow d(X_0, X_j) - min_{0 \in m_1, ..., m_i} d(X_0, 0)$
12: $\quad\quad$ **if** $\delta < 0$ **then** $\Delta TD \leftarrow \Delta TD + \delta$
13: $\quad\quad$ **if** $\Delta TD < \Delta TD^*$ **then** $(\Delta TD^*, x^*) \leftarrow (\Delta TD, X_j)$
14: $\quad (TD, m_{i+1}) \leftarrow (TD + \Delta TD^*, X^*)$
15: **return** $TD, \{m_1, ..., m_k\}$

### 4.4.2 SWAP Refinement Algorithm

SWAP is the name of the second algorithm of "Program PAM". To enhance the clustering, it considers any potential modification to the collection of k-medoids. Essentially, this entails changing some medoid with a non-medoid, resulting in $k(n-k)$ candidate swaps. In the spirit of a greedy steepest-descent technique, the best such change is implemented if it lowers TD, and this procedure is repeated until no more improvements are found. In Algorithm 5, we provide a pseudocode for this.

**Algorithm 5** PAM SWAP Algorithm Pseudocode

**repeat**
$\quad (\Delta TD^*, m^*, X^*) \leftarrow (0, null, null)$
$\quad$ **for each** $m_i \in \{m_1, ..., m_k\}$ **do**
$\quad\quad$ **for each** $X_i \notin \{m_1, ..., m_k\}$ **do**
$\quad\quad\quad \Delta TD \leftarrow 0$
$\quad\quad\quad$ **for each** $X_0 \notin \{m_1, ..., m_k\}$ **do** $\Delta TD \leftarrow \Delta TD + \Delta(x_0, m_i, X_j)$
$\quad\quad\quad$ **if** $\Delta TD < \Delta TD^*$ **then** $(\Delta TD^*, m^*, X^*) \leftarrow (\Delta TD, m_i, X_j)$
$\quad$ **break loop if** $\Delta TD \geq 0$
$\quad$ swap roles of medoid $m^*$ and $non-medoid$ $X^*$
$\quad TD \leftarrow TD + \Delta TD^*$
**return** $TD, M, C$

The pseudocode provided for BUILD and SWAP leaves out the specifics of handling the cached distances. The index of the closest medoid, $nearest(o)$, and its distance, $d_{nearest}(o)$, for every item must be stored, as must the distance to the second nearest medoid, $d_{second}(o)$. If we additionally store the index of the second nearest center, we might be able to avoid doing additional distance calculations.

PAM is not as successful for big data sets due to its complexity; nonetheless, it is effective for small data sets (e.g., 100 objects in 5 clusters). If we do a complexity study on PAM, this is not all that surprising. It follows that for large values of n and k, PAM becomes prohibitively expensive [38].

## 4.5 CLARA Algorithm

Ng and Han [39] created a sampling-based technique called Clustering LAR Applications (CLARA) to handle larger data sets. CLARA outputs the optimal clustering after taking several samples from the data set and applying PAM to each one. Now, each iteration's complexity is $O(ks^2 + k(n - k))$, where n is the total number of objects, $k$ is the number of clusters, and $s$ is the sample size. The sample size and sampling strategy determine how effective CLARA is. If any sampled medoid is not one of the top $k$ medoids, CLARA is unable to determine the optimal clustering. As such, figuring out the sample size is challenging. According to experiments published by Kaufman and Rousseeuw in 1999, five samples with a size of $40 + 2k$ produced acceptable outcomes. That being said, this is only true for tiny $k$. We will have hundreds of medoids in the road network extraction scenario (i.e., $k > 100$). The "optimal" sample size to ensure quality must be roughly ten times that of $k$, resulting in excessive computing time needed to conduct PAM on each sample set.

The implementation of the algorithm is described as follows:

1. For $i = 1 : 5$, repeat the following steps.
2. Draw a sample of $40 + 2k$ objects randomly from the entire data set and call the PAM algorithm to find the $k$ medoids of the sample.
3. For each object, $O_j$ in the entire data set, determine which of the k medoids is the most similar to $O_j$.
4. Calculate the average dissimilarity of the clustering obtained in the previous step. If this value is less than the current minimum, use this value as the current minimum, and retain the $k$ medoids found in Step 2 as the best set of medoids obtained so far.
5. Finish and start the next iteration.

In addition to PAM, CLARA functions well with big data sets. Remember that PAM iterations are $O(k(n - k)^2)$ in length. In the case of CLARA, however, each iteration is of $O(k(40 + k)^2 + k(n - k))$ when PAM is applied alone to the samples. This explains why, for large values of $n$, CLARA is more efficient than PAM [40].

## 4.6 CLARANS Algorithm

By employing distinct samples, CLARANS (clustering large applications based upon randomized search) outperforms CLARA. $Maxneighbor$ and $numlocal$ are two other parameters that CLARANS needs in addition to the standard input to PAM. The maximum number of neighbors to which a given node can be compared is known as $maxneighbor$. Because every node will be inspected, CLARANS increasingly resemble PAM as $maxneighbor$ rises. The number of samples to be taken is indicated by $numlocal$. Because every node will be inspected, CLARANS increasingly resemble PAM as $maxneighbor$ rises. The number of samples to be taken is indicated by $numlocal$. This also represents the number of clusterings that need to be produced because every sample undergoes a fresh clustering.

$Numlocal = 2$ and $maxneighbor = max((0.0125 * k(n-k)), 250)$ are good choices, according to performance studies. It has been proved that CLARANS is more efficient than PAM or CLARA for any dataset size. All data are assumed to be in main memory by CLARANS [31].

Similar to CLARA, the CLARANS algorithm does not examine each node's neighbors. However, in contrast to CLARA, it does not limit its search to a specific subgraph. The original graph is searched. The fact that CLARANS only examines a sample of a node's neighbors sets it apart from PAM. However, every sample is generated dynamically, in contrast to CLARA, meaning that nodes associated with specific items are not completely removed. Put differently, CLARANS draws a sample of neighbors at each search stage, whereas CLARA draws a sample of nodes at the start of a search. The advantage of this is that a search is not restricted to a specific location. Compared to CLARA, a CLARANS search yields clustering of greater quality and requires a relatively small number of searches.

The following steps govern this algorithm's implementation:

1. Input parameters $numlocal$ and $maxneighbor$. Initialize $i$ to 1, and $mincost$ to a large number.
2. Set current to an arbitrary node in $G_{n,k}$.
3. Set $j$ to 1.
4. Consider a random neighbor $S$ of $current$, and calculate the cost differential of the two nodes.
5. If $S$ has a lower cost, set current to $S$ and go to step 3.
6. Otherwise, increment $j$ by 1. If $j \leq maxneighbor$, go to step 4.
7. Otherwise, when $j > maxneighbor$, compare the cost of current with $mincost$. If the former is less than $mincost$, set $mincost$ to the cost of $current$, and set $bestnode$ to $current$.
8. Increment $i$ by 1. If $i > numlocal$, output $bestnode$ and halt. Otherwise, go to step 2.

In steps 3 to 6 above, look for nodes that cost progressively less. However, the current node is deemed to be a "local" minimum if it has already been compared to the maximum number of its neighbors and is still the least expensive. The cost of this local minimum is then compared to the lowest cost found thus far in step 7. *Mincost* contains the lowest of the two costs mentioned above. After that, the CLARANS algorithm searches for more local minima until a certain number is discovered [40].

## 5 Conclusions

This chapter addressed the problem of pattern clustering with a clear and detailed presentation. Various measures of proximity between patterns were explored, and the most prominent clustering validity functions in the literature were documented. In addition, a comprehensive analysis of various traditional clustering algorithms was performed, classifying them into four basic categories: hierarchical, partitional, density-based, and grid-based. The chapter presented an overview of the evolutionary family of algorithms and swarm intelligence approaches. The scope of these algorithms in designing dynamic, scalable, and robust clustering methods was further explored. This exploration provided valuable insights into how these approaches can be adapted and applied in different clustering contexts, offering effective solutions to complex clustering problems across diverse application domains.

## References

1. Abd Elaziz, M., Oliva, D., Xiong, S.: An improved opposition-based sine cosine algorithm for global optimization. Expert Syst. Appl. **90**, 484–500 (2017)
2. Yazan, E., Talu, M.F.: Comparison of the stochastic gradient descent based optimization techniques. In: International Artificial Intelligence and Data Processing Symposium (IDAP), pp. 1–5. IEEE (2017)
3. Hansen, N.: The cma evolution strategy: A tutorial (2016). arXiv:1604.00772
4. Storn, R., Price, K.: Differential evolution-a simple and efficient heuristic for global optimization over continuous spaces. J. Global Optim. **11**(4), 341–359 (1997)
5. Zhang, J., Sanderson, A.C.: Jade: Self-adaptive differential evolution with fast and reliable convergence performance. In: IEEE Congress on Evolutionary Computation, pp. 2251–2258. IEEE (2007)
6. Tanabe, R., Fukunaga, A.S.: Improving the search performance of shade using linear population size reduction. In: IEEE Congress on Evolutionary Computation (CEC), pp. 1658–1665. IEEE (2014)
7. Kennedy, J., Eberhart, R.: Particle swarm optimization. In: Proceedings of ICNN'95-International Conference on Neural Networks, vol. 4, 1942–1948. IEEE (1995)
8. Simon, D.: Biogeography-based optimization. IEEE Trans. Evol. Comput. **12**(6), 702–713 (2008)
9. Civicioglu, P.: Transforming geocentric cartesian coordinates to geodetic coordinates by using differential search algorithm. Comput. Geosci. **46**, 229–247 (2012)
10. Booker, L.B., Goldberg, D.E., Holland, J.H.: Classifier systems and genetic algorithms. Artif. Intell. **40**(1–3), 235–282 (1989)
11. Gao, W., Liu, S., Huang, L.: A global best artificial bee colony algorithm for global optimization. J. Comput. Appl. Math. **236**(11), 2741–2753 (2012)
12. Geem, Z.W., Kim, J.H., Loganathan, G.V.: A new heuristic optimization algorithm: harmony search. Simulation **76**(2), 60–68 (2001)

13. Morales-Castañeda, B., Zaldivar, D., Cuevas, E., Fausto, F., Rodríguez, A.: A better balance in metaheuristic algorithms: does it exist? Swarm Evol. Comput. **54**, 100671 (2020)
14. Van Laarhoven, P.J., Aarts, E.H., van Laarhoven, P.J., Aarts, E.H.: Simulated Annealing. Springer, Berlin (1987)
15. Glover, F., Laguna, M.: Tabu Search. Springer, Berlin (1998)
16. Velasco, L., Guerrero, H., Hospitaler, A.: A literature review and critical analysis of metaheuristics recently developed. Arch. Comput. Methods Eng. **31**(1), 125–146 (2024)
17. Alorf, A.: A survey of recently developed metaheuristics and their comparative analysis. Eng. Appl. Artif. Intell. **117**, 105622 (2023)
18. Khan, M.J., Mathew, L.: Comparative study of optimization techniques for renewable energy system. Arch. Comput. Methods Eng. **27**, 351–360 (2020)
19. Seo, J., Shneiderman, B.: Interactively exploring hierarchical clustering results [gene identification]. Computer **35**(7), 80–86 (2002)
20. Bouguettaya, A., Yu, Q., Liu, X., Zhou, X., Song, A.: Efficient agglomerative hierarchical clustering. Expert Syst. Appl. **42**(5), 2785–2797 (2015)
21. Zhang, T., Ramakrishnan, R., Livny, M.: Birch: an efficient data clustering method for very large databases. ACM Sigmod Record **25**(2), 103–114 (1996)
22. Aghdasifam, M., Izadkhah, H., Isazadeh, A.: A new metaheuristic-based hierarchical clustering algorithm for software modularization. Complexity **2020**, 1–25 (2020)
23. Consoli, S., Korst, J., Pauws, S., Geleijnse, G.: Improved metaheuristics for the quartet method of hierarchical clustering. J. Global Optim. **78**, 241–270 (2020)
24. Thaler, D.G., Ravishankar, C.V.: Distributed top-down hierarchy construction. In: Proceedings. IEEE INFOCOM'98, the Conference on Computer Communications. Seventeenth Annual Joint Conference of the IEEE Computer and Communications Societies. Gateway to the 21st Century (Cat. No. 98), vol. 2, pp. 693–701. IEEE (1998)
25. Renders, J.-M., Bersini, H.: Hybridizing genetic algorithms with hill-climbing methods for global optimization: two possible ways. In: Proceedings of the First IEEE Conference on Evolutionary Computation. IEEE World Congress on Computational Intelligence, pp. 312–317. IEEE (1994)
26. Holland, J.H.: Genetic algorithms. Sci. Am. **267**(1), 66–73 (1992)
27. Abobaker, R.A., Ayob, M., Hadwan, M.: Greedy constructive heuristic and local search algorithm for solving nurse rostering problems, in: 2011 3rd Conference on Data Mining and Optimization (DMO), IEEE, 2011, pp. 194–198
28. Hansen, P., Mladenović, N., Brimberg, J., Pérez, J.A.M.: Variable Neighborhood Search. Springer, Berlin (2019)
29. Kirkpatrick, S., Gelatt, C.D., Jr., Vecchi, M.P.: Optimization by simulated annealing. Science **220**(4598), 671–680 (1983)
30. Das, S., Abraham, A., Konar, A., Das, S., Abraham, A., Konar, A.: Metaheuristic pattern clustering–an overview. Metaheuristic Clustering 1–62 (2009)
31. Dunham, M.H.: Data Mining: Introductory and Advanced Topics. Pearson Education India (2006)
32. Yadav, J., Sharma, M.: A review of k-mean algorithm. Int. J. Eng. Trends Technol. **4**(7), 2972–2976 (2013)
33. Jain, A.K., Murty, M.N., Flynn, P.J.: Data clustering: a review. ACM Comput. Surv. (CSUR) **31**(3), 264–323 (1999)
34. Bezdek, J.C., Ehrlich, R., Full, W.: Fcm: the fuzzy c-means clustering algorithm. Comput. Geosci. **10**(2–3), 191–203 (1984)
35. Theodoridis, S., Koutroumbas, K.: Pattern Recognition. Elsevier (2006)
36. Kaufman, L.: Partitioning around medoids (program pam). Find. Groups T **344**, 68–125 (1990)
37. Schubert, E., Rousseeuw, P.J.: Faster k-medoids clustering: improving the pam, clara, and clarans algorithms. In: Similarity Search and Applications: 12th International Conference, SISAP 2019, Newark, NJ, USA, Proceedings 12, pp. 171–187. Springer, Berlin (2019)
38. Zhang, Q. Couloigner, I.: A new and efficient k-medoid algorithm for spatial clustering. In: International Conference on Computational Science and Its Applications, pp. 181–189. Springer, Berlin (2005)

39. Ng, R.T., Han, J.: Efficient and effective clustering methods for spatial data mining. In: Proceedings of VLDB, pp. 144–155. Citeseer (1994)
40. Ng, R.T., Han, J.: Clarans: a method for clustering objects for spatial data mining. IEEE Trans. Knowl. Data Eng. **14**(5), 1003–1016 (2002)

# Image Contrast Enhancement: Harnessing Metaheuristics and the Gauss Error Function

**Laritza Pérez-Enríquez, Miguel Jiménez-Domínguez, Néstor García-Rojas, Rafael Solar-Hernández, Saúl Zapotecas-Martínez, and Leopoldo Altamirano-Robles**

**Abstract** Contrast enhancement techniques play a vital role in image processing by amplifying features, brightness, and luminosity to create images tailored for diverse research applications. Addressing specific tasks in various domains necessitates an in-depth exploration of contrast enhancement, especially in computer vision systems where uncovering hidden features is crucial. This chapter delves into a contrast enhancement approach utilizing the Gauss error function, whose parameters can be fine-tuned through metaheuristics. Investigating the synergy of the Gauss error function with three widely recognized metaheuristics aims to optimize pixel values, enhancing features in low-contrast images effectively. Leveraging a dataset of public domain images, the presented method undergoes rigorous evaluation using standard performance indicators to measure the improvements in terms of contrast. Preliminary findings indicate that the Gauss error function enhances image contrast and exhibits rapid adaptability to various metaheuristics. The investigation presented in this chapter contributes to the advancement of contrast enhancement methodologies, offering a promising approach to enhance image quality in various computer vision applications.

---

L. Pérez-Enríquez · M. Jiménez-Domínguez · N. García-Rojas · R. Solar-Hernández · S. Zapotecas-Martínez (✉) · L. Altamirano-Robles
Instituto Nacional de Astrofísica, Óptica y Electrónica, Tonantzintla 72840, Puebla, Mexico
e-mail: szapotecas@inaoep.mx

L. Pérez-Enríquez
e-mail: laritza@inaoep.mx

M. Jiménez-Domínguez
e-mail: miguel.jimenez@inaoep.mx

N. García-Rojas
e-mail: nestor.garcia@inaoep.mx

R. Solar-Hernández
e-mail: rafael.solar@inaoep.mx

L. Altamirano-Robles
e-mail: robles@inaoep.mx

 
D. Oliva et al. (eds.), *Advances in Optimization Algorithms for Multidisciplinary Engineering Applications: From Classical Methods to AI-Enhanced Solutions*, Studies in Computational Intelligence 806, https://doi.org/10.1007/978-3-031-78440-8_4

## 1 Introduction

Contrast enhancement plays a crucial role in various fields, such as image processing, pattern recognition, and computer vision. It allows us to improve the clarity of images, making them more suitable for further analysis and processing [31].

Over the years, researchers have proposed various methods for modifying the histogram to enhance image contrast. Asamoah et al. [3] introduced an advanced approach that utilizes adaptive histogram equalization methods, such as Adaptive Histogram Equalization (AHE) [30] and Contrast Limited Adaptive Histogram Equalization (CLAHE) [42]. These methods directly improve the contrast measure in the images in contrast to conventional techniques like histogram equalization and global stretching, which may not always yield optimal results, especially for images with significant contrast variation.

In their review, Patel et al. [28] classify image contrast enhancement methods into two main categories. The first category encompasses techniques in the spatial domain that directly manipulate pixel values. Examples of methods belonging to this category include Gray Level Transformations [32], Contrast Stretching [26], and Histogram Processing [40]. The second category comprises methods in the frequency domain, which involve analyzing mathematical functions or signals in terms of frequency and operating on the image transform coefficients. Some techniques applied in this approach include Smoothing Filters and Sharpening Filters. These methods aim to enhance image contrast through frequency-based analysis and manipulation [28]. By categorizing these methods and understanding their underlying principles, researchers and practitioners in the field of image processing can effectively choose and apply the most suitable contrast enhancement techniques for their specific applications.

In addition to these techniques, other transformation functions have been proposed for contrast enhancement. For example, the Discrete Wavelet Transform with Singular Value Decomposition (DWT-SVD) and the Adaptive Gamma Correction (AGC) have been suggested. Sigmoid functions [18, 27], the adaptive local gamma function [41], and the incomplete Beta function [34] have been also proposed. On the other hand, metaheuristic algorithms, including Genetic Algorithm (GA) [11], Particle Swarm Optimization (PSO) [19], and Differential Evolution (DE) algorithm [36], have been employed for contrast enhancement. These algorithms offer alternative approaches to improving image contrast. Overall, these advancements in histogram modification techniques and utilizing various transformation functions and metaheuristic algorithms provide researchers with diverse tools for enhancing image contrast.

This chapter presents a transformation function based on the Gauss error function with adaptive parameters tailored for contrast enhancement. The effectiveness of this function in the context of contrast enhancement is systematically assessed by integrating it into three prominent bio-inspired metaheuristic algorithms: GA, DE, and PSO. The successful integration of the Gauss error function into these metaheuristics is highlighted, emphasizing its potential for contrast enhancement. Experimental findings validate the practicality and efficiency of integrating the Gauss error function into bio-inspired metaheuristics, demonstrating its adaptability and versatility. A comprehensive series of experiments is conducted using a diverse set of color images to evaluate the performance of the algorithms against standard quality indicators for digital images. The results offer empirical support for the effectiveness of the approach, showcasing its ability to enhance contrast across various scenarios. The adaptability of the Gauss error function, coupled with its seamless integration into established algorithms, suggests promising applications in real-world image processing tasks.

The structure of this chapter is outlined as follows. Section 2 reviews notable contributions in specialized literature dedicated to enhancing image contrast by applying metaheuristics and transformation functions. Section 3 introduces the overarching background providing essential context for understanding the study. Section 4 describes the transformation function detailing its integration within bio-inspired algorithms. Section 5 comprehensively presents the experimental study conducted to evaluate the proposed transformation function. Section 6 exposes a detailed analysis of results, incorporating a comparative performance study among the algorithms employed in the experimental investigation. Finally, Sect. 7 provides the conclusions and possible paths for future investigations.

## 2 Literature Review

Traditional methods for enhancing image contrast often resort to globally modifying pixel values across the entire image. However, this approach frequently results in abrupt highlights in certain areas, causing information loss and amplified noise. Advanced techniques have emerged to address these challenges, utilizing optimization algorithms to automatically determine optimal parameters.

Various methods leveraging transformation functions to enhance image contrast have been reported in the literature. For example, Hoseini et al. [13] proposed a hybrid algorithm integrating metaheuristic techniques such as Genetic Algorithm (GA) [11], Ant Colony Optimization (ACO) [8], and Simulated Annealing (SA) [21]. ACO generates transfer functions to map input intensities to output intensities, while SA identifies these functions. Finally, the GA manages the evolutionary process, yielding images with higher contrast while preserving natural appearance. Kallel et al. [15] introduced an algorithm dedicated to enhancing dark Computed Tomography (CT) scans, employing Discrete Wavelet Transform (DWT)-Singular Value Decomposition (SVD) [4] followed by Adaptive Gamma Correction (AGC) [39], suitable for low-contrast images. Chen et al. [7] utilized the Artificial Bee Colony (ABC) algorithm and introduced a new objective function for evaluating enhanced image quality, achieving improved image contrast. The improved image was created from the original image by applying a parametric image transformation function, and optimal parameter values are found through the ABC algorithm. Kanmani and Narasimhan [17], leveraging Particle Swarm Optimization (PSO), enhanced grayscale image contrast. This method was beneficial for identifying hidden details in an image and effective for satellite and nighttime imagery applications. Kamoona et al. [16] proposed an enhanced cuckoo search algorithm, facilitating automatic contrast enhancement within a new parameter search space for Local/Global Enhancement transformation (LGE).

Studies such as those conducted by Surya et al. [38] proposed a contrast enhancement algorithm with automated parameterization utilizing a GA and a fuzzy intensification operator. In another study, Guraksin et al. [12], a method involving wavelet transform and DE algorithm for image contrast enhancement was presented. Malik et al. [23] introduced a method that enhances images using the sigmoid function and employs PSO to maximize information content while preserving image intensity through gamma correction. Rao et al. [33] addressed the contrast enhancement of magnetic resonance imaging by employing modified histogram equalization through Fuzzy-Improved PSO (FIPSO) for dynamic histogram equalization. This technique captures image details through smoothing and utilizes the Gauss function to distribute pixel intensity to the nearest pixel. This approach has demonstrated good performance in terms of brightness and contrast. Ahmed et al. [2] proposed a method to enhance grayscale images, employing the Barnacles Mating Optimizer (BMO) [37] algorithm, which transforms input intensity into corresponding output intensity. The main objective of this study was to design an optimally equalized histogram by applying the BMO optimization algorithm. In another approach [41], a simple and novel correction method was proposed based on adaptive local gamma transformation and color compensation inspired by the illumination reflection model. The local gamma transformation function enhances image brightness by adapting parameters, reducing the influence of uneven illumination to avoid over-enhancement, and improving the visual effect of low-light images. Acharya et al. [1] proposed an adaptive image enhancement technique based on a GA, applying a novel approach of histogram subdivision using the exposure threshold and the optimal threshold to preserve brightness and reduce information loss. Khan et al. [20] employed a metaheuristic algorithm inspired by chimpanzee behavior for contrast enhancement, integrating Bilateral Gamma Correction (BGC) [5] to augment their approach. This selection was motivated by the shortcomings of the traditional beta function in effectively selecting parameters for image enhancement. The adjustable parameter range diminishes when extending into areas with low or high grayscale levels, leading to reduced effectiveness in contrast enhancement. This occurs due to expansion at both ends and compression in the middle of the image. Braik and Malik [5] presented the application of a hybrid approach that combines the Whale Optimization Algorithm (WOA) [25] with the Chameleon Swarm Algorithm (CSA) [5, 6], forming a method called Hybrid Whale Optimization Algorithm (HWOA). This approach merged CSA with WOA to enhance their exploration capability and achieve optimal balance. The resulting algorithm, HWOA, was used to adaptively select optimal parameters of the incomplete beta function through a predefined fitness function, thereby facilitating an adaptive image enhancement method.

In a more recent development, Pérez et al. [29] presented a method for improving image contrast utilizing a hyperbolic tangent sigmoid function. This transformation function was integrated with the Differential Evolution (DE), Genetic Algorithm (GA), and Particle Swarm Optimization (PSO) algorithms. The primary aim of this proposal is to elevate the intensity of pixels in low-contrast images through nature-inspired metaheuristics. These algorithms optimize the parameters of the sigmoid function specifically tailored for images with low contrast levels.

The main purpose of the aforementioned methods is to refine the adjustment of pixel intensities within an image, aiming to unveil hidden details in degraded images and amplify the contrast between light and dark areas. This ultimately enhances visual quality and facilitates image interpretation. Leveraging metaheuristics aids in efficiently configuring the parameters of these transformation functions, thereby achieving a notable improvement in image contrast through automated and adaptive processes.

## 3 General Background

### *3.1 Contrast*

The concept of contrast in an image is rooted in the discernible difference between its bright and dark regions, determined by the arrangement of pixels. It serves as a dynamic metric, reflecting the intensity levels exhibited by the pixels within the image. Higher contrast manifests as a pronounced divergence between bright and dark areas, engendering a visually striking image. Conversely, lower contrast points to a more subtle dissimilarity, resulting in a smoother and less conspicuously varied image in intensity. A judicious application of contrast in an image is instrumental in delineating objects from their surroundings and other elements within the visual composition. The ability to distinguish an object amid an image's complexity hinges on achieving an optimal contrast balance. Adequate contrast empowers viewers to identify and interpret the distinct features and nuances of the visual narrative. Furthermore, using contrast enhancement techniques is essential in elevating the overall visual quality of images. By refining the balance of bright and dark elements, these techniques create a more immersive and discernible visual experience. Consequently, observers can keenly perceive and appreciate the intricate details, textures, and characteristics embedded within the image, enriching their overall comprehension and enjoyment.

#### 3.1.1 Sum Edge Intensity

Edges in an image denote regions with sudden intensity changes between adjacent areas. The Sum of Edge Intensity (SEI) quantifies these intensity differences by summing them along identified edges. Edge detection algorithms such as Sobel or Canny aid in pinpointing these areas of significant intensity variation. SEI provides a measure of total intensity change within these regions by summing the intensity differences along edges. Consequently, the SEI value indicates the level of detail and features present at image edges. Higher SEI values correspond to increased image contrast, indicating a greater abundance of details and features along detected edges. This principle finds utility across various domains, including image contrast enhancement and feature identification.

#### 3.1.2 Count of Edge Pixels

Count of Edge Pixels (CEP) refers to the identification and quantification of pixels contributing to the edges in an image, offering a numerical gauge of the overall edge quantity. This procedure involves applying diverse edge detection techniques, ranging from convolution operators like Sobel or Roberts to thresholding algorithms and advanced methods such as the Canny algorithm. A higher pixel edge count implies a heightened prevalence of edges and intensity transitions within the image. This heightened count correlates with a more prominent presence of features and finer details. Beyond merely quantifying edges, pixel edge count is a valuable metric for evaluating the image's quality and sharpness. A heightened pixel edge count often indicates a clearer image, accentuating intricate details and structural nuances. Therefore, this metric serves a dual purpose by providing insights into the image's complexity and offering a quantitative measure for assessing its overall visual quality and sharpness.

#### 3.1.3 Image Entropy

The concept of Entropy (E) of an Image is a gauge for the distribution of intensity levels within an image. A higher entropy value indicates a more pronounced variation in intensity levels, signifying increased contrast in the image. Essentially, entropy measures the amount of information or uncertainty inherent in the distribution of intensity levels across the image. This term provides a quantifiable means to assess the richness of information present in the image. Calculating the entropy of an enhanced image involves evaluating the distribution of intensity levels. The higher the entropy value, the greater the variability and contrast, highlighting the richness and complexity of information captured in the enhanced image. The entropy of an enhanced image is calculated as follows.

$$E = \begin{cases} -\sum_{j=0}^{255} h_j log_2(h_j) & \text{if } h_j \neq 0 \\ 0 & \text{otherwise} \end{cases} \tag{1}$$

## 3.2 *Quality Indicators in Digital Images*

The evaluation of enhanced images involves the application of diverse metrics aimed at objectively gauging the quality of the processed results. These metrics are analytical tools to assess the fidelity and improvement achieved through the enhancement process. By comparing the processed image with a reference image, these metrics consider a spectrum of parameters outlined in this section. These parameters encompass critical aspects such as color accuracy, spatial resolution, noise reduction, and overall perceptual quality. This section presents a couple of standard metrics employed to assess the performance of the proposed enhanced contrast method.

### 3.2.1 Peak Signal-to-Noise Ratio

The Peak Signal-to-Noise Ratio (PSNR) [3] is a prevalent metric for assessing image reconstruction quality, comparing two images based on pixel information. A higher PSNR value signifies superior image quality. This metric is defined through the Root Mean Squared Error (RMSE) and its associated distortion metric, providing a quantitative insight into the fidelity of the reconstructed image. The PSNR is defined as:

$$PSNR = 20 \log \frac{255}{RMSE} \tag{2}$$

such that:

$$RMSE = \sqrt{\frac{1}{Ro \times Co} \sum_{i=1}^{Ro} \sum_{j=1}^{Co} (I_o(i, j) - I_e(i, j))} \tag{3}$$

where $Ro$ and $Co$ denote the height and weight of the image, $I_o$ is the original image, and $I_e$ represents the enhanced image.

### 3.2.2 Structural Similarity Index Measure

The Structural Similarity Index Measure (SSIM) [35] is a metric designed to evaluate the quality of an image by scrutinizing three essential components: luminance, contrast, and structure. Its primary purpose is to quantify the resemblance between a reference image and another one under comparison. By examining these key aspects, the SSIM provides a comprehensive measure of the likeness or dissimilarity between the two images, offering valuable insights into the overall perceptual quality and fidelity of the compared images. The definition of this metric is expressed in Eq. (4).

$$SSIM = \frac{(2\mu I_o \mu I_e + C_1)(2\sigma I_o I_e + C_2)}{((\mu^2 I_o + \mu^2 I_e) + C_1)((\sigma^2 I_o + \sigma^2 I_e) + C_2)} \tag{4}$$

such that:

$$\sigma_{(I_o I_e)} = \frac{1}{Co-1}\sum_{i=1}^{Co}(I_{o_i} + \mu_{I_o})(I_e + \mu_{I_e}) \tag{5}$$

### 3.2.3 Relative Enhancement Contrast

The Relative Enhancement Contrast (REC) [14] is a metric that indicates the contrast disparity between the original image and its enhanced counterpart. REC is quantified as the ratio of the contrast observed in the original image to the enhanced image. This metric offers a numerical assessment highlighting the proportional change in contrast brought about by the enhancement process, providing valuable information on how the visual distinctions within the image have been modified or intensified. The REC calculation is performed according to the Eq. (6).

$$REC = \frac{C_{I_o}}{C_{Ie}} \tag{6}$$

In Eq. (6), $C_{I_o}$ is computed using Eq. 7 and $C_{I_e}$ is computed in the same way.

$$C_{I_o} = 20\log\left[\frac{1}{Ro \times Co}\left(\sum_{i=1}^{Ro}\sum_{j=1}^{Co} {I_o}^2(i,j)\right) - \left(\sum_{i=1}^{Ro}\sum_{j=1}^{Co} I_o(i,j)\right)^2\right] \tag{7}$$

## 4 Contrast Enhancement Using the Gauss Error Function

This section introduces the proposed transformation function, which relies on the Gauss error function to enhance contrast. Additionally, it outlines the fitness function and the bio-inspired metaheuristics utilized in this investigation.

### *4.1 Transformation Function*

In the specialized literature, a common practice involves employing a transformation function to adjust the original intensity of each pixel within a low-contrast image. Various methodologies employ a piecewise linear transformation function, as documented in studies by Russo [34] and Ling et al. [22]. These approaches optimize the parameters within the transform functions to enhance the contrast of the underlying image to which they are applied.

The altered pixels may exhibit increased prominence upon applying the transformation function, potentially exceeding the permissible range of intensity levels or even taking on negative values, if applicable. Addressing this challenge necessitates replacing the piecewise curve with a continuous curve.

This investigation introduces a novel transformation function grounded in the Gauss error function with adaptive parameters designed explicitly for contrast enhancement.

The well-known Gauss error function spans a domain in $x \in [-\infty, \infty]$ and a codomain $(-1, 1)$. It is defined as:

$$f(x) = \frac{2}{\sqrt{\pi}} \int_{x}^{\infty} e^{-t^2} dt \tag{8}$$

To introduce adaptability, the Gauss error function can incorporate scaling parameter $\alpha$ and translation parameter $\beta$ adjusting Eq. (8) as shown below:

$$f(x, \alpha, \beta) = \frac{2}{\sqrt{\pi}} \int_{\frac{x-\beta}{\alpha}}^{\infty} e^{-t^2} dt \tag{9}$$

where $\alpha > 0$ and $\beta \in \mathbb{R}$. Figure 1 illustrates different shapes of the Gauss error function using various $\alpha$ values.

In the above function, the scaling parameter $\alpha$ accentuates the concave and convex aspects of the sigmoid drawn by the Gauss error function, while the translation parameter $\beta$ governs the inflection point's position in the sigmoid.

The Gauss error function can be scaled and translated to improve the contrast in digital images. Considering an image with $\Delta$ intensity levels, the Gauss error function for image contrast enhancement can be formulated as:

$$f(x, \alpha, \beta) = \frac{\Delta}{\sqrt{\pi}} \int_{\frac{x-\beta}{\alpha}}^{\infty} e^{-t^2} dt + \frac{\Delta}{2} \tag{10}$$

where $x$ denotes the pixel intensity, $\alpha > 0$ and $\beta \in [0, \Delta]$.

In this study, the transformation function is utilized to determine the new pixel values for the target image. The search space in this investigation was limited to $\alpha \in [10, 50]$ and $\beta \in \left[\frac{1}{3}\Delta, \frac{2}{3}\Delta\right]$.

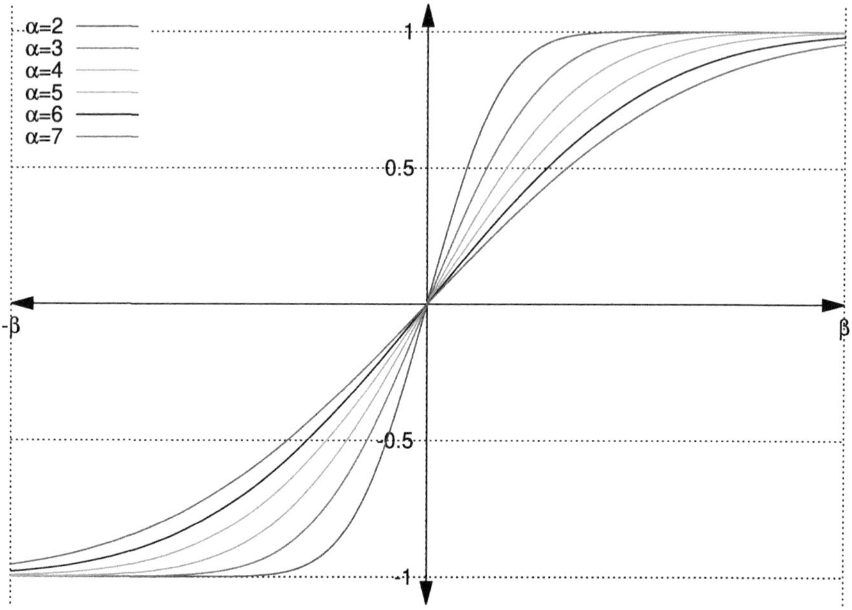


**Fig. 1** Gauss error function with different $\alpha$ values

## 4.2 Fitness Function

The methodology presented in this study employs a fitness function as a pivotal measure for assessing the quality of enhanced images. Existing literature demonstrates that researchers often leverage various image-related factors when utilizing optimization algorithms to evaluate the effectiveness of image enhancements. In this investigation, the fitness function is defined by considering several common image aspects, each contributing to the overall evaluation of enhanced image quality.[1] The key image-related factors integrated into the fitness function are as follows:

1. **Sum Edge Intensity (SEI) Levels of Pixels**: This aspect involves the cumulative intensity levels of pixels associated with edges.
2. **Count of Edge Pixels (CEP)**: This factor quantifies the number of pixels contributing to defining edges.
3. **Entropy of Image (E)**: Entropy, in the image context, measures the overall information or uncertainty present in the distribution of intensity levels.

These carefully chosen aspects encompass crucial dimensions of image quality, and their integration into the fitness function establishes a comprehensive approach for evaluating the effectiveness of image enhancement strategies. By systematically considering these factors, the proposed methodology aims to capture and quantify

[1] For a detailed understanding of these indicators, refer to Sect. 3.2.

various elements that collectively contribute to the enhanced image's visual fidelity and richness. These considerations collectively contribute to defining the following fitness function:

$$F(I_E) = \log(\log(SEI)) \times CEP \times E \tag{11}$$

where $I_E$ represents the image obtained after optimizing the $\alpha$ and $\beta$ parameters in Eq. (10).

Notably, this versatile fitness function allows any bio-inspired metaheuristic to adopt it. Subsequently, the proposed transformation function's performance is investigated, and its advantages in enhancing contrast in digital images are explored.

## 4.3 *Metaheuristics*

Metaheuristics are advanced strategies crafted to navigate the search for solutions in optimization problems. These versatile algorithms find utility across various optimization tasks where exact solutions are challenging or infeasible to obtain within reasonable timeframes. By adhering to relevant principles and tailoring the unique intricacies of each metaheuristic algorithm to the task of contrast enhancement, one can effectively leverage metaheuristics to enrich the visual quality of images. This section presents a comprehensive overview of the metaheuristic algorithms selected for addressing the contrast enhancement problem.

### 4.3.1 Genetic Algorithm

Genetic Algorithms (GAs) [11] constitute a well-established area of research inspired by the natural process of species. GAs work at the genotypic level and normally do not adopt a self-adaptation mechanism like evolutionary strategies. In GAs, solutions are encoded and can vary in different ways depending on the type of problem. A population consists of individuals encoded according to a certain criterion. The initial population is randomly chosen to cover the entire search space. In each generation, an objective function is evaluated for every solution in the population, assigning each individual a fitness value that guides the natural selection process. In the evolutionary process, a stochastic function, namely mating selection, is used to select parent individuals from the population based on their fitness for mating or recombination to produce offspring. Recombination between individuals involves generating one or more offspring solutions from two-parent solutions, ideally preserving and combining desirable features from both parents. The mutation phase is a variation operator that utilizes a single-parent solution to create a child solution by introducing a random change to the representation. In the final stage of a GA cycle, the selection of individuals to shape the succeeding generation is critical. Among the survival selection mechanisms, two commonly used schemes are $(\mu, \lambda)$- and $(\mu + \lambda)$-selection

**Algorithm 1** Genetic Algorithm with the $(\mu + \lambda)$-selection scheme

**Input:** A maximum number of generations $G_{max}$
**Output:** An evolved population $P$

$t = 0$
Generate an initial population $P_t$
evaluate($P_t$)
**while** $t < G_{max}$ **do**
    $Q$ = crossover($P_t$)
    $Q'$ = mutation($Q$)
    evaluate($Q'$)
    $T = P_t \cup Q'$
    $P_{t+1}$ = select_best($T$)
    $t = t + 1$
**end while**
**return** $P_t$

schemes [10]. In this investigation, the latter scheme was employed. In the $(\mu + \lambda)$-selection scheme, solutions generated in each generation compete against parental solutions, with only the finest individuals persevering. The general framework of GAs based on the $(\mu + \lambda)$-selection scheme is presented in Algorithm 1.

### 4.3.2 Differential Evolution

Differential Evolution (DE) [36] stands out as a robust population-based optimizer renowned for its ability to mitigate the challenge of selecting an optimal starting point in optimization tasks. This is achieved through a unique approach of sampling the objective function across multiple randomly chosen initial points. In the DE framework, the initial population selection is governed by preset parameter bounds, defining the domain from which vectors are chosen. Each vector within this initial population is assigned an index ranging from 0 to $N_p - 1$. Like other population-based methods, DE's hallmark is its capacity to generate new candidate solutions by perturbing existing ones. This perturbation is realized through the scaled difference of two randomly selected population vectors. The resultant perturbed vector, the trial vector, is crafted by adding this scaled, random vector difference to a third randomly selected population vector. In the subsequent selection stage, each trial vector competes against the population vector sharing the same index. The criterion for this competition is based on the objective function value, with the trial vector boasting the lower value earning its place as a member of the succeeding generation. This iterative process continues until all population vectors have engaged in competition against a randomly generated trial vector. Upon completion, the surviving vectors from these pairwise competitions serve as parents for the subsequent generation in the evolutionary cycle. DE's mutation strategy, denoted as DE/rand/1/bin, delineates its search mechanism. Here, "1" signifies the number of difference vectors involved, "rand" denotes the target vector considered during the mutation process, and "bin" indicates the type of crossover employed, specifically binomial in this context. The overarching methodology of DE, guided by the DE/rand/1/bin strategy, is encapsulated within Algorithm 2, offering a structured blueprint for its application and implementation.

### 4.3.3 Particle Swarm Optimization

Particle Swarm Optimization (PSO) [19] algorithm stands as a pioneering stochastic optimization technique inspired by the collective behaviors observed in animal swarms. Mimicking the collaborative quest for sustenance seen in various species, PSO orchestrates a cooperative search strategy where individual entities, akin to members of a swarm, dynamically adapt their search trajectories based on personal experiences and the collective knowledge gleaned from the swarm.

**Algorithm 2** Differential Evolution DE/rand/1/bin

**Input:** A maximum number of generations $G_{max}$, a population size $N$
**Output:** An evolved population $\{x_1, x_2, ..., x_N\}$

$C_r$ Crossover rate
$G_{max}$ Number of generations
$k = 0$
Generate an initial population $\{x_1, x_2, ..., x_N\} : x_i \in \mathbb{R}^n, \forall i, i = 1, ..., N$
**while** $k < G_{max}$ **do**
  **for** $i \in \{1, 2, ..., N\}$ **do**
    Select $i1, i2, i3 \in \{1, ..., N\} : i \neq i1 \neq i2 \neq i3$
    $v_i = x_{i1} + F(x_{i2} - x_{i3})$ //Mutation
    **if** $rand() < C_r, \forall j, j = 1, ..., n$ **then** //Trial vector
      $u_i = (u_i^{(1)}, u_i^{(2)}, ..., u_i^{(n)}) : u_i^{(1)} = v_i^{(j)}$
    **else**
      $u_i^{(j)} = x_i^{(j)}$
    **end if**
    **if** $f(u_i) \leq f(x_i)$ **then**
      $x_i = u_i$
    **end if**
  **end for**
  $k = k + 1$
**end while**
**return** $\{x_1, x_2, ..., x_N\}$

The foundation of PSO lies in the emulation of avian flocking patterns within a Cartesian coordinate framework. At the outset, each "bird" (representing a potential solution) assumes a randomized position and velocity. Adhering to the "nearest proximity velocity match rule", every bird continually adjusts its velocity to mirror that of its closest counterpart. However, to infuse a touch of realism, a stochastic element is injected into the process, introducing slight perturbations in velocity with each iteration. This stochasticity prevents premature convergence, ensuring that the swarm explores the solution space comprehensively and avoids premature convergence to suboptimal solutions.

Among the diverse array of PSO variants, the inertia approach emerges as the cornerstone, renowned for its widespread adoption and efficacy. This methodology introduces an inertia weight parameter into the velocity update equation, exerting a nuanced control over the influence of prior velocities on current movement dynamics. The judicious calibration of this parameter empowers practitioners to balance between exploration and exploitation delicately, endowing the algorithm with enhanced adaptability to traverse intricate solution landscapes. Consequently, the velocity update formula is succinctly articulated as follows:

$$v_{i,t+1}^{d} = \omega v_{i,t}^{d} + c_1 r_1 (p_{i,t}^{d} - x_{i,t}^{d}) + c_2 r_2 (p_{g,t}^{d} - x_{i,t}^{d}) \tag{12}$$

In Eq. (12), $\omega$ denotes the inertial weight, while $c_1$ and $c_2$ symbolize the respective weights assigned to the stochastic acceleration terms. These terms act as guiding forces, pulling the $i$-th particle towards its personal best ($p_i$) and the global best ($p_g$) positions within the solution space. The equation includes $r_1$ and $r_2$, representing uniformly distributed random numbers within the interval [0, 1].

**Algorithm 3** Particle Swarm Optimization

**Input:** A maximum number of iterations $G_{max}$, a population size $N$
**Output:** An evolved cluster $\{x_1(k), x_2(k), ..., x_N(k)\}$
$k = 0$
Generate an initial population of particle $\{x_1, x_2, ..., x_N\}$
Initialize personal bests $\{x_1^{pbest}, x_2^{pbest}, ..., x_N^{pbest}\}$
Initialize velocities $\{v_1, v_2, ..., v_N\}$
$x^{best}(k) =$ best particle($P$) //Update best global
**while** $k < G_{max}$ **do**
  **for** $i \in \{1, 2, ..., N\}$ **do**
    $v_i(k+1) =$ UpdateVelocity($x_i(k), v_i(k), x_i^{pbest}(k), x^{gbest}(k)$) //Update Velocity
    $x_i(k+1) = x_i(k) + v_i(k+1)$ //Update position
    **if** $f(x_i(k+1)) < f(x_i(k))$ **then**
      $x_i^{pbest}(k+1) = x_i(k+1)$ //Update personal best
    **end if**
    **if** $f(x_i(k+1)) < f(x^{gbest}(k))$ **then**
      $x^{gbest}(k) = x_i(k+1)$ //Update best global
    **end if**
  **end for**
  $k = k + 1$
**end while**
**return** $\{x_1(k), x_2(k), ..., x_N(k)\}$

To comprehensively understand the PSO algorithm's overarching structure and workflow, Algorithm 3 is presented, elucidating its procedural steps and operational mechanisms. Through this elucidation, practitioners can glean deeper insights into the intricate orchestration of PSO's iterative optimization process.

### 4.4 Contrast Enhancement Methodology

Figure 2 illustrates the process for enhancing image contrast using a transformation function based on the Gauss error function. This method involves several key steps to improve the input image's contrast systematically. First, we begin with the input image that requires contrast enhancement. A bio-inspired metaheuristic, such as GA, DE, or PSO, is employed to achieve this enhancement. The evolutionary algorithm operates with a representation of a vector of two real numbers $(\alpha, \beta)$, which are crucial parameters for the transformation function applied to the input image. The bio-inspired metaheuristic's objective is to determine the optimal values of $\alpha$ and $\beta$ that maximize the contrast of the input image. This optimization process is driven by a fitness function presented in Eq. (11), which evaluates the quality of the enhanced image using the quality indicators SEI, CEP, and E. As the bio-inspired metaheuristic iterates, it explores different $\alpha$ and $\beta$ values, guided by the fitness function to improve the image's contrast progressively. Once the algorithm converges to the optimal $\alpha$ and $\beta$ values, these parameters are used to generate the enhanced image. Thus, the enhanced image produced at the end of this process exhibits improved contrast significantly, demonstrating the effectiveness of using the Gauss error function in combination with bio-inspired metaheuristics.

## 5 Experimental Study

This section provides a comprehensive overview of the experiments carried out, presenting a detailed analysis of both the visual representations and numerical outcomes resulting from the application of three distinct optimization algorithms: Genetic Algorithm (GA), Differential Evolution (DE), and Particle Swarm Optimizer (PSO).

The primary objective of these experiments is to delve into the effectiveness of each optimization algorithm in fine-tuning the parameters of the Gauss error function to enhance image contrast. Through a systematic exploration, we aim to ascertain the algorithmic capabilities in optimizing this specific function, thereby shedding light on their respective strengths and weaknesses in the context of image enhancement tasks.

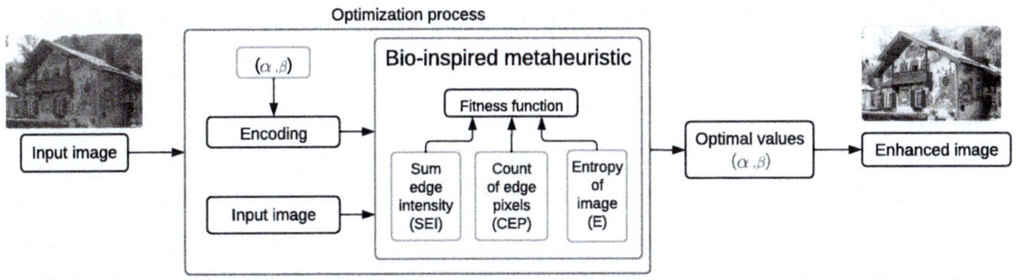


**Fig. 2** Contrast enhancement diagram

## 5.1 Experimental Setup

This study evaluates the efficacy of the proposed method through its application to images sourced from the Kodak dataset [9]. These images, depicting scenes in RGB color format and with a resolution of $756 \times 512$ pixels, serve as the basis for assessing the performance of three distinct algorithms. Key metrics utilized for gauging method effectiveness include PSNR, SSIM, and REC, enabling objective evaluation of image enhancement quality and facilitating comparisons with existing literature.

The following briefly describes the parameters utilized for each of the three bio-inspired metaheuristics.

**Genetic Algorithm**. The genetic algorithm integrates SBX and PBM operators [24] and employs the $(\mu + \lambda)$-selection scheme for survival selection. Performance evaluation of the genetic algorithm is conducted with a crossover probability ($P_c$) set to 0.9 and a mutation probability ($P_m$) set to 0.5.

**Differential Evolution**. The differential evolution algorithm employs a differential weight of $F = 0.9$ and a crossover probability of $C_r = 0.5$. This study adopted the scheme DE/rand/1/bin.

**Particle Swarm Optimization**. The Particle swarm optimizer adopts the inertia model, configured with acceleration coefficients ($c_1$ and $c_2$) set to 1.7 and an inertia weight ($w$) of 0.7289.

The algorithms were uniformly implemented with a population size of $N = 100$ and a maximum iteration limit of $G_{max} = 500$. These standardized parameters establish a fair and consistent evaluation framework for assessing the performance of each algorithm.

## 5.2 Performance Evaluation

A comprehensive comparative analysis was undertaken to assess the performance of bio-inspired algorithms, specifically Genetic Algorithm (GA), Differential Evolution (DE), and Particle Swarm Optimization (PSO), all leveraging the Gauss error function for image enhancement. The detailed comparison is presented in Tables 1, 2 and 3, which presents detailed statistics, including average and standard deviation values. These statistics were derived from 30 independent runs of each algorithm applied to the 24 images sourced from the Kodak Lossless True Color Image Suite [9].

In Tables 1, 2 and 3, the algorithm achieving the best average value for each performance indicator is boldly highlighted, facilitating straightforward identification of the algorithm with superior performance across a spectrum of metrics. This strategic highlighting approach enhances readability and enables swift comprehension of algorithmic prowess.

**Table 1** PSNR metric, obtained through the application of GA, DE, and PSO algorithms

| Image | PSNR | | |
|---|---|---|---|
| | GA | DE | PSO |
| kodim1 | 14.1747 (0.2757) | 14.3005 (0.2092) | **14.3815** (0.2051) |
| kodim2 | 15.8180 (0.0003) | 15.8181 (0.0000) | **15.8182** (0.0001) |
| kodim3 | 14.3632 (0.0043) | 14.3643 (0.0000) | **14.4663** (0.1771) |
| kodim4 | 15.0237 (0.2134) | 14.8929 (0.0004) | **15.3764** (0.0836) |
| kodim5 | 14.2227 (0.0066) | 14.2268 (0.0000) | **14.2301** (0.0216) |
| kodim6 | 14.7248 (0.0002) | 14.6956 (0.0881) | **14.7255** (0.0005) |
| kodim7 | 13.1273 (0.2455) | **13.1876** (0.1717) | 13.0561 (0.2244) |
| kodim8 | 15.9373 (0.0000) | 15.9373 (0.0000) | 15.9373 (0.0000) |
| kodim9 | 16.5879 (0.0508) | 16.5872 (0.0474) | **16.7127** (0.0186) |
| kodim10 | 16.3298 (0.0684) | **16.3924** (0.0000) | 16.3197 (0.0177) |
| kodim11 | 11.9131 (0.1670) | 11.9868 (0.0416) | **12.0250** (0.2972) |
| kodim12 | 16.6174 (0.0053) | 16.6165 (0.0000) | **16.6455** (0.0000) |
| kodim13 | **14.3346** (0.0000) | **14.3346** (0.0000) | 14.3159 (0.0785) |
| kodim14 | 13.0239 (0.0017) | 13.0233 (0.0000) | **13.0241** (0.0121) |
| kodim15 | 16.2121 (0.0029) | 16.2109 (0.0000) | **16.2151** (0.0147) |
| kodim16 | **14.9351** (0.0032) | 14.9295 (0.0013) | 14.8495 (0.0733) |
| kodim17 | **13.4761** (0.0012) | 13.4754 (0.0004) | 13.4482 (0.0125) |
| kodim18 | 16.2825 (0.0272) | **16.3288** (0.0115) | 16.2794 (0.0284) |
| kodim19 | 15.1944 (0.0621) | 15.0579 (0.0000) | **15.2228** (0.0005) |
| kodim20 | **14.9755** (0.0000) | **14.9755** (0.0000) | 14.9742 (0.0073) |
| kodim21 | 13.3927 (0.3607) | 13.1973 (0.0003) | **13.7491** (0.2527) |
| kodim22 | 15.3536 (0.0247) | 15.3381 (0.0000) | **15.4093** (0.0244) |
| kodim23 | 13.5433 (0.0137) | **13.5610** (0.0052) | 13.4940 (0.0503) |
| kodim24 | **13.8239** (0.2716) | 13.8237 (0.2717) | 13.5604 (0.1359) |

In conjunction with the quantitative assessment, Figs. 3 and 4 complement the analysis by offering visual depictions of the outcomes yielded by the Gauss error function for image contrast enhancement. These graphical representations vividly showcase the substantial enhancements in contrast and illumination achieved by the diverse algorithms. Such visual evidence serves to underscore the exceptional performance of the Gauss error function when synergized with the three adopted metaheuristics-GA, DE, and PSO.

The ensuing section embarks on a comprehensive comparative discussion, delving into the highlighted results obtained from the experimentation. This discussion endeavors to provide deeper insights into the strengths and weaknesses inherent in each algorithm, thereby illuminating their respective contributions to contrast

**Table 2** SSIM metric, obtained through the application of GA, DE, and PSO algorithms

| Image | SSIM | | |
|---|---|---|---|
| | GA | DE | PSO |
| kodim1 | 0.6848 (0.0071) | 0.6876 (0.0053) | **0.6904** (0.0055) |
| kodim2 | 0.2558 (0.0000) | 0.2558 (0.0000) | 0.2558 (0.0000) |
| kodim3 | 0.7608 (0.0020) | **0.7614** (0.0000) | 0.7592 (0.0035) |
| kodim4 | 0.6653 (0.0102) | 0.6647 (0.0001) | **0.6699** (0.0015) |
| kodim5 | 0.7406 (0.0001) | 0.7406 (0.0000) | **0.7412** (0.0002) |
| kodim6 | **0.6300** (0.0000) | 0.6282 (0.0055) | **0.6300** (0.0000) |
| kodim7 | **0.7266** (0.0067) | 0.7246 (0.0044) | 0.7238 (0.0051) |
| kodim8 | 0.7362 (0.0000) | 0.7362 (0.0000) | 0.7362 (0.0000) |
| kodim9 | 0.7540 (0.0059) | 0.7527 (0.0056) | **0.7614** (0.0003) |
| kodim10 | 0.7574 (0.0060) | 0.7519 (0.0000) | **0.7639** (0.0023) |
| kodim11 | 0.7144 (0.0012) | 0.7145 (0.0006) | **0.7234** (0.0058) |
| kodim12 | **0.7499** (0.0000) | **0.7499** (0.0000) | 0.7497 (0.0000) |
| kodim13 | 0.7235 (0.0000) | 0.7235 (0.0000) | **0.7236** (0.0001) |
| kodim14 | 0.7383 (0.0001) | 0.7383 (0.0000) | **0.7390** (0.0003) |
| kodim15 | **0.8095** (0.0005) | 0.8093 (0.0000) | 0.8093 (0.0004) |
| kodim16 | 0.7480 (0.0005) | 0.7471 (0.0002) | **0.7512** (0.0017) |
| kodim17 | 0.7387 (0.0000) | 0.7387 (0.0000) | **0.7400** (0.0017) |
| kodim18 | 0.7546 (0.0028) | **0.7594** (0.0012) | 0.7547 (0.0034) |
| kodim19 | 0.7313 (0.0005) | 0.7303 (0.0000) | **0.7318** (0.0001) |
| kodim20 | **0.6125** (0.0000) | **0.6125** (0.0000) | 0.6125 (0.0003) |
| kodim21 | 0.7765 (0.0017) | **0.7775** (0.0000) | **0.7775** (0.0017) |
| kodim22 | 0.6902 (0.0059) | **0.6952** (0.0000) | 0.6915 (0.0013) |
| kodim23 | 0.7984 (0.0003) | 0.7988 (0.0001) | **0.7991** (0.0006) |
| kodim24 | 0.7377 (0.0007) | 0.7376 (0.0007) | **0.7383** (0.0004) |

enhancement within the Gauss function framework. This elucidative discourse fosters a nuanced understanding of algorithmic efficacy in contrast enhancement, enriching the discourse surrounding image processing methodologies.

## 6 Analysis of Results

As mentioned earlier, the results generated by the three suggested algorithms were examined, taking into account the three indicators used to evaluate the contrast enhancement in the images. Additionally, a discussion is provided for each performance indicator.

**Table 3** REC metric, obtained through the application of GA, DE, and PSO algorithms

| Image | REC | | |
|---|---|---|---|
| | GA | DE | PSO |
| kodim1 | **1.0376** (0.0025) | 1.0364 (0.0017) | 1.0358 (0.0019) |
| kodim2 | 0.4872 (0.0001) | 0.4872 (0.0000) | 0.4872 (0.0000) |
| kodim3 | 1.0491 (0.0002) | **1.0492** (0.0000) | 1.0476 (0.0026) |
| kodim4 | 1.0148 (0.0024) | **1.0166** (0.0000) | 1.0105 (0.0016) |
| kodim5 | **1.0634** (0.0001) | 1.0633 (0.0000) | **1.0634** (0.0004) |
| kodim6 | **0.9630** (0.0000) | 0.9623 (0.0021) | 0.9629 (0.0000) |
| kodim7 | 1.0518 (0.0035) | 1.0506 (0.0021) | **1.0523** (0.0030) |
| kodim8 | 1.0151 (0.0000) | 1.0151 (0.0000) | 1.0151 (0.0000) |
| kodim9 | 1.0017 (0.0037) | 1.0009 (0.0037) | **1.0032** (0.0003) |
| kodim10 | 1.0090 (0.0049) | 1.0045 (0.0000) | **1.0128** (0.0016) |
| kodim11 | **1.0867** (0.0020) | 1.0858 (0.0004) | 1.0861 (0.0040) |
| kodim12 | 0.9987 (0.0001) | **0.9987** (0.0000) | 0.9979 (0.0000) |
| kodim13 | 1.0463 (0.0000) | 1.0463 (0.0000) | **1.0466** (0.0012) |
| kodim14 | 1.0698 (0.0000) | 1.0698 (0.0000) | **1.0699** (0.0002) |
| kodim15 | 1.0280 (0.0000) | 1.0280 (0.0000) | 1.0280 (0.0001) |
| kodim16 | 1.0360 (0.0001) | 1.0359 (0.0000) | **1.0378** (0.0013) |
| kodim17 | 1.0751 (0.0000) | 1.0752 (0.0000) | **1.0756** (0.0002) |
| kodim18 | 1.0403 (0.0004) | **1.0410** (0.0002) | 1.0403 (0.0005) |
| kodim19 | 1.0330 (0.0007) | **1.0344** (0.0000) | 1.0327 (0.0000) |
| kodim20 | 0.9703 (0.0000) | 0.9703 (0.0000) | 0.9703 (0.0000) |
| kodim21 | 1.0461 (0.0044) | **1.0484** (0.0000) | 1.0422 (0.0030) |
| kodim22 | 1.0053 (0.0022) | **1.0074** (0.0000) | 1.0046 (0.0009) |
| kodim23 | 1.0554 (0.0002) | 1.0553 (0.0000) | **1.0564** (0.0008) |
| kodim24 | 1.0518 (0.0041) | 1.0518 (0.0037) | **1.0558** (0.0021) |

## 6.1 *Peak Signal-to-Noise Ratio*

Table 1 presents the PSNR values derived from applying the Gauss error function within the GA, DE, and PSO algorithms for contrast enhancement in various images. Notably, the PSO algorithm consistently outperforms both DE and GA in terms of PSNR across most images within the Kodak dataset. This indicates the superior effectiveness of the PSO algorithm when employing the Gauss error function for enhancing image contrast. Figures 3 and 4 visually represent the results obtained from each image contrast enhancement algorithm. This graphical depiction allows for a clear observation of the changes in the contrast between the reference image and the enhanced image. The images improved using the Gauss error function, exhibiting a more visually appealing appearance and greater detail. The enhancements introduced by this function contribute to an overall improvement in image quality, as evidenced by the discernible contrast changes visible in the figures. This suggests that applying the Gauss error function is particularly effective in achieving enhanced visual outcomes in the context of image contrast enhancement.

### 6.2 Structural Similarity Index

The data obtained through the application of the SSIM metric utilizing GA, DE, and PSO algorithms is comprehensively presented in Table 2. Notably, a substantial proportion, at least 85%, of the gathered data consistently demonstrates SSIM values exceeding 0.70. This observation is indicative of commendable performance in terms of structural similarity when assessing the quality of the enhanced images. Table 2 implemented algorithms' efficacy in enhancing the images' visual quality PSO algorithm. Across most of the images examined, the PSO algorithm consistently outperforms the DE and GA algorithms, showcasing superior structural similarity values. This underscores the efficacy of the PSO algorithm in achieving enhanced image quality. To complement the qualitative analysis, a visual representation of the comparison between the results obtained by each algorithm and the reference image is provided in Figs. 3 and 4. The figures vividly illustrate the discernible improvements in both brightness and contrast within the enhanced images. This visual confirmation reinforces the quantitative findings and underscores the implemented algorithms' efficacy in enhancing the images' visual quality. The visual representation in Figs. 3 and 4 serves as a compelling visual testament to the efficacy of these algorithms in significantly improving the brightness and contrast of the images under consideration.

### 6.3 Relative Enhancement Contrast

The contrast improvement brought about by implementing the GA, DE, and PSO algorithms was quantitatively evaluated when calculating the REC metric. The outcomes, elucidated in Table 3, exhibit a range from 0.487 to 1.086. This range indicates that the algorithms have consistently enhanced the image contrast, with the lowest value representing a moderate improvement and the highest value indicating a significant enhancement. Table 3 provides a comprehensive breakdown of REC metric results for each algorithm—GA, DE, and PSO. After scrutinizing these findings, it becomes apparent that the DE and PSO algorithms have yielded comparable outcomes. This comparability underscores the reliability of our research, with a proportional distribution of images exhibiting standout REC values. While GA trails behind in prevalence compared to these algorithms, the attained values are not significantly inferior to those achieved by the DE and PSO algorithms. For a more intuitive understanding of the contrast enhancement, one can refer to Figs. 3 and 4. This visual representation not only facilitates a direct comparison between the

**Fig. 3** Improved images achieved through the use of GA, DE, and PSO algorithms applied to benchmark images kodim04, kodim08, kodim14, kodim16, and kodim17

reference and enhanced images but also serves as a robust validation of the quantitative REC metric results. It clearly showcases the discernible improvement in image contrast, reinforcing the applied algorithms' effectiveness in enhancing the images' contrast.

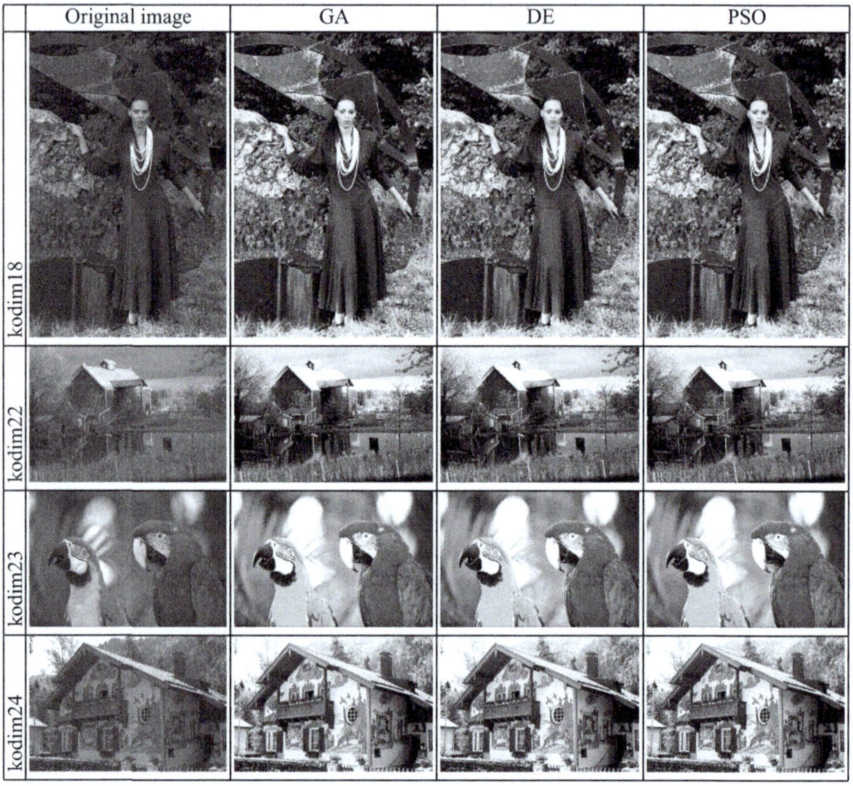


**Fig. 4** Improved images achieved through the use of GA, DE, and PSO algorithms applied to benchmark images kodim18, kodim22, kodim23, and kodim24

## 7 Conclusions and Future Work

This chapter has introduced an innovative technique for enhancing image contrast, leveraging the Gauss error function in conjunction with a Genetic Algorithm, a Differential Evolution algorithm, and a Particle Swarm Optimization algorithm. The method facilitates the refinement of intensity levels in each pixel, which is particularly beneficial for low-contrast images. Employing bio-inspired metaheuristics enables dynamic adjustment of the Gauss error function parameters, effectively addressing image contrast deficiencies. Furthermore, this contrast-enhancement approach is adaptable to the luminosity of an image, offering a means to elevate contrast levels according to the image's inherent characteristics. Initially applied to three traditional metaheuristics, the proposed transformation function demonstrates versatility, potentially extending to other bio-inspired approaches. Experimental findings confirmed the efficacy of the Gauss error function in contrast enhancement, with the PSO algorithm showcasing superior performance across various metrics when compared to

DE and GA. Notably, the PSO algorithm exhibited better performance in terms of the PSNR, SSIM, and REC metrics than DE and GA.

Future work in this research area is extensive, with several challenges requiring further investigation and experimentation. It is crucial to explore powerful bio-inspired metaheuristics that can effectively address the issue of local optima, currently caused by the characteristics of image luminosity. Additionally, an exhaustive exploration of quality indicators to quantify image contrast is necessary. This exploration will enable the development of new objective functions that can guide the behavior of metaheuristics, ultimately leading to improved results in terms of image contrast. Furthermore, there is a need to design new transformation functions for contrast enhancement that offer greater flexibility and freedom in transforming images. This contribution will significantly enhance the field by providing more options for improving image quality. By synergistically combining these research directions, we aim to achieve even more remarkable contrast enhancement results. These forward-looking initiatives, which we invite researchers to join, present interesting and promising avenues. Together, we can push the boundaries of image processing and contribute to continuously advancing image quality enhancement techniques.

## References


1. Acharya, U.K., Kumar, S.: Genetic algorithm based adaptive histogram equalization (gaahe) technique for medical image enhancement. Optik **230**, 166273 (2021)
2. Ahmed, S., Ghosh, K.K., Bera, S.K., Schwenker, F., Sarkar, R.: Gray level image contrast enhancement using barnacles mating optimizer. IEEE Access **8**, 169196–169214 (2020)
3. Asamoah, D., Ofori, E., Opoku, S., Danso, J.: Measuring the performance of image contrast enhancement technique. Int. J. Comput. Appl. **181**(22), 6–13 (2018)
4. Bhandari, A.K., Kumar, A., Singh, G.K., Soni, V.: Dark satellite image enhancement using knee transfer function and gamma correction based on DWT-SVD. Multidimension. Syst. Signal Process. **27**, 453–476 (2016)
5. Braik, M.: Hybrid enhanced whale optimization algorithm for contrast and detail enhancement of color images. Cluster Comput. 1–37 (2022)
6. Braik, M.S.: Chameleon swarm algorithm: a bio-inspired optimizer for solving engineering design problems. Expert Syst. Appl. **174**, 114685 (2021)
7. Chen, J., Yu, W., Tian, J., Chen, L., Zhou, Z.: Image contrast enhancement using an artificial bee colony algorithm. Swarm Evol. Comput. **38**, 287–294 (2018)
8. Dorigo, M., Di Caro, G.: Ant colony optimization: a new meta-heuristic. In: Proceedings of the 1999 Congress on Evolutionary Computation-CEC99 (Cat. No. 99TH8406), vol. 2, pp. 1470–1477. IEEE (1999)
9. Eastman Kodak Company: Kodak Lossless True Color Image Suite (2013). https://r0k.us/graphics/kodak/
10. García-Nájera, A., Zapotecas-Martínez, S., Bernal-Jaquez, R.: Selection schemes analysis in genetic algorithms for the maximum influence problem. In: Advances in Soft Computing: 19th Mexican International Conference on Artificial Intelligence, MICAI 2020, Mexico City, Mexico, October 12–17, 2020, Proceedings, Part I 19, pp. 211–222. Springer, Berlin (2020)
11. Goldenberg, D.: Genetic Algorithms in Search, Optimization and Machine Learning. Reading, Adisson Wesley (1989)

12. Guraksin, G.E., Deperlioglu, O., Kose, U.: A novel underwater image enhancement approach with wavelet transform supported by differential evolution algorithm. In: Nature Inspired Optimization Techniques for Image Processing Applications, pp. 255–278 (2019)
13. Hoseini, P., Shayesteh, M.G.: Efficient contrast enhancement of images using hybrid ant colony optimisation, genetic algorithm, and simulated annealing. Digital Signal Process **23**(3), 879–893 (2013)
14. Joseph, J., Periyasamy, R.: A fully customized enhancement scheme for controlling brightness error and contrast in magnetic resonance images. Biomed. Signal Process. Control **39**, 271–283 (2018)
15. Kallel, F., Sahnoun, M., Ben Hamida, A., Chtourou, K.: CT scan contrast enhancement using singular value decomposition and adaptive gamma correction. Signal, Image Video Process. **12**, 905–913 (2018)
16. Kamoona, A.M., Patra, J.C.: A novel enhanced cuckoo search algorithm for contrast enhancement of gray scale images. Appl. Soft Comput. **85**, 105749 (2019)
17. Kanmani, M., Narsimhan, V.: An image contrast enhancement algorithm for grayscale images using particle swarm optimization. Multimed. Tools Appl. **77**, 23371–23387 (2018)
18. Kannan, P., Deepa, S., Ramakrishnan, R.: Contrast enhancement of sports images using two comparative approaches. Am. J. Intell. Syst. **2**(6), 141–147 (2012)
19. Kennedy, J., Eberhart, R.: Particle swarm optimization. In: Proceedings of ICNN'95-International Conference on Neural Networks, vol. 4, pp. 1942–1948. IEEE (1995)
20. Khan, A.H., Ahmed, S., Bera, S.K., Mirjalili, S., Oliva, D., Sarkar, R.: Enhancing the contrast of the grey-scale image based on meta-heuristic optimization algorithm. Soft Comput. **26**(13), 6293–6315 (2022)
21. Kirkpatrick, S., Gelatt, C.D., Jr., Vecchi, M.P.: Optimization by simulated annealing. Science **220**(4598), 671–680 (1983)
22. Ling, Z., Liang, Y., Wang, Y., Shen, H., Lu, X.: Adaptive extended piecewise histogram equalisation for dark image enhancement. IET Image Process. **9**(11), 1012–1019 (2015)
23. Malik, R., Dhir, R., Mittal, S.K.: Remote sensing and landsat image enhancement using multiobjective PSO based local detail enhancement. J. Ambient Intell. Humanized Comput. **10**, 3563–3571 (2019)
24. Meyarivan, T., Deb, K., Pratap, A., Agarwal, S.: A fast and elitist multiobjective genetic algorithm: NSGA-II. IEEE Trans. Evol. Comput. **6**(2), 182–197 (2002)
25. Mirjalili, S., Lewis, A.: The whale optimization algorithm. Adv. Eng. Softw. **95**, 51–67 (2016)
26. Negi, S.S., Bhandari, Y.S.: A hybrid approach to image enhancement using contrast stretching on image sharpening and the analysis of various cases arising using histogram. In: International Conference on Recent Advances and Innovations in Engineering (ICRAIE-2014), pp. 1–6. IEEE (2014)
27. Park, S., Shin, Y.G., Ko, S.J.: Contrast enhancement using sensitivity model-based sigmoid function. IEEE Access **7**, 161573–161583 (2019)
28. Patel, P., Bhandari, A.: A review on image contrast enhancement techniques. Int. J. Online Sci. **5**(5), 14–18 (2019)
29. Pérez-Enríquez, L., Zapotecas-Martínez, S., Oliva, D., Altamirano-Robles, L.: Hyperbolic tangent sigmoid as a transformation function for image contrast enhancement. In: 2023 IEEE Symposium Series on Computational Intelligence (SSCI), pp. 282–287. IEEE (2023)
30. Pizer, S.M., Amburn, E.P., Austin, J.D., Cromartie, R., Geselowitz, A., Greer, T., ter Haar Romeny, B., Zimmerman, J.B., Zuiderveld, K.: Adaptive histogram equalization and its variations. Comput. Vis. Graph. Image Process. **39**(3), 355–368 (1987)
31. Qi, Y., Yang, Z., Sun, W., Lou, M., Lian, J., Zhao, W., Deng, X., Ma, Y.: A comprehensive overview of image enhancement techniques. Arch. Comput. Methods Eng. 1–25 (2021)
32. Raji, A., Thaibaoui, A., Petit, E., Bunel, P., Mimoun, G.: A gray-level transformation-based method for image enhancement. Pattern Recognit. Lett. **19**(13), 1207–1212 (1998)
33. Rao, B.S.: Dynamic histogram equalization for contrast enhancement for digital images. Appl. Soft Comput. **89**, 106114 (2020)

34. Russo, F.: Piecewise linear model-based image enhancement. EURASIP J. Adv. Signal Process. **2004**, 1–9 (2004)
35. Sara, U., Akter, M., Uddin, M.S.: Image quality assessment through FSIM, SSIM, MSE and PSNR-a comparative study. J. Comput. Commun. **7**(3), 8–18 (2019)
36. Storn, R., Price, K.: Differential evolution-a simple and efficient heuristic for global optimization over continuous spaces. J. Global Optim. **11**, 341–359 (1997)
37. Sulaiman, M.H., Mustaffa, Z., Saari, M.M., Daniyal, H.: Barnacles mating optimizer: a new bio-inspired algorithm for solving engineering optimization problems. Eng. Appl. Artif. Intell. **87**, 103330 (2020)
38. Surya Prabha, D., Satheesh Kumar, J.: An efficient image contrast enhancement algorithm using genetic algorithm and fuzzy intensification operator. Wirel. Pers. Commun. **93**, 223–244 (2017)
39. Tiwari, M., Gupta, B.: Brightness preserving contrast enhancement of medical images using adaptive gamma correction and homomorphic filtering. In: 2016 IEEE Students' Conference on Electrical, Electronics and Computer Science (SCEECS), pp. 1–4. IEEE (2016)
40. Vij, K., Singh, Y.: Enhancement of images using histogram processing techniques. Int. J. Comput. Tech. Appl. **2**(2), 309–313 (2009)
41. Wang, W., Yuan, X., Chen, Z., Wu, X., Gao, Z.: Weak-light image enhancement method based on adaptive local gamma transform and color compensation. J. Sens. **2021**, 1–18 (2021)
42. Zuiderveld, K.: Contrast Limited Adaptive Histogram Equalization, pp. 474–485. Academic Press Professional, Inc., USA (1994)

# Diversity Measurement in Different PSO Variants Applied to Global Optimization and Classical Engineering Problems

**Angel Casas-Ordaz, Jorge Ramos-Frutos, Mario A. Navarro, Eduardo H. Haro, Bernardo Morales-Castañeda, Marco Perez-Cisneros, Noé Ortega-S, and Daniel Zaldivar Navarro**

**Abstract** The Particle Swarm Optimization algorithm is one of the most popular swarm-based algorithms for solving global optimization problems, as evidenced by its prevalence in the literature. Although the PSO algorithm demonstrates satisfactory performance in terms of global optimization, it still exhibits shortcomings in terms of convergence and stagnation at local optima. Consequently, researchers have been exploring alternative implementations to address these limitations. As a result, the PSO algorithm has undergone numerous improvements and implementations since its original proposal. The improvements include different ways of initializing the swarm and self-configuration of the parameters, as well as hybridizations with other well-known algorithms. However, a comparative analysis of the population diversity behavior between the different variants has not been performed. As is known, population diversity within Evolutionary Algorithms can be used as an indication of the

A. Casas-Ordaz (✉) · M. A. Navarro · E. H. Haro · B. Morales-Castañeda · M. Perez-Cisneros · N. Ortega-S · D. Z. Navarro
Depto. de Ingeniería Elctro-Fotónica, Universidad de Guadalajara, CUCEI, Guadalajara, Mexico
e-mail: angel.casas5699@alumnos.udg.mx

M. A. Navarro
e-mail: mario.navarro@academicos.udg.mx

E. H. Haro
e-mail: eduardo.hernandezh@academicos.udg.mx

B. Morales-Castañeda
e-mail: juanbernardo.morales@academicos.udg.mx

M. Perez-Cisneros
e-mail: marco.perez@cucei.udg.mx

N. Ortega-S
e-mail: noe.ortega@academicos.udg.mx

D. Z. Navarro
e-mail: daniel.zaldivar@academicos.udg.mx

J. Ramos-Frutos
Posgrados, Centro de Innovación Aplicada en Tecnologías Competitivas, Guadalajara, Mexico
e-mail: jramos.estudiantepicyt@ciatec.mx

 
D. Oliva et al. (eds.), *Advances in Optimization Algorithms for Multidisciplinary Engineering Applications: From Classical Methods to AI-Enhanced Solutions*, Studies in Computational Intelligence 806, https://doi.org/10.1007/978-3-031-78440-8_5

distribution of solutions or population particles within the search space. This paper aims to analyze some of the PSO variants, focusing on the population diversity of each version. Subsequently, the behavior of the variants on engineering problems and the CEC 2017 dataset will be analyzed.

## 1 Introduction

The PSO algorithm is a stochastic technique based on swarm behavior. It was first introduced by Kennedy and Eberhart in 1995 [19]. The algorithm was based on the observation of the foraging behavior of a flock of birds. One of the peculiarities of this algorithm is that it is population-based and cooperative in nature. Each flock member changes its foraging pattern according to its own and other members' learning experiences. This social behavior of the birds has been translated into algorithmic operations, giving rise to the PSO algorithm as a global optimization algorithm, where the flock of birds is interpreted as a flock of particles, and each particle represents a candidate solution. The PSO algorithm is also characterized by a good balance between exploitation and exploration of its iterative search process. The PSO algorithm continues to attract the attention of researchers in recent decades due to its simplicity and few control parameters. Due to its advantages, PSO is a promising candidate for solving real-world optimization problems. Over the last few decades, we can find reviews in the PSO literature that provide a comprehensive review covering basic PSO concepts, binary PSO, and PSO topologies, among other things, as well as some remarkable engineering applications of PSO [31]. Some reviews cover aspects such as its application in specific areas [14, 16, 20]. As well as those that focus more on the existing variants of PSO [5, 28]. Only PSO of continuous space searches are considered, and PSO in binary form is excluded. In addition, [31] provides an informative table summarizing recent and important state-of-the-art survey papers. Another interesting and recent overview of PSO variants is the one published by [18], which covers basic concepts, several advanced versions, modifications, hybridizations, and theoretical analysis of PSO. Continuing with another interesting work, which studies the origin and background and performs the theoretical analysis of PSO, is the one done by [35]. This interesting work analyzes the current situation of research and application in algorithm structure, parameter selection, topology structure, discrete PSO algorithm and parallel PSO algorithm, multi-objective PSO optimization, and its engineering applications. Among the different works that make an investigation and overview and a general description of the PSO variants is the work of [17]. This research covers different variants of the PSO concerning aspects such as initialization, inertia weight, and mutation operators. Finally, some of the works related to a revision of the PSO algorithm, including one by the same author of the original PSO algorithm (Eberhart), are the following [30, 32, 33]. These studies are among the first that could be found in the literature that researched the different variants of the PSO algorithm that had been developed up to the time of publication of each article. Among each of these research papers, which are more related to overviews and revisions of the PSO algorithm, one can find a great variety of variant algorithms that have emerged from the original PSO algorithm. Studying each of them is a lot of hard work and an important task. In this work, the purpose is to study some of the variants of the PSO algorithm that can be found in the literature and to analyze them through a comparative study focused on two parts: the first part focuses on a comparative study of the implementation in single objective global optimization problems. At the same time, the second part focuses on engineering problems.

The remainder of the paper is organized as follows: Sect. 2 briefly introduces the algorithms used for the comparative analysis. Section 3 introduces the topic of diversity in metaheuristic algorithms. Then, Sect. 4 presents the implementation of the experimental analysis. Finally, Sect. 5 briefly mentions the conclusion and future work.

## 2 Particle Swarm Optimization Variants

This section analyzes the variants of the PSO algorithm, including this comparison. It is only intended to provide a brief introduction to each variant's proposed modification.

### *2.1 Particle Swarm Optimization*

Kennedy and Eberhart first presented PSO as a continuous real-valued algorithm [19]. Since their proposal, it has become one of the most widely used metaheuristic algorithms for single-objective optimization problems worldwide. The algorithm is inspired by the behavior of flocks of birds and shoals of fish. The behavior of the PSO algorithm starts with the exploration of the search space by randomly initializing the particles in the swarm, denoted as $i = \{1, 2, \ldots, N_p\}$, where $N_p$ is the maximum number of particles in the swarm. Each particle $i$ has a current position, denoted by $x$, and a current velocity $v$, which is adjusted throughout the iterative process. In the PSO algorithm, the motion of the particles is guided according to the best global and local positions. The best position per particle is denoted $p(t)$, while the best position $g(t)$ is identified as the best position of the whole swarm so far. The basic structure of the PSO algorithm can be seen in the flowchart in Fig. 2. The velocity and particle motion of the swarm in the PSO algorithm are mathematically represented as:

$$v_i(t+1) = w \cdot v_i(t) + c_1 r_1 (p_i(t) - x_i(t)) + c_2 r_2 (g(t) - x_i(t)) \quad (1)$$

$$x_i(t+1) = x_i(t) + v_i(t) \quad (2)$$

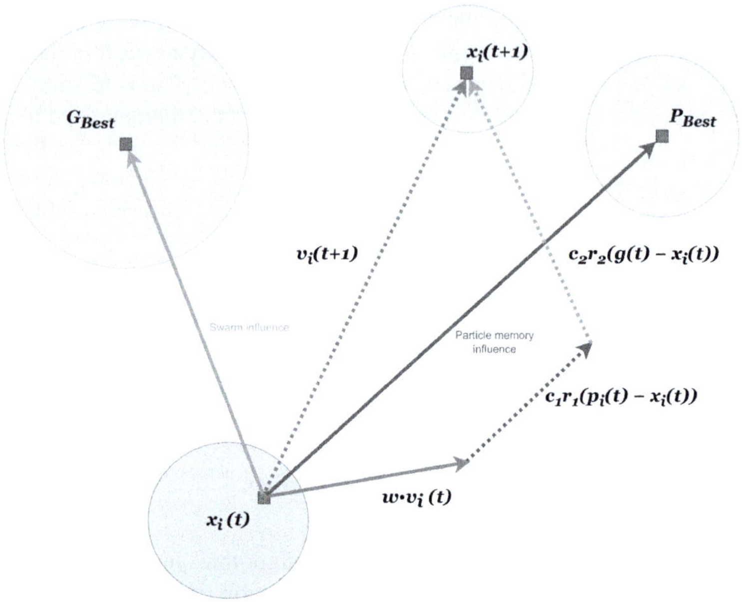


**Fig. 1** Representation of PSO velocity and particle locations

where $v_i(t+1)$ and $v_i(t)$ denote the velocity of the $i$th particle at the next iteration and the current iteration, respectively. $w$ is the inertia weight that controls the rate of change of the particle's position, and values range from $w_{max} = 0.9$ to $w_{min} = 0.4$, decreasing linearly. $c_1$ and $c_2$ are the cognitive and social coefficients, which are usually set to constant and are also called learning factors. $r_1$ and $r_2$ are random values in the range [0, 1]. $x_i(t+1)$ and $x_i(t)$ are the position vectors of the $i$th particle at iteration $t+1$ and $t$, respectively. $P_{Best} = p_i(t)$ is the best position of the $i$th particle at the current iteration, and $G_{Best} = g(t)$ is the global best position in the swarm. A graphical representation of the behavior of the PSO algorithm is shown in Fig. 1.

## 2.2 *Autonomous Groups Particles Swarm Optimization*

A modification to PSO was made by [23], called Autonomous Group Particle Swarm Particle Swarm Optimization (AGPSO). This algorithm was proposed to alleviate further the two problems of trapping in local minima and slow convergence rate in solving high-dimensional problems. The main idea of the AGPSO algorithm is inspired by the diversity of individuals in flocks of birds or swarms of insects. It is based on the premise that in natural colonies, individuals are basically not very similar regarding intelligence and ability. Still, they all fulfill their duties as members of a

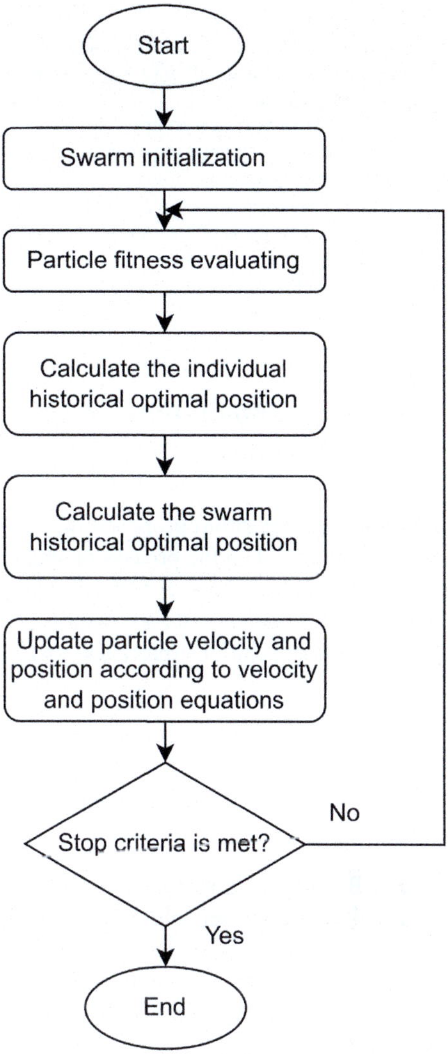


**Fig. 2** Flowchart of the PSO algorithm

colony. The ability of each individual can be useful in a particular situation. Different functions with different slopes, curvatures, and intercepts are proposed to adjust the social and cognitive parameters of the PSO algorithm to give the particles different behaviors as in natural colonies. The modifications concerning this proposal are basically described in the pseudo-code of Algorithm 1. A more detailed proposal description can be found in [23].

**Algorithm 1** Pseudo-code for AGPSO proposal [23]

```
1: Create and initialize a D-dimensional PSO.
2: Divide particles randomly into autonomous groups.
3: For Each particle:
4:     Extract the particle's group.
5:     Use its group strategy to update c1 and c2.
6:     Extract the particle's group.
7:     Use c1 and c2 to update velocities.
8: Use new velocities to define new positions
9: End for.
10: Until stopping condition is satisfied
```

### *2.3 Biogeography-Based Learning Particle Swarm Optimization*

BLPSO is an algorithm proposed in 2017 by Chen et al. [7]. The main features of BLPSO are based on the process of species migration in biogeography, where good species (particles) disperse to favorable habitats (regions of the search space with high solution quality) and bad species are eliminated. BLPSO uses two populations: a population of solutions (particles) and a population of habitats, and has three main operators:

- Migration: Particles migrate to better-quality habitats according to probability.
- Selection: The best particles in each habitat are selected to reproduce.
- Mutation: Particles mutate with a probability to explore new regions of the search space.

The BLPSO search strategy is exploratory and exploitative; migration and mutation enable exploration, while selection enables exploitation.

The advantages highlighted by the authors of this algorithm are:

**Good convergence capability**: BLPSO can find high-quality solutions with a high probability.

**Robustness**: BLPSO is robust to different types of problems and conditions.

**Scalability**: BLPSO can be scaled to large problem sizes.

On the other hand, the disadvantages are:

**Computational cost**: BLPSO can be more computationally expensive than other PSO algorithms.

**Parameters to be tuned**: BLPSO has several parameters that must be tuned to obtain good performance.

The main equations that make the search strategy of the BLPSO algorithm possible are summarized below.

$$E_i^t = 1 - S_i^t / H_i^t \tag{3}$$

where $E_i^t$ is the probability of emigration of particle $i$ at time $t$, $S_i^t$ is the number of "good" species in the habitat of particle $i$ at time $t$, and $H_i^t$ is the maximum number of species in the habitat of particle $i$ at time $t$.

$$I_i^t = \frac{S_i^t}{H_i^t} \sum_{j=1}^{N} E_j^t \tag{4}$$

where, $I_i^t$ is the probability of immigration of particle $i$ at time $t$, and $N$ is the particles number on the population.

The equation for updating the position of each particle is defined below:

$$x_i^{t+1} = x_i^t + v_i^{t+1} \tag{5}$$

where, $x_i^{t+1}$ is the position of particle $i$ at time $t+1$, $x_i^t$ is the position of particle $i$ at time $t$, and $v_i^{t+1}$ is the velocity of particle $i$ at time $t+1$.

Another equation of utmost importance for the method is the equation that defines the update of the velocity of each particle, and it is defined in Eq. 6:

$$v_i^{t+1} = wv_i^t + c_1 r_1 (p_i^t - x_i^t) + c_2 r_2 (g^t - x_i^t) \tag{6}$$

where:

- $v_i^{t+1}$: Velocity of particle $i$ at time $t+1$
- $v_i^t$: Velocity of particle $i$ at time $t$
- $w$: Inertia factor
- $c_1$ and $c_2$: Personal and social learning factor
- $r_1$ and $r_2$: Uniform random numbers between 0 and 1
- $p_i^t$: Personal best position of particle $i$ at time $t$
- $g^t$: Best overall population position at time $t$.

Finally, two indispensable equations for the BLPSO algorithm are shown; one represents the update of a habitat (Eq. 7), and the other is the method of selection of the best global position (Eq. 8).

$$H_i^{t+1} = H_i^t + \eta(E_i^t - I_i^t) \tag{7}$$

where $H_i^{t+1}$ is the number of species in the habitat of particle $i$ at time $t+1$, $H_i^t$ is the number of species in the habitat of particle $i$ at time $t$, and $\eta$, which is the habitat learning factor.

$$g^t = \arg\min_{i=1}^{N} f(x_i^t) \tag{8}$$

where $g^t$ is the best overall position of the population at time $t$, and $f(x_i^t)$ is the value of the objective function at position $x_i^t$.

These are the equations governing the BLPSO algorithm's search strategy. In practice, different variants of these equations can be used to improve the algorithm's performance.

## 2.4 Comprehensive Learning Particle Swarm Optimizer

In this variant of the PSO, a new strategy is proposed for calculating the speed using Eq. 9.

$$V_i^{t+1} = wV_i^t + cr_1(pbest_{f_i} - X_i^t) \tag{9}$$

where $c$ is the weighting factor, $w$ is a weighting function, $r_1$ is a random number between [0, 1], $f_i$ defines which particles of the $pbests$ maintain in the evolution of the algorithm, $pbest_{f_i}$ are the particles that should continue taking into account a probability vector $Pc$, from which a threshold and random numbers are defined to decide whether the particle continues or not, in the iterative process. After choosing these particles, a selection procedure is carried out as numbered below:

1. Two particles are chosen from the population.
2. The fitness of these two particles is compared, and the best one is chosen.
3. The $pbest$ winners are used as the exemplars to learn from them in the dimensions of the residents. If all exemplars are tied, another $pbest$ dimension in the particle is chosen randomly and interchanged into the other $pbest$ particle to maintain diversity.

## 2.5 An Improved PSO with Time-Varying Accelerator Coefficients

IPSO is an algorithm proposed in 2008 by Cui et al. [11]. IPSO is a variant of the Particle Swarm Optimization (PSO) algorithm that uses time-varying acceleration coefficients.

The acceleration coefficients are dynamically adjusted during algorithm execution to improve convergence and accuracy.

IPSO is executed in two phases: an exploration phase and an exploitation phase.

It is worth mentioning that IPSO uses the same operators as the standard PSO, i.e.: Particles update their velocity based on their personal best position, the global best position, and a random number. The position is updated based on its current velocity.

As for its heuristics, IPSO uses an exploratory and exploitative search strategy. The exploration phase uses higher acceleration coefficients to allow the particles to explore the search space. The exploitative phase uses lower acceleration coefficients to allow the particles to converge to an optimal solution.

The advantages highlighted by the authors of this algorithm are:

**Better convergence**: IPSO converges to optimal solutions faster than standard PSO.

**Higher accuracy**: IPSO generally finds higher quality solutions than standard PSO.

**Robustness**: IPSO is robust to different types of problems and conditions.

On the other hand, the disadvantages are:

**More complex**: IPSO is more complex than standard PSO due to the need to adjust acceleration coefficients.

**Computational cost**: IPSO can be more computationally expensive than standard PSO.

The equations of the IPSO algorithm are the same as those of the standard PSO, with the difference that the acceleration coefficients $c_1$ and $c_2$ are the same and are dynamically adjusted in time.

The equations for updating the acceleration coefficients are presented below:

$$c_1^t = c_1^{\min} + (c_1^{\max} - c_1^{min}) \cdot \frac{t}{T} \tag{10}$$

$$c_2^t = c_2^{\min} + (c_2^{\max} - c_2^{\min}) \cdot \frac{t}{T} \tag{11}$$

where $t$ is the current iteration, $c_1^{\min}$ and $c_1^{\max}$ are the minimum and maximum values of the coefficient $c_1$. In contrast, $c_2^{\min}$ and $c_2^{\max}$ are the minimum and maximum values of the coefficient $c_2$, and $T$ is the maximum number of iterations.

### *2.6 Particle Swarm Optimization Gravitational Search Algorithm*

Mirjalili and Mohd [22] proposed the Particle Swarm Optimization and Gravitational Search Algorithm (PSOGSA) in 2012 as a hybrid algorithm that combines the PSO and the Gravitational Search Algorithm (GSA). GSA is a metaheuristic based on the physics of Newton's Universal Law of Gravitation, proposed in 2009 by Rashedi et al., considering that each particle in the universe attracts the surrounding particles.

In PSOGSA, the first step consists of randomly generating individuals. After initializing the population with $N$ number of individuals, it is necessary to calculate the gravitational force (Eq. 12),

$$F_{ij}^{t} = G^{t}\left(\frac{M_{pi}^{t} M_{aj}^{t}}{R_{ij}^{t} + \varepsilon}\right)(X_{j}^{t} - X_{i}^{t}) \tag{12}$$

where $M_{aj}^{t}$ is the active gravitational mass related to $j$th agent, $M_{pi}^{t}$ is the passive gravitational mass related to $i$th agent, $\varepsilon$ is a small constant, $G^{t}$ is gravitational constant in the generation $t$ computed with Eq. 13, $R_{ij}^{t}$ is the Euclidean distance between particles $X_{i}^{t}$ and $X_{j}^{t}$.

$$G^{t} = G_{0}e^{-\alpha\frac{iter}{Maxiter}} \tag{13}$$

here $\alpha$ is the descending coefficient, $G_0$ is the initial value of $G$, $iter$ is the current iteration, and $Maxiter$ is the maximum number of iterations defined within the algorithm. The resulting forces acting on the $i$th agent are calculated with Eq. 14.

$$F_{i}^{t} = \sum_{j=1}^{N} r_{j} F_{ij}^{t} \tag{14}$$

where $F_i$ is the total force that acts in the $i$th agent, $r_j$ is a random number in the interval [0, 1]. Having the force that acts on each particle, the acceleration for them must be calculated using Eq. 15

$$ac_{i}^{t} = \frac{F_{i}^{t}}{M_{i}^{t}} \tag{15}$$

$ac_{i}^{t}$ is the acceleration in the $t$th iteration, $M_i$ is the mass of the object $i$. After calculating the acceleration for each particle, the velocity must be obtained with Eq. 16

$$V_{i}^{t+1} = wV_{i}^{t} + c_{1}r_{1}ac_{i}^{t} + c_{2}r_{2}(gbest - X_{i}^{t}) \tag{16}$$

where $V_{i}^{t}$ is the velocity of $i$th agent at iteration $t$, $c_1$ and $c_2$ are weighting factors, $w$ is a weighting function, $r_1$ and $r_2$ are random numbers between [0, 1], $gbest$ is the best solution so far. In each iteration, $i$th position is updated with equation.

$$X_{i}^{t+1} = X_{i}^{t} + V_{i}^{t+1} \tag{17}$$

This procedure is carried out until a stop criterion is reached.

## 2.7 A Modified Particle Swarm Optimization with an Adaptive Acceleration Coefficients

TACPSO is an algorithm proposed in 2009 by Ziyu et al. [39]. TACPSO is a variant of the Particle Swarm Optimization (PSO) algorithm that uses adaptive acceleration coefficients. The acceleration coefficients are automatically adjusted according to the algorithm's progress. TACPSO is executed in two phases: an exploration phase and an exploitation phase.

Like some of the improved versions of PSO presented in this chapter, TACPSO uses the same updated equations for the velocity and position of each particle as the standard PSO.

As with many current heuristics designed by the canonical PSO, TACPSO uses an exploratory and exploitative search strategy. The exploration phase uses higher acceleration coefficients to allow the particles to explore the search space. On the other hand, the exploitation phase uses lower acceleration coefficients to allow the particles to converge to an optimal solution.

The advantages highlighted by the authors of this algorithm are:

**Better convergence**: TACPSO converges to optimal solutions faster than the standard PSO.

**Higher accuracy**: TACPSO generally finds higher quality solutions than the standard PSO.

**Robustness**: TACPSO is robust to different types of problems and conditions.

On the other hand, the disadvantages are:

**More complex**: TACPSO is more complex than standard PSO due to the need to adjust acceleration coefficients.

**Computational cost**: TACPSO can be more computationally expensive than standard PSO.

The applications for which TACPSO has been used since its appearance are varied, some of which are listed below:

- Function optimization: TACPSO can be used to optimize continuous and discrete functions.
- Mechanical engineering: TACPSO has been used to optimize the design of structures, mechanisms, and other mechanical systems.
- Electrical engineering: TACPSO has been used to optimize the design of electrical circuits and power systems.
- Finance: TACPSO has been used to optimize investment portfolios and trading strategies.

The equations of the TACPSO algorithm are the same as those of the standard PSO, with the difference that the acceleration coefficients $c_1$ and $c_2$ are automatically adjusted according to the progress of the algorithm and are defined as follows:

$$c_1^t = c_1^{\min} + (c_1^{\max} - c_1^{\min}) \cdot \frac{f(p_i^t) - f(g^t)}{f(p_i^{t-1}) - f(g^{t-1})} \tag{18}$$

$$c_2^t = c_2^{\min} + (c_2^{\max} - c_2^{\min}) \cdot \frac{f(g^t) - f(x_i^t)}{f(p_i^t) - f(x_i^t)} \tag{19}$$

where, $c_1^{\min}$ and $c_1^{\max}$ are the minimum and maximum values of coefficient $c_1$, and $c_2^{\min}$ and $c_2^{\max}$ are the minimum and maximum values of coefficient $c_2$.

### 2.8 *Particle Swarm Optimization Algorithm with Asymmetric Time-Varying Acceleration Coefficients*

This variant of the PSO was proposed by Bao and Mao [4] in 2009. This algorithm aims to improve the original PSO's exploitation capacity to achieve faster convergence on the global optimum. The PSO variant proposes that the social and cognitive components be calculated differently to achieve convergence of the global optimum. Equations 20 and 21 shows the proposal of these two authors on the PSO variant.

$$c_1 = c_{1\max} - \frac{k(c_{1\max} - c_{1\min})}{k_{\max}} \tag{20}$$

$$c_2 = c_{2\min} - \frac{k(c_{2\max} - c_{2\min})}{k_{\max}} \tag{21}$$

here $k$ is the $k$th iteration, $k_{\max}$ is the maximum number of iterations defined. This method with a time-varying inertia weight factor is suggested to start from 1 and decrease towards 0.4.

## 3 Diversity in Metaheuristic Algorithms

This section introduces the notion of diversity in the field of metaheuristic algorithms. It briefly analyzes each of the most typical diversities in the state of the art. One of them was chosen to be the focus of the experimentation section. Through extensive study of optimization algorithms, it is well known that one of the most important factors affecting the performance of such algorithms is the ability to maintain a good balance between exploration and exploitation [10]. Starting from this premise, exploration and exploitation are considered two fundamental cornerstones of any metaheuristic algorithm. Therefore, a proper balance can lead the algorithm to a successful convergence [29]. Considering that exploration is referred to as the ability of an algorithm to explore the different areas within the search space, and exploitation is searching through the potential areas in a way that increases its probability of finding better solutions. As previously studied [2], the reflection of this balance can be seen in the diversity of the population. The diversity of the population is described as an attribute in the distribution of solutions in the search space [26]. Therefore, if the diversity is low, it can be understood as a high concentration of solutions in the same search space region. On the other hand, if the diversity of the population is high, it can refer to the fact that the solutions are scattered throughout the search space [24]. This motivates the research to adopt diversity measurement techniques. Among an exhaustive study of diversity measures, stands out one proposed in [37], called "True Diversity" ($D_{TD}$), which quantifies the average of the standard deviation observed for each individual gene across the population solutions. Subsequently, in [9], a normalization of the "True Diversity" is proposed by means of a Normalization with Maximum Diversity so Far (NMDF). A further well-known measure of diversity is proposed in [8], which is named dimension-wise diversity. This diversity measure calculates how far or close each agent is within the search space. It thus provides an idea of how much the algorithm is exploring or exploiting in its process of finding the best solution. This last measure of diversity will be the one used to make the comparative analysis of the different types of PSOs covered in this research.

### 3.1 Dimension-Wise Diversity Measurement

In order to find suitable solutions for optimization problems, metaheuristic algorithms use a set of possible solutions to traverse the search space. Better-performing agents usually steer the search, reducing the distance between agents and directing more attention toward exploitation. On the other hand, when the distance between agents increases, the exploration process gains prominence.

The dimension-wise diversity measurement [8] is a diversity metric used to estimate the variations in distance across search agents. This measurement defines population diversity as:

$$Div_j = \frac{1}{n}\sum_{i=1}^{n}\left|\text{median}(x^j) - x_i^j\right|$$

$$Div = \frac{1}{m}\sum_{j=1}^{m} Div_j$$

where median$(x^j)$ represents the median of dimension $j$ in the whole population. $x_i^j$ is the dimension $j$ of search agent $i$. $n$ corresponds to the number of search agents in the population, while $m$ symbolizes the number of design variables of the optimization problem. The diversity in each dimension $Div_j$ is defined as the distance between the dimension $j$ of every search agent and the median of that dimension averaged. The diversity of the entire population $Div$ is then calculated by averaging every $Div_j$ in each dimension. Both values are computed in every iteration.

Using the median value as a reference point when evaluating the balance response promotes consistency. However, one limitation of the dimension-wise diversity index is that it smooths out minor abrupt changes in diversity due to averaging across all dimensions, which may partially cause a tendency toward exploitation.

## 3.2 Average Population Diversity

Recently, work published by [27] has attracted the attention of researchers as it proposes a new mechanism of population size adaptation in the Differential Evolution algorithm. This adaptation mechanism is based on a linear reduction of population diversity. The diversity of the current population in this work is calculated by the $ADP$ criterion as follows:

$$ADP = \sqrt{\frac{1}{N_p}\sum_{i=1}^{N_p}\sum_{j=1}^{S}(x_{ij} - \overline{x}_j)^2} \tag{22}$$

where $\bar{x_j}$ represents the arithmetic mean of the entire population in dimension $j$.

$$\overline{x}_j = \frac{1}{N_p}\sum_{i=1}^{N_p} x_{ij} \tag{23}$$

In turn, $APD$ estimates the distance among a point in the population and centroid $\overline{x} = (\overline{x}_1, \overline{x}_2, ..., \overline{x}_D)$. $D$ is the maximum number of population dimensions. Evidently, $ADP \geq 0$ and $ADP$ are equal to zero when all population points are identical.

## 3.3 True Diversity

In a study by Wineberg and Oppacher [37], a comparative analysis of different diversities used in the field of metaheuristic algorithms is performed. Prominent among them is one that is referred to as "true diversity" (TD). The $TD$ depicted by

Eq. 24 represents the average standard deviation of each individual in the population and normalizes to the maximum diversity so far $NMD$.

$$TD = \frac{\frac{1}{S}\sqrt{\sum_{j=1}^{S}(\bar{x}_j^2 - (\bar{x}_j)^2)}}{NMD} \quad where \;\; \bar{x}_j^2 = \frac{1}{N}\sum_{i=1}^{N} x_{ij}^2 \tag{24}$$

## 4 Experiments and Results

This section analyzes the behavior of the different variants of the PSO algorithm used for this comparison. It is divided into two parts: the first compares the different variants covered in this manuscript regarding global optimization objective functions. In contrast, the second part focuses on engineering problems. The experimental tests were performed in the Matlab 2023B test environment with a computer equipped with an Intel(R) Core(TM) i7-10700 series processor, 16 GB of RAM, and a GPU (Radeon RX560 series). The algorithms used for the experimental tests were the AGPSO [23], BLPSO [7], CLPSO [21], IPSO [11], MPSO [4], PSO [19], PSOGSA [22], and TACPSO [39], the selection of these algorithms was based on the availability of the source codes by the authors. The remaining algorithms were transcribed based on the original articles.

To briefly describe the methodology used to incorporate diversity measurement into the algorithms, it is assumed that each algorithm starts with its initialization, a standard step of any MA, in which it starts from the initial parameters set by the user, such as the number of particles or agents in the population, the dimensions to be used, and the limits of the search space. The initialization process starts with a random process that generates particles in a uniform distribution. From this stage, we calculate the population's initial diversity for each algorithm using the wise diversity dimension. Then, the iterative process of each algorithm is performed to find the best solutions. For each iteration, the diversity measurement process is performed and stored for analysis. This allows us to observe and analyze how the diversity varies iteratively in each algorithm for each problem.

### *4.1 Experimental Results with CEC'2017 Benchmark*

The comparative analysis in this section includes the CEC 2017 benchmark test suite [3, 38], which is one of the most significant and important benchmark test suites designed to test the capability and performance of algorithms. This benchmark test suite includes 30 benchmark functions divided into unimodal, multimodal, multimodal, hybrid, and compositional functions. Table 1 lists the CEC 2017 benchmark functions with their respective name, search ranges ($D$ =dimensions), and global optimal values. The experimental tests for this test set were performed with

**Table 1** Summary of the CEC 2017 test functions

| | No. | Functions | Fi* = Fi(x*) |
|---|---|---|---|
| Unimodal functions | 1 | Shifted and rotated Bent Cigar function | 100 |
| | 2 | Shifted and rotated sum of different power function* | 200 |
| | 3 | Shifted and rotated Zakharov function | 300 |
| Simple multimodal functions | 4 | Shifted and rotated Rosenbrock's function | 400 |
| | 5 | Shifted and rotated Rastrigin's function | 500 |
| | 6 | Shifted and rotated Expanded Scaffer's F6 function | 600 |
| | 7 | Shifted and rotated Lunacek Bi_Rastrigin function | 700 |
| | 8 | Shifted and rotated Non-continuous Rastrigin's function | 800 |
| | 9 | Shifted and rotated Levy function | 900 |
| | 10 | Shifted and rotated Schwefel's function | 1000 |
| Hybrid functions | 11 | Hybrid function 1 (N = 3) | 1100 |
| | 12 | Hybrid function 2 (N = 3) | 1200 |
| | 13 | Hybrid function 3 (N = 3) | 1300 |
| | 14 | Hybrid function 4 (N = 4) | 1400 |
| | 15 | Hybrid function 5 (N = 4) | 1500 |
| | 16 | Hybrid function 6 (N = 4) | 1600 |
| | 17 | Hybrid function 6 (N = 5) | 1700 |
| | 18 | Hybrid function 6 (N = 5) | 1800 |
| | 19 | Hybrid function 6 (N = 5) | 1900 |
| | 20 | Hybrid function 6 (N = 6) | 2000 |
| Composition functions | 21 | Composition function 1 (N = 3) | 2100 |
| | 22 | Composition function 2 (N = 3) | 2200 |
| | 23 | Composition function 3 (N = 4) | 2300 |
| | 24 | Composition function 4 (N = 4) | 2400 |
| | 25 | Composition function 5 (N = 5) | 2500 |
| | 26 | Composition function 6 (N = 5) | 2600 |
| | 27 | Composition function 7 (N = 6) | 2700 |
| | 28 | Composition function 8 (N = 6) | 2800 |
| | 29 | Composition function 9 (N = 3) | 2900 |
| | 30 | Composition function 10 (N = 3) | 3000 |

Search range: $[-100, 100]^D$

50 dimensions as search space, considering 300,000 function calls as stopping criteria, and due to the stochastic behavior of the algorithms, all results were generated after 35 independent runs. The comparative analysis was performed using central tendency metrics such as Average Best (AB), Median (MD), and standard deviation (SD). Tables 2 and 3 show the statistical results used for the comparison of the PSO variant algorithms. Table 2 shows clearly statistically how the BLPSO algorithm achieves outstanding results for 11 of the 15 functions in the benchmark set (F1, F2, F4-F9, and F11-13). However, the exception that obtains remarkable results for the remaining functions is the original version of the PSO algorithm, which corroborates and confirms why the algorithm continues to be a promising option when considering it for solving global optimization problems. In the case of the second Table 3, the BLPSO algorithm continues to show remarkable results compared to the other algorithms considered. Again, the PSO algorithm, in its original version, is the one that obtains slightly favorable results for this second part of the functions.

The diversity analysis for the CEC2017 benchmark test set is performed using convergence curves. This analysis allows one to visualize how each algorithm's diversity behaves through the function accesses of the iterative process. It is worth mentioning that this analysis has not been considered in any of the manuscripts analyzed. However, analyzing this behavior could help improve the design of PSO variant proposals. Furthermore, the results of the comparative analysis of the diversity metric are presented in Table 4. In this table, the values that achieved the highest diversity are highlighted in bold black, while those that obtained the lowest diversity in the iterative process for each of the CEC2017 functions are highlighted in bold blue. This comparison differs from the results obtained for the fitness function, as it allows for the observation of how the algorithm is related to the highest diversity throughout the iterative process. In this case, the CLPSO algorithm exhibits the greatest diversity, while the PSOGSA algorithm exhibits the least. Conversely, the analysis reveals that the algorithm with the optimal fitness values is not necessarily the one with the highest or lowest diversity. Instead, it is the one that maintains a balance. Furthermore, the diversity of the population can be observed for each algorithm, allowing for a comparison between them. This analysis has not been performed to date, but it would be beneficial to include it in a future study or to consider it when designing new versions of the PSO algorithm (see Table 4).

## 4.2 *Experimental Results with Applied Optimization Problems*

In this subsection, the algorithms are tested on two different engineering optimization problems that are well-studied in the literature. These engineering applications are not the classical problems applied to understanding the performance of a novel proposal. Instead, these problems are characterized by their high level of complexity, which means that they are usually treated as independent optimization problems to be

**Table 2** Statistical results for the CEC 2017 benchmark (functions 1–15)

| Function | Metric | AGPSO | BLPSO | CLPSO | IPSO | MPSO | PSO | PSOGSA | TACPSO |
|---|---|---|---|---|---|---|---|---|---|
| F1 | AB | 5.39E+09 | **3.38E+03** | 2.44E+04 | 7.17E+09 | 3.56E+09 | 2.71E+08 | 1.01E+10 | 1.89E+08 |
| | MD | 3.89E+09 | **2.57E+03** | 1.78E+04 | 6.74E+09 | 2.52E+09 | 9.93E+03 | 7.91E+09 | 6.52E+03 |
| | SD | 5.38E+09 | **3.29E+03** | 2.15E+04 | 4.13E+09 | 3.05E+09 | 4.86E+08 | 8.29E+09 | 4.37E+08 |
| F2 | AB | 1.03E+53 | **4.19E+28** | 6.45E+35 | 1.23E+55 | 1.06E+53 | 3.75E+45 | 3.52E+72 | 1.23E+49 |
| | MD | 9.05E+36 | 3.93E+23 | 3.10E+33 | 1.37E+50 | 1.93E+46 | **5.19E+20** | 6.36E+46 | 2.67E+41 |
| | SD | 5.63E+53 | **2.06E+29** | 3.00E+36 | 5.79E+55 | 5.43E+53 | 1.44E+46 | 1.93E+73 | 3.79E+49 |
| F3 | AB | 5.49E+03 | 4.84E+04 | 1.44E+05 | 8.32E+03 | 2.70E+04 | **7.18E+02** | 5.25E+04 | 8.15E+03 |
| | MD | 3.92E+03 | 4.66E+04 | 1.47E+05 | 7.08E+03 | 2.35E+04 | **6.89E+02** | 3.82E+04 | 5.86E+03 |
| | SD | 5.09E+03 | 9.71E+03 | 1.89E+04 | 6.86E+03 | 1.19E+04 | **1.50E+02** | 4.99E+04 | 7.25E+03 |
| F4 | AB | 9.81E+02 | **5.46E+02** | 5.60E+02 | 1.04E+03 | 7.79E+02 | 5.76E+02 | 1.26E+03 | 5.81E+02 |
| | MD | 7.81E+02 | **5.46E+02** | 5.66E+02 | 9.93E+02 | 6.70E+02 | 5.70E+02 | 8.60E+02 | 5.50E+02 |
| | SD | 6.48E+02 | 4.98E+01 | **2.93E+01** | 3.92E+02 | 3.23E+02 | 8.94E+01 | 9.32E+02 | 1.25E+02 |
| F5 | AB | 7.22E+02 | **5.78E+02** | 6.72E+02 | 7.37E+02 | 7.15E+02 | 7.40E+02 | 9.89E+02 | 6.90E+02 |
| | MD | 7.20E+02 | **5.81E+02** | 6.75E+02 | 7.41E+02 | 7.16E+02 | 7.40E+02 | 9.81E+02 | 6.81E+02 |
| | SD | 4.88E+01 | **9.32E+00** | 1.95E+01 | 3.48E+01 | 3.37E+01 | 3.79E+01 | 9.22E+01 | 4.97E+01 |
| F6 | AB | 6.09E+02 | **6.00E+02** | 6.00E+02 | 6.19E+02 | 6.13E+02 | 6.33E+02 | 6.68E+02 | 6.11E+02 |
| | MD | 6.08E+02 | **6.00E+02** | 6.00E+02 | 6.18E+02 | 6.12E+02 | 6.32E+02 | 6.67E+02 | 6.10E+02 |
| | SD | 3.69E+00 | **5.30E-06** | 5.05E-04 | 6.20E+00 | 3.88E+00 | 8.53E+00 | 8.53E+00 | 4.17E+00 |
| F7 | AB | 1.03E+03 | **8.29E+02** | 9.27E+02 | 1.06E+03 | 1.03E+03 | 8.72E+02 | 2.33E+03 | 1.00E+03 |
| | MD | 1.02E+03 | **8.28E+02** | 9.24E+02 | 1.04E+03 | 1.02E+03 | 8.65E+02 | 2.24E+03 | 9.91E+02 |
| | SD | 6.17E+01 | **9.30E+00** | 1.78E+01 | 9.77E+01 | 6.54E+01 | 3.63E+01 | 3.62E+02 | 6.82E+01 |
| F8 | AB | 1.02E+03 | **8.78E+02** | 9.79E+02 | 1.05E+03 | 1.02E+03 | 1.06E+03 | 1.30E+03 | 1.00E+03 |
| | MD | 1.02E+03 | **8.77E+02** | 9.77E+02 | 1.05E+03 | 1.02E+03 | 1.06E+03 | 1.30E+03 | 1.00E+03 |
| | SD | 5.26E+01 | **9.56E+00** | 1.55E+01 | 3.48E+01 | 4.17E+01 | 3.22E+01 | 8.49E+01 | 4.35E+01 |
| F9 | AB | 5.78E+03 | **9.01E+02** | 3.24E+03 | 5.21E+03 | 3.86E+03 | 9.46E+03 | 1.95E+04 | 3.55E+03 |
| | MD | 4.73E+03 | **9.01E+02** | 3.01E+03 | 5.24E+03 | 3.85E+03 | 9.27E+03 | 1.99E+04 | 3.38E+03 |
| | SD | 3.15E+03 | **9.30E-01** | 1.03E+03 | 1.74E+03 | 1.28E+03 | 3.37E+03 | 4.36E+03 | 1.14E+03 |
| F10 | AB | 7.67E+03 | 6.73E+03 | 7.37E+03 | 8.23E+03 | 7.92E+03 | **6.47E+03** | 8.35E+03 | 8.07E+03 |
| | MD | 7.66E+03 | 6.77E+03 | 7.34E+03 | 8.32E+03 | 7.97E+03 | **6.38E+03** | 8.02E+03 | 8.18E+03 |
| | SD | 1.01E+03 | **3.60E+02** | 4.22E+02 | 9.51E+02 | 8.13E+02 | 8.99E+02 | 1.11E+03 | 1.00E+03 |
| F11 | AB | 1.49E+03 | **1.19E+03** | 1.34E+03 | 2.02E+03 | 1.75E+03 | 1.27E+03 | 4.07E+03 | 1.55E+03 |
| | MD | 1.49E+03 | **1.18E+03** | 1.34E+03 | 1.79E+03 | 1.70E+03 | 1.27E+03 | 1.62E+03 | 1.55E+03 |
| | SD | 1.12E+02 | 4.98E+01 | **2.77E+01** | 5.55E+02 | 1.61E+02 | 4.29E+01 | 6.19E+03 | 9.90E+01 |
| F12 | AB | 1.09E+09 | **1.42E+06** | 1.17E+07 | 5.76E+08 | 5.53E+08 | 3.77E+07 | 1.67E+09 | 6.15E+07 |
| | MD | 4.05E+08 | **1.31E+06** | 1.22E+07 | 2.03E+08 | 1.19E+08 | 2.89E+06 | 2.43E+08 | 2.15E+06 |
| | SD | 1.55E+09 | **5.31E+05** | 4.26E+06 | 8.65E+08 | 1.11E+09 | 1.60E+08 | 2.68E+09 | 1.75E+08 |
| F13 | AB | 1.13E+08 | **2.82E+03** | 2.77E+04 | 4.61E+07 | 1.13E+07 | 9.99E+05 | 1.53E+09 | 4.98E+04 |
| | MD | 9.27E+06 | **2.14E+03** | 2.28E+04 | 2.68E+05 | 9.93E+04 | 6.29E+03 | 9.27E+06 | 2.65E+04 |
| | SD | 2.75E+08 | **1.71E+03** | 1.32E+04 | 9.17E+07 | 4.51E+07 | 5.43E+06 | 2.51E+09 | 8.24E+04 |
| F14 | AB | 1.72E+05 | 1.45E+05 | 5.97E+05 | 1.35E+05 | 7.19E+04 | **6.52E+04** | 1.03E+05 | 8.49E+04 |
| | MD | 4.97E+04 | 1.27E+05 | 5.09E+05 | 6.74E+04 | 5.11E+04 | **2.69E+04** | 4.50E+04 | 4.26E+04 |
| | SD | 3.39E+05 | 9.19E+04 | 4.01E+05 | 1.98E+05 | **7.28E+04** | 1.70E+05 | 1.98E+05 | 1.06E+05 |
| F15 | AB | 2.40E+06 | 7.83E+03 | **6.12E+03** | 7.21E+04 | 3.47E+04 | 1.09E+04 | 3.68E+07 | 2.03E+04 |
| | MD | 2.10E+04 | 7.72E+03 | **5.56E+03** | 6.30E+04 | 3.23E+04 | 9.77E+03 | 2.21E+04 | 1.57E+04 |
| | SD | 1.30E+07 | 4.44E+03 | **2.64E+03** | 6.05E+04 | 2.05E+04 | 6.90E+03 | 1.62E+08 | 1.60E+04 |

**Table 3** Statistical results for the CEC 2017 benchmark (functions 16–30)

| Function | Metric | AGPSO | BLPSO | CLPSO | IPSO | MPSO | PSO | PSOGSA | TACPSO |
|---|---|---|---|---|---|---|---|---|---|
| F16 | AB | 3.36E+03 | **2.55E+03** | 2.77E+03 | 3.53E+03 | 3.38E+03 | 3.01E+03 | 4.03E+03 | 3.33E+03 |
| | MD | 3.26E+03 | **2.58E+03** | 2.79E+03 | 3.45E+03 | 3.27E+03 | 2.93E+03 | 3.98E+03 | 3.27E+03 |
| | SD | 3.58E+02 | **1.71E+02** | 2.38E+02 | 3.71E+02 | 4.24E+02 | 3.11E+02 | 6.50E+02 | 4.16E+02 |
| F17 | AB | 3.03E+03 | **2.24E+03** | 2.51E+03 | 3.39E+03 | 3.11E+03 | 2.85E+03 | 4.20E+03 | 3.14E+03 |
| | MD | 3.05E+03 | **2.22E+03** | 2.51E+03 | 3.33E+03 | 3.15E+03 | 2.87E+03 | 4.08E+03 | 3.19E+03 |
| | SD | 3.12E+02 | **1.16E+02** | 1.47E+02 | 3.72E+02 | 3.42E+02 | 2.61E+02 | 5.70E+02 | 3.48E+02 |
| F18 | AB | 3.18E+05 | 1.11E+06 | 1.50E+06 | 9.88E+05 | 6.92E+05 | 5.69E+05 | **2.78E+05** | 4.29E+05 |
| | MD | **1.51E+05** | 1.00E+06 | 1.44E+06 | 2.74E+05 | 3.87E+05 | 5.33E+05 | 1.73E+05 | 1.56E+05 |
| | SD | 4.39E+05 | 6.28E+05 | 6.65E+05 | 2.01E+06 | 1.05E+06 | **2.79E+05** | 3.13E+05 | 1.26E+06 |
| F19 | AB | 7.88E+05 | 1.71E+04 | **2.59E+03** | 9.83E+05 | 2.93E+05 | 2.08E+04 | 1.66E+08 | 3.91E+04 |
| | MD | 5.33E+04 | 1.84E+04 | **2.36E+03** | 2.11E+05 | 4.49E+04 | 1.46E+04 | 4.44E+04 | 1.10E+04 |
| | SD | 1.65E+06 | 7.74E+03 | **7.17E+02** | 2.19E+06 | 6.62E+05 | 3.16E+04 | 4.41E+08 | 7.99E+04 |
| F20 | AB | 3.14E+03 | **2.37E+03** | 2.65E+03 | 3.03E+03 | 3.03E+03 | 2.76E+03 | 3.69E+03 | 3.13E+03 |
| | MD | 3.15E+03 | **2.37E+03** | 2.67E+03 | 2.97E+03 | 3.05E+03 | 2.74E+03 | 3.74E+03 | 3.08E+03 |
| | SD | 3.35E+02 | **1.20E+02** | 1.22E+02 | 3.21E+02 | 1.88E+02 | 2.63E+02 | 3.54E+02 | 3.11E+02 |
| F21 | AB | 2.52E+03 | **2.38E+03** | 2.48E+03 | 2.56E+03 | 2.53E+03 | 2.55E+03 | 2.85E+03 | 2.50E+03 |
| | MD | 2.50E+03 | **2.38E+03** | 2.48E+03 | 2.56E+03 | 2.53E+03 | 2.54E+03 | 2.85E+03 | 2.49E+03 |
| | SD | 4.23E+01 | **9.78E+00** | 1.43E+01 | 3.08E+01 | 4.03E+01 | 5.64E+01 | 9.63E+01 | 4.16E+01 |
| F22 | AB | 9.36E+03 | 8.23E+03 | 8.70E+03 | 9.38E+03 | 9.08E+03 | **8.20E+03** | 1.02E+04 | 9.01E+03 |
| | MD | 9.27E+03 | **8.50E+03** | 9.29E+03 | 9.44E+03 | 9.24E+03 | 8.78E+03 | 1.02E+04 | 9.19E+03 |
| | SD | 9.48E+02 | 1.16E+03 | 1.80E+03 | 9.57E+02 | **7.41E+02** | 2.18E+03 | 8.96E+02 | 1.57E+03 |
| F23 | AB | 3.21E+03 | **2.82E+03** | 2.92E+03 | 3.14E+03 | 3.16E+03 | 3.17E+03 | 3.51E+03 | 3.01E+03 |
| | MD | 3.19E+03 | **2.81E+03** | 2.92E+03 | 3.13E+03 | 3.16E+03 | 3.17E+03 | 3.50E+03 | 3.00E+03 |
| | SD | 1.31E+02 | **1.43E+01** | 2.43E+01 | 7.40E+01 | 9.26E+01 | 8.60E+01 | 1.19E+02 | 7.18E+01 |
| F24 | AB | 3.50E+03 | **2.99E+03** | 3.22E+03 | 3.38E+03 | 3.39E+03 | 3.40E+03 | 3.59E+03 | 3.26E+03 |
| | MD | 3.46E+03 | **2.99E+03** | 3.22E+03 | 3.35E+03 | 3.39E+03 | 3.37E+03 | 3.59E+03 | 3.25E+03 |
| | SD | 1.76E+02 | **1.30E+01** | 3.85E+01 | 1.41E+02 | 1.33E+02 | 1.27E+02 | 1.64E+02 | 1.11E+02 |
| F25 | AB | 3.17E+03 | 3.05E+03 | 3.08E+03 | 3.33E+03 | 3.13E+03 | **3.04E+03** | 3.54E+03 | 3.06E+03 |
| | MD | 3.08E+03 | 3.05E+03 | 3.08E+03 | 3.26E+03 | 3.09E+03 | **3.04E+03** | 3.21E+03 | 3.06E+03 |
| | SD | 2.10E+02 | 2.03E+01 | **1.53E+01** | 2.60E+02 | 1.08E+02 | 3.89E+01 | 8.34E+02 | 3.08E+01 |
| F26 | AB | 7.21E+03 | **4.55E+03** | 5.41E+03 | 7.22E+03 | 6.60E+03 | 6.13E+03 | 1.22E+04 | 6.12E+03 |
| | MD | 7.24E+03 | **4.57E+03** | 5.54E+03 | 7.04E+03 | 6.67E+03 | 7.31E+03 | 1.23E+04 | 6.19E+03 |
| | SD | 8.32E+02 | **1.41E+02** | 3.74E+02 | 6.96E+02 | 8.20E+02 | 2.27E+03 | 1.60E+03 | 7.85E+02 |
| F27 | AB | 3.67E+03 | **3.31E+03** | 3.42E+03 | 3.63E+03 | 3.68E+03 | 3.52E+03 | 3.69E+03 | 3.53E+03 |
| | MD | 3.63E+03 | **3.30E+03** | 3.43E+03 | 3.56E+03 | 3.65E+03 | 3.50E+03 | 3.71E+03 | 3.52E+03 |
| | SD | 2.49E+02 | **2.91E+01** | 3.72E+01 | 2.09E+02 | 2.32E+02 | 1.82E+02 | 1.63E+02 | 1.40E+02 |
| F28 | AB | 3.88E+03 | **3.31E+03** | 3.38E+03 | 4.28E+03 | 3.77E+03 | 3.34E+03 | 5.30E+03 | 3.39E+03 |
| | MD | 3.50E+03 | **3.31E+03** | 3.38E+03 | 3.89E+03 | 3.54E+03 | 3.32E+03 | 5.48E+03 | 3.33E+03 |
| | SD | 8.73E+02 | 1.39E+01 | **1.20E+01** | 1.02E+03 | 6.12E+02 | 8.91E+01 | 1.27E+03 | 1.53E+02 |
| F29 | AB | 4.38E+03 | **3.42E+03** | 3.75E+03 | 4.52E+03 | 4.35E+03 | 4.38E+03 | 5.78E+03 | 4.24E+03 |
| | MD | 4.27E+03 | **3.42E+03** | 3.72E+03 | 4.51E+03 | 4.32E+03 | 4.40E+03 | 5.66E+03 | 4.24E+03 |
| | SD | 4.35E+02 | **8.13E+01** | 1.33E+02 | 4.23E+02 | 3.46E+02 | 2.86E+02 | 5.32E+02 | 3.61E+02 |
| F30 | AB | 3.57E+06 | **9.27E+05** | 1.18E+06 | 9.63E+06 | 5.73E+06 | 9.73E+05 | 6.23E+07 | 1.97E+06 |
| | MD | 2.47E+06 | 9.29E+05 | 1.17E+06 | 8.00E+06 | 4.14E+06 | **8.85E+05** | 2.71E+06 | 1.54E+06 |
| | SD | 3.05E+06 | **1.10E+05** | 1.88E+05 | 5.98E+06 | 4.99E+06 | 4.59E+05 | 2.80E+08 | 1.22E+06 |

**Table 4** Average diversity for the 2017 CEC benchmark test set over the entire iterative process

| Function | AGPSO (%) | BLPSO (%) | CLPSO (%) | IPSO (%) | MPSO (%) | PSO (%) | PSOGSA (%) | TACPSO (%) |
|---|---|---|---|---|---|---|---|---|
| F1 | 10.98 | 8.01 | **18.47** | 7.56 | 6.56 | 2.41 | **0.34** | 5.76 |
| F2 | 18.47 | 32.85 | **66.60** | 14.00 | 20.10 | 5.88 | **0.72** | 12.64 |
| F3 | 30.05 | 65.45 | **90.30** | 27.79 | 35.77 | 6.03 | **0.13** | 23.60 |
| F4 | 11.27 | 10.44 | **32.98** | 7.51 | 9.16 | 3.44 | **0.34** | 6.33 |
| F5 | 17.44 | 34.17 | **53.75** | 12.33 | 13.83 | 2.79 | **0.37** | 11.17 |
| F6 | 11.73 | 10.46 | **25.06** | 8.97 | 8.74 | 3.19 | **0.37** | 9.34 |
| F7 | 15.16 | 15.72 | **24.98** | 11.13 | 10.10 | 4.71 | **0.21** | 8.72 |
| F8 | 17.39 | 34.30 | **54.11** | 11.95 | 13.23 | 2.75 | **0.37** | 11.32 |
| F9 | 18.75 | 12.23 | **70.33** | 13.12 | 14.02 | 5.92 | **0.12** | 11.87 |
| F10 | 34.85 | 82.11 | **93.80** | 32.30 | 43.07 | 3.24 | **0.14** | 31.67 |
| F11 | 15.74 | 27.21 | **54.27** | 12.63 | 12.84 | 3.27 | **0.17** | 12.72 |
| F12 | 9.56 | 19.77 | **54.42** | 7.06 | 7.50 | 3.75 | **0.21** | 6.09 |
| F13 | 10.20 | 18.48 | **61.46** | 6.89 | 6.73% | 3.55 | **0.18** | 6.41 |
| F14 | 20.35 | 60.11 | **80.62** | 16.60 | 23.40 | 7.69 | **0.12** | 16.18 |
| F15 | 12.37 | 22.80 | **63.37** | 8.54 | 8.19 | 3.67 | **0.12** | 8.98 |
| F16 | 20.47 | 53.98 | **69.39** | 15.02 | 19.79 | 3.24 | **0.20** | 15.01 |
| F17 | 20.00 | 54.58 | **68.70** | 14.30 | 18.51 | 3.20 | **0.15** | 14.51 |
| F18 | 24.79 | 67.10 | **84.79** | 18.25 | 27.39 | 6.46 | **0.12** | 17.56 |
| F19 | 11.38 | 32.16 | **64.53** | 8.41 | 8.48 | 3.47 | **0.14** | 8.59 |
| F20 | 34.76 | 68.29 | **81.86** | 29.93 | 41.78 | 4.13 | **0.19** | 28.38 |
| F21 | 16.27 | 35.61 | **57.21** | 11.55 | 13.06 | 2.71 | **0.43** | 10.90 |
| F22 | 33.17 | 80.52 | **94.29** | 29.93 | 39.85 | 3.18 | **0.18** | 30.59 |
| F23 | 8.85 | 32.47 | **52.88** | 6.42 | 6.07 | 2.26 | **0.32** | 7.38 |
| F24 | 7.45 | 32.66 | **62.35** | 5.61 | 7.01 | 1.90 | **0.29** | 5.96 |
| F25 | 12.81 | 9.86 | **29.49** | 9.61 | 11.26 | 3.30 | **0.62** | 7.87 |
| F26 | 11.07 | 28.39 | **46.90** | 6.81 | 7.51 | 2.63 | **0.26** | 8.57 |
| F27 | 2.92 | 28.67 | **73.59** | 2.29 | 2.08 | 1.11 | **0.36** | 3.63 |
| F28 | 12.82 | 11.48 | **37.60** | 8.43 | 10.89 | 3.21 | **0.47** | 7.64 |
| F29 | 12.92 | 42.83 | **67.84** | 8.90 | 9.93 | 3.01 | **0.20** | 10.42 |
| F30 | 10.85 | 28.58 | **69.42** | 8.75 | 8.05 | 3.39 | **0.14** | 8.88 |

solved to enhance a specific research field. In addition, these two problems have been selected due to their difference between each other and the challenge they represent to be optimized by any proposed evolutionary approach.

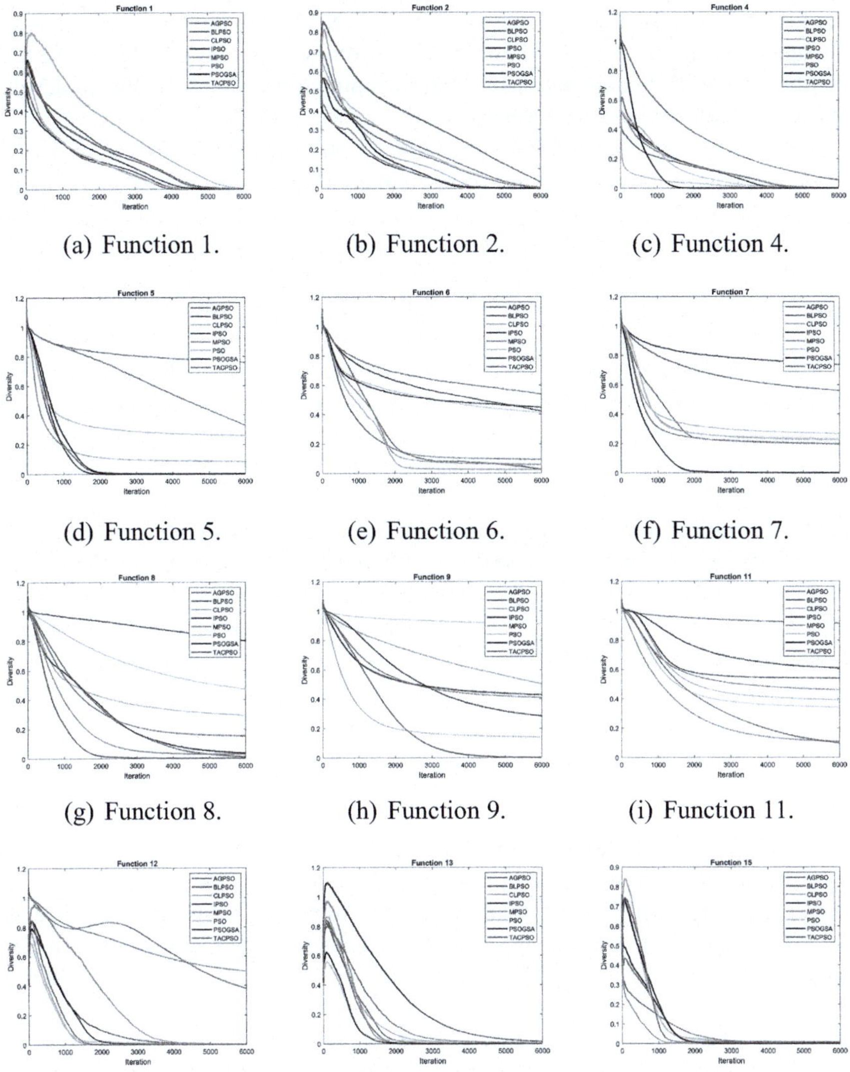


(a) Function 1. (b) Function 2. (c) Function 4.

(d) Function 5. (e) Function 6. (f) Function 7.

(g) Function 8. (h) Function 9. (i) Function 11.

(j) Function 12. (k) Function 13. (l) Function 15.

**Fig. 3** Diversity convergence for the 2017 CEC benchmark test suite

### 4.2.1 The Flexible Process Planning

The Flexible Process Planning (FPP) is an industrial optimization problem proposed and treated since 1980. Essentially, the FPP establishes that a manufacturing piece can be performed in different ways according to given specifications about the number of operations, the feasible machines to work, and many other variants [36]. In that sense, the FPP corresponds to a non-polynomial hard (NP-Hard) problem that belongs to the operations research field. The complexity of these types of tasks lies in the high number of possible combinations to find the optimal Process Plan (PP) to develop the aforementioned manufacturing part.

In this work, the treated FPP was originally proposed by Petrovic et al. [25] in 2016, and it has become one of the most complex FPP systems in modern literature. Figure 4 shows a graphical representation of the Petrovic FPP through an AND/OR network. As can be seen, the proposed FPP consists of 25 feasible operations, but each PP can be defined only by 10 operations, which involves finding the optimal way to manufacture the part. Moreover, it can be seen that each operation is defined by different feasible machines, tools, and Tool Access Directions (TADs). Specifically, Petrovic's model comprises 8 feasible machines, 12 possible tools, and 6 available TADs, which correspond to the viable angles of the tool's movements (+X, –X, +Y, –Y, +Z, and –Z). Therefore, each operation in a PP must define a specific machine, tool, and TAD, which in turn increases the possible combinations of PPs to more than two million possibilities. On the other hand, the optimization of this FPP lies in the minimization of the production time and production cost of the system given by Eqs. 25 and 26, respectively. These two functions are quite important since a feasible PP must consider a fast and cheap manufacturing process; otherwise, its implementation in a real-world environment is impractical. Hence, the aforementioned objective functions are described as follows:

$$TWPT = TW + TT + TCT + SCT \tag{25}$$

$$TWC = MC + TC + MCC + SCC + TCC \tag{26}$$

where TWPT is the Total Weighted Production Time of the PP, TW is the processing time defined for each tool used in a specific machine, TT is the transportation time between machines, TCT is the total tool change time required by each tool change, and SCT is the total setup change time that defines the time required by each TAD change. On the other hand, TWC corresponds to the Total Weighted Cost needed for each PP, MC, and TC, which are the machine cost and tool cost required by the use of each machine and tool, respectively. Meanwhile, MCC, SCC, and TCC define the total machine change cost, the total tool change cost, and the total setup change cost established by the change of each machine, tool, and TAD, respectively. It is worth mentioning that each objective function of this problem is much more complex and involves many elements to consider. Nonetheless, the entire explanation of the mathematical model of this FPP is not quite relevant to this study. Therefore, the author is referred to the Petrovic's work in order to have a better comprehension of the objective functions [25].

Based on the above, the studied algorithms have been tested on the Petrovic's FPP in order to optimize it through both objective functions. In that sense, all the approaches have been tested through 30 independent runs for each objective with an initial population of 100 individuals and 50,000 function accesses. In addition, all the algorithms have been tested with their original internal parameters defined in their

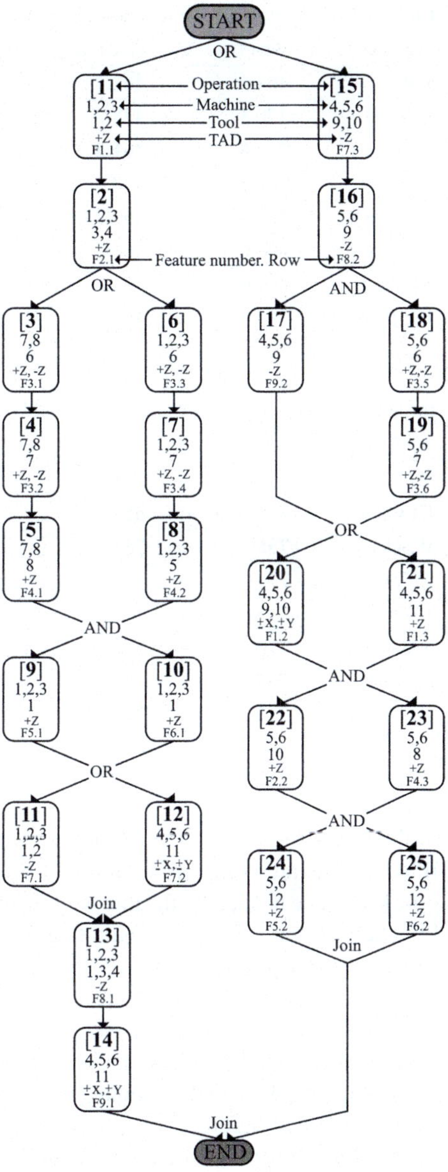


**Fig. 4** Flowchart of the FPP proposed system

respective articles in order to compete on equal terms. Under these specifications, Tables 5 and 6 present the obtained results of the algorithms for both objective functions F1 and F2, respectively. As can be noted, the IPSO reaches the best, average, and best worst results and best standard deviation for both functions. Furthermore, the IPSO obtains the optimal results for function F2 while its optimal best result is also reached by other techniques. This allows to understand the competitive qualities

**Table 5** Metric results for Production Time (TWPT)

| Approach | Best res. | Ave res. | Worst res. | STD | Time (s) |
|---|---|---|---|---|---|
| PSO | 464.4 | 467.1 | 475.3 | 0.932 | **179.01** |
| CLPSO | 470.3 | 473.2 | 481.1 | 1.951 | 185.77 |
| IPSO | **461.6** | **461.9** | **462.4** | **0.089** | 180.91 |
| MPSO | 465.3 | 470.3 | 475.2 | 1.132 | 186.55 |
| PSOGSA | 466.8 | 469.2 | 470.2 | 1.269 | 193.56 |
| TACPSO | **461.6** | 462.5 | 464.4 | 0.882 | 188.04 |
| AGPSO | **461.6** | 463.6 | 464.4 | 0.579 | 189.06 |
| BLPSO | 464.4 | 468.1 | 475.4 | 0.993 | 182.44 |

**Table 6** Metric results for Production Cost (TWC)

| Approach | Best res. | Ave res. | Worst res. | STD | Time (s) |
|---|---|---|---|---|---|
| PSO | 785 | 791 | 795 | 1.451 | **231.56** |
| CLPSO | 835 | 855 | 875 | 3.450 | 245.77 |
| IPSO | **755** | **757** | **765** | **1.083** | 240.01 |
| MPSO | 773 | 785 | 790 | 1.914 | 244.12 |
| PSOGSA | 775 | 779 | 785 | 1.099 | 265.01 |
| TACPSO | 775 | 782 | 788 | 1.105 | 240.92 |
| AGPSO | 765 | 785 | 792 | 2.371 | 243.67 |
| BLPSO | 765 | 770 | 785 | 2.018 | 238.66 |

of the IPSO for solving this manufacturing optimization problem. In addition, the IPSO probes are competitive in terms of execution time since they need a similar time to the one needed by the majority of approaches. On the other hand, the classical PSO demonstrates its simple but effective structure given the execution time results for the 30 independent runs for each function, making the PSO the best scheme compared with the rest of the algorithms. Finally, the TACPSO reports the highest execution time for both objectives, which corroborates its algorithmic complexity. At the same time, the CLPSO presents low stability, given the fact that it obtains the worst standard deviation for both functions.

As explained before, a PP is composed of 10 operations, and each operation is defined by a specific machine, tool, and TAD. Under that concept, Fig. 5 shows the optimal PPs reached by the IPSO based on the best metric results of Tables 5 and 6 for objective functions F1 and F2, respectively. About the optimal PP for F1 (Fig. 5a), the only machine used is M5, while there are only 6 used tools and 3 TADs, which in turn accelerates the processing time. In counterpart, the optimal PP for F2 (Fig. 5b) uses only 2 machines, 6 tools, and 3 TADs, which reduces the process cost. On the other hand, Fig. 6 shows the convergence rates of the algorithms for both functions through a logarithmic distribution. As can be seen, the IPSO reports the best convergence in

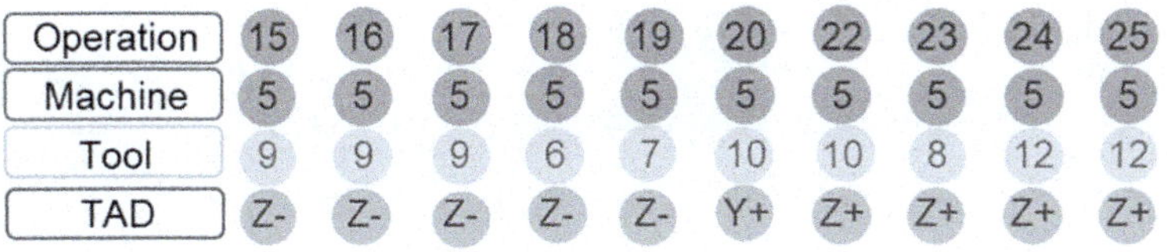


(a) Best PP for production time.

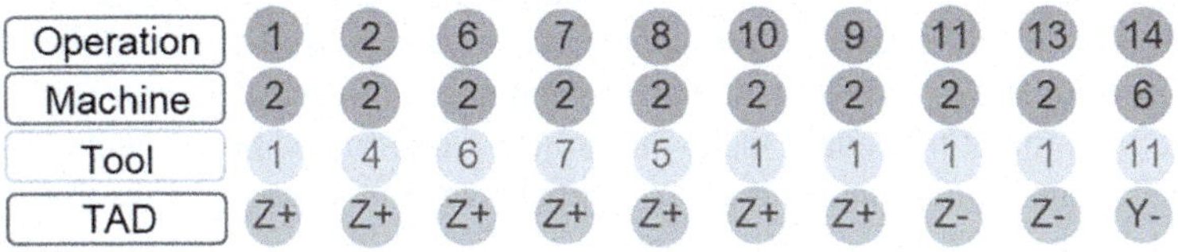


(b) Best PP for production cost.

**Fig. 5** Best process plans for both objective functions

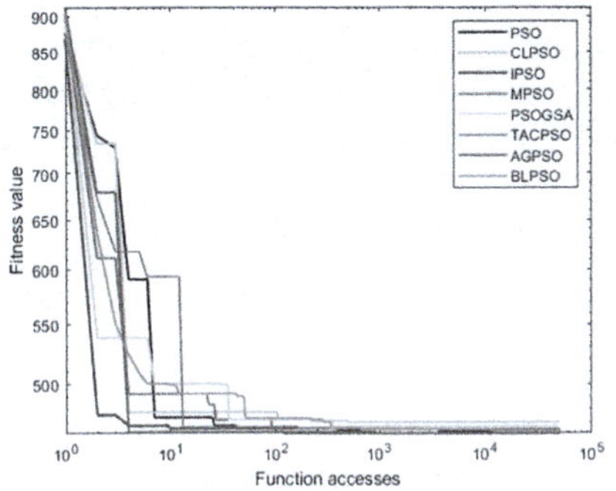


(a) Convergences for production time.

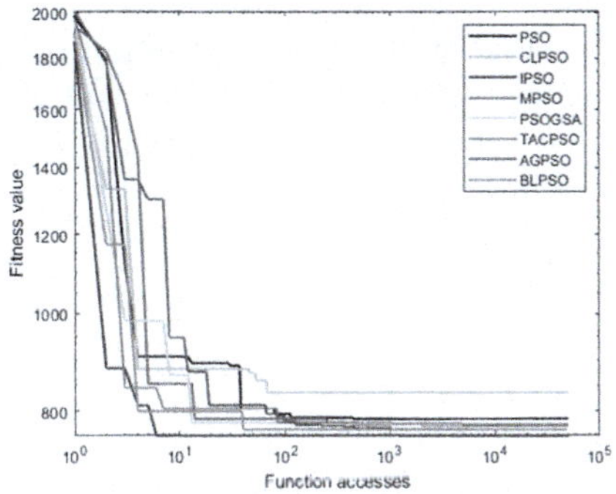


(b) Convergences for production cost.

**Fig. 6** Convergences for both objective functions

terms of speed and accuracy for both objectives, demonstrating the Robustness of the method for solving this engineering problem.

### 4.2.2 The Optimal Capacitor Placement

The Optimal Capacitor Placement (OCP) is one of the most relevant electrical engineering optimization problems in the state-of-the-art. Essentially, this problem involves the task of allocating specific capacitor banks in certain buses of a Radial Distribution Network (RDN). This process aims to reduce an electrical effect known as voltage decayment [34]. To better comprehend this concept, Fig. 7 shows an RDN of 10 buses. As can be seen, an RDN consists of a sub-station that supplies energy to the network buses. However, given the physics laws, the energy supplied by the sub-

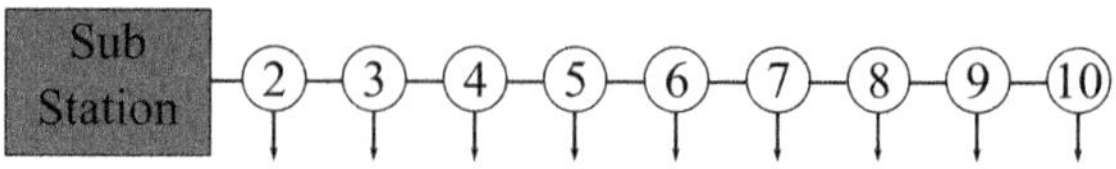


**Fig. 7** IEEE's 10-BUS RDN model

station is reduced for each bus due to the energy demands of the bus. Hence, the last bus receives the lowest voltage, which is counterproductive for optimal performance. Therefore, the solution to this problem lies in the fact of allocating capacitor banks in the buses to reduce voltage decay and preserve the best possible performance for each bus. Unfortunately, the installation and maintenance of each capacitor bank involve a huge investment that could make the project a non-viable investment [15]. Thus, the OCP consists of determining the optimal sizes of capacitor banks that must be used and their correct position through the network. In addition, the idea is to use the lowest possible number of capacitors with the lowest feasible sizes at the same time that the performance of the RDN is improved as best as possible. To have a better comprehension of this complex but interesting optimization problem, the reader is referred to the Boone's work [6].

Based on the above, the objective function of this OCP problem involves the minimization of the installation cost of the system at the same time that the performance of the voltage decayment is maximized. Hence, this model can be expressed as follows [13]:

$$C = K_p \times t \times P_{LOSS} + K_{ic} \times N_q + \sum_{j=1}^{n} k_j^c \times Q_j^c \tag{27}$$

where $C$ corresponds to the total cost (\$) of an RDN with a certain number of buses $n$, $K_p$ is the cost of power loss per hour for every unit ($\$/kW$), $P_{LOSS}$ is the total power loss ($kW$), while $k_j^c$ and $Q_j^c$ correspond to the cost per unit of reactive power ($\$/kVAR$) and the size of the capacitor ($kVAR$). On the other hand, $K_{ic}$ and $N_q$ are the installation cost of the system (\$) and its number of capacitors, respectively. Almost all the aforementioned parameters can be obtained from electrical analysis of the network. Nonetheless, the $P_{LOSS}$ corresponds to an estimation process that is always recommended to develop by the Newton-Raphson model [1]. It is worth mentioning that the capacitor sizes used in this work are related to 27 available commercial capacitors. However, given the table size to present these data, it has been decided not to include said information in this work. In any case, the reader is referred to Diaz et al. [12] to know this information. Respect to the rest of parameters, $K_p$ has a value of \$0.06, $t$ is defined as 8,760 h, $K_{ic}$ is \$1,000 while $K_c$ is \$3,000. About the algorithms, their respective optimal configuration proposed in their original articles are used for solving this optimization problem. Each algorithm is treated under 30 independent runs with an initial population of 100 individuals and 500 iterations. This last part is quite important since the nature of the problem does not allow for solving it by function accesses but by iterations. Furthermore, it is worth mentioning that many RDNs models have been proposed in the literature. In this case, the IEEE's

10-BUS system previously presented in Fig. 7 is used for finding its optimal solution, which supplies a voltage of 23 kV while the remaining buses generate a total active load of 12,368 kW and a total reactive load of 4,186 kVAR.

Table 7 presents the parameters obtained by each algorithm. The features used for this analysis are the Total Power Loses (TPL) expressed in kW, Power Losses Total Cost (PLTC) expressed in $ and Total Reactive Power (TRP) measured in kVAR. In addition the Capacitors Cost (CC), Total Annual Cost (TAC) and Network Savings (NS), all of them expressed in $. Meanwhile the minimum Voltage (mV) and Maximum Voltage (MV), both analyzed per unit. As mentioned before, the objective function of Eq. 27 is related to the system's installation cost. Therefore, the most important metrics are those directly related with this concept, which are the TAC and the NS in both senses. Under this principle, it can be noted that the PSO, IPSO, MPSO, PSOGSA, and BLPSO are the approaches capable of reaching the best RDN in terms of installation cost compared with the base case, which means the original RDN without capacitor banks. On the other hand, the CLPSO, TACPSO, and AGPSO obtained irregular values, which demonstrates their low capability for solving this energy-saving optimization problem. At the same time, it is worth remembering that the aim of this optimization process is not only to find the lower installation cost of the system but also the best improvement of the voltage decay through the network. This enhancement is appreciated by the voltage profile of each solution, and it is represented in Table 7 with the minimum and maximum voltage values of said profile. As can be noted, the BLPSO is the only scheme capable of reaching a minimum voltage of 0.8911 V, which in turn is the highest minimum voltage compared with the rest of the methods. Therefore, the BLPSO corresponds to the best approach, and its proposed capacitor distribution is seen in Fig. 8. In this proposed distribution the buses 3, 4, 5, 6, 7, and 10 are enhanced by the shunt capacitors of 4050 kVAR, 1200 kVAR, 2400 kVAR, 450 kVAR, 600 kVAR, and 350 kVAR, respectively.

When it is necessary to analyze the performance of the algorithms under conventional metrics, the TAC or voltage profile is not accurate since these values are provided by external measures of the fitness value of the objective function. Thus, since the objective function generates as fitness value the TPL, then the TPL is used to generate the metric results. This information is given in Table 8. The most remarkable information lies in the fact the BLPSO reaches the best values in the best and worst metrics, while the mean metric is led by the PSOGSA. However, the MPSO presents the best values regarding stability and resources since it reaches the best standard deviation and the lowest computational time needed for the 30 runs.

Finally, the obtained results in this experiment position the BLPSO as the best approach to solve this optimization problem. Hence, its voltage profile is shown graphically in Fig. 9. As can be seen, the BLPSO's profile is compared against the base case, and each bus of the RDN is represented as a $*$. It is worth mentioning that the rest of the profiles are allocated between the base case and the BLPSO's profile, but these profiles are not included in the figure to avoid saturating the graphic. In general terms, the BLPSO can significantly improve the original voltage decay, and it outperforms the rest of the swarm methodologies, positioning the BLPSO as a promising technique for future research in the field.

**Table 7** RDN parameters obtained by each algorithm

| Feature | Base | PSO | CLPSO | IPSO | MPSO | PSOGSA | TACPSO | AGPSO | BLPSO |
|---|---|---|---|---|---|---|---|---|---|
| TPL | 783.78 | 671.83 | 678.56 | 671.83 | 671.83 | 671.83 | 672.49 | 672.61 | 671.83 |
| PLTC | 411,953.95 | 353,112.60 | 356,653.71 | 353,112.60 | 353,112.60 | 353,112.60 | 353,459.42 | 353,521.26 | 353,112.60 |
| TRP | 0 | 9,250 | 6,800 | 9,250 | 9,250 | 9,250 | 9,050 | 9,050 | 9,250 |
| CC | 0 | 1,923.05 | 1,543.45 | 1,923.05 | 1,923.05 | 1,923.05 | 1,705.30 | 1,875.55 | 1,923.05 |
| TAC | 411,953.95 | **355,035.65** | 358,197.16 | **355,035.65** | **355,035.65** | **355,035.65** | 355,164.72 | 355,396.81 | **355,035.65** |
| NS | 0 | **56,918.30** | 53,756.79 | **56,918.30** | **56,918.30** | **56,918.30** | 56,789.24 | 56,557.15 | **56,918.30** |
| NS | 0 | **13.82** | 13.05 | **13.82** | **13.82** | **13.82** | 13.79 | 13.73 | **13.82** |
| mV | 0.8375 | 0.8910 | 0.8838 | 0.8910 | 0.8910 | 0.8910 | 0.8896 | 0.8870 | **0.8911** |
| MV | 0.9929 | 1.0055 | 1.0006 | 1.0055 | 1.0055 | 1.0055 | 1.0051 | 1.0051 | 1.0055 |

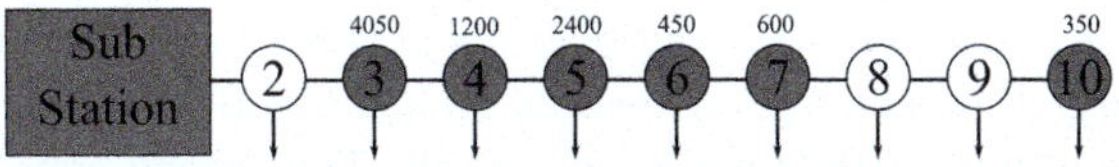


**Fig. 8** Optimal capacitors distribution

**Table 8** Metric results of the algorithms for the IEEE's 10-BUS model

| Method | Best | Mean | Worst | STD | Time |
|---|---|---|---|---|---|
| PSO | **671.83** | 672.69 | 697.60 | 0.232 | 6.32 |
| CLPSO | 678.56 | 696.15 | 735.80 | 10.611 | 6.02 |
| IPSO | **671.83** | 672.04 | 692.92 | 0.254 | 6.43 |
| MPSO | **671.83** | 671.86 | 692.06 | **0.021** | **5.38** |
| PSOGSA | **671.83** | **671.91** | 692.33 | 0.091 | 6.12 |
| TACPSO | 672.49 | 672.67 | 732.56 | 0.342 | 35.99 |
| AGPSO | 672.61 | 687.28 | 859.39 | 38.325 | 7.05 |
| BLPSO | **671.83** | 671.99 | **674.92** | 0.144 | 6.77 |

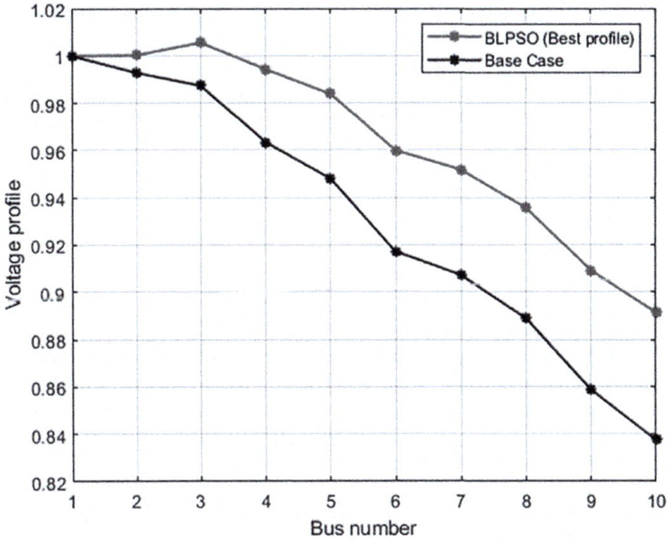


**Fig. 9** Voltage profile of the BLPSO against the Base case

## 5 Conclusions and Future Work

The present manuscript aims to provide an analysis of different PSO variants found in the literature. This analysis has two approaches: the first one aims to introduce a new approach to analyze the diversity in different PSO variants for single-objective global optimization problems. In this case, a widely known test suite such as CEC

2017 is used. It is worth mentioning that this analysis has not been considered in any of the analyzed manuscripts. However, an analysis of this behavior could help to improve the design of PSO variant proposals. The second approach starts with how these variants behave when applied to engineering problems and compares them. Comparing different variants of PSO, especially in terms of experimental tests, promises a wide field of research for further developing new methodologies that may arise from the original version of PSO. As a future work, it is intended to take this approach to new variants, as the development of the source code and the availability on the part of the authors will be obtained to expand this set of variants of the original algorithm.

## References

1. Akram, S., Ann, Q.U.: Newton raphson method. Int. J. Sci. Eng. Res. **6**(7), 1748–1752 (2015)
2. Aleti, A., Moser, I.: A systematic literature review of adaptive parameter control methods for evolutionary algorithms. ACM Comput. Surv. (CSUR) **49**(3), 1–35 (2016)
3. Awad, N., Ali, M., Suganthan, P., Liang, J., Qu, B.: Problem Definitions and Evaluation Criteria for the CEC 2017 Special Session and Competition on Single Objective Real-Parameter Numerical Optimization. Nanyang Technological University, Singapore, School of EEE (2016)
4. Bao, G., Mao, K.: Particle swarm optimization algorithm with asymmetric time varying acceleration coefficients. In: 2009 IEEE International Conference on Robotics and Biomimetics (ROBIO), pp. 2134–2139. IEEE (2009)
5. Bonyadi, M.R., Michalewicz, Z.: Particle swarm optimization for single objective continuous space problems: a review. Evol. Comput. **25**(1), 1–54 (2017)
6. Boone, G., Chiang, H.D.: Optimal capacitor placement in distribution systems by genetic algorithm. Int. J. Electri. Power Energy Syst. **15**, 155–161 (1993)
7. Chen, X., Tianfield, H., Mei, C., Du, W., Liu, G.: Biogeography-based learning particle swarm optimization. Soft Comput. **21**, 7519–7541 (2017)
8. Cheng, S., Shi, Y., Qin, Q., Zhang, Q., Bai, R.: Population diversity maintenance in brain storm optimization algorithm. J. Artif. Intell. Soft Comput. Res. **4**(2), 83–97 (2014)
9. Corriveau, G., Guilbault, R., Tahan, A., Sabourin, R.: Review and study of genotypic diversity measures for real-coded representations. IEEE Trans. Evol. Comput. **16**(5), 695–710 (2012)
10. Črepinšek, M., Liu, S.H., Mernik, M.: Exploration and exploitation in evolutionary algorithms: a survey. ACM Comput. Surv. (CSUR) **45**(3), 1–33 (2013)
11. Cui, Z., Zeng, J., Yin, Y.: An improved pso with time-varying accelerator coefficients. In: 2008 Eighth International Conference on Intelligent Systems Design and Applications, vol. 2, pp. 638–643. IEEE (2008)
12. Díaz, P., Perez-Cisneros, M., Cuevas, E., Camarena, O., Martinez, F.A.F., González, A.: A swarm approach for improving voltage profiles and reduce power loss on electrical distribution networks. IEEE Access **6**, 49498–49512 (2018)
13. El-Fergany, A.A.: Optimal capacitor allocations using evolutionary algorithms. IET Gener. Transm. Distrib. **7**(6), 593–601 (2013)
14. Elsheikh, A., Abd Elaziz, M.: Review on applications of particle swarm optimization in solar energy systems. Int. J. Environ. Sci. Technol. **16**, 1159–1170 (2019)
15. Gallego, R., Monticelli, A., Romero, R.: Optimal capacitor placement in radial distribution networks. IEEE Trans. Power Syst. **16**, 630–637 (2001)
16. Hajihassani, M., Jahed Armaghani, D., Kalatehjari, R.: Applications of particle swarm optimization in geotechnical engineering: a comprehensive review. Geotech. Geol. Eng. **36**, 705–722 (2018)

17. Imran, M., Hashim, R., Abd Khalid, N.E.: An overview of particle swarm optimization variants. Procedia Eng. **53**, 491–496 (2013)
18. Jain, M., Saihjpal, V., Singh, N., Singh, S.B.: An overview of variants and advancements of pso algorithm. Appl. Sci. **12**(17), 8392 (2022)
19. Kennedy, J., Eberhart, R.: Particle swarm optimization. In: Proceedings of ICNN'95-International Conference on Neural Networks, vol. 4, pp. 1942–1948. IEEE (1995)
20. Khare, A., Rangnekar, S.: A review of particle swarm optimization and its applications in solar photovoltaic system. Appl. Soft Comput. **13**(5), 2997–3006 (2013)
21. Liang, J.J., Qin, A.K., Suganthan, P.N., Baskar, S.: Comprehensive learning particle swarm optimizer for global optimization of multimodal functions. IEEE Trans. Evol. Comput. **10**(3), 281–295 (2006)
22. Mirjalili, S., Hashim, S.Z.M.: A new hybrid psogsa algorithm for function optimization. In: 2010 International Conference on Computer and Information Application, pp. 374–377. IEEE (2010)
23. Mirjalili, S., Lewis, A., Sadiq, A.S.: Autonomous particles groups for particle swarm optimization. Arab. J. Sci. Eng. **39**, 4683–4697 (2014)
24. Oliva, D., Rodriguez-Esparza, E., Martins, M.S., Abd Elaziz, M., Hinojosa, S., Ewees, A.A., Lu, S.: Balancing the influence of evolutionary operators for global optimization. In: 2020 IEEE Congress on Evolutionary Computation (CEC), pp. 1–8. IEEE (2020)
25. Petrovic, M., Mitic, M., Vukovic, N., Miljkovic, Z.: Chaotic particle swarm optimization algorithm for flexible process planning. Int. J. Adv. Manuf. Technol. **85**, 2535–2555 (2016)
26. Pluhacek, M., Viktorin, A., Kadavy, T., Kazikova, A.: On the common population diversity measures in metaheuristics and their limitations. In: 2021 IEEE Symposium Series on Computational Intelligence (SSCI), pp. 1–7. IEEE (2021)
27. Poláková, R., Tvrdík, J., Bujok, P.: Differential evolution with adaptive mechanism of population size according to current population diversity. Swarm Evol. Comput. **50**, 100519 (2019)
28. Rezaee Jordehi, A., Jasni, J.: Particle swarm optimisation for discrete optimisation problems: a review. Artif. Intell. Rev. **43**, 243–258 (2015)
29. Salleh, M.N.M., Hussain, K., Cheng, S., Shi, Y., Muhammad, A., Ullah, G., Naseem, R.: Exploration and exploitation measurement in swarm-based metaheuristic algorithms: An empirical analysis. In: Recent Advances on Soft Computing and Data Mining: Proceedings of the Third International Conference on Soft Computing and Data Mining (SCDM 2018), pp. 24–32. Springer, Berlin (2018)
30. Sedighizadeh, D., Masehian, E.: Particle swarm optimization methods, taxonomy and applications. Int. J. Comput. Theory Eng. **1**(5), 486–502 (2009)
31. Shami, T.M., El-Saleh, A.A., Alswaitti, M., Al-Tashi, Q., Summakieh, M.A., Mirjalili, S.: Particle swarm optimization: a comprehensive survey. IEEE Access **10**, 10031–10061 (2022)
32. Shi, Y., Eberhart, R.C.: Empirical study of particle swarm optimization. In: Proceedings of the 1999 Congress on Evolutionary Computation-CEC99 (Cat. No. 99TH8406), vol. 3, pp. 1945–1950. IEEE (1999)
33. Song, M.P., Gu, G.C.: Research on particle swarm optimization: a review. In: Proceedings of 2004 International Conference on Machine Learning and Cybernetics (IEEE Cat. No. 04EX826), vol. 4, pp. 2236–2241. IEEE (2004)
34. Sonwane, P.M., Kushare, B.E.: Optimal capacitor placement and sizing: an overview. Int. J. Innovative Res. Adv. Eng. (IJIRAE) **2**, 103–113 (2015)
35. Wang, D., Tan, D., Liu, L.: Particle swarm optimization algorithm: an overview. Soft Comput. **22**, 387–408 (2018)
36. Wang, S., Wang, X., ZhangView, J.: A review of flexible processes and operations. Prod. Oper. Manag. **30**(6), 1804–1824 (2021)
37. Wineberg, M., Oppacher, F.: The underlying similarity of diversity measures used in evolutionary computation. In: Genetic and Evolutionary Computation-GECCO 2003: Genetic and Evolutionary Computation Conference Chicago, IL, USA, July 12–16, 2003 Proceedings, Part II, pp. 1493–1504. Springer, Berlin (2003)

38. Wu, G., Mallipeddi, R., Suganthan, P.N.: Problem definitions and evaluation criteria for the cec 2017 competition on constrained real-parameter optimization. National University of Defense Technology, Changsha, Hunan, PR China and Kyungpook National University, Daegu, South Korea and Nanyang Technological University, Singapore, Technical Report (2017)
39. Ziyu, T., Dingxue, Z.: A modified particle swarm optimization with an adaptive acceleration coefficients. In: 2009 Asia-Pacific Conference on Information Processing, vol. 2, pp. 330–332. IEEE (2009)

# Pareto-Based Multiobjective Optimisation for JPEG Image Compression

**Seyed Jalaleddin Mousavirad, Davood Zabihzadeh, Seyyed Mohammad Tabatabaei, Mehran Pourvahab, Seyed Nooreddin Jafari, and Diego Oliva**

**Abstract** In this chapter, we focused on proposing a multi-objective strategy to address the conflicting objectives of enhancing image quality while simultaneously reducing image file size in the context of JPEG image compression, primarily through modifications to the quantisation table (QT). To this end, we employ two well-established methods: the non-dominated sorting genetic algorithm (NSGA-II), as a Pareto dominance-based approach, and the reference-point-based NSGA-II (NSGA-III), as a decomposition-based approach. By integrating our proposed strategy into these algorithms, we introduce two novel multi-objective techniques, namely NSGAII-JPEG and NSGAIII-JPEG. Our approach is designed to generate a diverse set of solutions, allowing users to select the most suitable solution based on their preferences after the optimisation process concludes. Our experimental results on different images show the effectiveness of our proposed algorithm.

S. J. Mousavirad (✉)
Department of Computer and Electrical Engineering, Mid Sweden University, Sundsvall, Sweden
e-mail: seyedjalaleddin.mousavirad@miun.se

D. Zabihzadeh
Computer Engineering Department, Hakim Sabzevari University, Sabzevar, Iran

S. M. Tabatabaei
Department of Medical Informatics, Faculty of Medicine, Mashhad University of Medical Sciences, Mashhad, Iran

M. Pourvahab
Departamento de Informática, Universidade da Beira Interior, Covilha, Portugal

NOVA LINCS, NOVA University of Lisbon, Lisbon, Portugal

S. N. Jafari
University of Beira Interior, Covilha, Portugal

D. Oliva
Depto. de Ingeniería Electro-Fotónica, Universidad de Guadalajara, CUCEI, Guadalajara, Mexico

 
D. Oliva et al. (eds.), *Advances in Optimization Algorithms for Multidisciplinary Engineering Applications: From Classical Methods to AI-Enhanced Solutions*, Studies in Computational Intelligence 806, https://doi.org/10.1007/978-3-031-78440-8_6

## 1 Introduction

The JPEG standard, established by the Joint Photographic Experts Group, is considered a cornerstone in digital image compression technologies, employing the Discrete Cosine Transform (DCT) [2]. This technique is initiated by transitioning the uncompressed image into the YCbCr colour space, delineating the image's luminance (Y) and chrominance components (Cb for blue and Cr for red). This transformation facilitates separate processing of each component, with a predominant emphasis on the luminance (Y) for its critical role in perceived image quality, although the method applies uniformly across the components.

In the JPEG workflow, the image's Y component is segmented into discrete $8 \times 8$ pixel blocks, subject to individual processing. A preparatory step involves zero-shifting each block by deducting 128 from all pixel values, setting the stage for the DCT application. The subsequent stage, quantisation, acts as the primary conduit for compression, albeit at the expense of information loss by abstracting DCT coefficients to a lower resolution. Despite the quantisation-induced data reduction, the ensuing entropy encoding stage [6] of each block secures the compression's efficiency, ensuring a lossless information encapsulation post-quantisation.

Quantisation's role in JPEG's compression schema is pivotal, effectively balancing between compressing the image data and maintaining its integrity. Although this step introduces irreversible information loss, it is meticulously calibrated to minimise perceptual degradation of the image. The entropy encoding, specifically designed to optimise data storage without further loss, leverages redundancy patterns within the quantised blocks, exemplifying JPEG's prowess in striking a balance between compression and quality retention.

Central to the efficacy of JPEG compression lies the utilisation of quantisation tables (QTs), as outlined in the influential Annex K variant [13] of the JPEG standard. This variant introduces two distinct types of QTs: one for luminance, known as the luminance quantisation table (LQT), and the other for chrominance, referred to as the chrominance quantisation table (CQT). These tables are instrumental in the quantisation process, applying discrete levels to the DCT coefficients for the luminance and chrominance components, thereby facilitating compression.

The challenge in optimising JPEG compression efficiency and image quality concurrently hinges on the precise calibration of these QTs. Ideally, each image would benefit from a uniquely tailored QT, reflecting its specific luminance and chrominance characteristics. Nonetheless, the complexity of deriving image-specific QTs has led to a prevalent reliance on standardised values. This approach simplifies the compression process but at the expense of potentially compromising the optimal balance between file size reduction and visual fidelity.

The task of determining the optimal values for LQT and CQT is complex, reflecting the nuanced trade-off between compressing image data effectively and preserving its quality. Despite the practical challenges in customising QTs for every image, the strategic selection of table values remains a critical aspect of JPEG compression, influencing both the efficiency of data reduction and the integrity of the resulting image.

To develop an optimal quantisation table, one can employ meta-heuristic algorithms (MAs) such as the genetic algorithm (GA) [38], particle swarm optimisation (PSO) [34]. MAs show considerable performance in solving different problems such as neural network training [22, 23, 27, 33], image segmentation [21, 28–30], lane-keeping systems in self-driving cars [18, 19], and image quantisation [24–26]. These MAs, being iterative, stochastic, and not specific to any particular problem, utilise various operators to navigate the search process in optimisation. Although they can yield solutions close to optimality, they do not guarantee a globally optimal solution. Recently, several MAs have been utilised in the creation of JPEG quantisation tables [8, 16, 32, 35, 37].

File size and image quality show two opposing objectives in JPEG image compression. To resolve this conflict, multi-objective meta-heuristic optimisation (MOMO) algorithms offer a viable solution. MOMO is designed to optimise problems with multiple conflicting objectives. Two primary methods for addressing multi-objective problems are scalarisation and Pareto-based approaches [14]. Scalarisation approaches aim to transform the multi-objective problem into a single-objective one, whereas Pareto-based approaches produce a set of optimal solutions.

MOMO algorithms are distinguished by their methodological execution strategies, which include a priori, interactive, and a posteriori approaches. In a priori strategies, the decision-makers' preferences are defined upfront, before the optimisation begins. This approach aims for a singular compromise solution, merging the multiple objectives into one via scalarisation techniques. Such methods convert multi-objective challenges into a singular objective landscape, tackled effectively with single-objective metaheuristic algorithms.

Contrastingly, a posteriori strategies unveil a spectrum of potential solutions post-optimisation, empowering users to select the option that best aligns with their preferences. A quintessential example within this domain is the Non-Dominated Sorting Genetic Algorithm II (NSGA-II) [11], celebrated for its efficacy in handling complex multi-objective problems through mechanisms like non-dominated sorting, elite preservation, and crowding distance assessment. NSGA-II's prowess is well-documented across diverse optimisation scenarios [1, 3, 39]. Building on this foundation, the NSGA-III [10] introduces a reference-point framework to enhance solution diversity, further broadening the algorithm's applicability and effectiveness [20, 39].

Within interactive methods, the optimisation process becomes a dynamic dialogue between the algorithm and the user. Here, the generation of solutions adapts in real time to evolving user preferences, offering a tailored optimisation experience. Techniques in this category range from the presentation of ideal and nadir vectors to stakeholders to the iterative refinement of Pareto optimal sets in response to newly specified preferences.

Notably, the application of NSGA-II to the optimisation of JPEG quantisation tables [17] underscores the potential of MOMO algorithms to transcend traditional boundaries. By comparing NSGA-II-based solutions with standard JPEG approaches, this study highlights the algorithm's capacity to innovate in the generation of efficient quantisation tables.

This chapter introduces a multi-objective approach aimed at addressing both image file size and image quality simultaneously. To achieve this goal, we integrate the proposed strategy into two established a posteriori Pareto-dominance and decomposition-based methodologies: NSGA-II [11] and NSGA-III [10]. Consequently, we present two novel Pareto-based techniques termed NSGAII-JPEG and NSGAIII-JPEG.

This chapter is organised as follows. Section 2 introduces the preliminaries for our study. Section 3 explains our NSGAII-JPEG and NSGAIII-JPEG algorithms, while Sect. 4 validates the algorithms. Eventually, Sect. 5 concludes the chapter and suggests some future works.

## 2 Preliminaries

### *2.1 Multi-objective Optimisation*

Multi-Objective Optimisation (MO) aims to identify the best possible solutions when dealing with several objectives that often conflict with one another. The primary goal in MO is to minimise these multiple objectives simultaneously. This means that rather than optimising a single objective, MO considers multiple criteria that need to be optimised together, acknowledging that improving one objective might lead to compromises in others. The mathematical formulation of an MO problem is encapsulated as:

$$\text{Minimise } F(x) = (f_1, f_2, \ldots, f_M), \text{ subject to } x \in \Omega, \tag{1}$$

where $\Omega$ signifies the set of feasible decisions, and the function $F : \Omega \rightarrow \mathbb{R}^M$ maps these decisions to outcomes within an $M$-dimensional objective space. The inherent objective conflicts preclude a singularly optimal solution; thus, Pareto optimality principles are employed to delineate a set of optimal solutions, constituting the Pareto optimal set, and visually demarcated by the Pareto front within the objective space.

Crucial to MO's evaluative framework is the concept of dominance, defined as follows:

**Definition 1** A solution $x_1$ is said to dominate another $x_2$ (notated as $x_1 \prec x_2$) if and only if it is equally good or superior in every objective while being strictly superior in at least one objective.

The notion of a non-dominated solution is then introduced:

**Definition 2** A solution $x_1$ achieves non-dominated or Pareto optimal status if no other solution $x_2$ within the decision space dominates it.

This concept is particularly illustrative in bi-objective minimisation scenarios, where certain solutions stand out as non-dominated by unmatched performance across the objectives.

The aggregation of these non-dominated solutions forms the Pareto optimal set, which is visually represented in the objective space as the Pareto front:

**Definition 3** The Pareto front (PF) serves as a graphical representation of the Pareto optimal set within the objective space, elucidating the inherent trade-offs between competing objectives.

Multi-objective optimisation (MO) involves finding the best solutions for two or more conflicting objectives, typically defined in terms of minimisation. The formal definition of a multi-objective optimization problem is given as follows:

$$\text{Minimise } F(x) = (f_1, f_2, \ldots, f_M) \text{ subject to } x \in \Omega \tag{2}$$

In this context, $\Omega$ denotes the decision space, and $F : \Omega \rightarrow \mathbb{R}^M$ represents the function that maps decision variables to outcomes in an $M$-dimensional objective space. Because these objectives can conflict, it is unlikely to find a single solution that optimises each objective simultaneously. Instead, the concept of Pareto optimality is introduced to identify a set of optimal solutions, represented by the Pareto optimal set, and visually depicted by the Pareto front in the objective space.

In the realm of multi-objective optimisation, the evaluation of solutions is more nuanced than in single-objective optimisation. Here, the concept of dominance is crucial.

**Definition 4** A solution $x_1$ dominates another solution $x_2$ (denoted $x_1 \prec x_2$) if it is as good or better in all objectives and strictly better in at least one objective.

Moreover, a non-dominated solution is characterised as follows:

**Definition 5** A solution $x_1$ is a non-dominated or Pareto optimal solution if there exists no other solution $x_2$ that dominates it across the entire decision space.

For instance, in a bi-objective minimisation scenario, specific solutions emerge as non-dominated because they are not bettered by any other across both objectives.

The Pareto optimal set, constituted of these non-dominated solutions, translates into a Pareto front in the objective space:

**Definition 6** The Pareto front (PF) is the representation of the Pareto optimal set in the objective space, illustrating the trade-offs and compromises between conflicting objectives.

## 2.2 *JPEG Image Compression*

### 2.2.1 DCT and IDCT Mechanisms

Initially, the input image is segmented into blocks of $8 \times 8$ pixels. The pixel values within these blocks are then normalised from the range $[0, 2^p - 1]$ to $[-2^{p-1}, 2^{p-1} - 1]$, where $p$ signifies the bit depth per pixel, typically set to 8 in standard JPEG compression. These $8 \times 8$ blocks are then treated as 64-dimensional vectors and are inputs for the Discrete Cosine Transform (DCT) process. DCT segregates the image data into 64 unique frequency components, known as DCT coefficients. The transformation is mathematically defined by:

$$F(u, v) = \frac{1}{4} c_u c_v \left[ \sum_{x=0}^{7} \sum_{y=0}^{7} f(x, y) \cos\left(\frac{(2x+1)u\pi}{16}\right) \cos\left(\frac{(2y+1)v\pi}{16}\right) \right] \tag{3}$$

where $F(u, v)$ represents the two-dimensional DCT of the image data, $c_u$ and $c_v$ are scaling factors that adjust the DCT coefficients to account for the DC component and to ensure orthogonality of the basis functions, and $f(x, y)$ denotes the pixel values of the image in the spatial domain. The double summation over $x$ and $y$ indicates that the DCT is applied to $8 \times 8$ blocks of pixel data. Also, $u$ and $v$ indicate representing the frequencies in the horizontal and vertical directions, respectively.

The Inverse DCT (IDCT) operates in reverse to DCT, reconstructing the approximate original image block from the DCT coefficients:

$$f(x, y) = \frac{1}{4} c_u c_v \left[ \sum_{u=0}^{7} \sum_{v=0}^{7} F(u, v) \cos\left(\frac{(2x+1)u\pi}{16}\right) \cos\left(\frac{(2y+1)v\pi}{16}\right) \right] \tag{4}$$

Bypassing quantisation ensures the exact reconstruction of the initial 64-point signal.

### 2.2.2 Quantisation and Dequantisation Procedures

During quantisation, a predefined 64-value matrix scales the DCT coefficients to reduce data, optimising the balance between compression and quality. Each matrix element dictates the quantiser's resolution for its associated DCT coefficient, falling within $[1, 255]$. The quantisation function is described as:

$$L(u, v) = \text{round}\left(\frac{F(u, v)}{Q(u, v)}\right), \tag{5}$$

where $L(u, v)$ denotes the quantised coefficients, and $Q(u, v)$ corresponds to the quantisation matrix values. Larger $Q(u, v)$ values indicate greater compression and data loss.

Dequantisation in the decoder reconstructs an estimate of $F(u, v)$ from $L(u, v)$, essential for approximating the original image after compression:

$$\bar{F}(u, v) = L(u, v) \times Q(u, v), \tag{6}$$

This stage is vital for JPEG compression, defining the trade-off between data reduction and image fidelity.

### 2.2.3 Symbol and Entropy Coding

Post-quantisation, the DC and AC coefficients are processed separately. The DC component uses Differential Pulse Code Modulation (DPCM) for encoding:

$$\text{DIFF}_i = DC_i - DC_{i-1} \tag{7}$$

where $DC_i$ and $DC_{i-1}$ represent the current and previous block's DC values. AC coefficients are reordered via a zigzag scan to facilitate subsequent entropy encoding, capitalising on their statistical properties to enhance compression. For entropy coding, JPEG typically employs Huffman coding, leveraging separate Huffman tables for luminance and chrominance components, optimising data reduction based on the coefficients' frequency distributions.

# 3 Proposed Approach

## *3.1 Encoding Scheme*

Our study introduces a novel methodology aimed at optimising quantisation tables for JPEG image compression, with a specific focus on both the LQT and the CQT. This optimisation process begins by linearising the conventional $8 \times 8$ matrix format of each quantisation table into a sequential $1 \times 64$ array. Subsequently, these arrays are amalgamated to form a comprehensive 128-dimensional vector, encapsulated as:

$$x = [LQT_{1,1}, \ldots, LQT_{8,8}, CQT_{1,1}, \ldots, CQT_{8,8}], \tag{8}$$

where the notation $LQT_{i,j}$ and $CQT_{i,j}$ represents the quantisation value at the $(i, j)$ position within the LQT and CQT, respectively. The vector's initial 64 components, derived from the LQT and constrained to integer values within the range $[0, 2^p - 1]$ (where $p = 8$ reflects the typical pixel bit depth), are followed by the 64 components corresponding to the CQT. To provide a visual comprehension of this methodological framework, Fig. 1 displays a schematic representation of a candidate solution's structure, showcasing the integration of LQT and CQT into a unified optimisation vector. This consolidated vector representation is pivotal, offering a streamlined approach to optimising the quantisation process within the JPEG compression standard.

The methodology introduced in our research is designed to identify optimal multi-objective quantisation tables, specifically targeting the luminance quantisation table (LQT) and chrominance quantisation table (CQT). For this purpose, we transform each $8 \times 8$ quantisation matrix into a linear $1 \times 64$ array before merging them. Consequently, our innovative approach results in a unified 128-dimensional vector represented as:

$$x = [LQT_{1,1}, \ldots, LQT_{8,8}, \ldots, CQT_{1,1}, \ldots, CQT_{8,8}] \tag{9}$$

where $LQT_{i,j}$ and $CQT_{i,j}$ denote the elements at position $(i, j)$ within the LQT and CQT matrices, respectively. Here, the initial 64 numbers, which are integral and fall within the range $[0, 2^p - 1]$ (considering $p = 8$ for pixel bit count), correspond to the LQT. The sequence that follows pertains to the CQT. Figure 1 illustrates an exemplar to elucidate the fundamental architecture of a typical candidate solution.

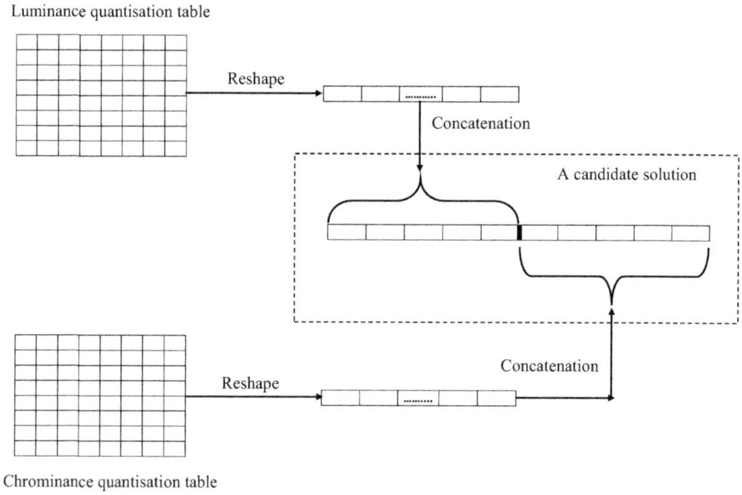


**Fig. 1** General structure of a candidate solution

### 3.2 *Objective Function*

In our research, we delineate two pivotal objective functions to navigate the complexities of image compression: minimising the file size and maximising the image quality. The evaluation begins with the creation of a JPEG image from a standard test image, employing a quantisation table derived from our optimisation algorithm. The effectiveness of compression is quantified through the file size ratio, mathematically articulated as:

$$\text{Compression Efficiency} = \frac{\text{Size of Compressed JPEG}}{\text{Size of Original Image}}, \tag{10}$$

where the numerator accounts for the compressed image's file size, and the denominator corresponds to the original image's file size. Herein, a diminished ratio is indicative of heightened compression performance.

Conversely, the integrity of the compressed image is assessed via the Peak Signal to Noise Ratio (PSNR), a widely recognised metric for evaluating image quality. The PSNR is calculated as follows:

$$\text{PSNR} = 20 \cdot \log_{10}\left(\frac{255}{\text{RMSE}}\right), \tag{11}$$

where the RMSE (Root Mean Square Error), representing the discrepancy between the original and compressed images, is computed by:

$$\text{RMSE} = \sqrt{\frac{\sum_{i=1}^{M}\sum_{j=1}^{N}(O(i,j) - C(i,j))^2}{M \cdot N}}, \tag{12}$$

with $M$ and $N$ denoting the image dimensions, while $O$ and $C$ symbolise the pixel values of the original and compressed images, respectively. An elevated PSNR value signals superior reconstructed image quality.

This dual-objective approach underscores the intrinsic balance between reducing file size and preserving image fidelity within JPEG compression. By employing these objective functions, our methodology seeks to pinpoint quantisation tables that offer an optimal compromise between compact file sizes and high-quality image reconstruction, addressing the fundamental challenge in image compression tasks.

This research defines two objective functions for the image compression: the file size and the image quality. The goal is to reduce the file size while enhancing the image quality. Initially, a JPEG image is generated using the designated candidate solution for a standard image. The objective function for file size is expressed as:

$$\text{File Size Ratio} = \frac{\text{Compressed File Size}}{\text{Original File Size}}, \tag{13}$$

where the numerator denotes the file size of the JPEG-compressed image, and the denominator represents the file size of the unaltered image. A lower ratio indicates more effective compression.

For image quality, the benchmark is the Peak Signal to Noise Ratio (PSNR), a prevalent criterion for image quality evaluation. It is defined as:

$$\text{PSNR} = 20 \cdot \log_{10} \left( \frac{255}{\text{RMSE}} \right), \tag{14}$$

where the RMSE (Root Mean Square Error) is calculated through:

$$\text{RMSE} = \sqrt{\frac{\sum_{i=1}^{M} \sum_{j=1}^{N} (O(i,j) - C(i,j))^2}{M \times N}}, \tag{15}$$

in which $M$ and $N$ signify the dimensions of the image, and $O$ and $C$ denote the original and compressed images, respectively. A superior PSNR reflects enhanced image quality.

## 3.3 *NSGAII-JPEG Algorithm*

The Non-dominated Sorting Genetic Algorithm II (NSGA-II) is a prominent method used in Pareto-based multi-objective optimisation. It employs four key mechanisms: Non-Dominated Sorting, Elite Preservation, Crowding Distance Assessment, and Selection, each detailed below.

**Non-Dominated Sorting**: This technique employs Pareto dominance principles to classify the population into different ranks. Initially, the best individuals (non-dominated) are identified, ranked highest, and segregated into a separate front. These individuals are then removed from consideration, and the process repeats to rank the remainder of the population, assigning them to successive fronts as illustrated in Fig. 2.

**Elite Preservation**: This mechanism ensures that top-tier solutions (elites) are carried over to subsequent generations without alteration, preserving high-quality candidates unless they are outperformed in future generations.

**Crowding Distance**: Crowding distance evaluates the proximity of solutions within the same front, estimating the density of solutions surrounding a particular point. A solution flanked by distant neighbours in objective space is considered superior, as defined by the crowding distance formula:

$$CD_i = \sum_{j=1}^{k} \frac{f_j^{i+1} - f_j^{i-1}}{f_j^{\max} - f_j^{\min}}, \tag{16}$$

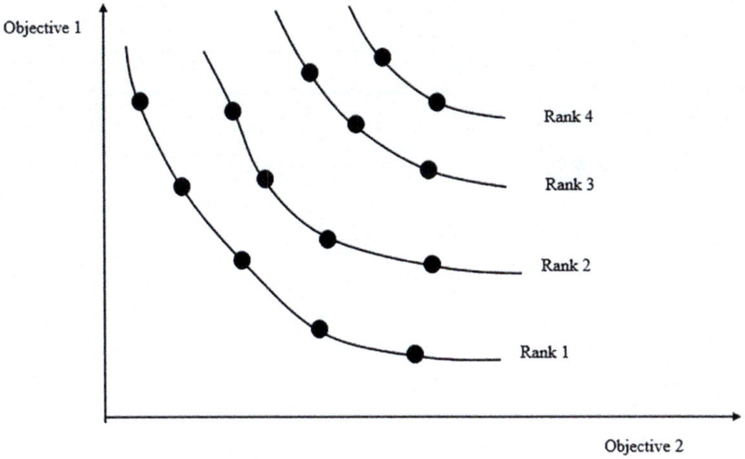


**Fig. 2** Illustration of the non-dominated sorting process

where $CD_i$ is the crowding distance for the $i$th solution, $f_j^i$ is the $j$th objective function value for that solution, with maximum and minimum values across the population denoted by $f_j^{\max}$ and $f_j^{\min}$, respectively. The process is depicted in Fig. 3.

**Selection Operator**: The population for the succeeding generation is chosen through a crowded tournament selection method, which considers both the rank and the crowding distance. This method prefers individuals of superior rank or, within the same rank, those with greater crowding distance.

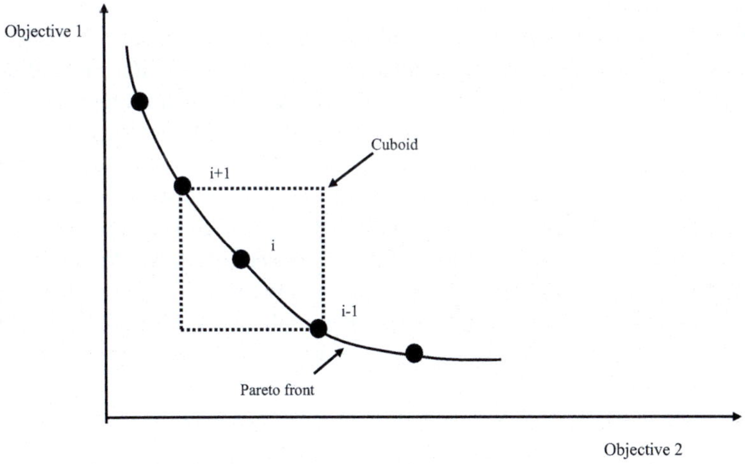


**Fig. 3** Visualization of crowding distance calculation

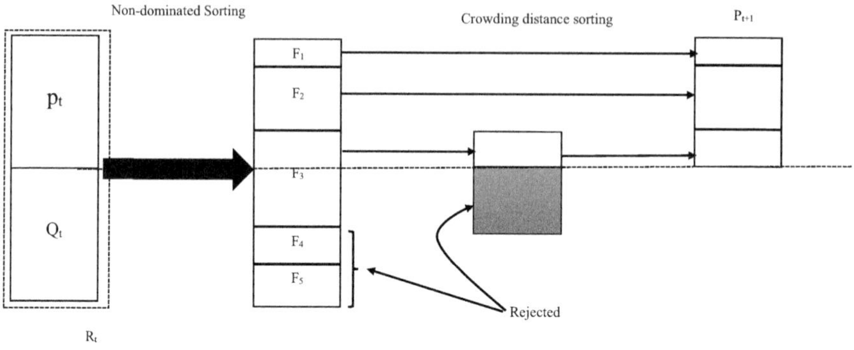


**Fig. 4** Depiction of the NSGA-II algorithm process

**Algorithm Process**: NSGA-II commences with an initial population $P_t$ and evolves it over generations. After applying crossover and mutation to $P_t$ to generate $Q_t$, both populations combine into $R_t$ for sorting. The selection of $N$ successors from $R_t$ depends on front rankings and crowding distances, ensuring diversity and quality. This selection process is visualised in Fig. 4, clearly outlining the iterative nature of NSGA-II across generations until a termination condition is achieved.

The developed NSGAII-JPEG methodology utilises the NSGA-II framework to optimise JPEG image compression. It incorporates standard NSGA-II components such as non-dominated sorting, an elitism mechanism, crowding distance calculation, and a selection process within the NSGAII-JPEG framework. The evolutionary phase is driven by tournament selection, SBX (Simulated Binary Crossover), and Polynomial mutation techniques. We employ tournament selection, which facilitates faster convergence. This method involves selecting the best candidate solutions from a randomly chosen subset of the population in each tournament. The size of the tournament is determined by the number of participants it contains. In addition, we employ SBX as the crossover operator [12]. Real values can be represented using binary notation, allowing for point crossover operations. SBX replicates this process by utilising a probability distribution model of the binary crossover. It leverages two primary parameters: the probability of crossover and the distribution index ($\eta$). In addition, Polynomial Mutation [12] is used in this paper, utilising the same probability distribution as the SBX operator within the parent's vicinity. Moreover, it shares identical parameters with the SBX operator. It is worthwhile to note that, although the representation is integer-based, during the implementation phase, we treat them as real-valued. In the postprocessing phase, we round them to the closest integer value.

The pseudocode for this NSGAII-JPEG approach is detailed in Algorithm 1.

**Algorithm 1** NSGAII-JPEG Procedure Outline

```
Inputs: Min/Max: minimum/maximum limits; PopSize: population size; VarCount: number
  of decision variables; MaxEvals: maximum allowable function evaluations; CrossOverProb:
  crossover likelihood; η_c: crossover distribution index.
Generate initial population of PopSize solutions based on the strategy discussed in Sect. 3.1.
Evaluate the fitness of each solution (Refer Sect. 3.2).
Apply Pareto ranking to categorize solutions.
Set Evals = PopSize
Initialize iteration = 0
while Evals ≤ MaxEvals do
    iteration = iteration + 1
    Conduct Tournament selection.
    Execute SBX crossover.
    Apply Polynomial mutation.
    Re-assess the fitness for all newly generated solutions (Sect. 3.2).
    R_t ← Merge current generation with offspring.
    Re-apply Pareto ranking for updated solution set.
    Identify and segregate non-dominated fronts.
    Populate the next generation with top-tier solutions until reaching PopSize.
    Compute and assign crowding distances.
    Prioritise and select individuals from the most sparse front.
    Increment Evals accordingly.
end while
```

### 3.4 NSGAIII-JPEG Algorithm

The Reference-point Based Non-dominated Sorting Genetic Algorithm (NSGA-III) distinguishes itself from NSGA-II through notable changes in its selection methodology, particularly in how it utilises adaptive reference points to enhance diversity within the solution population [10]. In contrast to NSGA-II, NSGA-III employs a systematic adjustment of multiple established reference points to foster population diversity.

NSGA-III utilises a predetermined array of reference points, as outlined by Das and Dennis [9], aligning them on a normalised hyperplane to ensure varied solution outputs. The algorithm calculates the number of reference points $H$ in a problem with $M$ objectives divided into $P$ parts as

$$H = \binom{M + P - 1}{P} \tag{17}$$

For example, in a problem with three objectives ($M = 3$), the algorithm configures reference points in a triangular layout, each corner representing a unique objective axis (1, 0, 0), (0, 1, 0), and (0, 0, 1), subdividing each axis into four parts ($P = 3$) to establish a total of 10 reference points as illustrated in Fig. 5.

**Fig. 5** Visualization of 10 reference points in 3D space with $P = 3$

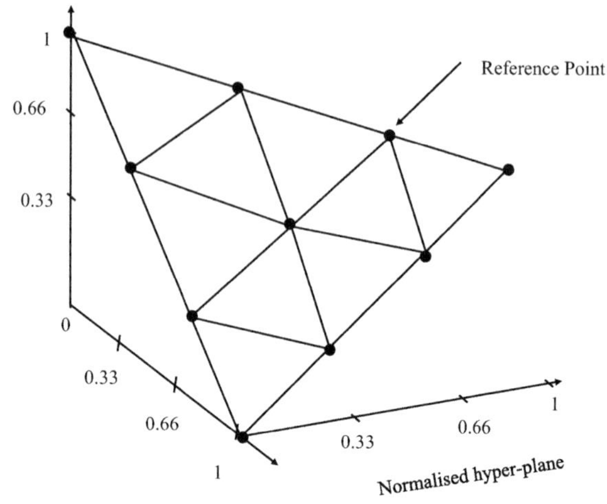


While NSGA-II applies crowding distance to select additional members post non-dominated sorting, NSGA-III leverages its reference points for this purpose. This necessitates normalising the objectives and reference points, followed by identifying the nearest population member to each reference line based on orthogonal distance.

Subsequently, the algorithm determines the niche count for every reference point to guide further selection, prioritising those with fewer connections. It then selects for inclusion in the next generation the population member linked to the least populated reference point. It is notable that some reference points might not be linked to any population members or could be connected to multiple members. The niche count $\rho_j$ for each reference point $j$ influences the selection: if $\rho_{j^*} = 0$, indicating no associated members, the algorithm responds based on the presence or absence of connected members in the current front. Conversely, if $\rho_j \geq 1$, indicating existing associations, a member from the corresponding front is chosen at random for inclusion. The process iteratively adjusts niche counts and selects members, repeating until the population size for $P_{t+1}$ meets the requisite number $N$.

**Algorithm Process:** NSGAIII-JPEG shares similarities with NSGAII-JPEG, albeit with a distinction in its utilisation of reference directions instead of crowding distance. Consequently, all operators employed in NSGAIII-JPEG mirror those utilised in NSGAII-JPEG. In other words, NSGAIII-JPEG utilises the non-dominated sorting, elite preservation, tournament selection, SBX operator, and Polynomial mutation for the optimisation process. The pseudocode for the NSGAIII-JPEG algorithm can be found in Algorithm 2.

**Algorithm 2** NSGAIII-JPEG Procudure Outline

**Inputs:** $L/U$: lower/upper bounds; $N_{pop}$: population size; $N_{var}$: number of variables; $NFE_{\max}$: maximum number of function evaluations; $prob$: crossover probability, $\eta$: distribution index.
Initialise a population of $N_{pop}$ candidate solutions using the encoding scheme described in Sect. 3.1.
Compute objective function values for all candidate solutions (see Sect. 3.2).
Rank the candidates based on Pareto sorting.
Set $NFE$ to $N_{pop}$.
Set $iter$ to 0.
**while** $NFE \leq NFE_{\max}$ **do**
    $iter \leftarrow iter + 1$
    Perform Tournament selection.
    Apply SBX mutation.
    Execute Polynomial mutation.
    Compute objective function values for all new candidate solutions.
    Combine parent and offspring populations to obtain $R_t$.
    Rank the individuals based on Pareto sorting.
    Generate sets of non-dominated solutions.
    Populate the next generation with solutions from the first front up to $N_{pop}$ individuals.
    Normalise objective functions and create a reference set.
    Assign each member to a reference point.
    Calculate the niche count for each reference point.
    Include new members in the population based on the niche count.
    Update the value of $NEF$.
**end while**

## 4 Experimental Results

To validate the effectiveness of our suggested approach, several experiments has been conducted. For this purpose, we have employed a diverse selection of images commonly used in image compression, comprising seven well-known benchmark images such as "Airplane," "Barbara," "Lena," "Mandrill," "Peppers," "Tiffany," and "Sailboat." Additionally, we included six images recommended in [31] for benchmarking image quantisation: "Snowman," "Beach," "Cathedrals beach," "Dessert," "Headbands," and "Landscape." These benchmark images are depicted in Fig. 6 (Following the new IEEE guidelines, we have removed the "Lena image" from this figure).

To ensure fairness, each algorithm is executed independently 30 times, and their statistical outcomes, including averages and standard deviations, are reported. The population size and the number of function evaluations are fixed at 50 and 1000, respectively, across all algorithms. Default parameters, as listed in Table 1, are utilised for other settings.

All algorithms are coded using Python 3.8 and the Pymoo framework [5], which is an open-source platform introduced in 2020. Pymoo encompasses cutting-edge single- and multi-objective algorithms along with functionalities tailored for multi-objective optimisation tasks, including visualisation capabilities. Also, the algorithms were executed on a computer equipped with Windows 10, 8GB of memory, and an AMD Quad Core FX-8800p processor.

(a) Airplane (b) Barbara (c) Mandrill

(d) Peppers (e) Sailboat (f) Snowman

(g) Tiffany (h) Beach (i) Cathedrals beach

(j) dessert (k) headbands (l) landscape

**Fig. 6** Benchmark images

**Table 1** Parameter settings

| Algorithm | Parameter | Value |
|---|---|---|
| NSGAII-JPEG | $Prob$ for crossover | 0.9 |
| | $\eta$ for crossover | 20 |
| | $Prob$ for mutation | 0.3 |
| | $\eta$ for mutation | 20 |
| NSGAIII-JPEG | $Prob$ for crossover | 0.9 |
| | $\eta$ for crossover | 20 |
| | $Prob$ for mutation | 0.3 |
| | $\eta$ for mutation | 20 |

As we know, the final solution is a set of solutions. There are two approaches for selecting only one solution. One approach is to give the responsibility to the user to select the option that best aligns with their preferences. The second approach is to select a solution based on specific criteria. For example, methods like ELECTRE compare solutions pairwise based on the decision-maker's preferences. Additionally, identifying the knee point or elbow point on the Pareto front, which is often seen as a good trade-off point, is another way to achieve this.

Various metrics exist to assess the outcomes of Pareto-based methodologies, such as Generational Distance (GD) [36] and Inverted Generational Distance (IGD) [7]. However, these metrics necessitate access to the true Pareto front, which isn not accessible in our context. Hence, we opted for an alternative metric called hyper-volume (HV) [15]. Unlike GD and IGD, HV does not rely on the true Pareto front for validation. It is considered a fair metric among alternatives [4] as it evaluates both proximity to the optimal solution and dispersion across the entire Pareto front. HV quantifies the region encompassed by solutions in relation to a reference point. In a minimisation scenario, a higher HV value indicates superior solution quality. The reference point selected for this paper is (4, 2).

Table 2 presents a comparison between two optimisation algorithms, NSGAII-JPEG and NSGAIII-JPEG, based on the hyper-volume (HV) measure across various benchmark images. Across most images, NSGAII-JPEG exhibits slightly higher mean HV values compared to NSGAII-JPEG, indicating its potential superiority in exploring and optimising the objective space. This consistency is observed in images such as "Barbara," "Sailboat," "Snowman," "Beach," "Cathedrals beach," "Dessert," "Heatbands," and "Landscape."

The algorithms' performance ranks are depicted for each image. NSGAII-JPEG achieves the top rank more frequently than NSGAIII-JPEG, as indicated by the higher number of top ranks across various images such as "Barbara," "Mandrill," "Peppers," "Sailboat," "Beach," "Cathedrals beach," "Dessert," "Heatbands," and "Landscape." In addition, the standard deviation values provide insights into the stability of each algorithm's performance. In some cases, NSGAII-JPEG demonstrates lower standard deviations compared to NSGAIII-JPEG, implying more consistent

**Table 2** Comparison of NSGAII-JPEG and NSGAIII-JPEG based on the HV measure

| Images | | NSGAII-JPEG | NSGAIII-JPEG |
|---|---|---|---|
| Airplane | Mean | 7.08 | 7.09 |
| | Std. | 0.18 | 0.21 |
| | Rank | 2 | 1 |
| Barbara | Mean | 7.12 | 7.04 |
| | Std. | 0.17 | 0.25 |
| | Rank | 1 | 2 |
| Lena | Mean | 7.12 | 7.14 |
| | Std. | 0.21 | 0.11 |
| | Rank | 2 | 1 |
| Mandrill | Mean | 7.06 | 7.06 |
| | Std. | 0.05 | 0.06 |
| | Rank | 1.5 | 1.5 |
| Peppers | Mean | 7.26 | 6.62 |
| | Std. | 0.06 | 0.55 |
| | Rank | 1 | 2 |
| Sailboat | Mean | 7.07 | 7.04 |
| | Std. | 0.20 | 0.10 |
| | Rank | 1 | 2 |
| Snowman | Mean | 7.19 | 7.09 |
| | Std. | 0.06 | 0.07 |
| | Rank | 1 | 2 |
| Tiffany | Mean | 6.96 | 7.10 |
| | Std. | 0.43 | 0.11 |
| | Rank | 2 | 1 |
| Beach | Mean | 7.08 | 6.96 |
| | Std. | 0.04 | 0.20 |
| | Rank | 1 | 2 |
| Cathedrals beach | Mean | 7.15 | 7.11 |
| | Std. | 0.12 | 0.06 |
| | Rank | 1 | 2 |
| Dessert | Mean | 7.12 | 7.01 |
| | Std. | 0.24 | 0.34 |
| | Rank | 1 | 2 |
| Heatbands | Mean | 7.23 | 7.18 |
| | Std. | 0.05 | 0.11 |
| | Rank | 1 | 2 |
| Landscape | Mean | 7.18 | 7.14 |
| | Std. | 0.08 | 0.08 |
| | Rank | 1 | 2 |
| Average rank | | 1.23 | 1.77 |
| Overall rank | | 1 | 2 |

performance across different runs. However, NSGAIII-JPEG shows higher standard deviations in images like "Peppers", and "Dessert," suggesting potential variability in its optimisation outcomes.

The average rank and overall rank provide a summary of the algorithms' performance across all images in the last rows. NSGAII-JPEG emerges as the overall winner with a lower average rank and securing the top overall rank compared to NSGAIII-JPEG. This suggests that while NSGAIII-JPEG may excel in certain scenarios, NSGAII-JPEG demonstrates more consistent and superior performance across the entire set of benchmark images.

For a more thorough examination, we additionally carried out a Wilcoxon signed rank test on the outcomes. The obtained $p$-value stands at 0.057, indicating a statistical distinction between the two algorithms solely at a 10% significance level. This outcome was anticipated since, in the majority of instances, reference points demonstrated superior performance compared to crowding distance when dealing with more than two objectives [10].

Due to the presence of two conflicting objective functions, plotting convergence curves based on objective functions is not feasible. Thus, in the subsequent experiment, we represented the convergence curves using the HV measure instead of objective functions. Figure 7 illustrates the convergence curves for our two proposed algorithms. It is evident that, in the majority of cases, NSGAII-JPEG exhibits a swifter convergence rate.

For a more encompassing perspective on the produced Pareto fronts, we depict them in Fig. 8. The visualisation reveals that NSGAII-JPEG yields a greater number of points in the Pareto front when compared to NSGAIII-JPEG.

## 5 Conclusion

This article introduces a multi-objective approach aimed at determining the optimal quantisation table values in JPEG image compression. We establish two conflicting objective functions, addressing both image quality and file size, while utilising a vector-based representation as the solution candidate. Subsequently, we integrate this proposed strategy into two metaheuristic algorithms, both of which are Pareto-based approaches: the non-dominated sorting genetic algorithm (NSGAII-JEPG) and the reference-based NSGA-II (NSGAIII-JPEG).

Furthermore, there are suggestions for extending this work in the future:

- Although this chapter employs established metaheuristic algorithms for embedding the strategy, further improvements could be achieved by incorporating more recent algorithms.
- This chapter employs default parameters for algorithms in the initial embedding stage, without focusing on parameter settings for all algorithms. Optimising these parameters using a self-adaptation approach could lead to improved performance.

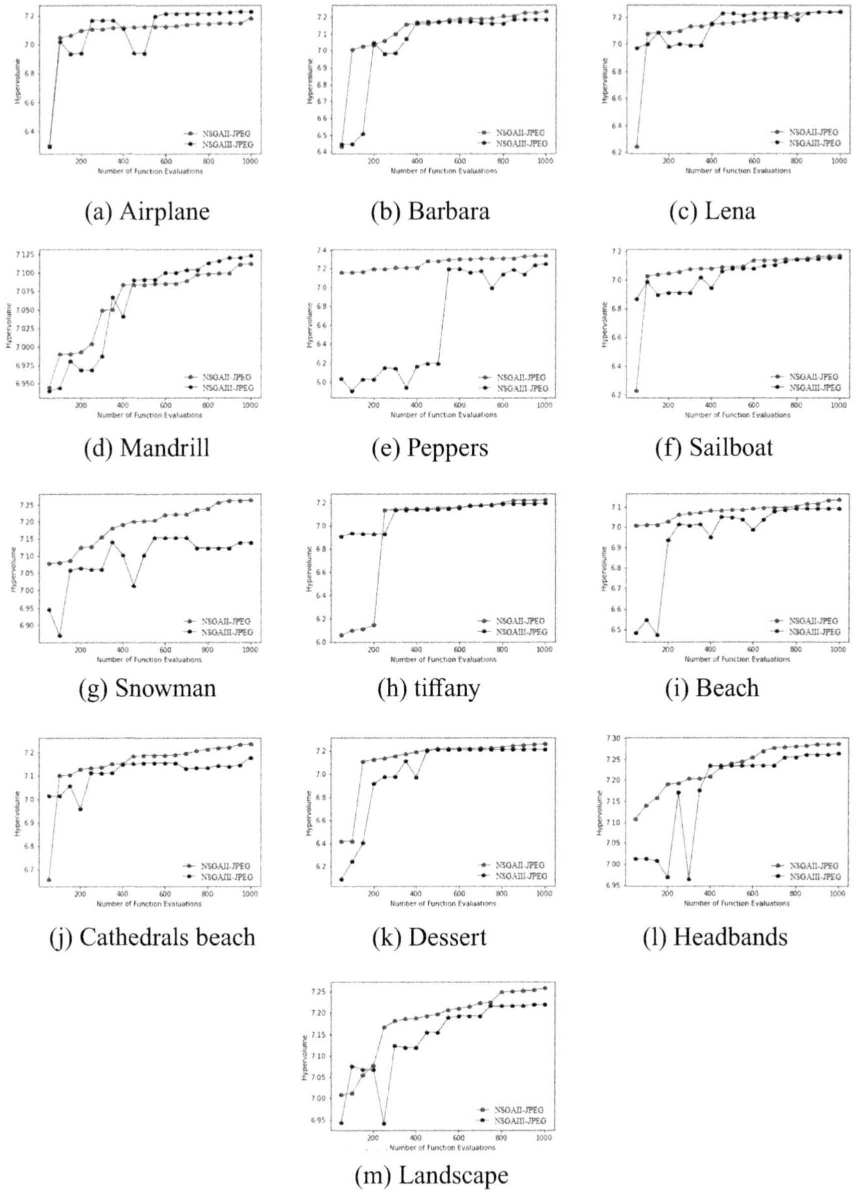


(a) Airplane (b) Barbara (c) Lena

(d) Mandrill (e) Peppers (f) Sailboat

(g) Snowman (h) tiffany (i) Beach

(j) Cathedrals beach (k) Dessert (l) Headbands

(m) Landscape

**Fig. 7** Convergence plots in terms of the HV metric

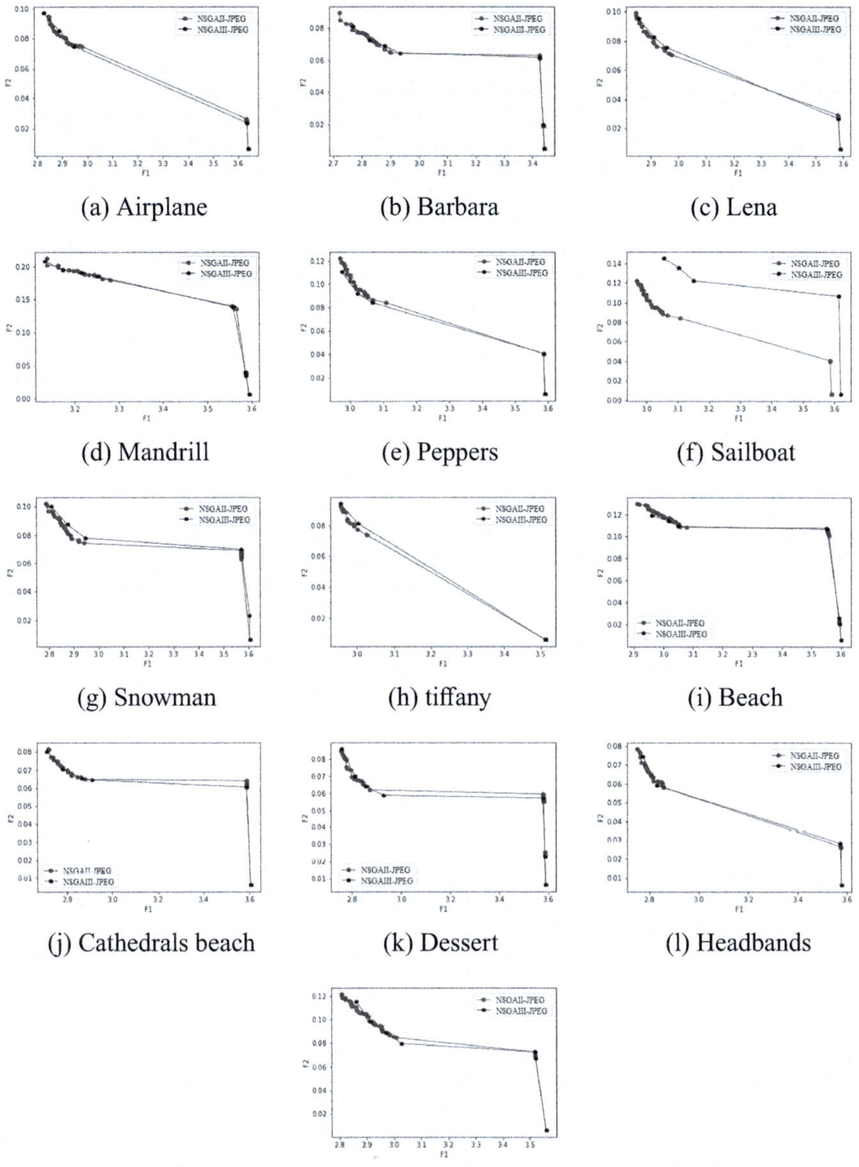

(a) Airplane (b) Barbara (c) Lena

(d) Mandrill (e) Peppers (f) Sailboat

(g) Snowman (h) tiffany (i) Beach

(j) Cathedrals beach (k) Dessert (l) Headbands

(m) Landscape

**Fig. 8** A comparison of the Pareto front of the proposed methods for each image

## References

1. Ahmed, F., Zhu, S., Yu, G., Luo, E.: A potent numerical model coupled with multi-objective NSGA-II algorithm for the optimal design of stirling engine. Energy **247**, 123468 (2022)
2. Andreadis, A., Benelli, G., Garzelli, A., Susini, S.: A DCT-based adaptive compression algorithm customized for radar imagery. In: IGARSS'97. 1997 IEEE International Geoscience and Remote Sensing Symposium Proceedings. Remote Sensing-A Scientific Vision for Sustainable Development, vol. 4, pp. 1993–1995. IEEE (1997)
3. Babajamali, Z., Aghadavoudi, F., Farhatnia, F., Eftekhari, S.A., Toghraie, D., et al.: Pareto multi-objective optimization of tandem cold rolling settings for reductions and inter stand tensions using NSGA-II. ISA Trans. **130**, 399–408 (2022)
4. Beume, N., Fonseca, C.M., Lopez-Ibanez, M., Paquete, L., Vahrenhold, J.: On the complexity of computing the hypervolume indicator. IEEE Trans. Evol. Comput. **13**(5), 1075–1082 (2009)
5. Blank, J., Deb, K.: pymoo: multi-objective optimization in python. IEEE Access **8**, 89497–89509 (2020)
6. Choi, J., Han, B.: Task-aware quantization network for JPEG image compression. In: European Conference on Computer Vision, pp. 309–324. Springer, Berlin (2020)
7. Coello Coello, C.A., Reyes Sierra, M.: A study of the parallelization of a coevolutionary multi-objective evolutionary algorithm. In: Mexican International Conference on Artificial Intelligence, pp. 688–697. Springer, Berlin (2004)
8. Costa, L.F., Veiga, A.C.P.: Identification of the best quantization table using genetic algorithms. In: PACRIM. 2005 IEEE Pacific Rim Conference on Communications, Computers and signal Processing, pp. 570–573. IEEE (2005)
9. Das, I., Dennis, J.E.: Normal-boundary intersection: a new method for generating the pareto surface in nonlinear multicriteria optimization problems. SIAM J. Optim. **8**(3), 631–657 (1998)
10. Deb, K., Jain, H.: An evolutionary many-objective optimization algorithm using reference-point-based nondominated sorting approach, part i: solving problems with box constraints. IEEE Trans. Evol. Comput. **18**(4), 577–601 (2013)
11. Deb, K., Pratap, A., Agarwal, S., Meyarivan, T.: A fast and elitist multiobjective genetic algorithm: NSGA-II. IEEE Trans. Evol. Comput. **6**(2), 182–197 (2002)
12. Deb, K., Sindhya, K., Okabe, T.: Self-adaptive simulated binary crossover for real-parameter optimization. In: Proceedings of the 9th Annual Conference on Genetic and Evolutionary Computation, pp. 1187–1194 (2007)
13. Ferrer-Roca, O., Rodriguez, R., Sousa Pereira, A.: Annex x: image formats. In: Handbook of Telemedicine, pp. 252–261. IOS Press (1998)
14. Gunantara, N.: A review of multi-objective optimization: methods and its applications. Cogent Eng. **5**(1), 1502242 (2018)
15. Knowles, J., Corne, D.: Properties of an adaptive archiving algorithm for storing nondominated vectors. IEEE Trans. Evol. Comput. **7**(2), 100–116 (2003)
16. Kumar, B.V., Karpagam, M.: Differential evolution versus genetic algorithm in optimising the quantisation table for JPEG baseline algorithm. Int. J. Adv. Intell. Paradigms **7**(2), 111–135 (2015)
17. Lazzerini, B., Marcelloni, F., Vecchio, M.: A multi-objective evolutionary approach to image quality/compression trade-off in JPEG baseline algorithm. Appl. Soft Comput. **10**(2), 548–561 (2010)
18. Moghadam, M.H., Borg, M., Mousavirad, S.J.: Deeper at the sbst 2021 tool competition: Adas testing using multi-objective search. In: 2021 IEEE/ACM 14th International Workshop on Search-Based Software Testing (SBST), pp. 40–41. IEEE (2021)
19. Moghadam, M.H., Borg, M., Saadatmand, M., Mousavirad, S.J., Bohlin, M., Lisper, B.: Machine learning testing in an adas case study using simulation-integrated bio-inspired search-based testing. J. Softw. Evol. Process e2591 (2022)
20. Mostafazadeh, F., Eirdmousa, S.J., Tavakolan, M.: Energy, economic and comfort optimization of building retrofits considering climate change: a simulation-based NSGA-III approach. Energy Build. **280**, 112721 (2023)

21. Mousavirad, S.J., Oliva, D., Chakrabortty, R.K., Zabihzadeh, D., Hinojosa, S.: Population-based self-adaptive generalised masi entropy for image segmentation: a novel representation. Knowl.-Based Syst. **245**, 108610 (2022)
22. Mousavirad, S.J., Oliva, D., Hinojosa, S., Schaefer, G.: Differential evolution-based neural network training incorporating a centroid-based strategy and dynamic opposition-based learning. In: 2021 IEEE Congress on Evolutionary Computation (CEC), pp. 1233–1240. IEEE (2021)
23. Mousavirad, S.J., Oliva, D., Schaefer, G., Moghadam, M.H., El-Abd, M.: A novel two-level clustering-based differential evolution algorithm for training neural networks. In: International Conference on the Applications of Evolutionary Computation (Part of EvoStar), pp. 259–272. Springer, Berlin (2024)
24. Mousavirad, S.J., Schaefer, G., Celebi, M.E., Fang, H., Liu, X.: Colour quantisation using human mental search and local refinement. In: IEEE International Conference on Systems, Man, and Cybernetics, pp. 3045–3050. IEEE (2020)
25. Mousavirad, S.J., Schaefer, G., Fang, H., Liu, X., Korovin, I.: Colour quantisation by human mental search. In: International Conference on Swarm Intelligence, pp. 130–141 (2020)
26. Mousavirad, S.J., Schaefer, G., Korovin, I.: Colour quantisation using self-organizing migrating algorithm. In: Genetic and Evolutionary Computation Conference Companion, pp. 1448–1453 (2020)
27. Mousavirad, S.J., Schaefer, G., Korovin, I.: An effective approach for neural network training based on comprehensive learning. In: International Conference on Pattern Recognition (2020)
28. Mousavirad, S.J., Schaefer, G., Moghadam, M.H., Saadatmand, M., Pedram, M.: A population-based automatic clustering algorithm for image segmentation. In: Proceedings of the Genetic and Evolutionary Computation Conference Companion, pp. 1931–1936 (2021)
29. Mousavirad, S.J., Schaefer, G., Zhou, H., Moghadam, M.H.: How effective are current population-based metaheuristic algorithms for variance-based multi-level image thresholding? Knowl.-Based Syst. **272**, 110587 (2023)
30. Mousavirad, S.J., Zabihzadeh, D., Oliva, D., Perez-Cisneros, M., Schaefer, G.: A grouping differential evolution algorithm boosted by attraction and repulsion strategies for masi entropy-based multi-level image segmentation. Entropy **24**(1), 8 (2022)
31. Pérez-Delgado, M.L.: Color image quantization using the shuffled-frog leaping algorithm. Eng. Appl. Artif. Intell. **79**, 142–158 (2019)
32. Preethi, D., Loganathan, D.: Quantization table selection using firefly with teaching and learning based optimization algorithm for image compression. In: Handbook of Multimedia Information Security: Techniques and Applications, pp. 473–499. Springer, Berlin (2019)
33. Rahmani, S., Mousavirad, S.J., El-Abd, M., Schaefer, G., Oliva, D.: Centroid-based differential evolution with composite trial vector generation strategies for neural network training. In: International Conference on the Applications of Evolutionary Computation (Part of EvoStar), pp. 608–622. Springer, Berlin (2023)
34. Shi, Y., Eberhart, R.: A modified particle swarm optimizer. In: IEEE International Conference on Evolutionary Computation, pp. 69–73 (1998)
35. Tuba, M., Bacanin, N.: JPEG quantization tables selection by the firefly algorithm. In: 2014 International Conference on Multimedia Computing and Systems (ICMCS), pp. 153–158. IEEE (2014)
36. Van Veldhuizen, D.A.: Multiobjective Evolutionary Algorithms: Classifications, Analyses, and New Innovations. Air Force Institute of Technology (1999)
37. Wang, Q., Liu, P., Zhang, L., Cheng, F., Qiu, J., Zhang, X.: Rate-distortion optimal evolutionary algorithm for JPEG quantization with multiple rates. Knowl.-Based Syst. **244**, 108500 (2022)
38. Whitley, D.: A genetic algorithm tutorial. Stat. Comput. **4**(2), 65–85 (1994)
39. Xue, Y., Zhu, H., Neri, F.: A feature selection approach based on NSGA-II with relieff. Appl. Soft Comput. 109987 (2023)

# Evolutionary Self-Adjusting Masi Entropy Thresholding

**Seyed Jalaleddin Mousavirad, Diego Oliva, Seyyed Mohammad Tabatabaei, Davood Zabihzadeh, and Mehran Pourvahab**

**Abstract** Image segmentation is crucial in image analysis and computer vision applications. Image thresholding, a popular approach due to its simplicity and robustness, becomes less efficient with an increase in thresholds due to an exhaustive search. Evolutionary algorithms like differential evolution (DE) can address this. Generalised Masi entropy (GME) leverages entropic measure (r) to explore additive/non-extensive information for thresholding. However, determining a proper value for $r$ is vital for GME's performance. This paper proposes a self-adjusting method to find the $r$ value without prior knowledge of histogram distribution. Introducing a new representation compatible with any evolutionary algorithm, the approach retains efficiency without additional function evaluations. Using DE for optimisation, our experiments on benchmark images show significant improvements in the objective function, peak signal-to-noise ratio (PSNR), feature similarity index (FSIM), and structural similarity index (SSIM). This demonstrates the proposed approach's ability to automatically find the optimal $r$ value and enhance GME-based thresholding efficacy.

S. J. Mousavirad (✉)
Department of Computer and Electrical Engineering, Mid Sweden University, Sundsvall, Sweden
e-mail: seyedjalaleddin.mousavirad@miun.se

D. Oliva
Depto. de Ingeniería Electro-Fotónica, Universidad de Guadalajara, CUCEI, Guadalajara, Mexico

S. M. Tabatabaei
Faculty of Medicine, Department of Medical Informatics, Mashhad University of Medical Sciences, Mashhad, Iran

D. Zabihzadeh
Computer Engineering Department, Hakim Sabzevari University, Sabzevar, Iran

M. Pourvahab
Departamento de Informática, Universidade da Beira Interior, Covilha, Portugal

NOVA LINCS, NOVA University of Lisbon, Lisbon, Portugal



D. Oliva et al. (eds.), *Advances in Optimization Algorithms for Multidisciplinary Engineering Applications: From Classical Methods to AI-Enhanced Solutions*, Studies in Computational Intelligence 806, https://doi.org/10.1007/978-3-031-78440-8_7

# 1 Introduction

Image segmentation is a preliminary step in image processing and machine vision applications. In a broad context, it entails the division of an image into distinct regions based on characteristics such as colour and texture. Its applications span various domains, including medicine [5, 27], crowd sensing platforms [26], thermographies [25] and food quality assessment [23]. The literature encompasses a diverse array of image segmentation methodologies [1, 2, 33], with image thresholding being particularly acknowledged for its simplicity and robustness [4, 21].

The primary objective of thresholding is to determine one or more threshold values by leveraging information in the histogram. Generally, image thresholding methods fall into two categories: bi-level image thresholding (BLIT) and multi-level image thresholding (MLIT). BLIT seeks a single threshold to segregate an image into two regions (e.g., foreground and background), while MLIT involves the selection of multiple threshold values. Identifying appropriate threshold values in MLIT poses a challenging task, attracting substantial research attention in recent years [18, 20, 22].

In recent years, entropy has gained prominence as a widely adopted technique for image thresholding. Entropy is a quantitative measure of randomness, wherein homogeneous regions exhibit minimal unpredictability, while non-homogeneous regions manifest maximal unpredictability [30]. A heightened entropy value indicates improved differentiation between background and foreground. Renyi entropy [28], Shannon entropy [7], and Tsallis entropy [3] represent distinct types of entropy commonly employed in this context.

The intrinsic characteristics of pixel information within an image, exhibiting either additive or non-additive traits, are well recognised. The fundamental purpose of entropy-based image segmentation lies in leveraging this information. Renyi entropy effectively addresses additivity concerns by incorporating an entropic parameter [28]. Conversely, Tsallis entropy, utilised in image segmentation, constitutes a generalisation of Shannon entropy [3]. The pseudo-additive nature of Tsallis entropy, modulated by an entropic parameter denoted as $r$, accommodates non-extensive information about statistically independent subsystems. A limitation common to both Renyi's and Tsallis's entropies is their inability to account for additive and non-additive information simultaneously.

To overcome the limitations associated with Renyi and Tsallis entropies, Masi proposes a novel entropy concept outlined in [9]. This innovative approach amalgamates the additivity characteristic of Renyi entropy with the non-extensivity trait of Tsallis entropy, incorporating an entropic parameter denoted as $r$. When $r$ equals 1, the Masi entropy aligns with the characteristics of Shannon entropy. Despite its efficacy in bi-level image thresholding, Masi entropy experiences a significant reduction in efficiency for multi-level image thresholding, primarily attributed to the escalating time complexity resulting from an increased number of thresholds.

To address this challenge, population-based evolutionary algorithms (PBEAs), including differential evolution (DE) [16, 31], particle swarm optimisation (PSO)

[15, 29], artificial bee colony (ABC) [6], and human mental search (HMS) [12, 17], are suggested. PBEAs, classified as stochastic optimisation algorithms, utilise a population set to exchange information and show excellent performance in solving different problems [10, 11, 14]. Despite the efficacy of PBEAs, it is essential to acknowledge the No Free Lunch (NFL) theorem [34], which posits that no PBEA algorithm can universally solve all optimisation problems effectively.

A crucial aspect of Masi entropy lies in selecting the appropriate value for the entropic parameter $r$. In contrast to probability functions associated with Renyi and Tsallis entropy, where each state-probability is raised to the power of their entropic parameters, Masi entropy involves raising the entire probability function to the power $r$ [9, 24]. The parameter $r$ indicates the degree of extensivity or non-extensivity within the system (image). The entropic measure introduces flexibility, yielding diverse outcomes based on specific demands [30]. Consequently, determining a suitable value for $r$ is imperative depending on the image type, although existing literature commonly employs a fixed value.

This research presents an innovative method for Masi-based image thresholding, aiming for adaptive determination of the entropic measure tailored to each image. In pursuit of this objective, a unique representation is proposed, characterised by simplicity and effectiveness. The emphasis of our approach lies in the modification of representation, a straightforward task yielding significant enhancements in Masi entropy-based image thresholding.

Noteworthy attributes of the proposed representation include its simplicity and effectiveness, as altering the representation is a straightforward task leading to substantial improvements in Masi entropy-based image thresholding. Moreover, our approach obviates the need for prior knowledge about the image type or the histogram distribution to ascertain the entropic parameter. Being representation-based, it is independent of PBEAs and can seamlessly integrate with all PBEAs. The proposed method does not necessitate additional function evaluations, ensuring efficiency without altering the time complexity.

The remainder of this paper is organised as follows. Section 2 provides the background knowledge on DE and Masi entropy, while Sect. 3 explains the proposed algorithm. Section 4 validates the proposed algorithm, and Sect. 5 concludes the paper and provides the future work for our proposed algorithm.

## 2 Background

### 2.1 *Differential Evolution*

Differential Evolution (DE), introduced by Storn and Price [31], is a proficient PBEA algorithm featuring four primary operators: initialisation, mutation, crossover, and selection. Its effectiveness in addressing complex optimisation problems has been demonstrated in various studies [13, 19, 32]. In the following, the leading operators are explained.

### 2.1.1 Initialisation

Like other PBEA algorithms, DE initiates the process by generating a set of random individuals $N_{pop}$. In a $D$-dimensional space, an individual is represented as $X = [x_1, x_2, ..., x_D]$, initialized as $X_{i,j} = x_j^{min} + rand \times (x_j^{max} - x_j^{min})$, where $i = 1, 2, ..., N_{pop}$, $j = 1, 2, ..., D$, $rand$ is a random number between 0 and 1, and $x_j^{min}$ and $x_j^{max}$ are the lower and upper bounds, respectively.

Individuals are evaluated using an objective function $f()$:

$$f\,(individual) = f(x_1, x_2, ..., x_D) \tag{1}$$

### 2.1.2 Mutation

The mutation operator generates a mutant vector based on scaled individual differences, with several variants available. One of the most famous mutation operators is called DE/rand/1 defined as

$$v_i = x_{r1} + F \times (x_{r2} - x_{r3}) \tag{2}$$

where $x_{r1}$, $x_{r2}$, and $x_{r3}$ are three randomly selected individuals, and $F$ is the scaling factor controlling the amplification of differential variation $(x_{r2} - x_{r3})$.

### 2.1.3 Crossover

Crossover operators integrate the mutant and parent vectors, aiming to increase the diversity of the population. A popular crossover operator is binomial crossover, defined as:

$$u_{i,j} = \begin{cases} v_{i,j} & \text{if rand}(0,1) \leq CR \text{ or } j == j_{\text{rand}} \\ x_{i,j} & \text{otherwise} \end{cases} \tag{3}$$

where $u$ is the trial vector, $i = 1, ..., N_{pop}$, $j = 1, ..., D$, $CR$ is the crossover rate, and $j_{\text{rand}}$ is a random number between 1 and $N_{pop}$.

### 2.2 Selection

The selection operator is responsible for choosing the better individual between the trial vector and the parent vector. If the trial vector is superior, the parent vector is updated with the trial vector.

### 2.3 Masi Entropy

In 2005, Masi introduced a novel entropy metric known as Generalised Masi Entropy (GME) [9]. This entropy measure is defined as

$$S_r = \frac{1}{1-r} \log \left[ 1 - (1-r) \sum_{i=1}^{n} p_i \times \log p_i \right] \tag{4}$$

Here, $r > 0$ serves as a measure of the degree of non-extensivity. When $r = 1$, the Masi entropy reduces to the well-known Shannon entropy ($S = -\sum_{i=1}^{n} p_i \times \log p_i$). Masi entropy amalgamates the non-extensive characteristics of Tsallis entropy with the additive properties of Renyi entropy [9]. The parameter $r$ governs the extent of non-extensivity and extensivity within a given system.

## 3 DE-Based Self-Adjusting Masi Entropy

This research introduces a self-adjusting generalised Masi entropy tailored for image segmentation, where the parameter $r$ is autonomously selected without requiring prior knowledge about the image. A DE-based approach is presented to adjust this parameter dynamically. When employing an evolutionary algorithm for optimisation tasks, two critical considerations involve the encoding scheme, dictating the structure of each individual, and the objective function, indicating an individual's quality.

### 3.1 Encoding Scheme

The customary encoding schehme employed in multi-level image thresholding tasks is a vector denoted as

$$x = [th_1, th_2, ..., th_m] \tag{5}$$

where $m$ signifies the number of threshold values, and $th_i$ denotes the $i$-th threshold value. This prevalent encoding schehme is widely adopted in the literature, with most research employing it for optimisation purposes. It exclusively encodes threshold values, and the optimisations objective is centred around finding optimal threshold values. An illustrative example of the conventional encoding schehme is depicted in Fig. 1.

This paper introduces a novel encoding scheme that incorporates the threshold values and the $r$ parameter. The encoding scheme is realised as a vector of real and integer numbers, with a dimensionality of $m + 1$ (where $m$ is the number of threshold values), expressed as

$$x = [r, th_1, th_2, ..., th_m] \tag{6}$$

where $r$ serves as the entropic parameter.

The initial entry is a positive floating-point number within the range [0, 1], controlling the value of the $r$ parameter. The subsequent entries are allocated for the $m$ threshold values, which are integer numbers ranging from 0 to $2^n + 1$ (where $n$ signifies the number of bits representing a pixel). Figure 2 provides an example of a problem featuring $m$ threshold values in an 8-bit image.

The new encoding scheme incorporates both real and integer numbers with distinct ranges. While the first entry falls within 0 and 1, the subsequent entries span 0 to 255. This discrepancy can pose challenges during the evolution process, such as calculating the distance between two individuals, where threshold values might dominate the entropic measure, resulting in an insignificant impact on the distance

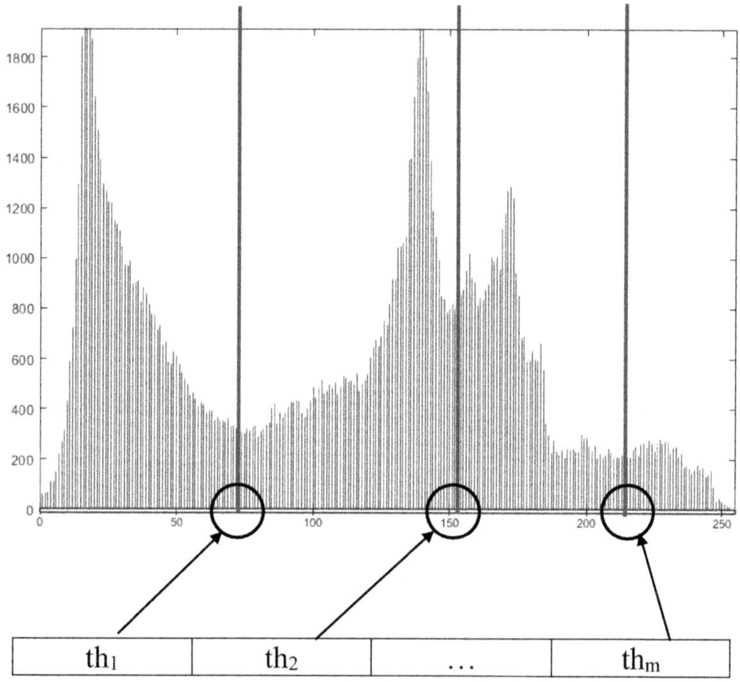


**Fig. 1** Encoding schehme for conventional Masi Entropy. $th_i$ denotes the $i$-th threshold value

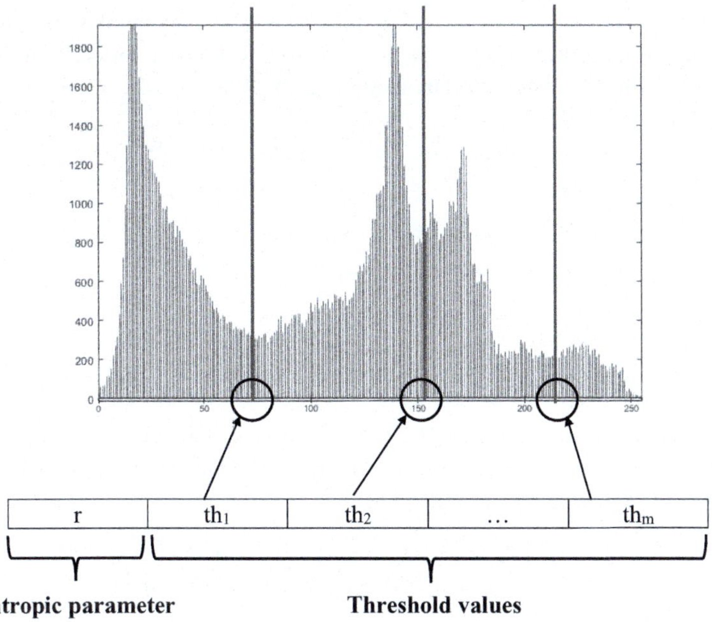


**Fig. 2** The proposed encoding scheme for Masi Entropy

measure. To address this issue, all variables' lower and upper bounds are constrained between 0 and 1. Consequently, when calculating the objective function, the elements corresponding to threshold values should be mapped to the range 0 to 255 using the equation:

$$th_{new} = round(th \times 2^{(n-1)} - 1) \tag{7}$$

where $n$ represents the number of bits representing a pixel. Specifically, for $n = 8$, the new thresholds are computed as

$$th_{new} = round(th \times 255) \tag{8}$$

## 3.2 *Objective Function*

The formulation of bi-level image thresholding can be expressed as follows:

$$\begin{aligned} M_0 &= f(x, y) \in I \mid 0 \leq f(x, y) \leq th - 1 \\ M_1 &= g(x, y) \in I \mid th \leq f(x, y) \leq L - 1 \end{aligned} \tag{9}$$

where $th$ is the threshold value, and $L$ represents the number of grey levels. Multi-level image thresholding extends bi-level thresholding to determine $m(m \geq 2)$ threshold values to obtain a segmented image as follows:

$$\begin{aligned} M_0 &= f(x, y) \in I \mid 0 \leq f(x, y) \leq th_1 - 1 \\ M_1 &= f(x, y) \in I \mid th_1 \leq f(x, y) \leq th_2 - 1 \\ M_i &= f(x, y) \in I \mid th_i \leq f(x, y) \leq th_{i+1} - 1 \\ M_m &= f(x, y) \in I \mid th_m \leq f(x, y) \leq L - 1 \end{aligned} \tag{10}$$

where $th_i, i = 1, ..., m$ is the $i$-th threshold value, and $m$ is the total number of thresholds.

If $f$ denotes an image of size $M \times N$ with intensity levels in the range $[0, \ldots, L - 1]$, and the number of pixels at level $i$ is denoted as $n_i$, then the probability of the occurrence of gray level $i$ is defined as:

$$h_i = \frac{n_i}{M \times N}, \quad h_i \geq 0, \quad \sum_{i=0}^{L-1} h_i = 1 \tag{11}$$

For multi-level image thresholding, class probabilities are defined as:

$$w_1 = \sum_{i=0}^{th_1} h_i, \quad w_2 = \sum_{i=th_1+1}^{th_2} h_i, \quad \ldots, \quad w_k = \sum_{i=th_k-1}^{L-1} h_i \tag{12}$$

Multi-level Masi entropy (MME) for each class is defined as:

$$\begin{aligned} H_1 &= \frac{1}{1-r} \log \left[ 1 - (1-r) \sum_{i=0}^{th_1} \left( \frac{h_i}{w_1} \right) \log \left( \frac{h_i}{w_1} \right) \right] \\ H_2 &= \frac{1}{1-r} \log \left[ 1 - (1-r) \sum_{th_1+1}^{th_2} \left( \frac{h_i}{w_2} \right) \log \left( \frac{h_i}{w_2} \right) \right] \\ H_m &= \frac{1}{1-r} \log \left[ 1 - (1-r) \sum_{th_m+1}^{L} \left( \frac{h_i}{w_m} \right) \log \left( \frac{h_i}{w_m} \right) \right] \end{aligned} \tag{13}$$

where the entropy is denoted by $H$, and $r$ is the value of the entropic parameter.

Finally, the objective function is mathematically defined as:

$$f(t_1, t_2, ..., t_m, r) = H_1 + H_2 + ... + H_m \tag{14}$$

Therefore, the optimal threshold values can be formulated as:

$$t_1^*, t_2^*, ..., t_m^* = \underset{r, 0<t_1<t_2<\cdots<t_m<L-1}{\arg\max} f(r, t_1, t_2, ..., t_m, r) \tag{15}$$

### 3.3 *Evolutionary Operators*

This research introduces an innovative encoding scehme for Masi entropy, enabling the adaptive determination of the entropic parameter ($r$). This representation is algorithm-agnostic and suitable for various evolutionary algorithms. While we employ the DE algorithm for optimisation in this instance due to its simplicity and efficiency, the approach remains flexible for utilising other PBEAs. The proposed algorithm's primary steps, SAME-DE, are delineated below.

1. Initialise each individual to encompass $m$ threshold values and one entropic parameter, all within the range of 0 and 1. The individual encoding scheme is expounded in Sect. 3.1, which constitutes the main contribution of this paper.
2. For each individual:
   Map the elements corresponding to threshold values to the new search space using Eq. 7.
   Calculate the objective function using Eq. 14 and the individual in the new search space.
3. Perform the mutation operator.
4. Perform the crossover operator.
5. For each individual:
   Map the elements corresponding to threshold values to the new search space using Eq. 7.
   Calculate the objective function using Eq. 14 and the individual in the new search space.
6. Perform the selection operator.
7. If the stopping condition is not satisfied, return to Step (3).

## 4 Experiments

An extensive series of experiments has been conducted to assess the effectiveness of our proposed individual representation. We utilised five well-known benchmark images, namely 'Boats,' 'Peppers,' 'Goldhill,' 'Lenna,' and 'House,' alongside seven images sourced from the Berkeley segmentation dataset [8]-specifically '12003,' '181079,' '175043,' '101085,' '147091,' '101087,' and '253027.'

In Fig. 3, all images are depicted along with their corresponding histograms. The selected images exhibit diverse histogram distributions, providing a robust benchmark for evaluating algorithms.

(a) Boats (b) Peppers (c) Goldhill (d) Lenna

(e) House (f) 12003 (g) 181079 (h) 175043

(i) 101085 (j) 147091 (k) 101087 (l) 253027

**Fig. 3** Test images and their histograms

For a fair comparison, each algorithm undergoes 25 independent runs, with reported statistical results, including average and standard deviation over the 25 runs. The population size for all algorithms is fixed at 50, and the stopping criterion is set at 10,000 function evaluations.

## 4.1 Evaluation Criteria

This section outlines the criteria for assessing the proposed algorithm, described in detail below.

### 4.1.1 Objective Function

The primary criterion for evaluating algorithms is the objective function. The self-adaptive approach is utilised to determine entropic parameters specific to an image. This criterion aims to demonstrate the efficacy of the proposed method in achieving higher Masi entropy values by appropriately adjusting the $r$ parameter.

### 4.1.2 Peak Signal to Noise Ratio (PSNR)

PSNR, a widely recognised metric for evaluating image quality post-thresholding, is formulated as:

$$PSNR = 20 \log_{10}(255/RMSE) \tag{16}$$

Here, $RMSE$ represents the root mean square error, calculated by:

$$RMSE = \sqrt{\frac{\sum_{i=1}^{M} \sum_{j=1}^{N} (I(i,j) - \hat{I}(i,j))^2}{MN}} \tag{17}$$

where $M$ and $N$ denote the image dimensions, and $I$ and $\hat{I}$ are the original and thresholded images, respectively. A higher PSNR value indicates superior image quality.

### 4.1.3 Feature Similarity Index Measure (FSIM)

FSIM [35], based on the human visual system, is another criterion employed in this paper. It calculates similarity using high phase congruency (PC) and gradient magnitude (GM) as:

$$FSIM = \frac{\sum_{x\in\Omega} S_L(X) \cdot PC_m(X)}{\sum_{x\in\Omega} PC_m(X)} \tag{18}$$

Here, $\Omega$ represents the entire image, $S_L(X)$ is the similarity measure, and $PC$ denotes the high phase congruency ratio. The similarity and PC values are determined by:

$$S_L(X) = \frac{2PC_1(X)PC_2(X) + T_1}{PC_1^2(X) + PC_2^2(X) + T_1} \tag{19}$$

$$S_G(X) = \frac{2G_1(X)G_2(X) + T_2}{G_1^2(X) + G_2^2(X) + T_2} \tag{20}$$

where $T_1$ and $T_2$ are coefficients, and $G$ represents the gradient magnitude. Higher FSIM values indicate more effective performance.

#### 4.1.4 Structural Similarity Index Measure (SSIM)

SSIM, integrating luminance, contrast, and structure, is the fourth evaluation criterion:

$$SSIM(x, y) = \frac{(2\mu_x\mu_y + C_1)(2\sigma_{xy} + C_2)}{(\mu_x^2 + \mu_y^2 + C_1)(\sigma_x^2 + \sigma_y^2 + C_2)} \tag{21}$$

where $\mu_x$ and $\mu_y$ are the average intensities, $\sigma_x$ and $\sigma_y$ are the variances, and $\sigma_{xy}$ is the covariance of images $X$ and $Y$, with $C_1$ and $C_2$ being constants. Higher SSIM values signify better image quality.

### *4.2 Results*

Table 1 presents a comprehensive comparison between the proposed and DE (as our baseline) algorithms, focusing on the objective function for varying threshold values (m = 2, 3, 4, 5, and 10). The results, as depicted in the last column of the table, consistently show that the Improved Rate (IR) for all images and threshold values is greater than 1, ranging between 1.17 and 7.01. This unequivocally indicates that for every case studied, the proposed algorithm achieved higher objective function values than the DE algorithm.

Such results are significant as they demonstrate the superiority of the proposed encoding scheme in our proposed algorithm over the conventional encoding scheme used in the DE algorithm. Particularly noteworthy is the case of the 'Goldhill' image at a threshold value of 2, where the proposed algorithm achieves an IR of 7.01, significantly outperforming the DE algorithm. This trend is consistently observed across various images and threshold values, reinforcing the effectiveness of the individual encoding scheme in the proposed algorithm.

**Table 1** Comparison of Masi Entropy with Self-Adaptive Masi Entropy in terms of objective function measure

| Images | M | Masi entropy | | Self-adaptive masi entropy | | IR value |
|---|---|---|---|---|---|---|
| | | AVG | Std | AVG | Std | |
| Boats | 2 | 33.8085 | 0.0000 | 81.9137 | 5.8184 | **2.42** |
| | 3 | 35.2333 | 0.0250 | 96.1013 | 7.3906 | **2.73** |
| | 4 | 35.3906 | 0.1952 | 102.9289 | 8.6680 | **2.91** |
| | 5 | 37.7555 | 0.2165 | 106.9565 | 7.0567 | **2.83** |
| | 10 | 50.6134 | 0.1559 | 120.4249 | 7.2057 | **2.38** |
| Peppers | 2 | 61.4491 | 8.1325 | 84.6408 | 7.0526 | **1.38** |
| | 3 | 66.1996 | 8.7457 | 92.3636 | 7.3795 | **1.40** |
| | 4 | 66.4486 | 6.4888 | 96.3773 | 5.6290 | **1.45** |
| | 5 | 68.2616 | 7.7628 | 102.2122 | 6.5273 | **1.50** |
| | 10 | 51.2683 | 3.2975 | 116.8918 | 4.9612 | **2.28** |
| Goldhill | 2 | 13.9091 | 0.0000 | 97.5567 | 34.9021 | **7.01** |
| | 3 | 15.5554 | 0.2077 | 79.9875 | 8.6250 | **5.14** |
| | 4 | 16.7997 | 0.4703 | 59.6882 | 7.3422 | **3.55** |
| | 5 | 17.4773 | 0.6361 | 61.6198 | 8.3377 | **3.53** |
| | 10 | 24.2124 | 0.4509 | 68.0360 | 5.8051 | **2.81** |
| Lenna | 2 | 66.1115 | 1.4811 | 84.2458 | 6.7056 | **1.27** |
| | 3 | 70.7372 | 2.1710 | 89.9052 | 6.8297 | **1.27** |
| | 4 | 72.5674 | 2.0647 | 94.7045 | 6.8787 | **1.31** |
| | 5 | 73.1578 | 2.5071 | 100.6330 | 7.0391 | **1.38** |
| | 10 | 58.8415 | 7.2076 | 115.4231 | 6.3712 | **1.96** |
| House | 2 | 62.9681 | 3.2058 | 83.2243 | 7.0778 | **1.32** |
| | 3 | 64.7518 | 2.9153 | 91.4188 | 6.1981 | **1.41** |
| | 4 | 64.9181 | 2.4136 | 97.7577 | 6.8026 | **1.51** |
| | 5 | 67.6959 | 3.3574 | 104.6733 | 5.2237 | **1.55** |
| | 10 | 49.5071 | 0.2145 | 116.8489 | 5.2397 | **2.36** |
| 12003 | 2 | 67.7324 | 4.0981 | 89.6073 | 7.2095 | **1.32** |
| | 3 | 66.3032 | 6.6072 | 95.1857 | 6.3433 | **1.44** |
| | 4 | 65.8359 | 7.1440 | 101.4224 | 7.0517 | **1.54** |
| | 5 | 68.5491 | 4.6451 | 105.4893 | 6.4691 | **1.54** |
| | 10 | 61.1679 | 5.8045 | 122.4221 | 7.0039 | **2.00** |
| 181079 | 2 | 62.7319 | 4.4973 | 83.9473 | 4.2645 | **1.34** |
| | 3 | 66.2420 | 3.4382 | 91.9735 | 6.7032 | **1.39** |
| | 4 | 68.2890 | 2.7933 | 96.7852 | 11.4986 | **1.42** |
| | 5 | 70.1628 | 3.0208 | 99.3471 | 5.3884 | **1.42** |
| | 10 | 50.2968 | 0.3559 | 118.3056 | 7.4391 | **2.35** |
| 175043 | 2 | 60.4357 | 4.1206 | 88.6626 | 7.8079 | **1.47** |
| | 3 | 63.1564 | 3.4971 | 92.7214 | 6.1854 | **1.47** |
| | 4 | 62.8356 | 3.6252 | 100.7232 | 6.3669 | **1.60** |
| | 5 | 63.9570 | 4.7977 | 105.9257 | 5.3544 | **1.66** |
| | 10 | 50.8374 | 0.1999 | 119.7741 | 5.1867 | **2.36** |
| 101085 | 2 | 62.2314 | 5.6364 | 87.7891 | 7.3550 | **1.41** |
| | 3 | 63.9595 | 4.8575 | 94.5298 | 7.9036 | **1.48** |
| | 4 | 64.2398 | 4.2241 | 104.3580 | 7.0467 | **1.62** |
| | 5 | 67.4625 | 4.3901 | 104.8397 | 5.8789 | **1.55** |
| | 10 | 60.0898 | 7.0116 | 122.0190 | 5.2272 | **2.03** |

(continued)

**Table 1** (continued)

| Images | M | Masi entropy | | Self-adaptive masi entropy | | IR value |
|---|---|---|---|---|---|---|
| | | AVG | Std | AVG | Std | |
| 147091 | 2 | 65.6337 | 1.2243 | 89.9635 | 8.1575 | **1.37** |
| | 3 | 67.8831 | 1.5645 | 97.9851 | 6.7131 | **1.44** |
| | 4 | 70.1127 | 2.2803 | 105.2204 | 6.1622 | **1.50** |
| | 5 | 70.6823 | 1.6655 | 107.8314 | 6.9576 | **1.53** |
| | 10 | 56.5639 | 6.6437 | 125.9096 | 6.5130 | **2.23** |
| 101087 | 2 | 61.2515 | 6.2463 | 71.8026 | 5.1852 | **1.17** |
| | 3 | 59.4643 | 7.1952 | 80.6284 | 5.2781 | **1.36** |
| | 4 | 62.2509 | 5.9372 | 90.8687 | 8.3988 | **1.46** |
| | 5 | 65.9350 | 7.5047 | 99.3601 | 8.4006 | **1.51** |
| | 10 | 55.6898 | 6.8016 | 118.2269 | 6.5213 | **2.12** |
| 253027 | 2 | 26.7864 | 0.0000 | 83.0778 | 4.8062 | **3.10** |
| | 3 | 29.9862 | 0.0698 | 89.0026 | 6.1107 | **2.97** |
| | 4 | 32.9903 | 0.0863 | 93.7718 | 5.4209 | **2.84** |
| | 5 | 35.8719 | 0.1498 | 100.1583 | 6.7445 | **2.79** |
| | 10 | 48.8008 | 0.2181 | 117.0751 | 6.1896 | **2.40** |

Furthermore, the consistent improvement across different images and threshold values suggests that the advantage of our proposed encoding scheme is not confined to specific conditions or image types but is broadly applicable. This versatility is crucial for practical applications with typical scenarios and varying conditions.

Table 2 provides a detailed comparison of the proposed algorithm and DE algorithms regarding the PSNR measure across various images and threshold values. The IR values in the table offer insights into the relative performance of the two algorithms.

A key observation from the table is that in most cases, the IR values are more significant than 1, indicating that the proposed encoding scheme generally achieves higher PSNR values than CR-DE. This trend is particularly pronounced in images like 'Boats,' 'Peppers,' 'Lenna,' and 'House,' where the IR values across multiple threshold values are consistently above 1, sometimes reaching as high as 1.69 (as seen in the 'Boats' image with a threshold of 3). These results suggest that the self-adaptive approach of the proposed algorithm is more effective in optimising image quality in terms of PSNR.

However, there are instances, particularly at higher threshold values (like 10 in the case of the 'Boats,' 'Goldhill,' 'House,' and '253027' images), where the IR values fall below 1. This indicates that in these specific scenarios, DE outperforms the proposed approach. It highlights that the advantage of the proposed approach in terms of PSNR is not absolute and may vary depending on the image and the threshold value.

Interestingly, for the 'Goldhill' image, the performance of the proposed approach could be more consistent, with IR values both above and below one another across different threshold values. This suggests that specific characteristics of the images may influence the effectiveness of the proposed algorithm in enhancing PSNR.

**Table 2** Comparison of Masi Entropy with Self-Adaptive Masi Entropy in terms of PSNR measure

| Images | M | Masi entropy | | Self-adaptive masi entropy | | IR value |
|---|---|---|---|---|---|---|
| | | AVG | Std | AVG | Std | |
| Boats | 2 | 14.7583 | 0.0000 | 21.9196 | 0.8218 | **1.49** |
| | 3 | 14.7665 | 0.0033 | 24.9660 | 0.3429 | **1.69** |
| | 4 | 21.1537 | 1.9951 | 25.1173 | 0.7486 | **1.19** |
| | 5 | 24.1281 | 1.6868 | 25.9170 | 0.9856 | **1.07** |
| | 10 | 30.6378 | 1.1935 | 29.1210 | 1.3385 | 0.95 |
| Peppers | 2 | 15.9571 | 0.0512 | 21.6902 | 1.1951 | **1.36** |
| | 3 | 15.9804 | 0.0578 | 23.6149 | 1.0184 | **1.48** |
| | 4 | 15.9687 | 0.0568 | 24.2968 | 1.2995 | **1.52** |
| | 5 | 15.9899 | 0.0452 | 25.5439 | 1.2510 | **1.60** |
| | 10 | 25.7010 | 7.4698 | 29.4703 | 1.0571 | **1.15** |
| Goldhill | 2 | 14.3017 | 0.0000 | 17.3628 | 2.0482 | **1.21** |
| | 3 | 18.2403 | 0.5343 | 17.6931 | 1.4535 | 0.97 |
| | 4 | 19.7408 | 1.7250 | 19.0102 | 2.1876 | 0.96 |
| | 5 | 21.5380 | 2.0669 | 18.6290 | 1.8349 | 0.86 |
| | 10 | 24.7976 | 1.6491 | 22.0777 | 2.5131 | 0.89 |
| Lenna | 2 | 16.9620 | 0.1980 | 21.9957 | 1.0918 | **1.30** |
| | 3 | 16.9833 | 0.0943 | 23.7498 | 1.7016 | **1.40** |
| | 4 | 16.8460 | 0.3704 | 24.5557 | 1.7432 | **1.46** |
| | 5 | 16.9674 | 0.1980 | 26.3399 | 1.7802 | **1.55** |
| | 10 | 17.6003 | 4.1614 | 29.4793 | 1.3528 | **1.67** |
| House | 2 | 15.6324 | 0.0736 | 21.6474 | 0.8128 | **1.38** |
| | 3 | 15.6581 | 0.0925 | 22.9470 | 1.2840 | **1.47** |
| | 4 | 15.6186 | 0.0626 | 24.8415 | 1.1193 | **1.59** |
| | 5 | 15.6678 | 0.0945 | 25.3892 | 1.2753 | **1.62** |
| | 10 | 30.2118 | 1.1396 | 28.5671 | 1.3658 | 0.95 |
| 12003 | 2 | 16.6783 | 0.3629 | 20.7166 | 0.7720 | **1.24** |
| | 3 | 16.6025 | 0.3166 | 22.1386 | 1.2695 | **1.33** |
| | 4 | 16.7209 | 0.3404 | 23.7583 | 1.3957 | **1.42** |
| | 5 | 16.8061 | 0.2671 | 25.1458 | 1.5356 | **1.50** |
| | 10 | 18.3339 | 4.7384 | 28.8249 | 1.1562 | **1.57** |
| 181079 | 2 | 16.1842 | 0.0320 | 21.6453 | 1.0946 | **1.34** |
| | 3 | 16.1244 | 0.2889 | 23.4321 | 1.0112 | **1.45** |
| | 4 | 16.1886 | 0.0310 | 24.4759 | 1.1346 | **1.51** |
| | 5 | 16.1905 | 0.0337 | 24.8274 | 1.4885 | **1.53** |
| | 10 | 28.5708 | 3.9444 | 28.8641 | 1.3833 | **1.01** |
| 175043 | 2 | 16.8524 | 0.0190 | 22.9002 | 0.5622 | **1.36** |
| | 3 | 16.8495 | 0.0188 | 24.0016 | 1.1054 | **1.42** |
| | 4 | 16.8521 | 0.0200 | 25.4340 | 1.0314 | **1.51** |
| | 5 | 16.8501 | 0.0229 | 26.1023 | 1.1830 | **1.55** |
| | 10 | 30.0862 | 1.1518 | 28.4958 | 1.7233 | 0.95 |
| 101085 | 2 | 16.3399 | 0.6974 | 20.0245 | 0.9216 | **1.23** |
| | 3 | 16.6037 | 0.8569 | 21.8110 | 1.0953 | **1.31** |
| | 4 | 16.4483 | 0.7995 | 23.3634 | 1.1137 | **1.42** |
| | 5 | 16.9517 | 1.1630 | 24.8175 | 1.3589 | **1.46** |
| | 10 | 21.7081 | 5.7472 | 28.0942 | 1.6369 | **1.29** |

(continued)

**Table 2** (continued)

| Images | M | Masi entropy | | Self-adaptive masi entropy | | IR value |
|---|---|---|---|---|---|---|
| | | AVG | Std | AVG | Std | |
| 147091 | 2 | 13.8490 | 0.4648 | 21.0999 | 0.8073 | **1.52** |
| | 3 | 13.8004 | 0.3728 | 22.9297 | 0.6920 | **1.66** |
| | 4 | 13.9544 | 0.6179 | 23.6874 | 1.6676 | **1.70** |
| | 5 | 13.7331 | 0.0037 | 25.1044 | 1.2560 | **1.83** |
| | 10 | 24.0450 | 8.0590 | 28.6006 | 1.1434 | **1.19** |
| 101087 | 2 | 17.8501 | 1.0810 | 15.1165 | 3.1843 | 0.85 |
| | 3 | 16.8152 | 1.4715 | 20.4673 | 3.3977 | **1.22** |
| | 4 | 17.7472 | 1.1903 | 23.2322 | 2.9891 | **1.31** |
| | 5 | 17.8029 | 1.2045 | 25.5356 | 1.9341 | **1.43** |
| | 10 | 21.8835 | 5.7075 | 29.3977 | 1.3339 | **1.34** |
| 253027 | 2 | 19.3974 | 0.0000 | 21.7490 | 1.5896 | **1.12** |
| | 3 | 19.6239 | 0.0962 | 23.0486 | 2.0046 | **1.17** |
| | 4 | 20.3754 | 0.4778 | 25.2586 | 1.7915 | **1.24** |
| | 5 | 25.5548 | 1.6095 | 26.2924 | 1.6903 | **1.03** |
| | 10 | 31.1360 | 0.9813 | 29.2952 | 1.2060 | 0.94 |

Table 3 compares the Masi Entropy and Self-Adaptive Masi Entropy algorithms using the FSIM across various images and threshold values.

A notable trend observed in the table is that the Self-Adaptive Masi Entropy often achieves higher FSIM values than the Masi Entropy, as evidenced by the IR values greater than 1 in many cases. For instance, in the 'Boats' image at thresholds 2 and 3, the IR values are 1.55 and 1.75, respectively, suggesting a significant improvement in the feature similarity index with the self-adaptive approach. This improvement is consistent across other images like 'Peppers', 'Lenna', and 'House', where the IR values for lower threshold values are considerably above 1, implying better performance by the Self-Adaptive Masi Entropy in preserving image features.

However, it is essential to note that this trend is not universal. For some images, particularly at higher threshold values, the IR dips below 1. For example, in the 'Boats' image at a threshold value of 10, the IR is 0.98, indicating a slight underperformance of the Self-Adaptive Masi Entropy compared to the standard Masi Entropy. This pattern is also observed in other images, such as 'Goldhill', 'House', and '253027' at higher thresholds.

Furthermore, in some instances, the performance of the Self-Adaptive Masi Entropy is relatively close to that of the Masi Entropy, as indicated by IR values hovering around 1. This is observed in the '181079' and '253027' images at a threshold 10, suggesting comparable performance between the two algorithms under these specific conditions.

The variations in IR values across different images and thresholds suggest that while the Self-Adaptive Masi Entropy generally improves FSIM, its efficacy can be influenced by the specific characteristics of the image and the threshold value.

Table 4 compares the Masi Entropy and Self-Adaptive Masi Entropy algorithms in terms of the SSIM across various images and threshold values.

**Table 3** Comparison of Masi Entropy with Self-Adaptive Masi Entropy in terms of FSIM measure

| Images | M | Masi entropy | | Self-adaptive masi entropy | | IR value |
|---|---|---|---|---|---|---|
| | | AVG | Std | AVG | Std | |
| Boats | 2 | 0.4793 | 0.0000 | 0.7443 | 0.0306 | **1.55** |
| | 3 | 0.4784 | 0.0006 | 0.8389 | 0.0084 | **1.75** |
| | 4 | 0.7608 | 0.0870 | 0.8474 | 0.0204 | **1.11** |
| | 5 | 0.8391 | 0.0302 | 0.8667 | 0.0278 | **1.03** |
| | 10 | 0.9521 | 0.0157 | 0.9321 | 0.0321 | 0.98 |
| Peppers | 2 | 0.6088 | 0.0109 | 0.6871 | 0.0224 | **1.13** |
| | 3 | 0.6064 | 0.0196 | 0.7302 | 0.0242 | **1.20** |
| | 4 | 0.6094 | 0.0161 | 0.7474 | 0.0270 | **1.23** |
| | 5 | 0.6098 | 0.0169 | 0.7764 | 0.0273 | **1.27** |
| | 10 | 0.7943 | 0.1394 | 0.8601 | 0.0202 | **1.08** |
| Goldhill | 2 | 0.4421 | 0.0000 | 0.6491 | 0.0673 | **1.47** |
| | 3 | 0.6152 | 0.0565 | 0.6015 | 0.0357 | 0.98 |
| | 4 | 0.6856 | 0.0779 | 0.6560 | 0.0826 | 0.96 |
| | 5 | 0.7417 | 0.0879 | 0.6427 | 0.0591 | 0.87 |
| | 10 | 0.8447 | 0.0481 | 0.7609 | 0.0926 | 0.90 |
| Lenna | 2 | 0.6381 | 0.0103 | 0.7372 | 0.0148 | **1.16** |
| | 3 | 0.6381 | 0.0109 | 0.7756 | 0.0304 | **1.22** |
| | 4 | 0.6329 | 0.0193 | 0.7988 | 0.0337 | **1.26** |
| | 5 | 0.6400 | 0.0105 | 0.8338 | 0.0311 | **1.30** |
| | 10 | 0.6295 | 0.0902 | 0.8985 | 0.0189 | **1.43** |
| House | 2 | 0.4502 | 0.0117 | 0.7371 | 0.0236 | **1.64** |
| | 3 | 0.4519 | 0.0137 | 0.7844 | 0.0394 | **1.74** |
| | 4 | 0.4461 | 0.0076 | 0.8454 | 0.0301 | **1.90** |
| | 5 | 0.4565 | 0.0149 | 0.8598 | 0.0361 | **1.88** |
| | 10 | 0.9525 | 0.0131 | 0.9257 | 0.0321 | 0.97 |
| 12003 | 2 | 0.5042 | 0.0172 | 0.6179 | 0.0212 | **1.23** |
| | 3 | 0.5288 | 0.0214 | 0.6583 | 0.0349 | **1.24** |
| | 4 | 0.5347 | 0.0221 | 0.7051 | 0.0421 | **1.32** |
| | 5 | 0.5221 | 0.0200 | 0.7455 | 0.0442 | **1.43** |
| | 10 | 0.5950 | 0.1161 | 0.8505 | 0.0328 | **1.43** |
| 181079 | 2 | 0.5115 | 0.0016 | 0.6414 | 0.0294 | **1.25** |
| | 3 | 0.5123 | 0.0029 | 0.6886 | 0.0234 | **1.34** |
| | 4 | 0.5124 | 0.0022 | 0.7213 | 0.0246 | **1.41** |
| | 5 | 0.5129 | 0.0021 | 0.7298 | 0.0351 | **1.42** |
| | 10 | 0.8247 | 0.0950 | 0.8290 | 0.0299 | **1.01** |
| 175043 | 2 | 0.2925 | 0.0025 | 0.6988 | 0.0249 | **2.39** |
| | 3 | 0.2920 | 0.0033 | 0.7473 | 0.0464 | **2.56** |
| | 4 | 0.2925 | 0.0028 | 0.8083 | 0.0369 | **2.76** |
| | 5 | 0.2919 | 0.0033 | 0.8290 | 0.0374 | **2.84** |
| | 10 | 0.9327 | 0.0168 | 0.8950 | 0.0430 | 0.96 |
| 101085 | 2 | 0.5321 | 0.0164 | 0.6343 | 0.0591 | **1.19** |
| | 3 | 0.5475 | 0.0294 | 0.7141 | 0.0630 | **1.30** |
| | 4 | 0.5573 | 0.0354 | 0.7741 | 0.0475 | **1.39** |
| | 5 | 0.5916 | 0.0549 | 0.8111 | 0.0647 | **1.37** |
| | 10 | 0.7335 | 0.1426 | 0.8922 | 0.0619 | **1.22** |

(continued)

**Table 3** (continued)

| Images | M | Masi entropy | | Self-adaptive masi entropy | | IR value |
|---|---|---|---|---|---|---|
| | | AVG | Std | AVG | Std | |
| 147091 | 2 | 0.5973 | 0.0147 | 0.7215 | 0.0212 | **1.21** |
| | 3 | 0.5974 | 0.0126 | 0.7535 | 0.0261 | **1.26** |
| | 4 | 0.6045 | 0.0210 | 0.7683 | 0.0416 | **1.27** |
| | 5 | 0.5981 | 0.0016 | 0.8001 | 0.0313 | **1.34** |
| | 10 | 0.7907 | 0.1192 | 0.8494 | 0.0213 | **1.07** |
| 101087 | 2 | 0.6339 | 0.0041 | 0.6954 | 0.0320 | **1.10** |
| | 3 | 0.6353 | 0.0076 | 0.7321 | 0.0346 | **1.15** |
| | 4 | 0.6398 | 0.0082 | 0.7645 | 0.0395 | **1.19** |
| | 5 | 0.6419 | 0.0051 | 0.8037 | 0.0329 | **1.25** |
| | 10 | 0.7122 | 0.1178 | 0.8722 | 0.0184 | **1.22** |
| 253027 | 2 | 0.6128 | 0.0000 | 0.7006 | 0.0356 | **1.14** |
| | 3 | 0.6052 | 0.0113 | 0.7322 | 0.0468 | **1.21** |
| | 4 | 0.6512 | 0.0233 | 0.7884 | 0.0402 | **1.21** |
| | 5 | 0.7938 | 0.0358 | 0.8121 | 0.0433 | **1.02** |
| | 10 | 0.9057 | 0.0168 | 0.8720 | 0.0284 | 0.96 |

A significant observation from the table is that the Self-Adaptive Masi Entropy generally outperforms the standard Masi Entropy in SSIM, indicated by IR values greater than 1 in most cases. For example, in the 'Boats' image at thresholds 2 and 3, the IR values are notably high at 1.54 and 1.70, respectively. This suggests a substantial enhancement in structural similarity when employing the self-adaptive approach. Similar trends are observed in other images, such as 'Peppers,' 'Lenna,' and 'House,' where the Self-Adaptive Masi Entropy consistently shows superior performance in maintaining structural integrity per the SSIM metric.

However, it is important to note variations in performance at higher threshold values. In some instances, such as with the 'Boats' image at a threshold of 10, the IR value falls below 1 (0.96), indicating a slightly lower performance of the Self-Adaptive Masi Entropy compared to the standard approach. This pattern is observed in several other images, suggesting that the advantage of the self-adaptive method in SSIM is more pronounced at lower thresholds.

Interestingly, for specific images like 'Goldhill,' the performance of the Self-Adaptive Masi Entropy varies, with some threshold levels showing IR values below 1. This highlights that the effectiveness of the self-adaptive method in enhancing SSIM may depend on the specific characteristics of the images.

Moreover, the performance difference between the two methods is marginal in certain cases, as indicated by IR values close to 1. This is seen in images such as '181079' and '253027' at higher thresholds, suggesting that both algorithms perform similarly under these conditions. Lastly, we presented the image thresholding results for a representative image, "101087," in Fig. 4.

**Table 4** Comparison of Masi Entropy with Self-Adaptive Masi Entropy in terms of SSIM measure

| Images | M | Masi entropy | | Self-adaptive masi entropy | | Improved rate |
|---|---|---|---|---|---|---|
| | | AVG | Std | AVG | Std | |
| Boats | 2 | 0.4118 | 0.0000 | 0.6330 | 0.0306 | **1.54** |
| | 3 | 0.4130 | 0.0004 | 0.7014 | 0.0084 | **1.70** |
| | 4 | 0.6433 | 0.0726 | 0.7144 | 0.0204 | **1.11** |
| | 5 | 0.7207 | 0.0306 | 0.7328 | 0.0278 | **1.02** |
| | 10 | 0.8443 | 0.0294 | 0.8146 | 0.0321 | 0.96 |
| Peppers | 2 | 0.5692 | 0.0095 | 0.6833 | 0.0224 | **1.20** |
| | 3 | 0.5624 | 0.0205 | 0.7271 | 0.0242 | **1.29** |
| | 4 | 0.5679 | 0.0152 | 0.7403 | 0.0270 | **1.30** |
| | 5 | 0.5663 | 0.0174 | 0.7681 | 0.0273 | **1.36** |
| | 10 | 0.7650 | 0.1471 | 0.8423 | 0.0202 | **1.10** |
| Goldhill | 2 | 0.3759 | 0.0000 | 0.4706 | 0.0673 | **1.25** |
| | 3 | 0.4794 | 0.0482 | 0.4611 | 0.0357 | 0.96 |
| | 4 | 0.5378 | 0.0715 | 0.5094 | 0.0826 | 0.95 |
| | 5 | 0.5914 | 0.0853 | 0.4999 | 0.0591 | 0.85 |
| | 10 | 0.6955 | 0.0599 | 0.6146 | 0.0926 | 0.88 |
| Lenna | 2 | 0.6345 | 0.0128 | 0.6923 | 0.0148 | **1.09** |
| | 3 | 0.6355 | 0.0085 | 0.7267 | 0.0304 | **1.14** |
| | 4 | 0.6275 | 0.0243 | 0.7421 | 0.0337 | **1.18** |
| | 5 | 0.6364 | 0.0134 | 0.7737 | 0.0311 | **1.22** |
| | 10 | 0.6184 | 0.0764 | 0.8296 | 0.0189 | **1.34** |
| House | 2 | 0.4046 | 0.0035 | 0.6211 | 0.0236 | **1.54** |
| | 3 | 0.4060 | 0.0043 | 0.6591 | 0.0394 | **1.62** |
| | 4 | 0.4043 | 0.0029 | 0.7091 | 0.0301 | **1.75** |
| | 5 | 0.4066 | 0.0045 | 0.7254 | 0.0361 | **1.78** |
| | 10 | 0.8441 | 0.0230 | 0.8073 | 0.0321 | 0.96 |
| 12003 | 2 | 0.4441 | 0.0110 | 0.5954 | 0.0212 | **1.34** |
| | 3 | 0.4589 | 0.0125 | 0.6417 | 0.0349 | **1.40** |
| | 4 | 0.4607 | 0.0127 | 0.6926 | 0.0421 | **1.50** |
| | 5 | 0.4542 | 0.0126 | 0.7322 | 0.0442 | **1.61** |
| | 10 | 0.5175 | 0.1367 | 0.8357 | 0.0328 | **1.61** |
| 181079 | 2 | 0.5039 | 0.0011 | 0.6413 | 0.0294 | **1.27** |
| | 3 | 0.5033 | 0.0026 | 0.6832 | 0.0234 | **1.36** |
| | 4 | 0.5041 | 0.0021 | 0.7097 | 0.0246 | **1.41** |
| | 5 | 0.5048 | 0.0020 | 0.7156 | 0.0351 | **1.42** |
| | 10 | 0.7954 | 0.0912 | 0.8019 | 0.0299 | **1.01** |
| 175043 | 2 | 0.1982 | 0.0012 | 0.6994 | 0.0249 | **3.53** |
| | 3 | 0.1980 | 0.0012 | 0.7398 | 0.0464 | **3.74** |
| | 4 | 0.1981 | 0.0014 | 0.7934 | 0.0369 | **4.01** |
| | 5 | 0.1980 | 0.0015 | 0.8143 | 0.0374 | **4.11** |
| | 10 | 0.9075 | 0.0221 | 0.8727 | 0.0430 | 0.96 |
| 101085 | 2 | 0.4208 | 0.0268 | 0.5553 | 0.0591 | **1.32** |
| | 3 | 0.4347 | 0.0399 | 0.6465 | 0.0630 | **1.49** |
| | 4 | 0.4377 | 0.0501 | 0.7186 | 0.0475 | **1.64** |
| | 5 | 0.4861 | 0.0823 | 0.7623 | 0.0647 | **1.57** |
| | 10 | 0.6754 | 0.1750 | 0.8522 | 0.0619 | **1.26** |

(continued)

**Table 4** (continued)

| Images | M | Masi entropy | | Self-adaptive masi entropy | | Improved rate |
|---|---|---|---|---|---|---|
| | | AVG | Std | AVG | Std | |
| 147091 | 2 | 0.5384 | 0.0265 | 0.7135 | 0.0212 | **1.33** |
| | 3 | 0.5362 | 0.0208 | 0.7494 | 0.0261 | **1.40** |
| | 4 | 0.5472 | 0.0341 | 0.7623 | 0.0416 | **1.39** |
| | 5 | 0.5363 | 0.0021 | 0.7952 | 0.0313 | **1.48** |
| | 10 | 0.7648 | 0.1349 | 0.8407 | 0.0213 | **1.10** |
| 101087 | 2 | 0.6529 | 0.0237 | 0.6977 | 0.0320 | **1.07** |
| | 3 | 0.6322 | 0.0309 | 0.7435 | 0.0346 | **1.18** |
| | 4 | 0.6516 | 0.0249 | 0.7785 | 0.0395 | **1.19** |
| | 5 | 0.6546 | 0.0258 | 0.8196 | 0.0329 | **1.25** |
| | 10 | 0.7325 | 0.1058 | 0.8799 | 0.0184 | **1.20** |
| 253027 | 2 | 0.5841 | 0.0000 | 0.6758 | 0.0356 | **1.16** |
| | 3 | 0.5792 | 0.0060 | 0.6994 | 0.0468 | **1.21** |
| | 4 | 0.6153 | 0.0246 | 0.7516 | 0.0402 | **1.22** |
| | 5 | 0.7512 | 0.0404 | 0.7726 | 0.0433 | **1.03** |
| | 10 | 0.8738 | 0.0227 | 0.8355 | 0.0284 | 0.96 |

(a) Masi Entropy

(b) Self-adaptive Masi Entropy

**Fig. 4** Thresholded *101087* images

## 5 Conclusion

Masi entropy is a technique for image thresholding designed to improve upon the limitations observed in Renyi and Tsallis entropies. It integrates Renyi entropy's additive property with the non-additive characteristic of Tsallis entropy. Notably, the effectiveness of Masi entropy diminishes with an increase in threshold values. To address this issue, Population-based evolutionary algorithms (PBEAs) are recommended as a robust solution. A critical aspect of Masi entropy is its entropic parameter, $r$, which significantly influences its performance in image thresholding applications. Although existing studies typically utilise a constant value for this parameter, it is suggested that the histogram distribution could inform the optimal value of $r$. The aim of this study is to introduce an innovative encoding scheme for PBEAs that enables the autonomous identification of the most suitable $r$ value without prior knowledge of image distribution. This novel approach is algorithm-agnostic, focusing on the encoding scheme. In the optimisation phase, the differential evolution algorithm, a proven PBEA, was employed. Comparative experiments were conducted using this method and the DE baseline algorithm across various complex images. The results, evaluated based on objective function value, peak signal-to-noise ratio (PSNR), feature similarity index (FSIM), and structural similarity index (SSIM), demonstrate that this new encoding scheme effectively determines both the ideal $r$ value and threshold levels. Future work could apply this encoding scheme across diverse PBEAs to enhance their efficacy, expand its application to satellite and medical imagery, and adapt it for colour image processing.

## References


1. Adhikari, S.K., Sing, J.K., Basu, D.K., Nasipuri, M.: Conditional spatial fuzzy C-means clustering algorithm for segmentation of MRI images. Appl. Soft Comput. **34**, 758–769 (2015)
2. Basar, S., Ali, M., Ochoa-Ruiz, G., Zareei, M., Waheed, A., Adnan, A.: Unsupervised color image segmentation: a case of RGB histogram based k-means clustering initialization. Plos one **15**(10), e0240015 (2020)
3. De Albuquerque, M.P., Esquef, I.A., Mello, A.G.: Image thresholding using Tsallis entropy. Pattern Recognit. Lett. **25**(9), 1059–1065 (2004)
4. Esmaeili, L., Mousavirad, S.J., Shahidinejad, A.: An efficient method to minimize cross-entropy for selecting multi-level threshold values using an improved human mental search algorithm. Expert. Syst. Appl. **182**, 115106 (2021)
5. Jamil, U., Sajid, A., Hussain, M., Aldabbas, O., Alam, A., Shafiq, M.U.: Melanoma segmentation using bio-medical image analysis for smarter mobile healthcare. J. Ambient Intell. Humaniz. Comput. 1–22 (2019)
6. Karaboga, D., Basturk, B.: A powerful and efficient algorithm for numerical function optimization: artificial bee colony (ABC) algorithm. J. Glob. Optim. **39**(3), 459–471 (2007)
7. Lin, J.: Divergence measures based on the Shannon entropy. IEEE Trans. Inf. Theory **37**(1), 145–151 (1991)
8. Martin, D., Fowlkes, C., Tal, D., Malik, J.: A database of human segmented natural images and its application to evaluating segmentation algorithms and measuring ecological statistics. In: 8th International Conference on Computer Vision, vol. 2, pp. 416–423 (2001)

9. Masi, M.: A step beyond tsallis and rényi entropies. Phys. Lett. A **338**(3–5), 217–224 (2005)
10. Moghadam, M.H., Borg, M., Mousavirad, S.J.: Deeper at the SBST 2021 tool competition: ADAS testing using multi-objective search. In: 2021 IEEE/ACM 14th International Workshop on Search-Based Software Testing (SBST), pp. 40–41. IEEE (2021)
11. Moghadam, M.H., Borg, M., Saadatmand, M., Mousavirad, S.J., Bohlin, M., Lisper, B.: Machine learning testing in an ADAS case study using simulation-integrated bio-inspired search-based testing. J. Softw. Evol. Process. e2591 (2022)
12. Mousavirad, S.J., Ebrahimpour-Komleh, H.: Human mental search: a new population-based metaheuristic optimization algorithm. Appl. Intell. **47**(3), 850–887 (2017)
13. Mousavirad, S.J., Oliva, D., Hinojosa, S., Schaefer, G.: Differential evolution-based neural network training incorporating a centroid-based strategy and dynamic opposition-based learning. In: 2021 IEEE Congress on Evolutionary Computation (CEC), pp. 1233–1240. IEEE (2021)
14. Mousavirad, S.J., Oliva, D., Schaefer, G., Moghadam, M.H., El-Abd, M.: A novel two-level clustering-based differential evolution algorithm for training neural networks. In: International Conference on the Applications of Evolutionary Computation (Part of EvoStar), pp. 259–272. Springer (2024)
15. Mousavirad, S.J., Rahnamayan, S.: CenPSO: A novel center-based particle swarm optimization algorithm for large-scale optimization. In: International Conference on Systems, Man, and Cybernetics (2020)
16. Mousavirad, S.J., Rahnamayan, S.: Enhancing SHADE and L-SHADE algorithms using ordered mutation. In: 2020 IEEE Symposium Series on Computational Intelligence (SSCI), pp. 337–344. IEEE (2020)
17. Mousavirad, S.J., Schaefer, G., Ebrahimpour-Komleh, H.: The human mental search algorithm for solving optimisation problems. In: Enabling AI Applications in Data Science, pp. 27–47. Springer (2020)
18. Mousavirad, S.J., Schaefer, G., Korovin, I.: High-dimensional multi-level image thresholding using self-organizing migrating algorithm. In: Genetic and Evolutionary Computation Conference Companion, pp. 1454–1459 (2020)
19. Mousavirad, S.J., Schaefer, G., Korovin, I., Oliva, D.: RDE-OP: A region-based differential evolution algorithm incorporation opposition-based learning for optimising the learning process of multi-layer neural networks. In: 24th International Conference on the Applications of Evolutionary Computation (2021)
20. Mousavirad, S.J., Schaefer, G., Oliva, D., Hinojosa, S.: HCS-BBD: an effective population-based approach for multi-level thresholding. In: Proceedings of the Genetic and Evolutionary Computation Conference Companion, pp. 1923–1930 (2021)
21. Mousavirad, S.J., Schaefer, G., Zhou, H., Moghadam, M.H.: How effective are current population-based metaheuristic algorithms for variance-based multi-level image thresholding? Knowl. Based Syst. **272**, 110587 (2023)
22. Mousavirad, S.J., Zabihzadeh, D., Oliva, D., Perez-Cisneros, M., Schaefer, G.: A grouping differential evolution algorithm boosted by attraction and repulsion strategies for masi entropy-based multi-level image segmentation. Entropy **24**(1), 8 (2022)
23. Mousavirad, S., Akhlaghian, F., Mollazade, K.: Classification of rice varieties using optimal color and texture features and BP neural networks. In: 7th Iranian Conference on Machine Vision and Image Processing, pp. 1–5 (2011)
24. Nie, F., Zhang, P., Li, J., Ding, D.: A novel generalized entropy and its application in image thresholding. Signal Process. **134**, 23–34 (2017)
25. Oliva, D., Ortega-Sanchez, N., Navarro, M.A., Ramos-Michel, A., El-Abd, M., Mousavirad, S.J., Nadimi-Shahraki, M.H.: Segmentation of thermographies from electronic systems by using the global-best brain storm optimization algorithm. Multimed. Tools Appl. 1–31 (2023)
26. Rezaee, K., Mousavirad, S.J., Khosravi, M.R., Moghimi, M.K., Heidari, M.: Distance-aware crowd sensing platform using deep shufflenet transfer learning. IEEE Trans. Intell. Transp. Syst. (2021)
27. Rose, R.A., Annadhason, A.: GHT based automatic kidney image segmentation using modified AAM and GBDT. Health Technol. 1–10 (2019)

28. Sahoo, P., Wilkins, C., Yeager, J.: Threshold selection using Renyi's entropy. Pattern Recognit. **30**(1), 71–84 (1997)
29. Shi, Y., Eberhart, R.: A modified particle swarm optimizer. In: IEEE International Conference on Evolutionary Computation, pp. 69–73 (1998)
30. Shubham, S., Bhandari, A.K.: A generalized masi entropy based efficient multilevel thresholding method for color image segmentation. Multimed. Tools Appl. **78**(12), 17197–17238 (2019)
31. Storn, R., Price, K.: Differential evolution-a simple and efficient heuristic for global optimization over continuous spaces. J. Glob. Optim. **11**(4), 341–359 (1997)
32. Tang, Y., Ji, J., Zhu, Y., Gao, S., Tang, Z., Todo, Y.: A differential evolution-oriented pruning neural network model for bankruptcy prediction. Complexity **2019**, (2019)
33. Wang, G., Zuluaga, M.A., Li, W., Pratt, R., Patel, P.A., Aertsen, M., Doel, T., David, A.L., Deprest, J., Ourselin, S., et al.: Deepigeos: a deep interactive geodesic framework for medical image segmentation. IEEE Trans. Pattern Anal. Mach. Intell. **41**(7), 1559–1572 (2018)
34. Wolpert, D., Macready, G.: No free lunch theorems for optimization. IEEE Trans. Evol. Comput. **1**(1), 67–82 (1997)
35. Zhang, L., Zhang, L., Mou, X., Zhang, D.: FSIM: a feature similarity index for image quality assessment. IEEE Trans. Image Process. **20**(8), 2378–2386 (2011)

# Machine Learning

# A Comparative Study of Bird-Based Metaphor Algorithms for Feature Selection Problems

**Kanak Kalita, G. Shanmugasundar, and Jasgurpreet Singh Chohan**

**Abstract** Data must be interpreted correctly to understand the relevant information it carries. Especially today, data has massive content that needs to be retrieved by a specialist to be evaluated and validated before deciding on a solution. These massive data are commonly executed by computers trained through various specialized algorithms. Feature selection (FS) is critical in modern machine learning frameworks. Metaheuristic techniques can efficiently carry out FS to reduce the data dimension. In this paper, feature selection is carried out by using a K-Nearest Neighbors (KNN) wrapper with bird-based metaphor algorithms. Five different bird-based optimizers, namely, Cuckoo Search, Harris Hawks, Crow Search, Stain Bowerbird and Emperor Penguin, are considered for the study. For analyzing the different algorithms, six different types of datasets are used. The performance of an average number of features selected (AFS), accuracy, fitness, convergence capabilities and computational cost is compared.

**Keywords** Feature selection · Metaheuristics · Optimization algorithms · Swarm intelligence

K. Kalita (✉)
Department of Mechanical Engineering, Vel Tech Rangarajan Dr. Sagunthala R&D Institute of Science and Technology, Avadi 600062, India
e-mail: drkanakkalita@veltech.edu.in; kanakkalita02@gmail.com

Department of Machining, Assembly and Engineering Metrology, Faculty of Mechanical Engineering, VSB-Technical University of Ostrava, 70800 Ostrava, Czechia

G. Shanmugasundar
Department of Mechanical Engineering, Sri Sai Ram Institute of Technology, Chennai 600044, India
e-mail: shanmugasundar.mech@sairamit.edu.in

J. S. Chohan
Department of Mechanical Engineering and University Centre for Research & Development, Chandigarh University, Mohali 140413, India
e-mail: jasgurpreet.me@cumail.in

 
D. Oliva et al. (eds.), *Advances in Optimization Algorithms for Multidisciplinary Engineering Applications: From Classical Methods to AI-Enhanced Solutions*, Studies in Computational Intelligence 806, https://doi.org/10.1007/978-3-031-78440-8_8

# 1 Introduction

In real-world problems, we have to deal with large amounts of data, handling which is a very expensive, prominent and complex task. In a dataset, there are many attributes, but for fetching useful information, not always, there a need for all the attributes because there can be some redundancy or irrelevancy that can reduce the performance of the algorithm or model. Thus, to deal with such problems, there are feature reduction approaches that not only aim to reduce the dataset's original size but also help maintain and increase the model's performance and accuracy. Feature construction and feature selections are the two main parts of the feature reduction problem. In the feature construction part, we create new pairs of attributes using the provided dataset, but on the other hand, feature selection is concerned with selecting only the relevant attributes or features from the dataset that we can further use for processing and finding results. It is one of the most important elements of machine learning and data mining, which helps select the best feature subset according to the given data and thus removes irrelevant data. However, on the other hand, selecting an optimal set of features from a dataset with the help of traditional methods is very complex work. Today, some applications, like social media, continuously generate new features in a specific period. Optimization methods allow decision makers of different fields, such as engineers, managers and scientists, to develop the best or near-best decision that will help them improve the overall performance of their system or project., these optimization techniques are divided into two groups, metaheuristic and heuristic algorithms. The main categories for the feature selection problem-solving techniques are hybrid-based, filter and wrapper. The main advantage of traditional or filter-based feature selection problems is that they not only have the ability to deal with and scale large datasets but also are speedier than other techniques. In cases such as Information Gain (IG), the wrapper technique is even more useful when dealing with feature selection. Information gain is very helpful in reducing the number of features that can make the classification models even more accurate. It also lets us know how much information we can retrieve from the class.

In machine learning, feature selection is one of the most important and challenging tasks. According to this, if there are some attributes in a set, you can select the best subset from a total of subsets. However, if the value is very large, it will be challenging because performance evaluation of the model at each subset is impossible. Thus, different methodologies have been proposed for dealing with such situations. Greedy search, exhaustive search, etc., are different methodologies used to find the best subset according to feature selection problems, but all these techniques have to suffer the same issues like enormous complexity, premature convergence and high computational cost. Thus, metaphor-based metaheuristic algorithms have been developed to avoid these issues that gain much attention while dealing with such circumstances. These algorithms are the most effective and efficient algorithms that cannot only find the best subsets of attributes but also maintain the performance and accuracy of the model. For the feature selection problem, De Souza et al. [1] proposed using the crow search method. Because this is a binary problem, a transfer function with a V shape was used. To get over its shortcomings, Arora et al. [2] combined the grey wolf algorithm with the crow search algorithm. Binomial cuckoo search, developed by Pandey et al. [3], is a new approach to selecting features. For the stability problem, a hybrid data transformation approach was used, combining principal component analysis with quick independent component analysis. Binary cuckoo search, binary bat algorithm, binary gravitational search algorithm, binary whale optimization using simulated annealing and binary grey wolf optimization were used to compare their findings. A streaming feature selection approach for multi-label data was developed by Paul et al. [4]. When choosing the best features, a multi-objective variation of the cuckoo search-based method was utilized. Datasets from various fields, including text, biology and audio, were used for evaluation. Alzaqebah et al. [5] demonstrated a Cuckoo search technique based on a memory-based mechanism to save the most useful features found by the best solutions. The selected characteristics were remembered at each iteration so that the features that improve classification accuracy could be located. To address the issue of early convergence, Turabieh et al. [6] presented a variant of the HHO algorithm in which population diversity is regulated. When all solutions fall into one cluster and get stuck at a local optimum, population diversity is controlled by injecting new solutions into the existing population. Specifically, they used it as a feature selection technique to find the most informative characteristics for solving the challenging task of predicting students' performance. Prediction accuracy was measured using several different machine learning classifiers, including k-nearest neighbors (kNN), layered recurrent neural networks (LRNN), naive Bayes and artificial neural networks (ANN).

Abdel-Basset et al. [7] hybridized an HHO using simulated annealing (SA) for feature selection tasks. K-nearest neighbors, a common wrapping method using the Euclidean distance metric, was used to evaluate the new approaches. Elgamal et al. [8] too used hybridized Chaotic HHO and SA for feature selection of medical data. They used chaotic maps at the initialization phase of HHO to enhance the population diversity in the search space and SA was used to improve the current best solution to improve HHO exploitation. Similarly, Hussain et al. [9] amalgamated an HHO with a sine cosine algorithm for low and high-dimensional feature selection problems. Along with improved convergence, the hybrid algorithm demonstrated an 87% feature reduction and 92% accuracy. Al-Waji et al. [10] demonstrated that a hybrid Grey Wolf Optimizer and HHO can outperform its counterparts.

The No-Free-Lunch (NFL) theorem [11] states that no unique optimization algorithm can outperform all other optimization strategies and solve every optimization problem. This means that one optimizer may work better than the others for some problems but not for others. Therefore, to discover the best answer to the feature selection problem, it is essential to compare multiple optimization techniques on different datasets. This research only uses a small subset of the hundreds of optimization algorithms described in the literature; instead, it focuses on a small subset of the most popular and widely-cited bird-based metaheuristic algorithms. Six different classification datasets ranging in size from somewhat small to quite large are utilized to assess the efficacy of various methods. Various algorithms' convergence rates, average fitness and accuracy are contrasted. In this study, we make the following contributions:

- Five widely used and extensively referenced bird-based optimization techniques are used to solve the feature-selection test problems.
- Using classification datasets of varying sizes to evaluate metaheuristics for the feature-selection problem.
- Feature selection as a benchmark problem for comparing bird-based metaheuristic algorithms.

The rest of the paper is organized as explained below. In Sect. 2, we discuss the various algorithms considered in this study. Section 3 discusses the experimental settings. The outcomes and comparative analysis of all five algorithms are presented in Sect. 4, titled Results and Discussion, while Sect. 5 provides the concluding remarks.

## 2 Materials and Methods

For the feature selection problems, we have made a comparison by using different bird-based metaphor algorithms for several datasets. A detailed description of each of the algorithms used in this research is presented in the following subsections.

### *2.1 Cuckoo Search*

In simple words, Cuckoo Search work on the basis of three idealized rules [12] followed by cuckoo birds in real life.

- At a time, the cuckoo lays only one egg and randomly chooses nests to dump its egg.
- The nest having high-quality eggs will be carried over to the next generation.
- Available host nests are fixed where the egg is laid by a cuckoo and then the host bird discovers the laid eggs with the probability of $p_a \in |0,1|$. This is because sometimes the host bird builds a completely new nest by either destroying the nest or throwing the eggs away.

Thus, we can simply represent that every egg in a nest represents a solution and thus, every new egg of the cuckoo will represent a new solution. In this way, we can consider better and new solutions for replacing such solutions that are not so good in the nest. Although, we can use this algorithm to extend the most complex cases in which every nest consists of multiple eggs that would represent a set of different solutions. In this case, where we need to maximize a problem, the fitness or quality of a solution and the values of objective functions would be proportional to each other. We can also define other forms of fitness similarly to the fitness functions in a genetic algorithm. While dealing with feature selection problems, most of the time, we use a simple case in which each nest has only one egg [12]. On the basis of the above-mentioned rules, we can summarize the cuckoo search algorithm in the following way. While developing new solutions $x^{[t+1]}$, for every cuckoo $i$, we perform L'evy flight

$$x_i(t+1) = x_i^{[t]} + \alpha \oplus L'evy(\lambda) \quad (1)$$

where in step size $\alpha > 0$, relates with the scales of interesting problem. On the other hand, in many of the cases, we use $\alpha = 1$. Basically, the above-mentioned equation is stochastic in the case of random walks. Generally, a random walk is a Markov chain in which the next location depends on transition probability and the current location. The current location is the first term in the above-mentioned equation while transition probability is the second term. Also, in the above-mentioned equation, the term $\oplus$ stands for entry-wise multiplications that are similar to those which we used in PSO, while here random walk along with *L'evy* flight is much more efficient while exploring the search space because its lengths of steps are longer in the case of the long run. In actuality, *L'evy* flight provides a random walk while on the other hand, the random length of steps is developed from a *L'evy* distribution [12].

$$L'evy \sim u = t^{-\lambda}, (1 < \lambda \leq 3) \quad (2)$$

There is an infinite variance with an infinite mean. In this equation, the steps at the end show a stochastic process that has a distribution along the force law step with a heavy tail. You can also create new solutions with L'evy by going through the best solutions obtained so far, which will speed up your local search. Although many remote domain randomizations can be generated for new solutions, any site should be far enough away from the existing best solutions to ensure that the system is not invaded for the local best solution [12]. The flowchart for the cuckoo search is shown in Fig. 1a.

## 2.2 *Crow Search Algorithm*

Crows are known as one the most intelligent birds as they contain large brains compared to their size. Along with this, recent research evidence that their cleverness is mind-blowing. They can remember unfriendly faces and can warn each other in case any of those faces approaches. Besides this, they can communicate in a more sophisticated way as compared to other birds as well as they are also able to use tools. They are also so clever that even after several months, they can recall their hidden food [13].

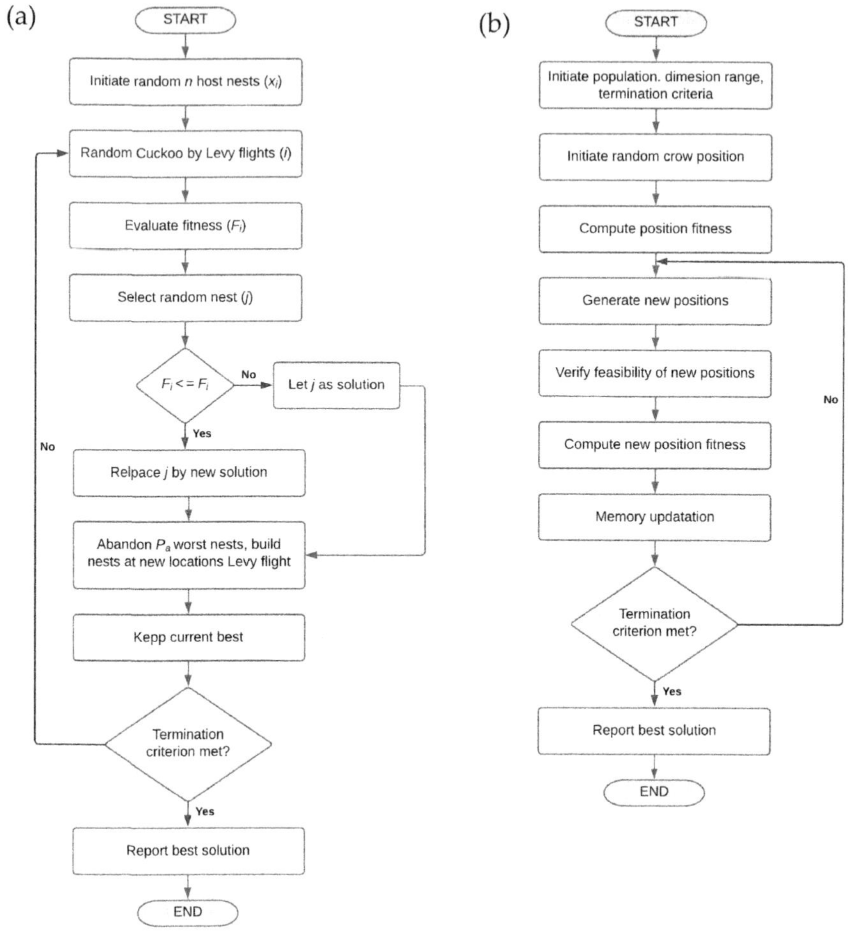


**Fig. 1** Flowchart for **a** Cuckoo Search **b** Crow Search Algorithm

Based on their intelligent behavior, the crow search algorithm was developed and thus, the principles of the crow search algorithm are as follows [13]

- Crows always live in groups.
- Crows follow each other to steal.
- Crows remember the hiding places of their food.
- Crows protect their nests from being robbed.

Let us assume that in an environment, there is a $d$-dimension that is followed by a number of crows. Thus, $N$ is used for representing a number of crows, while for the position of crow $i$ at every iteration there is always a vector $X^{i \cdot iter}(i = 1,2,3\ldots, N; iter = 1,2,3\ldots, iter_{max})$ when we are having a search space and

$iter_{max}$ represents the number of maximum iterations. Along with this, in every iteration, the hiding position of the crow is always specified by $m^{i \cdot iter}$. Thus, by keeping in view that, crows move in the environment in the search of better food and for this purpose, they also follow other crows, CSA works in eight steps that are as follows [13]. The flowchart for crow search algorithm is shown in Fig. 1b.

Step 1: Problem initialization and problem adjustment.

Step 2: Position initialization and crow's memory.

Step 3: Fitness evaluation function.

Step 4: New position generation.

Step 5: Checking the new position's feasibility.

Step 6: Fitness function evaluation of the new position.

Step 7: Memory updating.

Step 8: Termination criterion checking.

## 2.3 *Stain Bowerbird Optimization*

In polygynous species and birds, complex and multifaceted displays in terms of mating are very common. These displays contain visual, chemical, tactile and auditory components that give the females different information, such as sex, species relatedness, quality and individual identity. Bowerbirds are closely relative to the birds of paradise. Satin is the bowerbird that feeds on fruits and insects of the rainforests or mesic forests in eastern Australia. These birds live in a particular area but are mainly found in forests. Apart from that, in the winter and autumn seasons, these birds have to leave their forest habitat and move toward the open woodlands in search of insects and forests. While with the onset of the spring season, in the form of small groups, they collect together, inhabiting the land they seem to occupy year after year. Bower is built by the male satin bowerbirds are specialized stick structures where courtship and copulation occur. For the choice of female and to carry the successful mal mating, they must decorate the bowers with different kinds of feathers, flowers and barriers. To complete decoration, male bowerbirds have to steal the decoration of other males and destroy their neighbor bowers. Males have to attract their mates with the construction of their bower as well as also by decorating their surrounding areas. Thus, when the bowers are complete, females visit different bowers to choose a partner and then return to their bower [14].

In the Stain Bowerbird Optimization algorithm, during the mating season, adult male bowerbirds construct a bower by using different kinds of materials such as flowers, sparkling objects, branches, dramatic gestures and fruits that are very important for attracting females. Females are attracted to the bower because of the male's dramatic gestures and the bower's beauty. It is also important to notice that imitation

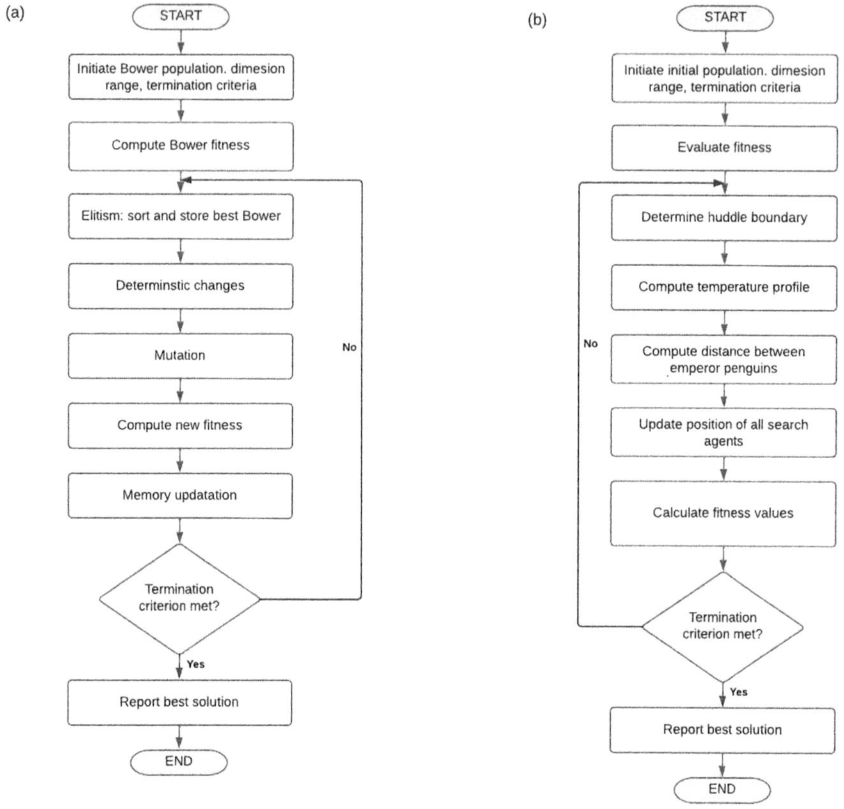


**Fig. 2** Flowchart for **a** Stain Bowerbird Optimization **b** Emperor Penguin Optimizer

of other males and natural instincts is only used in constructing the bower. Thus, by considering the strain bower bird life, this algorithm is systemized in different stages discussed below [14]. The flowchart for the Stain Bowerbird optimization algorithm is shown in Fig. 2a.

- **Generation of a random bower set**: Starts with creating a random initial population. These initial populations also contain a set of positions. Then, $n$-dimensional vectors are defined for each position that has to be optimized. Upper and lower limit parameters are set for the normal distribution of these randomly initialized values. For determining the bower's attractiveness, a combination of parameters is used.
- **Population member probability calculation**: Probability can be calculated by the following equation:

$$Prob_i = \frac{fit_i}{\sum_{n=1}^{N_b} fit_n} \tag{3}$$

where $N_b$ stands for the number of bowers, $f_{it_i}$ is the $i$ th fitness.

- **Elitism**: It is one of the most important factors for all evolutionary algorithms that help in finding the best solution that needs to be preserved at every stage while dealing with optimization problems.
- **Determination of new changes in any position**: At every stage of the algorithm, with the below-mentioned equation, you can determine any new change in the position of the bower:

$$X_{ik}^{new} = x_{ik}^{old} + \lambda_k \left( \left( \frac{X_{jk} + x_{eLite,k}}{2} \right) - x_{ik}^{old} \right) \tag{4}$$

where xi represents bower, $x_j$ represents the target solution in the current iteration, $x_{ik}$ determines $k$ th member of $x_i$ vector.

- **Combining new and old populations**: At the end of every complete cycle, in this algorithm, both old and new populations are combined together and thus are sorted. In this way, a new population is generated and the previous ones are deleted.

## 2.4 *Emperor Penguin Optimizer*

Emperor penguins breed in winter and spend their whole life on open ice. In the breeding season, they come ashore in huge colonies home to hundreds of thousands of emperor penguins. The females lay one egg and can travel 50 miles to the ocean to hunt. The emperor penguin is a social animal whose behavior is similar to foraging and hunting. At sea, emperor penguins can dive to depths of up to 1,900 feet and stay there for more than 25 min. Emperor penguins do not fly like other species with hard, flat wings. At sea, emperor penguins can dive to depths of up to 1,900 feet and stay underwater for more than 25 min. Emperor penguins do not fly like other species with hard, flat wings [15]. For them, it is known that this is the only specie that can survive even in the Antarctic winter. This behavior of the emperor penguin is decomposed into the following four phases [15].

- Determining and creating huddle boundaries for the emperor penguin.
- Calculation of temperature around the huddle.
- Distance calculation between emperor penguins.
- Relocation of the effective movement.

All the mentioned steps are further explained in mathematical notations as follows. The flowchart for Emperor Penguin Optimizer is shown in Fig. 2b.

- **Determining and creating huddle boundaries of the emperor penguin**: While huddling, emperor penguins are usually placed on a polygon-shaped web border. Emperor penguins have at least two randomly chosen neighbors in the group. The wind flows around the group and is determined to find the group boundary around the polygon. However, the flow of the wind is faster than the movement of the emperor penguin. Let's assume the variables that specify a specific notation. Thus, $\Phi$ represents the velocity of the wind while $\Psi$ is the gradient of $\Phi$. So,

$$\Psi = \nabla \Phi \tag{5}$$

- Also, the $\Omega$ vector is combined with wind velocity for generating complex potential.

$$F = \Phi + i\Omega \tag{6}$$

where $F$ stands for analytical function and $i$ represents the imaginary constant on the polygon.
- **Temperature calculation**: The temperature around the huddle can be determined as,

$$\begin{aligned} T' &= (T - \max \textit{iteration}/x - \max \textit{iteration}) \\ T &= 0, \textit{if}\, R > 1 \text{ and } 1, \textit{if}\, R < 1 \end{aligned} \tag{7}$$

While $R$ determines the radius, max iteration represents the maximum iterations, $T$ is the time period for finding the best and optimal solution and $x$ shows the current iteration.
- **Distance calculation**: The distance between an emperor penguin and the best solution obtained is calculated after the meeting limit is generated. The current best solution is one whose value is appropriately approaching the optimum level.
- **Relocation of the movement**: For updating the positions of the penguins we need the best optimal solution that is obtained in the previous steps such as a mover who is responsible for changing and relocating the positions of other agents within a provided search space to vacate the current position.

## 2.5 *Harris Hawks Optimization*

HHO is a crowd-based incremental optimization algorithm with many active and time-varying exploration and exploitation phases. This algorithm was first published in the Journal of Future Generation Computer Systems in 2019 and since then, it has attracted much attention among researchers due to its flexible structure, high performance and high-quality results. The basic logic of the HHO method is based on the cooperative behavior and stalking technique of Harris's hawks in the forest called the "surprise jump." There are currently many proposals on how to improve the functionality of HHO [16]. HHO works in the following stages. The flowchart for HHO is shown in Fig. 3.

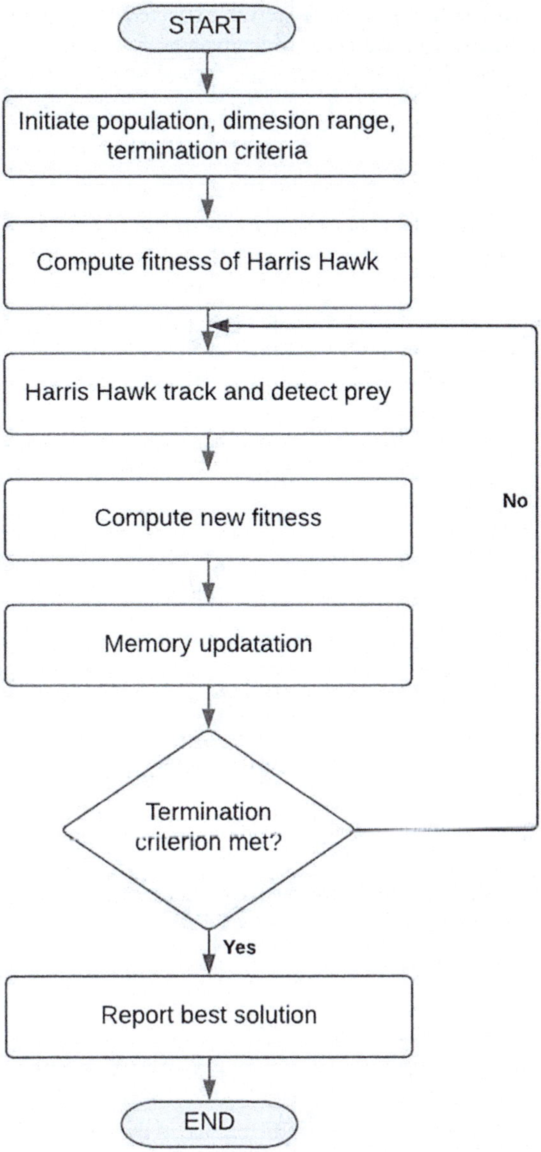


**Fig. 3** Flowchart for Harris Hawks optimization

- **Exploration phase**: In this phase, the mechanism of HHO exploration is proposed. In this, we consider the nature of Harris Hawks because they are able to track and detect the prey with their strong eyes, although, sometimes, they can't prey because it can be seen easily. Therefore, during the whole cycle, Harris hawks are waiting and watching a specific desert location to discover prey perhaps after several hours. In HHO, Harris Hawks are the candidate and best solutions. Then the solution at each step is expected to be almost optimum. The mathematical structure for this phase is given as,

$$X(t+1) = \begin{cases} X_{rand}(t) - r_1|X_{rand}(t) - 2r_2X(t)| & q \geq 0.5 \\ (X_{rabbit}(t) - X_m(t)) - r_3(LB + r_4(UB - LB)) & q < 0.5 \end{cases} \quad (8)$$

where $X(t+1)$ represents the position vector of Hawks in subsequent iteration, $X_{rabbit}$ shows the position of the rabbit in $t$th iteration, $X(t)$ determines the position vector of Hawks in $t$th iteration, $X_{rand}$ (t) is the randomly selected hawk in $t$th iteration, $X_m$ is the mean of all the hawk positions in $t$th iteration. $r_1$, $r_2$, $r_3$ and $r_4$ are random numbers between [0, 1].

- **Transition from exploration to exploitation**: The HHO algorithm can be moved from exploration to exploitation and then switched between different exploitation behaviors. Based on the escape energy of the prey. Prey energy decreases significantly during escape behavior. To show this fact, the prey capacity is designed as follows,

$$E = 2E_0(1 - t/T) \quad (9)$$

Here, $E$ represents the prey escaping energy, $T$ shows the maximum number of iterations, while $E_0$ is the initial state of its energy.

- **Exploitation phase**: The exploitation phase is modeled by the following rules,
- **Hard Besiege**: In this case ($|E| < 0.5$ and $r \geq 0.5$), you can update the current position by the following equation,

$$X(t+1) = X_{rabbit}(t) - E|\Delta X(t)| \quad (10)$$

- **Soft Besiege**: In the case of soft besiege ($|E| \geq 0.5$ and $r \geq 0.5$), the behavior is modeled as:

$$X(t+1) = \Delta X(t) - E|JX_{rabbit}(t) - X(t)| \quad (11)$$

$$\Delta X(t) = X_{rabbit}(t) - X(t) \quad (12)$$

Thus, $\Delta X(t)$ is the difference in the position vector between the position of the rabbit as well as the current location in iteration $t$. On the other hand, the value of $J$ changes after every iteration so that it can simulate the nature of the motion of the rabbit.

- **Soft besiege with progressive rapid dives**: For performing this ($|E| \geq 0.5$ and $r < 0.5$), we suppose that the hawks, always evaluate their next move on the basis of the following equation,

$$Y = X_{rabbit}(t) - E|JX_{rabbit}(t) - X(t)| \quad (13)$$

Thus, in hawk position in soft besiege with progressive rapid dives are updated as,

$$X(t+1) = \begin{cases} Y & \text{if} F(Y) < F(X(t)) \\ Z & \text{if} F(Z) < F(X(t)) \end{cases} \tag{14}$$

where $Z$ is given as,

$$Z = Y + S \times LF(D) \tag{15}$$

- **Hard besiege with progressive rapid dives**: In this condition ($|E| < 0.5$ and $r < 0.5$), we perform the following rule,

$$X(t+1) = \begin{cases} Y & \text{if} F(Y) < F(X(t)) \\ Z & \text{if} F(Z) < F(X(t)) \end{cases} \tag{16}$$

Although $Y$ remains the same as the for the case of soft besiege while for $Z$,

$$Z = Y + S \times LF(D) \tag{17}$$

## 3 Experimental Details

We have applied the algorithms to six different datasets to analyze the working of the algorithms. The collected datasets are diverse regarding dataset domain, number of features, data points etc. Table 1 provides the details about the dataset, number of instances as well as number of features.

Since all the considered metaheuristics are stochastic, it is important to properly tune each algorithm so that the best performance is achieved [17, 18]. Pilot studies are carried out to tune the respective parameters of each of the five considered algorithms (i.e., CS, CSA, SBO, EPO and HHO) and reported in Table 2. However, for proper comparison between the algorithms, the common parameters like the number of search agents (i.e., population size) and number of iterations (i.e., generation limit) are

**Table 1** Information about the datasets considered in this study

| Dataset | Number of Instances | Number of features |
|---|---|---|
| Heart | 303 | 13 |
| Ionosphere | 351 | 34 |
| German | 1000 | 24 |
| Breast cancer | 699 | 9 |
| Sonar | 206 | 60 |
| Ovarian cancer | 216 | 4000 |

**Table 2** Parameter settings for all algorithms

| Algorithm | Parameter | Value |
|---|---|---|
| Common for all algorithms | K | 5 |
| | Maximum number of iterations | 200 |
| | Number of search agents | 30 |
| | Number of independent runs | 10 |
| | Ratio of validation data | 0.2 |
| Cuckoo Search (CS) | Discovery rate | 0.25 |
| | $\alpha$ | 1 |
| | $\beta$ | 1.5 |
| Crow Search Algorithm (CSA) | Awareness probability | 0.1 |
| | Flight length | 1.5 |
| Satin Bowerbird Optimization (SBO) | $\alpha$ | 0.94 |
| | $z$ | 0.02 |
| | Mutation rate | 0.05 |
| Emperor Penguin Optimizer (EPO) | $M$ | 2 |
| | $f$ | 3 |
| | $l$ | 2 |
| Harris Hawks Optimization (HHO) | $\beta$ | 1.5 |

fixed for all algorithms as 30 and 200, respectively. Thus, the total function evaluation is limited to 6000. Moreover, to account for the stochasticity of the algorithms, each algorithm is independently evaluated 10 times [19, 20].

## 4 Results and Discussion

In context of this paper, the best fitness value of an algorithm is the best solution obtained for the problem among all 6000 function evaluations. However, in Table 3, the best solution obtained among the 10 repeated trials (each of 6000 function evaluations) is reported. The rank (out of five algorithms) assigned to each algorithm for each dataset is also reported in Table 3. The rank is reported in parentheses in the tables. On a cumulative basis of comparison for the six datasets the algorithms can be ranked as CSA > EPO > CS ~ HHO > SBO.

The mean best fitness value of 10 independent trials is reported in Table 4. This is necessary to indicate that the performance of an algorithm is not one-off but a repetitive event. Based on Table 4, the algorithms can be ranked as CS > CSA > EPO > HHO > SBO.

**Table 3** Best fitness value for all algorithms

| Dataset | CS | CSA | SBO | EPO | HHO |
|---|---|---|---|---|---|
| Heart | 0.11781 (3) | 0.08558 (1) | 0.11858 (5) | 0.10208 (2) | 0.11781 (3) |
| Ionosphere | 0.01738 (1) | 0.04566 (2) | 0.07630 (5) | 0.05745 (3) | 0.05775 (4) |
| German | 0.19805 (3) | 0.19638 (2) | 0.20712 (4) | 0.23432 (5) | 0.19597 (1) |
| Breast cancer | 0.01268 (3) | 0.00333 (1) | 0.01268 (3) | 0.00444 (2) | 0.01869 (5) |
| Sonar | 0.02665 (4) | 0.00333 (2) | 0.02681 (5) | 0.00133 (1) | 0.02498 (3) |
| Ovarian cancer | 0.00440 (4) | 0.00466 (5) | 0.00432 (3) | 0.00001 (1) | 0.00002 (2) |
| Average rank | 3.00 | 2.17 | 4.17 | 2.33 | 3.00 |

**Table 4** Mean best fitness value of all algorithm

| Dataset | CS | CSA | SBO | EPO | HHO |
|---|---|---|---|---|---|
| Heart | 0.1412 (3) | 0.1062 (1) | 0.1520 (4) | 0.1216 (2) | 0.1623 (5) |
| Ionosphere | 0.0648 (1) | 0.0801 (4) | 0.1056 (5) | 0.0801 (3) | 0.0776 (2) |
| German | 0.2079 (1) | 0.213 (2) | 0.2268 (3) | 0.2404 (5) | 0.2278 (4) |
| Breast cancer | 0.0208 (3) | 0.0149 (1) | 0.0255 (4) | 0.0158 (2) | 0.0296 (5) |
| Sonar | 0.0319 (1) | 0.0327 (2) | 0.0758 (3) | 0.0879 (5) | 0.0842 (4) |
| Ovarian cancer | 0.0092 (3) | 0.0139 (4) | 0.0366 (5) | 0.0001 (1) | 0.0047 (2) |
| Average rank | 2.00 | 2.33 | 4.00 | 3.00 | 3.67 |

The standard deviation represents the dispersion (for 10 repeated trials) of the obtained solutions concerning the mean best fitness value. So, if the standard deviation is higher, the solutions are more spread, but in the case of low value, the obtained solutions are clustered around the mean best value. In other words, a lower standard deviation ensures better repeatability of obtained solution in successive independent trials. The standard deviation of the fitness values for all algorithms is presented in Table 5.

The convergence curve for all the algorithms for all six test cases for trial number 1 is shown in Fig. 4. The convergence of HHO and EPO is very rapid, whereas CS

**Table 5** Standard deviation of fitness values for all algorithms

| Dataset | CS | CSA | SBO | EPO | HHO |
|---|---|---|---|---|---|
| Heart | 0.019 (3) | 0.0138 (1) | 0.0331 (5) | 0.0139 (2) | 0.0329 (4) |
| Ionosphere | 0.0338 (5) | 0.0259 (3) | 0.0301 (4) | 0.0161 (1) | 0.0213 (2) |
| German | 0.0135 (2) | 0.0178 (4) | 0.014 (3) | 0.0062 (1) | 0.0223 (5) |
| Breast cancer | 0.006 (1) | 0.0085 (3) | 0.0104 (5) | 0.0082 (2) | 0.0102 (4) |
| Sonar | 0.0108 (1) | 0.0266 (2) | 0.0345 (3) | 0.0527 (5) | 0.0373 (4) |
| Ovarian cancer | 0.0103 (3) | 0.0207 (5) | 0.0206 (4) | 0.0001 (1) | 0.0103 (2) |
| Average rank | 2.50 | 3.00 | 4.00 | 2.00 | 3.50 |

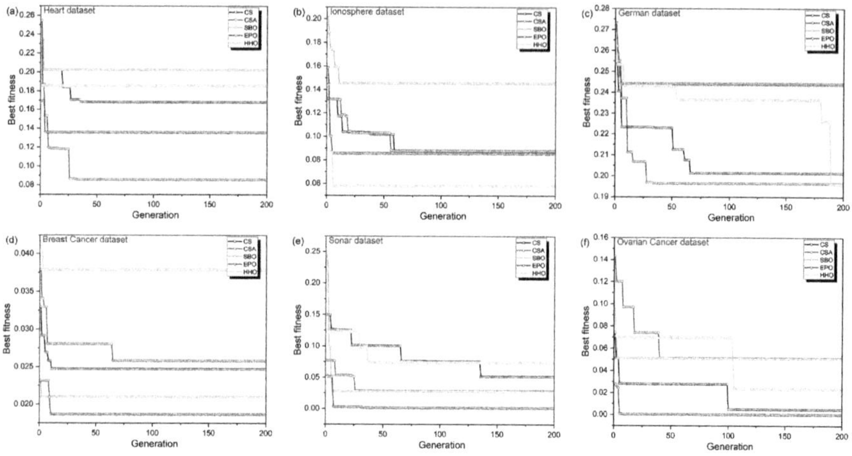


**Fig. 4** Convergence curve of all algorithms

and CSA algorithms are slightly sluggish. Also, the trend followed by CS and CSA is very similar. On the other hand, notable similarity in HHO and EPO converge trends is seen.

In feature selection problems, classification accuracy is one of the main tasks. It is desired to have classification accuracy as close to 1 as possible. A classification accuracy of 1 indicates that the algorithms can correctly predict all the instances into the correct classes. The classification accuracy will only improve if irrelevant features clog the classifier or significant features are included in the feature selection step. Thus, in the context of this study, classification accuracy is an indirect measure of the ability of the metaheuristics to leave out insignificant features while retaining the significant ones. The best classification accuracy is presented in Table 6. The algorithms are ranked as CSA > CS > EPO > HHO > SBO based on the best classification accuracy. Similarly, Table 7 contains the mean (of 10 trials) best classification accuracy for the algorithms, which infer that CS ~ CSA > EPO > HHO > SBO. The standard deviation of the classification accuracy for all algorithms is presented in Table 8.

The number of features each algorithm selects for each dataset is also studied. The mean number of selected features is reported in Table 9 and the standard deviation of the number selected is in Table 10. The algorithms are ranked as EPO > HHO > CS > SBO > CSA regarding the mean features selected. Interestingly, in large feature datasets like the Ovarian cancer dataset, EPO and HHO achieve approximately 99% reduction in features. This, combined with the fact that EPO and HHO achieved a mean best classification accuracy of 1 and 0.9953 on the corresponding large feature dataset, makes their performance even more remarkable. Additionally, EPO showed zero deviation in classification accuracy on the 10 repeated trials for the ovarian cancer dataset. It should be noted that for the ovarian cancer dataset, the standard deviation of the number of features selected for EPO is slightly larger

**Table 6** Best classification accuracy for all algorithms

| Dataset | CS | CSA | SBO | EPO | HHO |
|---|---|---|---|---|---|
| Heart | 0.8833 (3) | 0.9167 (1) | 0.8833 (3) | 0.9000 (2) | 0.8833 (3) |
| Ionosphere | 0.9857 (1) | 0.9571 (2) | 0.9286 (5) | 0.9429 (3) | 0.9429 (3) |
| German | 0.8050 (1) | 0.8050 (1) | 0.7950 (4) | 0.765 (5) | 0.8050 (1) |
| Breast cancer | 0.9928 (3) | 1 (1) | 0.9928 (3) | 1 (1) | 0.9856 (5) |
| Sonar | 0.9756 (3) | 1 (1) | 0.9756 (3) | 1 (1) | 0.9756 (3) |
| Ovarian cancer | 1 (1) | 1 (1) | 1 (1) | 1 (1) | 1 (1) |
| Average rank | 2.00 | 1.17 | 3.17 | 2.17 | 2.67 |

**Table 7** Mean best classification accuracy for all algorithms

| Dataset | CS | CSA | SBO | EPO | HHO |
|---|---|---|---|---|---|
| Heart | 0.8600 (3) | 0.8967 (1) | 0.8500 (4) | 0.8800 (2) | 0.8400 (5) |
| Ionosphere | 0.9371 (1) | 0.9229 (3) | 0.8971 (5) | 0.9200 (4) | 0.9229 (2) |
| German | 0.7940 (1) | 0.7900 (2) | 0.7760 (3) | 0.7590 (5) | 0.7740 (4) |
| Breast cancer | 0.9842 (3) | 0.9899 (1) | 0.9799 (4) | 0.9885 (2) | 0.9741 (5) |
| Sonar | 0.9707 (2) | 0.9707 (1) | 0.9268 (3) | 0.9122 (5) | 0.9171 (4) |
| Ovarian cancer | 0.9953 (2) | 0.9907 (4) | 0.9674 (5) | 1 (1) | 0.9953 (3) |
| Average rank | 2.00 | 2.00 | 4.00 | 3.17 | 3.83 |

**Table 8** Standard deviation of classification accuracy for all algorithms

| Dataset | CS | CSA | SBO | EPO | HHO |
|---|---|---|---|---|---|
| Heart | 0.0190 (3) | 0.0139 (2) | 0.0333 (5) | 0.0139 (1) | 0.0325 (4) |
| Ionosphere | 0.0344 (5) | 0.0260 (3) | 0.0310 (4) | 0.0163 (1) | 0.0217 (2) |
| German | 0.0139 (2) | 0.0177 (4) | 0.0139 (2) | 0.0065 (1) | 0.0225 (5) |
| Breast cancer | 0.0060 (1) | 0.0082 (3) | 0.0107 (4) | 0.0082 (2) | 0.0109 (5) |
| Sonar | 0.0109 (1) | 0.0267 (2) | 0.0345 (3) | 0.0534 (5) | 0.0370 (4) |
| Ovarian cancer | 0.0104 (2) | 0.0208 (5) | 0.0208 (4) | 0 (1) | 0.0104 (2) |
| Average rank | 2.33 | 3.17 | 3.67 | 1.83 | 3.67 |

than its corresponding mean number of features selected, indicating some extreme solutions in the 10 trials. The simulation data on the ovarian cancer dataset (not presented here) shows that the best and worst solution (among 10 trials) regarding the number of features selected for EPO is 4 and 113. Nevertheless, in both these cases, an ideal classification accuracy of 1 is obtained.

Figure 5 shows the boxplot analysis of computational time of all algorithms. It shows that in general the time taken to analyze large feature dataset like ovarian cancer dataset is much more compared to other smaller datasets like Ionosphere

**Table 9** Mean number of features selected and percentage reduction in features for each dataset

| Dataset | CS | | CSA | | SBO | | EPO | | HHO | |
|---|---|---|---|---|---|---|---|---|---|---|
| | Mean (rank) | % | Mean (rank) | % | Mean (rank) | % | Mean (rank) | % | Mean (rank) | % |
| Heart | 3.4 (1) | 74% | 5.0 (4) | 62% | 4.6 (3) | 65% | 3.6 (2) | 72% | 5 (4) | 62% |
| Ionosphere | 9.8 (3) | 71% | 12.3 (4) | 64% | 12.4 (5) | 64% | 3.5 (1) | 90% | 4.3 (2) | 87% |
| German | 9.6 (2) | 60% | 12.2 (5) | 49% | 12 (4) | 50% | 4.4 (1) | 82% | 9.8 (3) | 59% |
| Breast cancer | 4.6 (4) | 49% | 4.4 (3) | 51% | 5 (5) | 44% | 4 (2) | 56% | 3.6 (1) | 60% |
| Sonar | 17.8 (3) | 70% | 22.4 (5) | 63% | 20 (4) | 67% | 5.8 (1) | 90% | 12.4 (2) | 79% |
| Ovarian cancer | 1819.6 (4) | 55% | 1894.0 (5) | 53% | 1738.8 (3) | 57% | 41.6 (1) | 99% | 49.8 (2) | 99% |
| Average rank | 2.83 | | 4.33 | | 4.00 | | 1.33 | | 2.33 | |

**Table 10** Standard deviation of number of features selected

| Algorithm | CS | CSA | SBO | EPO | HHO |
|---|---|---|---|---|---|
| Heart | 1.1402 (4) | 1 (3) | 0.5477 (1) | 0.8944 (2) | 1.2247 (5) |
| Ionosphere | 1.6866 (2) | 1.8886 (3) | 2.9136 (5) | 0.7071 (1) | 2.0028 (4) |
| German | 1.5166 (1) | 2.6833 (4) | 2.3452 (3) | 1.5166 (1) | 4.2071 (5) |
| Breast cancer | 1.5166 (5) | 1.1402 (2) | 0.7071 (1) | 1.2247 (4) | 1.1402 (2) |
| Sonar | 3.1145 (4) | 1.8166 (2) | 2.4495 (3) | 1.4832 (1) | 8.1731 (5) |
| Ovarian cancer | 38.2727 (3) | 34.8855 (2) | 18.5122 (1) | 48.8293 (5) | 44.8185 (4) |
| Average rank | 3.17 | 2.67 | 2.33 | 2.33 | 4.17 |

dataset. In general, in terms of computational time the algorithms can be ranked as EPO > HHO > CSA > SBO > CS.

## 5 Conclusions

In this paper, five different bird-based metaphor algorithms for feature selection problems were analyzed. Six different datasets varying in domain type, number of features and instances were selected to evaluate the algorithms. The algorithms were comprehensively evaluated based on fitness value, convergence, classification accuracy, number of features selected and computational time. The following conclusions are drawn based on the analysis.

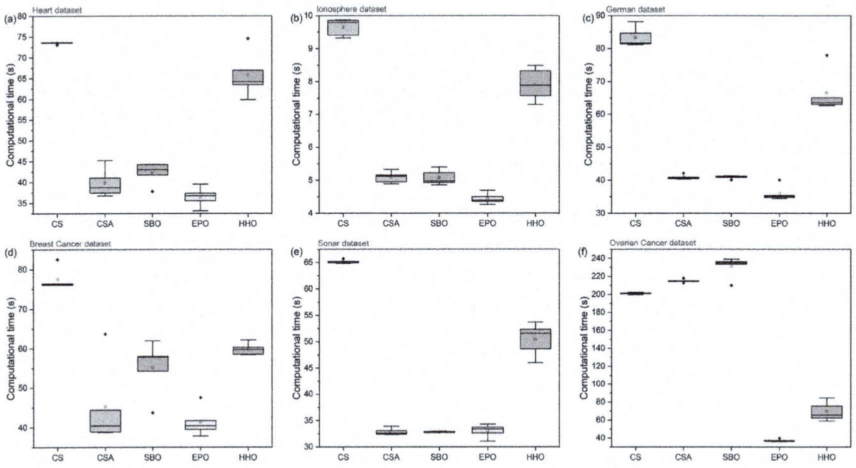


**Fig. 5** Boxplot of the computational time of all algorithms

- Based on best fitness, the tested bird-based metaphor algorithms can be ranked as CSA > EPO > CS ~ HHO > SBO. The ranking of the mean best fitness is CS > CSA > EPO > HHO > SBO. However, EPO is seen to have better reproducibility of results on repeated independent trials and thus, in terms of the standard deviation of best fitness, EPO > CS > CSA > HHO > SBO ranking is obtained.
- The convergence of EPO and HHO is superior to the other three algorithms.
- Regarding best classification accuracy, the algorithms are ranked as CSA > CS > EPO > HHO > SBO. However, the algorithms are ranked as CS ~ CSA > EPO > HHO > SBO for the mean best classification accuracy. EPO > CS > CSA > SBO ~ HHO ranking is obtained for a standard deviation for classification accuracy.
- For the mean number of features selected, the ranking is EPO > HHO > CS > SBO > CSA, whereas, for the standard deviation of the number of features selected, the ranking is EPO ~ SBO > CSA > CS > HHO.
- EPO and HHO are found to be computationally faster than the other algorithms and thus, as per computation time, the algorithms are ranked as EPO > HHO > CSA > SBO > CS.

From the study, it can be concluded that EPO and HHO are highly suitable for feature selection problems where, for large feature datasets, a 99% reduction in features can be achieved without compromising the classification accuracy.

Despite the comprehensiveness, this study is limited by analysis of only bird-cased metaphor algorithms. Other metaphor algorithms like human behavior-inspired, mammal-inspired, hunting behavior inspired etc., can also be studied in feature. The study can be further extended to include hybrid variants of these algorithms.

## References


1. Souza, D., Thom, R.C., Coelho, L.D.S., Macedo, C.A.D., Pierezan, J.: A V-shaped binary crow search algorithm for feature selection. In: IEEE Congress on Evolutionary Computation (CEC). Rio de Janeiro, Brazil (2018)
2. Arora, S., Singh, H., Sharma, M., Sharma, S., Anand, P.: A new hybrid algorithm based on grey wolf optimization and crow search algorithm for unconstrained function optimization and feature selection. IEEE Access **7**, 26343–26361 (2019). https://doi.org/10.1109/ACCESS.2019.2897325
3. Avinash Chandra, P., Dharmveer Singh, R.: Feature selection method based on hybrid data transformation and binary binomial cuckoo search. J. Ambient. Intell. Humaniz. Comput. **11**(2), 719–738 (2020). https://doi.org/10.1007/s12652-019-01330-1
4. Paul, D., Kumar, R., Saha, S., Mathew, J.: Multi-objective cuckoo search-based streaming feature selection for multi-label dataset. ACM Trans. Knowl. Discov. Data (TKDD) **15**(6), 1–24 (2021). https://doi.org/10.1145/3447586
5. Alzaqebah, M., Briki, K., Alrefai, N., Brini, S., Jawarneh, S., Alsmadi, M.K., Mohammad, R.M.A., et al.: Memory based cuckoo search algorithm for feature selection of gene expression dataset. Inf. Med. Unlocked **24**, 100572 (2021). https://doi.org/10.1016/j.imu.2021.100572
6. Turabieh, H., Azwari, S.A., Rokaya, M., Alosaimi, W., Alharbi, A., Alhakami, W., Alnfiai, M.: Enhanced Harris Hawks optimization as a feature selection for the prediction of student performance. Computing **103**(7), 1417–1438 (2021). https://doi.org/10.1007/s00607-020-00894-7
7. Abdel-Basset, M., Ding, W., El-Shahat, D.: A hybrid Harris Hawks optimization algorithm with simulated annealing for feature selection. Artif. Intell. Rev. **54**(1), 593–637 (2021). https://doi.org/10.1007/s10462-020-09860-3
8. Elgamal, Z.M., Yasin, N.B.M., Tubishat, M.A.M., Mirjalili, S.: An improved Harris Hawks optimization algorithm with simulated annealing for feature selection in the medical field. IEEE Access **8**, 186638–186652 (2020). https://doi.org/10.1109/ACCESS.2020.3029728
9. Hussain, K., Neggaz, N., Zhu, W., Houssein, E.: An efficient hybrid sine-cosine Harris Hawks optimization for low and high-dimensional feature selection. Expert Syst. Appl. **176**, 114778 (2021). https://doi.org/10.1016/j.eswa.2021.114778
10. Al-Wajih, R., Abdulkadir, S., Aziz, N., Al-Tashi, Q., Talpur, N.: Hybrid binary grey wolf with Harris Hawks optimizer for feature selection. IEEE Access **9**, 31662–31677 (2021). https://doi.org/10.1109/ACCESS.2021.3060096
11. Wolpert, D.H., Macready, W.G.: No free lunch theorems for optimization. IEEE Trans. Evol. Comput. **1**(1), 67–82 (1997). https://doi.org/10.1109/4235.585893
12. Yang, X.-S., Deb, S.: Cuckoo search via Levy flights. In: 2009 World Congress on Nature and Biologically Inspired Computing (NaBIC). Coimbatore, India (2009). https://doi.org/10.1109/NABIC.2009.5393690
13. Askarzadeh, A.: A novel metaheuristic method for solving constrained engineering optimization problems: crow search algorithm. Comput. Struct. **169**, 1–12 (2016). https://doi.org/10.1016/j.compstruc.2016.03.001
14. Seyyed Hamid Samareh, M., Vahid Khatibi, B.: Satin bowerbird optimizer: a new optimization algorithm to optimize ANFIS for software development effort estimation. Int. J. Intell. Real-Time Autom. **60**, 1–15 (2017). https://doi.org/10.1016/j.engappai.2017.01.006
15. Dhiman, G., Kumar, V.: Emperor Penguin optimizer: a bio-inspired algorithm for engineering problems. Knowl.-Based Syst. **159**, 20–50 (2018). https://doi.org/10.1016/j.knosys.2018.06.001
16. Heidari, A.A., Seyedali, M., Hossam, F., Ibrahim, A.M.M., Huiling, C.: Harris Hawks optimization: algorithm and applications. Futur. Gener. Comput. Syst. **97**, 849–872 (2019). https://doi.org/10.1016/j.future.2019.02.028
17. Premalatha, M., Jayasudha, M., Čep, R., Priyadarshini, J., Kalita, K., Chatterjee, P.: A comparative evaluation of nature-inspired algorithms for feature selection problems. Heliyon **10**(1), e23571 (2024). https://doi.org/10.1016/j.heliyon.2023.e23571

18. Priyadarshini, J., Premalatha, M., Čep, R., Jayasudha, M., Kalita, K.: Analyzing physics-inspired metaheuristic algorithms in feature selection with K-nearest-neighbor. Appl. Sci. **13**(2), 906 (2023). https://doi.org/10.3390/app13020906
19. Ganesh, N., Shankar, R., Čep, R., Chakraborty, S., Kalita, K.: Efficient feature selection using weighted superposition attraction optimization algorithm. Appl. Sci. **13**(5), 3223 (2023). https://doi.org/10.3390/app13053223
20. Ali, M.Z., Abdullah, A., Zaki, A.M., Rizk, F.H., Eid, M.M., El-Kenway, E.M.: Advances and challenges in feature selection methods: a comprehensive review. J. Artif. Intell. Metaheuristics **7**(1), 67–77 (2024). https://doi.org/10.54216/JAIM.070105

# Metaheuristic Algorithms for Data Clustering in Multivariate Data Sets: A Comparative Analysis

**Mario A. Navarro-Velázquez, Beatriz A. Rivera-Aguilar, Alfonso Ramos-Michel, Luis A. Beltran, and Noé Ortega-Sánchez**

**Abstract** This chapter explores the application of various metaheuristic optimization algorithms to improve the clustering process, addressing the challenge of finding optimal centroids for the clusters. We present an in-depth analysis of the problem and how to tune the algorithms. In addition, we compare 10 metaheuristic algorithms using a MATLAB implementation on 18 different cluster tests distributed over 3 other datasets with different variations in their data sets with different variations in their functions. The chapter investigates the convergence behavior of these algorithms when considering diverse databases, which include Shapes, Synthetic, and UCI datasets. Emphasis is placed on selecting algorithms based on their performance regarding computational speed and function access. The iterative nature of clustering is discussed, with stopping criteria such as a minimum change in centroids or reaching a maximum number of iterations. Importantly, each algorithm is rigorously tested with 50,000 function accesses per test, repeated 30 times to account for stochastic variations. Unlike previous works focusing on a single algorithm, this chapter aims to identify the most suitable algorithm for centroid determination in clustering problems. The comparison covers classical and contemporary algorithms, considering their various inspirations, and seeks to understand their effectiveness in solving clustering problems.

M. A. Navarro-Velázquez (✉) · B. A. Rivera-Aguilar · L. A. Beltran · N. Ortega-Sánchez
División de Tecnologías para la integración ciber-humana, CUCEI, Universidad de Guadalajara, Guadalajara, Mexico
e-mail: mario.navarro@academicos.udg.mx

B. A. Rivera-Aguilar
e-mail: beatriz.rivera@academicos.udg.mx

L. A. Beltran
e-mail: luis.beltran7882@alumnos.udg.mx

N. Ortega-Sánchez
e-mail: noe.ortega@academicos.udg.mx

A. Ramos-Michel
Departamento de ingenierías, CUCSUR, Universidad de Guadalajara, Autlán de Navarro, Mexico
e-mail: alfonso.rmichel@academicos.udg.mx

 
D. Oliva et al. (eds.), *Advances in Optimization Algorithms for Multidisciplinary Engineering Applications: From Classical Methods to AI-Enhanced Solutions*, Studies in Computational Intelligence 806, https://doi.org/10.1007/978-3-031-78440-8_9

**Keywords** Metaheuristics · Machine learning · Clustering · Square euclidean distance

# 1 Introduction

Data clustering is an unsupervised machine learning technique used in data mining. This process classifies data with one or more characteristics in common into bins, works by looking for the position of centroids and making groups depending on the Euclidean distance between centroids and each data point. Each point has the distance measured with each centroid and is assigned to the nearest one (see Fig. 1).

Clustering is currently used in a wide variety of areas, such as production, traffic safety, and biomedical. Applications span across multi-period planning to solve production and inventory problems [1], data segmentation for analyzing potential hazard factors in cyclist crashes [2], automatic identification of significant groups within biomedical and gene expression datasets without manual parameter selection [3], or understanding infectious disease spread in clustered networks to design epidemic prevention strategies [4].

Real-world problems present significant challenges for their solution due to the number of variables involved in their operation. The greater the number of variables present, the greater the complexity of the computation to solve the given problem. Some problems can be simplified by idealizing parameters to establish their mathematical definition with few variables. However, there are situations and issues where we obtain wrong and uncertain results by ignoring some of the variables that define them. However, computing many variables implies a barrier to algorithmic efficiency

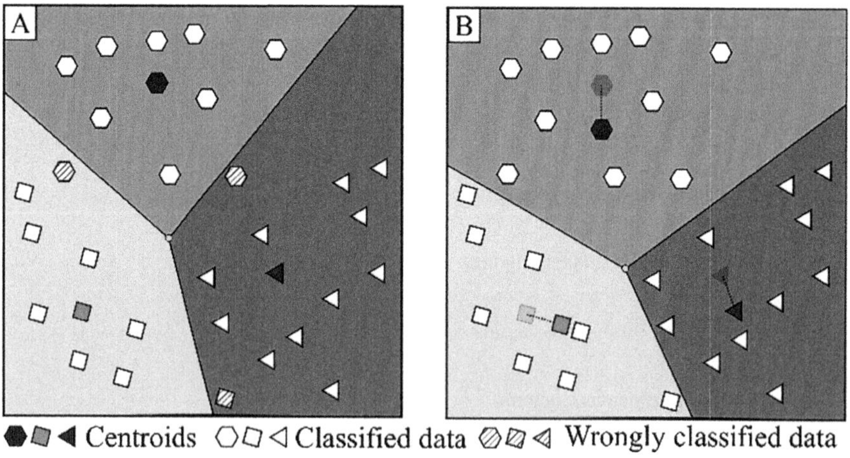


**Fig. 1** Example of clustering data

due to the high computational cost of the solving process and the high time taken to do it [5].

With an appropriate application of clustering task on the variables that compose a problem, the high dimensionality of it may be reduced, facilitating the work with it, solving the lacks before mentioned. By reducing problem dimensionality, Balakrishnan et al. improved the accuracy of infrastructure resilience prediction models, grouping component-level features based on their characteristics [6]. Ismkhan and Izaldi recently proposed in [7] a dimensionality reduction technique for the process of finding the nearest neighbors of a point inside a high-dimensional dataset, reducing the time of computation, and increasing the accuracy and efficiency of the results.

Applying clustering helps to optimize tasks and solve complex problems. However, to think that the algorithms that help us to perform clustering are perfect is not valid. The process of automatically clustering data could be better. Different artificial intelligence techniques have been used to improve clustering results, such as convolutional neural networks (CNN), support vector machines (SVM), extreme learning machines (ELM), and Metaheuristic Algorithms (MA). All these techniques, along with many others not mentioned, help to optimize the number of clusters and the parameters of each cluster [8–14].

This chapter explores the improvement of clustering by using metaheuristic algorithms. For this purpose, we have selected some of the most representative metaheuristic algorithms from the various techniques presented in the optimization field. When working with MA to improve clustering, the algorithm can stop the process when there is little change in the centroid vector, do not exists significant changes in the distance, or when the algorithm exceeds the minimum number of iterations or functions access to evaluate the fitness. Working with these algorithms, we found that even in their original versions, some are excellent options for finding the most appropriate centroids for clustering in the selected data sets, such as Shapes benchmark, Synthetic s-sets, and UCI datasets. We will reserve the study of the other AI techniques for future work.

We present in the rest of this chapter an explanation of the basics of metaheuristics and their classification in Sect. 2; the objective function considered for clustering and the process of metaheuristics to optimize the solution in Sect. 3; a comparative analysis, the convergence curves and the results in Sect. 4; and draw the conclusions and comment about the future work which continues the present one, in Sect. 5.

## 2 Metaheuristic Algorithms

A metaheuristic algorithm is a high-level algorithmic framework that provides a set of strategies for developing heuristic optimization algorithms. Due to their stochastic nature, which prevents them from easily being stuck on local optima [15], MAs represent a better way to explore the search space than exhaustive search algorithms, reducing computational cost and search time. For this reason, applying metaheuristics has become countless, and mentioning every single paper in existence has become an overwhelming task [16]. However, the advantages of the MAs do not guarantee to find the optimal solution to the problem. MAs are used to find approximate solutions to complex combinatorial, continuous, or mixed optimization problems, better than classical optimization techniques. In addition, the ease of scaling and adapting the MAs for specific applications and the known robustness convert this optimization algorithm into a top-rated and valuable search tool.

As mentioned, metaheuristic optimization algorithms have been widely used for many applications. Sometimes, due to the natural lack of MAs, combining algorithms is a common task to reinforce their performance in applications like solving engineering problems [17], the calculation of the size for dynamic rail car fleet [18] or the balance of robotic assembly lines [19]. The reason to combine them is that there is no perfect algorithm [20], and their combination results in a better performance. Nowadays, the number of different optimization proposals makes it possible to select an algorithm for a specific task after vast experimentation, comparing several algorithms before their final application in solving a particular problem. The results obtained from an appropriate comparison serve as a base to choose an algorithm in particular or, probably, readjust its parameters to improve its results. Sometimes, the results indicate that there is not just one best algorithm, and two or more must be hybridized to obtain the better part of each one and get the best possible solution from their cooperative participation.

In this paper, we compare classical algorithms like Particle Swarm Optimization (PSO—1995), a method based on heuristics that draws inspiration from the social interactions observed in animals, such as flocks of birds or swarms of insects, and was proposed by Kennedy and Eberhart [21]; the Differential Evolution algorithm (DE—1997) from Storn and Price [22], represents an alternative optimization approach focused on identifying solutions for optimization challenges within multidimensional spaces. Its primary feature lies in working with groups of potential solutions and employing mutation, recombination, and selection operators to progress toward optimal solutions. The Harmony Search algorithm (HS—2001) draws inspiration from the musical concept of harmony search. A sequence of musical notes is fused to simulate this concept, with the resulting harmonies serving as potential solutions for the optimization problem. This algorithm was presented by Geem et al. [23]. D. Karaboga and B. Basturk presented The Artificial Bee Colony algorithm (ABC—2007) [24]. The origin of this optimization algorithm lies in the social dynamics observed in bee colonies. It aims to efficiently replicate the interaction patterns among worker bees and their adeptness in locating food sources.

We consider the Bat Algorithm (BA—2010) from the modern algorithms. BA is a metaheuristic that relies on the natural behavior of bats. The procedure encompasses using sounds emitted by bats to locate their prey. Additionally, it replicates the random exploratory movements bats undertake in search of food areas. Ultimately, it adapts the amplitude and frequency of the sounds they emit based on the degree of success in locating prey. This algorithm was developed for Yang [25]. The Social Spider Optimizer (2013—SSO), whose inspiration comes from emulating the conduct of social spiders, involves utilizing diverse evolutionary operators based on the gender of the spiders, whether male or female, to replicate their behaviors. It was proposed by Cuevas et al. [26]. Another modern algorithm we considered is the Grey Wolf Optimizer (GWO—2014). The basis of this algorithm lies in replicating the social and hierarchical conduct observed in grey wolves in their natural environment. The objective is to mimic the leadership and hierarchical structure existing within a group of grey wolves to improve the exploration and exploitation of the search space in optimization problems and was developed by Mirjalili et al. [27].

Lastly, we compare some more recent algorithms like the Moth-Flame Optimization algorithm (MFO—2015). The inspiration for this algorithm comes from observing how moths navigate in their natural environment. The primary focus lies in their nighttime flight, where they consistently hold a set angle about the moon. MFO was presented by Mirjalili [28]. The Whale Optimization Algorithm (WOA—2016) is considered too. The algorithm draws inspiration from whales' natural behaviors, primarily focusing on the hunting and migration strategies these marine creatures employ; it was designed by Mirjalili and Lewis [29]. The last considered algorithm is the Selfish Herd Optimizer (SHO—2017), which draws inspiration from the self-serving behavior observed in herds, mimicking the predatory interactions between clusters of prey and a group of ravenous predators; made by Fausto et al. [30]. There is another classification for the algorithms that is extensively accepted, in which the algorithms are classified by their inspiration and divided into Nature-inspired, which gets their search methods by coding animal behavior; Evolution-based, which combines their candidate solutions to get similar solutions on the next generation and can mutate their solutions to find a new one; Physic based that emulate physics phenomena and tries to limit themselves to the laws of physics and Human-based that are inspired in activities, social interaction or lifestyles of the people. Even if there are different classifications, every algorithm has in common the search for the global minima or maxima for a problem by exploring and exploiting an objective function that is used to model the problem at hand.

As mentioned before, metaheuristic algorithms are frameworks that, through a general structure, offer different heuristic strategies for the optimization (minimization or maximization) of mathematical functions. To understand how they overcome the difficulties presented by complex problems without increasing the computational cost, it is necessary to understand how the search and evaluation process of possible solutions is performed during its execution. Figure 2 shows the general flowchart that all metaheuristic algorithms follow. Some techniques make minor variations in the sequence or repetition of some steps without technically departing from the general process shown.

The general operation can be divided into two main stages: the initialization process and the iterative process. Both stages are critical and directly impact the results obtained at the end of the algorithm execution. The initialization process is fundamental because the main variables of the algorithm receive their initial value here. This value determines the starting point for the algorithm's operation. At this stage, the parameters are also tuned, defining the behavior the algorithm will follow during the iterative process. The iterative process is the stage where the algorithm executes the search for possible solutions, the feasibility of each of them is evaluated and determined, and the final answer of the algorithm is chosen for the problem posed.

As shown in Fig. 2, everything starts with the initialization of parameters. These control the percentage of exploration and exploitation that the algorithm will perform. With the parameters defined, the algorithm proceeds to generate the first population of possible solutions, which, depending on the type of algorithm, can be called

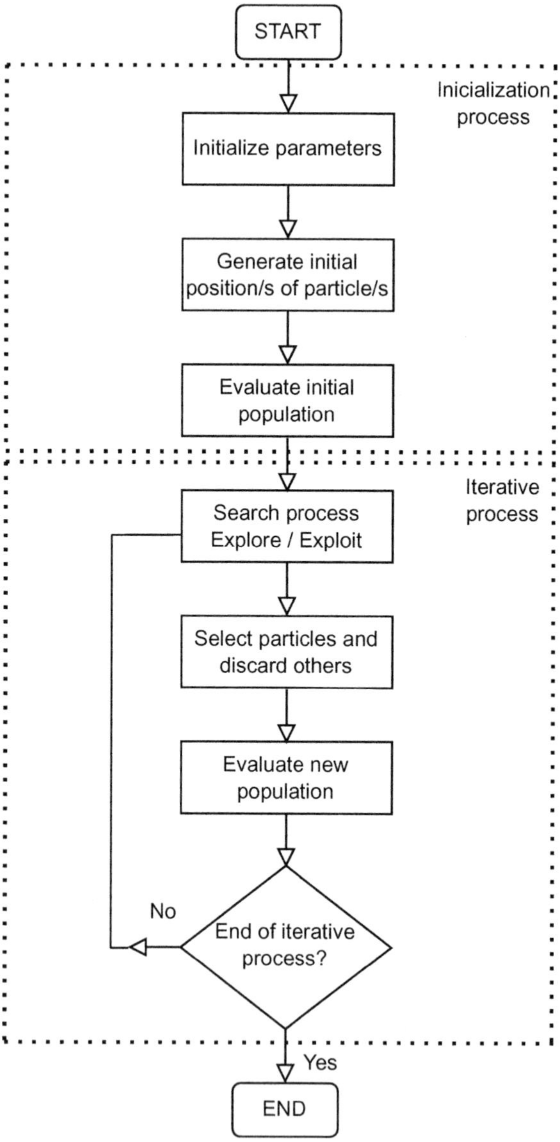


**Fig. 2** General process of metaheuristics

candidate solutions, individuals, or search agents. There are many methods to start the first population: Stochastic, such as randomly inserting candidate solutions in the search area without distance restriction; stochastic-deterministic methods, such as the Latin hypercube that divides the search area into smaller areas and then randomly sets one candidate solutions per area, two-step methods such as opposition-based learning in which half of the population is randomly inserted and the other half is calculated with its opposite.

Once the initial population is generated, the quality of the possible solution represented by each element is evaluated. With this first evaluation of the objective function, the algorithm has the necessary components and information to enter the second stage of its operation.

In the iterative process, new solutions are searched, starting from the candidate solutions obtained during the initialization process. The search consists of performing exploration and exploitation of the search space. During the exploration phase, the algorithm searches for new solutions independent of those previously obtained. The exploitation phase refers to refining the known solutions, where the algorithm performs an exhaustive search near the current position of the candidate solutions that the algorithm considers. This search task is where the significant difference in the performance of the different metaheuristic algorithms lies.

The next block represents a selection process to avoid growing the initial population and not to make the process slower. At this stage, many algorithms can generate or remove elements from their population as defined in their strategy. Each algorithm has its selection process because selecting only the candidate solutions with the best fitness could make the algorithm exploit a local solution, and selecting just bad candidate solutions could result in exploring a larger area and finding the global solution but not being able to refine it or even losing it. The search and candidate solutions selection strategies mainly define each algorithm's personality.

The last step is to evaluate whether to continue or terminate the iterative process. The iterative process goes on until a condition is met. It is common to have a certain number of iterations as stop criteria. In this case, when the iterative process reaches the maximum iteration number, the algorithm decides to finish its job. Another stop criterion is based on its iterative process, counting the number of evaluations done to the objective function. Although these are the main ways of establishing the stop criteria, each algorithm, and in particular, the problem to be solved, can determine some other stop criteria by evaluating any other variables handled within the algorithm or the problem itself. Once the stop criterion is reached, the iterative process ends, and the algorithm delivers the best solution according to the established criteria.

## 3 Clustering as an Objective Function of Metaheuristics

Supervised learning is an unsupervised artificial intelligence technique, these techniques infer results from an unlabeled input data. Their goal is to determine hidden patterns or grouping data that has characteristics in common, mostly used in exploratory data analysis denoting that both input and output are unknown.

Data clustering has the aim of make unlabeled bins and select data points with common characteristics to fill these bins. There are many applications in which it can be applied, clustering is vastly used in data mining, information retrieval and image segmentation. The data in clusters can represent a great variety of objects, sizes, degrees of complexity, distances, etc. The clustering technique does not have any knowledge of the meaning of the characteristics, instead looks for similarities.

K-means is a clustering technique based on centroids (see Fig. 3), it is a popular clustering approach due to its simplicity and efficiency, However the final solution for the centroids position can converge to a local optimum, this is highly dependant of the initial random positions. Because of the already mentioned problem we took the metaheuristics algorithms ability to escape local optima by exploration and to improve the solutions by exploitation of the search space, the algorithms treat each cluster axis as a dimension for the problem and each position will represent a solution that comprise all the positions for every cluster.

Figure 3 shows a clustering example by K-means with metaheuristics, in image A, we can observe that the data is classified as a square as the minimum Euclidean distance between the centroids is 25 which is the square centroid, and at the image B the centroids move with the search strategy of an algorithm and the data between them changes at triangle because the minimal distance is now 31 and comes from the triangle centroid. Both are candidate solutions and in most cases the solution with the better fitness is the function to continue on the iterative process, the total fitness of the clustering process is calculated as the sum of the minimal distance of every data point, the minimum distance of the data point between the centroids got higher but almost every centroid got closer to most of their already classified data which turns to be a better solution for the problem.

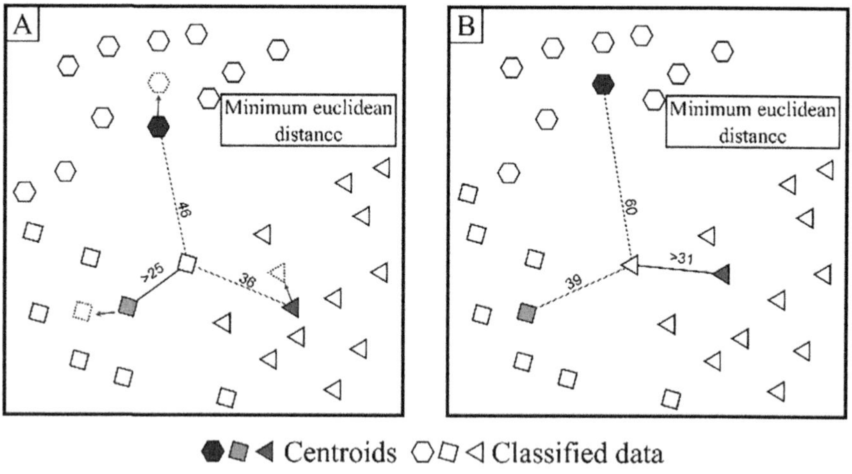


**Fig. 3** Clustering by minimum Euclidean distance

To get the fitness for the clustering problem, we use the square Euclidean distance between the data point x and the cluster centroid Cc which is calculated as:

$$\sqrt{(Cc_i - x)^2}^{\,2} = (Cc_i - x)^2 \tag{1}$$

This process is made for each of the cluster centroids and it selects the minimum result, so the formula is:

$$X_{min_j} = \min\left(\sum_{i=1}^{K} (Cc_i - x_j)^2\right) \tag{2}$$

And it obtains the fitness by making a sum of every minimal distance of each candidate solution with each cluster.

$$\sum_{j=1}^{n} X_{minj} \tag{3}$$

Figure 4 represents the final diagram of the evaluation method of clustering for the metaheuristic, each iteration comprises of several function access equal to itself multiplied by the population.

The diagram is called each time the algorithm needs to obtain the new fitness of the population and it can be inserted in any metaheuristic algorithms that follows the Fig. 5.

In this paper we adapted 10 algorithms to test in several clustering problems, we explain experimental results in Sect. 4.

## 4 Experimental Results

We divided our experimental results in three parts depending on the clustering benchmark: In Sect. 4.1; we present the Shape benchmark, that usually consist on different shapes to be clustered by density, but we test it by distance, for Sect. 4.2; we show the synthetic s-sets which are gaussian clusters with different degrees of overlapping, Sect. 4.3; we present the UCI datasets that works with real world-data, in Sect. 4.4; we show convergence tables to display the algorithms process over the iterations. The main objective of this type of experiment is to cover several clustering problems to evaluate and perform a robust and comprehensive comparative analysis with experimentation to set a benchmark for researchers exploring these areas. Currently, several old and new metaheuristics are state of the art, some more popular than oth-

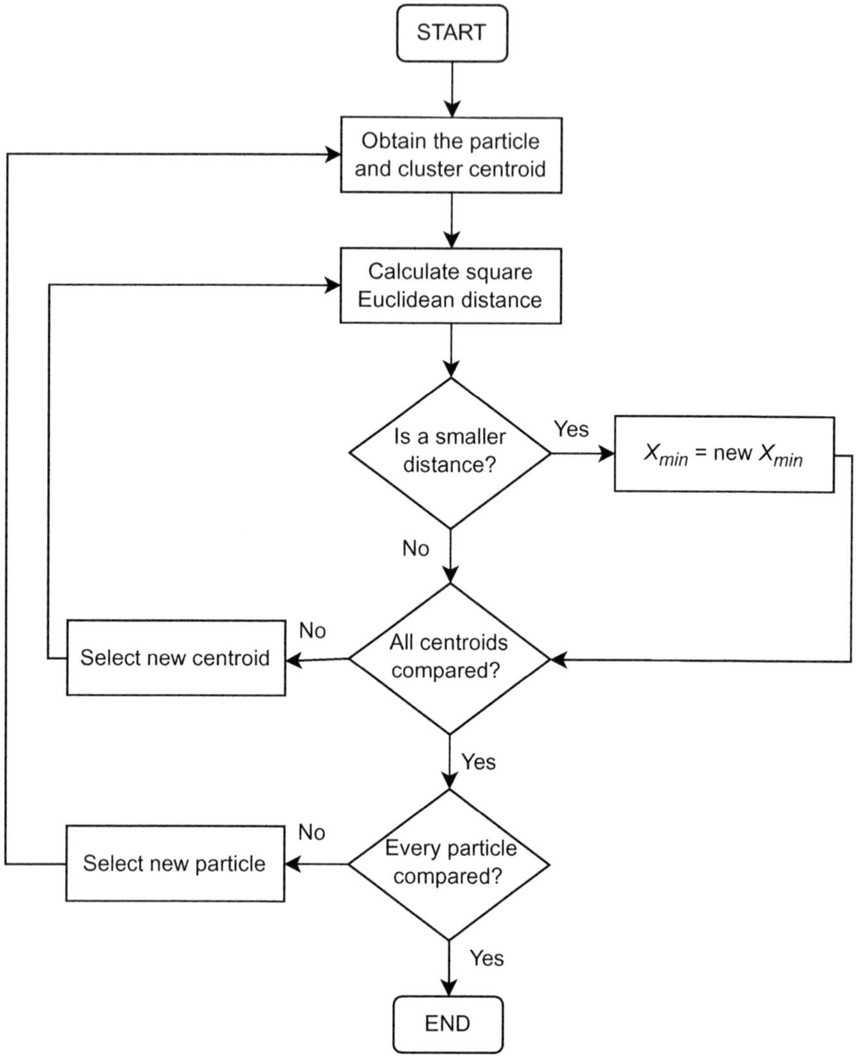


**Fig. 4** Evaluation method of clustering for the metaheuristic

ers. In that sense, very popular and relatively new metaheuristics were chosen for comparative analysis.

The algorithms compared are; Artificial bee colony (ABC) [24], Bat Algorithm (BA) [25], Differential evolution (DE) [22], Grew wolf optimizer (GWO) [27], Harmony search (HS) [23], Moth-flame optimization (MFO) [28], Particle swarm optimization (PSO) [13], Selfish herd optimizer (SHO) [30], Social spider optimizer (SSO) [26], and Whale optimization algorithm (WOA) [29]. And for each algorithm

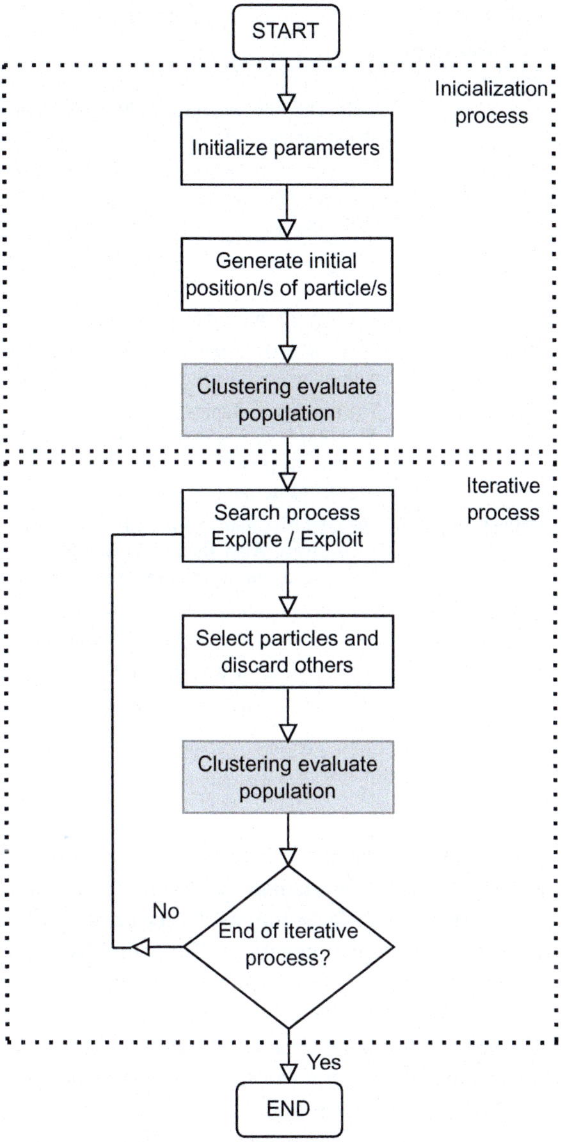


**Fig. 5** Procedure of clustering with metaheuristics algorithms

the total population has been set to 50 individuals, and they run for 50,000 function access (Faccs) and it is performed 30 times per experiment.

For the comparison analysis, the parameter configuration for each algorithm has been set according to its reported guidelines. Such configurations represent the best general performance according to its author. We show the configuration of these settings in Table 1.

**Table 1** Parameter settings of the algorithms chosen for the analysis

| Parameter configuration | |
|---|---|
| ABC [24] | Onlooker 50%, employees 50%, acceleration coefficient upper bound = 1, abandonment limit $L$ = round (0.6*dimensions*population) |
| BA [25] | The parameters where set as follows: Initial loudness rate $A = 2$, pulse emission rate $r = 0.9$, minimum frequency $f\_min = 0$ and maximum frequency $f\_max = 1$, respectively |
| DE [22] | The crossover rate is set to $CR = 0.5$, while the differential weight is given as $F = 0.2$ |
| GWO [27] | The algorithm's parameter $a$ is set to decrease linearly from 2 to 0 |
| HS [23] | $HMCR = 0.7$ Harmony Memory Considering Rate finds notes randomly within the possible playable range. $PAR = 0.3$ Pitch Adjusting Rate |
| MFO [28] | The number of flames is set as $N\_flames = round((N\_pop - k) * (N\_pop - 1)/k\_max)$, where N_pop denotes the population size, k the current iteration and k_max the maximum number of iterations |
| PSO [13] | *The cognitive and social coefficients are set to c_1=2.0 and c_2=2.0, respectively.* Also, the inertia weight factor $\omega$ is set to decreases linearly from 0.9 to 0.2 as the search process evolves |
| SHO [30] | *The herd population proportion randomly chosen from between a 70% and 90%,* with the remaining individuals assigned as predators |
| SSO [26] | The female attraction probability parameter is set as $PF = 0.7$ |
| WOA [29] | *The internal parameters A and C are set to decreases linearly from 2 to 0, and –1 to –2, respectively* |

For comparisons, three different indicators considered are: The Average Best-so-far (AB) solution, the Median Best-so-far (MB) solution and the Standard Deviation (SD) of the best-so-far solutions. Every benchmark function is tested on each algorithm 30 times which generates 30 results that represent the best-found solutions. The mean of these 30 outcomes is calculated to obtain AB. Similarly, the median of these 30 results is determined to generate MB, and the standard deviation of the 30 results is estimated to acquire SD of the best-so-far solutions. Indicators AB and MB evaluate the accuracy of the solutions while SD the dispersion and therefore the robustness of the algorithm.

## *4.1 Shapes Sets*

The shape dataset has 8 functions with 2 dimensions each, but each of them has a different distribution and number of data Dn and different number of clusters K; Flame: Dn = 240 K = 2 [31], Jain: Dn = 373 K = 2 [32], PathBased: Dn = 300 K = 3 [33], Spiral: Dn = 312 K = 3 [33], Aggregation: Dn = 788 K = 7 [34], Compound:

Dn = 399 K = 6 [35], R15: Dn = 600 K = 15 [36] and D31: Dn = 3100 K = 31 [36], see Fig. 6.

In Tables 2 and 3 we can see that SSO gets the best result in 4 of the clustering tests, follow by DE with 2 and lastly ABC tied with GWO with one. We can say that SSO is a reliable algorithm to make clustering in 2D even if the number of clusters or data changes, as it got the best result in Spiral with k = 3 and D31 that has 31 clusters to place.

## 4.2 Synthetic Sets

The Synthetic 2D dataset comprise 4 functions with 2 dimensions, with 5000 data points and 15 clusters [37], the thing that changes between them is the degree of cluster overlapping, which varies spatial complexity.
As it can be seen at the Fig. 7, the clusters expand and overlap more from S1 to S4.

The results of the Synthetic 2D table had SSO as the best performing algorithm, with this result it is clearly that SSO works well in the case of normally distributed data.

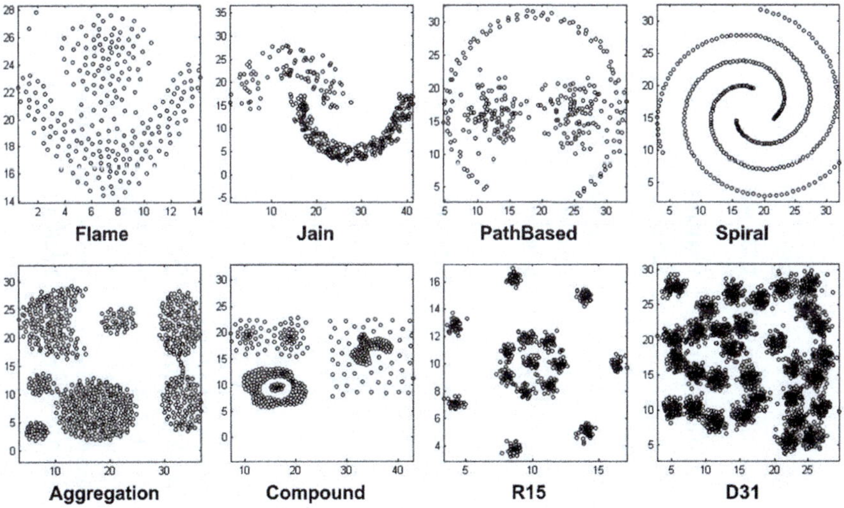


**Fig. 6** Distribution of data of the shape set

**Table 2** Results from shape dataset

| Fn | Fitness | ABC | BA | DE | GWO | HS | MFO | PSO | SHO | SSO | WOA |
|---|---|---|---|---|---|---|---|---|---|---|---|
| Shapes flame | $f_{mean}$ | 7.73E+03 | 8.95E+03 | **7.73E+03** | 7.73E+03 | 7.73E+03 | 7.73E+03 | 7.73E+03 | 1.34E+04 | 7.74E+03 | 7.73E+03 |
| | $f_{md}$ | 7.73E+03 | 8.23E+03 | **7.73E+03** | 7.73E+03 | 7.73E+03 | 7.73E+03 | 7.73E+03 | 1.34E+04 | 7.74E+03 | 7.73E+03 |
| | $f_{std}$ | 3.43E-08 | 1.91E+03 | **1.28E-12** | 5.73E-04 | 3.50E-06 | 4.06E-12 | 3.01E-12 | 2.14E+03 | 3.41E+00 | 8.70E-07 |
| Shapes jain | $f_{mean}$ | **2.22E+04** | 2.26E+04 | 2.22E+04 | 2.25E+04 | 2.22E+04 | 2.24E+04 | 2.24E+04 | 2.40E+04 | 2.23E+04 | 2.26E+04 |
| | $f_{md}$ | 2.22E+04 | 2.22E+04 | 2.22E+04 | 2.22E+04 | **2.22E+04** | 2.22E+04 | 2.22E+04 | 2.41E+04 | 2.22E+04 | 2.22E+04 |
| | $f_{std}$ | **1.01E-11** | 8.11E+02 | 8.99E+00 | 7.01E+02 | 3.23E-04 | 5.64E+02 | 5.64E+02 | 2.23E+02 | 3.38E+02 | 7.52E+02 |
| Shapes pathbased | $f_{mean}$ | 8.96E+03 | 8.96E+03 | **8.96E+03** | 8.96E+03 | 8.96E+03 | 8.97E+03 | 8.99E+03 | 8.96E+03 | 8.96E+03 | 8.97E+03 |
| | $f_{md}$ | 8.96E+03 | 8.96E+03 | 8.96E+03 | 8.96E+03 | 8.96E+03 | **8.96E+03** | 8.96E+03 | 8.96E+03 | 8.96E+03 | 8.96E+03 |
| | $f_{std}$ | 4.63E-02 | 7.75E+00 | **1.26E-03** | 4.77E-01 | 4.42E-03 | 5.28E+01 | 7.19E+01 | 4.76E-01 | 3.83E-01 | 1.15E+01 |
| Shapes spiral | $f_{mean}$ | 1.23E+04 | 1.23E+04 | 1.23E+04 | 1.23E+04 | 1.23E+04 | 1.23E+04 | 1.23E+04 | 1.23E+04 | **1.23E+04** | 1.23E+04 |
| | $f_{md}$ | 1.23E+04 | 1.23E+04 | 1.23E+04 | **1.23E+04** | 1.23E+04 | 1.23E+04 | 1.23E+04 | 1.23E+04 | 1.23E+04 | 1.23E+04 |
| | $f_{std}$ | 7.44E-01 | 1.84E+02 | 1.74E+00 | 3.11E-01 | 7.83E-01 | 1.83E+00 | 3.33E-01 | 2.55E+00 | **2.14E-01** | 1.01E+02 |
| Shapes aggregation | $f_{mean}$ | 1.32E+04 | 1.26E+04 | 1.21E+04 | 1.17E+04 | 1.19E+04 | 1.18E+04 | 1.21E+04 | 1.32E+04 | **1.12E+04** | 1.21E+04 |
| | $f_{md}$ | 1.31E+04 | 1.13E+04 | 1.21E+04 | 1.13E+04 | 1.20E+04 | 1.15E+04 | 1.19E+04 | 1.32E+04 | **1.12E+04** | 1.16E+04 |
| | $f_{std}$ | 7.08E+02 | 2.02E+03 | 1.93E+02 | 8.24E+02 | 2.36E+02 | 1.05E+03 | 8.50E+02 | 3.15E+02 | **1.85E+02** | 9.22E+02 |
| Shapes compound | $f_{mean}$ | 4.06E+03 | 4.79E+03 | 3.95E+03 | **3.89E+03** | 3.92E+03 | 4.14E+03 | 4.23E+03 | 8.56E+03 | 3.97E+03 | 4.55E+03 |
| | $f_{md}$ | 4.06E+03 | 4.96E+03 | 3.95E+03 | **3.87E+03** | 3.89E+03 | 3.87E+03 | 4.00E+03 | 8.65E+03 | 3.87E+03 | 4.59E+03 |
| | $f_{std}$ | 1.01E+02 | 4.91E+02 | **2.52E+01** | 1.27E+02 | 6.77E+01 | 3.49E+02 | 4.44E+02 | 1.27E+03 | 2.67E+02 | 6.12E+02 |

(continued)

**Table 2** (continued)

| Fn | Fitness | ABC | BA | DE | GWO | HS | MFO | PSO | SHO | SSO | WOA |
|---|---|---|---|---|---|---|---|---|---|---|---|
| Shapes R15 | $f_{mean}$ | 1.10E+03 | 3.21E+02 | 5.86E+02 | 2.23E+02 | 7.44E+02 | 3.06E+02 | 4.56E+02 | 2.27E+02 | **1.65E+02** | 3.79E+02 |
| | $f_{md}$ | 1.11E+03 | 3.24E+02 | 5.84E+02 | 2.22E+02 | 7.54E+02 | 2.87E+02 | 4.40E+02 | 2.33E+02 | **1.69E+02** | 3.69E+02 |
| | $f_{std}$ | 1.23E+02 | 9.85E+01 | **2.75E+01** | 6.45E+01 | 6.14E+01 | 9.96E+01 | 1.07E+02 | 7.00E+01 | 4.40E+01 | 9.08E+01 |
| Shapes D31 | $f_{mean}$ | 1.72E+04 | 5.58E+03 | 1.29E+04 | 5.41E+03 | 1.19E+04 | 5.14E+03 | 6.87E+03 | 7.23E+03 | **4.65E+03** | 7.44E+03 |
| | $f_{md}$ | 1.73E+04 | 5.71E+03 | 1.30E+04 | 5.50E+03 | 1.19E+04 | 5.01E+03 | 6.75E+03 | 6.60E+03 | **4.70E+03** | 7.56E+03 |
| | $f_{std}$ | 1.07E+03 | 7.30E+02 | 5.35E+02 | 5.68E+02 | 4.21E+02 | 6.36E+02 | 9.38E+02 | 2.27E+03 | **3.92E+02** | 7.48E+02 |

**Table 3** Results from the synthetic dataset

| Fn | Fitness | ABC | BA | DE | GWO | HS | MFO | PSO | SHO | SSO | WOA |
|---|---|---|---|---|---|---|---|---|---|---|---|
| Sset S1 | $f_{mean}$ | 4.67E+13 | 5.18E+13 | 3.30E+13 | 1.40E+13 | 3.53E+13 | 1.49E+13 | 2.07E+13 | 1.44E+13 | **1.30E+13** | 2.06E+13 |
| | $f_{md}$ | 4.65E+13 | 5.21E+13 | 3.32E+13 | **1.37E+13** | 3.55E+13 | 1.40E+13 | 2.09E+13 | 1.38E+13 | 1.38E+13 | 2.03E+13 |
| | $f_{std}$ | 3.19E+12 | 6.67E+12 | **2.05E+12** | 3.31E+12 | 2.19E+12 | 4.75E+12 | 4.76E+12 | 3.26E+12 | 3.37E+12 | 3.72E+12 |
| Sset S2 | $f_{mean}$ | 4.46E+13 | 5.11E+13 | 3.35E+13 | 1.84E+13 | 3.54E+13 | 1.77E+13 | 2.08E+13 | 1.70E+13 | **1.51E+13** | 2.26E+13 |
| | $f_{md}$ | 4.48E+13 | 5.09E+13 | 3.35E+13 | 1.72E+13 | 3.55E+13 | 1.72E+13 | 2.02E+13 | 1.68E+13 | 1.59E+13 | 2.20E+13 |
| | $f_{std}$ | 2.52E+12 | 5.65E+12 | 1.52E+12 | 2.17E+12 | 1.52E+12 | 3.02E+12 | 4.18E+12 | 2.49E+12 | 1.97E+12 | 5.21E+12 |
| Sset S3 | $f_{mean}$ | 3.59E+13 | 4.29E+13 | 2.90E+13 | 1.92E+13 | 3.07E+13 | 2.08E+13 | 2.26E+13 | 1.89E+13 | **1.83E+13** | 2.26E+13 |
| | $f_{md}$ | 3.56E+13 | 4.30E+13 | 2.90E+13 | 1.91E+13 | 3.09E+13 | 2.10E+13 | 2.22E+13 | 1.89E+13 | **1.88E+13** | 2.22E+13 |
| | $f_{std}$ | 2.25E+12 | 5.29E+12 | 8.86E+11 | 2.04E+12 | **1.18E+12** | 2.20E+12 | 2.27E+12 | 1.08E+12 | 9.75E+11 | 2.07E+12 |
| Sset S4 | $f_{mean}$ | 3.11E+13 | 3.61E+13 | 2.52E+13 | 1.75E+13 | 2.75E+13 | 1.73E+13 | 2.03E+13 | 1.66E+13 | **1.65E+13** | 2.09E+13 |
| | $f_{md}$ | 3.11E+13 | 3.56E+13 | 2.54E+13 | 1.73E+13 | 2.74E+13 | 1.72E+13 | 2.01E+13 | 1.66E+13 | **1.66E+13** | 2.03E+13 |
| | $f_{std}$ | 1.27E+12 | 3.83E+12 | **9.32E+11** | 1.40E+12 | 1.01E+12 | 1.11E+12 | 2.04E+12 | 8.72E+11 | 8.06E+11 | 2.50E+12 |

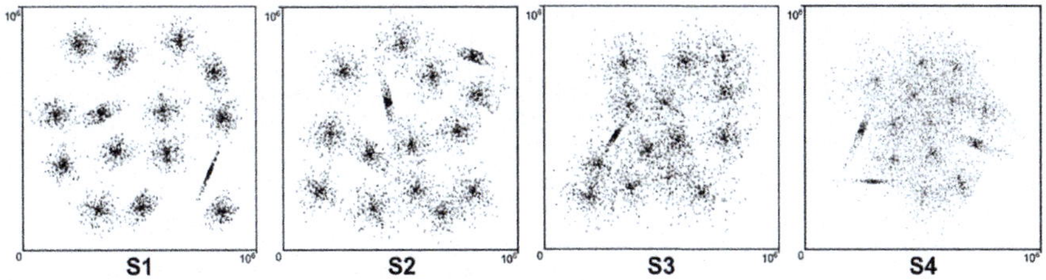


**Fig. 7** Distribution of the data of the synthetic set

## *4.3 UCI Datasets*

UCI machine learning repository is a collection of databases, we selected collections made with real data; we present 6 datasets for clustering that have different numbers of data Dn, clusters K, and dimensions D, the datasets compared are; Breast: Dn = 699, K = 2, D= 9, Glass: Dn = 214, K = 7, D = 9, Iris: Dn = 150, K = 3, D = 4, Thyroid: Dn = 215, K = 2, D = 5, Wine: Dn = 178, K = 3, D = 13 and Yeast: Dn = 1484, K = 10, D = 9.
As it can be seen, each function has a different dimensionality, and as every function has a dimension over 3 it makes difficult to show an image of the cluster density.

The results of the UCI dataset table had several algorithms performing better than the other in one test, this includes DE, GWO, HS, and SSO and SHO outperforms the algorithms in 2 of 6 functions, this make it the best algorithm for multidimensional clustering.

## *4.4 Convergence Analysis*

The algorithms improve their solutions though the iterative process, the convergence analysis has the values for the best solution each iteration, the performance results of each algorithm in a single execution have been plotted to illustrate the speed of convergence. The axis of the graph represents the fitness obtained in the function and the function access used.

The convergence analysis considers 16 clustering tests, the selection of the functions was made considering 4 tests from each database, Shape database; aggregation, compound, D31 and R15, Synthetic database; S1, S2, S3 and S4, and UCI dataset; breast, glass, wine and yeast. The test evaluates the performance of algorithms through time in different dimensions. The graphs with the results of convergence are shown in Figs. 8, 9, 11, 12, 13, 13.

From Fig. 8 in the Aggregation test, we can see that SSO, BA and GWO obtains similar results, but SSO only took about 2 x $10^4$ function access, for the Compound test SSO and MFO obtained the best results in similar times.

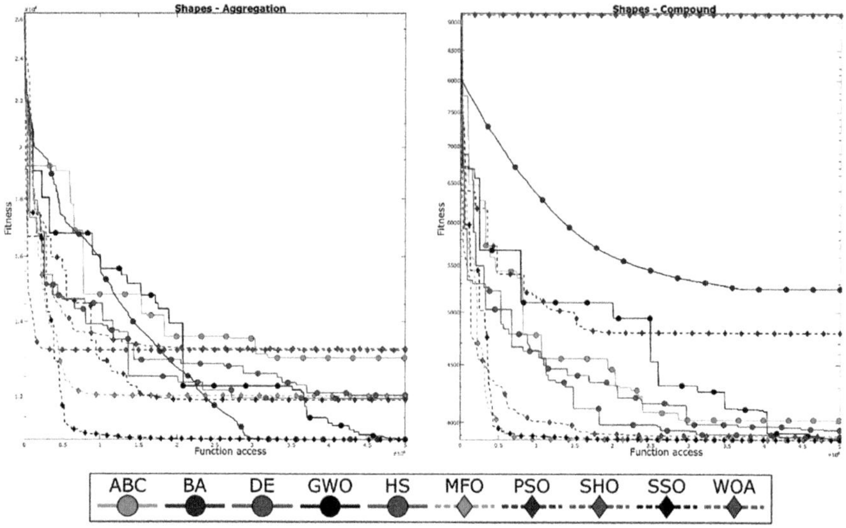


**Fig. 8** Convergence graph for Shape dataset Aggregation (Left) and Compound (Right)

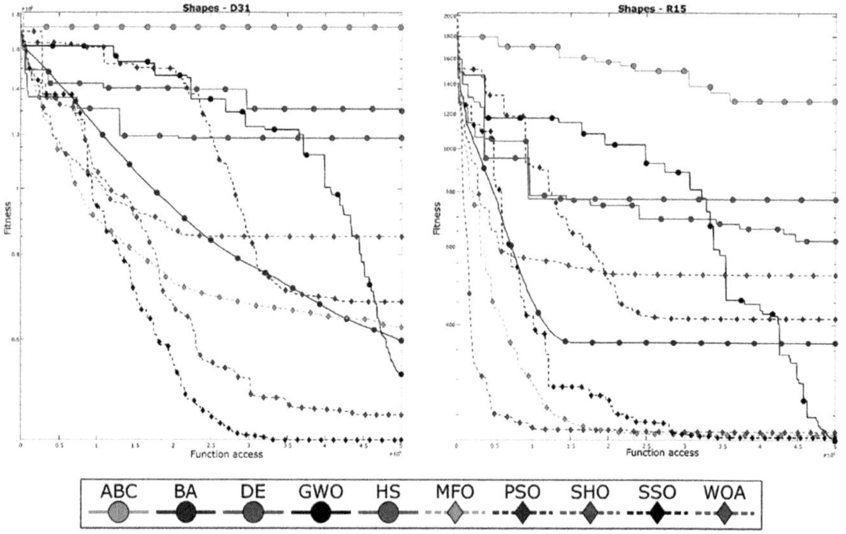


**Fig. 9** Convergence graph for Shape dataset D31 (Left) and R15 (Right)

In Fig. 9 the test D31 has a clearly best algorithm which is the SSO, and on the test R15 the best algorithm is GWO, but SSO got a really similar result 20,000 function access before, so the best algorithm for this test is debatable (Fig. 9 and Table 2).

From synthetic test S1 we observed that SSO obtains the best solution in 1.2 x $10^4$ function access, this means that it found a great solution in very little time, and

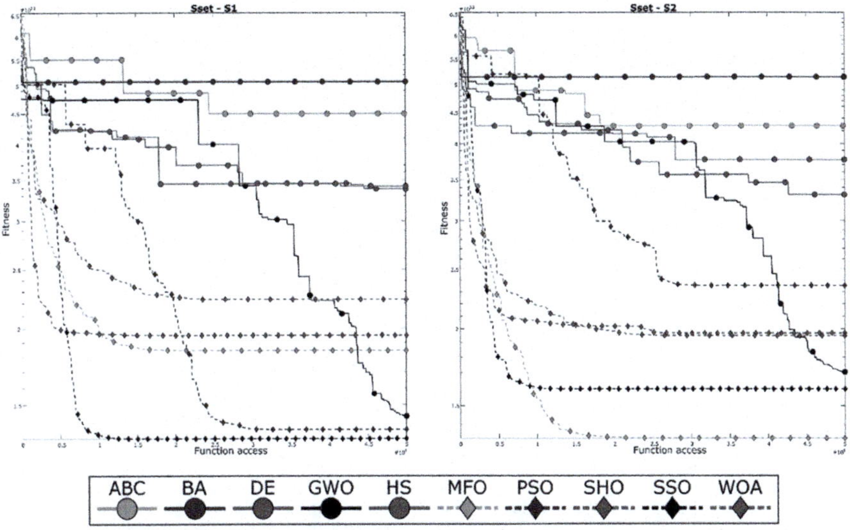


**Fig. 10** Convergence graph for synthetic dataset S1 (Left) and S2 (Right)

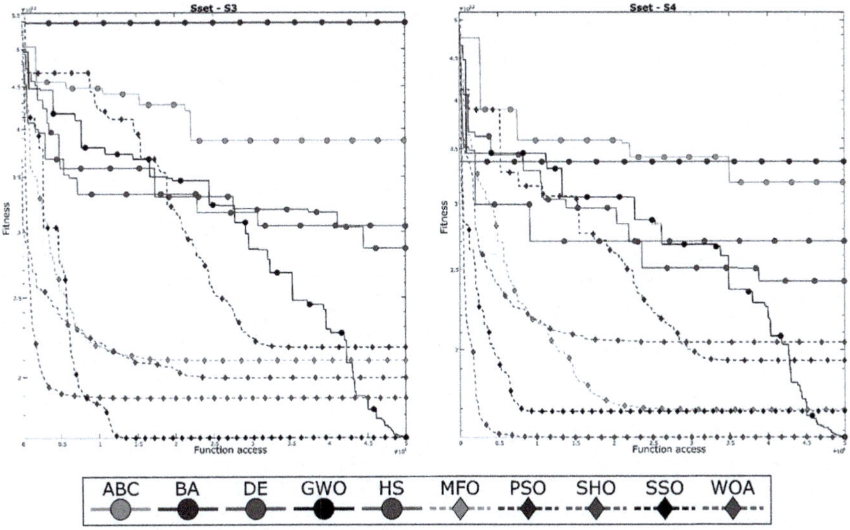


**Fig. 11** Convergence graph for synthetic dataset S3 (Left) and S4 (Right)

in function S2 SSO find a good solution but it gets stuck in a local minimum, MFO in the other hand obtains the best result in only 2.3 x $10^4$ Faccs.

Synthetic test S3 and S4 are better solved by GWO, tied with SSO in S3 and with SHO in S4, however GWO got the solutions using almost the 50,000 Faccs, therefore sometimes could be better to use the other algorithms.

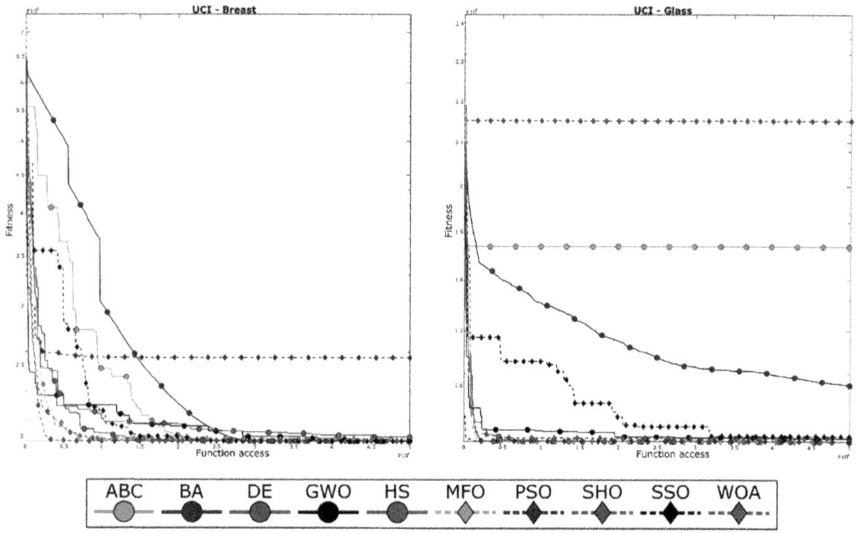


**Fig. 12** Convergence graph for UCI dataset Breast (Left) and Glass (Right)

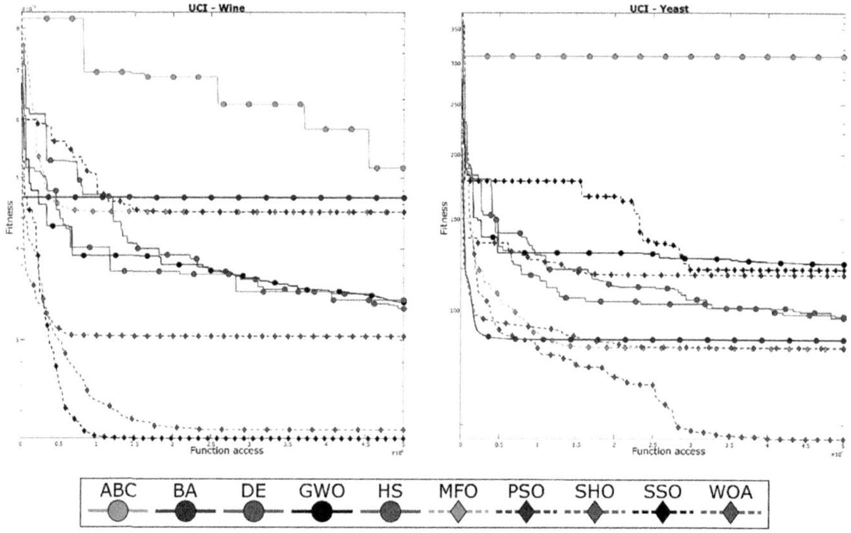


**Fig. 13** Convergence graph for UCI dataset Wine (Left) and Yeast (Right)

**Table 4** Results from the UCI dataset

| Fn | Fitness | ABC | BA | DE | GWO | HS | MFO | PSO | SHO | SSO | WOA |
|---|---|---|---|---|---|---|---|---|---|---|---|
| UCI breast | $f_{mean}$ | 1.97E+04 | 1.97E+04 | 1.97E+04 | 1.97E+04 | 2.00E+04 | 2.02E+04 | 2.31E+04 | **1.97E+04** | **1.97E+04** | 1.97E+04 |
| | $f_{md}$ | 1.97E+04 | 1.97E+04 | 1.97E+04 | 1.97E+04 | 2.00E+04 | 1.99E+04 | 2.34E+04 | 1.97E+04 | 1.97E+04 | 1.97E+04 |
| | $f_{std}$ | 2.77E+00 | 3.23E-04 | 1.58E-01 | 9.00E-02 | 6.65E+01 | 9.64E+02 | 1.78E+03 | 4.56E-11 | 5.50E-01 | 7.57E+00 |
| UCI glass | $f_{mean}$ | 2.16E+04 | 1.61E+04 | 1.50E+04 | 1.52E+04 | **1.50E+04** | 1.50E+04 | 1.51E+04 | 2.22E+04 | 1.52E+04 | 1.55E+04 |
| | $f_{md}$ | 2.16E+04 | 1.61E+04 | 1.50E+04 | 1.50E+04 | 1.50E+04 | 1.50E+04 | 1.51E+04 | 2.18E+04 | 1.52E+04 | 1.54E+04 |
| | $f_{std}$ | 1.55E+03 | 3.58E+02 | 2.43E-01 | 3.20E+02 | 1.54E-02 | 2.61E+00 | 8.67E+01 | 1.78E+03 | 6.70E+01 | 4.72E+02 |
| UCI Iris | $f_{mean}$ | 1.40E+14 | 3.15E+14 | 1.42E+14 | **1.40E+14** | 1.40E+14 | 1.49E+14 | 1.76E+14 | 1.42E+14 | 1.40E+14 | 1.57E+14 |
| | $f_{md}$ | 1.40E+14 | 3.19E+14 | 1.42E+14 | 1.40E+14 | 1.40E+14 | 1.42E+14 | 1.52E+14 | 1.40E+14 | 1.40E+14 | 1.40E+14 |
| | $f_{std}$ | 1.40E+11 | 5.43E+13 | 6.70E+11 | 1.15E+09 | 1.83E+11 | 1.48E+13 | 3.71E+13 | 9.26E+12 | 2.00E+10 | 2.88E+13 |
| UCI thyroid | $f_{mean}$ | 7.53E+14 | 1.48E+15 | **7.42E+14** | 7.58E+14 | 7.56E+14 | 7.61E+14 | 8.29E+14 | 7.72E+14 | 7.58E+14 | 7.68E+14 |
| | $f_{md}$ | 7.57E+14 | 1.40E+15 | 7.40E+14 | 7.57E+14 | 7.60E+14 | 7.60E+14 | 7.93E+14 | 7.62E+14 | 7.60E+14 | 7.62E+14 |
| | $f_{std}$ | 6.41E+12 | 2.87E+14 | 5.72E+12 | 9.80E+12 | 6.69E+12 | 1.63E+13 | 9.98E+13 | 1.33E+13 | 5.04E+12 | 3.46E+13 |
| UCI wine | $f_{mean}$ | 6.05E+11 | 5.46E+11 | 3.46E+11 | 3.17E+11 | 3.45E+11 | 4.03E+11 | 4.27E+11 | 2.26E+11 | **2.23E+11** | 2.85E+11 |
| | $f_{md}$ | 6.02E+11 | 5.38E+11 | 3.46E+11 | 3.13E+11 | 3.45E+11 | 4.41E+11 | 4.49E+11 | 2.20E+11 | 2.20E+11 | 2.96E+11 |
| | $f_{std}$ | 6.27E+10 | 5.23E+10 | 1.60E+10 | 4.32E+10 | 1.18E+10 | 6.23E+10 | 4.46E+10 | 2.00E+10 | 1.52E+10 | 5.59E+10 |
| UCI yeast | $f_{mean}$ | 3.27E+02 | 7.47E+01 | 9.88E+01 | 1.08E+02 | 9.88E+01 | 1.06E+02 | 1.21E+02 | **6.20E+01** | 1.10E+02 | 1.05E+02 |
| | $f_{md}$ | 3.25E+02 | 7.48E+01 | 9.96E+01 | 1.15E+02 | 9.89E+01 | 1.12E+02 | 1.23E+02 | 6.29E+01 | 1.08E+02 | 1.09E+02 |
| | $f_{std}$ | 5.54E+01 | 1.19E+01 | 2.87E+00 | 1.25E+01 | 2.44E+00 | 1.29E+01 | 5.10E+00 | 4.53E+00 | 7.54E+00 | 1.26E+01 |

In UCI tests breast and glass every algorithm has similar results except for PSO in breast and both BA and ABC in glass that got stuck in a local solution.

For UCI wine SSO obtains the best solution among the algorithms in very short time which proves that it is a great algorithm to make clusters and in the test UCI yeast SHO, an algorithm that has been in the top three places in most of the tests gets the best solution by far.

## 5 Conclusions and Future Work

In this paper, we compared multiple metaheuristics algorithms to solve the clustering problem of finding the best position for the centroids. The extensive experimental results over 3 different datasets comprising 18 unique functions showed that:

- The algorithms SSO, SHO and GWO had the top 3 best results over the clustering functions.
- SSO search strategy got great results over the experiments, obtaining the best results on 10 out of 18 functions compared against 10 algorithms. Its attraction or repulsion movement of the female spiders could have been the reason of their achievement.
- Convergence tables shows that SSO has a great exploration as the curve goes the furthest down than other algorithms, and his exploitation also is remarkable as even with good solutions it stills keep improving the solution.
- Looking at the data of convergences tables shows that SSO did not exploit much after 35,000 function access is most of the functions, late exploitation must improve to do so.
- As future work, it is envisioned to use this method for a more extensive comparative study that includes several recent generation algorithms and to show graphical results of the clustering performed with this methodology, in addition to adding non-parametric statistics to validate the results.

**Competing Interests** Please declare any competing interests in the context of your chapter. The following sentences can be regarded as examples.

This study was funded by [X] [grant number X].

[Author A] has a received research grant from [Company W].

[Author B] has received a speaker honorarium from [Company X] and owns stock in [Company Y].

[Author C] is a member of [committee Z].

The authors have no conflicts of interest to declare that are relevant to the content of this chapter.

**Ethics Approval** If your chapter includes primary studies with humans please declare the adherence of ethical standards. Example text: This study was performed in line with the principles of the Declaration of Helsinki. Approval was granted by the Ethics Committee of University B (Date.../No. ...).

In addition, for human participants, authors are required to include a statement that informed consent (to participate and/or to publish) was obtained from individual participants or parents/guardians if the participant is minor or incapable.

If animals are studied, authors should make sure that the legal requirements or guidelines in the country and/or state or province for the care and use of animals have been followed or specify that no ethics approval was required.

**Acknowledgements** If you want to include acknowledgments of assistance and the like at the end of an individual chapter please use the `acknowledgement` environment—it will automatically be rendered in line with the preferred layout.

## References


1. Nananukul, N.: Clustering model and algorithm for production inventory and distribution problem. Appl. Math. Model. **37**(24), 9846–9857 (2013)
2. Sivasankaran, S.K., Balasubramanian, V.: Exploring the severity of bicycle-vehicle crashes using latent class clustering approach in India. J. Saf. Res. **72**, 127–138 (2020)
3. Rahman, M.A., Ang, L.-M., Seng, K.P.: Clustering biomedical and gene expression datasets with kernel density and unique neighborhood set based vein detection. Inf. Syst. **91**, 101490 (2020)
4. Wu, Q., Hadzibeganovic, T.: An individual-based modeling framework for infectious disease spreading in clustered complex networks. Appl. Math. Model. **83**, 1–12 (2020)
5. Dimension reduction for projective clustering (2022)
6. Application of clustering algorithms for dimensionality reduction in infrastructure resilience prediction models (2022). arXiv:2205.03316
7. Proposing a dimensionality reduction technique with an inequality for unsupervised learning from high-dimensional big data. **53**, 3880–3889 (2023)
8. Yang, J., Parikh, D., Batra, D.: Joint unsupervised learning of deep representations and image clusters. In: Proceedings of the IEEE Conference on Computer Vision and Pattern Recognition, pp. 5147–5156 (2016)
9. Zhang, Y., Tang, Y., Liu, Y., Liang, Z.: Fault diagnosis of transformer using artificial intelligence: a review. Front. Energy Res. **10**, 1006474 (2022)
10. Lin, X., et al.: A road network traffic state identification method based on macroscopic fundamental diagram and spectral clustering and support vector machine. Math. Probl. Eng. **2019**, (2019)
11. Hu, X.-F., Zhao, Z., Wang, S., Wang, F.-L., He, D.-K., Wu, S.-K.: Multi-stage extreme learning machine for fault diagnosis on hydraulic tube tester. Neural Comput. Appl. **17**, 399–403 (2008)
12. Nasiri, J., Khiyabani, F.M.: A whale optimization algorithm (WOA) approach for clustering. Cogent Math. & Stat. **5**(1), 1483565 (2018)
13. Janani, R., Vijayarani, S.: Text document clustering using spectral clustering algorithm with particle swarm optimization. Expert. Syst. Appl. **134**, 192–200 (2019)
14. Dutta, D., Sil, J., Dutta, P.: Automatic clustering by multi-objective genetic algorithm with numeric and categorical features. Expert. Syst. Appl. **137**, 357–379 (2019)
15. Cuevas, E., Osuna, V., Oliva, D., et al.: Evolutionary Computation Techniques: a Comparative Perspective, vol. 686. Springer (2017)
16. Fausto, F., Reyna-Orta, A., Cuevas, E., Andrade, Á.G., Perez-Cisneros, M.: From ants to whales: metaheuristics for all tastes. Artif. Intell. Rev. **53**, 753–810 (2020)
17. Zhang, J., Xiao, M., Gao, L., Pan, Q.: Queuing search algorithm: a novel metaheuristic algorithm for solving engineering optimization problems. Appl. Math. Model. **63**, 464–490 (2018)

18. Yaghini, M., Khandaghabadi, Z.: A hybrid metaheuristic algorithm for dynamic rail car fleet sizing problem. Appl. Math. Model. **37**(6), 4127–4138 (2013)
19. Janardhanan, M.N., Li, Z., Bocewicz, G., Banaszak, Z., Nielsen, P.: Metaheuristic algorithms for balancing robotic assembly lines with sequence-dependent robot setup times. Appl. Math. Model. **65**, 256–270 (2019)
20. Wolpert, D.H., Macready, W.G.: No free lunch theorems for optimization. IEEE Trans. Evol. Comput. **1**(1), 67–82 (1997)
21. Kennedy, J., Eberhart, R.: Particle swarm optimization. In: Proceedings of ICNN'95-International Conference on Neural Networks, vol. 4, pp. 1942–1948. IEEE (1995)
22. Storn, R., Price, K.: Differential evolution-a simple and efficient heuristic for global optimization over continuous spaces. J. Glob. Optim. **11**, 341–359 (1997)
23. Geem, Z.W., Kim, J.H., Loganathan, G.V.: A new heuristic optimization algorithm: harmony search. Simulation **76**(2), 60–68 (2001)
24. Karaboga, D., Basturk, B.: A powerful and efficient algorithm for numerical function optimization: artificial bee colony (ABC) algorithm. J. Glob. Optim. **39**, 459–471 (2007)
25. Yang, X.-S.: A new metaheuristic bat-inspired algorithm. In: Nature Inspired Cooperative Strategies for Optimization (NICSO 2010), pp. 65–74. Springer (2010)
26. Cuevas, E., Cienfuegos, M., Zaldívar, D., Pérez-Cisneros, M.: A swarm optimization algorithm inspired in the behavior of the social-spider. Expert. Syst. Appl. **40**(16), 6374–6384 (2013)
27. Mirjalili, S., Mirjalili, S.M., Lewis, A.: Grey wolf optimizer. Adv. Eng. Softw. **69**, 46–61 (2014)
28. Mirjalili, S.: Moth-flame optimization algorithm: a novel nature-inspired heuristic paradigm. Knowl. Based Syst. **89**, 228–249 (2015)
29. Mirjalili, S., Lewis, A.: The whale optimization algorithm. Adv. Eng. Softw. **95**, 51–67 (2016)
30. Fausto, F., Cuevas, E., Valdivia, A., González, A.: A global optimization algorithm inspired in the behavior of selfish herds. Biosystems **160**, 39–55 (2017)
31. Fu, L., Medico, E.: Flame, a novel fuzzy clustering method for the analysis of DNA microarray data. BMC Bioinform. **8**(1), 1–15 (2007)
32. Jain, A.K., Law, M.H.: Data clustering: a user's dilemma. In: Pattern Recognition and Machine Intelligence: First International Conference, PReMI 2005, Kolkata, India, December 20–22, 2005. Proceedings 1, pp. 1–10. Springer (2005)
33. Chang, H., Yeung, D.-Y.: Robust path-based spectral clustering. Pattern Recognit. **41**(1), 191–203 (2008)
34. Gionis, A., Mannila, H., Tsaparas, P.: Clustering aggregation. ACM Trans. Knowl. Discov. Data (Tkdd) **1**(1), 4–es (2007)
35. Zahn, C.T.: Graph-theoretical methods for detecting and describing gestalt clusters. IEEE Trans. Comput. **100**(1), 68–86 (1971)
36. Veenman, C.J., Reinders, M.J.T., Backer, E.: A maximum variance cluster algorithm. IEEE Trans. Pattern Anal. Mach. Intell. **24**(9), 1273–1280 (2002)
37. Fränti, P., Virmajoki, O.: Iterative shrinking method for clustering problems. Pattern Recognit. **39**(5), 761–775 (2006)

# Innovative Machine Learning Techniques for Pedestrian Detection in Autonomous Vehicles

**G. Shanmugasundar**

**Abstract** Due to the numerous societal advantages of autonomous vehicles, which include improved safety for passengers and comfort, decreased fuel expenditure, reduced travel congestion and accident, cost savings, and increased dependability, their acceptance has steadily increased in recent years. Nonetheless, there exist several functionality issues in autonomous vehicles that necessitate rectification prior to their widespread adoption on public roads. One of the most formidable challenges in developing reliable autonomous vehicles is pedestrian detection, primarily due to two factors: (i) Deformation and occlusion and (ii) the level of visual quality and spectral diversity. However, recent advancements in machine learning technology have demonstrated potential in ameliorating the ability of autonomous vehicles to accurately recognize pedestrians. In this research, a novel hybrid approach —K-Nearest Neighbour (KNN) integrated with Gaussian naive Bayes (GNB) is proposed. Autonomous cars must deal with a variety of environmental factors, such as changing weather, illumination, and pedestrian appearances. By using both the statistical modelling of GNB and the flexibility of K-NN, the hybrid technique is able to adapt to these changes. For autonomous driving to be safe, this adaptability is essential. The Waymo Open Dataset, the largest of its kind, served as the benchmark for the experimental investigation. The highly precise data that 3D LIDAR technology offers gets utilized to identify people in any lighting situation. The proposed method displays superior accuracy in comparison to conventional techniques, thereby underscoring its efficacy.

**Keywords** Pedestrian detection · Autonomous vehicles · Waymo open dataset · KNN · Gaussian Naive Bayes

---

G. Shanmugasundar (✉)
Department of Mechanical Engineering, Sri Sai Ram Institute of Technology, Chennai 600044, India
e-mail: shanmugasundar.mech@sairamit.edu.in



D. Oliva et al. (eds.), *Advances in Optimization Algorithms for Multidisciplinary Engineering Applications: From Classical Methods to AI-Enhanced Solutions*, Studies in Computational Intelligence 806, https://doi.org/10.1007/978-3-031-78440-8_10

# 1 Introduction

Autonomous vehicles have the potential to boost the safety and effectiveness of the transportation system, enhance the quality and productivity of time spent in automobiles, and turn transportation into a service that is accessible to all people at all times. To achieve this, improvements must be made in a variety of vehicle autonomy-related areas, including vehicle design, control, perception, planning, coordination, and human contact. Using sensors like cameras, lidar, and radar, an autonomous vehicle's computer system's ability to identify and track nearby pedestrians is known as pedestrian detection. Pedestrian detection is a crucial technology for autonomous cars to function safely in urban environments where pedestrians are often present. Computer vision (CV) and machine learning (ML) are only two of the methods used by pedestrian detection systems to identify and track people. Through the use of CV techniques, pedestrians can be recognized based on their size, shape and movement patterns. Using supervised or unsupervised learning approaches, ML techniques entail training algorithms to distinguish pedestrians based on a huge collection of pictures [1]. In autonomous vehicles (AVs) and connected and automated vehicles (CAVs), vehicle control is one of the most important technical issues since it is essential to vehicle safety, passenger comfort, transit efficiency, and energy conservation. This study emphasizes the transition from microscopic vehicle state estimation and trajectory tracking control in AVs, where each vehicle individually manages its path with high precision, to macroscopic collaborative control in CAVs. In this collaborative framework, multiple vehicles communicate and coordinate their movements to optimize overall traffic flow and safety, leveraging real-time data exchange and advanced algorithms for collective decision-making.

To offer more details about the surroundings and increase detection precision, pedestrian detection systems may additionally use other sensors, such as lidar and radar. Lidar, for instance, may be used to gauge an object's depth, whereas radar can gauge a moving object's velocity and direction. To offer more details about the surroundings and increase detection precision, pedestrian detection systems may additionally use other sensors, such as lidar and radar. Both LiDAR and radar are sensor technologies that are often used in conjunction with cameras in autonomous cars and other applications to offer a thorough sense of the surrounding environment. Combining diverse sorts of data from each sensor technology may improve the vehicle's ability to understand its environment and make decisions [2].

Although autonomous vehicles have many potential advantages such as safety, traffic flow and congestion, impacts of travel behaviour, effects on safety; there are several issues that need to be overcome, including legislative and legal issues, technological constraints, and ethical choices. Continuous research and development will be required to solve these issues and bring autonomous cars to market as the technology develops further. A variety of navigational systems, including GPS, mapping, and vehicle-to-vehicle communication, are used by autonomous cars. With the fast advancement of autonomous driving, the research focus has shifted more and more to the creation of emergency anti-collision systems, which have a significant influence on road safety. While several anti-collision techniques have been developed recently, the majority of them simply take steering or braking into account. The complexity and dynamic nature of the driving environment make it difficult to create robust collision avoidance algorithms for emergency situations [3, 4].

The ability of an autonomous vehicle to identify and react correctly to nearby pedestrians makes pedestrian detection essential work. CV and ML techniques are only two of the methods that may be utilized to identify pedestrians in autonomous vehicles. ML techniques are used to train a model on a large dataset of visual information to detect and classify pedestrians. Two methods may be used for this purpose: supervised learning, in which the model is taught by seeing and interacting with labelled examples and unsupervised learning, in which the model learns to detect patterns in the data without any labelling. It is possible to enhance pedestrian identification by combining several approaches, such as using both CV and ML technologies. In addition, sensors like lidar and radar may be utilized in combination with cameras to offer more details about the surroundings and enhance pedestrian recognition. Pedestrian detection can also be improved by combining multiple confronts, such as using both ML and CV techniques [5].

One of the most important topics in autonomous vehicles (AVs) is obstacle detection. The Advanced Driver Assistant System (ADAS) and collision avoidance systems make an effort to have a thorough understanding of the surroundings. Due to safety concerns that have developed in recent years, pedestrians are one of the most significant obstacles. There are several difficulties in detecting pedestrians, such as variations in position, clothing, occlusion, movement, lighting, size, and weather. Additionally, coverage and surround view encourage researchers to include additional cameras to cover blind areas around the vehicle [6]. The autonomous vehicle, a significant intelligent transportation systems participant, can operate on its own. Driving safely without endangering other drivers' property, pedestrians, or passengers is the autonomous vehicle's primary goal. When a vehicle crashes, pedestrians are far more susceptible than the occupants of the vehicles. Therefore, it is crucial to take pedestrian safety into account. The position of the pedestrians must be communicated to the vehicle as soon as possible so that it can determine whether or not to stop to properly protect them. They focused on the field of pedestrian detection which involves locating each pedestrian in the picture and indicating their whereabouts using bounding boxes that are near to the pedestrians [7].

Cui et al. [8] provided a concise overview of the deep learning (DL) that is used to analyse data from point clouds and images. Analyses of camera-LiDAR fusion systems are provided, organized by the various fusion levels they support, such as object identification, tracking, online cross-sensor calibration and semantic segmentation. An emerging research subject is LiDAR fusion. However, methods for fusing camera and LiDAR data using DL are still being given much consideration. Fayyad et al. [9] offered a thorough analysis of the cutting-edge techniques used to enhance the effectiveness of AV systems nearby or in regional vehicle scenarios. Their major focus was on DL sensor fusion techniques for perception, localization and projection. They argue that the present state of ground transportation is anticipated to be improved, transformed and revolutionized by autonomous vehicles.

It is predicted that one-day intelligent automobiles will replace conventional ones, capable of autonomous decision-making and driving activities. Autonomous vehicles are outfitted having sensors for seeing and understanding the immediate environment and the environment in the distance utilizing more advanced communication technologies, such as 5G, to accomplish this goal. Chen and Huang [10] discussed developments in DL technology for autonomous vehicles. The DL learning paradigm, convolutional and recurrent neural networks and AI-based self-driving architectures are introduced initially. The algorithms for driving scene recognition, route planning, behaviour arbitration and motion control are based on these approaches. Ranganayaki et al. [11] explained the choice to avoid a person must be made as swiftly and with as little delay as feasible in autonomous driving, because pedestrian detection is a safety–critical task. SVM, Naive Bayes, and KNN are a few examples of machine learning classifiers that are used to predict pedestrians on the road. Feng et al. [12] offered a survey and comparative analysis of current probabilistic object identification techniques for autonomous driving applications to address this issue. The crowd first provides a general overview of useful DL algorithms for uncertainty estimates before doing a thorough analysis of the current approaches and performance measures for probabilistic object identification. It is crucial for safe autonomous driving to capture ambiguity in object recognition. Despite substantial developments in the field of autonomous vehicle pedestrian detection, there are several challenges that remain unresolved in existing studies. Many of the current ML methods, though successful to a degree, still struggle with precision and efficiency in complex, real-world scenarios. Certain techniques falter in the face of unforeseen situations, non-linear data, or unstructured environments that exhibit high variability. In the extant literature, it is noticeable that there is an underrepresentation of studies focusing on the efficient combination of different Machine Learning techniques to achieve higher accuracy and precision. Moreover, the majority of the existing works do not adequately address the variations in lighting conditions and obstacles such as buildings and moving cars that can significantly impede the effectiveness of pedestrian detection systems. Furthermore, identifying smaller or non-standard subjects such as children or animals presents an additional layer of complexity that has not been satisfactorily addressed in current research.

This paper seeks to address these gaps by presenting a novel Hybrid K-Nearest Neighbour (KNN) Integrated Gaussian Naive Bayes (HKIGNBA) methodology for pedestrian detection in autonomous vehicles. The present method specifically addresses the aforementioned challenges and demonstrates a superior ability to handle non-linear data, thereby increasing classification accuracy and recall in complex environments. Further notable strides are made in enhancing precision and F1-score, as substantiated by the comparative analysis with traditional methods such as SSD-Mobile Net, R-CNN, and VGG-16. Through the implementation of this methodology, substantial contribution is made towards improving the safety of autonomous vehicles. This study thus serves as a critical step forward in the field of machine learning for pedestrian detection, promising a new paradigm of safety and reliability in autonomous vehicles. This paper explores and validates the practical utility of the HKIGNBA approach which will inspire and enable further research and development in this area.

The introduction of the Advanced Driving Assistance technology (ADS) technology aids in the reduction of unexpected crashes. One of its well-known characteristics is the ability to recognize pedestrians. Engineers later added this capability to autonomous vehicles. But even with this capability, there are still a number of problems with pedestrian recognition that need to be fixed. Numerous researchers have attempted to resolve these problems via various inventions. Poor pedestrian tracking and identification, poor obstacle recognition in a variety of lighting settings, including clear visibility problems at night, shadowing conditions, inadequate resolution, and the occurrence of microscopic size are some of these challenging problems.

The rest of the paper is structured as follows: A more detailed description of the methodologies is given in Sect. 2 with focused discussion of the proposed method in Sect. 3. The results and the discussion are covered in Sect. 4, and the conclusion is provided with future scope in Sect. 5.

## 2 Methodology

### *2.1 Algorithm Details*

Hybrid KNN integrated Gaussian Naive Bayes approach (HKIGNBA) is an ML algorithm that integrates the K-Nearest Neighbours and Gaussian Naive Bayes algorithms for pedestrian detection in autonomous vehicles. A hybrid KNN-integrated GNB method combines the advantages of two separate algorithms to increase the accuracy of pedestrian identification in autonomous vehicles. The GNB algorithm's presumptions are accurate and it handles characteristics that are complicated or strongly linked and be useful. For each unique dataset and occupation, the implementation's particulars and its associated parameters may need to be adjusted and confirmed. For the algorithm to be secure and reliable in autonomous vehicles, it is also essential to make sure that it has been validated and assessed in real-world scenarios. Hence, for pedestrian detection in autonomous vehicles, a HKIGNBA modelling is suggested. Recent improvements in ML technology have shown encouraging results in enhancing the capacity of autonomous vehicles to recognize pedestrians.

### *2.2 Data Set*

The newly published Waymo Open Dataset [13], this has over 1,000 driving actions recorded in diverse urban locations, is for the investigation. In terms of various local

**Table 1** Instances of different items in both the training and evaluation sets

| Images | Objects | | | Total |
|---|---|---|---|---|
| | 74,420 | 24,770 | | 99,190 |
| | Vehicle | Pedestrian | Cyclist | |
| Training | 589,583 | 77,569 | 3842 | 670,994 |
| Validation | 256,076 | 21,678 | 1332 | 279,086 |
| Total objects | 845,659 | 99,247 | 5174 | 950,080 |

and meteorological conditions (Clear sky, rain, fog, snow, and other weather conditions) it is the most varied dataset in this subject to date. Data is gathered during the day, at night, and in between the hours of day and night. The driverless cars made by Waymo are fitted with a variety of LiDAR sensors, including mechanical and solid-state spinning LiDAR units. These sensors produce laser beams and track the time it takes for the laser pulses to return, building a 3D point cloud that is very comprehensive. Additionally, Waymo recently ran an online contest for dataset detection. However, the competition only assesses the accuracy, disregarding the models' efficacy. 3D LiDAR and 2D camera information have labels that were separately produced in the dataset. A range of cameras, including high-resolution cameras with RGB pixels and grayscale cameras, are included in Waymo's cars. The visible and near-infrared spectrums are both used by these cameras to record pictures. A Waymo car has three frontal cameras with a resolution of 1920 * 1280, as well as two lateral cameras with a resolution of 1920 * 886.

In this dataset, there are three main sorts of things that must be detected is automobiles, people and bicycles. Table 1 lists the number of photos in each collection as well as the distribution of labels among the three groups. Waymo explicitly provides the training and testing division when you download the dataset. Nearly one million objects have been personally annotated across approximately 100,000 total images in the dataset. As can be observed, there is a severe imbalance in the distribution of the items. The number of bikes and pedestrians is far less than that of automobiles. The 2D bounding box labels in this dataset are precise and only include the visible portions of the items.

## *2.3 Data Pre-processing*

Pre-processing refers to a set of techniques applied to the input data before running the pedestrian detection algorithm in autonomous vehicles. The goal of pre-processing is to enhance the input data and prepare it for efficient and accurate detection of pedestrians.

The pre-processing strategies for pedestrian detection in autonomous vehicles typically include image resizing to a standard size, image normalization to remove variations in lighting conditions and image filtering to reduce noise and smooth the image. Feature extraction is also a critical pre-processing step where relevant features such as edge detection, colour histograms and Histograms of Oriented Gradients are extracted from the input image.

Data augmentation methods like flipping, rotating and scaling are used to create additional training data to improve the robustness of the algorithm and reduce over-fitting. Image segmentation techniques can also be applied to separate the image into different regions and identify potential areas where pedestrians might be present. We use Waymo Open dataset in our proposed method is to implement advanced pedestrian detection in autonomous vehicles which includes parameters such as vehicle, pedestrian and cyclist. Object detection model training requires a dataset that have been annotated. In photos or frames, human annotators to identify things of interest (such as vehicle, pedestrian and cyclist), giving limits or segmentation masks to denote object positions. Metrics used in validation techniques include recall, accuracy, precision, and f1-score. To address safety and reliability problems and to ensure that the model is ready for real-world deployment, testing techniques for object detection in autonomous cars are essential. The number of instances of different items in both the training and evaluation sets was determined by Table 1.

## 3 Hybrid KNN Integrated Gaussian Naive Bayes Approach

The HKIGNBA is an innovative approach created to greatly improve pedestrian detection within the autonomous vehicle framework. K-Nearest Neighbour (KNN) and Gaussian Naive Bayes (GNB), two essential machine learning algorithms, are combined in HKIGNBA to take use of their complimentary characteristics. One of the major problems with autonomous cars is pedestrian identification, and the HKIGNBA provides an in-depth solution.

### *3.1 K-Nearest Neighbour (KNN)*

The KNN method is an instance-based learning technique employed in the recognition of patterns to categorize objects according to the training instances that are closest to them in the attribute space. A substance is categorized by a large number of its neighbours, who assign it to the group that comprises the bulk of its k-nearest neighbours.

In this case, $k$ is a positive value. In the KNN approach, a novel test for the feature vector's classification is based on the classes of the feature vector's k-nearest neighbors.

Euclidean distance measurements are used to construct the KNN method and find the closest neighbour. Equation (1) is used to get the Euclidean distance metrics $d(x, y)$ between two points $x$ and $y$.

$$d(x, y) = \sum_{i=1}^{N} \sqrt{x_i^2} - y_i^2 \tag{1}$$

The KNN classifier is one of several methods that aims to categorize an observation into the class with the greatest estimated probability after attempting to calculate the conditional distribution of $Y$ given $X$.

A positive number is provided $k$ and a test monitoring, $x_0$ the KNN classifier initially determines the $k$ indicates in the most similar training data to $x_0$, represented by $N_0$. It then predicts the conditional chance for class $j$ as the fraction of points in $N_0$ whose reaction values equal $j$ as suggested in Eq. (2),

$$Pr(Y = j|X = x_0) = \frac{1}{K} \sum_{i \in N_0} I(y_i = j) \tag{2}$$

where, $I(y_i = j)$ is a metric constant that corresponds to 1 if $y_i = j$ and zero if $y_i \neq j$.

KNN works well with plenty of training examples and is resilient to noisy training data. However, it is necessary to decide on the type of distance to be used as well as the value of the parameter $k$ for this algorithm. The calculation may take a while since it is necessary to calculate the separation between every query example and each instruction sample and it becomes noticeably the number slower of instances, predictors and separate components rises.

Due to its inherent flaw of getting substantially slower as the quantity of input increases, KNN is an undesirable alternative when speedy predictions are needed. There are also speedier algorithms that provide classification and regression results that are more accurate. KNN may still help address issues when finding comparable items is necessary for an answer, assuming there are enough computational resources to process the data quickly for producing forecasts.

The KNN algorithm is iterated with various $k$ values to choose the $k$ that is best for a dataset. The $k$ that decreases mistakes while preserving the technique's capacity to make accurate predictions when used with data it has never seen before is selected. Other methods of calculating distance exist and depending on the issue at hand, one method may be better. However, a well-liked and recognized option is the straight-line distance, also known as the Euclidean distance.

In contrast, as $k$ rises, the predictions develop more reliable because the majority casting averaging and are hence more likely. As time passes, more and more errors become apparent. One realizes that the correct value of $k$ has now been surpassed. When determining the mode in a classification problem, for instance, if a majority vote among labels is required, the value of $k$ is often an odd integer to serve as a tiebreaker. The KNN approach may be used to address classification, regression and search-related issues.

### 3.2 Gaussian Naive Bayes

The GNB classification algorithm is among the simplest ones. It follows a Gaussian distribution; it involves choosing a class designation that maximizes the iterative probability of all of the samples.

The GNB decision rule is stated in terms of the discriminant value for every class k for every searchlight s. According to the Bayes rule, the discriminant value is calculated as the total of the average distances from every class' centroid across all searchlight voxels, scaled by variation and using the learning groups of a-priori probability logarithm.

The predicted class, for instance, $i$ within the test set is allocated by choosing the category of the discriminatory function's label $\delta_i^k$ having the highest value across all discriminant functions for all classes, which denotes the maximum posterior probability.

$$\widehat{c_i} \arg\max_k = \{a, b\}\{\delta_i^k\} \tag{3}$$

$$\delta_i^k = -\sum_{j=1}^{v}\left(\frac{\left(x_{ij}^{Te} - \widehat{\mu_J^k}\right)}{\widehat{2\sigma_j^2}}\right) + \log(P_k) \tag{4}$$

where,

| | |
|---|---|
| $\delta_i^k$ | Data point discriminate function |
| $x_{ij}^{Te}$ | Value of the jth feature |
| $\mu_J^k$ | Estimated mean of the jth feature |
| $\widehat{2\sigma_j^2}$ | Estimated variance of the jth feature |
| $\log(P_k)$ | Natural logarithm of the prior probability of class k. |

This formula reflects a binary categorization issue using classes. Assigning data points to the class with the higher score in this binary classification involves calculating the discriminate score (0 or 1) for each class based on the formula. $\{a, b\}$ for several classes' classification issues, it is readily generalized. The average deviation $\mu_j^k$; $\widehat{\sigma_j}$ every voxel $j$ are calculated throughout the practice set. The initial equation is stated in one matrix product when GNB is applied to numerous searchlights, utilizing the sparse binary matrix $S$ to choose and total the contributions from each searchlight's voxels in a single operation.

$$D_{m,s}^k = F_{m,v}^k S_{v,s} + \log(P_k) \tag{5}$$

The elements of the matrix $F_{m,v}^k$ are the voxel contributing to discrimination for each instance $i$ within the parameters of the test set $f_j^k = \frac{\left(x_{ij}^{Te} - \widehat{\mu_j^k}\right)}{\widehat{2\sigma_j^2}}$, which match the contributions made by Eq. (4) voxel-wise. The matrix $D^k_{m,s}$ includes the discriminatory operates for the class $k$, additionally, for every sample at each searchlight. The GNB technique yields outcomes identical to those obtained by progressively executing the MATLAB categorize program and is numerically stable. The expectation of a diagonal covariance matrix throughout factors and similar across-class standard deviations makes the GNB classifier equivalent to linear discriminant analysis. Assuming once again a diagonal covariance matrix, the assumption of different standard deviations across classes leads to a quadratic discriminant analysis. QDA calculations are simple and produce a function of discrimination that is similar to the linear case.

A variation of the Naive Bayes method that is often used in ML for problems with categorization is called Gaussian Naive Bayes. It is a probabilistic model that relies on expectations of how the data will be distributed. The simple and quick Gaussian Naive Bayes technique can handle high-dimensional datasets with many characteristics. However, it might not work well if the data distribution significantly deviates from a normal distribution or if the assumption of feature independence is broken.

## 4 Result and Discussion

Pedestrian detection is an essential element of autonomous vehicles as it assists the vehicle to detect and avoid pedestrians in its surroundings, lowering the likelihood of accidents. These techniques are combined with autonomous vehicles to improve the accuracy of pedestrian detection. A strong and dependable pedestrian detection system that can operate in a variety of weather and lighting conditions and identify pedestrians even if they are partly obscured by surrounding objects is also important. As a result, the HKIGNBA is implemented for efficient pedestrian detection in autonomous vehicles. The performance indicators used for evaluation are accuracy, precision, recall and F1-score. The conventional techniques used for SSD-Mobile Net [13], R-CNN [13] and VGG-16 [14].

### *4.1 Accuracy*

This accuracy metric quantifies the number of correctly classified instances, establishing a correlation between the predictions and the ground-truth labels [15]. It is observed from the graph that the novel method exhibits superior precision compared to its predecessor. The accuracy of pedestrian detection systems in autonomous vehicles can be influenced by several factors, including the quality of sensors deployed, the algorithms implemented for object detection, and the operational conditions of the system. A discrepancy between the predicted outcome and the true value may arise from inadequate precision. The percentage of accurate results provides insight into the overall data balance. The accuracy is evaluated using Eq. (6).

$$\text{Accuracy} = \frac{\text{TP} + \text{TN}}{\text{TP} + \text{TN} + \text{FP} + \text{FN}} \tag{6}$$

where,

TP True positive
TN True negative
FP False positive
FN False negative.

Figure 1 illustrates the accuracy of the proposed method with respect to SSD (Single Shot MultiBox) Detector-Mobile Net [13], R-CNN [13] and VGG-16 [14]. The proposed HKIGNBA method outperforms the other techniques in terms of accuracy. With an accuracy of 95.10%, it improves upon SSD-Mobile Net, R-CNN, and VGG-16 by 13, 5, and 53% respectively. This demonstrates that HKIGNBA correctly identifies a greater percentage of instances, indicating a strong correspondence between predictions and ground-truth labels.

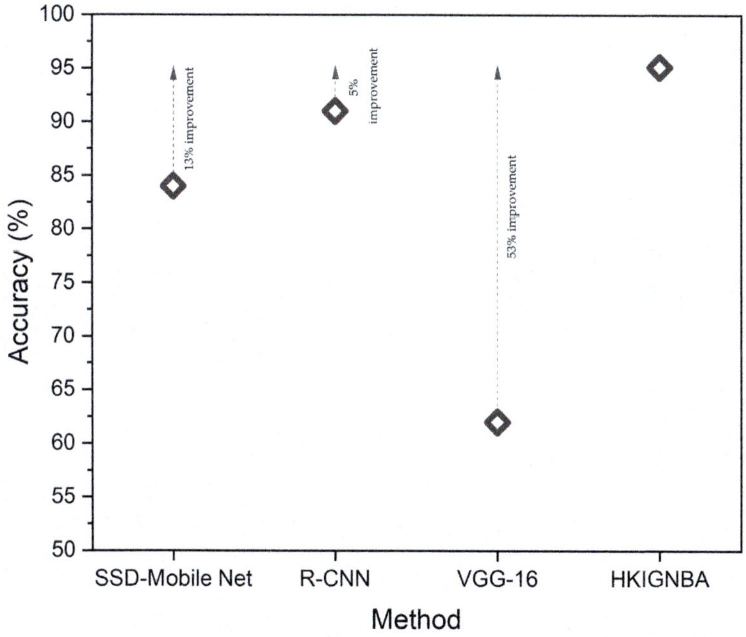


**Fig. 1** Accuracy of proposed HKIGNBA with respect to SSD-Mobile Net, R-CNN and VGG-16

## 4.2 Precision

A crucial measure of accuracy is precision, which is defined as the fraction of accurately classified instances among all instances that are predicted as positive [16]. Precision is computed using Eq. (7).

$$\text{Precision} = \frac{\text{TP}}{\text{TP} + \text{FP}} \tag{7}$$

The performance of pedestrian detection systems in autonomous vehicles can be affected by various factors, such as the quality of the employed sensors, the complexity of the surrounding environment, and the specific object detection algorithm in use. As these systems typically generate more false positives than false negatives, their precision tends to be lower than their recall.

Figure 2 illustrates the precision of the proposed method with respect to SSD-Mobile Net [13], R-CNN [13] and VGG-16 [14]. The proposed HKIGNBA method registers a precision score of 98.22%, demonstrating superior precision over SSD-Mobile Net and R-CNN by 6 and 24% over VGG-16. This high precision score signifies that when HKIGNBA predicts an instance as positive, it is highly likely to be a true positive, thus reducing false positives.

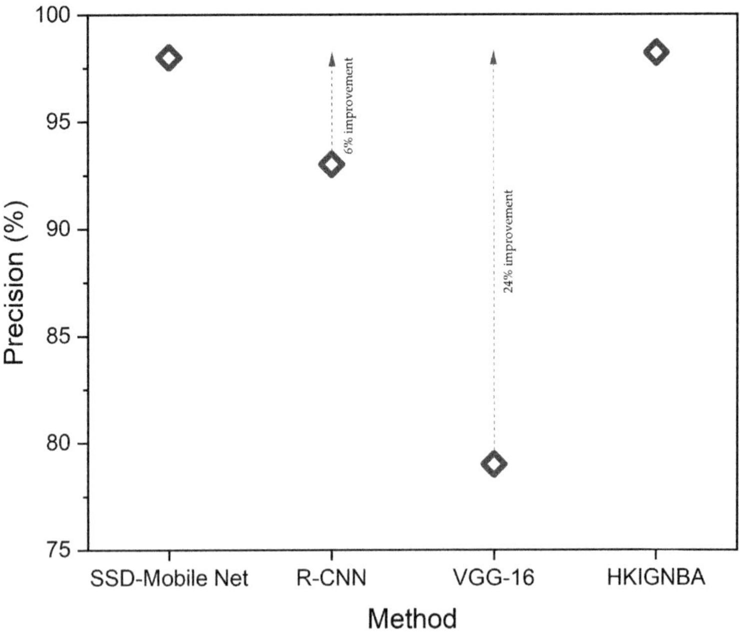


**Fig. 2** Precision of proposed HKIGNBA with respect to SSD-Mobile Net, R-CNN and VGG-16

## 4.3 Recall

Recall represents a model's ability to identify all pertinent instances within a dataset [17]. It is statistically defined as the ratio of True Positives (TPs) to the sum of True Positives and False Negatives (FNs). Recall is calculated using Eq. (8).

$$\text{Recall} = \frac{\text{TP}}{\text{TP} + \text{FN}} \tag{8}$$

The recall of pedestrian detection systems in autonomous vehicles may be influenced by multiple factors, including the quality of the deployed sensors, the environmental complexity, and the specific object detection algorithm. In the context of pedestrian detection systems, recall is often prioritized over accuracy as false negatives can have more serious implications than false positives.

Figure 3 illustrates the recall of the proposed method with respect to SSD-Mobile Net [13], R-CNN [13] and VGG-16 [14]. HKIGNBA achieves the highest recall value of 98.72%, marking improvements of 18, 2, and 25% over SSD-Mobile Net, R-CNN, and VGG-16, respectively. The high recall value implies that the HKIGNBA method is proficient in detecting all pertinent instances within the data set, making it less prone to false negatives.

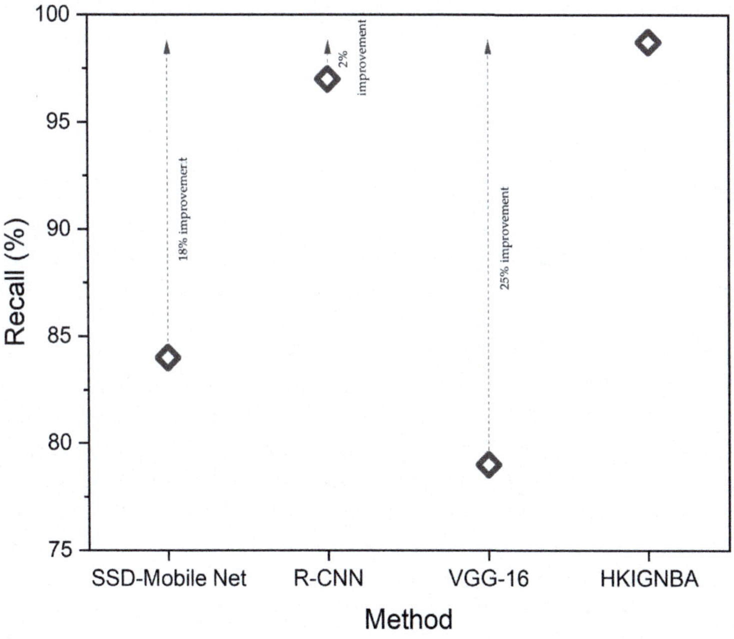


**Fig. 3** Recall of proposed HKIGNBA with respect to SSD-Mobile Net, R-CNN and VGG-16

### 4.4 F1-Score

The F1-score denotes a harmonic mean of precision and recall, serving as a balanced measure of these two metrics. By considering both true and false outcomes, it facilitates a comprehensive evaluation of a model's performance. This metric, also known as the F-measure, is often employed in information retrieval scenarios. The F1-score can be adjusted to prioritize precision over recall or vice versa, as needed. F1-score is calculated using Eq. (9).

$$F1 - \text{score} = \frac{2 \times \text{precision} \times \text{recall}}{\text{precision} + \text{recall}} \tag{9}$$

The F1-score of pedestrian detection systems in autonomous vehicles may be influenced by various factors such as the quality of the sensors used, the environmental complexity, and the specific object detection algorithm. It should be noted that a system demonstrating both high precision and recall will invariably yield a higher F1-score.

Figure 4 illustrates theF1-score of the proposed method with respect to SSD-Mobile Net [13], R-CNN [13] and VGG-16 [14]. HKIGNBA again leads with a score of 98.47%. This indicates an improvement of 8, 4, and 20% over SSD-Mobile Net, R-CNN, and VGG-16, respectively. The high F1-score of HKIGNBA signifies that it maintains an effective balance between precision and recall. Therefore, even in scenarios with uneven class distribution, HKIGNBA can be relied upon to provide robust performance.

### 4.5 Discussion

Small or poorly contrasted objects may be difficult for SSD-Mobile Net [13] to recognize properly in an image. As pedestrians, bicycles, and other vehicles can appear little or far away in the camera's range of vision, this is essential to autonomous cars in particular. The model's ability to recognize and track objects accurately was hampered by shadows, glare, or quickly changing lighting conditions. For real-time applications like autonomous driving, R-CNNs' [13] high computing requirements might be something to consider. R-CNNs have comparatively slow inference times because of their computing requirements. Real-time decision-making is essential in autonomous cars, and slower inference may pose a safety risk. VGG [14] architecture has been used extensively in picture classification tasks and computer vision applications. Input pictures for VGG networks should often be of a particular size. Cameras are capable of taking pictures with different resolutions and aspect ratios in actual driving situations. With VGG, managing varying input sizes is potentially difficult. HKIGNBA excels at being straightforward and flexible, making it reliable for making decisions in real-time based on adjacent objects. It makes use of spatial

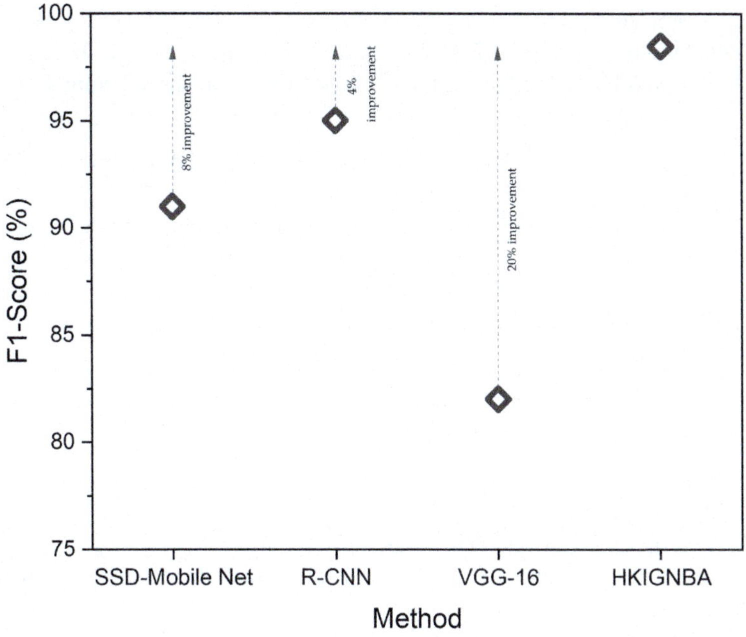


**Fig. 4** F1-score of proposed HKIGNBA with respect to SSD-Mobile Net, R-CNN and VGG-16

affection to classify objects. It also offers probabilistic reasoning that helps with managing uncertainty and decision confidence. It makes autonomous driving safer and more dependable by managing sensor data ambiguity and intricate environment modelling.

## 5 Conclusion

Modern sensor technology and machine learning techniques are used in advanced pedestrian detection in autonomous vehicles to improve pedestrian safety and identification. By taking into account a number of variables, including pedestrian behaviour, occlusion, and complicated metropolitan surroundings, accuracy is increased. The main achievement of this research is the development and validation of a novel HKIGNBA methodology. HKIGNBA improves the identification of items in real time based on proximity to one another, adds deterministic insights for dealing with complicated situations and ambiguity, resulting in more reliable and secure pedestrian detection in dynamic settings. It exhibits profound capability in increasing the classification accuracy and handling non-linear data, thereby, making it highly suitable for practical applications. The results show the superior performance of HKIGNBA across various metrics, namely accuracy (95.10%), recall

(98.22%), precision (98.72%), and F1-score (98.47%). The HKIGNBA outperform established models such as SSD-Mobile Net, R-CNN, and VGG-16. This reinforces the robustness of HKIGNBA in correctly classifying instances and minimizing false predictions. The key results are:

- In comparison to benchmark techniques, the proposed method significantly increased the accuracy of pedestrian detection.
- Our system performed well under a variety of environmental situations, including bad weather and low light situations.
- The system demonstrated real-time processing capabilities, which are essential for applications involving autonomous vehicles and guarantee rapid pedestrian detection and reaction.

However, there are considerable challenges that need to be addressed to enhance the precision and reliability of these technologies:

- Unexpected movements and obstructions caused by factors like other vehicles or buildings.
- The diversity in pedestrian behaviour.
- Shifting lighting conditions which can compromise the precision of detection technologies.
- Difficulty in accurately detecting smaller entities, like small children or animals.

While these hurdles exist, continued research and development aimed at resolving these limitations holds immense promise. The creation of ML algorithms that can better recognize and track pedestrians, even under challenging conditions such as low brightness or occlusion, remains an area of focus. In light of these findings and the ongoing research efforts, the future of pedestrian detection in autonomous vehicles appears promising. As autonomous vehicles become increasingly prevalent, it is anticipated that significant advancements in the precision and reliability of pedestrian detection algorithms, further bolstering the safety and efficiency of this transformative technology.

The combination of these two techniques may be improved in the future by enhancing precision and real-time processing capabilities. For effective pedestrian detection systems, it is also essential to investigate how well they can adapt to diverse environmental situations, such as adverse weather or complicated metropolitan environments. The performance of this hybrid method is possibly further improved by using deep learning techniques for feature extraction and improving, eventually leading to safer autonomous driving.

## References

1. Chen, S., Dong, J., Ha, P.J., Li, Y., Labi, S.: Graph neural network and reinforcement learning for multi-agent cooperative control of connected autonomous vehicles. Comput.-Aided Civil Infrastruct. Eng. **36**, 838–857 (2021). https://doi.org/10.1111/mice.12702

2. Daniel, J.A., Vignesh, C.C., Muthu, B.A., Senthil Kumar, R., Sivaparthipan, C.B., Marin, C.E.M.: Fully convolutional neural networks for LIDAR-camera fusion for pedestrian detection in autonomous vehicles. Multimed. Tools Appl. (2023). https://doi.org/10.1007/s11042-023-14417-x
3. Chen, L., Ma, N., Wang, P., Li, J., Wang, P., Pang, G., Shi, X.: Survey of pedestrian action recognition techniques for autonomous driving. Tsinghua Sci. Technol. **25**, 458–470 (2020). https://doi.org/10.26599/TST.2019.9010018
4. Galvao, L.G., Abbod, M., Kalganova, T., Palade, V., Huda, M.N.: Pedestrian and vehicle detection in autonomous vehicle perception systems: a review. Sensors (Basel) **21**, 7267 (2021). https://doi.org/10.3390/s21217267
5. Liu, W., Liao, S., Ren, W., Hu, W., Yu, Y.: High-Level semantic feature detection: A new perspective for pedestrian detection. In: 2019 IEEE/CVF Conference on Computer Vision and Pattern Recognition (CVPR). Long Beach, CA, USA (2019). https://doi.org/10.1109/CVPR.2019.00533
6. Boukerche, A., Sha, M.: Design guidelines on deep learning-based pedestrian detection methods for supporting autonomous vehicles. ACM Comput. Surv. **54**, 1–36 (2022). https://doi.org/10.1145/3460770
7. Sha, M., Boukerche, A.: A novel visibility semantic feature-aided pedestrian detection scheme for autonomous vehicles. Comput. Commun. **179**, 50–61 (2021). https://doi.org/10.1016/j.comcom.2021.06.009
8. Cui, Y., Chen, R., Chu, W., Chen, L., Tian, D., Li, Y., Cao, D.: Deep learning for image and point cloud fusion in autonomous driving: a review. IEEE Trans. Intell. Transp. Syst. **23**, 722–739 (2022). https://doi.org/10.1109/TITS.2020.3023541
9. Fayyad, J., Jaradat, M.A., Gruyer, D., Najjaran, H.: Deep learning sensor fusion for autonomous vehicle perception and localization: a review. Sensors (Basel) **20**, 4220 (2020). https://doi.org/10.3390/s20154220
10. Chen, Z., Huang, X.: Pedestrian detection for autonomous vehicle using multi-spectral cameras. IEEE Trans. Intell. Veh. **4**, 211–219 (2019). https://doi.org/10.1109/TIV.2019.2904389
11. Ranganayaki, V., Ramesh, P.S., Rammohan, K., Lincy, R.B., Deepak, A.: Machine learning approaches on pedestrian detection in an autonomous vehicle. In: 2023 Fifth International Conference on Electrical, Computer and Communication Technologies (ICECCT), pp. 1–8. IEEE (2023). https://doi.org/10.1109/ICECCT56650.2023.10179836
12. Feng, D., Harakeh, A., Waslander, S.L., Dietmayer, K.: A review and comparative study on probabilistic object detection in autonomous driving. IEEE Trans. Intell. Transp. Syst. **23**, 9961–9980 (2022). https://doi.org/10.1109/TITS.2021.3096854
13. Iftikhar, S., Zhang, Z., Asim, M., Muthanna, A., Koucheryavy, A., Abd El-Latif, A.A.: Deep learning-based pedestrian detection in autonomous vehicles: Substantial issues and challenges. Electronics (Basel) **11**, 3551 (2022). https://doi.org/10.3390/electronics11213551
14. Tarchoun, B., Jegham, I., Khalifa, A.B., Alouani, I., Mahjoub, M.A.: Deep CNN-based pedestrian detection for intelligent infrastructure. In: 2020 5th International Conference on Advanced Technologies for Signal and Image Processing (ATSIP). Sousse (2020). https://doi.org/10.1109/ATSIP49331.2020.9231712
15. Ganesh, N., Shankar, R., Čep, R., Chakraborty, S., Kalita, K.: Efficient feature selection using weighted superposition attraction optimization algorithm. Appl. Sci. **13**, 3223 (2023). https://doi.org/10.3390/app13053223
16. Shaik, K., Ramesh, J.V.N., Mahdal, M., Rahman, M.Z.U., Khasim, S., Kalita, K.: Big data analytics framework using squirrel search optimized gradient boosted decision tree for heart disease diagnosis. Appl. Sci. **13**, 5236 (2023). https://doi.org/10.3390/app13095236
17. Priyadarshini, J., Premalatha, M., Čep, R., Jayasudha, M., Kalita, K.: Analyzing physics-inspired metaheuristic algorithms in feature selection with K-nearest-neighbour. Appl. Sci. **13**, 906 (2023). https://doi.org/10.3390/app13020906

# Machine Learning-Enhanced Dynamic Routing for Internet of Things Energy Efficiency

**G. Shanmugasundar, Janjhyam Venkata Naga Ramesh, S. Suthagar, K. Karthik, and V. Ramesh**

**Abstract** Since Internet of Things (IoT) gadgets are restricted in energy resources, energy consumption is a serious problem. In IoT networks, conventional routing protocols cannot take energy consumption into account, which results in early network downtime and a shorter lifespan of the network. Because of this, the article suggests a dynamic energy-efficient routing protocol that can react to the varying energy levels of network nodes by using machine learning (ML) to forecast the amount of energy consumed and modify routing patterns to reduce energy consumption. Using cluster- based technologies to shorten the distance between sensor nodes is one common way to lower the energy consumption of wireless sensor networks (WSNs). The Variance Ratio (VR) technique is used initially to determine the ideal number of clusters. To accomplish the major objective of this work, this improvement will occur through two stages that pass using better ways to enhance the chain-greedy hierarchical routing protocol (CGHRP). The fuzzy c-means (FCM) technique was then used to increase the network lifetime. As a consequence, data is carried via several shorter parallel lines instead of a single lengthy route. The Python tool

G. Shanmugasundar
Department of Mechanical Engineering, Sri Sai Ram Institute of Technology, Chennai 600044, India
e-mail: shanmugasundar.mech@sairamit.edu.in

J. V. N. Ramesh
Department of Computer Science and Engineering, Koneru Lakshmaiah Education Foundation, Vaddeswaram, Andhra Pradesh 522502, India

Department of CSE, Graphic Era Deemed To Be University, Dehradun, Uttarakhand 248002, India

Department of CSE, Graphic Era Hill University, Dehradun 248002, India

S. Suthagar
Department of Aeronautical Engineering, Vel Tech Rangarajan Dr. Sagunthala R&D Institute of Science and Technology, Chennai, Tamilnadu 600062, India
e-mail: suthagars@veltech.edu.in

K. Karthik (✉) · V. Ramesh
Department of Mechanical Engineering, Vel Tech Rangarajan Dr. Sagunthala R&D Institute of Science and Technology, Chennai, Tamilnadu 600062, India
e-mail: karthikmeed@gmail.com

 
D. Oliva et al. (eds.), *Advances in Optimization Algorithms for Multidisciplinary Engineering Applications: From Classical Methods to AI-Enhanced Solutions*, Studies in Computational Intelligence 806, https://doi.org/10.1007/978-3-031-78440-8_11

is used for simulating the protocol, and the findings, especially regarding energy savings, are clear and useful. The findings from the simulation demonstrate that in terms of network lifespan and energy consumption, the suggested protocol (CGHRP-FCM) performs better than the state-of-the-art energy-efficient routing protocols. The protocol is suited for IoT networks wherein nodes are portable and have restricted resources for energy because of its dynamic nature, which enables it to adapt to modifications in the network architecture and energy levels of nodes.

**Keywords** Internet of Things (IoT) · Energy consumption · Clustering · Variance Ratio (VR) · Chain-Greedy Hierarchical Routing Protocol (CGHRP) · Fuzzy C-Means (FCM)

## 1 Introduction

To create a system with full network coverage and to increase the number of intelligent objects that are associated with the deployed system, the contemporary world has adopted new communication technologies. By linking users and contemporary technologies, one of the useful patterns known as the IoT can increase the amount of information created via the Internet. In addition, there are several problems with the IoT when connecting to people and equipment, particularly in terms of position and time [1]. Additionally, the Quality of Service (QoS), which includes power, latency, productivity, etc., is a goal of the IoT. To achieve the greatest QoS, effective management of bandwidth, one of the key sources in the platform for IoT, is required. Additionally, there is a strong increase in the demand for multimedia capabilities in IoT platforms [2]. It finds use in smart homes, smart grids, etc. [3]. The IoT is a network of disparate devices for communication and sharing data to provide customers with better services. There has been a rise in research and development in the IoT space across several application sectors [4]. Among the use cases for IoT networks are clever design for homes and gadgets, automated surveillance and safety, intelligent transport control, and medical evacuation plans for emergencies. Radio Frequency Identification (RFID) tags, telephones, and sensors that collect environmental data are all part of the vast and highly sophisticated IoT networks [5]. These devices, which are also known as sensor nodes, have a modest computing capacity and a short battery life [6]. WSN is a core piece of IoT technology. It is the foundation of the IoT system. WSN encourages the development and evolution of the Internet of Things by allowing inexpensive gadgets with restricted funds and providing game-changing services [7]. It makes use of many thousands of sensors connected wirelessly. With the development of sensor technology, it is now feasible to create wireless IoT-enabled sensors that are both inexpensive and compact, adding intelligence to small- to large-scale products. A typical WSN is made up of several sensor nodes that are capable of sensing, communicating, and computing [8]. The assessment of significant environmental variables (such as humidity, light, temperature, strain, and others) in smart agriculture, safe and trustworthy interaction, military purposes and tracking, health care, and medicine, various types of industries, traffic observation, and other domains are just a few examples where WSN can be used as a platform [9]. IoT is entering a new era as a result of the development of Wi-Fi technology, data systems, and mobile devices [10]. Through the acknowledgment of identifying objects, and data creation, transfer, and extraction, the latter has developed into the foundation of ubiquitous computing. With the (IoT), physical objects and people may connect to and participate in the online world through wireless connections [11]. To first enable things to learn about and interact with their surroundings, IoT initially concentrated on the network of RFID tags before being substantially expanded to a variety of gadgets and applications. These items can communicate wirelessly with one another, allowing the system to be remotely controlled through the Internet [12]. The MEC has become a key factor in the provision of low latency, high bandwidth real-time radio network resources in the ongoing development of the internet [13]. MEC completes the task by allowing cloud computing at the network's edge. MEC is a crucial instrument in the IoT concept for ensuring a good network level of service (QoS). Wireless Sensor Networks (WSNs) have also become a substantial innovation in the area of wireless interaction that directly affects daily life. Miniature, inexpensive sensors may now be produced because of ongoing advances in semiconductor manufacturing. Thus, hundreds or thousands of sensor nodes make up the entire WSN. To assess or identify certain occurrences, these nodes are placed widely over an area. These sensors may do many tasks or sense a variety of things, including noise, moisture, temperature, light, etc. [14].

The proposed protocol (CGHRP-FCM) outperforms the most recent energy-effective protocols for routing in terms of performance. The protocol's dynamic nature, which enables it to adapt to changes in the network's design and energy levels of nodes, makes it suitable for IoT networks where nodes are portable and have limited energy resources.

The structure of the paper is as follows. The associated work is shown in Sect. 2. Algorithms for the proposed protocol are described in Sect. 3. The simulations and results that the algorithms have produced are shown in Sect. 4, and the conclusion is presented in Sect. 5.

## 2 Related Works

The suggested architecture allows IoT devices to use energy effectively to increase network coverage. The GEER-DLAD methodology employs the error lossy compression (ELC) technique to lessen the quantity of information transmitted across the network. The moth flame swarm optimization (MSO) technique is also used to determine the best possible pathways inside the network [15]. Study [16] presents an optimization problem model for the energy limitation issue faced by IoT device applications. The protocol for routing messages initially groups electronics into clusters depending on a variety of variables, like proximity to the primary location, info/message length, and information currently detected from the environment, to save the electrical power of gadget nodes. A directed acyclic graph (DAG) is then made with each cluster head acting as a node after a cluster head has been selected for each cluster. Study [17] suggests an energy-effective opportunistic navigation system for sustainable IoT applications that depends on fuzzy ranking and ARFOR. The proposed methodology extends the network's longevity during the entire transmission cycle by using a volunteer node (VN) as a sender to transfer packets with threshold energy constraints to the parent node (PN). PN serves as the cluster's head node and aggregates signals to the DODAG root. In the study [18], a statistical concurrent navigation framework-based software-defined network (SDN) is presented by utilizing the enormous computing capacity for interactively improving multi constrained QOS variables in the IIoT using a graphics processing unit (GPU). Delay-sensitive, jitter-sensitive, and loss-sensitive QoS applications are taken into account by the structure for intelligent medical traffic. The suggested approach uses a three-tier clustering technique with an embedded security system to identify and flag sensor nodes that are engaging in malicious activity, placing them on a blacklist. A sink node chooses the cluster head and grid head depending on the value of its cost function in this center-based clustering protocol. Hardware-based link quality estimators are more utilized to evaluate links' performance and boost routing efficacy [19]. According to the study [20], mobile sinks in energy-effective green IoT networks with tree-based tracking algorithms have been proposed to reduce end-to-end delay. The suggested protocol introduces two brand-new, different methods for managing routing in networks. They suggest a novel data aggregation-aware, energy-efficient routing method built on Q-learning. To find the best path, the suggested strategy employs reinforcement learning for maximizing benefits at each sensor node as measured by the effectiveness of the sensing-type reliant data gathering, interaction power, and node residue energy. They made use of incentives that depended on the type of sensor [21]. The study [22] aims to improve the end-to-end latency and packet obtaining speed of network functioning, as well as the power consumption of wireless sensor networks (WSN). The main focuses of the study were the adaptable surroundings detecting technique known as WSN and the powerless-energy data routing approach. The study [23], provides a structure for transmitting information in enormous networks for emergencies, where numerous wireless sensors are spread across transportation facilities parking lots at retail establishments, airports, etc. IoT applications that send information collected by numerous sensor nodes to a gateway connected to a vast network like the Internet are theoretically supported by our design.

The main problem in IoT networks and assumes that other issues, such as security and reliability, are already addressed. While energy consumption is a significant concern for IoT devices, other challenges must also be considered in the design of a robust routing protocol, including data transmission delays, packet loss, and network congestion.

## 3 Proposed Model

This study proposes the flexible and energy-effective chain-greedy hierarchical routing protocol fuzzy C-Means optimization (CGHRP-FCM). Consider that a group of sensor nodes is scattered randomly to assess and keep an eye on the environment of the network. According to the suggested protocol, the network's region is divided into clusters of varying sizes, each with a disparate number of randomized nodes. Every cluster uses a greedy method with a minimum range-based next hop, and the centers of every cluster change the sojourn places. Data is received by the terminal nodes for every cluster, after which BS is sent to a neighboring sojourn location. As a consequence, shorter paths are used to convey data.

### 3.1 Energy Model

The suggested approach makes use of a first-order radio model to calculate the amount of energy consumed during sensor data transfer in Fig. 1. Equations (1) and (2) may be used to determine *Ex*(*l*, *c*), which stands for the quantity of power required to convey a between the transmitter and recipient nodes, a k-packet of bits [24, 25]. A sensor node in a WSN needs energy each time it sends a data packet to different sensor nodes, and the amount of power used is inversely associated with the number of communication hops. It is possible to measure *FSu*(*l*), which is utilized in Eq. (3) to represent the quantity of power required to transmit and receive nodes each getting a k-packet of bits.

$$FSu(l, c) = l \times F_{elec} + l \times \varepsilon_{et} \times c^2, \text{if}\, c < c_0 \tag{1}$$

$$FSu(l, c) = l \times Felec + l \times \varepsilon amp \times c^2, \;\text{if}\, c \geq c_0 \tag{2}$$

$$FSu(l) = l \times Felec \tag{3}$$

$$c_0 = \sqrt{\frac{set}{S_{amp}}} \tag{4}$$

$$EDA(l) = EDA - Felec \tag{5}$$

where *Felec* (60 nJ/bit) is the quantity of power required the transmission device is powered by the radio or reception electronics. The multiple paths absorption algorithm's dissipations of energy factor, or $\varepsilon amp$, is 0.0014 pJ/bit/m$^4$. $l$ is the number of bits. $c$ is the distance and *EDA* is the quantity of power (6 nJ/bit) used by data collected child nodes are merged via the transmitters circuit. $c_0$ is a cutoff range

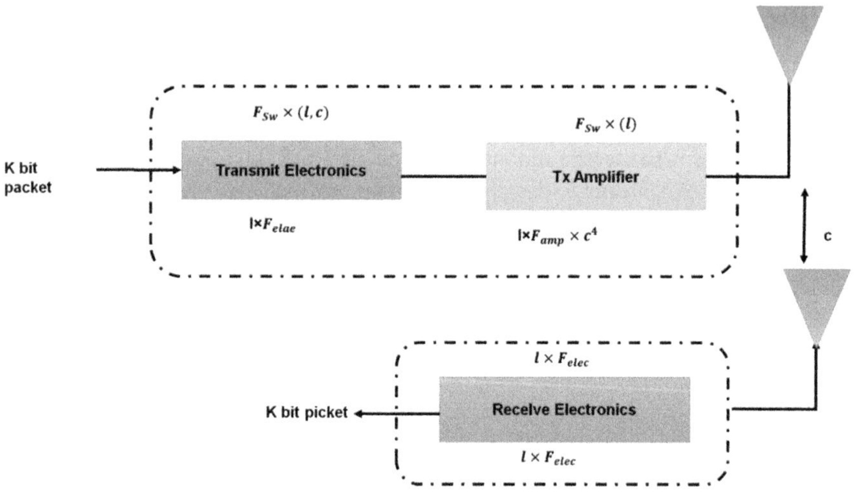


**Fig. 1** Model of energy

established by Eq. (4). Utilization of energy is measured using Eq. (1) if there is a lesser distance among two nodes than $c_0$ $(c < c_0)$ or using Eq. (2) if $(c \geq c_0)$. Equation (5) is employed to calculate the consumption of energy.

## *3.2 Network Model*

A sensor area with 120 arbitrarily positioned nodes and a size of 120 m × 120 m is considered in the network model. In the sensor industry, clusters have developed. Additionally, the clustering of these 120 nodes is asymmetric. The greedy approach has now been repeatedly enforced in clusters to produce chains. To further increase energy savings, mobile BS has been used in place of fixed BS. The following are the basic presumptions of the network model—Assuming that there is a homogenous network of sensors with all nodes with sensors having.

1. Equivalent capacities to process and detect.
2. Energy and processing power in the sink are infinite.
3. Nodes deliver rounds, or kbit data packets, at regular intervals.
4. In the range of the network, there is only one BS.
5. Data can be sent to the BS from any node.
6. Every node's position is known to the BS during installation.

### 3.3 *The Chain-Greedy Hierarchical Routing Protocol Based Fuzzy C-Means Algorithm (CGHRP-FCM)*

The shortcomings prompted the investigators to suggest an enhanced CGHRP procedure known as CGHRPFCM, which will be covered in this paper. This will then allow us to discuss why we believe utilizing FCM will enhance the CGHRP procedure.

The single chain lead has a significant load on it because of the separation between the sink and it, which requires a lot more energy for CGHRP to build a chain.

Due to the lengthy single chain, there is an enormous lag in the transmission of data, making it unsuitable for applications that require rapid response.

The five steps of the suggested CGHRP-FCM protocol, include machine learning (ML), formation of a multi-chain, clustering validation, data transfer, and chain leader choice.

**Validate Clustering-Step 1**

The correct number of clusters must be selected for cluster analysis for it to be successful in a variety of situations. As indicated in Fig. 2, we applied the variance ratio, referred to as the variance ratio criteria. Equation (6) defines the Variance Ratio index.

$$VRC = \frac{TT_a}{TT_v} \times \frac{(\mathrm{M} - \mathrm{l})}{l - 1} \tag{6}$$

Well-defined clusters have a significant among-cluster variation ($TT_a$) and a small within-cluster variance ($TT_v$). There are the most clusters in the solution with the highest variance ratio index value.

Equations (7) and (8) illustrate how the total *TTa* and *TTv* and are defined.

$$TT_a = \sum_{j=1}^{l} m_j||n_j - \mathrm{n}|| \tag{7}$$

$$TT_v = \sum_{j=1}^{l} \sum W \in d_j U - n^2 \tag{8}$$

where

| | |
|---|---|
| *TTa* | is there a general among-cluster variance. |
| *TTv* | is the total inside-of-cluster variation. |
| $l$ | The number of clusters. |
| *mj* | the quantity of evaluation in cluster j. |
| *M* | is the number of evaluations. |
| *nj* | is the cluster's centroid. |
| $n$ | is the collection of information total average. |
| $mj\|\|nj - n\|\|$ | are distances among both vectors in Euclidean space. |

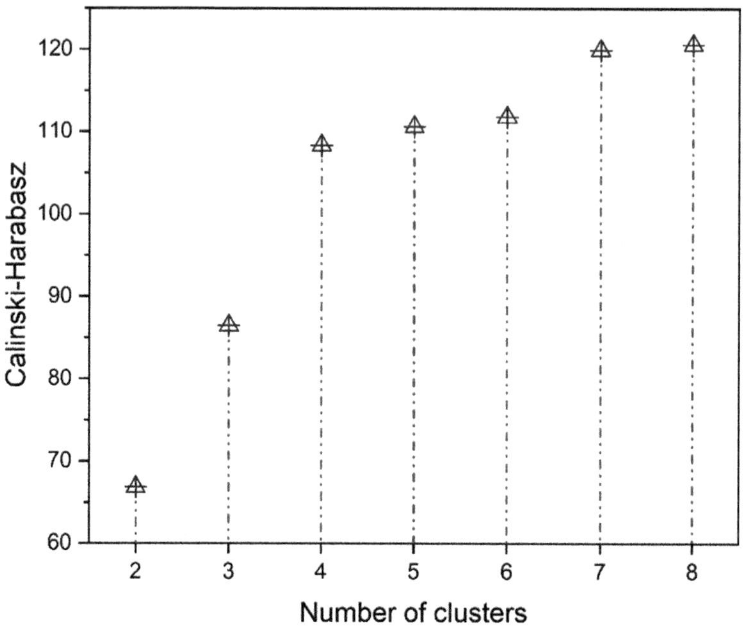


**Fig. 2** When $n = 100$, the Calinski-Harabasz values are compared to the values that indicate the number of clusters

| | |
|---|---|
| $U$ | is a data point. |
| $dj$ | is the cluster's centroid. |
| $\|\|U - n\|\|$ | are distances among both vectors in Euclidean space. |

**Fuzzy C-means (FCM)-Step 2**

A popular clustering method that works well for building clusters in networks of sensors is the fuzzy c-means (FCM). The FCM algorithm allocates each data point to each cluster with a degree of membership [26, 27]. It is a gentle clustering technique. In sensor networks, data points can represent data from different sensors, and clusters can reflect multiple geographic or event-related locations that the sensors have observed.

The following actions are part of the FCM algorithm:

Initialization: Randomly initialize the membership degrees for each data point and define the number of clusters.

Cluster centroids calculation: Based on the degrees of membership of the data points, calculate the centroids of each cluster.

Recalculate each data point's membership degrees following its separation from the cluster centroids.

Termination: Keep going back and forth between stages 2 and 3 until the membership degrees converge or a predetermined number of iterations are attained.

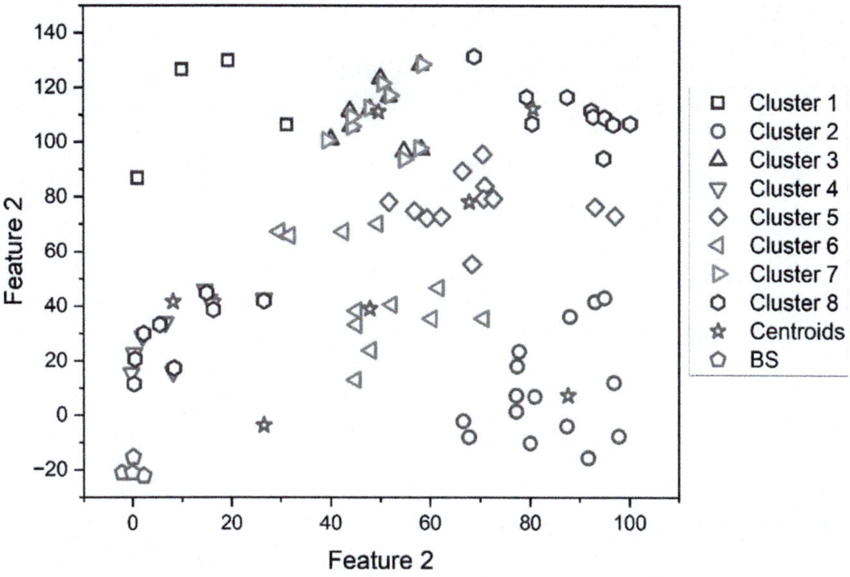


**Fig. 3** Fuzzy C-means

The sensor data may be thought of as the input data points for the FCM algorithm when applied to a sensor network. The intended degree of resolution or the estimated number of unique regions or events that the sensor is anticipated to observe can be used to calculate the number of clusters (Fig. 3).

**Multi-chain Formation-Step 3**

The sink communicates with each node to say hello to find out more about them. Using Eq. (9), determine the node that is the furthest from all sinks by calculating the sink distance from each node.

$$C = \sqrt{(V \cdot uc - A \cdot uc)^2 + (B \cdot zc - A \cdot zc)^2} \tag{9}$$

The number sensor nodes A and B are located in the $X$ and $Y$ values, correspondingly; $uc$ and $zc$ stand for the corresponding values.

Every node that is not an element of the chain must find the nearest node and connect with it by following the identical steps. As shown in Eq. (10).

$$CM = \sqrt{(V(j-1) \cdot uc - A(i) \cdot uc)^2 + (B(j-1) \cdot zc - A(i) \cdot zc)^2} \tag{10}$$

where $j$ and $i$ indicate the number of sensor nodes that remain due to the chain being created. Each node $j$ in the chain that gets data from node $i$ serves as the node's parent, while node $i$ serves as a node's child. Figure 4 illustrates how the same chain formation process is repeated in each section.

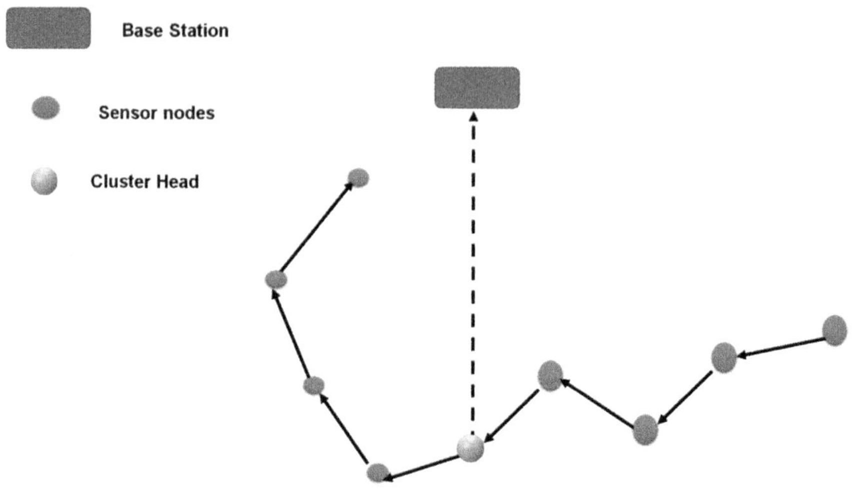


**Fig. 4** Formation of chain

### Select Chain Leader

The head of the group is going to be in charge of giving the BS the data that has been gathered. The most effective node will be chosen as the node that leads in the CGHRP-FCM protocol using the weight mechanism outlined in Eq. (11).

$$Xm = Fm/Cm \tag{11}$$

where,

$Xm$ The separation between the sink and the sensor node.

The most weighted node is chosen by the network to serve as the crucial chain leader. Every node calculates its separation from its parent node $dp$ following chain formation and contrasts it with its separation from the sink$dbs$. If ($dp \geq dbs$), the node serves as a second chain head, forwarding the collected data to the sink rather than the parent node.

### Transmit Data

Figure 5 illustrates the token transfer strategy, which starts at the end node and moves to the chain head before arriving at the sink. The 4 chains go through this procedure again. The duration- division multi-access approach has been used to prevent collisions when a round consists of numerous chains. A data aggregating factor compressed the data using a data density strategy (DCS).

### Sink Mobility

Static sinks are heavily loaded with big data volumes and have limited energy. As a result, network mobility is used to enhance network lifespan. In the WSN, Sink

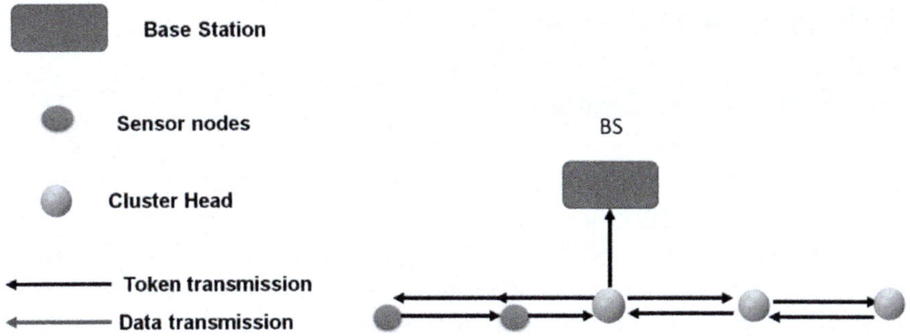


**Fig. 5** The chain-based routing methods use an authorization passing of tokens technique

moves along a predetermined path, stopping at each point to wait for a certain period. In the WSN, Sink moves along a predetermined path, stopping at each point to wait for a certain period. Sojourn time is the period a sink remains at a particular location to gather information from the chain leaders. The sojourn place is the transient home of the sink for data collection. The mobile sink will move between clusters in the suggested scenario along a defined path chosen by the centers of each area using a fuzzy C-mean technique.

## *3.4 Pseudocode for CGHRP-FCM Protocol*

The pseudocode for CGHRP-FCM Protocol is shown below.

```
Algorithm: CGHRP-FCM Protocol
Initialize clusters using Fuzzy C-means
For each cluster:
  | Calculate cluster centroid
  | Assign nodes to clusters based on proximity
For each cluster:
  | Form multi-chains using greedy approach
  | Select chain leader based on energy levels and distance

  | Transmit data from end nodes to chain leader
  | Chain leader transmits aggregated data to sink
```

# 4 Results and Discussion

## *4.1 Experimental Details*

This section compares the suggested protocol to the LEACH [28], TEEN [29], PEGASIS [10], and SEP [30] protocols through simulation using Python 3.11.3.120 nodes are randomly dispersed around the network region (120 × 120 m). The starting energy of each node is the same (0.6 J). Table 1 displays the virtual parameters that were utilized to calculate the recommended methodology. The simulation parameters were used to evaluate the recommended approach.

The protocols LEACH, TEEN, PEGASIS, and SEP were chosen for comparison due to their widespread acceptance and extensive use in the field of energy-efficient routing for WSNs. LEACH is one of the earliest hierarchical routing approaches, known for its simplicity and efficiency. TEEN and PEGASIS build upon LEACH, introducing modifications to improve performance under different scenarios. SEP provides a foundation for understanding energy efficiency in heterogeneous networks. Comparing the proposed protocol with these well-known techniques allows for a comprehensive evaluation of its effectiveness and innovation. The performance of the suggested approach is compared and examined using the following metrics shown in Table 2.

**Table 1** Parameters of network

| Parameters | Value |
|---|---|
| Size of environment | 120 × 120 |
| Position of BS | (0,0) |
| No of normal nodes | 120 |
| Size of packet | 3000 bits |
| Initial energy for normal node | 0.6 J |
| No of rounds | 5000 |
| Eelec | 60 nJ/bit |
| Efs | 20 pJ/bit/m$^2$ |
| Emp | 0.0014 pJ/bit/m$^4$ |
| EDA | 6 nJ/bit |

**Table 2** Energy consumption metrics

| Metric | Description |
|---|---|
| FND | Power supply expiry on the first node |
| HND | Half nodes died |
| LND | Last node died |
| Residual energy | Energy left after network operation |

## 4.2 Lifetime Network

The network lifespan, stability, and instability duration of the suggested protocols are contrasted to those of the LEACH, PEGASIS, TEEN, and SEP protocols, correspondingly. Figure 6 displays the number of nodes that are still operational after a predetermined no of rounds, whereas Fig. 7 displays the network's stability and instability (FND, HND, and LND). It is demonstrated that the recommended protocol (CGHRP-FCM) outperforms conventional protocols due to decreased stress on the chain heads, which results in load balancing over the whole network. Since there are fewer distances between connected nodes rather than in a single chain, in a multi-chain idea, less energy is needed to transmit data. Consequently, the improved development outcomes demonstrate the average life expectancy.

$$Enhancement\ gain = \frac{CGHRP\ FCM\ - Other\ protocol}{Other\ protocol} \times 100\%$$

The percentages of enhancement for FND, HND, and LND were determined by comparing the suggested protocol for CGHRP-FCM to older protocols.

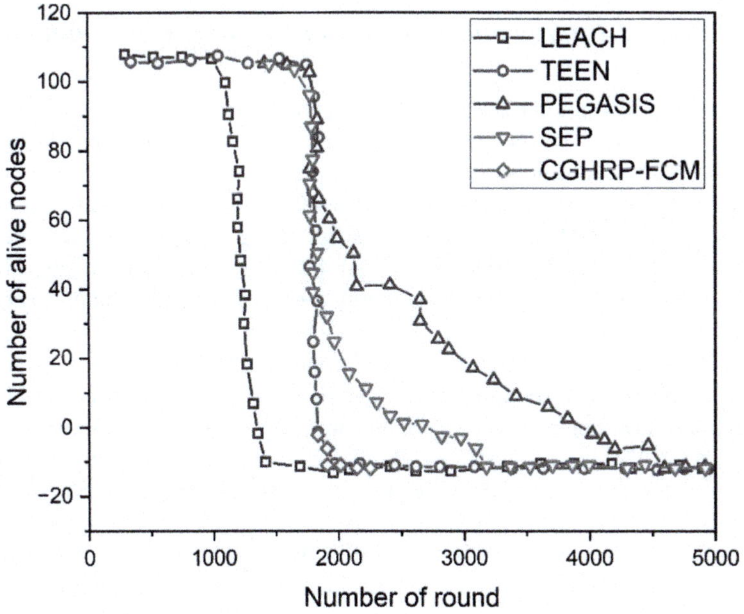


**Fig. 6** Nodes that are alive in each cycle

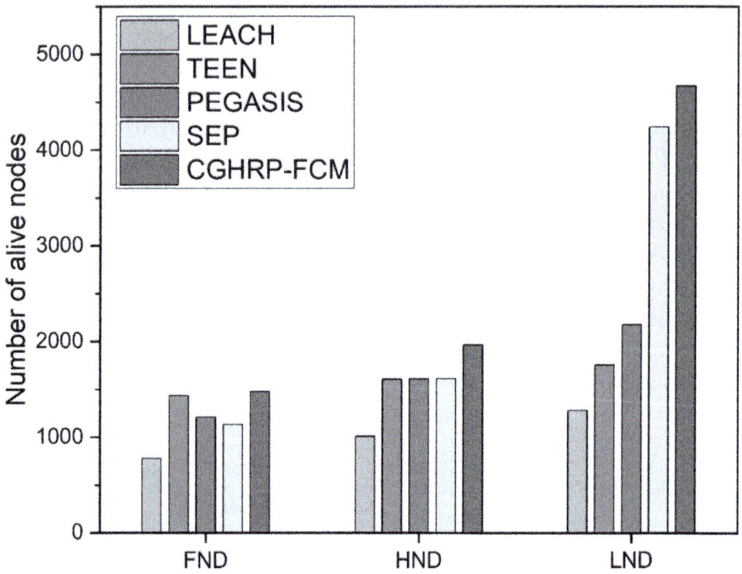


**Fig. 7** Lifetime of the network comparison

## *4.3 Residual Energy and Typical Node Energy Consumption*

The SEP and FCM are the least energy-intensive procedures use all the protocols. LEACH, PEGASIS, TEEN, and SEP protocols all perished at rounds 1275, 1751, 2176, and 4246, accordingly, and the networks that are using them have also perished following all protocols. As a result, the network lifespan is increased while energy usage is minimized. Demonstrates the concept of Fig. 8, which displays the quantity of energy saved after 1000, 2500, and 4500 rotations, respectively. Figure 9 illustrates that a network's remaining energy for CGHRPFCM exhibits an apparent enhancement in the behavior of the network lifecycle.

The altered and basic routing protocols' chain lengths are summarized. It shows that the CGHRP- FCM routing protocol significantly lowers the problem of lengthy connections. The single PEGASIS chains are (887.4 m) long, while the four SEP chains are (9197.9 m), (221 m), (182.4 m), and (243.8 m, respectively. In CGHRP-FCM, these lengths are reduced to (96.3 m), (49.0 m), (96.9 m), (135.8 m), (79.7 m), (139.8 m), (94.3 m), and (9104.5 m).

Finally, it can be stated that the CGHRP-FCM procedures correct the basic problems with the main enhancement, which eliminated the CHs being arbitrarily selected without considering the duration or present power use of nodes, and the CHs communicating with the sink using one hop, which makes LEACH inapplicable for big networks. Large loads on single chains provide a difficulty that PEGASIS and TEEN

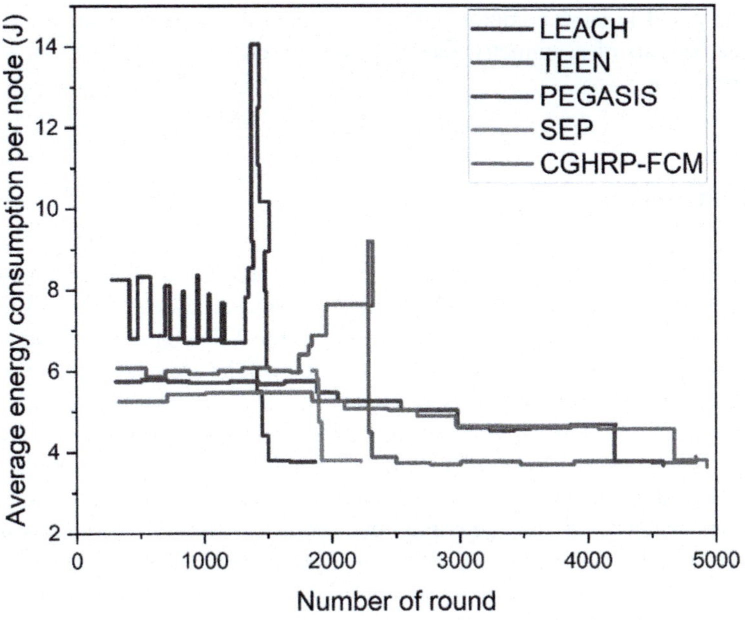


**Fig. 8** Average energy consumption per node for the number of rounds

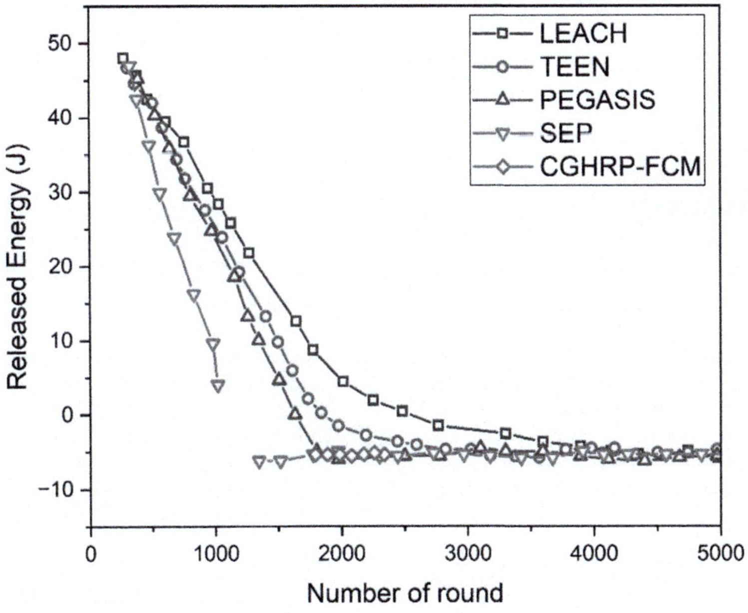


**Fig. 9** Round-by-round residual energy

can address. The more complex linkages that consume greater amounts of energy and the benefits of asymmetry above equality in clustering employed in SEP are explored.

## 5 Conclusion

The purpose of this effort is to increase the sensor network's lifespan. Energy efficiency is the main concern while building an enhanced power-efficient PEGASIS-based protocol for routing FCM (CGHRP-FCM) for WSNs since sensor nodes have limited resources for energy. The WSN's lifespan will be extended if the sensors are kept functioning for as long as is practical. The primary energy utilization of the sensor nodes is for tasking-hop connections inside a specific cluster that transfers data. Long chains are divided into smaller chains, each of which is a cluster and has two head clusters that are in communication with the BS. The variance ratio approach, which is used to determine the optimal number of clusters, is utilized to determine the number of these clusters. The outcomes demonstrate a decrease in transmission of data interruptions, a prolongation of the lifespan of the network through decreasing the power consumed by sensor nodes, a decline in the separation among nodes connected by smaller chains, and a decline in the stress put on the sparse nodes through the employment of sink movement. In future, look at how throughput, delay, and loss of packets affect the quality of service (QoS) when energy usage is reduced. Either an algorithm based on genetics or a particle swarm optimization technique may be used to choose the optimal cluster heads for every cluster.

## References

1. Gupta, B.B., Quamara, M.: An overview of Internet of Things (IoT): architectural aspects, challenges, and protocols. Concurr. Comput. Practice Exp. **32**(21), e4946 (2020). https://doi.org/10.1002/cpe.4946
2. Kumar, S.V., Mary, G.A.A., Chohan, J.S., Kalita, K.: Original research article efficient sensor anomaly detection using Markov-LSTM architecture for methane sensing. J. Auton. Intell. **7**(3) (2024). https://doi.org/10.32629/jai.v7i3.1285
3. Khanh, Q.V.: An energy-efficient routing protocol for MANET in the Internet of Things environment. iJOE **17**(07), 89 (2021). https://doi.org/10.3991/ijoe.v17i07.23273
4. Kalita, K., Ganesh, N., Balamurugan, S. (eds.).: Metaheuristics for Machine Learning: Algorithms and Applications. Wiley, New York (2024). https://doi.org/10.1002/9781394233953
5. Kumar, S., Tiwari, P., Zymbler, M.: Internet of Things is a revolutionary approach for future technology enhancement: a review. J. Big Data **6**, 111 (2019). https://doi.org/10.1186/s40537-019-0268-2
6. Mutombo, V.K., Lee, S., Lee, J., Hong, J.: EER-RL: energy-efficient routing based on reinforcement learning. Mob. Inf. Syst. **2021**, 1–12 (2021). https://doi.org/10.1155/2021/5589145

7. Gupta, A., Gulati, T., Bindal, A.K.: WSN based IoT applications: A review. In: 2022 10th International Conference on Emerging Trends in Engineering and Technology—Signal and Information Processing (ICETET-SIP-22), pp. 1–6. Nagpur, India (2022). https://doi.org/10.1109/ICETET-SIP-2254415.2022.9791495
8. Majid, M., Habib, S., Javed, A.R., Rizwan, M., Srivastava, G., Gadekallu, T.R., Lin, J.C.-W.: Applications of wireless sensor networks and Internet of Things frameworks in the Industry Revolution 4.0: a systematic literature review. Sensors **22**, 2087 (2022). https://doi.org/10.3390/s22062087
9. Santiago, S., Arockiam, L.: A novel fuzzy-based energy-efficient routing for the Internet of Things. In: 2017 International Conference on Algorithms, Methodology, Models and Applications in Emerging Technologies (ICAMMAET), pp. 1–4. IEEE (2017). https://doi.org/10.1109/ICAMMAET.2017.8186645
10. Khedr, A.M., Aziz, A., Osamy, W.: Successors of PEGASIS protocol: a comprehensive survey. Comput. Sci. Rev. **39**, 100368 (2021). https://doi.org/10.1016/j.cosrev.2021.100368
11. Shabnam, Pandey, M.: A review on IoT based wireless sensor network and security. In: Garg, L., et al. (eds.) Key Digital Trends Shaping the Future of Information and Management Science. ISMS 2022. Lecture Notes in Networks and Systems, vol. 671. Springer, Cham (2023). https://doi.org/10.1007/978-3-031-31153-6_13
12. Lei, T., Jia, X., Zhang, Y., He, L., Meng, H., Nandi, A.K.: Significantly fast and robust fuzzy c-means clustering algorithm based on morphological reconstruction and membership filtering. IEEE Trans. Fuzzy Syst. **26**(5), 3027–3041 (2018). https://doi.org/10.1109/TFUZZ.2018.2796074
13. Randhawa, S., Jain, S.: Data Aggregation in Wireless Sensor Networks: Previous Research, Current Status and Future Directions. Springer, US (2017). https://doi.org/10.1007/s11277-017-4674-5
14. Li, K., Zhao, J., Hu, J., Chen, Y.: Dynamic energy efficient task offloading and resource allocation for noma-enabled IoT in smart buildings and environment. Build. Environ. **226**, 109513 (2022). https://doi.org/10.1016/j.buildenv.2022.109513
15. Lydia, E.L., Jovith, A.A., Devaraj, A.F.S., Seo, C., Joshi, G.P.: Green energy efficient routing with deep learning-based anomaly detection for Internet of Things (IoT) communications. Mathematics **9**(5), 500 (2021). https://doi.org/10.3390/math9050500
16. Vashishth, V., Chhabra, A., Khanna, A., Sharma, D.K., Singh, J.: An energy-efficient routing protocol for wireless Internet-of-Things sensor networks (2018). arXiv:1808.01039
17. Chithaluru, P., Kumar, S., Singh, A., Benslimane, A., Jangir, S.K.: An energy-efficient routing scheduling based on a fuzzy ranking scheme for the Internet of Things. IEEE Internet Things J. **9**(10), 7251–7260 (2021). https://doi.org/10.1109/JIOT.2021.3098430
18. Naeem, F., Tariq, M., Poor, H.V.: SDN-enabled energy-efficient routing optimization framework for industrial Internet of Things. IEEE Trans. Industr. Inf. **17**(8), 5660–5667 (2020). https://doi.org/10.1109/TII.2020.3006885
19. Ilyas, M., Ullah, Z., Khan, F.A., Chaudary, M.H., Malik, M.S.A., Zaheer, Z., Durrani, H.U.R.: Trust-based energy-efficient routing protocol for Internet of Things-based sensor networks. Int. J. Distrib. Sens. Netw. **16**(10), 1550147720964358 (2020). https://doi.org/10.1177/1550147720964358
20. Yarinezhad, R., Azizi, S.: An energy-efficient routing protocol for the Internet of Things networks based on geographical location and link quality. Comput. Netw. **193**, 108116 (2021). https://doi.org/10.1016/j.comnet.2021.108116
21. Yun, W.K., Yoo, S.J.: Q-learning-based data-aggregation-aware energy-efficient routing protocol for wireless sensor networks. IEEE Access **9**, 10737–10750 (2021). https://doi.org/10.1109/ACCESS.2021.3051360
22. Ju, X.: Energy-efficient routing sensing technology of wireless sensor networks based on the Internet of Things. J. High-Speed Netw. **27**(3), 225–235 (2021). https://doi.org/10.3233/JHS-210663
23. Al-Turjman, F.: Cognitive routing protocol for disaster-inspired Internet of Things. Futur. Gener. Comput. Syst. **92**, 1103–1115 (2019). https://doi.org/10.1016/j.future.2017.03.014

24. Shahzad, M.K., Islam, S.R., Kwak, K.S., Nkenyereye, L.: AEF: adaptive en-route filtering to extend network lifetime in wireless sensor networks. Sensors **19**(18), 4036 (2019). https://doi.org/10.3390/s19184036
25. Bouguera, T., Diouris, J.F., Chaillout, J.J., Jaouadi, R., Andrieux, G.: Energy consumption model for sensor nodes based on LoRa and LoRaWAN. Sensors **18**(7), 2104 (2018). https://doi.org/10.3390/s18072104
26. Hashemi, S.E., Gholian-Jouybari, F., Hajiaghaei-Keshteli, M.: A fuzzy C-means algorithm for optimizing data clustering. Expert Syst. Appl. **227**, 120377 (2023). https://doi.org/10.1016/j.eswa.2023.120377
27. Dhal, K.G., Das, A., Sasmal, B., Ray, S., Rai, R., Garai, A.: Fuzzy C-means for image segmentation: challenges and solutions. Multimed. Tools Appl. **83**(9), 27935–27971 (2024). https://doi.org/10.1007/s11042-023-16569-2
28. Lonkar, B.B., Karmore, S.: Recent advancements on energy-saving LEACH protocol in wireless sensor network-review. In: Security, Privacy and Data Analytics: Select Proceedings of the 2nd International Conference, ISPDA 2022, vol. 1049, p. 1. Springer Nature (2023). https://doi.org/10.1007/978-981-99-3569-7_1
29. Chauhan, T., Nayyer, M.: Review on energy efficient protocol based on LEACH, PEGASIS and TEEN. In: 2016 International Conference on Emerging Trends in Communication Technologies (ETCT), pp. 1–5. IEEE (2016). https://doi.org/10.1109/ETCT.2016.7882929
30. Tanwar, S., Kumar, N., Rodrigues, J.J.: A systematic review on heterogeneous routing protocols for wireless sensor network. J. Netw. Comput. Appl.Comput. Appl. **53**, 39–56 (2015). https://doi.org/10.1016/j.jnca.2015.03.004

# Adaptive Multigrid Long Short-Term Memory Algorithm for Improved Air Quality Forecasting

**G. Shanmugasundar, Janjhyam Venkata Naga Ramesh, Banu Murali, Krishnasamy Karthik, and Velumayil Ramesh**

**Abstract** As urbanization and industrialization accelerate, the imperative for monitoring air quality intensifies due to its significant impact on health and the environment. This paper presents an innovative air pollution prediction model employing the adaptive multigrid LSTM (Long Short-Term Memory) algorithm. Our methodology begins with the preprocessing of a substantial dataset, comprising approximately 450,000 records across India, with a focus on 100,000 randomly drawn records. These records encompass 13 distinct features such as $SO_2$, $NO_2$, PM2.5, location etc. Through logistic gradient clustering for feature extraction, the data is prepared for subsequent analysis. The core of our approach, the adaptive multigrid LSTM algorithm, enables efficient and accurate prediction of air pollution levels. Evaluations indicate a significant improvement over conventional models, achieving an impressive accuracy of 93%. These metrics decisively outperform traditional techniques like AdaBoost, MLP, Gaussian NB, and SVM classifiers. This study demonstrates the feasibility of using advanced machine learning techniques for air quality forecasting, highlighting their potential for enhancing environmental health management and policy planning, especially in rapidly developing regions like India.

G. Shanmugasundar
Department of Mechanical Engineering, Sri Sai Ram Institute of Technology, Chennai 600044, India
e-mail: shanmugasundar.mech@sairamit.edu.in

J. V. N. Ramesh
Department of Computer Science and Engineering, Koneru Lakshmaiah Education Foundation, Vaddeswaram 522502, Andhra Pradesh, India

Department of Computer Science and Engineering, Graphic Era Deemed To Be University, Dehradun 248002, Uttarakhand, India

Department of CSE, Graphic Era Hill University, Dehradun 248002, India

B. Murali · K. Karthik (✉) · V. Ramesh
Department of Mechanical Engineering, Vel Tech Rangarajan Dr. Sagunthala R&D Institute of Science and Technology, Chennai 600062, Tamilnadu, India
e-mail: karthikmeed@gmail.com



D. Oliva et al. (eds.), *Advances in Optimization Algorithms for Multidisciplinary Engineering Applications: From Classical Methods to AI-Enhanced Solutions*, Studies in Computational Intelligence 806, https://doi.org/10.1007/978-3-031-78440-8_12

**Keywords** Air pollution · Pollutant forecasting model · LSTM algorithm · Deep learning · Machine learning

## 1 Introduction

Air pollution, a critical environmental challenge, has profound effects on both ecological and human health. As urbanization and industrialization accelerate, particularly in rapidly developing countries, the urgency to forecast air pollution levels becomes paramount. The detrimental impacts of air pollutants, notably particulate matter (PM) and ozone ($O_3$), on respiratory, cardiovascular, immunological, and neurological systems have been well documented [1]. Predictive models for air pollution are essential in guiding mitigation strategies and informing public health policies.

The global burden of air pollution, especially in densely populated megacities, is aggravated by industrial emissions, vehicle exhausts, and coal combustion. The World Health Organization's alarming statistic of approximately seven million premature deaths in 2012 due to atmospheric aerosols underscores the urgent need for effective air quality prediction models [2]. These concerns are amplified by the fact that the environmental issue of air pollution has significantly escalated throughout the twenty-first century.

Recent research endeavors have focused on developing various forecasting models for air pollution, with a significant emphasis on the health impacts of pollutants, urban exposure methods, and land use regression approaches [3]. The advent of deep learning (DL) techniques and remote sensing data has shifted the focus towards long-term projections of fine particulates, ozone, and nitrogen oxide, while also examining the associated mortality and morbidity rates [4].

The evolution of forecasting algorithms, from basic predictors to advanced deep learning models, has marked a significant advancement in the field of air pollution prediction [5]. The application of deep learning frameworks, particularly for large-scale data modeling, has shown promise but also faces challenges in outcome consistency [6]. These models have been instrumental in predicting concentrations of various air pollutants, such as PM2.5, PM10, and nitrogen dioxide ($NO_2$), using techniques like recurrent neural networks, thereby aiding in strategic planning for pollution control and health risk mitigation [7].

The accuracy of air pollution forecasting is not only crucial for immediate public health responses but also for long-term environmental policy-making. Despite extensive research, long-term forecasting models for specific air pollutants often overlook the potential of integrating deep learning and statistical forecasting models [8]. Learning-based models, such as transferrable bidirectional LSTM models, have emerged as vital tools in environmental quality forecasting [9].

Urban planning strategies also play a pivotal role in controlling air pollution. Studies suggest that air pollution hotspots are not exclusively linked to heavy traffic, and smart urban planning can effectively reduce pollution levels [10]. Additionally, the health impacts of air pollution extend to various physiological systems, pregnancies, and overall life expectancy [11]. Advanced technological interventions, like simulation analyses of air circulation and developed engine technologies, are being explored to reduce vehicular air pollution [12].

Comprehensive studies using content analysis have revealed the extensive effects of air pollution on human life, encompassing physical health conditions and broader socioeconomic impacts [13]. The relationship between air pollution and increased population morbidity, along with its economic implications, emphasizes the necessity of accurate and efficient predictive models [14].

Given the complex nature of air pollution and its multifaceted impacts, this paper proposes an innovative approach to enhance air pollution prediction using an adaptive multigrid LSTM (Long Short-Term Memory) algorithm. The objective is to provide an in-depth comparative analysis of various predictive models, evaluating their effectiveness in air quality prediction.

## 2 Methodology

### *2.1 Problem Statement*

Air pollution is a serious issue throughout the world, but it is especially difficult for emerging nations like India. This is because India needs more rapid and long-term growth. To properly address this problem, continuous measuring and monitoring of air pollution levels throughout cities is essential. By deciphering the concealed characteristics in regulated traditional pollution records, administrators may predict pollutant levels specific to a given particular region. Thus, the fundamental purpose of our research is to evaluate the rising level of air pollution by predicting future values based on data from the gathered parameters throughout time. Furthermore, time series forecasting using machine learning methods is required to complete the task. This forecasting system can provide predicted values regularly, allowing for better management of air pollution issues.

### *2.2 Proposed Work*

The direction of the proposed methodology is explained in this section. The process of analyzing air pollution CSV datasets, preprocessing using normalization, feature extraction using logistic gradient clustering, dividing the data into training and testing datasets, adaptive multigrid long short-term memory (AM-LSTM), and performance analysis are all included in the schematic representation of the suggested technique. The suggested strategy is schematically shown in Fig. 1.

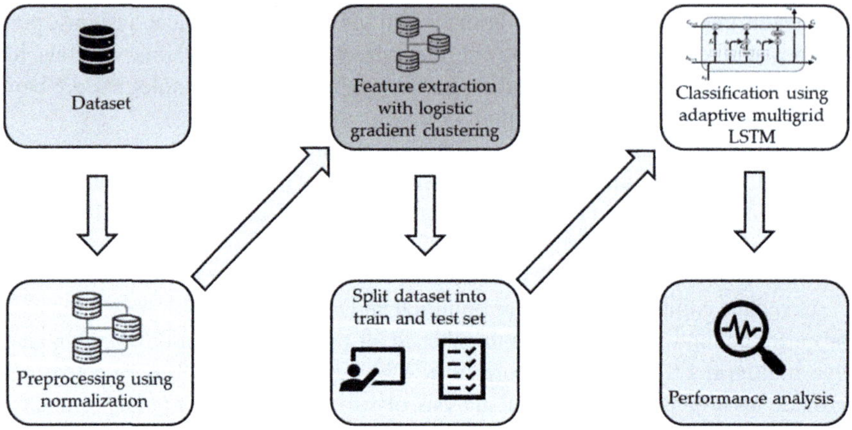


**Fig. 1** Flowchart of the present methodology

## 2.3 Dataset

The dataset utilized in this study is sourced from the comprehensive and cleaned version available on Kaggle (https://www.kaggle.com/datasets/shrutibhargava94/india-air-quality-data), which consolidates the Historical Daily Ambient Air Quality Data (https://data.gov.in/catalog/historical-daily-ambient-air-quality-data) originally released by the Ministry of Environment and Forests and Central Pollution Control Board of India. This dataset is a part of the National Data Sharing and Accessibility Policy (NDSAP) initiative and represents a wide-ranging compilation of air quality metrics across different states and years in India [15].

The dataset encompasses approximately 450,000 entries throughout India, from which after removing the incomplete or missing value rows, randomly 100,000 entries are selected in the sample set. However, it was ascertained that each class has equal amount of representation in the sample set. The data spans several years, offering a longitudinal view of air quality variations. The dataset is structured into multiple columns, each representing a specific aspect of the air quality measurements:

- stn_code: Denotes the station code, a unique identifier for each monitoring station.
- sampling_date: The date when the air quality data was collected.
- state: The name of the Indian state where the monitoring station is located.
- location: The specific location of the air monitoring station within the state.
- agency: The name of the agency responsible for data collection.
- type: Classifies the environment of the monitoring station (e.g., industrial, residential).
- $so_2$: Measures the concentration of sulfur dioxide ($SO_2$) in the air.
- $no_2$: Measures the concentration of nitrogen dioxide ($NO_2$) in the air.
- rspm: Refers to Respirable Suspended Particulate Matter in the air.
- spm: Denotes Suspended Particulate Matter in the air.

- location_monitoring_station: Provides additional details about the monitoring station's location.
- pm2_5: Concentration of particulate matter with a diameter of less than 2.5 μm.
- date: The exact date of the air quality measurement.

This robust dataset forms the foundation for our predictive modeling, enabling a comprehensive analysis of air quality trends over time and across various geographic locations. The diverse range of features, including key pollutants like $SO_2$, $NO_2$, PM2.5, and overall air quality indices, provides a rich basis for our Adaptive Multigrid LSTM algorithm to effectively forecast air pollution levels.

## 2.4 Preprocessing

The preprocessing stage is critical for preparing the air quality dataset for predictive analysis. Given the dataset's diverse origins and the complexity of the data, our preprocessing efforts focused on ensuring data quality and consistency. This stage involved several key steps:

1. **Data Segregation**: Each monitoring station's dataset was individually identified and stored in separate files, named after the station's unique code. This step ensured that the data from different locations could be analyzed separately, allowing for more nuanced insights.
2. **Handling Missing Values**: A common challenge in real-world datasets is the presence of missing values. To address this, we adopted a stringent approach, removing any row in the dataset that contained one or more missing values. This decision was made to maintain the integrity and reliability of our analysis.
3. **Normalization**: The air quality data, especially pollutant concentrations like $SO_2$, $NO_2$, PM2.5, and others, vary greatly in range. To standardize these values and ensure they fall within a comparable scale, we applied the MinMaxScaler technique. This normalization process adjusts each feature to a range between 0 and 1, using the following equation:

$$v' = \frac{v - \min_A}{\max_A - \min_A} \tag{1}$$

   Here, $v'$ is the normalized value, $v$ is the original value, $\min_A$ and $\max_A$ represent the minimum and maximum values of feature $A$, respectively. This normalization is essential for the effective training of machine learning models, as it ensures that all features contribute equally to the predictive process.
4. **Data Splitting**: After preprocessing, the dataset was divided into two subsets: 80% of the data was allocated for training the model, and the remaining 20% was reserved for testing. This split was crucial for evaluating the performance of the model on unseen data, thereby ensuring the robustness and generalizability of our predictive algorithm.

Through these preprocessing steps, we ensured that the data was clean, consistent, and optimally formatted for feeding into the Adaptive Multigrid LSTM model for further analysis and air quality prediction.

### 2.5 Feature Extraction

In general, three sorts of characteristics may be used to solve a typical recommendation problem: user feature, item feature, and cross-feature extraction. As a result of the dataset's size, the generation of features should be as basic and efficient as feasible. Logistic gradient clustering is the focus of this research. To effectively describe input sequences for data mining methods such as logistic gradients clustering or classification, one must first perform feature extraction. Because of this, the logistic gradient method was used for this problem. The Logistic Regression was chosen because of the ease with which it can be implemented and the wide range of applications it has seen.

$$i_\theta(y_j) = i_\theta(y_{j1}, y_{j2}, \ldots, y_{jE}) = \frac{1}{1 + \exp(-\eta_j)} = \frac{1}{1 + \exp(-\theta^U y_j)} \tag{2}$$

where, $\eta_j = \theta_j + \theta_1 y_{j1} + \theta_2 \theta_{j2} \cdots + \theta_E y_{jE} = \theta^U y_j$. $h(\eta_j) = \frac{1}{1+\exp(-\eta_j)}$ is known as the logistic gradient function and its derivative is given in the following manner.

$$h'(\eta_j) = h(\eta_j) \times (1 - h(\eta_j)) \tag{3}$$

For the traveling (z = 1) and non-raveling (z = 0) classes, the probabilities are given in the following way,

$$Q(z_j = 1|y_{j,}\theta) = i_\theta(y_j) \tag{4}$$

$$Q(z_j = 0|y_{j,}\theta) = 1 - i_\theta(y_j) \tag{5}$$

The logistic gradient may be used to optimize the model parameter. A training dataset with ground—truth given to each data instance is required for logistic gradient clustering.

### 2.6 Adaptive Multigrid Long Short-Term Memory

Adaptive multigrid long short-term memory (AM-LSTM) is a kind of recurrent neural network-based structure that uses feedback connections to interpret sequence data. The four components of long short-term memory are the cells, which are capable of recollecting data with arbitrary time intermission; which determines whether data from the prior cell state is maintained (if value 1) or deleted or forgotten if the value is 0; the current step's data is saved or discarded by the input gate [16, 17]. Moreover, there is an output gate, which transmits data to the next concealed state. The LSTM's greatest asset is its ability to retain information for extended periods [18, 19]. The output gates information is multiplied by the cell states, and the output is transferred to the hidden state (Fig. 2).

$$j_u = \sigma\left(X_j y_u + V_j i_{u-1} + W_j d_{u-1}\right) \tag{6}$$

$$g_u = \sigma\left(X_g y_u + V_g i_{u-1} + W_g d_{u-1}\right) \tag{7}$$

$$p_u = \sigma\left(X_p y_u + V_p i_{u-1} + W_p d_u\right) \tag{8}$$

$$\widetilde{D_u} = \tan i(X_d y_u + V_d i_{u-1}) \tag{9}$$

$$d_u = g_u^j \odot d_{u-1} + j_u \odot \widetilde{D_u} \tag{10}$$

$$i_u = p_u \odot \tan i(d_u) \tag{11}$$

where $j_u$ represents the input, $p_u$ represents the output and $g_u$ represents the forget gates respectively, $d_u$ represents the memory cell, $andi_u$ is hidden state the same is represented by the below formula in which $d_u$ represents the cell state vector, $j_u$ represents the input gate vector, $\sigma_h$ represents the sigmoid function, and $y_u$ represents the input vector respectively. Mappings of parameter values such as $X_j$, $V_j$. $c_j$ is denoted by bias vector.

$$g_u = \sigma_h\left(X_g y_u + V_g i_{u-1} + c_g\right) \tag{12}$$

$$j_u = \sigma_h\left(X_j y_u + V_j i_{u-1} + c_j\right) \tag{13}$$

$$p_u = \sigma_h\left(X_j y_u + V_j i_{u-1} + c_p\right) \tag{14}$$

$$\widetilde{D_u} = \tan i(X_d y_u + V_d i_{u-1} + c_d) \tag{15}$$

$$d_u = g_u * d_{u-1} + j_u * \widetilde{D_u} \tag{16}$$

$$I_u = p_u * \tan i(D_u) \tag{17}$$

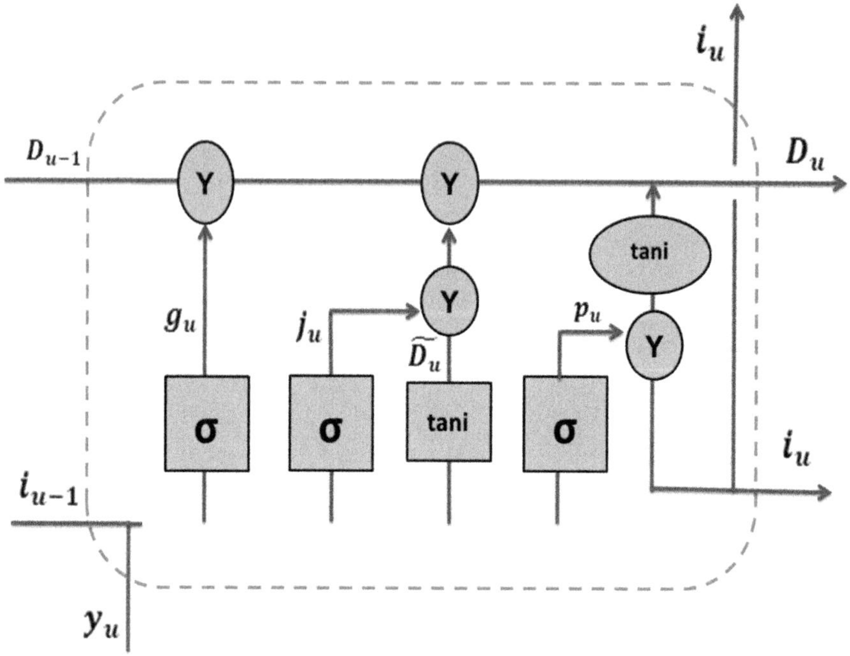


**Fig. 2** Adaptive LSTM architecture

It is obvious from Eqs. (11), and (12) how the input, forget, and output gates are used to produce cell state and output vector (ct, ht). Θ is a Hadamard product.

Algorithm 1: AM-LSTM ALGORITHM

**Step 1:**
All fine particle matter & meteorological data are preprocessed
AM- LSTM pre-training
**Step 2:**
$y_{(u)}$ is an element-wise input that ignores bias
To translate information from cells to gate vectors, create a weight matrix X
For each member of the gate vector, define n and get input from the cell state
Set d as an internal memory of the cell state Fine-tuning
**Step 3:**
Use the cell's state as input for the tan *i* method to build a vector
Filter out the value of iu-1 and $y_{(u)}$ with the help of the sigmoid function
**Step 4:**
Obtain the predicted outcomes

## 3 Results and Discussion

In this section, the suggested framework's is critically analyzed and compared with other existing machine learning algorithms. The approaches include the AdaBoost (ADB) classifier [20, 21], multilayer perceptron (MLP) classifier [22, 23], Gaussian Naive Bayes (GNB) classifier [24, 25], support vector machine (SVM) classifier [26, 27], Random Forest (RF) classifier [28, 29], Gradient Boosting Machines (GBM) classifier [30, 31], Extreme Gradient Boosting (XGBoost) classifier [23, 32], and adaptive multigrid-long short-term memory (AM-LSTM). Since the tuning parameters play an important role in the performance of the algorithms [33], a pilot study is carried out to fine tune the parameters for the study. Table 1 contains the various parameters and their values considered in this study.

Figure 3 shows the normalized confusion matrix for all class labels. The AQI is distributed into six classes i.e., 'Good', 'Satisfactory', 'Moderate', 'Poor', 'Very poor' and 'Severe'. From Fig. 3 it is evident that all the algorithms classify the labels 'Very poor' and 'Severe' with 100% accuracy while for other classes misclassification occurs.

Using these confusion matrix, four different metrics, namely accuracy, precision, recall and F1-score are computed based on the following equations [33],

$$Accuracy = \frac{(TP + TN)}{TP + TN + FP + FN} \tag{18}$$

$$Precision = \frac{(TP)}{(TP + FP)} \tag{19}$$

$$Recall = \frac{(TP)}{(TP + FN)} \tag{20}$$

$$\text{F1 score} = \frac{2 \times (precision \times recall)}{precision + recall} \tag{21}$$

where *TP* is the number of true positives and *FP* is the number of false negatives, respectively, the number of false negatives is denoted by *FN* and true negatives are denoted by *TN* respectively.

Figure 4 presents the performance comparison of the eight algorithms based on accuracy, precision, recall and F1-score. AM-LSTM showed the most promising results with uniform scores across accuracy, precision, recall and F1-score at 93%. The high performance of AM-LSTM can be attributed to its ability to capture temporal dependencies and complex patterns in air quality data, which are characteristic of environmental datasets. Its balanced precision and recall indicate a strong ability to correctly identify true positives while minimizing false positives and negatives, a crucial aspect in air pollution forecasting where both overestimation and underestimation can have significant implications.

**Table 1** Tuning parameters and their values of ML algorithms

| Algorithm | Parameter | Value | Description |
|---|---|---|---|
| AM-LSTM | Learning rate | 0.01 | Controls the step size at each iteration while moving toward a minimum of a loss function |
| | Batch size | 32 | Number of training examples utilized in one iteration |
| | Number of layers | 3 | The depth of the LSTM network |
| | Neurons per layer | 100 | Number of neurons in each LSTM layer |
| | Dropout rate | 0.2 | Fraction of the input units to drop to prevent overfitting |
| ADB | Number of estimators | 100 | The maximum number of estimators at which boosting is terminated |
| | Learning rate | 1 | Weight applied to each classifier at each boosting iteration |
| MLP | Hidden layer sizes | 100 | The number of neurons in the $i^{th}$ hidden layer |
| | Activation function | ReLU | The activation function for the hidden layers |
| | Solver | Adam | The solver for weight optimization |
| | Learning rate | Adaptive | Schedule for weight updates |
| GNB | Var_ smoothing | 1.00E-09 | Portion of the largest variance of all features to be added to variances for stability |
| SVM | Kernel | RBF | Specifies the kernel type to be used in the algorithm |
| | C | 1 | Regularization parameter, controlling the trade-off between smooth decision boundary and classifying correctly |
| | Gamma | Scale | Kernel coefficient for 'rbf', 'poly' and 'sigmoid' |
| RF | Number of trees | 100 | The number of trees in the forest |
| | Min samples split | 2 | The minimum number of samples required to split an internal node |
| | Max features | Auto | The number of features to consider when looking for the best split |
| GBM | Learning rate | 0.1 | Shrinks the contribution of each tree by this value |
| | Number of estimators | 100 | The number of boosting stages to be run |
| | Max depth | 3 | Maximum depth of the individual regression estimators |
| | Subsample | 1 | The fraction of samples to be used for fitting the individual base learners |
| XGBoost | Learning rate | 0.1 | Step size shrinkage used to prevent overfitting. Represents the "eta" value |

(continued)

**Table 1** (continued)

| Algorithm | Parameter | Value | Description |
|---|---|---|---|
| | Number of estimators | 100 | Number of gradient boosted trees. Equivalent to the number of boosting rounds |
| | Max depth | 6 | Maximum tree depth for base learners |
| | Subsample | 0.8 | Subsample ratio of the training instances |
| | Colsample_bytree | 0.8 | Subsample ratio of columns when constructing each tree |

AdaBoost demonstrated moderate performance with an accuracy of 78% and an F1 Score of 77%. While AdaBoost is known for its ability to enhance the performance of weak learners, its relatively lower scores might indicate challenges in handling the complexity and noise inherent in air quality data. The equal precision and recall suggest that it maintains a balance between sensitivity and specificity, yet it does not excel in either.

The MLP classifier, with an accuracy of 83% and a slightly higher precision of 84%, indicates a reasonable capability to model the non-linear relationships in the data. The higher precision relative to recall suggests it might be more conservative in predicting high pollution levels, prioritizing minimizing false positives over capturing all true high pollution instances.

GNB showed the weakest performance among the algorithms with an accuracy of 68% and an F1 Score of 66%. The lower scores could be a result of the algorithm's assumption of feature independence, which is often violated in environmental datasets where factors influencing air quality are interrelated. Its lower precision and recall indicate a struggle in effectively differentiating between pollution levels.

SVM, with an accuracy and precision of 77%, performed moderately. Its performance might be limited by the choice of kernel or hyperparameters, suggesting that while SVMs are powerful in high-dimensional spaces, they require careful tuning to effectively capture the complexities of air pollution data.

Random Forest achieved an 84% score across all metrics, reflecting its strength in handling diverse datasets and robustness against overfitting. Its balanced performance indicates a good overall ability to capture the variance in air quality data without being biased towards overestimating or underestimating pollution levels.

GBM scored 89% in accuracy and precision, and 88% in the F1 Score, demonstrating its effectiveness in modeling complex datasets. The high scores suggest that GBM's stage-wise additive modeling is suitable for capturing the nuanced patterns in air quality data, although it falls slightly short of the best-performing models.

XGBoost, with scores of 92% across all metrics, was one of the top-performing algorithms. Its high scores can be attributed to its advanced handling of underfitting/overfitting and capability to process large and complex datasets effectively. The consistency across all metrics suggests it is well-tuned and capable of accurately predicting air pollution levels with a balanced approach towards precision and recall.

**Fig. 3** Normalized confusion matrix of **a** AM-LSTM **b** ADB **c** MLP **d** GNB **e** SVM **f** RF **g** GBM **h** XGBoost algorithms

The results clearly indicate that deep learning models like AM-LSTM and advanced ensemble methods like XGBoost and GBM are more suited for predicting air pollution, likely due to their ability to handle complex, non-linear relationships and temporal dependencies in the data. Traditional models like GNB and SVM, while useful in certain contexts, seem to underperform in this specific application,

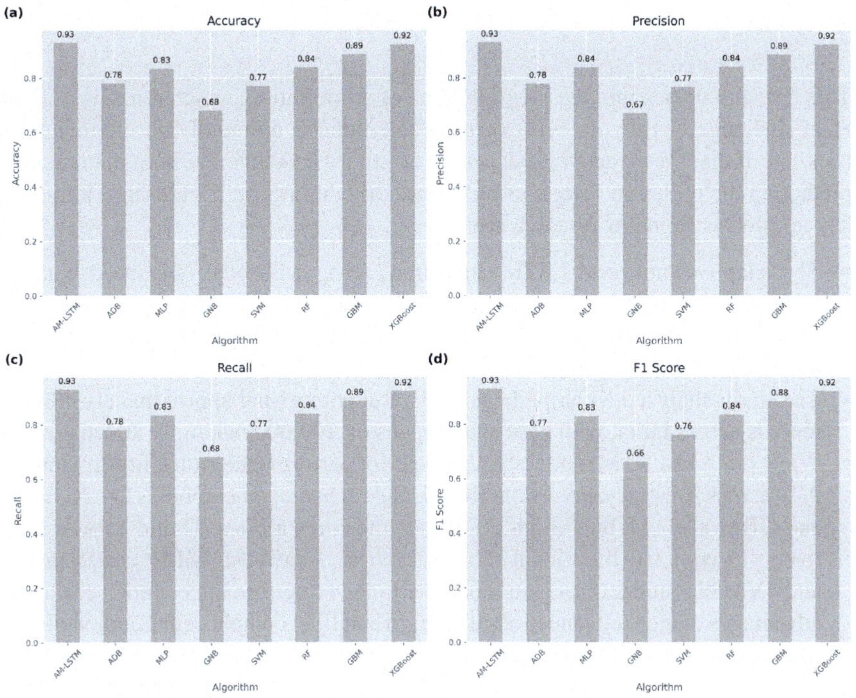


**Fig. 4** Performance comparison of the algorithms based on **a** Accuracy **b** Precision **c** Recall **d** F1 Score

possibly due to the multifaceted nature of environmental data and the need for more sophisticated feature handling and pattern recognition.

The uniform high performance of AM-LSTM across all metrics notably sets it apart, suggesting its potential as a reliable tool in environmental monitoring and policy-making. However, the computational complexity and the need for extensive data for training are factors that should be considered in practical applications.

In contrast, models like RF and GBM, though slightly less accurate than AM-LSTM and XGBoost, offer a good balance between performance and computational efficiency, making them suitable alternatives for scenarios where resources are limited. The moderate performance of algorithms like ADB and MLP indicates their potential utility in certain contexts, especially when combined with other methods or specific feature engineering techniques.

Overall, these results underscore the importance of choosing the right algorithm based on the specific characteristics of the environmental data and the objectives of the air pollution prediction task. The choice of model should balance accuracy, computational efficiency, and the ability to handle the complexity and nuances of environmental data.

## 4 Conclusions

This research delves into the pressing issue of air pollution, which remains a significant challenge worldwide, with a particular emphasis on rapidly developing countries like India. Our study introduces an advanced machine learning approach for predicting air pollution levels, using the adaptive multigrid LSTM algorithm. The key takeaways from our research are:

- The adaptive multigrid LSTM algorithm, along with other advanced machine learning models, has demonstrated substantial effectiveness in predicting air pollution levels. This highlights the potential of sophisticated algorithms in environmental monitoring.
- The study analyzed a comprehensive dataset comprising approximately 450,000 records across India, with a particular focus on 100,000 randomly drawn records.
- While our AM-LSTM model showed superior performance with uniform scores of 93% across Accuracy, Precision, Recall, and F1 Score, other models like XGBoost and GBM also exhibited high efficacy with scores around 92 and 89%, respectively. In contrast, traditional models such as AdaBoost, MLP, Gaussian NB, and SVM classifiers displayed moderate to lower performance, underscoring the advantages of more advanced algorithms in handling complex environmental data.
- The application of logistic gradient clustering for feature extraction was a key factor in enhancing the accuracy of predictions, demonstrating its effectiveness in preparing complex environmental data for analysis.

The predictive capability of our model can significantly aid in proactive health and policy measures, especially in anticipating high pollution days and formulating appropriate responses. Since recognizing air pollution as a global issue, the methodology and findings of this study can be applied in various geographical contexts, making it relevant for global environmental health management.

**Acknowledgements** The authors would like to thank their respective institutes for their kind encouragement.

**Funding Statement** This research received no specific grant from any funding agency in the public, commercial, or not-for-profit sectors.

**Data Availability** All the data or models analyzed during the study are included in this published article.

**Conflicts of Interest** The authors declare that they have no conflicts of interest to report regarding the present study.

## References


1. Ma, X., Chen, T., Ge, R., Xv, F., Cui, C., Li, J.: Prediction of PM2.5 concentration using spatiotemporal data with machine learning models. Atmosphere **14**, 1517 (2023). https://doi.org/10.3390/atmos14101517
2. Liao, Q., et al.: Deep learning for air quality forecasts: a review. Curr. Pollut. Rep. **6**, 399–409 (2020). https://doi.org/10.1007/s40726-020-00159-z
3. Liao, K., et al.: Statistical approaches for forecasting primary air pollutants: a review. Atmosphere **12**, 686 (2021). https://doi.org/10.3390/atmos12060686
4. Usharani, B., Sreedevi, M.: Deep learning techniques for air pollution prediction using remote sensing data. In: Smart Technologies in Data Science and Communication, pp. 107–123. Springer, Singapore (2021). https://doi.org/10.1007/978-981-16-1773-7_9
5. Liu, H., Yan, G., Duan, Z., Chen, C.: Intelligent modeling strategies for forecasting air quality time series: a review. Appl. Soft Comput. **102**, 106957 (2021). https://doi.org/10.1016/j.asoc.2020.106957
6. Gugnani, V., Singh, R.K.: Analysis of deep learning approaches for air pollution prediction. Multimed. Tools Appl. **81**, 1 (2022). https://doi.org/10.1007/s11042-021-11836-6
7. Al-Janabi, S., Mohammad, M., Al-Sultan, A.: A new method for prediction of air pollution based on intelligent computation. Soft. Comput. **24**, 661–680 (2020). https://doi.org/10.1007/s00500-019-04495-1
8. Middya, A.I., Roy, S.: Pollutant-specific optimal deep learning and statistical model building for air quality forecasting. Environ. Pollut. **301**, 118972 (2022). https://doi.org/10.1016/j.envpol.2022.118972
9. Guo, R., Zhang, Q., Yu, X., Qi, Y., Zhao, B.: A deep spatio-temporal learning network for continuous citywide air quality forecast based on dense monitoring data. J. Clean. Prod. **414**, 137568 (2023). https://doi.org/10.1016/j.jclepro.2023.137568
10. Huang, Y., et al.: A review of strategies for mitigating roadside air pollution in urban street canyons. Environ. Pollut. **280**, 116971 (2021). https://doi.org/10.1016/j.envpol.2021.116971
11. Hu, F., Guo, Y.: Health impacts of air pollution in China. Front. Environ. Sci. Eng. **15**, 1–18 (2021). https://doi.org/10.1007/s11783-020-1367-1
12. Yu, H., Zahidi, I.: Environmental hazards posed by mine dust, and monitoring method of mine dust pollution using remote sensing technologies: an overview. Sci. Total. Environ. **864**, 161135 (2023). https://doi.org/10.1016/j.scitotenv.2022.161135
13. Gautam, D., Bolia, N.B.: Air pollution: impact and interventions. Air Qual. Atmos. Health **13**, 209–223 (2020). https://doi.org/10.1007/s11869-019-00784-8
14. Ye, Y., Tao, Q.: The dynamic relationship among economic development, air pollution, and health production in China: the DNSBM efficiency model. Front. Environ. Sci. **11** (2023). https://doi.org/10.3389/fenvs.2023.1205712
15. Narasimhan, D., Vanitha, M.: Ambient air quality assessment using ensemble techniques. Soft. Comput. **25**, 9943–9956 (2021). https://doi.org/10.1007/s00500-020-05470-x
16. Hochreiter, S., Schmidhuber, J.: Long short-term memory. Neural Comput. **9**, 1735–1780 (1997). https://doi.org/10.1162/neco.1997.9.8.1735
17. Dey, K., Kalita, K., Chakraborty, S.: Prediction performance analysis of neural network models for an electrical discharge turning process. Int. J. Interact. Des. Manuf. (IJIDeM) **17**(2), 827–845 (2023). https://doi.org/10.1007/s12008-022-01003-y
18. Xu, H., Wu, H.: Accurate tsunami wave prediction using long short-term memory based neural networks. Ocean Model **186**, 102259 (2023). https://doi.org/10.1016/j.ocemod.2023.102259
19. Ganesh, N., Shankar, R., Mahdal, M., Murugan, J.S., Chohan, J.S., Kalita, K.: Exploring deep learning methods for computer vision applications across multiple sectors: challenges and future trends. CMES-Comput. Model. Eng. Sci. **139**(1) (2024). https://doi.org/10.32604/cmes.2023.028018
20. Ferreira, J.M., Pires, I.M., Marques, G., Garcia, N.M., Zdravevski, E., Lameski, P., Flórez-Revuelta, F., Spinsante, S.: Identification of daily activities and environments based on the

adaboost method using mobile device data: a systematic review. Electronics **9**(1), 192 (2020). https://doi.org/10.3390/electronics9010192

21. Burande, D.V., Kalita, K., Gupta, R., Kumar, A., Chohan, J.S., Kumar, D.: Machine learning metamodels for thermo-mechanical analysis of friction stir welding. Int. J. Interact. Des. Manuf. (IJIDeM) 1–19 (2024). https://doi.org/10.1007/s12008-024-01871-6
22. Sardar, A.A.M., Sanzidul Islam, M., Bhuiyan, T.: A review on automatic speech emotion recognition with an experiment using multilayer perceptron classifier. In: Soft Computing Techniques and Applications: Proceeding of the International Conference on Computing and Communication (IC3 2020), pp. 381–388. Springer, Singapore (2021). https://doi.org/10.1007/978-981-15-7394-1_36
23. Kalita, K., Ganesh, N., Jayalakshmi, S., Chohan, J.S., Mallik, S., Qin, H.: Multi-objective artificial bee colony optimized hybrid deep belief network and XGBoost algorithm for heart disease prediction. Front. Digital Health **5**, 1279644 (2023). https://doi.org/10.3389/fdgth.2023.1279644
24. Reddy, E.M.K., Gurrala, A., Hasitha, V.B., Kumar, K.V.R.: Introduction to Naive Bayes and a review on its subtypes with applications. In: Bayesian Reasoning and Gaussian Processes for Machine Learning Applications, pp. 1–14 (2022). https://doi.org/10.1201/9781003164265-1
25. Ganesh, N., Balamurugan, M., Chohan, J.S., Kalita, K.: Development of a grey wolf optimized-gradient boosted decision tree metamodel for heart disease prediction. Int. J. Intell. Syst. Appl. Eng. **12**(8s), 515–522 (2024)
26. Pisner, D.A., Schnyer, D.M.: Support vector machine. In: Machine Learning, pp. 101–121. Academic (2020). https://doi.org/10.1016/B978-0-12-815739-8.00006-7
27. Narayanan, G., Joshi, M., Dutta, P., Kalita, K.: PSO-tuned support vector machine metamodels for assessment of turbulent flows in pipe bends. Eng. Comput. **37**(3), 981–1001 (2020). https://doi.org/10.1108/EC-05-2019-0244
28. Sheykhmousa, M., Mahdianpari, M., Ghanbari, H., Mohammadimanesh, F., Ghamisi, P., Homayouni, S.: Support vector machine versus random forest for remote sensing image classification: a meta-analysis and systematic review. IEEE J. Sel. Top. Appl. Earth Obs. Remote Sens. **13**, 6308–6325 (2020). https://doi.org/10.1109/JSTARS.2020.3026724
29. Jain, P., Choudhury, A., Dutta, P., Kalita, K., Barsocchi, P.: Random forest regression-based machine learning model for accurate estimation of fluid flow in curved pipes. Processes **9**(11), 2095 (2021). https://doi.org/10.3390/pr9112095
30. Konstantinov, A.V., Utkin, L.V.: Interpretable machine learning with an ensemble of gradient boosting machines. Knowl.-Based Syst. **222**, 106993 (2021). https://doi.org/10.1016/j.knosys.2021.106993
31. Kalita, K., Haldar, S., Chakraborty, S.: A comprehensive review on high-fidelity and metamodel-based optimization of composite laminates. Arch. Comput. Methods Eng. 1–36 (2022). https://doi.org/10.1007/s11831-021-09699-z
32. Ali, Z.A., Abduljabbar, Z.H., Taher, H.A., Sallow, A.B., Almufti, S.M.: Exploring the power of eXtreme gradient boosting algorithm in machine learning: a review. Acad. J. Nawroz Univ. **12**(2), 320–334 (2023). https://doi.org/10.25007/ajnu.v12n2a1612
33. Kalita, K., Ganesh, N., Balamurugan, S., (eds.).: Metaheuristics for Machine Learning: Algorithms and Applications. Wiley, New York (2024). https://doi.org/10.1002/9781394233953

# A Hybrid Approach for Optic Disc Localization in Eye Fundus Images

**Oscar Ramos-Soto, Erick Rodríguez-Esparza, Francisco Carrasco-Hernández, Gonzalo Pajares, and Sandra E. Balderas-Mata**

**Abstract** Visual examination of eye structures like the optic disc (OD) allows precise detection of conditions such as glaucoma, diabetic retinopathy, and other abnormalities. Medical image processing, with a focus on ophthalmic imaging, plays a crucial role in the diagnosis and grading of retinal pathologies. In this chapter, an automated approach for OD localization in eye fundus images is introduced. The method involves a vessel diminishing stage utilizing the Optimized Top-Hat transformation, with parameter optimization performed by the RUNge Kutta metaheuristic optimizer (RUN). Subsequently, OD segmentation is carried out using the Minimum Cross-Entropy Thresholding-Harris Hawks Optimization (MCET-HHO) method, followed by centroid calculation and image cropping for final OD detection. The proposed method is assessed across three datasets, encompassing a variety of eye conditions and imaging settings. A numerical evaluation employing key performance metrics compares the calculated OD center coordinates and segmentations with manual expert marks, where accuracy values of 99.69% and 99.77% were achieved for

O. Ramos-Soto (✉) · S. E. Balderas-Mata
Depto. de Ingeniería Electro-Fotónica, CUCEI, Universidad de Guadalajara, Guadalajara, Jalisco, Mexico
e-mail: oscar.ramos9279@alumnos.udg.mx

S. E. Balderas-Mata
e-mail: sandra.balderas@academicos.udg.mx

E. Rodríguez-Esparza
Faculty of Engineering, DeustoTech, University of Deusto, Av. Universidades, 24, 48007 Bilbao, Spain
e-mail: erick.rodriguez@deusto.es

F. Carrasco-Hernández
Engineering Department in Mechatronics and Renewable Energy, Technological University of Durango, Carr. Durango - Mezquital, Km. 4.5 S/N, Durango, Durango, Mexico
e-mail: francisco.carrasco@utd.edu.mx

G. Pajares
Depto. de Ingeniería de Software e Inteligencia Artificial, Universidad Complutense de Madrid, Madrid, Spain
e-mail: pajares@ucm.es

 
D. Oliva et al. (eds.), *Advances in Optimization Algorithms for Multidisciplinary Engineering Applications: From Classical Methods to AI-Enhanced Solutions*, Studies in Computational Intelligence 806, https://doi.org/10.1007/978-3-031-78440-8_13

normal and glaucomatous images, respectively. This methodology serves as a robust pre-processing tool for OD analysis, demonstrating great robustness in a wide range of fundus images.

## 1 Introduction

In the last decades, the medical domain has experienced a transformative synergy with technological advances, bringing medical specialists useful tools for enhanced diagnostics and disease management, making clinical work easier and more efficient. In particular, medical image analysis has served as an element that leads to more effective interventions and improved patient outcomes by achieving early disease detection, pathology classification, and planning a medical follow-up [1].

In this context, the ophthalmic field has had a significant impact, derived from the increased use of medical images in its development. The human eye, a marvel of biological engineering, works as the gateway to our visual perception and plays a crucial role by providing information from our surroundings. Anatomically, it comprises several elements with specific functions that collectively contribute to its optimal operation. With the aid of ophthalmic imaging devices, such as the ophthalmoscope, it becomes possible to visually inspect the inner part of the eye [2], known as the retina.

In a general way, three main structures within the eye are notably observed: blood vessels, the macular zone, and the optic disc (OD), with the latter being particularly significant for the focus of this study. Figure 1 illustrates an example of an eye fundus image captured using an ophthalmoscope, highlighting the most relevant ophthalmic structures.

Ophthalmic imaging analysis, particularly the OD localization in eye fundus images, is essential for early detection and monitoring of eye diseases. For instance, in conditions like glaucoma, precisely analyzing the OD is crucial for identifying early signs and preventing vision loss [3]. Glaucoma is characterized by a rise in intraocular pressure, which affects the optic nerve and causes a distortion in the

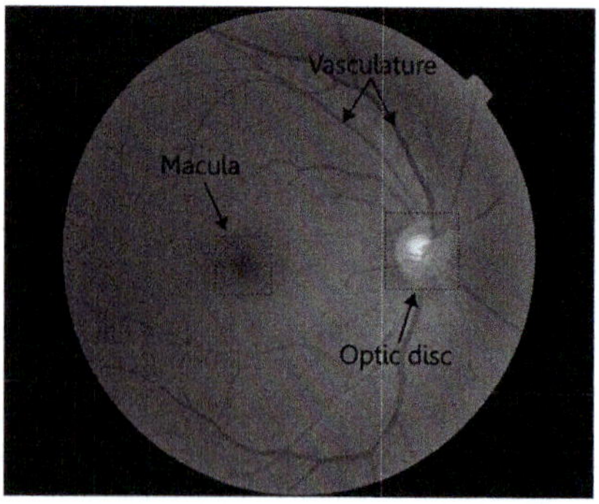


**Fig. 1** Main retinal elements in the eye fundus image

cup-to-disc ratio [4]. Moreover, in diabetic retinopathy, the OD localization is vital for determining the severity of the disease and its impact on the retinal vasculature. In proliferative diabetic retinopathy (PDR), abnormal neovascularization can extend into the OD, causing structural changes and potential complications [5]. Also, diabetic macular edema, prevalent in diabetic retinopathy, may lead to fluid accumulation in the macula, with ramifications extending to the OD, causing OD edema [6]. Beyond these, comprehensive OD analysis proves to be crucial in identifying retinal abnormalities such as exudates, drusen, microaneurysms, and hemorrhages, offering valuable insights into diverse retinal conditions [7–9].

As depicted in Fig. 2, the characteristics of the image vary significantly based on the OD condition. This underscores the imperative for a robust OD detection method that remains effective irrespective of the condition of this region.

Regular screenings and comprehensive assessments of fundus images not only contribute to the early detection of the above-mentioned conditions but also pave the way for timely interventions. In this context, the development of methodologies for accurate and efficient localization of OD holds great promise for the advancement of ophthalmologic diagnosis.

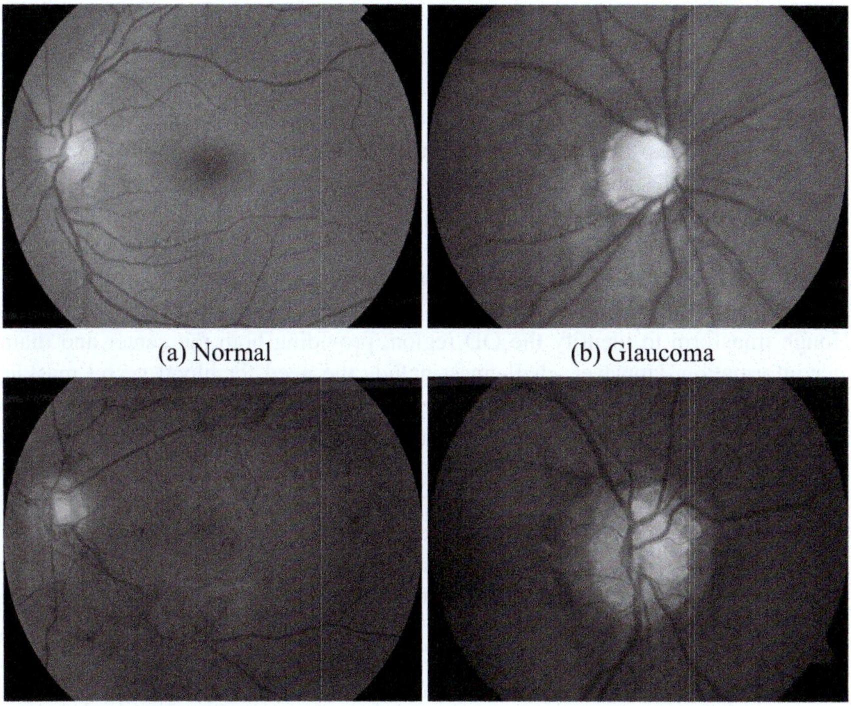

(a) Normal (b) Glaucoma

(c) PDR (d) Drusen

**Fig. 2** Different OD conditions

The integration of metaheuristic techniques and medical image processing creates a powerful synergy between computational intelligence and clinical applications. This proposed methodology aims to contribute to the broader landscape of ophthalmic diagnostics. By merging technological innovation and medical insight, this chapter explores a new approach to automated OD detection, which holds promise as a pre-processing tool for early disease diagnosis and monitoring through ophthalmic imaging analysis.

The remaining sections are structured as follows: Sect. 2 provides a comprehensive review of previously proposed methods for ophthalmic image processing, focusing on OD segmentation. Following this, each processing stage is individually detailed in Sect. 3, providing a comprehensive insight to the reader. Section 4 elucidates the integration of all stages, encompassing considerations in the hybridization of these steps. The experimental framework is outlined in Sect. 5, featuring a concise description of other optimizers considered for comparison, the performance metric used, and the criteria for selecting the dataset. Subsequently, Sect. 6 presents both numerical and visual analyses, emphasizing the strengths of the proposed method. Finally, the conclusions drawn from this work are summarized in Sect. 7.

## 2 Related Work

In the current literature, methods for OD segmentation have been broadly divided into two categories: unsupervised and supervised algorithms [10, 11]. Shanthi et al. have further classified these methods into various types, including edge detection, threshold-based, region-based, template-based, pattern recognition, and deep learning techniques [12].

Edge-based methods are employed to identify each point in the fundus image as background or foreground. Recently, Ali et al. [13] used techniques like the Hough transform to identify the OD region, providing both the center and diameter information. However, challenges include the need for blood vessel masking and the dependence on specific image operators, leading to potential discontinuities. Although edge detection methods use multiple operators to identify and separate various objects in retinal fundus images, they lack a universal operator suitable for all images. Moreover, generating a discontinuous edge is a potential limitation.

Regarding threshold-based methods are commonly employed to convert multi-level color fundus images into binary images by selecting an appropriate threshold to categorize the fundus images into distinct regions. Shanthamalar et al. adopted the dynamic gray-level thresholding approach, adjusting to varying intensities, and employed morphological operations with two disk structuring elements to eliminate the OD region [14]. It is worth mentioning that while thresholding techniques offer advantages in terms of simplicity, cost-effectiveness, and real-time interaction, they come with notable limitations. These include a critical dependence on the threshold value, sensitivity to noise, and a tendency to overlook spatial information in retinal fundus images.

Region-based methods segment the OD region by dividing the fundus image into distinct regions based on its features, like intensity and color. In the work of Kim et al., a morphological operation is employed to remove the blood vessels and substituted based on intensity values from adjacent non-vessel pixels. Subsequently, the OD is extracted using image features such as shape, intensity, and size [15]. In 2022, Yi et al. enhanced the OD segmentation based on modified particle swarm optimization, applying iterative updates to optimize the contour of the segmented OD [16]. Region-based techniques generally provide accurate OD boundaries but may suffer from challenges in defining stopping conditions and choosing initial seeds.

Template-based methods serve the purpose of identifying the optimal match between models and candidates. In a recent work, Shalchi et al. presented an innovative OD detection approach employing the grasshopper optimization algorithm. This method entails the evaluation of a fitness function and the iterative adjustment of the optimal positions of individual grasshoppers to enhance the results progressively. The procedural steps involve the removal of vessels from grayscale images through morphological operations and subsequent utilization of the bat algorithm to optimize the image's threshold value for precise OD localization. Finally, the authors used ellipse fitting to refine and smooth the obtained results [17]. Template-based methods are concise but sensitive to noise.

On the other hand, pattern recognition techniques involve two primary steps: formulating decision rules and classifying fundus images based on these rules. Naqvi et al. proposed constructing the OD boundary using local Laplacian filtering in regions affected by major vascular structures and utilizing gradient-independent active contour estimation for unconstrained OD boundary detection [18]. Pattern recognition techniques establish input-output relationships but are limited by shape features and complexity.

In recent times, medical images have been analyzed using deep learning techniques. The convolutional neural network is trained to identify and isolate the OD to prevent false positive detection. Bright regions like exudates and blood vessels are the two primary causes of false positive detection. Carvalho et al. used a hybrid machine learning model with a conditional generative adversarial network for rapid and accurate results [19]. Moving to 2022, Joshi et al. applied a CNN architecture for precise OD segmentation [20], while Latif et al. employed a Shallow CNN to detect the OD [21]. Yu et al. proposed a combination of deep CNN and generative adversarial networks for OD and optic cup segmentation [22]. In the same year, Hervella et al. presented a multi-task architecture for simultaneous OD segmentation and glaucoma classification [23]. Xiong et al. introduced a weak label-based Bayesian UNet for pixel-level OD annotation [24].

The algorithms discussed above present diverse limitations, particularly in cases where reliance on the OD's appearance leads to false region detection. Difficulties in specifying a threshold value and using reliable edge detection methods contribute to the identification of exudates or blood vessels instead of accurately detecting the OD. Similar challenges emerge in methods utilizing these approaches during the preprocessing step. Additionally, algorithms based on deep learning face obstacles such as lengthy training times and the requirement for high-end computing resources. These limitations hinder their practicality in various clinical applications and pose challenges for real-time diagnosis. To address these weaknesses, this study proposes an innovative and novel method.

## 3 Preliminary Concepts

This section introduces the most relevant stages of the study, to provide a thorough understanding of the method by detailing each step. Specifically, the approaches used in this study involve the utilization of the Optimized top-hat for vessel diminishing, the description of the RUNge-Kutta optimizer (RUN), and the integration of the Harris Hawks Optimization (HHO) algorithm with the Minimum Cross-Entropy Thresholding (MCET-HHO) with multilevel image segmentation purposes.

### *3.1 Optimized Top-Hat*

Morphological operations are a type of image processing that relies on the shape or morphology of features in an image. These operations involve the use of a structuring element, which is a small shape or template, to analyze or modify the spatial arrangement of pixels in an image [25]. The two fundamental morphological operations are dilation and erosion, and they serve as building blocks for more complex operations. Morphological operations have been widely used in image processing, including in biomedical applications [26, 27].

In morphological transformations, one of the widely recognized operations is the Top-hat transform, also known as the White top-hat transform. This transformative process involves two fundamental morphological operations: opening and subtraction. Its application is extensive, particularly in enhancing small details, especially in images characterized by uneven illumination or subtle edges. Various adaptations of this transform have been employed in the analysis of the eye fundus images [28, 29]. Let $I$ represent the input image, and $S_o$ denote the structural element for closing. The original top-hat transform is formally defined in Eq. 1:

$$TH = I - (I \circ S_o) \tag{1}$$

where ($\circ$) denotes the opening operator.

Nevertheless, when applying the classical top-hat transform to eye fundus images, some small edges may go undetected, leading to a lack of detailed structure enhancement. To overcome these issues, the authors have developed an improved modification of the top-hat transform [11]. This modification involves a sequence of eroding, dilating, and image complement operations. The optimized version, referred to as the optimized top-hat, exhibits enhanced vessel-background contrast compared to classical implementations. The formulation of the modified top-hat transform is presented in Eq. 2:

$$TH_{opt} = I^c - (I^c \circ S_o) \bullet S_c \tag{2}$$

where $S_o$ and $S_c$ represent the structuring elements corresponding to the opening ($\circ$) and closing ($\bullet$) operators, respectively. On the other hand, $I^c$ denotes the complement of $I$, which represents the difference between the maximum intensity value and the corresponding pixel value of $I$. For this scenario, vessels have to be diminished (especially on the OD zone), so a slight modification is done by adding up $I$ and $TH_{opt}$ to obtain a vessel-diminished image $I_{dim}$. The latter is expressed in Eq. 3:

$$I_{dim} = I + TH_{opt} \tag{3}$$

As mathematically represented above, the response of this transformation relies totally on the $S_o$ and $S_c$ values, hence the importance of optimal parameter selection.

### 3.2 RUNge Kutta Optimizer (RUN)

The RUN optimizer [30] is a metaphor-free metaheuristic introduced in 2021, leveraging slight variations from other metaheuristic proposals. It draws inspiration from the Runge Kutta method for solving ordinary differential equations [31, 32]. The optimizer comprises two main operators: the runge-kutta root (solution) search mechanism and the enhanced solution quality operator. The efficiency of RUN has been demonstrated in diverse applications, including renewable energy [33], image segmentation [34], and gene expression for cancer classification [35].

In metaheuristic algorithms, the typical initial step involves generating an initial population. For initializing search agents, the initial swarm distribution of $N$ positions in the $D$-dimensional search space is determined by Eq. 4:

$$x_{n_d} = L_d + \text{rand}(U_d - L_d) \tag{4}$$

where rand is a random number within [0, 1], and $L_d$ and $U_d$ are the lower and upper bounds of the $d$th variable.

To update solutions, the RUN algorithm employs the fourth-order Runge-Kutta (RK4) method for optimization [30]. The update of each agent's position is defined in Eq. 5. This expression distinguishes between two scenarios: when rand $< 0.5$, the algorithm emphasizes exploration, introducing randomness to the search process. In contrast, the second case occurs when rand $\geq 0.5$ and represents exploitation, focusing on refining solutions based on current knowledge. This dual approach enhances the algorithm's adaptability to both exploration and exploitation aspects.

$$x_{n+1} = \begin{cases} (x_c + \kappa \cdot x_c) + SF \cdot SM + \mu \cdot (\text{rand} \cdot (x_m - x_c)) & \text{rand} < 0.5 \\ (x_m + \kappa \cdot x_m) + SF \cdot SM + \mu \cdot (\text{rand} \cdot (x_{r1} - x_{r2})) & \text{otherwise} \end{cases} \quad (5)$$

where $\kappa = r \cdot SF \cdot g$, $r$ takes values of -1 or 1, $\mu = 0.5 + 0.1 \cdot \text{rand}$, and $g$ is a random number within [0, 2]. The variables $x_c, x_m, \kappa$, and $SF$ are defined in Eqs. 6–8.

$$x_c = \phi \cdot x_n + (1 - \phi) \cdot x_{r1} \quad (6)$$

$$x_m = \phi \cdot x_{best} + (1 - \phi) \cdot x_{dbest} \quad (7)$$

$$SF = 2 \cdot (0.5 - \text{rand}) \cdot a \cdot exp\Big( - b \cdot \text{rand} \cdot \Big(\frac{It}{MaxIt}\Big)\Big) \quad (8)$$

Here, $\phi$ is a random number within the range [0, 2], $x_{best}$ and $x_{dbest}$ represent the best solution so far and the best solution obtained at each iteration, $It$, respectively. Additionally, $a$ and $b$ are control parameters of $SF$, and $SM$ is defined in the mathematical analysis of the Runge-Kutta method in this chapter's appendix.

To improve solution quality, the enhanced solution quality operator enhances the quality of solution candidates. The expression in Eq. 9 is used to generate the solution ($x_{new2}$). The enhanced solution quality block is executed when rand < 0.5; if this condition is true, a criterion based on $w$ is evaluated. The variables are defined as $w = \text{rand}(0, 2) \cdot exp(-c(\frac{i}{MaxIt}))$, $x_{new1} = \eta \cdot x_{avg} + (1 - \eta) \cdot x_{best}$, and $x_{avg}$ is the average of three solutions. Here, $c$ is defined as $5 \cdot \text{rand}$, while $\eta$ is a random number constrained within the range [0, 1]. Similar to Eq. 5, $r$ is an integer that can take values of 1, 0, or -1.

$$x_{new2} = \begin{cases} x_{new1} + r \cdot w \cdot |(x_{new1} - x_{avg}) + \text{rand}| & w < 1 \\ (x_{\text{new1}} - x_{avg}) + r \cdot w \cdot |(u \cdot x_{new1} - x_{\text{avg}}) + \text{rand}| & \text{otherwise} \end{cases} \quad (9)$$

Furthermore, a refinement step, as described in Eq. 10, has been incorporated to steer $x_{new2}$ towards an enhanced position when the condition rand < $w$ is true. This additional step helps to improve the fitness of the candidate. Also, it's worth mentioning that although $x_{best2}$ represents an improved solution among all candidates, it doesn't necessarily guarantee the best fitness level.

$$x_{new3} = (x_{new2} - \text{rand} \cdot x_{new2}) + SF \cdot (\text{rand} \cdot x_{RK} + (v \cdot x_b - x_{new2})) \quad (10)$$

Here, $v$ represents a random number with a value defined as $2 \cdot \text{rand}$. As mentioned above, the objective of Eq. 10 is to guide the solution $x_{new2}$ towards an improved position. The coefficient $v$ is introduced to underscore the significance of the best solution. It is important to highlight that, in calculating $x_{RK}$ (formally detailed in the appendix), the solutions $x_b$ and $x_w$ are substituted with $x_k$ and $x_{new2}$, respectively, due to the lower fitness value of $x_n$ compared to $x_{new2}$, so $f(x_{new2}) > f(x_n)$.

### 3.3 Harris Hawks Optimization (HHO)

The HHO, introduced by Heidari et al. in 2019 [36], emulates the collaborative hunting behavior of Harris Hawks. Inspired by their *seven kills* approach, HHO features exploration and exploitation phases. In the HHO algorithm, a population of hawks represents candidate solutions, and the best solution found is referred to as the *rabbit*. The adaptability to different scenarios is intrinsic to this algorithm, reflecting the dynamic hunting styles of hawks based on prey behavior.

In exploration, hawks adapt using random and guided strategies ($X(t+1)$) based on $q$, as described in Eq. 11. The transition to exploitation is controlled by the dynamic energy parameter $E$. The energy value is updated using Eq. 12, where $E_0$ is a randomly generated parameter within the range of $(-1, 1)$. When $|E| \geq 1$, it corresponds to the exploration phase, while $|E| < 1$ indicates the exploitation phase.

$$X(t+1) = \begin{cases} X_{rand}(t) - r_1 \left|X_{rand}(t) - 2r_2 X(t)\right| & \text{if } q \geq 0.5 \\ (X_{rabbit}(t) - X_m(t)) - r_3(LB + r_4(UB - LB)) & \text{if } q < 0.5 \end{cases} \tag{11}$$

$$E = 2E_0(1 - \frac{t}{T}) \tag{12}$$

In the exploitation phase, strategies like soft besiege (Eq. 13), hard besiege (Eq. 14), and progressive rapid dives refine solutions, outlined in Eq. 15 for the soft besiege with progressive rapid dives strategy and in Eq. 16 the hard besiege with progressive rapid dives strategy. These strategies adapt dynamically based on the rabbit's energy state.

$$X(t+1) = \Delta X(t) - E \left|J X_{rabbit}(t) - X(t)\right| \tag{13}$$

$$X(t+1) = X_{rabbit}(t) - E \left|\Delta X(t)\right| \tag{14}$$

where the displacement vector $\Delta X(t)$ is defined as the difference between the rabbit's position $X_{rabbit}(t)$ and the current position of the hawk $X(t)$. The parameter $J$ is intricately tied to the jump strength of the rabbit and is given by $J = 2(1 - r_5)$.

$$X(t+1) = \begin{cases} Y \ if \ F(Y) < F(X(t)) \\ Z \ if \ F(Z) < F(X(t)) \end{cases} \tag{15}$$

Here, $Y$ and $Z$ are calculated by Eqs. 17 and 18, respectively.

$$X(t+1) = \begin{cases} Y' \quad if \ F(Y') < F(X(t)) \\ Z' \ if \ F(Y + Z') < F(X(t)) \end{cases} \tag{16}$$

Variables $Y'$ and $Z'$ are computed using the Eqs. 19 and 20.

$$Y = X_{rabbit}(t) - E \left|J X_{rabbit}(t) - X(t)\right| \tag{17}$$

$$Z = Y + S \times LF(D) \tag{18}$$

$$Y' = X_{rabbit}(t) - E\,|JX_{rabbit}(t) - X_m(t)| \tag{19}$$

$$Z' = Y' + S \times LF(D) \tag{20}$$

where $S$ is a vector by size $1 \times$ D and $LF$ shows the levy flight function, which is computed by Eq. 21:

$$LF(x) = \frac{u \times \sigma}{|v|^{\frac{1}{\beta}}}, \sigma = \left( \frac{\Gamma(1+\beta) \times sin(\frac{\pi\beta}{2})}{\Gamma(\frac{1+\beta}{2}) \times \beta \times 2^{(\frac{\beta-1}{2})}} \right)^{\frac{1}{\beta}} \tag{21}$$

where $u$, $v$ are values of the $LF$ randomly defined and $\beta$ is 1.5.

The algorithm dynamically adapts its exploration and exploitation strategies according to the energy levels of the "rabbit", allowing for efficient and effective convergence. The HHO algorithm requires two main input parameters: the swarm size ($N$) and the maximum number of iterations ($T$). Its goal is to find the best solution ($X_{rabbit}$) by emulating the cooperative hunting behavior of Harris hawks.

### 3.4 Minimum Cross-Entropy Thresholding (MCET)

The MCET algorithm is a technique designed to determine the optimal threshold for image segmentation. It achieves this by minimizing the statistical criterion of cross-entropy between the input grayscale image and its histogram [37]. The grayscale image ($I_{Gr}$) is then thresholded using the determined threshold, resulting in a binary segmented image with foreground and background areas.

For bilevel thresholding, the segmented image ($I_{th}$) is defined Eq. 22 and $\mu$ is defined in Eq. 23 [38]:

$$I_{th}(x, y) = \begin{cases} \mu(1, th), & \text{if } I_{Gr}(x, y) < th \\ \mu(th, L+1), & \text{if } I_{Gr}(x, y) \geq th \end{cases} \tag{22}$$

$$\mu(a, b) = \sum_{i=a}^{b-1} ih(i) \Bigg/ \sum_{i=a}^{b-1} h(i) \tag{23}$$

The cross-entropy is calculated as the objective function $f_{MCET}(th)$ by evaluating the entropy values for the segmented image. This function considers both sides of the threshold, capturing the divergence between the original and segmented images [39].

For multilevel thresholding, a vector $th = [th_1, th_2, \ldots, th_{nt}]$ is introduced. However, the computational cost can be high. To address this, Yin [40] proposed a faster recursive programming technique, reducing $f_{MCET}(th)$ to Eq. 24:

$$f_{MCET}(th) = \sum_{i=1}^{L} ih(i)log(i) - \sum_{i=1}^{nt} H_i \tag{24}$$

where $nt$ is the total number of thresholds, and $H_i$ is defined in Eq. 25:

$$\begin{aligned} H_1 &= \sum_{i=1}^{th_1-1} ih(i)\log(\mu(1, th_1)) \\ H_k &= \sum_{i=th_{k-1}}^{th_k-1} ih(i)\log(\mu(th_{k-1}, th_k)), \qquad 1 < k < nt \\ H_{nt} &= \sum_{i=th_{nt}}^{L} ih(i)\log(\mu(th_{nt}, L+1)) \end{aligned} \tag{25}$$

In practical terms, the MCET, implemented within the HHO algorithm, iteratively refines threshold values to improve image segmentation. The algorithm seeks to minimize cross-entropy, promoting higher homogeneity and lower uncertainty in the resulting segmented image. The MCET-HHO approach offers a powerful tool for image segmentation tasks, balancing computational efficiency with segmentation quality [41].

## 4 Proposed Method

Broadly speaking, in the context of image processing, a metaheuristic approach enhances the performance of a technique by iteratively searching for optimal values within a specific transformation function. This process is evaluated using a fitness function. The metaheuristic algorithm explores the solution space, seeking the combination of parameters that maximizes or minimizes the fitness function. In other words, the metaheuristic fine-tunes the transformation function to enhance its effectiveness in a specific image processing task.

In Fig. 3, a graphical workflow of this proposal is shown. Initially, the input RGB retinal image undergoes channel separation to extract the red channel. Subsequently, the image is subjected to an iterative vessel-diminishing process, employing the optimized top-hat as the transformation function. This process involves the assessment of a fitness function, and the input parameters of the optimized top-hat are further refined using the RUN metaheuristic algorithm. Upon reaching the stop criterion, the vessel-removed image is thresholded for multilevel segmentation via the MCET-HHO to delineate the OD. In the final step, the centroid of the OD segmentation is calculated to facilitate subsequent cropping of the region of interest.

Next, all steps involving the OD detection proposal are detailed below, including all the considerations for its design.

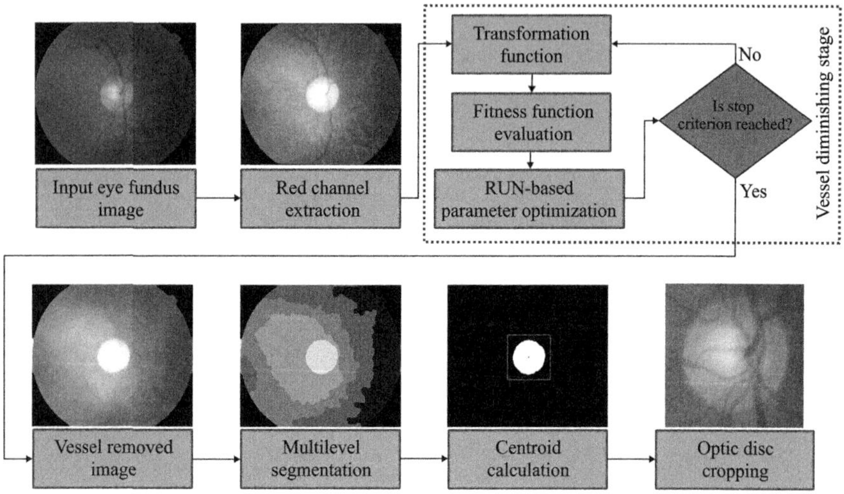


**Fig. 3** General diagram of the proposed method

## 4.1 Color Channel Selection

In a RGB eye fundus image, each channel (red, green, and blue) encapsulates unique characteristics that collectively contribute to the overall composition of retinal elements. Analyzing these channels individually provides valuable insights into the specific features of each color component, and further processing is tailored based on the intended approach. Visual representations of the red, green, and blue channels are depicted in Fig. 4.

The red channel proves to be a valuable tool for emphasizing the OD due to the absence of melanin in this specific area [42], facilitating efficient localization, as illustrated in Fig. 4a, d. Moving on to the green channel provides an overall view of the retinal anatomy, and its balanced composition ensures that details related to both blood vessels and non-vascular structures are discernible, owing to the hemoglobin absorption spectral range [43, 44], as demonstrated in Fig. 4b, e. On the other hand, the blue channel usually exhibits lower contrast for blood vessels or the OD compared to the red and green channels, as presented in Fig. 4c, f.

Considering the above description of each channel, the red one is taken out to the next stage since the OD analysis is intended.

## 4.2 Transformation Function

As described in Sect. 3, the Optimized top-hat was initially defined for vessel enhancement purposes as described in Eq. 2. However, this approach is modified for vessel reduction and mathematically defined in Eq. 3. This function requires the

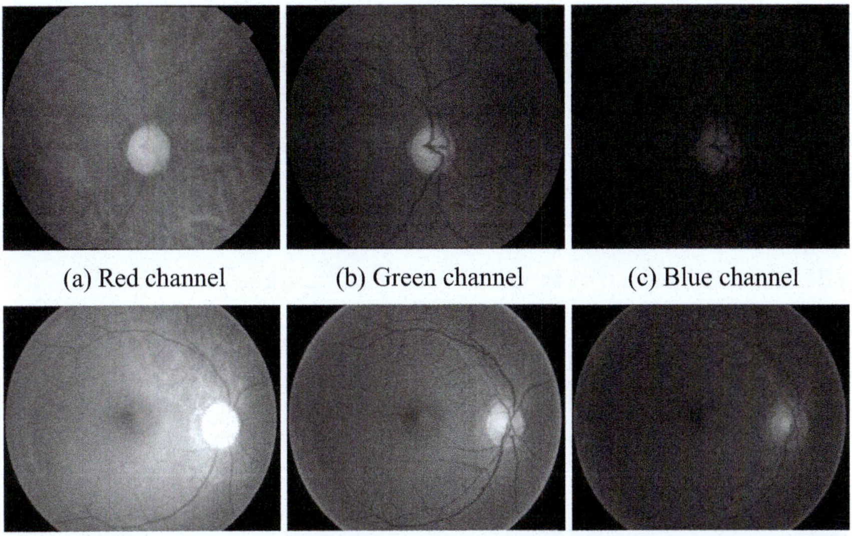

(a) Red channel (b) Green channel (c) Blue channel

(d) Red channel (e) Green channel (f) Blue channel

**Fig. 4** Visual representation of red, green, and blue channels of two different eye fundus images: **a**–**c** and **d**–**f**

configuration of two values ($S_o$ and $S_c$), whose performance depends on the parameters value combination. As a visual example, a comparison of different $S_o$ and $S_c$ values combination applied to vessel diminishing is shown in Fig. 5.

In Fig. 5a, the red channel of an example eye fundus image is displayed, still revealing noticeable vasculature. Moving to Fig. 5b, a subtle diminishing effect is observed, with thick vessels remaining visible. Increasing the $S_c$ value in Fig. 5c leads to further vasculature reduction. However, elevating both $S_o$ and $S_c$ values, as seen in Fig. 5d, e, results in a darker image, even when these values are substantially apart or the same.

The manual identification of optimal values for vessel diminishing in each image represents a challenge and diminishes the automation efficiency of this approach. The RUN metaheuristic algorithm becomes useful to overcome this issue by identifying the best value combinations for each image, representing a diverse search space. In Fig. 5f, vessel removal is executed using the optimal values $S_o = 18.2$ and $S_c = 7.4$ determined by the RUN algorithm. This process effectively eliminates both thick and thin vessels from the OD and retina while preserving the eye fundus boundary and OD limits. Noteworthy is the rounding of $S_o$ and $S_c$ to structural elements (technically represented as an array of pixels), denoted as 18 and 7, respectively, when the transformation function is applied. It is noteworthy to mention that vessels initially appearing in the OD zone are also removed, which makes further steps more accurate.

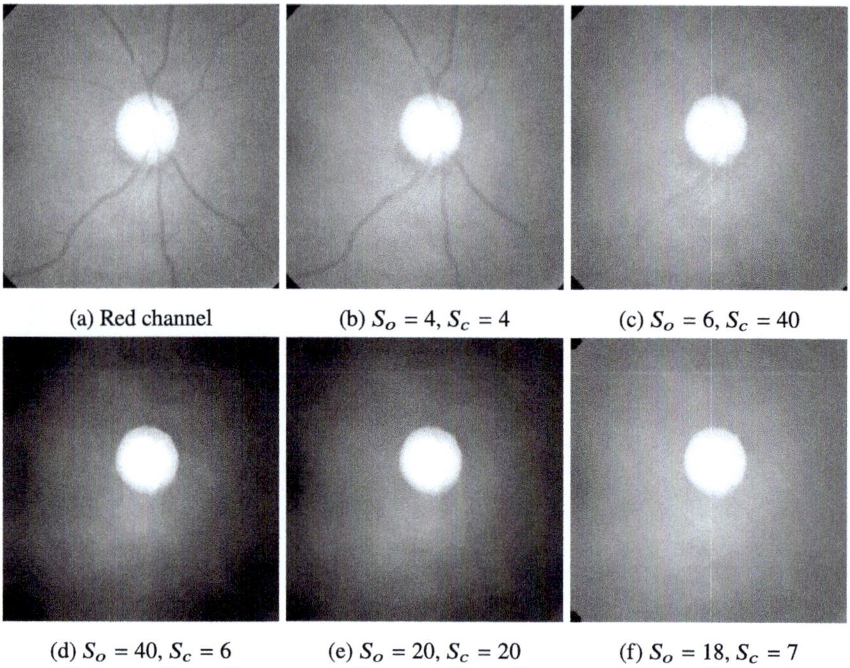


**Fig. 5** Comparison of different $S_o$ and $S_c$ values for optimized top hat

## 4.3 Fitness Function

To iteratively evaluate the performance of the parameter optimization for the transformation function, a new fitness function specifically designed for vessel diminishing and artifact removal is proposed. This design is based on the following premise: keep OD edges delimited while smoothing any elements on the retinography, including vessels on the OD.

Taking into account the above, the first element considered is the Standard Deviation (SD), which measures the amount of variation or dispersion of pixel values in an image. It provides information about the distribution of pixel intensities and the level of detail or contrast in the image. Considering an image $I$ with elements $a_{ij}$, where $1 \leq i \leq m$ and $1 \leq j \leq n$, the SD is calculated as shown in Eq. 26:

$$SD = \sqrt{\frac{1}{mn}\sum_{i=1}^{m}\sum_{j=1}^{n}(a_{ij} - \bar{I})^2} \tag{26}$$

where $\bar{I}$ is the mean of the matrix $I$, given by:

$$\bar{I} = \frac{1}{mn}\sum_{i=1}^{m}\sum_{j=1}^{n} a_{ij} \tag{27}$$

On the other hand, it is necessary to quantify a level of smoothing to strike a balance between the smoothness and contrast of the image. The smoothing score of an image can be calculated based on its gradient information, using the mean value of the gradient magnitude of the image $I$ [45]. To assess the smoothness of the image, the mean value of the gradient magnitude of the image $I$ is chosen, as it provides numerical information about edges in all directions. The partial derivatives with respect to the horizontal ($x$) and vertical ($y$) directions are denoted as $\frac{\partial I}{\partial x}$ and $\frac{\partial I}{\partial y}$, respectively, where the gradient vector $\nabla I$ is defined in Eq. 28:

$$\nabla I = \begin{bmatrix} \frac{\partial I}{\partial x} \\ \frac{\partial I}{\partial y} \end{bmatrix} \tag{28}$$

The numerical gradient is an approximation of the gradient based on finite differences. Using central differences, the partial derivatives are approximated in Eq. 29 and Eq. 30:

$$\frac{\partial I}{\partial x} \approx \frac{I(x+\Delta x, y) - I(x-\Delta x, y)}{2\Delta x} \tag{29}$$

$$\frac{\partial I}{\partial y} \approx \frac{I(x, y+\Delta y) - I(x, y-\Delta y)}{2\Delta y} \tag{30}$$

where $\Delta x$ and $\Delta y$ are the spacings between grid points in the $x$ and $y$ directions, respectively. The magnitude of the gradient is then calculated as defined in Eq. 31:

$$||\nabla I|| = \sqrt{\nabla I_x^2 + \nabla I_y^2} \tag{31}$$

Finally, the smoothing score is computed as the negative mean (since this is a minimization process) of the gradient magnitude is presented in Eq. 32:

$$SS = -\frac{1}{T}\sum_{i=1}^{T} ||\nabla I|| \tag{32}$$

where $T$ is the total number of elements in the gradient magnitude matrix. The negative sign is applied to minimize the smoothing score. The final fitness function, $FF$, which defined in Eq. 33, is an equally balanced merge of the standard deviation (Eq. 26) and the negative mean value of the gradient magnitude (Eq. 32):

$$FF = SD + SS \tag{33}$$

The lower the value obtained through this fitness function in Eq. 33, the better the performance of a determined transformation function in vessel diminishing, in this case, the optimized top-hat. The RUN algorithm, detailed in Sect. 3, fine-tunes the input parameters of the transformation function in each iteration, whose stop criterion is reached when the 20th iteration is completed.

## 4.4 Multilevel Segmentation

After extracting the vessels from the image, a segmentation process is performed using the MCET-HHO detailed in the previous section. For this scenario, the parameter configuration comprises 50 hawks and a maximum iteration value of 500, with six thresholds. This setup ensures a precise histogram division for accurate OD delineation.

In Fig. 6, two examples of this process are illustrated. Figure 6a, d display two vessel-removed images, each accompanied by their respective histograms shown in Fig. 6b, e. The optimal threshold values determined by the MCET-HHO are marked with a vertical red line on the histograms. Notably, the calculated threshold values

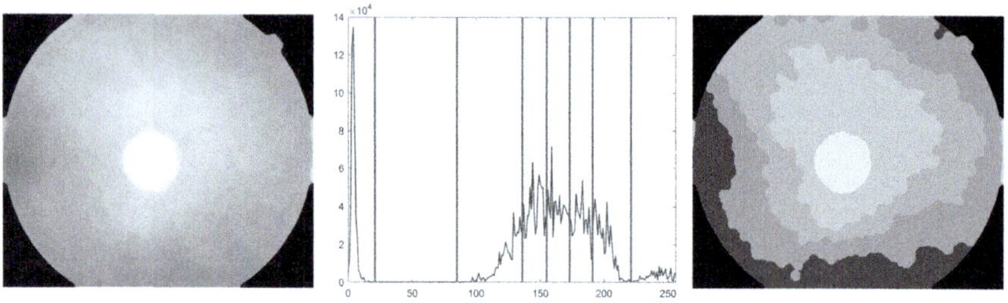


(a) Vessel removed image 042 from DRISHTI dataset (b) Image histogram and threshold values (c) Multilevel image segmentation

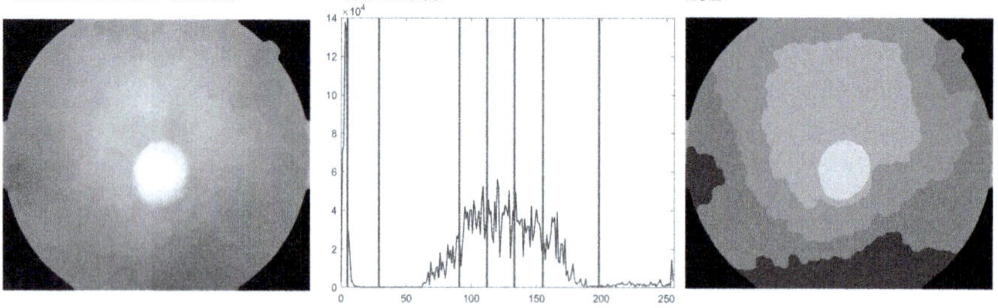


(d) Vessel removed image 077 from DRISHTI-GS dataset (e) Image histogram and threshold values (f) Multilevel image segmentation

**Fig. 6** Vessel-diminished images with their corresponding histograms and multilevel segmentations: **a** and **d** vessel-removed images, **b** and **e** present their respective histograms with optimal threshold, and **c**, and **f** display the multilevel segmentation vary depending on the image histogram, emphasizing the adaptability of the approach to different image characteristics and enhancing the precision of OD delimitation. Finally, in Fig. 6c, e, the multilevel segmentation of each image is presented, clearly demonstrating the accurate delineation of the OD.

## 4.5 Centroid Calculation

The highest threshold value calculated in the previous subsection corresponds to the pixels representing the OD since it is the brightest object in the vessel-diminished image. Based on this observation, the thresholded image is binarized, assigning a pixel value of 1 to those belonging to the OD and 0 to the remaining pixels. Once the OD is binary-segmented, the next step is to obtain the coordinates of this element to do a proportional crop of the original image to complete the detection. This is achieved by calculating the centroid of the OD binary segmentation image, essentially representing the average position of all pixels in that object. This process involves finding the center of mass for each labeled region, where the center of mass is determined by averaging the positions of all pixels in that region. The centroid calculation involves two general steps: first, the calculation of the total area $A$, and second, determining the $x$ and $y$ centroid coordinates.

The total area $A$ of a region in an image is calculated as the sum of all the pixels in that region. Mathematically, it can be represented as presented in Eq. 34:

$$A = \sum_{i} 1 \tag{34}$$

where the sum is taken over all pixels $i$ in the region. Each pixel contributes a value of 1 to the sum, this way counting the number of pixels in the region.

Regarding the $x$ and $y$ coordinates, they are obtained by using Eq. 35 and Eq. 36:

$$C_x = \frac{\sum_i x_i}{A} \tag{35}$$

$$C_y = \frac{\sum_i y_i}{A} \tag{36}$$

where $x_i$ and $y_i$ are the $x$ and $y$ coordinates of each pixel in the region.

In other words, $C_x$ is the average of the x-coordinates of all pixels in the region, and $C_y$ is the average of the y-coordinates. For all images, a squared crop of a size proportional to the binary area is done over the original RGB image, considering the centroid as the square center to ensure a complete OD visualization. Figure 7 illustrates the described steps applied to two eye fundus images, extracted from the DRISHTI-GS dataset [46], image collection that is detailed in the following section. In the first

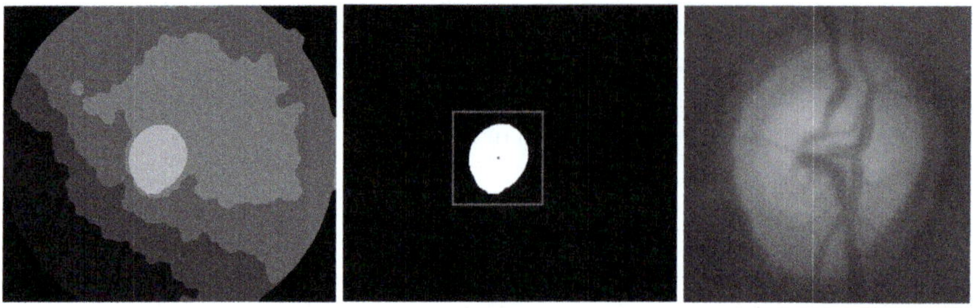

(a) Thresholded image 024 from DRISHTI-GS dataset (b) OD centroid detection and crop region delimited (c) OD cropped from original RGB image

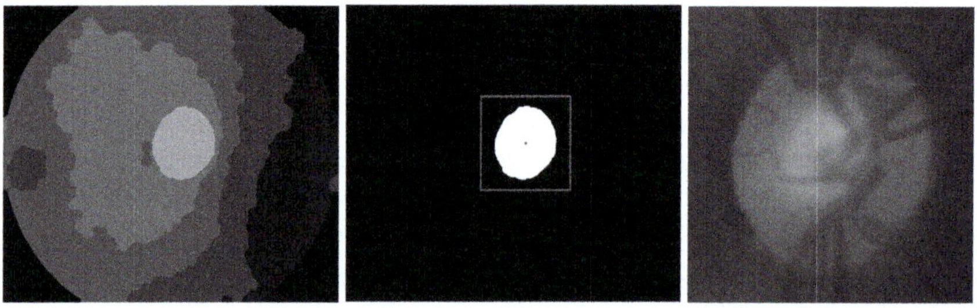

(d) Thresholded image 099 from DRISHTI-GS dataset (e) OD centroid detection and crop region delimited (f) OD cropped from original RGB image

**Fig. 7** Centroid calculation and cropping process: **a** and **d** thresholded images, **b** and **e** centroid and crop region marked on red dot and square, respectively, and **c** and **f** display the cropped OD

row, Fig. 7a shows a multi-level segmented image (thresholded), while Fig. 7b visually depicts the OD binary image, where the calculated centroid coordinate and the crop region are shown, marked in a red color dot and square, respectively. Similarly, the second row of images Fig. 7d–f correspond to the same process applied to a different image. These visual representations showcase the correct centroid calculation and OD localization, the latter being the final result of this approach.

## 5 Experimental Framework

This section encompasses a brief description of optimizers considered for comparison, including their respective parameters. Additionally, it details the performance metrics employed to assess the coordinates calculation and the effectiveness of segmentation, providing a detailed understanding of the evaluation criteria. Furthermore, the selected datasets are presented, including their peculiarities and characteristics.

## 5.1 Parameter Optimization Comparison

The performance of the parameter optimization stage for the transformation function was evaluated, along with the RUN algorithm, using three state-of-the-art metaheuristic algorithms: Particle Swarm Optimization (PSO) [47], Differential Evolution (DE) [48], and Sine-Cosine Algorithm (SCA) [49].

The PSO algorithm was introduced in 1995 [47] and takes inspiration from the social behavior of birds and fish. It models potential solutions as particles in a multidimensional space and adjusts their positions based on individual and global best-known positions. On the other hand, DE was proposed in 1996 [48] as an evolutionary algorithm that iteratively improves a population of candidate solutions using the difference between them to create new trials. Lastly, the SCA developed in 2016 [49], takes inspiration from sine and cosine functions to balance exploration and exploitation in the search space. The SCA updates solution positions using trigonometric functions and has demonstrated promising results across various optimization problems.

In Table 1, general and individual parameters for each algorithm are presented.

## 5.2 Performance Metrics

To numerically assess this method's performance, different performance metrics are used. In this scenario, the Root Mean Squared Error (RMSE) is selected to measure the difference between the calculated OD center coordinates and the ground truth coordinates. Regarding the segmentation performance, three metrics to evaluate the OD segmentation are selected [50], namely Accuracy (Acc), Intersection over Union (IoU), and Dice Similarity Coefficient (DSC). All four metrics are briefly detailed below.

**Table 1** Parameter settings for algorithms

| Algorithm | Parameter settings |
|---|---|
| General settings | Population size = 10, |
| | Maximum iterations = 20 |
| | $S_o$ boundaries = [2, 40] |
| | $S_c$ boundaries = [2, 40] |
| PSO [47] | Inertia weight = 0.7, |
| | Personal acceleration coefficient = 1.5, |
| | Social acceleration coefficient = 1.5 |
| DE [48] | Scaling factor = 0.7, |
| | Crossover rate = 0.9 |
| SCA [49] | Range in [-2 ,2] (A = 2) |

### 5.2.1 Root Mean Squared Error

The Root Mean Squared Error (RMSE) [51] is a widely used metric to quantify the difference between predicted and observed values. It is calculated by taking the square root of the mean of the squared differences between corresponding predicted ($y_i$) and observed ($\hat{y}$) values. The formula for RMSE is expressed Eq. 37:

$$\text{RMSE} = \sqrt{\frac{\sum_{i=1}^{n}(y_i - \hat{y}_i)^2}{n}} \tag{37}$$

Here, $n$ represents the total number of observations. The RMSE provides a measure of the average magnitude of errors, allowing for the assessment of the overall accuracy of a predictive model or algorithm. Lower RMSE values indicate better agreement between predicted and observed values, while higher values suggest greater discrepancies.

In the context of image segmentation, evaluating the performance of segmentation algorithms is crucial. Three commonly used metrics for this purpose are Accuracy (Acc), Intersection over Union (IoU), and Dice Similarity Coefficient (DSC). These metrics rely on key components to quantify the algorithm's ability to delineate between the Region of Interest (ROI) and background pixels (boolean data): True Positives ($T_P$), True Negatives ($T_N$), False Positives ($F_P$), and False Negatives ($F_N$). $T_P$ represents correctly identified ROI pixels, while $T_N$ denotes accurately identified background pixels. Conversely, $F_P$ encloses ROI pixels erroneously labeled as background, and $F_N$ encompasses ROI pixels overlooked or misclassified as background. The Acc, IoU, and DSC metrics are briefly detailed below.

### 5.2.2 Accuracy

Accuracy (Acc) provides a global measure of the correctness of the segmentation results. It takes into account both $T_P$ and $F_P$ pixels. The formula for Accuracy is given by Eq. 38:

$$\text{Acc} = \frac{T_P + T_N}{T_P + T_N + F_P + F_N} \tag{38}$$

### 5.2.3 Intersection over Union

The Intersection over Union (IoU), also known as the Jaccard Index, measures the spatial overlap between the predicted segmentation and the ground truth. It considers the intersection of the predicted and ground truth regions divided by their union. The IoU is calculated as shown in Eq. 39:

$$\text{IoU} = \frac{T_P}{T_P + F_P + F_N} \tag{39}$$

#### 5.2.4 Dice Similarity Coefficient

The DSC, also known as the F1 score, evaluates the spatial agreement between predicted and actual segments. It is defined as twice the ratio between the intersection and the sum of their areas, obtained in Eq. 40:

$$\mathrm{DSC} = \frac{2 \cdot T_P}{2 \cdot T_P + F_P + F_N} \tag{40}$$

Similar to IoU metric, the DSC provides a measure of segmentation accuracy, focusing on balancing precision and recall [52].

These metrics collectively offer valuable insights into the performance of image segmentation algorithms, considering aspects such as correctness, spatial overlap, and spatial agreement.

### *5.3 Dataset Description*

To test the robustness of this approach, three datasets of eye fundus images were selected, encompassing a diverse spectrum of lighting conditions, spatial resolutions, and OD contrast variations. These datasets are briefly described below.

#### 5.3.1 DRISHTI-GS

DRISHTI-GS [46] is a widely used dataset in fields related to the analysis of retinal images. It is designed for research in glaucoma analysis and includes retinal fundus images with annotated OD and optic cup manual segmentations, cup-to-disc ratio, and OD center coordinates, all done by glaucoma experts. The dataset comprises a total of 101 images, with a division of 50 images for training and 51 images for testing. The entire image set was acquired using the same equipment, centered on OD, with a field of view of 30 degrees and final resolution after non-fundus black region cropping of $\sim 2047 \times 1760$ pixels.

This dataset is incorporated into the experimental framework since it provides ground truth segmentation and OD center to numerically evaluate the proposal performance. In Fig. 8 eye fundus images from the DRISHTI-GS dataset are presented.

#### 5.3.2 EyePACS

The EyePACS dataset [53] consists of assorted smaller datasets acquired using diverse cameras in different hospitals worldwide. It encompasses distinct settings, varying sizes, and notable differences in brightness and illumination. These images

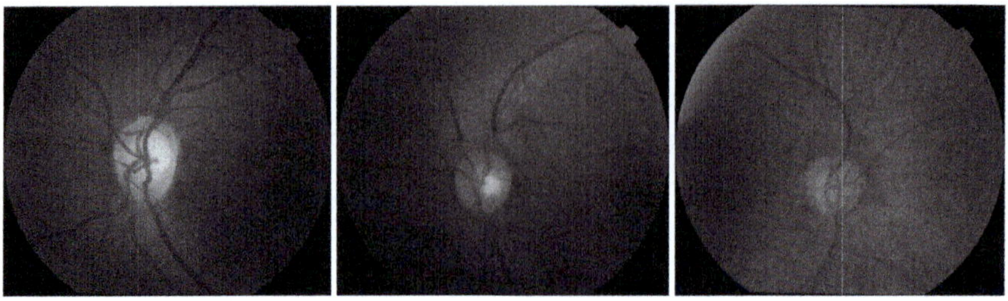

**Fig. 8** Examples of DRISHTI-GS dataset

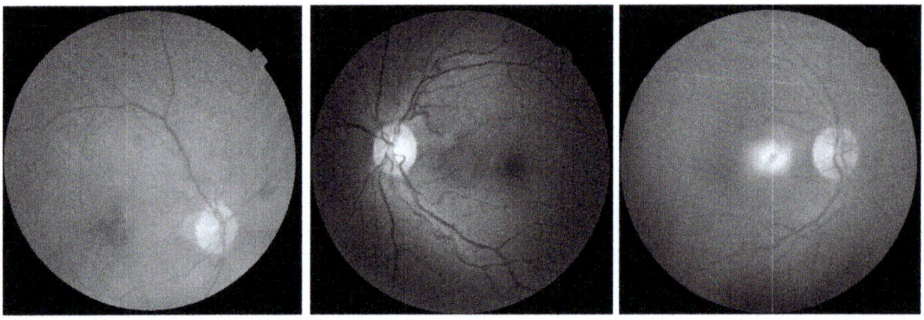

**Fig. 9** Examples of EyePACS dataset

were acquired as a diabetic retinopathy screening program for disease grading, whose publicly available version comprises 35,126 scans. In contrast with Fig. 8, images presented in Fig. 9 show significant variation in terms of contrast of retinal elements and illumination of the images.

### 5.3.3 Large-Scale Attention Based Glaucoma

With the aim of testing this approach on images acquired under different settings, it is also consider the Large-scale Attention-based Glaucoma (LAG) dataset [54]. The retinographies were captured at Beijing Tongren Hospital and encompassed a total of 5,824 images categorized into glaucoma and non-glaucoma classes, with 2,392 and 3,432 images, respectively. All images underwent classification by glaucoma experts based on clinical evaluations (including signs and symptoms, intraocular pressure, vision loss, etc.). They were subsequently cropped to a resolution of 500 $\times$ 500 pixels to retain glaucoma-related information and standardize them for classification purposes. In Fig. 10, image examples of this dataset are presented.

While EyePACS and LAG datasets lack ground truth, they were selected to evaluate OD detection efficiency in real-world scenarios with varying image conditions.

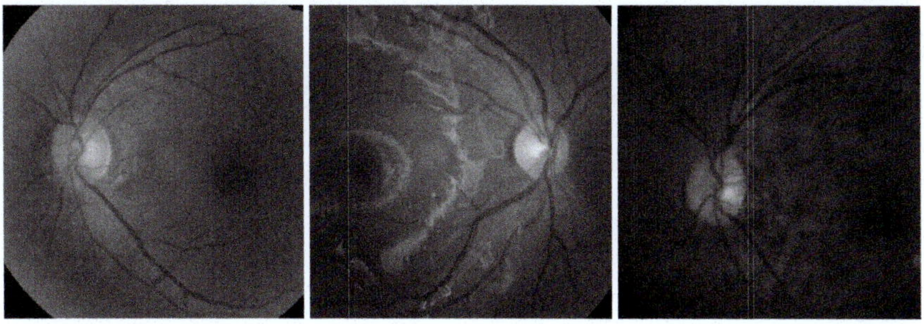

**Fig. 10** Examples of LAG dataset

# 6 Results

In this section, the results of our proposed method using both quantitative and qualitative analysis are introduced. First, the performance of the RUN against other metaheuristic algorithms for vessel diminishing is presented through convergence curves. Then, a numerical assessment of the performance in OD segmentation is shown using the previously detailed performance metrics. Finally, an OD center coordinates comparison against the ground truth is presented, and a final visual evaluation of this proposal in eye fundus images with various conditions.

## *6.1 Optimization Analysis*

To assess the efficiency of the RUN algorithm in this specific scenario, it was compared against three state-of-the-art metaheuristic algorithms (PSO, DE, and SCA) using the optimized top-hat as a transformation function and the proposed fitness function, both described in Sect. 4. Figure 11 illustrates the convergence behavior of all four tested algorithms when applied to images from all selected datasets.

It is noteworthy that across all convergence curves, the RUN algorithm (depicted by the solid purple line) outperforms the other metaheuristic approaches. In specific behaviors, as illustrated in Fig. 12a, the PSO (blue dotted line), DE (orange dash-dotted line), and RUN algorithms converge to the same fitness value. However, the RUN algorithm achieves convergence before the fifth iteration, whereas the SCA (yellow dashed line) encounters difficulties in escaping a local minimum. Conversely, in scenarios depicted in Fig. 12b–d, each algorithm obtains a distinct best fitness value, yet the RUN algorithm consistently surpasses them all. Additionally, as showcased in Fig. 12e, the effectiveness of the RUN algorithm persists, reaffirming its robust performance across diverse convergence scenarios.

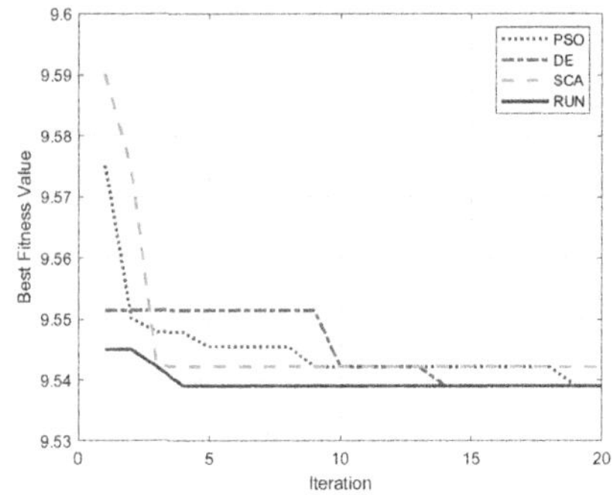


(a) Convergence curve for image 84 _right from EyePACS dataset

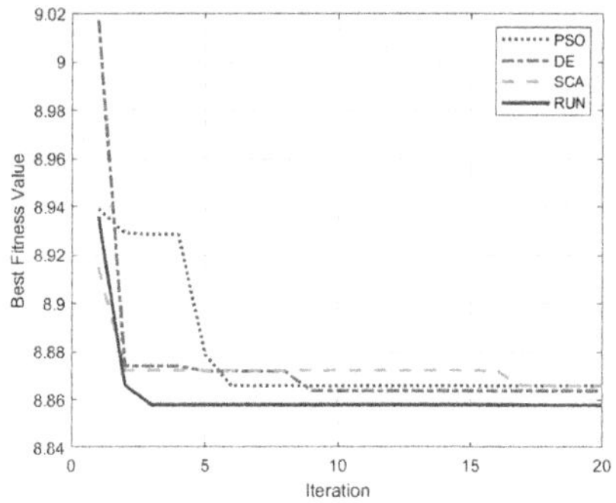


(b) Convergence curve for image g.0507 from LAG dataset

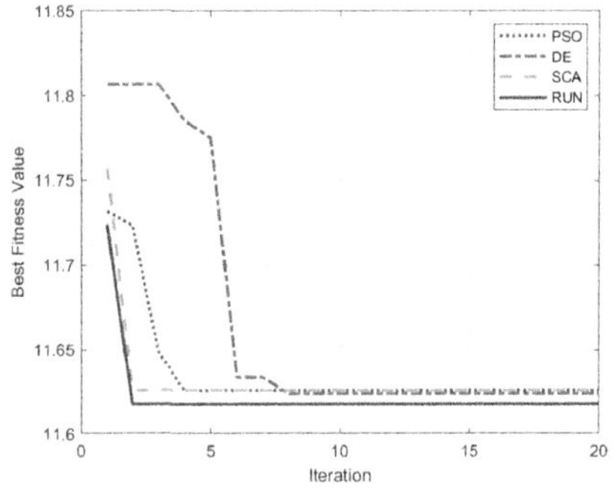


(c) Convergence curve for image 007 from DRISHTI-GS dataset

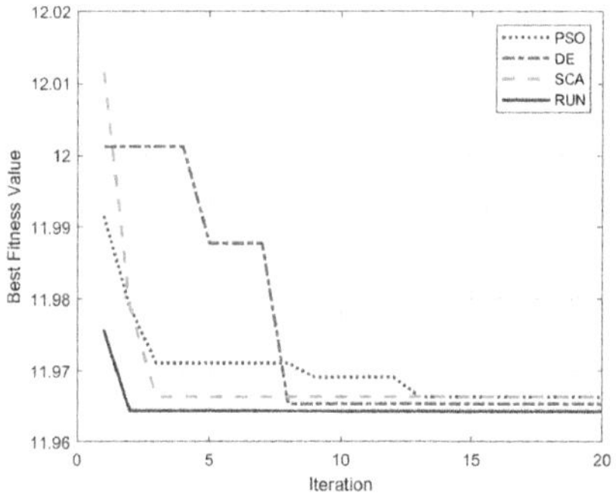


(d) Convergence curve for image 100 from DRISHTI-GS dataset

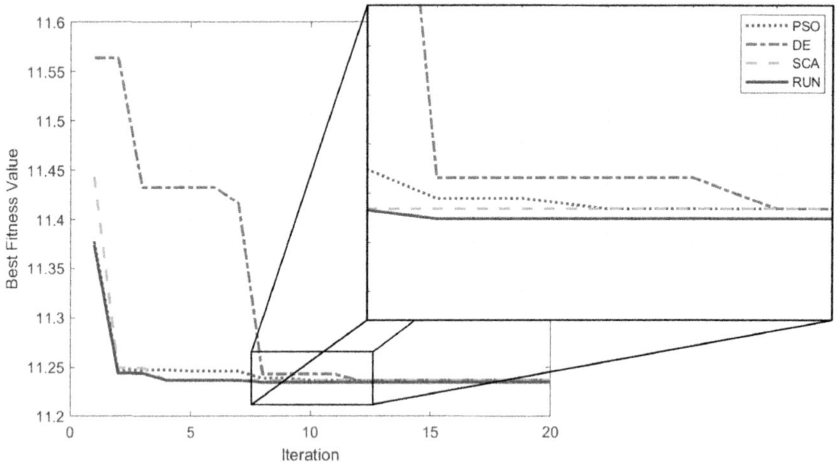


(e) Convergence curve for image ng.0001 from LAG dataset

**Fig. 11** Convergence curves of different dataset images

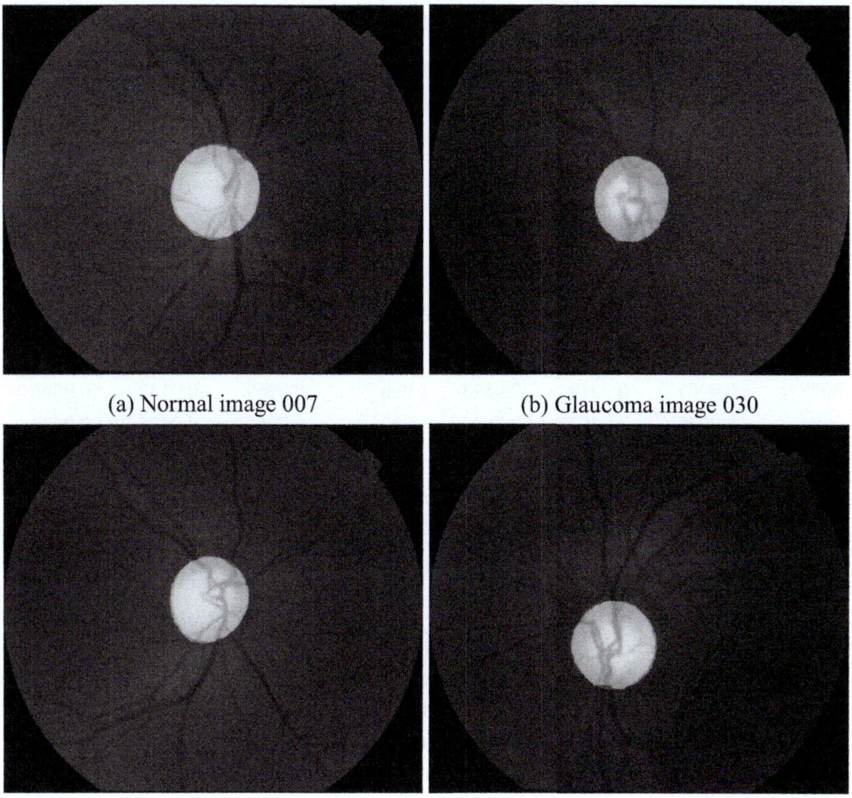

(a) Normal image 007 (b) Glaucoma image 030

(c) Normal image 085 (d) Glaucoma image 052

**Fig. 12** Manual and automated segmentation comparison on normal and glaucomatous images. Red, green, and yellow pixels correspond to manual, automated, and matching segmentation, respectively

## *6.2 Segmentation Performance*

As mentioned in Sect. 5, the DRISHTI-GS dataset provides manually marked ground truths of the optic disc (OD), which are used for the segmentation performance analysis. Ten images were randomly selected for both "Normal" and "Glaucoma" classes to test the segmentation efficiency. The segmentation performance is evaluated using three metrics detailed in Sect. 5: Acc, IoU, and DSC.

For comparison analysis, Table 2 provides a comprehensive analysis of segmentation metrics for both normal and glaucomatous eye fundus images. The "Normal" category represents healthy eye fundus images, and the segmentation metrics exhibit consistently high values. The average Acc is 99.69%, indicating an accurate delineation of the OD in normal images. The IoU and DSC values are also notably high, with an average of 90.67% and 95.00%, respectively. Moving to the "Glaucoma"

**Table 2** Segmentation performance on DRISHTI-GS images

| Image | Normal | | | Image | Glaucoma | | |
|---|---|---|---|---|---|---|---|
| | Acc | IoU | DSC | | Acc | IoU | DSC |
| 007 | 0.9975 | 0.9431 | 0.9707 | 020 | 0.9969 | 0.9195 | 0.9581 |
| 009 | 0.9981 | 0.9404 | 0.9693 | 025 | 0.9976 | 0.9199 | 0.9582 |
| 013 | 0.9934 | 0.7703 | 0.8702 | 029 | 0.9983 | 0.9491 | 0.9739 |
| 072 | 0.9983 | 0.9440 | 0.9712 | 030 | 0.9986 | 0.9587 | 0.9789 |
| 078 | 0.9983 | 0.9460 | 0.9722 | 052 | 0.9979 | 0.9453 | 0.9719 |
| 085 | 0.9982 | 0.9486 | 0.9736 | 054 | 0.9985 | 0.9359 | 0.9669 |
| 091 | 0.9945 | 0.8172 | 0.8994 | 073 | 0.9965 | 0.8901 | 0.9418 |
| 096 | 0.9969 | 0.8949 | 0.9445 | 077 | 0.9965 | 0.9037 | 0.9494 |
| 097 | 0.9967 | 0.9219 | 0.9593 | 086 | 0.9969 | 0.9132 | 0.9546 |
| 099 | 0.9976 | 0.9403 | 0.9692 | 087 | 0.9960 | 0.9675 | 0.9834 |
| **Mean** | **0.9969** | **0.9067** | **0.9500** | **Mean** | **0.9977** | **0.9303** | **0.9637** |

category, representing eye fundus images with glaucomatous conditions, the average Acc is 99.77%, indicating a high level of accuracy in identifying the OD in glaucomatous images. The IoU and DSC metrics also exhibit strong performance, with an average of 93.03% and 96.37%, respectively. This implies that the segmentation algorithm is robust and effective, even in the presence of glaucomatous abnormalities.

In summary, Table 2 provides a quantitative assessment of the segmentation performance, demonstrating the reliability and accuracy of the algorithm in detecting the OD in both normal and glaucomatous eye fundus images. These findings underscore the potential clinical utility of the segmentation method for automated analysis in ophthalmic applications.

To visually assess segmentation efficiency, a graphical comparison between the obtained segmentation and ground truth is presented in Fig. 12. In the representation, green pixels correspond to the obtained segmentation, red pixels to the ground truth, and yellow pixels indicate matching segmentations and are placed over the corresponding gray-scale eye fundus image. While slight differences between manual and automated OD delimitation are observable, the overall shapes and total area exhibit remarkable similarity, affirming the effectiveness of the segmentation method.

## *6.3 Centroid Calculation Performance*

As mentioned in Sect. 5, the DRISHTI-GS dataset includes the OD center coordinates determined by medical experts. These coordinates play a crucial role in assessing the algorithm's accuracy in detecting the OD positions. A numerical analysis from the obtained OD center is employed to corroborate a correct OD detection. The aim

**Table 3** Ground truth and predicted OD center coordinates comparison

| Image | GTx | PRx | RMSE (x) | GTy | PRy | RMSE (y) |
|---|---|---|---|---|---|---|
| 007 | 1043 | 1037 | 6.08 | 886 | 892 | 6.08 |
| 009 | 900 | 897 | 3.74 | 875 | 872 | 3.74 |
| 013 | 1630 | 1598 | 39.58 | 906 | 895 | 11.40 |
| 020 | 1007 | 997 | 10.63 | 921 | 915 | 6.00 |
| 025 | 1328 | 1333 | 4.47 | 795 | 794 | 1.00 |
| 029 | 1102 | 1102 | 0.00 | 898 | 900 | 2.00 |
| 030 | 989 | 982 | 6.08 | 921 | 922 | 1.00 |
| 052 | 894 | 912 | 13.42 | 1052 | 1050 | 2.00 |
| 054 | 937 | 937 | 0.00 | 1019 | 1019 | 0.00 |
| 072 | 1120 | 1117 | 3.00 | 806 | 803 | 3.00 |
| 073 | 890 | 872 | 18.97 | 920 | 913 | 8.06 |
| 077 | 1063 | 1053 | 10.05 | 996 | 990 | 5.07 |
| 078 | 993 | 994 | 1.00 | 964 | 965 | 1.00 |
| 085 | 1006 | 1003 | 3.00 | 833 | 835 | 2.00 |
| 086 | 1007 | 1001 | 6.08 | 1006 | 1005 | 1.00 |
| 087 | 1049 | 1049 | 0.00 | 1027 | 1025 | 2.00 |
| 091 | 1035 | 1033 | 2.00 | 941 | 940 | 1.00 |
| 096 | 1018 | 1016 | 2.00 | 1128 | 1119 | 8.06 |
| 097 | 1283 | 1283 | 0.00 | 837 | 843 | 6.08 |
| 099 | 1097 | 1105 | 8.06 | 816 | 811 | 5.07 |
| **Mean** | | | **10.29** | | | **4.79** |

is to compare the predicted coordinates with the ground truth values, evaluating the algorithm's performance in terms of both precision and reliability.

This analysis is conducted over the same ten glaucomatous and ten normal images randomly selected for segmentation performance analysis, and it is presented in Table 3. The columns GTx and GTy represent the manually annotated ground truth x and y coordinates, respectively, of the OD centroids in retinal images. Conversely, the columns PRx and PRy denote the corresponding predicted x and y coordinates generated by the automated algorithm. The RMSE metric is employed to quantify the spatial disparities between the predicted and ground truth centroid positions, as shown in Table 3.

The mean RMSE for x and y coordinates is 10.29 and 4.79, respectively, as indicated by the results. In some images, the absolute coordinate difference is zero, resulting in a RMSE value of zero. However, the biggest RMSE value is 39.58, which corresponds to an absolute difference of 32 pixels. It is worth noting that the images from the DRISHTI-GS dataset have a resolution of $2047 \times 1760$, representing a minor difference that does not pose a performance issue when cropping the OD.

### 6.4 Visual Analysis

To validate the effectiveness of this method in real clinical scenarios, comprehensive testing was conducted on the LAG and EyePACS datasets, encompassing diverse retinal images with varying imaging settings and eye conditions, as elaborated in Sect. 5. Figures 13 and 14 present the results of the OD detection. The first and third columns present the original images, while the second and fourth columns exhibit the automatically cropped OD regions corresponding to the respective left images. This set of images serves to visualize different scenarios where the method consistently demonstrates exceptional performance.

As illustrated in Fig. 13, the proposed method consistently detects the OD across various eye fundus image conditions. Notably, in Fig. 13a, c, k, where different eye diseases like peripapillary atrophy and pathologic myopia are present, the OD is accurately identified. Furthermore, the method demonstrates robust performance in images with overall low contrast, as evidenced in Fig. 13g, i

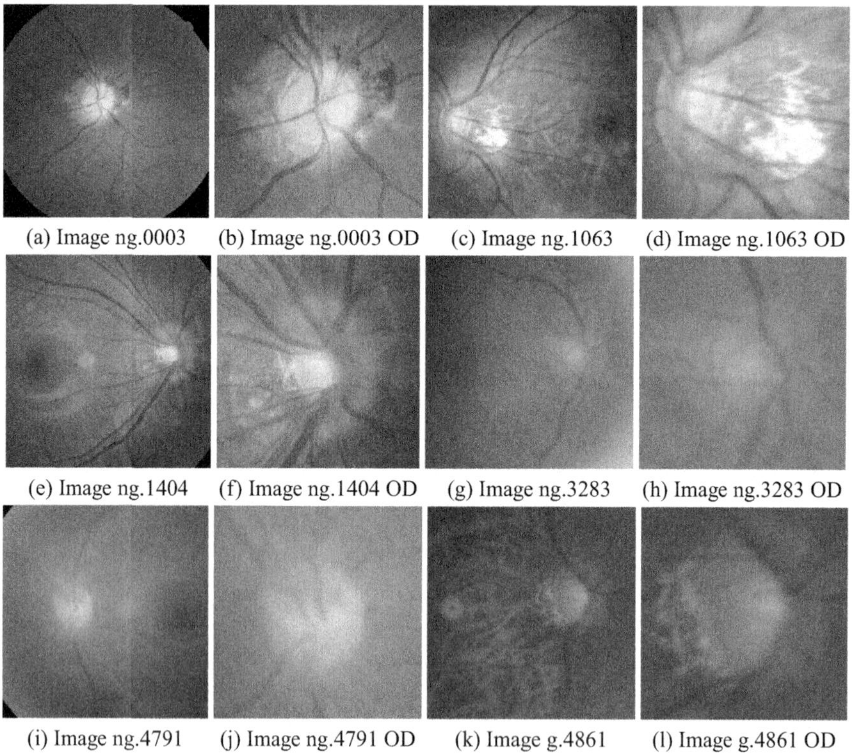


**Fig. 13** Detected OD on LAG dataset

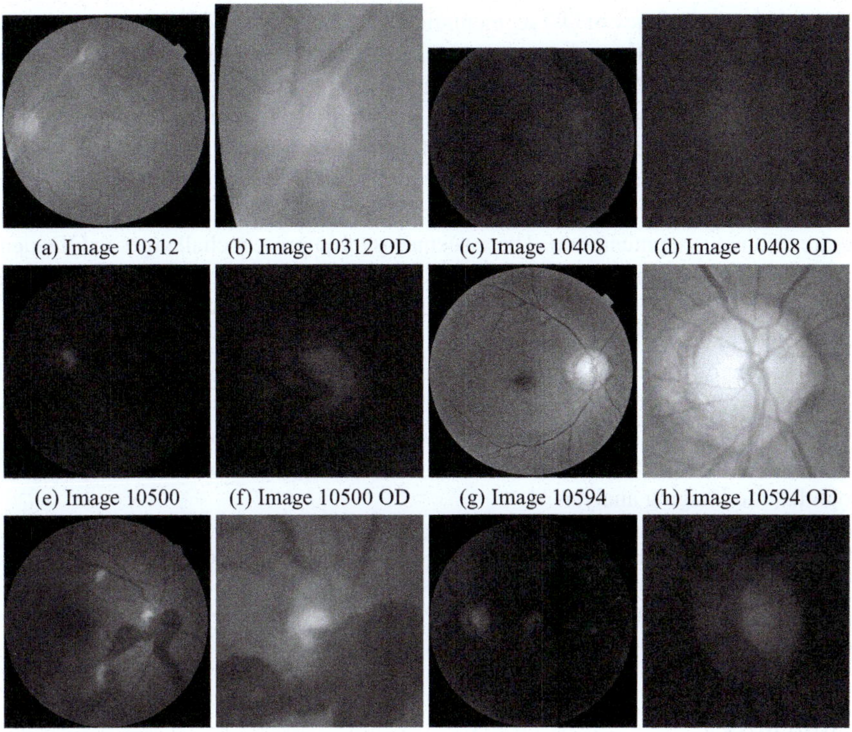

(a) Image 10312 (b) Image 10312 OD (c) Image 10408 (d) Image 10408 OD

(e) Image 10500 (f) Image 10500 OD (g) Image 10594 (h) Image 10594 OD

(i) Image 10580 (j) Image 10580 OD (k) Image 13180 (l) Image 13180 OD

**Fig. 14** Detected OD on EyePACS dataset

In the images from the EyePACS dataset showcased in Fig. 14, the proposed method consistently achieves accurate OD detection, even under challenging conditions. Notably, in Fig. 14c, e, k, where illumination is poor, the OD is correctly identified. The method also proves effective in cases where the OD is blurred, as seen in Fig. 14a. Furthermore, it demonstrates great performance in scenarios where the OD is partially obscured by disease elements, as depicted in Fig. 14i.

## 7 Conclusions

This chapter delves into the analysis of ophthalmic imaging, with a particular focus on localizing the OD in eye fundus images. Given the pivotal role of retinography analysis in early disease detection and pathological condition monitoring, an OD detection approach is proposed.

The efficacy of the RUN metaheuristic algorithm has been highlighted, revealing its ability to optimize parameters for vessel diminishing and, consequently, easing the accuracy and efficiency of OD segmentation through the MCET-HHO approach. The subsequent presentation of segmentation performance on both normal and glaucomatous DRISHTI-GS images, along with the comparison of various metrics, added a quantitative dimension to the evaluation process thus providing an insight into how well the proposed method performed. Lastly, the chapter delved into the visual analysis of OD detection in a wide range of eye fundus images, including several diseased eye fundus, where the OD presents several pathological elements that tend to generate erroneous segmentation. These scenarios often present challenges in OD identification due to pathological elements. Remarkably, this proposed method demonstrated high performance in overcoming these challenges, showcasing its robustness in real-world clinical applications.

In summary, this chapter introduces a robust automated OD segmentation method. The optimization of a previously proposed transformation function, coupled with the evaluation of its performance using a specially designed fitness function, signifies a notable advancement in the realm of early disease diagnosis and monitoring through ophthalmic imaging analysis.

**Acknowledgements** The authors express their sincere gratitude to Dr. Omar Rodríguez-Esparza for his invaluable assistance in emphasizing the significance of our study in medicine. He provided expert guidance on the accurate use of medical terminology and helped validate the visual results.

## Appendix

The RKM approach is primarily used to solve ordinary differential equations (ODE) by defining $f(x, y)$ as a slope (S) of a curve plot in a determined $(x, y)$ point.

Let's consider Eq. 41 as a first-order ODE:

$$\frac{dy}{dx} = f(x, y), y(x_0) = y_0 \tag{41}$$

By evaluating the slope at the point $(x_0, y_0)$, an additional point can be derived using the best-fitted straight line: $(x_1, y_1) = (x_0 + \Delta x, y_0 + S_0\Delta x)$, where $S_0 = f(x_0, y_0)$. Following a similar pattern, $(x_2, y_2) = (x_1 + \Delta x, y_1 + S_1\Delta x)$ can be determined. This iterative process is repeated $m$ times, resulting in an approximate solution within the range $[x_0, x_0 + m\Delta x]$. The RKM is derived from the Taylor series, which is defined as:

$$y(x + \Delta x) = y(x) + y'(x)\Delta x + y''(x)\frac{(\Delta x)^2}{2!} + ... \tag{42}$$

By neglecting high-order terms and applying the Euler method (or the first-order Runge Kutta), Eq. 42 can be expressed as:

$$y(x + \Delta x) = y(x) + k_1\Delta x \tag{43}$$

where $k_1 = y'(x) = f(x, y)$ and $\Delta x = x_{n+1} - x_n$.

This first-order derivative of Eq. 43 [55], $y'(x)$, is approximated as:

$$y'(x) = \frac{y(x + \Delta x) - y(x - \Delta x)}{2\Delta x} \tag{44}$$

Hence, Eq. 43 can be expressed as:

$$y'(x) = y(x) + \frac{y(x + \Delta x) - y(x - \Delta x)}{2} \tag{45}$$

The Runge Kutta (RK4) method [56] (which is based on Eq. 42 and works with a fourth-order solution), is used for the optimization process. The RK4 can be formally written as:

$$y(x + \Delta x) = y(x) + \frac{1}{6}(k_1 + 2 \cdot k_2 + 2 \cdot k_3 + k_4)\Delta x \tag{46}$$

where weighted factors are defined as:

$$\begin{aligned} k_1 &= y'(x) = f(x, y) \\ k_2 &= f\left(x + \frac{\Delta x}{2}, y + \frac{\Delta x}{2} \cdot k_1\right) \\ k_3 &= f\left(x + \frac{\Delta x}{2}, y + \frac{\Delta x}{2} \cdot k_2\right) \\ k_4 &= f(x + \Delta x, y + \Delta x \cdot k_3) \end{aligned} \tag{47}$$

The parameter $k_1$ represents the initial increment, determining the slope at the beginning of the interval $[x, x + \Delta x]$. $k_2$ represents the second increment and is computed considering the slope at the midpoint, incorporating $y$ and $k_1$. Similarly, $k_3$ denotes the third increment and $k_4$ corresponds to the fourth increment.

By applying some considerations widely explained by the RUN optimizer author [30], the $\Delta x$ is rewritten as:

$$\Delta x = 2 \cdot \text{rand} \cdot |St| \tag{48}$$

where $|St|$ controls the increment parameter mathematically defined as:

$$|St| = \text{rand} \cdot ((x_b - \text{rand} \cdot x_{avg}) + \zeta) \tag{49}$$

and the scaling factor $\zeta$ used in Eq. 49 as:

$$\zeta = \text{rand}(x_n - \text{rand} \cdot (u - d))exp\left(-4 \cdot \frac{It}{MaxIt}\right) \tag{50}$$

where $u$ is a random parameter to control the best solution $x_b$, which is defined as $u = round(1 + \text{rand}) \cdot (1 - \text{rand})$, and $d$ the $d - th$ variable of the problem ($d = 1, 2, ..., D$).

Considering the worst solution $x_w$ and all mathematical definitions detailed above, the parameters $k_1$ to $k_4$ defined in Eq. 47, modified for swarm-based optimization purposes, are rewritten as follows:

$$\begin{aligned} k_1 &= \frac{1}{2\Delta x}(\text{rand} \cdot x_w - u \cdot x_b) \\ k_2 &= \frac{1}{2\Delta x}\left(\text{rand} \cdot (x_w + \text{rand}_1 \cdot k_1 \cdot \Delta x) - (u \cdot x_b + \text{rand}_2 \cdot k_1 \cdot \Delta x)\right) \\ k_3 &= \frac{1}{2\Delta x}\left(\text{rand} \cdot (x_w + \text{rand}_1 \cdot \frac{1}{2}k_2 \cdot \Delta x) - (u \cdot x_b + \text{rand}_2 \cdot \frac{1}{2}k_2 \cdot \Delta x)\right) \\ k_4 &= \frac{1}{2\Delta x}\left(\text{rand} \cdot (x_w + \text{rand}_1 \cdot k_3 \cdot \Delta x) - (u \cdot x_b + \text{rand}_2 \cdot k_3 \cdot \Delta x)\right) \end{aligned} \tag{51}$$

where $\text{rand}_1$ and $\text{rand}_2$ are random numbers whose values are contained in [0, 1]. Finally, the SM is defined as:

$$SM = \frac{1}{6}(x_{RK})\Delta x \tag{52}$$

where:

$$x_{RK} = k_1 + 2 \cdot k_2 + 2 \cdot k_3 + k_4 \tag{53}$$

**Competing Interests**The authors have no conflicts of interest to declare that are relevant to the content of this chapter.

## References


1. Scholl, I., Aach, T., Deserno, T.M., Kuhlen, T.: Challenges of medical image processing. Computer science-Research and development **26**, 5–13 (2011)
2. Timberlake, G.T., Kennedy, M.: The direct ophthalmoscope how it works and how to use it, University of Kansas 39 (2005)
3. Medeiros, F.A., Alencar, L.M., Zangwill, L.M., Bowd, C., Sample, P.A., Weinreb, R.N.: Prediction of functional loss in glaucoma from progressive optic disc damage. Arch. Ophthalmol. **127**(10), 1250 (2009)
4. Artes, P.H., Chauhan, B.C.: Longitudinal changes in the visual field and optic disc in glaucoma. Prog. Retin. Eye Res. **24**(3), 333–354 (2005)
5. Jansson, R.W., Frøystein, T., Krohn, J.: Topographical distribution of retinal and optic disc neovascularization in early stages of proliferative diabetic retinopathy. Investig. Ophthalmol. Vis. Sci. **53**(13), 8246–8252 (2012)
6. Shinohara, Y., Akiyama, H.: Relationship between neovascularization of the optic disc and diabetic macular edema. Clin. Ophthalmol. 1203–1205 (2023)

7. Zohora, S.E., Chakraborty, S., Khan, A., Dey, N.: Detection of exudates in diabetic retinopathy: a review. In: 2016 International Conference on Electrical, Electronics, and Optimization Techniques (ICEEOT), pp. 2063–2068. IEEE (2016)
8. Sanders, T., Gay, A., Newman, M.: Drüsen of the optic disk-hemorrhagic complications. Trans. Am. Ophthalmol. Soc. **68**, 186 (1970)
9. Niemeijer, M., van Ginneken, B., Russell, S.R., Suttorp-Schulten, M.S., Abramoff, M.D.: Automated detection and differentiation of drusen, exudates, and cotton-wool spots in digital color fundus photographs for diabetic retinopathy diagnosis. Investig. ophthalmol. Vis. Sci. **48**(5), 2260–2267 (2007)
10. Rehman, Z.U., Naqvi, S.S., Khan, T.M., Arsalan, M., Khan, M.A., Khalil, M.: Multi-parametric optic disc segmentation using superpixel based feature classification. Expert Syst. Appl. **120**, 461–473 (2019)
11. Ramos-Soto, O., Rodríguez-Esparza, E., Balderas-Mata, S.E., Oliva, D., Hassanien, A.E., Meleppat, R.K., Zawadzki, R.J.: An efficient retinal blood vessel segmentation in eye fundus images by using optimized top-hat and homomorphic filtering. Comput. Methods Prog. Biomed. **201**, 105949 (2021)
12. Shanthi, R., Prabakaran, S.: A review paper on an automatic localization of optic disc and segmentation approaches for glaucoma diagnosis. Int. J. Appl. Eng. Res. **10**(75), 2015 (2015)
13. Ali, A., Wan Zaki, W.M.D., Hussain, A., Hashim, N., Mohd Isa, W.N.: Vessel masking and hough transform for optic disc localisation from retinal images. F1000Research **11**, 181 (2022)
14. Shanthamalar, J.J., Ramani, R.G.: Automatic blood vessel segmentation in retinal fundus images using image enhancement and dynamic gray-level thresholding. In: Proceedings of International Conference on Computational Intelligence and Data Engineering: ICCIDE 2021, pp. 337–348. Springer (2022)
15. Kim, G.Y., Lee, S.H., Kim, S.M.: Automated segmentation and quantitative analysis of optic disc and fovea in fundus images. Multimed. Tools Appl. **80**, 24205–24220 (2021)
16. Yi, J., Ran, Y., Yang, G.: Particle swarm optimization-based approach for optic disc segmentation. Entropy **24**(6), 796 (2022)
17. Al Shalchi, N.F.A., Rahebi, J.: Human retinal optic disc detection with grasshopper optimization algorithm. Multimed. Tools Appl. **81**(17), 24937–24955 (2022)
18. Naqvi, S.S., Fatima, N., Khan, T.M., Rehman, Z.U., Khan, M.A.: Automatic optic disk detection and segmentation by variational active contour estimation in retinal fundus images. SIViP **13**, 1191–1198 (2019)
19. Carvalho, T.H., Moraes, C.H., Almeida, R.C., Spadoti, D.H.: Application of conditional gan models in optic disc/optic cup segmentation of retinal fundus images. In: 17th International Symposium on Medical Information Processing and Analysis, Vol. 12088, pp. 389–395. SPIE (2021)
20. Joshi, A., Sharma, K.: Graph deep network for optic disc and optic cup segmentation for glaucoma disease using retinal imaging. Phys. Eng. Sci. Med. **45**(3), 847–858 (2022)
21. Latif, J., Tu, S., Xiao, C., Ur Rehman, S., Imran, A., Latif, Y.: Odgnet: a deep learning model for automated optic disc localization and glaucoma classification using fundus images. SN Appl. Sci. **4**(4), 98 (2022)
22. Yu, L.: Joint segmentation of optic cup and optic disc using deep convolutional generative adversarial network. In: Journal of Physics: Conference Series, Vol. 2234, p. 012008. IOP Publishing (2022)
23. Hervella, Á.S., Rouco, J., Novo, J., Ortega, M.: End-to-end multi-task learning for simultaneous optic disc and cup segmentation and glaucoma classification in eye fundus images. Appl. Soft Comput. **116**, 108347 (2022)
24. Xiong, H., Liu, S., Sharan, R.V., Coiera, E., Berkovsky, S.: Weak label based bayesian u-net for optic disc segmentation in fundus images. Artif. Intell. Med. **126**, 102261 (2022)
25. Gonzalez, R.C.: Digital Image Processing. Pearson education India (2009)
26. Pastore, J., Bouchet, A., Brun, M., Ballarin, V.: New windows based color morphological operators for biomedical image processing. In: Journal of Physics: Conference Series, Vol. 705, p. 012023. IOP Publishing (2016)

27. Cuevas, E., Osuna, V., Oliva, D.: White blood cells detection in images, pp. 155–180. A Comparative Perspective, Evolutionary Computation Techniques (2017)
28. BahadarKhan, K., Khaliq, A.A., Shahid, M.: A morphological hessian based approach for retinal blood vessels segmentation and denoising using region based otsu thresholding. PLoS ONE **11**(7), e0158996 (2016)
29. Wang, W., Wang, W., Hu, Z.: Segmenting retinal vessels with revised top-bottom-hat transformation and flattening of minimum circumscribed ellipse. Med. Biol. Eng. Comput. **57**, 1481–1496 (2019)
30. Ahmadianfar, I., Heidari, A.A., Gandomi, A.H., Chu, X., Chen, H.: Run beyond the metaphor: an efficient optimization algorithm based on Runge Kutta method. Expert Syst. Appl. **181**, 115079 (2021)
31. Runge, C.: Über die numerische auflösung von differentialgleichungen. Math. Ann. **46**(2), 167–178 (1895)
32. Kutta, W.: Beitrag zur näherungsweisen Integration totaler Differentialgleichungen, Teubner (1901)
33. Shaban, H., Houssein, E.H., Pérez-Cisneros, M., Oliva, D., Hassan, A.Y., Ismaeel, A.A., AbdElminaam, D.S., Deb, S., Said, M.: Identification of parameters in photovoltaic models through a Runge Kutta optimizer. Mathematics **9**(18), 2313 (2021)
34. Casas-Ordaz, A., Oliva, D., Navarro, M.A., Ramos-Michel, A., Pérez-Cisneros, M.: An improved opposition-based Runge Kutta optimizer for multilevel image thresholding. J. Supercomput. 1–108 (2023)
35. Houssein, E.H., Hassan, H.N., Samee, N.A., Jamjoom, M.M.: A novel hybrid Runge Kutta optimizer with support vector machine on gene expression data for cancer classification. Diagnostics **13**(9), 1621 (2023)
36. Heidari, A.A., Mirjalili, S., Faris, H., Aljarah, I., Mafarja, M., Chen, H.: Harris hawks optimization: algorithm and applications. Futur. Gener. Comput. Syst. **97**, 849–872 (2019)
37. Li, C.H., Lee, C.: Minimum cross entropy thresholding. Pattern Recognit. **26**(4), 617–625 (1993)
38. Rodríguez-Esparza, E., Zanella-Calzada, L.A., Oliva, D., Hinojosa, S., Pérez-Cisneros, M.: Multilevel segmentation for automatic detection of malignant masses in digital mammograms based on threshold comparison. In: 2019 IEEE Latin American Conference on Computational Intelligence (LA-CCI), pp. 1–6. IEEE (2019)
39. Rodríguez-Esparza, E., Zanella-Calzada, L.A., Zaldivar, D., Galván-Tejada, C.E.: Automatic detection of malignant masses in digital mammograms based on a mcet-hho approach. Applications of Hybrid Metaheuristic Algorithms for Image Processing, pp. 351–374 (2020)
40. Yin, P.-Y.: Multilevel minimum cross entropy threshold selection based on particle swarm optimization. Appl. Math. Comput. **184**(2), 503–513 (2007)
41. Rodríguez-Esparza, E., Zanella-Calzada, L.A., Oliva, D., Heidari, A.A., Zaldivar, D., Pérez-Cisneros, M., Foong, L.K.: An efficient Harris hawks-inspired image segmentation method. Expert Syst. Appl. **155**, 113428 (2020)
42. Fadzil, M.A., Nugroho, H.A., Nugroho, H., Iznita, I.L.: Contrast enhancement of retinal vasculature in digital fundus image. In: 2009 International Conference on Digital Image Processing, pp. 137–141. IEEE (2009)
43. Nakano, K., Hirofuji, R., Ohnishi, T., Hauta-Kasari, M., Nishidate, I., Haneishi, H.: RGB camera-based imaging of oxygen saturation and hemoglobin concentration in ocular fundus. IEEE Access **7**, 56469–56479 (2019)
44. Delori, F.C., Pflibsen, K.P.: Spectral reflectance of the human ocular fundus. Appl. Opt. **28**(6), 1061–1077 (1989)
45. Acton, S.T.: Diffusion partial differential equations for edge detection. The Essential Guide to Image Processing, pp. 525–552. Elsevier (2009)
46. Sivaswamy, J., Krishnadas, S., Joshi, G.D., Jain, M., Tabish, A.U.S., Drishti-gs: Retinal image dataset for optic nerve head (onh) segmentation. In: IEEE 11th International Symposium On Biomedical Imaging (ISBI), pp. 53–56. IEEE (2014)

47. Kennedy, J., Eberhart, R.: Particle swarm optimization. In: Proceedings of ICNN'95-International Conference on Neural Networks, Vol. 4, pp. 1942–1948. IEEE (1995)
48. Price, K.V.: Differential evolution: a fast and simple numerical optimizer. In: Proceedings of North American Fuzzy Information Processing, pp. 524–527. IEEE (1996)
49. Mirjalili, S.: SCA: a sine cosine algorithm for solving optimization problems. Knowl.-Based Syst. **96**, 120–133 (2016)
50. Setiawan, A.W.: Image segmentation metrics in skin lesion: accuracy, sensitivity, specificity, dice coefficient, Jaccard index, and Matthews correlation coefficient. In: 2020 International Conference on Computer Engineering, Network, and Intelligent Multimedia (CENIM), pp. 97–102. IEEE (2020)
51. Willmott, C.J., Matsuura, K.: Advantages of the mean absolute error (MAE) over the root mean square error (RMSE) in assessing average model performance. Climate Res. **30**(1), 79–82 (2005)
52. Ma, T., Zhang, H., Ong, H., Vora, A., Nguyen, T.D., Gupta, A., Wang, Y., Sabuncu, M.R.: Ensembling low precision models for binary biomedical image segmentation. In: Proceedings of the IEEE/CVF Winter Conference on Applications of Computer Vision, pp. 325–334 (2021)
53. Cuadros, J., Bresnick, G.: Eyepacs: an adaptable telemedicine system for diabetic retinopathy screening. J. Diabetes Sci. Technol. **3**(3), 509–516 (2009)
54. Li, L., Xu, M., Wang, X., Jiang, L., Liu, H.: Attention based glaucoma detection: a large-scale database and CNN model. In: Proceedings of the IEEE/CVF Conference on Computer Vision And Pattern Recognition, pp. 10571–10580 (2019)
55. Patil, P.B., Verma, U.: Numerical Computational Methods. Alpha Science International, Ltd (2006)
56. England, R.: Error estimates for Runge-Kutta type solutions to systems of ordinary differential equations. Comput. J. **12**(2), 166–170 (1969)

# Classification of University Students Using Feature Selection and Wrapping Methods in a Pattern Recognition System

**Cynthia Cristina Martinez Padilla**

**Abstract** Dropping out of school is a problem that affects students around the world, with consequences for society. Therefore, in this study we will try to analyze this situation in higher education that affects not only students, but also different institutions training courses. By applying a pattern recognition system, feature selection methods are implemented to find those factors that could significantly influence the school career of university students at higher education institution in Portugal. These methods correspond to the so-called Features Selection Methods and Wrapper Methods, which allow analyzing sets with large volumes of data and which in turn correspond to mixed data types (numeric, categorical, object, etc.). The complexity of each of these methods used mostly involves all the features of the set, which will allow more accurate results to be obtained. After analyzing the results in the six methods used for Feature Selection, we can observe that in the Fisher Index, Pearson Correlation and Forward Selection, the characteristics of previous grades are more relevant to determine the success or failure of the student. For its part, in Variance Selection, Backward Selection and Exhaustive Selection, the characteristics that had the greatest weight were those related to the background and previous courses of the father and mother. With the training results, the resulting features with the Nayve Bayes and Nayve Bayes with Neyman Pearson models have an accuracy rate of around 70%. The case of the Hidden Markov model with its accuracy rate of 32.5% does not imply a Gaussian criterion, which gives us a lower accuracy, but no less reliable.

**Keywords** Feature selection · Wrapper methods · Pattern recognition · Academic dropout

C. C. M. Padilla (✉)
Universidad Politécnica de Aguascalientes, Aguascalientes, Ags., México
e-mail: mc230019@alumnos.upa.edu.mx



D. Oliva et al. (eds.), *Advances in Optimization Algorithms for Multidisciplinary Engineering Applications: From Classical Methods to AI-Enhanced Solutions*, Studies in Computational Intelligence 806, https://doi.org/10.1007/978-3-031-78440-8_14

# 1 Introduction

One of the challenges facing universities around the world is how to appropriately address the diverse learning styles of their students [1] and the diverse educational outcomes [2] they will experience while attending those institutions. The ability to predict and anticipate difficulties that students may face [3] is useful for universities seeking to develop support and guidance strategies for students who may be at risk of academic failure [4] based on information collected annually [5].

Today, the concept of acquiring knowledge from data is prevalent in almost every facet of contemporary technology. In these cases, as in all others, learning theory refers to methodologies that enable computer systems to effectively address complex questions by improving their performance through adaptive means using data samples from the respective domain. Learning processes are the fundamental component of machine learning techniques. These approaches meet the growing demand for automated methodologies to analyze large amounts of data in digital format [6].

Machine learning is about statistics, a theory that examines how to estimate functional dependencies from a collection of data [7]. Patter recognition is one of those common statistical problems. Patterns can be defined as regular, recurring patterns that help identify a situation or problem and can be used as indicators of future behavior [6]. From a statistical perspective, pattern recognition is one of the simplest forms of statistical (or inferential) modeling [8].

Predictive modeling using data and machine learning has been successfully applied in other areas, including business and public health, and plays an important role in creating early warning systems to define school dropouts. In this study, data related to attendance, student performance, parent feedback, and classroom activity are used as machine learning techniques to predict school dropout. Trained predictive models can help you identify at-risk students early and provide appropriate interventions to prevent them from dropping out of school [9]. Depriving students of their education has greater social costs for society. Furthermore, it can cause great harm to the lives of individual students. While it is important to help youth who do not attend and drop out of school, identifying those who do not attend school and preventing migration is a more important step. Predictive modeling using machine learning has the potential to identify at-risk students [9].

Talking about Dropping out of school, it is a problem that affects not only students but also different institutions around the world, with consequences for society. It can be defined as a person who withdrew from an educational institution before receiving a degree or a person who enrolled in the previous year and decided not to transfer to another school and not continue in their classes [10]. In principle, they see this as a learning program until the student leaves school. On the other hand, the spatial connection of concepts occurs when students change subjects in the same institution, when they change universities, even when they leave the educational system, but continue in the future. Whether it is the same house or different houses [11].

A review of the literature shows that students drop out of school for many reasons, including personal factors, issues related to lack of motivation, interpersonal relationships, teenage pregnancy, etc., and financial issues related to family income level; Know strategies, activities, resources, and educational evaluation. Educational institutions and parents must participate. This is a great support for children, because it has a positive effect on motivation, behavior and even learning [11].

Some studies related to dropping out of school have found it efficient to take courses during their stay at the academic institution to improve their grades and reinforce what they have learned, thus students have greater chances of completing their studies [12].

### *1.1 Contribution*

This work seeks to contribute to the prediction of the level of academic failure or success in higher education students using feature selection and machine learning techniques [13]. We seek to identify those factors that intervene in their studies at an early stage and that help school authorities and parents to respond to these questions and pay more attention and courses of action to those that concern them through courses, workshops, etc. It will help society to create more awareness about what is happening to our young people and help them in their professional and personal training [14]. A more exhaustive investigation is necessary to find the reason why these factors are of great relevance, and we hope that this work will be the basis for future work.

## 2 Background

In relation to works that have implemented classification and pattern recognition techniques for the analysis of school dropouts, we find the following within the literature.

A study showed that receiving comprehensive support in all economic, social and educational areas during the years of basic and secondary education had approximately half the probability of completing or advancing to another level than those who did not receive any support [15]. Other studies indicate that one of the main factors for dropping out of school in students is stress due to very important circumstances to which they have been subjected at some stage of their life [16].

In the research presented by Samuel Robin in 2020 the points to risk and protective factors for school failure. Using a 4-year longitudinal design, he focuses on the role of fixed and time-varying factors in predicting dropout and future dropout. This allowed him to show that negative life events were significantly associated with increased intentions to drop out of school and increased dropout. Furthermore, average levels of self-efficacy and social support were associated with lower levels of intention to quit, but did not prevent the dropout [17].

In the 2020 paper, it was analyzed that one of the reasons why students leave is the attractiveness of the labor market, because the working day does not coincide with the day of graduation, and the income is more than that. People with degrees. The qualitative and quantitative results show that financial problems are the main cause of student behavior and that students enter a negative cycle. These results indicate that employment during or after university is an unavoidable area in the analysis of dropouts [18].

Diane Rodriguez's work in 2022 explored the perspectives of teachers, counselors, and administrators on issues related to school services and the educational and academic problems of students in New York City. This study attempted to identify the factors that increase dropout among high school students. In addition, we seek to identify school services that improve academic and learning success [19].

Another study in Pakistan sought to examine how students drop out of school through the perspective of their parents. Data were collected through interviews with parents of children who had dropped out of school. Family poverty, poor academic performance and teachers' loyalty to the school are the main reasons [20].

Other studies of socioeconomic factors provide a comprehensive assessment of higher education choices in dynamic environments where informational and relational factors are important in predicting student achievement. This review suggests that persistence/absence depends on a combination of individual, institutional and economic factors, and that effectiveness depends on the student's ability to integrate into the system [21].

Other studies have shown that online learning is becoming more popular in higher education because it allows for more flexibility and flexibility in student schedules. However, there are also challenges in ensuring student retention and continued in the course. This has led to the need for educators and researchers to analyze student behavior and personal and academic information to provide models and strategies to reduce dropout in these courses and make accurate predictions to reduce. In Mubarak's work [22] on online course dropouts, he showed that the model used achieved 84% accuracy in predicting the risk of dropping out.

Low student performance in so-called massive open online courses (MOOCs) is a major concern for the higher education community. Aldowah's work in 2020 used multi-criteria decision and techniques to identify the root and potential causes of high dropout rates in MOOCs. The results identified six main factors: academic ability, prior experience, course design, feedback, socialization, and social support. Other factors such as interaction, course difficulty, time, effort, motivation, and family/work situation play a secondary role when it comes to student attrition [23].

# 3 Theoretical Framework

## *3.1 Pattern Recognition System*

A discipline that studies the description and classification of objects, it is also a collection of mathematical, statistical, heuristic and heuristic methods that play an important role in human computing [8].

Pattern recognition has grown in popularity and importance to us since the 1960s and is generating impressive interest in a wide range of areas. The main objective is to examine how objects are described and classified. Finally, a pattern recognition system based on pattern recognition methods has three different and interrelated processes: one is data creation. Another is pattern analysis. Finally, classification of the patterns [24].

## *3.2 Preprocessing Stage*

A database obtained at the time of the student's registration in the different courses of the four different schools belonging to it (such as agronomy, design, education, nursing, journalism, administration, social service and technologies) is used as a reference and contains information such as day/night attendance, age of enrollment, educational level of the father and mother, etc. enrolled and the academic performance of the students at the end of the first and second semester [5].

To facilitate the capture of information, the information must be analyzed and examined from various perspectives and the data processed to be useful. This research deals with data **transformation** [25], converting raw data into numerical data. Some of the variables in the analyzed data set are clustered variables, that is, grouped and unordered variables. Therefore, you need a data transformation method that can convert categorical data to numeric data and use coding techniques to assign numbers to different categories for each variable in the data set. It can be illustrated by the following function (Fig. 1):

Once you have converted the segmented data into numbers, the next step is to organize the data. **Normalization** is a number assigned to a different category for

```
data_int = data.copy()
for i in range (data.shape[1]);
   if data_int.iloc[0:,i].dtype=='0':

data_int.iloc[0:,i]=data_int.iloc[0:,i].astype('category').cat.codes
   return data_int.astype('int')
```

**Fig. 1** Function in Python for transforming

each variable in the data set from 0 to 1, which calculates the trigonometric function [13]. This is necessary when there are large differences in the scope of the various activities. This scaling method is useful when your data set has no outliers [26]. It can be illustrated by the following function (Fig. 2) and for the following formulas (Eqs. 1 and 2) used for normalization correspond to the standard deviation and the scaling of variables [27].

$$X_{std} = \frac{x - X_{\min}}{X_{\max} - X_{\min}} \tag{1}$$

$$X_{scaled} = X_{std} \times (X_{\max} - X_{\min} + X_{\min}) \tag{2}$$

Finally, we must identify outliers in the data that deviate so much from other observations that they raise suspicions that they were generated by a different mechanism than the bulk of the data. **Outliers** can be erroneous or real and can exert undue influence on the results of statistical analysis, so they should be identified using reliable detection methods before performing data analysis. Some noisy points may be far from the data points, while others may be close. The distant noisy points would affect the result more significantly because they are more different from the data points [28]. In this data set we are going to identify outliers with respect to number of times the standard deviation and calculating the mean value of the data as shown in the following function (Fig. 3):

```
data_norm = data.copy()
for i in range (data.shape[1]);
   npvalues = np.asarray(data_norm.iloc[0:,i])
   xmin = np.min(npvalues)
   data_pre = npvalues - xmin
   xmaxdata = np.max(data_pre)
   data_nomr.iloc[0:,i] = data_pre/xmaxdata
   return data_norm
```

**Fig. 2** Function in Python for Normalizer_linear

```
miu = feaure.mean()
sigma = feature.std()
features_mantained = [i for i in range(len(feature)) if
(abs(feature[i].miu) < sigma*threshold)]
```

**Fig. 3** Function in Python for outlier removal

### 3.3 Feature Selection Stage

In this section, we included the different methods for feature selection used in this work, including filter and wrapper methods, as shown below:

1. **Fisher Index**: Is an efficient way that reduces the dimension of features in a data set. The main goal is to find a subset of features in the data space that covers the selected features, maximizes the distance between data points of different classes and minimizes the distance between data points of the same class [29]. Use the following formula (Eq. 3).

$$\mathrm{F} = \frac{\sum_{j=1}^{k} P_j \cdot \left(\mu_j - \mu\right)^2}{\sum_{j=1}^{k} P_j \cdot \sigma_j^2} \tag{3}$$

2. **Pearson's Correlation**: Measures whether two sets of data lie on a straight line and is used to measure the linear relationship between distant variables. The coefficient is calculated as the Pearson correlation coefficient between the selected variables. For a sample with sample size n, we transform the n original data into master data and the coefficient r is [30]. The formula (Eq. 4) to determine the coefficient for this method [27] is:

$$\rho(X, Y) = \frac{\sum_{\mathrm{i}=1}^{\mathrm{n}} \left(\mathrm{X_i} - \overline{\mathrm{X}}\right)\left(\mathrm{Y_i} - \overline{\mathrm{Y}}\right)}{\sum_{\mathrm{i}=1}^{\mathrm{n}} \left(\mathrm{X_i} - \overline{\mathrm{X}}\right)^2 \sum_{\mathrm{i}=1}^{\mathrm{n}} \left(\mathrm{Y_i} - \overline{\mathrm{Y}}\right)} \tag{4}$$

3. **Variance Threshold**: Classifies the features by calculating the variance of each feature in the data set. A characteristic is given a high variance. This is a simple selection method used to select text mining features. Is obtained by the following formula (Eq. 5) [31]. The equation to determine the margin of variation between the samples and the average value [27] is:

$$\sigma^2 = \frac{\sum_{\mathrm{i}=1}^{\mathrm{n}} \left(\mathrm{X_i} - \overline{\mathrm{X}}\right)^2}{\mathrm{n} - 1} \tag{5}$$

4. **Forward Feature Selection**: Part of an empty set with only a small number of features. Start with the first best-performing variable, then compare the other best-performing variables in the tree until you get the set with the best features [32].
5. **Backward Feature Selection**: Operates in the opposite direction to the sequence selection. We create a model that starts with the full set of features and compares them by removing values that do not meet the criteria until we reach the final set [32].
6. **Exhaustive Selection**: After quantitatively verifying the predictive power of all possible sets of variables, the combination with the highest predictive power is taken as the best model equation. When testing all possible combinations of

variables, each variable has two aspects: used and obsolete. Expanding this to all variables, the total number of possible states is 2n. n is the number of variables [32].

### 3.4 Data Training

The objective of the classification process is to know the model parameters that best adjust the characteristics of the explanatory variables to the response variables. A discriminant model is a classifier that learns the decision boundaries that are used to distinguish between classes and decide which class is given input features. Cluster models, model the probability distribution of features and classes. This is like modeling the probability of a class and the probability of a given feature in that class [33]. In this section, we included the different methods for classification process used in this work, as shown below:

1. **Gaussian Naïve Bayes Classifier**: Is a simple probabilistic classifier that calculates a series of probabilities using the frequencies and combinations of values in a data set. The theorem considers the values of the class variables, assuming that all variables are independent although this independence is very rarely valid in real world applications and is therefore considered naive, the algorithm constantly learns for a variety of operational classification problems [33]. Baye's mathematical formula used is (Eq. 6):

$$P(Y|X_1, X_2, \ldots, X_n)\alpha P(Y)\prod_{i=1}^{n} P(X_i|Y) \tag{6}$$

   In pattern recognition, a common way in training data to evaluate classification quality is through predictive decision classes and a confusion matrix, also known as an error matrix. The matrix is a tool that contains only the information contained in the data. The sum of the main diagonal elements of a confusion matrix is widely used to measure the success of classification based on algorithms or human observations. An entry hYˆ i,Yji = ni j in the matrix is the number of elements of Yj which are predicted to be in Yi; in particular the number of correctly classified elements is defined by the next formula (Eq 7) [33]:

$$\sum\{n_{ii} : 1 \leq i \leq k\} \tag{7}$$

2. **Bayesian Theory Classifiers**: Can be applied to reproductive samples to estimate the probability of a class based on its characteristics. A formula (Eq. 8) that calculates the probability of an event using prior knowledge of the conditions associated with the event [34].

$$P(A|B) = \frac{P(B|A)P(A)}{P(B)} \tag{8}$$

3. **Hidden Markov Models Classifier (HMM)**: The construction of an HMM relies on a representing state space from which states are drawn. Is a method that trains probabilistic data to represent hidden sequences for observations in different scenarios and provides a basis and method for finding the best sequence that best explains the observations [35]. We start with a state and go through a sequence of states based on transitions that are characterized probabilistically. We use these observations to learn. We need calculate (Eqs. 9 and 10):

$$\pi_i = \frac{\text{Number of times the first observation is assigned to cluster } i}{\text{Total number of patterns}} \tag{9}$$

$$\alpha_{ij} = \frac{\text{Number of times } O_t \text{ is in cluster } i \text{ and } O_{t+1} \text{ is in cluster } j}{\text{Number of times } O_t \text{ is in cluster } i} \tag{10}$$

## 4 Methods

The review and selection of the articles was done based on 2 phases. These phases correspond to planning the review and carrying out the review in which the following activities were carried out:

1. **Review planning**: Research activities were carried out focused on finding the different problems that arise in the educational field as a result of the recent pandemic, several situations of abandonment were detected and with this the question arose of how much they can affect factors such as social, economic, cultural, in the success or failure of students and how these factors can be related to determine it. An investigation was also carried out into the methods and techniques that supported this study and whose main objective was the classification of data.
2. **Review articles**: In this stage of the process, the search, evaluation and selection of previous works in which selection methods were applied and had satisfactory results was carried out. It began by detecting keywords such as pattern recognition, feature selection methods, fisher selection, pearson correlation, wrapper methods, data normalization, outlier removal. These keywords were searched in combination with each other and as individual terms [36]. After the selection of the keywords, the search was carried out in some online article databases such as IEEE Xplore, Google Schoolar and Elsevier, the search was restricted to the title of the article and the English language. Afterwards, a selection of some articles was made by initially reading their title and summary and, if it contained relevant information, the conclusion was read, seeking to ensure that they met the area of our interest. As the articles were read, some were classified as primary, secondary and others were discarded because they did not meet expectations.

3. **Data Set**: The data covers a variety of majors including agriculture, design, education, nursing, journalism, business, social work and technology from the 2008–2009 school year to 2018–2019 [5]. The data includes features such as school age, marital status, nationality, special needs, parents' occupations, admission grades, previous semesters, and several more [5]. The final data set is composed of 4424 records, 36 independent variables, and 1 final classification variable.
4. **Data Analysis**: Data analysis was carried out using Python language, using the following libraries: Numpy, Pandas, Matplotlib, Seaborn and the Scikit-learn package. In this way, we were able to standardize categorical data to numerical data and normalize the information based on similar ranges from 0 to 1, as described in Sect. 2.3 (Preprocessing Stage) using the Numpy library that is for mathematical operations, computational arrays and linear algebra operation; the results of the standardization were: of the 3 data types contained in our data set (int64, float64, object), the 37 variables were converted to int32 type.

Once the data was standardized, the call to the outlier_removal function was made as mentioned in Sect. 3.2 Preprocessing Stage to eliminate those records that, because they are further away from the standard mean, would cause a poor selection of features. The test was carried out using several values for threshold (number of times the variance), leaving 4 as the final value, to eliminate a good percentage of elements without sacrificing the rest for selection; the result was from 4424 we obtained 3455 records. In all the methods applied for data visualization, the matplotlib library was used, which in Pytohn is very popular for the design of 2D graphs.

6 selection methods were applied to determine which features are the most relevant and to determine the success or failure of the students. The 6 methods correspond to mixed functions that involve both numerical and categorical features. Some libraries and their utilities were used to apply each of the selection methods, as described below:

- For the Fisher's Index, Variance Threshold, Forward and Bakcward Select methods, the scikit-learn utilities were used. These utilities include pooling, reduction, addition, and dimensionality reduction transformations.
- For the Pearson's Correlation method, the seaborn utility was used, which provides us with a high-level interface to create quality statistical graphs with a few lines of code; and finally
- For the Exaustive Selection method, the utilities from the mlxtend library were used, which is responsible for providing all the tools for generate the association rules.

3 training methods were also used to analyze the behavior of the features of the evaluated data set and obtain the subset with the best characteristics that will help us determine the success or failure of the students in their studies. The methods used are:

- Bayesian Theory Classifiers
- Naïve Bayes Classifier with Neyman Pearson
- Hidden Markov Models Classifier.

# 5 Results and Discussion

## *5.1 Fisher Index*

This method is applied for both numerical and categorical features, therefore, here we use all the features of the data set. The library we used for Python was the skfeature from which we imported the fisher_score method.

First, we calculate the Fisher score of each feature using the training set (Eq. 3). The features are sorted in descending order based on their Fisher score and the ranking index is generated. To select the best features, we display the results in a graph showing the score for each feature (Fig. 4).

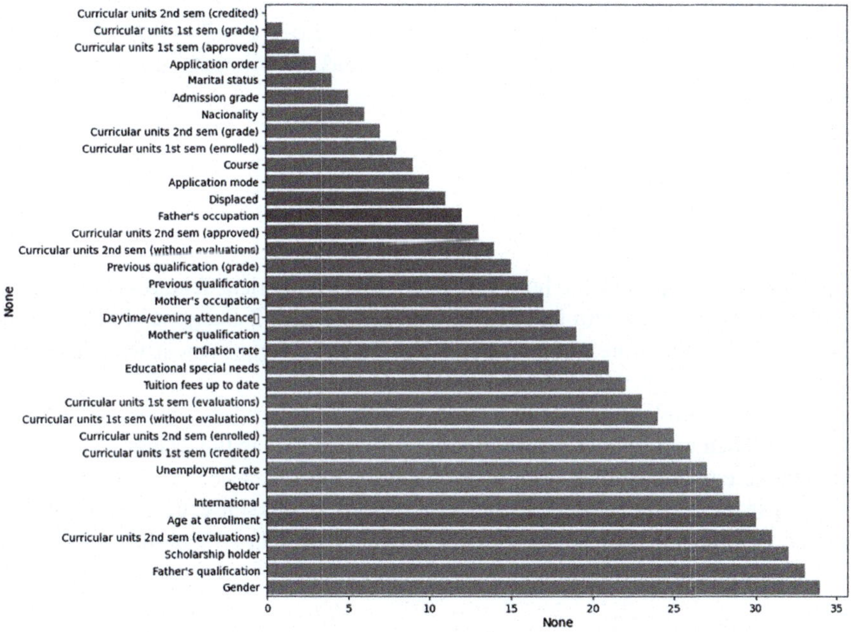


**Fig. 4** Results of student classification using the feature selection model

In the function fisher_ranks, we specified the number of features selected for evaluation. In this case, we need 7, which represents 20% of all features. The final score for features obtained were:

1. `Curricular units 2nd sem (credited),`
2. `Curricular units 1st sem (grade),`
3. `Curricular units 1st sem (approved),`
4. `Application order,`
5. `Marital status,`
6. `Admission grade,`
7. `Nacionality.`

`The first 3 features indicate that students decide to leave school based on the grades obtained in the first and second semester of their careers, which indicates a possible demotivation due to feeling that they are low compared to the rest of their classmates or with your personal academic history. The rest of the factors, such as nationality and marital status, also indicate that they are important elements to continue studying, perhaps because they do not feel included in the group or have other family and economic commitments corresponding to their marital status.`

## *5.2 Pearson's Correlation*

This method is applied for both numerical and categorical features, however its functionality is better with features that have more than 2 categories in each of the data variables, therefore, those features that meet this requirement were analyzed, reducing our variables from 37 to 28 (Table 1).

Using the correlation formula (Eq. 4), we obtain a score for each of the evaluated features. Then we use the seaborn library for Python to map the results obtained and compare the relationship of each of the features (Fig. 5).

We use these estimates as selection criteria [38]. In this sense, the results obtained for the 7 features that we are looking for and that represent 20% of the total features of the dataset are the following:

1. `Marital status,`
2. `Application mode,`
3. `Curricular units 2nd sem (credited),`
4. `Curricular units 2nd sem (evaluations),`
5. `Curricular units 2nd sem (grade),`
6. `Age at enrollment,`
7. `Course.`

**Table 1** Features used in Pearson's correlation

| | |
|---|---|
| Marital status | Curricular units 1st sem (credited) |
| Application mode | Curricular units 1st sem (enrolled) |
| Application order | Curricular units 1st sem (evaluations) |
| Course | Curricular units 1st sem (approved) |
| Previous qualification | Curricular units 1st sem (grade) |
| Previous qualification (grade) | Curricular units 1st sem (without evaluations) |
| Nacionality | Curricular units 2st sem (credited) |
| Mother's qualification | Curricular units 2st sem (enrolled) |
| Father's qualification | Curricular units 2st sem (evaluations) |
| Mother's occupation | Curricular units 2st sem (approved) |
| Father's occupation | Curricular units 2st sem (grade) |
| Admission grade | Curricular units 2st sem (without evaluations) |
| Aget at enrollment | Unemployment rate |
| GDP | Inflation rate |

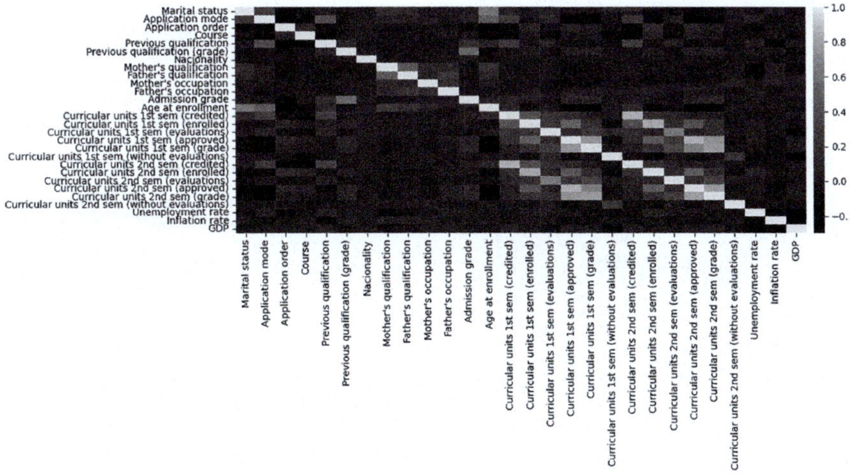


**Fig. 5** Results of student classification using Pearson's correlation model

With this analysis, the first apparent cause of dropping out of school is marital status, as mentioned in the case of the Fisher Index method, having other family and economic commitments corresponding to their marital status in some cases affects compliance with the corresponding tasks and studies. The way in which face-to-face or online classes are carried out also has a negative impact, beyond the advantages of either. Age at the time of entry and the selected career are not considered risk factors or with high impact as they are in sixth and seventh place respectively.

## 5.3 Variance Threshold

This method is applied for both numerical and categorical features, therefore, here we use all the features of the data set. Analyze the characteristics, not the desired outcome. Specify a training value of 0.095, which removes features with a training set variance less than this threshold. The goal is to keep all features constant with non-zero variance, that is, for the data to have the same value in all samples.

Finally, we define the number of features that we want to obtain as a result (Fig. 6), 7 in this case are the desired optimal ones. The final score for each features obtained (Eq. 5) were:

All features with zero variance are eliminated, and we consider those that have a higher threshold as those that may contain useful information. So, the 7 best features obtained were:

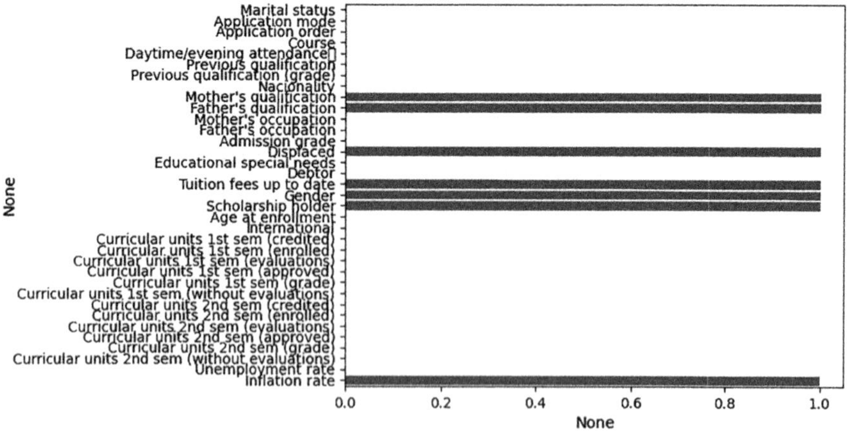


**Fig. 6** Results of student classification using the Variance Threshold model

1. Mother's qualification,
2. Father's qualification,
3. Displaced,
4. Tuition fees up to date,
5. Gender,
6. Scholarship holder,
7. Inflation rate.

These results show that the educational level of the parents has a high impact on the students, due to the pressure to reach the same level as the parents or surpass them and/or because in some cases the parents do not have studies and their position is the of expecting the same and continuing with that pattern, making excuses for not studying.

## 5.4 *Forward Feature Selection*

This method is applied for both numerical and categorical features, therefore, here we use all the features of the data set. We start with an empty set with no variables, and the quantity with the highest correlation coefficient is added as the first variable. Variables that have larger partial coefficients with the dependent variables specified in the model are then included in the model. The process ends when the increase in the coefficient of determination due to the inclusion of new variables is no longer significant. The final score for features obtained were (Fig. 7):

From there we take the 7 features with the highest score:

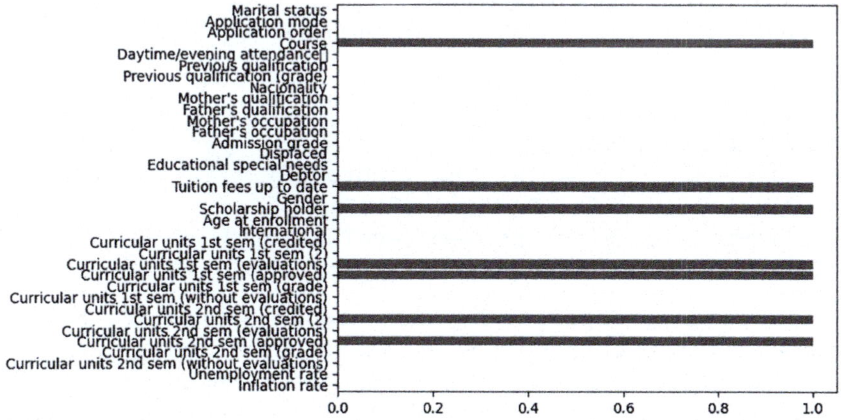


**Fig. 7** Results of student classification using the Forward model

1. Course,
2. Tuition fees up to date,
3. Scholarship holder,
4. Curricular units 1st sem (evaluations),
5. Curricular units 1st sem (approved),
6. Curricular units 2nd sem (2),
7. Curricular units 2nd sem (approved).

The results obtained show that the selected career may in certain cases not be appropriate or desired by the students, causing them to change to another that is of interest to them or that they can cover with their expenses. Being a scholarship holder, working and studying at the same time for some students is also a high reason for drop out because they are not able to organize their activities in the best way. In the case of grades in the first semesters it seems to be a less significant factor.

## 5.5 Backward Feature Selection

This method is applied for both numerical and categorical features, therefore, here we use all the features of the data set. We first consider all available variables in the data set and exclude them one by one, the first variable to exclude is the variable with the lowest partial correlation coefficient with the dependent variable and so on until we reach a situation where there is one more variable. Eliminated Variables reflect the decrease in the representativeness of the coefficient of determination. The final score for features obtained were (Fig. 8):

From there we take the 7 features with the highest score:

1. Course,
2. Tuition fees up to date,
3. Scholarship holder,
4. Curricular units 1st sem (approved),
5. Curricular units 2nd sem (2),
6. Curricular units 2nd sem (evaluations),
7. Curricular units 2nd sem (approved).

The results obtained by Backward selection are the same as those obtained by Fordward selection, both selection methods show that the career selected by the students is not the appropriate or desired one, causing them to change to another that is of interest or that they can cover with their expenses. Being a scholarship

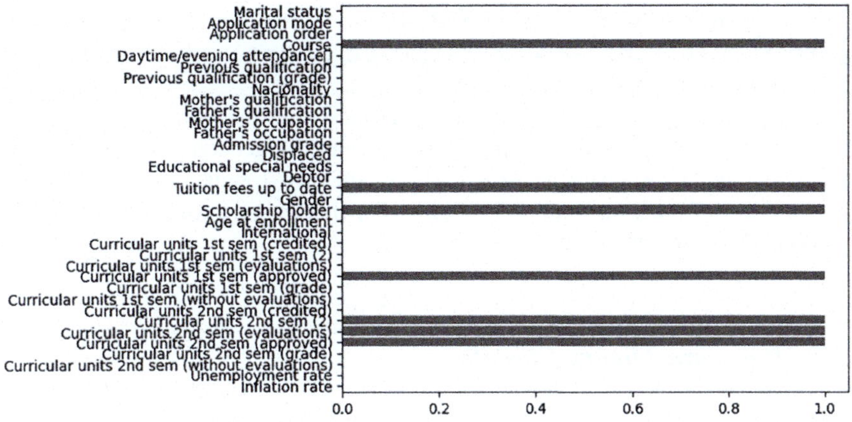


**Fig. 8** Results of student classification using the Backward model

holder, working and studying at the same time affect the academic and personal performance of some students, significantly affecting their continuity. In the case of grades in the first semesters it seems to be a less significant factor.

## 5.6 *Exhaustive Selection*

This method is applied for both numerical and categorical features, therefore, all features of the data set whether numerical or categorical are used here. As our data set is coarse, a smaller search range was defined, only 4 characteristics of the 7 that were searched in the other methods in order to optimize process times and we used a function to evaluate all possible combinations of the characteristics. functions. This means that we evaluate each feature in the data set to obtain the best subset that provides the best estimate of those features. The final score of the characteristics obtained by the best four was:

1. Application mode,
2. Course,
3. Mother's qualification,
4. Father's qualification.

As in the results obtained by Fisher Index, using Exhaustive selection, the way in which face-to-face or online classes are carried out also has a negative impact, beyond the advantages of either of the two. In the same way, career selection, as already mentioned in the Forward, Backward and Pearson correlation methods, career selection also significantly influences the decision to continue studying.

After applying the selection methods, and based on the results of the Pearson Correlation, we train the data taking as reference the Target column from which we will obtain the 5 best characteristics, and subsequently from these columns we will obtain their information to do the training. For this training we use the following criteria, as shown below:

## 5.7 Bayesian Theory Classifiers

From the data set, we separate the features and Target columns into arrays. Later we divided our entire set into 4 parts; The first 2 correspond to the features used for testing and training, the last 2 correspond to the Target column for testing and training. Finally, create our model using the training data from features and Target and apply a function to evaluate the data (Fig. 9).

The results obtained from some of the training characteristics with the highest and lowest ratings were:

| | |
|---|---|
| Target | 1.000000 |
| Curricular units 2nd sem (approved) | 0.740580 |
| Curricular units 1st sem (approved) | 0.667954 |
| Curricular units 2nd sem (grade) | 0.616634 |
| Curricular units 1st sem (grade) | 0.536621 |
| Age at enrollment | 0.294247 |
| Application mode | 0.265989 |
| Curricular units 2nd sem (enrolled) | 0.260226 |

```
X = data_int;
Y = data_int.Target;
X_train, X_test, y_train, y_test = function_split(X, Y);
clf.fit(X_train, y_train);
```

**Fig. 9** Function used to calculate the Bayesian classification

| | |
|---|---|
| Curricular units 1st sem (enrolled) | 0.225899 |
| Admission grade | 0.111897 |
| Previous qualification (grade) | 0.110424 |
| Marital status | 0.105112 |
| Application order | 0.097845 |
| Previous qualification | 0.067806 |
| Curricular units 2nd sem (evaluations) | 0.061842 |
| Father's occupation | 0.060874 |
| Curricular units 1st sem (without evaluations) | 0.046280 |
| GDP | 0.043375 |
| Inflation rate | 0.041972 |
| Mother's qualification | 0.037656 |
| Curricular units 2nd sem (without evaluations) | 0.036264 |
| Nacionality | 0.025070 |
| Mother's occupation | 0.024921 |
| Curricular units 1st sem (credited) | 0.014914 |
| Course | 0.012438 |
| | … |
| Father's qualification | 0.002732 |
| Unemployment rate | 0.000343 |

The final score for features obtained with an effectiveness accuracy of 72.3% were:

1. Curricular units 2nd sem (approved),
2. Curricular units 1st sem (approved),
3. Curricular units 2nd sem (grade),
4. Curricular units 1st sem (grade),
5. Age at enrollment',
6. Application mode,
7. Curricular units 2nd sem (enrolled).

Applying the Bayesian method, the best trained and best evaluated characteristics to predict the behavior of students are those that correspond mostly to the grades obtained during the first and second semester of their studies, which gives rise to the fact that in many In most cases, students are given the opportunity to get to know each other and, after this first year, decide whether to continue studying or not. This training was 72.3% effective in guaranteeing the correct classification of students.

## 5.8 *Naïve Bayes Classifier with Neyman Pearson*

From the data set, we separate the features and Target columns into arrays. Later we divided our entire set into 4 parts; The first 2 correspond to the features used for testing and training, the last 2 correspond to the Target column for testing and training. A range is defined to evaluate the features based on a selection rule. Finally, create our model using the training data from features and Target, the sample_weight and apply a function to evaluate the data (Fig. 10).

The results obtained from some of the training characteristics with the highest and lowest ratings were:

| | |
|---|---|
| Target | 1.000000 |
| Curricular units 2nd sem (approved) | 0.740580 |
| Curricular units 1st sem (approved) | 0.667954 |
| Curricular units 2nd sem (grade) | 0.616634 |
| Curricular units 1st sem (grade) | 0.536621 |
| Age at enrollment | 0.294247 |
| Application mode | 0.265989 |
| Curricular units 2nd sem (enrolled) | 0.260226 |
| Curricular units 1st sem (enrolled) | 0.225899 |

```
X = data int;
Y = data int.Target;
X train, X test, y train, y test = function split(X, Y);
sample weights = [1 if i == 0 else 20 for i in Y];
clf.fit(X train, Y train, sample weight);
```

**Fig. 10** Function used to calculate the classification with Naïve bayes model

| | |
|---|---|
| Admission grade | 0.111897 |
| Previous qualification (grade) | 0.110424 |
| Marital status | 0.105112 |
| Application order | 0.097845 |
| Previous qualification | 0.067806 |
| Curricular units 2nd sem (evaluations) | 0.061842 |
| Father's occupation | 0.060874 |
| Curricular units 1st sem (without evaluations) | 0.046280 |
| GDP | 0.043375 |
| Inflation rate | 0.041972 |
| Mother's qualification | 0.037656 |
| Curricular units 2nd sem (without evaluations) | 0.036264 |
| Nacionality | 0.025070 |
| Mother's occupation | 0.024921 |
| Curricular units 1st sem (credited) | 0.014914 |
| Course | 0.012438 |
| | … |
| Father's qualification | 0.002732 |
| Unemployment rate | 0.000343 |

The final score for features obtained with an effectiveness accuracy of 68.7% were:

1. Curricular units 2nd sem (approved),
2. Curricular units 1st sem (approved),
3. Curricular units 2nd sem (grade),
4. Curricular units 1st sem (grade),
5. Age at enrollment,
6. Application mode,
7. Curricular units 2nd sem (enrolled).

As with the Bayes method, in the case of the Nayve Bayes with Neyman Pearson method, the best trained and best evaluated characteristics to predict student behavior are mostly the grades obtained during the first and second semester of their studies. . In both cases, age at the time of enrollment is also considered as a key factor, in fifth place, for the success or

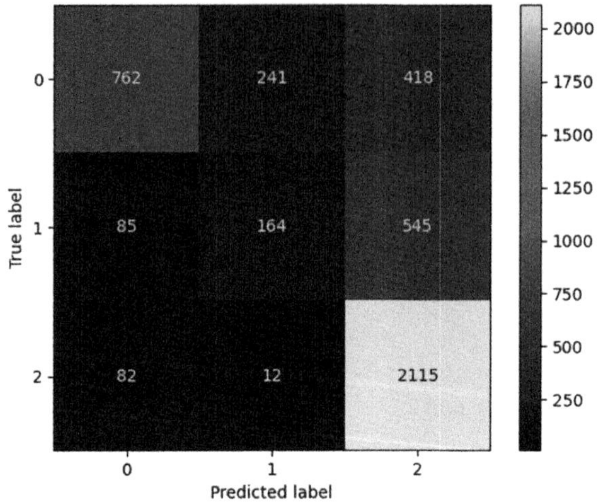


**Fig. 11** Confusion matrix showing the ratio of correctly classified results to those that were not correctly classified

failure of their studies. This training was 68.7% effective in guaranteeing the correct classification of students, placing it below the Bayes method.

The respective confussion matrix is (Fig. 11). The problem is formulated as a three-category classification task (dropout, enrolled, and graduate) at the end of the normal duration of the course [5]. Based on the $3 \times 3$ confusion matrix, the main diagonal (762, 164, 2115) gives the correct predictions. That is, the cases in which the actual values and the model predictions are the same.

TP real predicted as dropout (762)
TN real enrolled predicted as enrolled and real graduated predicted as graduated (164 + 2115)
FP real enrolled and graduated predicted as dropout (85 + 82)
FN real dropout predicted as enrolled or graduated (241 + 418)

Where TP is (True Positive), TN is (True Negative), FP is (False Positive) and FN is (False Negative).

## 5.9 *Hidden Markov Models Classifier*

From the data set, we separate the features and Target columns into arrays. Later we divided our entire set into 4 parts; The first 2 correspond to the features used for testing and training, the last 2 correspond to the Target column for testing and training. Using the training data, we categorize the Target column with relative and absolute values, thus generating 2 new arrays. Subsequently, we categorize the features columns into

```
X = data int
Y = data int.Target
X train, X test, y train, y test = function split(X, Y)
X1 = X train[y train]
X2 = X train[abs(y train-1)]
clf1.fit(X1)
clf2.fit(X2)
```

**Fig. 12** Function used to calculate the classification with Hidden markov model

2 new arrays, one corresponds to the evaluation of the training data with the relative Target values, and the other corresponds to the training data with the absolute Target values. Finally, create our model using the training data from features X1 and X2 and apply a function to evaluate the data (Fig. 12).

The results obtained from some of the training characteristics with the highest and lowest ratings were:

| | |
|---|---|
| Target | 1.000000 |
| Curricular units 2nd sem (approved) | 0.740580 |
| Curricular units 1st sem (approved) | 0.667954 |
| Curricular units 2nd sem (grade) | 0.616634 |
| Curricular units 1st sem (grade) | 0.536621 |
| Age at enrollment | 0.294247 |
| Application mode | 0.265989 |
| Curricular units 2nd sem (enrolled) | 0.260226 |
| Curricular units 1st sem (enrolled) | 0.225899 |
| Admission grade | 0.111897 |
| Previous qualification (grade) | 0.110424 |
| Marital status | 0.105112 |
| Application order | 0.097845 |
| Previous qualification | 0.067806 |
| Curricular units 2nd sem (evaluations) | 0.061842 |
| Father's occupation | 0.060874 |
| Curricular units 1st sem (without evaluations) | 0.046280 |

| | |
|---|---|
| GDP | 0.043375 |
| Inflation rate | 0.041972 |
| Mother's qualification | 0.037656 |
| Curricular units 2nd sem (without evaluations) | 0.036264 |
| Nacionality | 0.025070 |
| Mother's occupation | 0.024921 |
| Curricular units 1st sem (credited) | 0.014914 |
| Course | 0.012438 |
| | ... |
| Father's qualification | 0.002732 |
| Unemployment rate | 0.000343 |

The final score for features obtained with an effectiveness accuracy of 32.5% with:

1. Mother's occupation,
2. Father's occupation,
3. Mother's qualification,
4. Previous qualification (grade),
5. Course
6. Father's qualification,
7. Application mode.

In the case of the Hiden Markov Model method, the results of the best-trained features for classification correspond to the data obtained from both the father and the mother, for the most part. As mentioned previously in the Variance Threshold selection method, the educational level of the parents impacts the students, either by reaching that same level, exceeding it and/or stagnating and choosing to help with family expenses. The grade obtained in the immediately preceding grade and the career selected are also considered important.

The degree of reliability of this method is 32.5% compared to the previous 2 methods, which makes it not as effective for classification. However, if we consider that the first two training methods follow a Gaussian model and the latter follows a linear model, the accuracy obtained by HMM does not mean that it is not suitable for classification.

## 6 Conclusion

After analyzing the results in the six methods used, we can observe that in the Fisher's Index, Pearson Correlation and Forward Select method, the characteristics of previous grades are more relevant to determine the success or failure of the student. For its part, in Variance Select, Backward Select and Exhaustive Select, the characteristics that had the greatest weight were those involved with the father's and mother's background and previous courses. Therefore, the range of characteristics that we can consider determining in greater detail the expected result are the characteristics in the following table (Table 2).

By summarizing the characteristics obtained by each of the selection and training methods, we can see how those related to the academic field are more likely to be a cause of dropout, this is because the grades obtained in the first and second semester demotivate students. students, making them assume that these careers are not for

**Table 2** Table of results best characteristics and total repetitions

| Best feature | Total |
|---|---|
| Curricular units 2nd sem (credited) | 2 |
| Curricular units 1st sem (grade) | 1 |
| Curricular units 1st sem (approved) | 3 |
| Application order | 1 |
| Marital status | 2 |
| Admission grade | 1 |
| Nacionality | 1 |
| Application mode | 1 |
| Curricular units 2nd sem (evaluations) | 2 |
| Curricular units 2nd sem (grade) | 1 |
| Age at enrollment | 1 |
| Mother's qualification | 2 |
| Father's qualification | 2 |
| Displaced | 1 |
| Tuition fees up to date | 3 |
| Gender | 1 |
| Scholarship holder | 3 |
| Inflation rate | 1 |
| Course | 3 |
| Curricular units 1st sem (evaluations) | 1 |
| Curricular units 2nd sem (2) | 2 |
| Curricular units 2nd sem (approved) | 2 |
| Application mode | 1 |

them or that the further they advance they will be more difficult or motivate them to continue; The course in which they are enrolled is also a common factor, since this makes students and parents think that they have low demand and therefore will have a lot of job competition upon graduation or high demand, therefore the acceptance rate is low and it is very difficult to stay in them due to academic and work demands. and the current registration fees are also another factor as there is not enough financial support (scholarships) from the institutions.

## References

1. Cimermanová, I.: The effect of learning styles on academic achievement in different forms of teaching. Int. J. Instr. **11**, 219–232 (2018). https://doi.org/10.12973/iji.2018.11316a
2. Okubo, F., Shimada, A., Yamashita, T., Ogata, H.: A neural network approach for students' performance prediction. In: ACM International Conference Proceeding Series, pp. 598–599. Association for Computing Machinery (2017). https://doi.org/10.1145/3027385.3029479
3. Ifenthaler, D., Yau, J.Y.K.: Utilising learning analytics to support study success in higher education: a systematic review. Educ. Tech. Res. Dev. **68**, 1961–1990 (2020). https://doi.org/10.1007/s11423-020-09788-z
4. Glewwe, P.W., Hanushek, E.A., Humpage, S.D., Ravina, R., Mn, S.P.: School resources and educational outcomes in developing countries (2021). https://doi.org/10.3386/w17554
5. Rocha, Á., Adeli, H., Dzemyda, G., Moreira, F., Maria, A., Correia, R.: Early prediction of student's performance in higher education: a case study. Trends Appl. Inf. Syst. Technol. **1** (2021). https://doi.org/10.1007/978-3-030-72657-7_16
6. Antonio Giraldi, G.: Machine learning and pattern recognition. In: Lecture Notes-Graduate Program in Nanobiosystems (2021)
7. Mahesh, B.: Machine learning algorithms-a review. Int. J. Sci. Res. (2018). https://doi.org/10.21275/ART20203995
8. Fazilov, S., Mamatov, N., Samijonov, A., Abdullaev, S.: Reducing the dimensionality of feature space in pattern recognition tasks. J. Phys. Conf. Ser. Institute of Physics Publishing (2020). https://doi.org/10.1088/1742-6596/1441/1/012139
9. Chung, J.Y., Lee, S.: Dropout early warning systems for high school students using machine learning. Child Youth Serv. Rev. **96**, 346–353 (2019). https://doi.org/10.1016/j.childyouth.2018.11.030
10. Moscoviz, L., Evans, D.K.: Learning Loss and Student Dropouts during the COVID-19 Pandemic: A Review of the Evidence Two Years after Schools Shut Down Abstract. Center for Global Development. Working Paper. 609 (2022)
11. Gottfried, M.A., Plasman, J.S.: Linking the timing of career and technical education course-taking with high school dropout and college-going behavior. Am. Educ. Res. J. **55**, 325–361 (2018). https://doi.org/10.3102/0002831217734805
12. Terrence, T.J., Walsh, M.E., Raczek, A.E., Vuilleumier, C.E., Foley, C., Heberle, A., Sibley, E., Dearing, E.: The long-term impact of systemic student support in elementary school: reducing high school dropout. AERA Open **4** (2018). https://doi.org/10.1177/2332858418799085
13. Rivera, M.M., Escalante-Garcia, N., Dena-Aguilar, J.A., Olvera-Gonzalez, E., Vacas-Jacques, P.: Feature selection to predict LED light energy consumption with specific light recipes in closed plant production systems. Appl. Sci. (Switzerland) **12** (2022). https://doi.org/10.3390/app12125901
14. Neuman, A., Oz, G.: Different solutions to similar problems: parents' reasons for choosing to homeschool and social criticism of the education system. Int. J. Qual. Stud. Educ. **34**, 37–54 (2021). https://doi.org/10.1080/09518398.2020.1720852

15. Dupéré, V., Dion, E., Leventhal, T., Archambault, I., Crosnoe, R., Janosz, M.: High school dropout in proximal context: the triggering role of stressful life events. Child Dev. **89**, e107–e122 (2018). https://doi.org/10.1111/cdev.12792
16. Oviedo, B., Zambrano-Vega, C., Gómez Gómez, J., Kovalenko, I.D.: Clasificador Bayesiano Simple aplicado al aprendizaje. Revista Ibérica de Sistemas e Tecnologias de Informação **18**, 34–46 (2019)
17. Samuel, R., Burger, K.: Negative life events, self-efficacy, and social support: risk and protective factors for school dropout intentions and dropout. J. Educ. Psychol. **112**, 973–986 (2020). https://doi.org/10.1037/edu0000406
18. Kocsis, Z., Pusztai, G.: Student employment as a possible factor of dropout. Acta Polytechnica Hungarica **17** (2020)
19. Rodriguez, D., Carrasquillo, A., Garcia, E., Howitt, D.: Factors that challenge English learners and increase their dropout rates: recommendations from the field. Int. J. Biling. Educ. Biling. **25**, 878–894 (2022). https://doi.org/10.1080/13670050.2020.1722059
20. Mughal, A.W.: Secondary school students who drop out of school in rural Pakistan: the perspectives of fathers. Educ. Res. **62**, 199–215 (2020). https://doi.org/10.1080/00131881.2020.1755604
21. Aina, C., Baici, E., Casalone, G., Pastore, F.: The determinants of university dropout: a review of the socio-economic literature (2022). https://doi.org/10.1016/j.seps.2021.101102
22. Mubarak, A.A., Cao, H., Zhang, W.: Prediction of students' early dropout based on their interaction logs in online learning environment. Interact. Learn. Environ. (2020). https://doi.org/10.1080/10494820.2020.1727529
23. Aldowah, H., Al-Samarraie, H., Alzahrani, A.I., Alalwan, N.: Factors affecting student dropout in MOOCs: a cause and effect decision-making model. J. Comput. High. Educ. **32**, 429–454 (2020). https://doi.org/10.1007/s12528-019-09241-y
24. Joshi, A.: Convolutional neural networks: an overview and its applications in pattern recognition. Inf. Commun. Technol. Intell. Syst. **195**, 21–30 (2021)
25. Wang, H., Xing, G., Chen, K.: Categorical Data Transformation Methods for Neural Networks (2015)
26. Singh, D., Singh, B.: Investigating the impact of data normalization on classification performance. Appl. Soft. Comput. **97** (2020). https://doi.org/10.1016/j.asoc.2019.105524
27. Sun, L., Wang, T., Ding, W., Xu, J., Lin, Y.: Feature selection using fisher score and multilabel neighborhood rough sets for multilabel classification. Inf Sci (N Y) **578**, 887–912 (2021). https://doi.org/10.1016/j.ins.2021.08.032
28. Chen, P., Li, F., Wu, C.: Research on intrusion detection method based on pearson correlation coefficient feature selection algorithm. J. Phys. Conf. Ser. IOP Publishing Ltd. (2021). https://doi.org/10.1088/1742-6596/1757/1/012054
29. Wang, H., Hong, M.: Distance variance score: an efficient feature selection method in text classification. Math. Probl. Eng. **2015** (2015). https://doi.org/10.1155/2015/695720
30. Rahman, S.A., Fong, S.: Feature selection methods: case of filter and wrapper approaches for maximising classification accuracy (2018)
31. Ueki, K., Kuwatani, T., Okamoto, A., Akaho, S., Iwamori, H.: Thermodynamic modeling of hydrous-melt–olivine equilibrium using exhaustive variable selection. Phys. Earth Planet. Inter. **300** (2020). https://doi.org/10.1016/j.pepi.2020.106430
32. Rivera Zárate, I., Bolaños, M.H., Romero, P.P.: Clasificador Bayesiano del Síndrome Post COVID-19 usando Python (2022)
33. Sociedad, U.Y., Oviedo Bayas, B., Zambrano-Vega, C., Bayas, O., Zambrano.: Nuevo clasificador bayesiano simple para el análisis de Datos Educativos. Revista Científica de la Universidad de Cienfuegos **11**(2), 278–285 (2019)
34. Saritas, M.M., Yasar, A.: International journal of intelligent systems and applications in engineering performance analysis of ANN and Naive Bayes classification algorithm for data classification. Orig. Res. Paper Int. J. Intell. Syst. Appl. Eng. IJISAE. **7**, 88–91 (2019). https://doi.org/10.1039/b000000x

35. Khalifa, Y., Mandic, D., Sejdii, E.: A Review of Hidden Markov Models and Recurrent Neural Networks for Event Detection and Localization in Biomedical Signals. Elsevier (2020)
36. Wang, H., Xing, G., Chen, K.: Categorical Data Transformation Methods for Neural Networks (2008)
37. Rochin Berumen, F.L.: Deserción escolar en la educación superior en México: revisión de literatura. RIDE Revista Iberoamericana para la Investigación y el Desarrollo Educativo **11** (2021). https://doi.org/10.23913/ride.v11i22.821.
38. Akoglu, H.: User's guide to correlation coefficients. Turk. J. Emerg. Med. **18**, 91–93 (2018). https://doi.org/10.1016/j.tjem.2018.08.001

# Prediction of Heart Disease Using a Pattern Recognition Approach with Feature Selection and Naïve Bayesian Classifier

**David Alonso Carranza Escobar and Martín Montes Rivera**

**Abstract** Heart disease is a major health risk for people worldwide. It is a complex medical condition that affects the cardiovascular system. Predictive techniques have been developed and implemented to detect heart disease. Several factors, including unhealthy lifestyle and lack of physical activity, play a critical role in diagnosing the disease. There are two approaches for predicting heart disease: physical and chemical modeling and artificial intelligence (AI). Mathematical models rely on complex physics and chemistry concepts to create intricate models that increase in complexity as the level of detail increases. AI techniques, on the other hand, can automatically generate simpler models based on raw measurements using both deterministic and heuristic methods. However, physical modeling must strike a balance between complexity and assumptions to create simplified yet representative models. Feature selection involves identifying the most significant variables in a model, known as features. The procedure is supported by statistical and search tools that facilitate experts in examining extensive datasets and identifying the most significant characteristics. This research utilizes a public dataset derived from four databases, which originally contained 76 attributes. Including the predicted attribute. However, we focus on 14 attributes including the target as most published experiments have done. The method we propose is to use pattern recognition to improve the accuracy of a model by identifying important correlations and eliminating features with insufficient variance. This simplifies the training process and reduces complexity, resulting in better results than models without feature selection. To strengthen our findings, we employed several filtering methods based on statistical metrics for feature selection, such as mutual information gain, Chi-square, Fisher score, Pearson correlation, and variance thresholding. In addition, we evaluate wrapper methods based on greedy search, namely forward search, backward search and exhaustive search, with the

D. A. C. Escobar · M. M. Rivera (✉)
Research and Postgraduate Studies Department, Universidad Politécnica de Aguascalientes, Aguascalientes, Ags., México
e-mail: martin.montes@upa.edu.mx

D. A. C. Escobar
e-mail: mc230011@alumnos.upa.edu.mx

 
D. Oliva et al. (eds.), *Advances in Optimization Algorithms for Multidisciplinary Engineering Applications: From Classical Methods to AI-Enhanced Solutions*, Studies in Computational Intelligence 806, https://doi.org/10.1007/978-3-031-78440-8_15

possibility of selecting up to six features. We evaluate the accuracy of the models by using Naive Bayes classifiers, Neyman Pearson Criterion and Hidden Markov Models and thus perform a detailed comparison of the models' accuracies. Finally, after performing the analysis using the filter, wrapper and classifier methods we have found that the most effective way to predict heart disease is with a Naïve Bayesian classifier, using three specific features, age, chest pain and gender. This particular configuration has resulted in an accuracy rate of 87%.

**Keywords** Feature selection · Heart disease prediction · Pattern recognition · Python

## 1 Introduction

The development and implementation of predictive techniques for detecting heart disease have been a critical research area.

The heart is a vital organ for the body, and its malfunction can directly affect the organism, making it a significant health risk for people [1]. This is since cardiovascular diseases are a leading cause of mortality worldwide, with an estimated 17.9 million deaths annually, according to the World Health Organization (WHO) [1].

Heart disease is a complex medical condition that encompasses a wide range of disorders affecting the cardiovascular system. It can involve conditions related to the blood vessels, such as atherosclerosis and peripheral artery disease, as well as heart rhythm abnormalities, such as arrhythmias and conduction disorders. Understanding the underlying mechanisms of heart disease is critical for developing effective diagnostic and treatment strategies [2].

Several factors play a significant role in diagnosing the disease, including a history of diabetes, smoking, alcohol consumption, as well as an unhealthy lifestyle and lack of physical activity. These factors have been extensively researched and documented in various investigations like those in [1–5].

Predicting heart disease imply using physical modeling approaches to predict system behaviors; however, they rely on the description of complex physics concepts. As a result, the models produced tend to be intricate, increasing in complexity as the level of detail increases [6].

Alternatively, Artificial Intelligence (AI) researchers have put forth various techniques to automatically generate models and equations based on datasets containing measurements. Furthermore, Machine Learning (ML), a subset of AI, employs both deterministic and heuristic methods to create models with reduced complexity based on raw measurements. However, adhering to the principle of Occam's razor, physical modeling must strike a balance between complexity and assumptions to create simplified yet representative models [6, 7].

Pattern recognition offers the feature selection stage useful statistical and search tools to the field of ML for finding the more important variables in a model denominated as features [7, 8].

In machine learning, feature selection is a critical stage as in pattern recognition that involves identifying the most significant variables in a model, known as features. This process is facilitated by a range of statistical and search tools that enable experts to analyze large datasets and identify the most relevant features to include in their models. By leveraging these tools, experts can significantly enhance the accuracy and efficiency of their machine learning models, enabling them to achieve superior results in a wide range of applications [7, 8].

This study employs a machine learning (ML) approach inspired by pattern recognition to analyze a publicly available dataset in Kaggle [9]. The dataset comprises 303 samples and dates back to 1988. The dataset consists of four databases generated in Cleveland, Long Beach V, Hungary and Switzerland, containing 76 attributes, and one attribute of what was being predicted. However, in this work we use 14 attributes including the target as most published experiments have done.

Our proposal under pertains to utilizing pattern recognition methods to augment a model's efficacy by identifying the most prominent correlations and discarding the features that exhibit insufficient variance. This procedure will streamline the system's training process and curtails its complexity, thereby producing superior results compared to those obtained without a feature selection stage.

To reinforce our findings, we employed various filter methods based on statistical metrics for feature selection such as mutual information gain, Chi-square, Fisher score, Pearson's correlation, and Variance threshold. Also, we evaluated wrapper methods based on greedy search, namely Forward search, Backward search, and Exhaustive search, with the possibility of selecting up to six features.

Likewise, three machine learning classifiers were implemented to address cardiac disease detection. These classifiers are the Naïve Bayes Gaussian Classifier, the Neyman-Pearson criterion and the Hidden Markov Model (HMM) Classifier. Each of them has its own methodology which allows a wider exploration of the data and a higher probability of accuracy.

After performing an analysis of filtering methods, wrapping and classifiers, we found that the most effective way to predict heart disease with a naive Bayesian classifier resulted in an accuracy rate of 87% for predicting heart disease. The rest of the paper is structured into five sections. In Sect. 2, we present the background along with similar works to our proposal. In Sect. 3, we discuss the relevant materials and methods. In Sect. 4, we present our results and provide a detailed discussion. Finally, in Sect. 5, we describe our conclusions and findings.

## 2 Backgrounds

The papers discussed in this section primarily deal with the application of machine learning in predicting heart disease. They focus on the use of data preprocessing and feature selection techniques to improve the accuracy of predictions.

In [1] mentions deep learning with hyperparameter adjustment, and also suggests that the errors presented in the data are attributed to the irreducible error of the problem; he concludes with an accuracy of 78%, being the lowest percentage of the research.

Kavitha et al. [5] presents a different approach by making a friendly visual interface guide, in which the physician can enter the patient's data and generate the prediction immediately, in this research data mining techniques are used, and performs the use of feature selection at the end of the study concludes with the experimental results with a percentage of 88.7%.

In [10] the objective of the research is to reduce the size of the dimensions and find the most relevant features through feature selection, in addition, his dataset has categorical variables they employ the use of chi-2 in combination with PCA and conclude with an accuracy of 99% indicating that they obtain better results in comparation to other classifiers.

Also, [11] looks for to predict early diagnosis of heart disease, using data science and feature selection techniques such as Decision Tree, Logistic Regression, Logistic Regression SVM, Naïve Bayes and Random to obtain a minimum error rate in the research, concludes with an average error rate of 84.80% in his system.

Ayon et al. [12], focuses mainly on coronary heart artery disease, compares various computational intelligence techniques for a Cleveland data set obtained from the UCI repository, the author concludes that the highest accuracy obtained was 98.15% by using deep neural network and points out that his analysis has a minimum percentage error and can be used in clinical applications for this specific condition.

Finally, [13], seeking to predict cardiac diseases, from the database of Clevland, Swiss Hungary and Longbeach, uses algorithms such as naive Bayes, decision tree, stochastic gradient descent to predict cardiac diseases, ending with an average of 97.77% accuracy among the most prominent algorithms used.

Table 1 presents a brief comparison of similar papers identified in the research, ordered from highest to lowest accuracy according to the models presented. It summarizes the accuracy achieved by the different models used in the compared studies.

**Table 1** Comparison of the different investigations

| Authors | Methods | Accuracy (%) |
|---|---|---|
| Dun et al. [1] | Random Forest, Logistic Regression and Neural Network | 78 |
| Kavitha et al. [5] | Logistic Regression, Support Vector Machine, Deep Neural Network, Decision Tree, Naïve Bayes, Random Forest | 88.7 |
| Gárate-Escamila et al. [10] | Chi-2 combined with Principal Component Analysis (PCA) | 99 |
| Zafar-uz-Zaman [11] | Decision Tree, Logistic Regression, Support Vector Machine, Naïve Bayes and Random Forest | 84.80 |
| Ayon et al. [12] | Logistic Regression, Support Vector Machine, Deep Neural Network, Decision Tree, Naïve Bayes, Random Forest | 98.15 |
| Almustafa [13] | K-Nearest Neighbor, Naive Bayes, JRip, SVM (Support Vector Machine), Decision Table (DT) | 97.77 |

## 3 Materials and Methods

The use of pattern recognition for condition prediction is necessary for disease prognosis as it removes irrelevant features and highlights those that are essential for prediction using preprocessing and feature selection methods. In addition, the use of classifiers is essential, as it allows the application of models that recognize distinct patterns within the data, which improves the accuracy and reliability of predictions.

### *3.1 Preprocessing Stage*

This phase seeks to improve the quality of the information in the database before being used for feature selection, this preparation phase is necessary to optimize the performance of the system and reduce the error to the probability of model [12, 14].

- Outlier removal: it is defined as a data that looks graphically distanced from the data set, these points known as outliers produce errors during the training process, so removing them ensures a more accurate training. Equation (1) provides a criterion for identifying outliers in the samples.

$$|x - \mu| > k\sigma \tag{1}$$

  where $x$ is the original value to be processed, $\mu$ is the mean, $\sigma$ is the standard deviation and $k$ is the threshold. If this equation is satisfied then x is considered an outlier and is removed, this process checks all the values in the database.
- Data normalization: This technique seeks that the analyzed data remain normalized in the same range between 0 and 1. This method standardize the data to make easy the comparison, by performing the normalization process it is secured that all the analyzed variables collaborate with the model. The Eq. (2) represents the standard deviation required in the transformation as $x_{\text{norm}}$ and describe the stage of preprocessing normalization.

$$x_{norm} = \frac{x - x_{\min}}{x_{\max} - x_{\min}} \tag{2}$$

  where $x_{\text{norm}}$ is the normalized value, $x$ represent the original value, and $x_{\min}$ and $x_{\max}$ are the minimum and maximum values of the data, respectively.

### *3.2 Feature Selections*

In this stage the most important characteristics of the system are searched and through the filter methods eliminate those that are not important for the model, only using the properties that improve the performance, this reduces the time for the training of the system [10].

- Mutual information: The use of this technique facilitates the identification of characteristics and the evaluation of their degree of interaction by listing them from the best characteristic to the worst one, in addition, this filtering method is effective for both linear models and nonlinear.
- Chi-square: Chi-square is used for databases with categorical characteristics, the process of this filter starts with the calculation of Chi-square between the characteristics by selecting the desired number of characteristics therefore allowing to obtain key characteristics.
- Fisher score: This technique reviews all the features and returns in ascending order the variables according to the Fisher score, which evaluates the capacity of the properties, that is, the higher the score, the more relevant the analyzed feature is considered.
- Pearson's correlation coefficient: This method measures the linear relationship between variables and looks for a relationship between the characteristics; if both are related, one can be predicted from the other so that one can be discarded.
- Variance threshold: This is a selection method that doesn't consider the relationships between variables and eliminates all characteristics with zero variance.

### *3.3 Wrapper Methods*

These methods are based on an algorithm that follows a search measure and evaluates all the characteristics within an evaluation criterion; this achieves more concrete results compared to filtering methods [7].

- Forward feature selection: For the operation of this method, it starts with the best performing variable and then selects another variable that gives a better performance combined with the first one; this process combines the variables until reaching an established parameter.
- Backward function elimination: This method is the opposite of the forward feature selection method, starting with all available functions and discarding the variable that contributes most to improving the value of the evaluation measure until its objective is reached.
- Exhaustive function selection: This method tests all possible combinations of the variables; this obtains the best performance of the system ensuring the evaluation of all the characteristics of the model giving as a result the percentage of prediction of the system.

### 3.4 Classifier Methods

These methods involve the use of classification algorithms to assign the observations into different classes; they are used to identify patterns in specific data or modes and generate a more accurate and reliable prediction, these classifiers are fundamental for machine learning and data analysis applications.

- Naïve Bayes is a classification technique based on the Bayes theorem, which assumes that the effect of a particular feature on a class is independent of other features. This classifier determines the likelihood that a pattern fits into a given class, and is a good choice when speed and accuracy are sought, since it achieves these characteristics by working with a large amount of data [15].
- Neyman Pearson Criterior: This particular classifier is designed with the goal of reducing type II error, also known as false negatives, while controlling for type I error, or false positives. To achieve this, the classifier employs hypothesis testing as a method of distinguishing between two options while regulating the probability of making false positive conclusions. In addition, this classifier is versatile, as it can fit both parametric and non-parametric models, making it a valuable tool in a variety of fields, such as medical diagnostics, computer security, search engine optimization and many more applications [16].
- Hidden Markov Models: These tools are useful in applications in which the observations are noisy or incomplete, they allow determining the real states of a model from a series of observations. Their versatility allows modeling discrete systems with finite or continuous state spaces [17].

To better understand the steps to be followed in the methodology employed in the use of the various classifiers, Fig. 1 describes the process to be followed to obtain the corresponding results according to the selected model.

To carry out this process, we made use of the Python programming language, which is one of the most used programming languages and has shown a major growth in recent years [18, 19]. Python is a high-level language that has the ability to perform complex tasks with short code compared to other languages. The versatility of Python in the scientific and engineering community highlights this programming language as an optimization tool. In addition, it has a lot of libraries and packages that developers update, which makes it a good choice for approach technical challenges. Also, Python is considered one of the most accessible languages to start programming due to its coding style. It is worth noting that its open-source license encourages the development and expansion of diverse programs for the community [20].

The sk-learn library is a functional open-source library for automatic learning in the Python programming language, it has different tools that facilitate the study of various topics and allows to obtain among its various uses the selection of models, the preprocessing of information, the selection of features mentioned in the background, as well as some other functions that can be used in various branches of science [21].

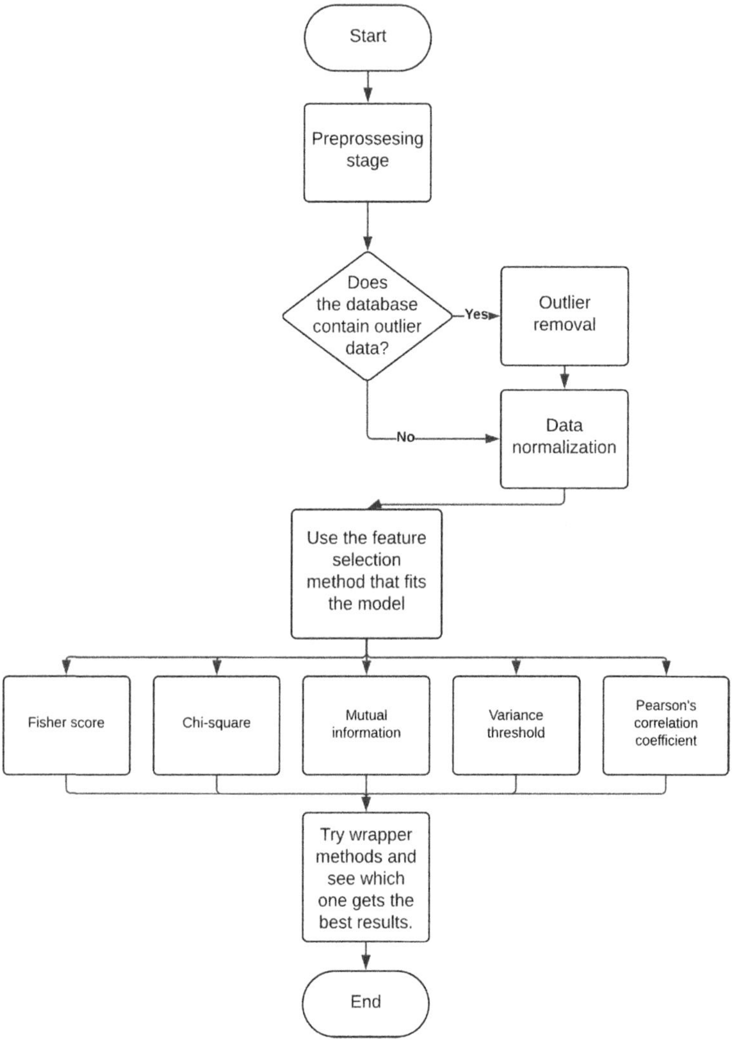


**Fig. 1** Flow diagram that describes the methodology

## 4 Results and Discussion

In the present study, a database with an approximate size of 303 collected data related to the prediction of heart disease available in [12] is used. In which the variables of age (Age), gender (sex), type of chest pain (cp), blood pressure (trestbps), cholesterol (chol), blood sugar level (fbs), electrocardiographic findings (restecg), maximum heart rate (thalach), exercise-induced angina (exang) are also considered, ST segment changes during exercise compared to rest (oldpeak), ST segment slope during exercise (slope), number of major vessels (ca), type of thalach (thal), and finally, the output (target) which indicates a value of zero if the patient does not have the disease and one if has heart disease [12]. Table 2 shows a part of the database without the data preprocessing provided by [12] and also Table 3 shows the normalized data set in a range of zero and one in order to facilitate the training of the program.

After having normalized the data set, it is necessary to eliminate the outliers to avoid malfunctions. Table 4 shows the elimination of the outliers, obtaining as a result only the points that are within the range.

Once the database has been prepared, it is necessary to apply the filtering methods to determine those characteristics that present the most significant information for the analysis of the model. Figure 2 shows the results obtained from the mutual information method, showing those characteristics that have a greater relationship with the analyzed variable.

Likewise, the Chi-2 filtering method was not used because it is only for databases with categorical characteristics; the next method used was the Fisher Score, which ranks the most important characteristics in ascending order. Figure 3, shows that the characteristic with the highest Fisher score is the changes in the ST segment during exercise compared to rest (oldpeak), while the lowest is the patient's age (Age).

The next method used is Pearson's correlation in Fig. 4 shows a heat map with the input variables, if the relationship between two variables shows a dark tone indicates that there is no correlation, on the other hand if they show a lighter tone there is a relationship between the two variables.

**Table 2** Database for heart disease prediction before data preprocessing

| | Age | Sex | Cp | Trestbps | Chol | Restecg | Thalach | Exang | Oldpeak | Slope | Ca | Thal | Target |
|---|---|---|---|---|---|---|---|---|---|---|---|---|---|
| 0 | 63 | 1 | 3 | 145 | 233 | 0 | 150 | 0 | 2.3 | 0 | 0 | 1 | 1 |
| 1 | 37 | 1 | 2 | 130 | 250 | 1 | 187 | 0 | 3.5 | 0 | 0 | 2 | 1 |
| 2 | 41 | 0 | 1 | 130 | 204 | 0 | 172 | 0 | 1.4 | 2 | 0 | 2 | 1 |
| 3 | 56 | 1 | 1 | 120 | 236 | 1 | 178 | 0 | 0.8 | 2 | 0 | 2 | 1 |
| 4 | 57 | 0 | 0 | 120 | 354 | 1 | 163 | 1 | 0.6 | 2 | 0 | 2 | 1 |
| 5 | 57 | 1 | 0 | 140 | 192 | 1 | 148 | 0 | 0.4 | 1 | 0 | 1 | 1 |
| 6 | 56 | 0 | 1 | 140 | 294 | 0 | 153 | 0 | 1.3 | 1 | 0 | 2 | 1 |
| 7 | 44 | 1 | 1 | 120 | 263 | 1 | 173 | 0 | 0 | 2 | 0 | 3 | 1 |
| 8 | 52 | 1 | 2 | 172 | 199 | 1 | 162 | 0 | 0.5 | 2 | 0 | 3 | 1 |
| 9 | 57 | 1 | 2 | 150 | 168 | 1 | 174 | 0 | 1.6 | 2 | 0 | 2 | 1 |
| 10 | 54 | 1 | 0 | 140 | 239 | 1 | 160 | 0 | 1.2 | 2 | 0 | 2 | 1 |
| 11 | 48 | 0 | 2 | 130 | 275 | 1 | 139 | 0 | 0.2 | 2 | 0 | 2 | 1 |
| 12 | 49 | 1 | 1 | 130 | 266 | 1 | 171 | 0 | 0.6 | 2 | 0 | 2 | 1 |
| 13 | 64 | 1 | 3 | 110 | 211 | 0 | 144 | 1 | 1.8 | 1 | 0 | 2 | 1 |
| 14 | 58 | 0 | 3 | 150 | 283 | 0 | 162 | 0 | 1 | 2 | 0 | 2 | 1 |
| 15 | 50 | 0 | 2 | 120 | 219 | 1 | 158 | 0 | 1.6 | 1 | 0 | 2 | 1 |

**Table 3** Database with the normalization phase completed

| | Age | Sex | Cp | Trestbps | Chol | Restecg | Thalach | Exang | Oldpeak | Slope | Ca | Thal | Target |
|---|---|---|---|---|---|---|---|---|---|---|---|---|---|
| 0 | 0.71 | 1.00 | 1.00 | 0.48 | 0.24 | 0.00 | 0.60 | 0.00 | 0.37 | 0.00 | 0.00 | 0.33 | 1.00 |
| 1 | 0.17 | 1.00 | 0.67 | 0.34 | 0.28 | 0.50 | 0.89 | 0.00 | 0.56 | 0.00 | 0.00 | 0.67 | 1.00 |
| 2 | 0.25 | 0.00 | 0.33 | 0.34 | 0.18 | 0.00 | 0.77 | 0.00 | 0.23 | 1.00 | 0.00 | 0.67 | 1.00 |
| 3 | 0.56 | 1.00 | 0.33 | 0.25 | 0.25 | 0.50 | 0.82 | 0.00 | 0.13 | 1.00 | 0.00 | 0.67 | 1.00 |
| 4 | 0.58 | 0.00 | 0.00 | 0.25 | 0.52 | 0.50 | 0.70 | 1.00 | 0.10 | 1.00 | 0.00 | 0.67 | 1.00 |
| 5 | 0.58 | 1.00 | 0.00 | 0.43 | 0.15 | 0.50 | 0.59 | 0.00 | 0.06 | 0.50 | 0.00 | 0.33 | 1.00 |
| 6 | 0.56 | 0.00 | 0.33 | 0.43 | 0.38 | 0.00 | 0.63 | 0.00 | 0.21 | 0.50 | 0.00 | 0.67 | 1.00 |
| 7 | 0.31 | 1.00 | 0.33 | 0.25 | 0.31 | 0.50 | 0.78 | 0.00 | 0.00 | 1.00 | 0.00 | 1.00 | 1.00 |
| 8 | 0.48 | 1.00 | 0.67 | 0.74 | 0.17 | 0.50 | 0.69 | 0.00 | 0.08 | 1.00 | 0.00 | 1.00 | 1.00 |
| 9 | 0.58 | 1.00 | 0.67 | 0.53 | 0.10 | 0.50 | 0.79 | 0.00 | 0.26 | 1.00 | 0.00 | 0.67 | 1.00 |
| 10 | 0.52 | 1.00 | 0.00 | 0.43 | 0.26 | 0.50 | 0.68 | 0.00 | 0.19 | 1.00 | 0.00 | 0.67 | 1.00 |
| 11 | 0.40 | 0.00 | 0.67 | 0.34 | 0.34 | 0.50 | 0.52 | 0.00 | 0.03 | 1.00 | 0.00 | 0.67 | 1.00 |
| 12 | 0.42 | 1.00 | 0.33 | 0.34 | 0.32 | 0.50 | 0.76 | 0.00 | 0.10 | 1.00 | 0.00 | 0.67 | 1.00 |
| 13 | 0.73 | 1.00 | 1.00 | 0.15 | 0.19 | 0.00 | 0.56 | 1.00 | 0.29 | 0.50 | 0.00 | 0.67 | 1.00 |
| 14 | 0.60 | 0.00 | 1.00 | 0.53 | 0.36 | 0.00 | 0.69 | 0.00 | 0.16 | 1.00 | 0.00 | 0.67 | 1.00 |
| 15 | 0.44 | 0.00 | 0.67 | 0.25 | 0.21 | 0.50 | 0.66 | 0.00 | 0.26 | 0.50 | 0.00 | 0.67 | 1.00 |

**Table 4** Process for the elimination of outliers

| Initial | After removal |
|---|---|
| 303 | 293 |
| 293 | 293 |
| 293 | 293 |
| 293 | 278 |
| 278 | 267 |
| 267 | 265 |
| 265 | 257 |
| 257 | 257 |
| 257 | 246 |
| 246 | 234 |
| 234 | 214 |
| 214 | 200 |
| Final value | 200 |

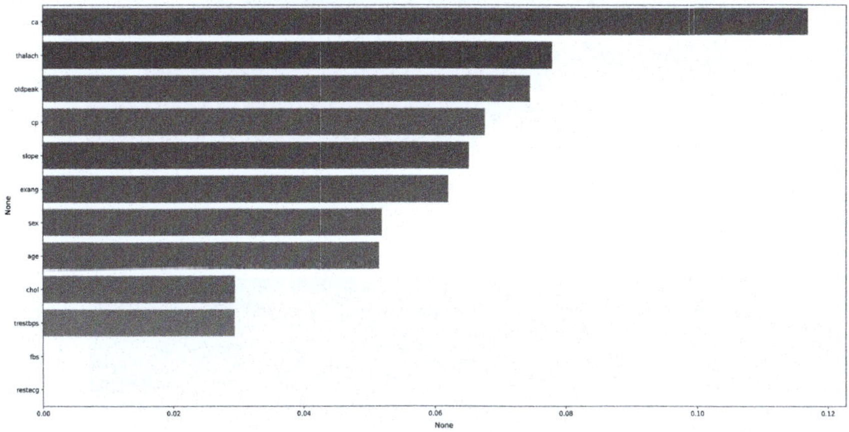


**Fig. 2** Results obtained from the mutual information method

To finish with the selection of characteristics, the last filtering method called variance threshold is used, which discards the variables that present the same value in all the samples. Figure 5 shows the results obtained from the feature selection method in which the variables that were eliminated as well as those that obtained a variance different from zero are displayed.

Wrapper methods are necessary to test all possible combinations between features, these methods are applied to classes by using machine learning algorithms, so these techniques are usually more precision compared to filter methods. The technique of wrapper methods is based on a greedy search that explores all the combinations between the characteristics obtaining the best results [5, 14].

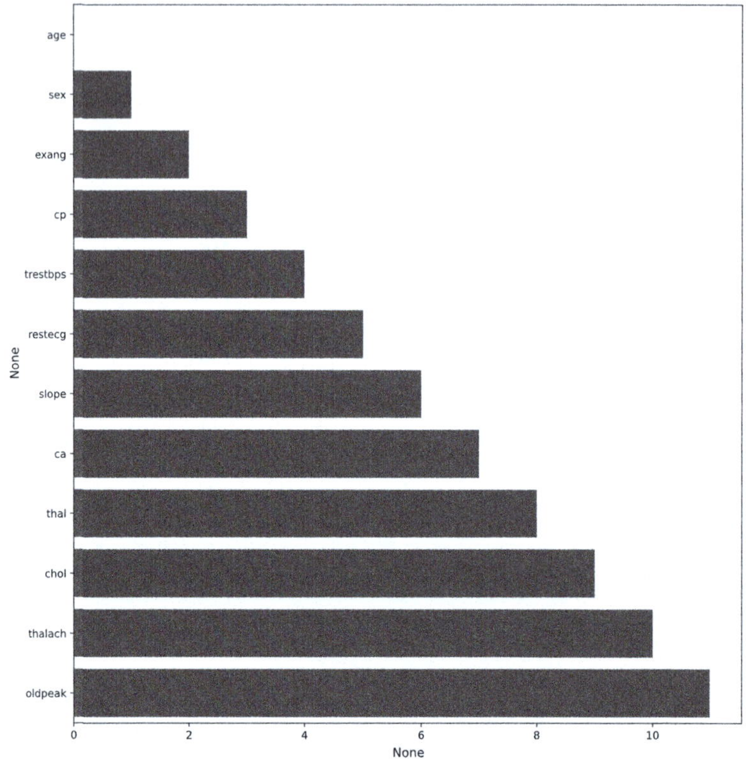


**Fig. 3** Results obtained using the fisher score method

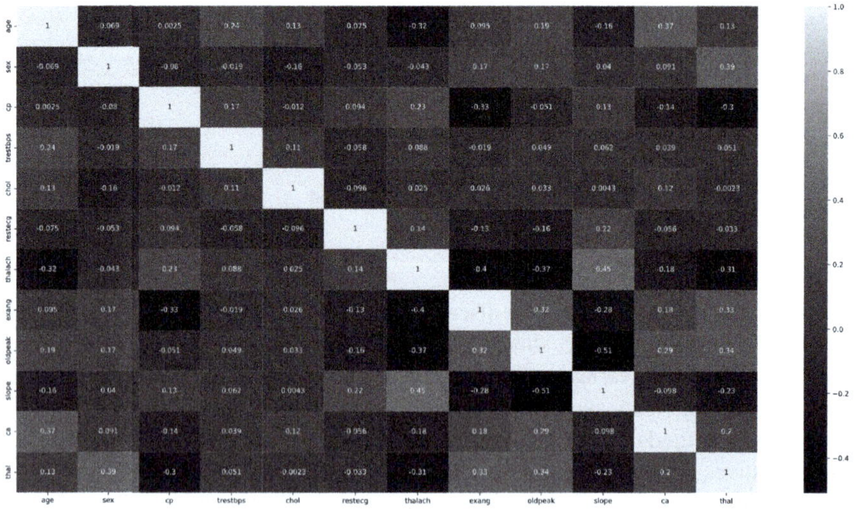


**Fig. 4** Results obtained from Pearson's correlation

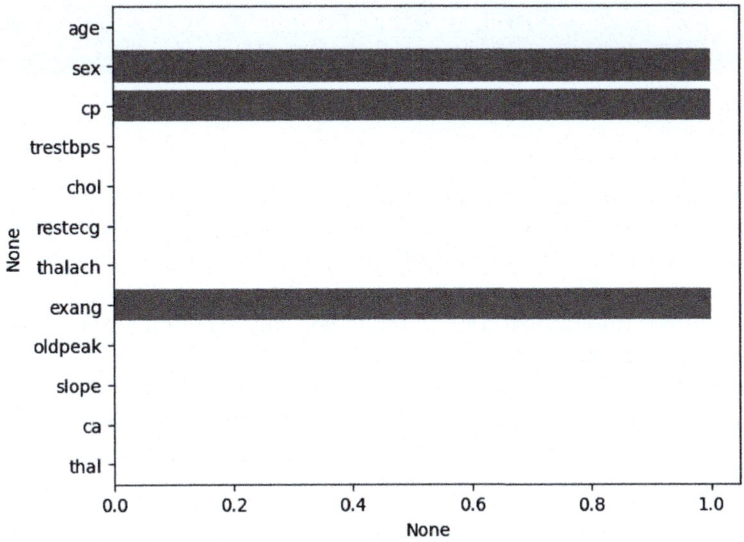


**Fig. 5** Results of the variance threshold method

1. The first wrapper method is forward selection, which selects a characteristic from the database that has the best relationship with the output characteristic, then combines it with another characteristic that correlates with the first variable and continues the process to reach a previously defined objective.
2. The second method is backward selection, this method is the reverse of forward selection, it starts with all the variables in the database to be analyzed, compares and relates them to each other and discards those that increase the efficiency of the evaluation mean until the target is reached.
3. Finally, the exhaustive search method tests all possible combinations of database characteristics, in order to obtain the best characteristics of the system.

It is necessary to say that for this specific problem the exhaustive search method was used because it tests all possible combinations between characteristics. Also, worked with a database of moderate sample size, which means that the computational demand is not high.

The first classifier used for the prediction of heart disease is the Naive Bayes classifier, which assumes independence among the characteristics [15]. Figure 6 shows the accuracy obtained after applying the classifier to the database analyzed.

After applying the Neyman Pearson criterion, a confusion matrix is obtained, which describes the predictions obtained by means of a 2 $\times$ 2 matrix; in position

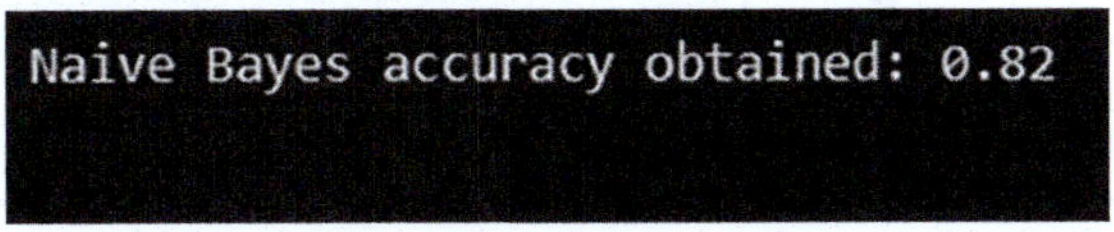


**Fig. 6** Accuracy obtained applying Naïve Bayes classifier

**Fig. 7** Percentage of success applying Neyman Pearson criterior

[0, 0] are shown the correctly classified instances or true positives (VP), in [0, 1] are shown those classified incorrectly as positive or false positives, in the position [0, 1] are shown those cases classified correctly as negative or true negatives (VN), finally, the position [1] represents the cases classified incorrectly as negative or false negatives (FN) [16] as shown in Fig. 7. In addition, Fig. 8 shows the accuracy obtained by applying the above-mentioned model.

Finally, the Hidden Markov model classifier is applied which in the context of the problem is used to predict the probability of having heart disease based on the patient's symptoms. The accuracy of the model within the database is shown in Fig. 9, which shows the level of accuracy achieved.

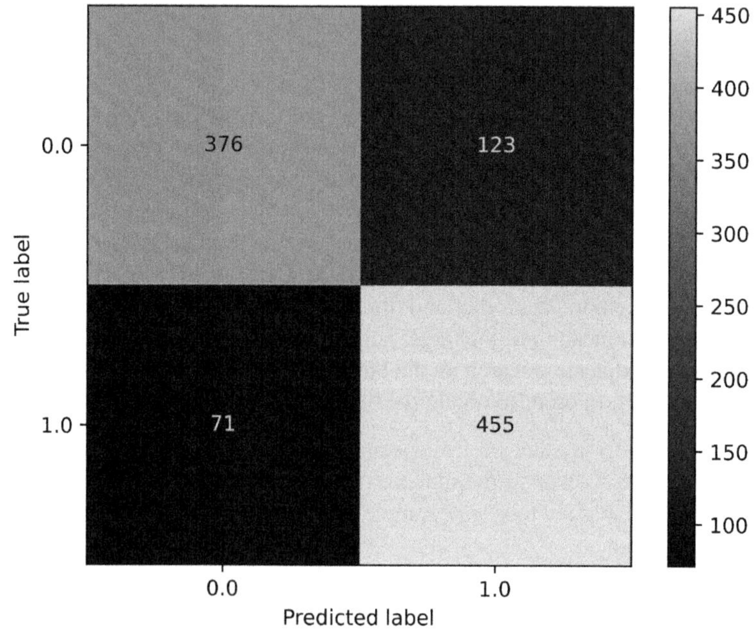


**Fig. 8** Results obtained in the Neyman Pearson criterion, represented by the confusion matrix

**Fig. 9** Results obtained from the Hidden Markov model

```
Best Accuracy score: 0.87
Best subset(indices):   (0, 1, 2)
Best subset(Corresponding Names): ('age', 'sex', 'cp')
```

**Fig. 10** Results obtained from the exhaustive search method

The results section shows the best accuracy obtained after training the model, performing the preprocessing steps and applying the filtering methods, compared to the accuracy obtained from the classifiers. In addition, Fig. 10 shows that the exhaustive search model has an effectiveness of about 87%, showing that the best sets of the database in this case are the patient's age (Age), gender (sex) and chest pain (cp). While in the Naive Bayes classification method it obtained an effectiveness of 82% being the highest of the 3 classifiers. Therefore, it can be said that the model in most cases is functional for the prediction of heart disease, but based on the ethical aspect, I consider that it cannot be an applicable method because we are talking about human lives, and the prediction of heart disease obtained by the model should be supported by health professionals with experience in this sector.

## 5 Conclusion

The following study uses the preprocessing stage, filtering methods for the selection of the best features and classifiers in the prediction of heart disease using a database and concludes that this model can solve this type of problem by using the techniques and can also be applied to a wide variety of situations, always with the help of a health professional. This provides an effective tool for predicting, according to the patient's symptoms and measurements, whether a cardiovascular disease may be present, improving early decision making and reducing risks at all times.

After analyzing the information, with the different filtering methods, it was found that the mutual information and Fisher score methods repeated the characteristics of thalamic type (thal) and ST-segment changes during exercise compared to rest (oldpeak). Moreover, in variance thresholding and exhaustive search we obtain that the best features of the model are gender (sex), chest pain type (cp) and exercise-induced angina (exang).

In conclusion, the best features to consider for the analysis of heart disease are those shown in Table 5, which shows the frequency of occurrence of each feature in the different methods. Likewise, the classifier that obtains the highest accuracy in this specific problem is the Naive Bayes classifier, but if we compare the accuracies between the classifiers and the wrapper methods, the exhaustive method obtained a higher accuracy compared to the others, so I consider that it would be a better method for this specific application.

**Table 5** Frequency of features with different models

| Best feature | Repetition |
|---|---|
| Age (Age) | 1 |
| Gender (sex) | 2 |
| Type of chest pain (cp) | 2 |
| Cholesterol (chol) | 1 |
| Maximum heart rate (thalach) | 2 |
| Exercise-induced angina (exang) | 1 |
| ST segment changes during exercise compared to rest (oldpeak) | 2 |
| Number of major vessels (ca) | 1 |

*Future works*

Although various classifiers were used to obtain the above results, to further advance the research, better results can be obtained by employing other models such as neural networks, decision trees, and more complex ensemble methods. These approaches have the potential to capture more complex relationships between variables and improve the predictive capability of the model. In addition, the exploration of more advanced data preprocessing techniques and the expansion of the data set used may contribute to improving the accuracy and capability of models in the prediction of cardiac disease.

## References

1. Dun, B., Wang, E., Majumder, S.: Heart Disease Diagnosis on Medical Data Using Ensemble Learning
2. Ozcan, M., Peker, S.: A classification and regression tree algorithm for heart disease modeling and prediction. Healthcare Anal. **3** (2023). https://doi.org/10.1016/j.health.2022.100130
3. Chandrasekhar, N., Peddakrishna, S.: Enhancing heart disease prediction accuracy through machine learning techniques and optimization. Processes **11** (2023). https://doi.org/10.3390/pr11041210
4. Nandy, S., Adhikari, M., Balasubramanian, V., Menon, V.G., Li, X., Zakarya, M.: An intelligent heart disease prediction system based on swarm-artificial neural network. Neural Comput. Appl. **35**, 14723–14737 (2023). https://doi.org/10.1007/s00521-021-06124-1
5. Kavitha, M., Gnaneswar, G., Dinesh, R., Sai, Y.R., Suraj, R.S.: Heart disease prediction using hybrid machine learning model. In: Proceedings of the 6th International Conference on Inventive Computation Technologies, ICICT 2021, pp. 1329–1333. Institute of Electrical and Electronics Engineers Inc. (2021). https://doi.org/10.1109/ICICT50816.2021.9358597
6. Hosseinzadeh, A., Zhou, J.L., Altaee, A., Li, D.: Machine learning modeling and analysis of biohydrogen production from wastewater by dark fermentation process. Bioresour. Technol. **343**, 126111 (2022). https://doi.org/10.1016/J.BIORTECH.2021.126111
7. Rivera, M.M., Escalante-Garcia, N., Dena-Aguilar, J.A., Olvera-Gonzalez, E., Vacas-Jacques, P.: Feature selection to predict LED light energy consumption with specific light recipes in closed plant production systems. Appl. Sci. **12**, 5901 (2022). https://doi.org/10.3390/APP12125901

8. Mosavi, A., Ozturk, P., Chau, K.W.: Flood prediction using machine learning models: literature review. Water **10**, 1536 (2018). https://doi.org/10.3390/W10111536
9. Heart disease prediction, https://www.kaggle.com/datasets/moazeldsokyx/heart-disease. Last accessed 29 Feb 2024
10. Gárate-Escamila, A.K., Hajjam El Hassani, A., Andrès, E.: Classification models for heart disease prediction using feature selection and PCA. Inform. Med. Unlocked **19** (2020). https://doi.org/10.1016/j.imu.2020.100330
11. Zafar-uz-Zaman, M.: National Centre for Physics, Centres of Excellence in Science and Applied Technologies, Institute of Electrical and Electronics Engineers. Islamabad Section, Institute of Electrical and Electronics Engineers: Proceedings of 2019 16th International Bhurban Conference on Applied Sciences & Technology (IBCAST) (2019)
12. Ayon, S.I., Islam, M.M., Hossain, M.R.: Coronary artery heart disease prediction: a comparative study of computational intelligence techniques. IETE J. Res. **68**, 2488–2507 (2022). https://doi.org/10.1080/03772063.2020.1713916
13. Almustafa, K.M.: Prediction of heart disease and classifiers' sensitivity analysis. BMC Bioinform. **21** (2020). https://doi.org/10.1186/s12859-020-03626-y
14. Rivera, M.M., Escalante-Garcia, N., Dena-Aguilar, J.A., Olvera-Gonzalez, E., Vacas-Jacques, P.: Feature selection to predict LED light energy consumption with specific light recipes in closed plant production systems. Appl. Sci. (Switzerland) **12** (2022). https://doi.org/10.3390/app12125901
15. Patel, T.S., Daxesh, Patel, P., Sanyal, M., Pranav, Shrivastav, S.: Prediction of Heart Disease and Survivability Using Support Vector Machine and Naive Bayes Algorithm. https://doi.org/10.1101/2023.06.09.543776
16. Tong, X., Xia, L., Wang, J., Feng, Y.: Neyman-Pearson classification: parametrics and sample size requirement (2020)
17. Xia, X.: Hidden Markov models and protein secondary structure prediction. In: Bioinformatics and the Cell, pp. 145–172. Springer International Publishing (2018). https://doi.org/10.1007/978-3-319-90684-3_7
18. Oberoi, A., Chauhan, R.: Visualizing data using Matplotlib and Seaborn libraries in Python for data science. Int. J. Sci. Res. Publ. (IJSRP) **9**, 8733 (2019). https://doi.org/10.29322/ijsrp.9.03.2019.p8733
19. Stančin, I., Jović, A.: An overview and comparison of free Python libraries for data mining and big data analysis
20. Cutting, V., Stephen, N.: A review on using python as a preferred programming language for beginners 4 publications 3 citations see profile a review on using python as a preferred programming language for beginners. Int. Res. J. Eng. Technol. (2021)
21. Buitinck, L., Louppe, G., Blondel, M., Pedregosa, F., Mueller, A., Grisel, O., Niculae, V., Prettenhofer, P., Gramfort, A., Grobler, J., Layton, R., Vanderplas, J., Joly, A., Holt, B., Varoquaux, G.: API design for machine learning software: experiences from the scikit-learn project (2013)

# Machine Learning and Data Analysis in the Prevention of Complications Derived from Diabetes

Víctor E. Moreno, Itzel Aranguren, and Marco A. Pérez

**Abstract** Diabetes is an overall global health problem that impacts millions of people worldwide, with a disproportionate impact on low- and middle-income countries. Mexico, in particular, has a high prevalence of diabetes among its population. In 2021, Mexico was ranked 7th in the world for the highest number of diabetic adults by the International Diabetes Federation. This has led to a country's health system crisis and is a cause for concern. In the research, machine learning techniques and data analysis were used to establish the specific relationship between the habits of Mexican diabetic patients and the development of complications associated with this condition. Arising from the increasing incidence of diabetes in Mexico and its association with health complications in patients, understanding how specific habits can influence the progression of the disease provides a valuable tool to promote changes in patient behavior to prevent adverse events. The results obtained provide a clear understanding of how certain habits of diabetic patients are directly linked to the development of complications such as heart attacks, vision loss, and amputations, among others.

**Keywords** Diabetes · Patient habits · Prevention of complications · Machine learning · Support vector machine

## 1 Introduction

Diabetes is a persistent condition characterized by insufficient insulin production or ineffective insulin utilization by the pancreas. Insulin is crucial in regulating glucose levels in the bloodstream, enabling cells to absorb and utilize it for energy while

V. E. Moreno (✉) · I. Aranguren · M. A. Pérez
Depto. de Innovación Basada en la Información y el Conocimiento, Universidad de Guadalajara, CUCEI, Av. Revolución 1500, Guadalajara, Jalisco, Mexico
e-mail: victor.moreno@academicos.udg.mx

I. Aranguren
e-mail: itzel.aranguren@academicos.udg.mx

M. A. Pérez
e-mail: marco.perez@cucei.udg.mx

 
D. Oliva et al. (eds.), *Advances in Optimization Algorithms for Multidisciplinary Engineering Applications: From Classical Methods to AI-Enhanced Solutions*, Studies in Computational Intelligence 806, https://doi.org/10.1007/978-3-031-78440-8_16

maintaining blood sugar levels. This imbalance results in a notable elevation in blood glucose (hyperglycemia), which can, over time, significantly damage various organs and systems in individuals affected by this condition [1]. In turn, diabetes has a strong impact on the economy of patients and their families. Additionally, it is one of the four leading causes of death from non-communicable diseases worldwide. In Mexico, diabetes is one of the most prevalent chronic diseases in the population and is associated with high rates of morbidity and mortality. There are two main types of diabetes: Type 1 Diabetes and Type 2 Diabetes. The latter, which accounts for 85–90% of cases, is the most common in the population and mainly occurs in adults [2].

Type 2 diabetes is characterized by the pancreas's inability to regulate blood sugar levels using insulin effectively. It primarily develops due to patients' poor habits, which, over time, can lead to further complications and contribute to various health issues [3]. Fortunately, with the help of emerging technologies and data digitization, we now have access to artificial intelligence tools that allow us to analyze data on a large scale. This technology enables us to develop machine learning models that provide information that may be impossible to obtain using traditional data analysis methods. Given that Type 2 diabetes develops largely due to poor habits, and that these habits can contribute to the development of other diseases and complications arising from the same condition, we have formulated the following hypothesis: "By analyzing data on the habits of diabetic patients, a machine learning model can be developed to analyze how these habits contribute to the development of complications derived from the disease." The goal is to provide patients with a tool that helps them understand how their habits influence the prevention of complications arising from their condition.

For this research, was used the database of the National Health and Nutrition Survey (ENSANUT), conducted by the National Institute of Geography and Statistics (INEGI) in Mexico [4]. This study aims to understand the health status and nutritional conditions of various population groups in Mexico. We managed to obtain information about the habits of patients with diabetes and data on complications derived from the disease. With this information, we decided to use a machine learning model based on a Support Vector Machine to determine to what extent habits can affect the development of complications. Support Vector Machine (SVM) is a widely used machine learning algorithm based on statistical learning theory [5, 6]. Initially designed for binary classification problems, it was later extended for multiclass problems. SVMs are particularly effective in handling complex medical data within the healthcare industry. Unlike simpler classification methods, SVM can manage nonlinear relationships between features (such as symptoms or test results) and disease states. This makes SVM ideal for two main purposes: diagnosing diseases by determining the presence or absence of a condition in binary classification and predicting disease progression by forecasting how a disease might develop over time in multi-class classification [7]. The rest of the paper is arranged as follows:

**Section** 2 depicts the related works.

**Section** 3 describes the impact of diabetes in Mexico.

**Section** 4 defines the main complications associated with diabetes.

**Section** 5 presents the relevance and advantages of implementing machine learning models in healthcare.

**Section** 6 displays the results of the proposed SVM model for the prediction of complications derived from diabetes.

**Section** 7 closes the study with the conclusions and future work.

## 2 Related Works

One of the main challenges in applying machine learning techniques for the early detection of diseases focuses on identifying and modeling the data to obtain accurate information that provides the markers for the prediction of a particular disease. Below, we conduct an analysis of the most recent studies on data mining and machine learning for early disease detection, focusing our research on diabetes. Below is a summary of the results of the studies analyzed in this review of the state of the art:

In October 2023, authors Toni Mora, David Roche, and Beatriz Rodriguez published the results of their research in the article titled *"Predicting the onset of diabetes-related complications after a diabetes diagnosis with machine learning algorithms"*. In this study, the authors aimed to predict the risk of developing certain complications one, two, and three years after being diagnosed with Type 2 Diabetes. For the development of the research, they utilized clinical data from over 600,000 patients diagnosed with diabetes in Catalonia, Spain. Among the variables used to predict the risk of complications were data such as age, gender, copayment levels, and other comorbidities. They employed four machine learning algorithms (Logistic Regression, Decision Tree, Random Forest, and Extreme Gradient Boosting), with the Random Forest algorithm being the most relevant for predicting the risk of hypertension, myocardial infarction, and retinopathy, while Logistic Regression was more relevant for predicting the risk of other comorbidities, with precision rates ranging from 60 to 75%. Age was the variable most associated with an elevated risk for all complications except retinopathy [8]. This research provides a methodology for predicting the risk of diabetes-related complications using personal data and machine learning algorithms. However, our proposal focuses on analyzing how the habits of patients with diabetes can influence the development of complications derived from the disease, addressing a different dimension based on patient behavior.

In the article titled *"Evaluation of Machine Learning Methods Developed for Prediction of Diabetes Complications,"* authors Kuo Ren Tan, Jun Jie Benjamin Seng,

Yu Heng Kwan, Ying Jie Chen, Zueziani Binte Zainudin, Dionne Hui Fang Loh, Nan Liu, and Lian Leng Low, published in 2023, conducted a study to assess the performance of 32 studies and 87 machine learning models in predicting microvascular and macrovascular complications in adults with Type 2 Diabetes. They found that 36% of these models demonstrated good performance in predicting complications, with the Random Forest model yielding the best results. However, most of the models showed a high risk of bias, leading the authors to note that they are still in an exploratory stage and require external validation before clinical implementation [9]. While this study focuses on evaluating machine learning models, our proposal, based on patient habits, will allow for a more comprehensive understanding of the factors contributing to the onset of diabetes complications. This is aimed at improving strategies for the prevention and management of the disease.

In the article *"Prediction of Complications of Type 2 Diabetes: A Machine Learning Approach,"* published in 2022, authors Antonio Nicolucci, Luca Romeo, Michele Bernardini, Marco Vespasiani, Maria Chiara Rossi, Massimiliano Petrelli, Antonio Ceriello, Paolo Di Bartolo, Emanuele Frontoni, and Giacomodo Vespasiani utilized a supervised learning approach based on tree-based learning (XGBoost) and leveraging large amounts of electronic medical records data to develop a model for predicting diabetes complications within 5 years of diagnosis. They considered ocular complications, cardiovascular diseases, cerebrovascular diseases, peripheral vascular diseases, nephropathy, and diabetic neuropathy. For all complications considered, the model showed a precision greater than 70%, and the AUC significantly exceeded 0.80. Additionally, they conducted external validation in five additional centers [10]. The results demonstrate good precision in predicting diabetes complications, underscoring the potential utility of these tools for improving the care of patients with diabetes. While this study is based on clinical data, our project modifies the approach by adding a perspective based on patient habits.

The reviewed studies demonstrate the growing interest in using machine learning techniques to predict complications related to type 2 diabetes. Our research proposal, focused on the influence of patient habits on the development of diabetic complications, aims to complement these existing approaches and contribute to a more comprehensive understanding of the factors influencing the onset of complications. By considering patient habits, we can advance toward more effective strategies for diabetes prevention and management, thereby improving the quality of life of patients affected by this disease.

With a data-driven approach utilizing electronic medical records (Cern-er Health Facts EMR Data), authors Abu Saleh Mohammad Mosa, Chalermpon Thongmotai, Humayera Islam, Tanmoy Paul, K. S. M. Tozammel Hossain, and Vasanthi Mandhadi developed a model in July 2022 to predict which patients are at higher risk of developing three types of diabetes-related complications (eye diseases, renal failure, and neuropathy). For the study, they employed seven classification techniques (Decision tree, Random Forest, Support Vector Machine, Multi-layer Perceptron, Logistic regression, Adaptive Boosting, Gradient Boosting) to predict these three complications using diagnostic comorbidity data. The authors note that, as a result, through the performance metrics of the machine learning models, the specificity ranged between

94% and 99%, with the SVM model achieving an F1 score of 0.759, which was the highest [11]. In contrast, our project considers information from the 2018 ENSANUT survey conducted by INEGI instead of medical records and comorbidity data, providing a more detailed perspective on the patient's habits and behavior.

## 3 Impact of Diabetes in Mexico

In Mexico, diabetes mellitus has become one of the main public health concerns. The prevalence of this disease has shown an alarming upward trend in recent decades, leading to an increase in the incidence of other conditions because of complications arising from the disease itself. According to the International Diabetes Federation, in 2019, there were 12.8 million diabetic patients in Mexico, with a rising trend predicting that by 2045, the number of people with this disease in the country will increase to 22.9 million. This places Mexico in sixth place in diabetes prevalence worldwide, following India, China, the United States, Pakistan, and Brazil [12].

For the year 2020, in Mexico, 151,000 people died due to diabetes. This represents 14% of the total deaths that occurred in the country that year. The mortality rate for diabetes in 2020 was 11.95 deaths per 10,000 inhabitants. These figures were the highest reported in the last 10 years. According to death statistics recorded by INEGI, in 2021, 13% of the deaths registered in the country were due to diabetes [13]. Given that this disease affects different organs and systems of patients, a significant number of complications can arise from it, which can significantly affect the health of the patients. Thanks to the data collected in the 2018 ENSANUT survey by INEGI, we can get an approximation of the prevalence rate of the various conditions developed by patients as a consequence of the disease [4]. Figure 1 presents the prevalence of different complications in diabetic patients.

A trend has been observed indicating that as people's age increases in our country, the incidence of disease also rises. According to the 2018 National Health and Nutrition Survey (ENSANUT) conducted by the National Institute of Geography and Statistics (INEGI), 28 5% of the population between 60 and 69 years old reported being diagnosed with diabetes nationwide, which accounts for 2.3 million individuals [4]. Figure 2 illustrates the prevalence of diabetes among the population aged 20 years and above, categorized by age group and gender in 2018.

## 4 Diabetes Complications

Before continuing with the project development, we will address diabetes complications. It is important to understand how this disease can affect various organs and systems of the body, which could even endanger the lives of patients if not treated promptly. The next subsections explore some of the main complications that arise from diabetes, as well as their treatment and prevention.

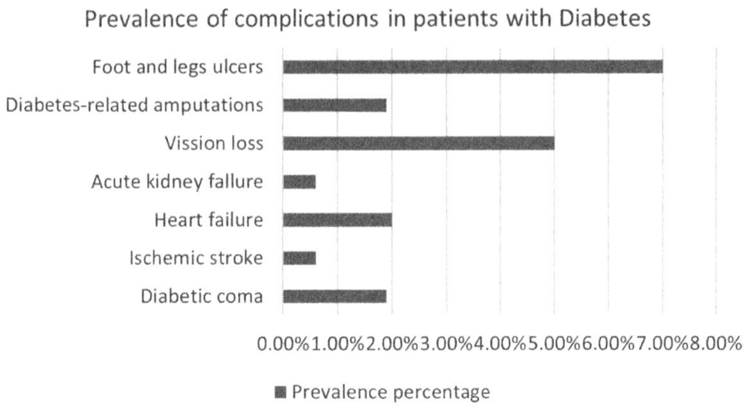


**Fig. 1** Prevalence rate of Diabetes complications. Mexico (INEGI)

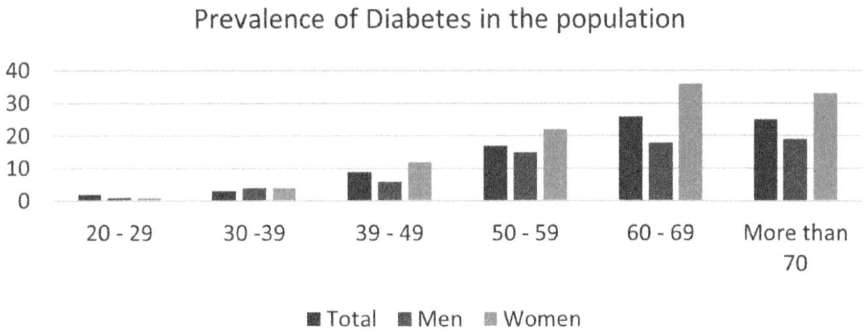


**Fig. 2** Prevalence Diabetes in the population. Mexico (INEGI)

## *4.1 Hypoglycemia*

Can be defined as a concentration of glucose in venous blood below 60 mg/dl. It is closely related to the pharmacological treatment of diabetes using oral medications that help control blood glucose levels or the administration of insulin. Any patient undergoing treatment can experience it, although it is more common in patients with a long history of diabetes mellitus who are undergoing intensive insulin therapy.

Hyperglycemia can occur mildly with symptoms such as anxiety, restlessness, tachycardia, palpitations, tremors, decreased concentration, dizziness, hunger, or blurred vision, among others, which do not interfere with normal activities. However, it can also present moderately with symptoms such as evident deterioration of the patient's motor function, preventing them from addressing the symptoms autonomously, confusion, or inappropriate behavior, and severely with seizures and coma. The main complications of hypoglycemia range from acute cardiovascular or cerebrovascular accidents to hypoglycemic encephalopathy or permanent damage to the cerebral cortex resulting from frequent episodes of severe hypoglycemia.

In order to prevent episodes of severe hypoglycemia, patients are advised to pay attention to any symptoms of mild hypoglycemia so that they can receive timely attention. Self-monitoring of blood glucose levels and proper control of the patient's habits are the best protection against complications arising from this metabolic disorder [14].

## 4.2 *Hyperglycemia*

The total or partial deficit of insulin can cause an increase in blood glucose levels, which is called hyperglycemia, and can lead to various metabolic complications for the patient. One of the main complications of hyperglycemia is diabetic ketoacidosis. This complication can develop mainly in type 1 diabetes, although cases have also been reported in patients with type 2 diabetes, commonly when they are in stressful situations. Diabetic ketoacidosis occurs primarily when there is an absolute or relative deficit of insulin that results in glucose concentrations exceeding 300 mg/dl. Among the factors that more frequently trigger ketoacidosis are infectious processes and errors in insulin administration during treatment, most commonly resulting from the patient omitting a dose or an inappropriate prescription of a treatment regimen.

Through the underlying metabolic disorder, the clinical manifestations associated with this complication are easy to recognize for diagnosis. In the initial phase, the patient may experience polyuria, polydipsia, weight loss, weakness, and anorexia. As the condition advances, nausea, vomiting, abdominal pain, alterations in consciousness, and, in a small percentage of patients, coma may occur. This complication mainly occurs in patients with type 2 diabetes in advanced age, which leads to a higher mortality rate than that caused by diabetic ketoacidosis. The clinical picture starts with acute impairment of the central nervous system function and severe dehydration and can occur in patients commonly not previously diagnosed with diabetes (in 35% of cases, this is the first manifestation of diabetes in the patient). Self-monitoring by the patient allows for control over blood glucose and ketone levels, enabling early diagnosis of ketoacidosis for better treatment. Furthermore, proper control of treatment and patient habits are the best tools for preventing such complications [15].

## 4.3 *Macroangiopathy*

It is the atherosclerotic condition of medium and large-caliber vessels. Unlike in non-diabetic patients, arteriosclerosis can commonly occur in younger patients with greater severity and extension and a worse prognosis, with cardiovascular diseases being the leading cause of morbidity and mortality among people with diabetes mellitus. Approximately 70–80% of diabetic patients die as a result of cardiovascular diseases. The presence of microalbuminuria or proteinuria is an important precursor or determinant factor of cardiovascular disease.

Properly controlling risk factors, with particular emphasis on lipid abnormalities, high blood pressure, obesity, and smoking, through lifestyle changes in patients is crucial to preventing serious complications arising from vascular disease, such as cardiovascular or cerebrovascular accidents or death. Motivating them to improve their habits and maintain proper body weight control through an appropriate diet and exercise plan is essential. The American Diabetes Association recommends the use of antiplatelet drugs, such as aspirin, as a primary prevention measure in diabetic patients with a high-risk cardiovascular profile. This is based on studies that have shown significant effectiveness in reducing the risk of myocardial infarction [16].

## *4.4 Ischemic Heart Disease*

In diabetic patients, ischemic heart disease may be present prior to its detection, as the risk of developing this condition in these patients is 2–5 times higher than in non-diabetic patients. The way this condition typically presents itself shows certain peculiarities in diabetic patients, in addition to common symptoms such as angina pectoris, acute myocardial infarction, heart failure, or sudden death. For example, in the case of angina pectoris or acute myocardial infarction, patients may experience, in addition to the classic symptoms, sweating, weakness, nausea, vomiting, dyspnea, or syncope, with an incidence three times higher in these patients than in the general population. Moreover, they have a significantly higher risk of cardiogenic shock and heart failure.

Because patients often do not present symptoms and detection is only possible through studies such as electrocardiograms, Holter monitoring, or stress tests, periodic electrocardiograms, at least once a year, are necessary. Furthermore, atherosclerotic cardiovascular disease (ASCVD) is the leading cause of morbidity and mortality in individuals with diabetes, representing a significant financial burden associated with this condition. Hypertension and dyslipidemia, common in patients with type 2 diabetes, are clear risk factors for ASCVD, which also presents an independent risk due to diabetes. Additionally, heart failure emerges as a significant cause of cardiovascular complications in diabetic patients, with significantly higher hospitalization rates compared to those without diabetes [17].

## *4.5 Heart Failure*

Diabetic patients have a 5 times higher risk than the general population of developing heart failure as a complication of the disease. This condition shows a greater tendency to occur in female patients. To prevent cardiovascular diseases in diabetic patients, it is necessary to establish strict control of blood glucose and dyslipidemia. In addition to this, the consumption of acetylsalicylic acid can help reduce the possibility of death. Furthermore, in smokers, the risk increases considerably, so quitting smoking completely is of vital importance [14].

### *4.6 Peripheral Artery Disease*

Another cardiovascular complication in diabetic patients is peripheral artery disease, which tends to occur 4 times more often in men and up to 8 times more often in women compared to the prevalence in the general population. The condition primarily affects the lower limbs, although it can sometimes occur in the upper limbs.

This condition can cause leg pain that may hinder the patient from walking, worsening as the patient travels certain distances and worsening over time to the point where intense pain may persist even at rest. If the disease continues to progress, it can lead to injuries such as ulcers or gangrene. The treatment for this type of complication consists of endovascular techniques such as balloon angioplasty, atherectomy, laser angioplasty, and revascularization. The best way to prevent this type of complication is to properly control risk factors, such as maintaining a healthy weight, controlling blood lipid levels, constantly monitoring blood pressure, following a proper diet and exercise program, quitting smoking, and maintaining proper blood glucose control [18].

### *4.7 Ischemic Stroke*

Ischemic stroke which occurs 2 times more often in diabetic patients than in non-diabetics. One of the main risk factors for ischemic stroke in diabetic patients is hypertension. 50% of cases of sudden death in diabetic and hypertensive patients are due to these types of complications.

Stroke is one of the leading causes of disability worldwide, with ischemic stroke being one of its most prevalent subtypes. In 2016 alone, there were 80.1 million cases of stroke globally, and ischemic stroke accounted for 84.4% of them. Individuals who have survived a stroke are at a high risk of experiencing another stroke within the first 30 d, and recurrent strokes tend to have more severe consequences than the initial event. Diabetes, a prevalent chronic condition affecting millions worldwide, is widely recognized as a significant risk factor for neurovascular diseases, including ischemic stroke. A considerable proportion of ischemic stroke patients also have concurrent diabetes. Given the substantial health and economic burden associated with stroke recurrence and diabetes, especially as the global population ages, it is crucial to explore the impact of diabetes on stroke recurrence and improve outcomes for patients with both conditions [19].

## 4.8 Acute Kidney Failure

Another vascular complication that can arise as a consequence of poor blood glucose control is renal artery stenosis and damage to blood vessels and the kidneys' nephrons. This can lead to alterations and asymmetries in kidney size, ultimately leading to acute renal failure.

Understanding the epidemiology of diabetes and chronic kidney disease sheds light on the significant health implications associated with this condition, characterized by persistent elevation of urinary albumin excretion, low estimated glomerular filtration rate, or other signs of kidney damage. This condition, known as diabetic kidney disease, affects a substantial portion of individuals with diabetes, between 20% and 40%. It typically develops approximately ten years after the onset of Type 1 Diabetes, although it may manifest earlier, even at the diagnosis of Type 2 Diabetes. As chronic kidney disease progresses, it can lead to end-stage renal disease, which requires dialysis or kidney transplantation. It is important to note that the presence of this condition significantly increases the risk of cardiovascular complications and healthcare costs for patients. Therefore, it is crucial to understand and effectively manage diabetic kidney disease to mitigate its impact on both patient health and healthcare system burden [20].

## 4.9 Diabetic Retinopathy

Diabetic retinopathy is a specific vascular complication that particularly affects patients with diabetes and constitutes a significant health issue due to its high prevalence. It is the leading cause of new cases of blindness in adults, affecting a wide range of patients from 20 to 74 years old. Additionally, it can trigger other vision disorders such as glaucoma and cataracts. It is known that factors such as chronic hyperglycemia, nephropathy, hypertension, and dyslipidemia increase the risk of retinopathies. However, the long-term impact of glycemic control can improve the prognosis of developing this complication [21].

## 4.10 Diabetic Foot

Derived from prolonged hyperglycemia, diabetic foot is a clinical condition with a neuropathic etiology that results in lesions or ulcers on the feet of individuals with diabetes. These lesions may occur with or without ischemia in the extremity's vessels and arteries. This complication stands as one of the primary causes of non-traumatic lower extremity amputation in diabetic patients.

To determine appropriate treatment, such complications are categorized into risk levels ranging from 0 to 5 based on the severity of the injury. Grade 0 denotes a foot at risk without any lesions, while grade 1 indicates a minor lesion such as a superficial ulcer. Grade 2 signifies a more significant lesion, like a deep ulcer penetrating the subcutaneous cellular tissue and affecting tendons and ligaments, but without infection or bone involvement. Grade 3 represents a deep ulcer accompanied by inflammation, abscess, or osteitis. Grade 4 refers to a localized lesion manifesting gangrene, typically in the heel, toes, or distal areas of the foot. Grade 5 indicates severe gangrenous lesions affecting a substantial portion of the limb.

Preventing such complications requires identifying patients at higher risk of neuropathy or artery disease, such as smokers, those with more than 10 years of diabetes duration, and those with poor glycemic control. Moreover, timely detection of the diabetic foot at grade 0 through regular self-inspection by the patient and examinations by the attending physician, along with the use of appropriate footwear and regular podiatrist visits, are crucial measures for preventing more severe complications and avoiding limb amputation. It concerns that up to half of the diabetic peripheral neuropathy cases may be asymptomatic, posing a significant risk if left undetected and if preventive foot care measures are not implemented, potentially leading to injuries, diabetic foot ulcers, and amputations. Recognizing and treating autonomic neuropathy can improve symptom management, reduce complications, and enhance overall quality of life. Although there is currently no specific treatment to reverse the underlying nerve damage, maintaining adequate glucose control plays a crucial role in preventing diabetic peripheral neuropathy (DPN) and cardiac autonomic neuropathy (CAN) in Type 1 diabetes and may modestly slow their progression in type 2 diabetes [22].

## 5 Machine Learning in Healthcare

In the medical field, copious amounts of data are generated daily. Before the era of computers, this data was predominantly recorded on paper. However, with the rapid trend towards digitalization and the emergence of new technologies, massive amounts of data, commonly referred to as Big Data, are now generated. The digital health dataset has become so vast and intricate that traditional data management tools and methods struggle to process it effectively. The sheer volume and complexity of health data today are overwhelming, compounded by the diversity of data and the urgency with which it needs to be managed.

Through the analysis and mining of large datasets, healthcare organizations can derive various benefits, including effective population health management, fraud detection in healthcare, and optimization of hospital resources, among others. One of the primary advantages is early disease detection, leading to more efficient treatment and the prevention of complications that may arise if the disease is not detected promptly [23].

## 5.1 *Advantages of Machine Learning in Healthcare*

The digitization of information has led to the accumulation of vast amounts of data, presenting numerous opportunities for leveraging machine learning techniques across various aspects of healthcare. One significant area of opportunity lies in the early detection of diseases and their complications. By analyzing extensive historical databases, machine learning algorithms can predict events that may occur, leading to improved patient care across different disease spectrums. This predictive capability enables healthcare providers to offer patients tools for better disease management, ultimately reducing the risk of complications [24]. While early disease detection is a crucial objective in large-scale data mining, its benefits represent only a fraction of the extensive advantages achievable in healthcare. Below are outline of several applications of large-scale data mining and machine learning in healthcare:

**Applications of machine learning in healthcare**

- **Clinical operations:** Through large-scale data classification and analysis, research can be conducted to compare existing medical treatments, aiming to determine procedures that are both clinically relevant and cost-effective in diagnosing and treating patients.

- **Research and development:** Various prediction and classification models can improve clinical trials by categorizing patients into personalized treatments, reducing trial failures, and fostering the development of new and improved medical treatments. Additionally, analyzing trials and clinical histories can identify indicators of adverse drug effects before market release.

- **Public health:** Large-scale disease data analysis can identify patterns to track outbreaks and transmission, enhancing public health surveillance for faster response and contributing to the rapid development of vaccines. Analyzing vast amounts of data also informs needs identification, service delivery, prediction, and prevention of public health crises.

- **Evidence-based medicine:** Integrating financial, operational, clinical, and genomics data, both structured and unstructured, enables treatment standardization, prediction of disease or complication risks, and delivery of more efficient healthcare.

- **Genomic analysis:** Improved efficiency and cost-effectiveness in gene sequencing make genomics analysis a routine part of patient care decision-making and medical history management.

- **Analytics for fraud prevention:** Analyzing a multitude of incidents provides valuable information for fraud prevention, reducing unnecessary expenses, and minimizing abuses.
- **Remote monitoring:** Real-time and remote analysis of large datasets allows for enhanced monitoring systems, ensuring safety and predicting adverse events.
- **Patient profile analysis:** Advanced analysis models on patient profiles can identify disease indicators or complications. By analyzing extensive data, predictive or classification models enable early disease or complication detection, leading to better preventive treatments.

Figure 3 presents a diagram illustrating the influence of different machine learning techniques applied to healthcare sciences.

## 5.2 *The Attributes of Data Analytics in Healthcare*

In line with other fields, the analysis of large datasets (Big Data) in healthcare is characterized by four main attributes: volume, velocity, variety, and veracity [25].

**Big data attributes**

**Volume:** The proliferation of new technologies has led to an exponential growth in health-related data. Each day, an immense amount of information is generated, encompassing medical records, images, clinical trial data, genetic sequences, sensor readings, and patient-specific data. Moreover, the continuous emergence of new technologies further augments this volume.

**Velocity:** Advancements in data management, such as virtualization and cloud computing, have facilitated the development of platforms capable of capturing, storing, and processing data in real-time and at an unprecedented scale and speed.

**Variety:** With the capacity to perform real-time analysis on constantly evolving datasets, a multitude of applications can now be generated, offering

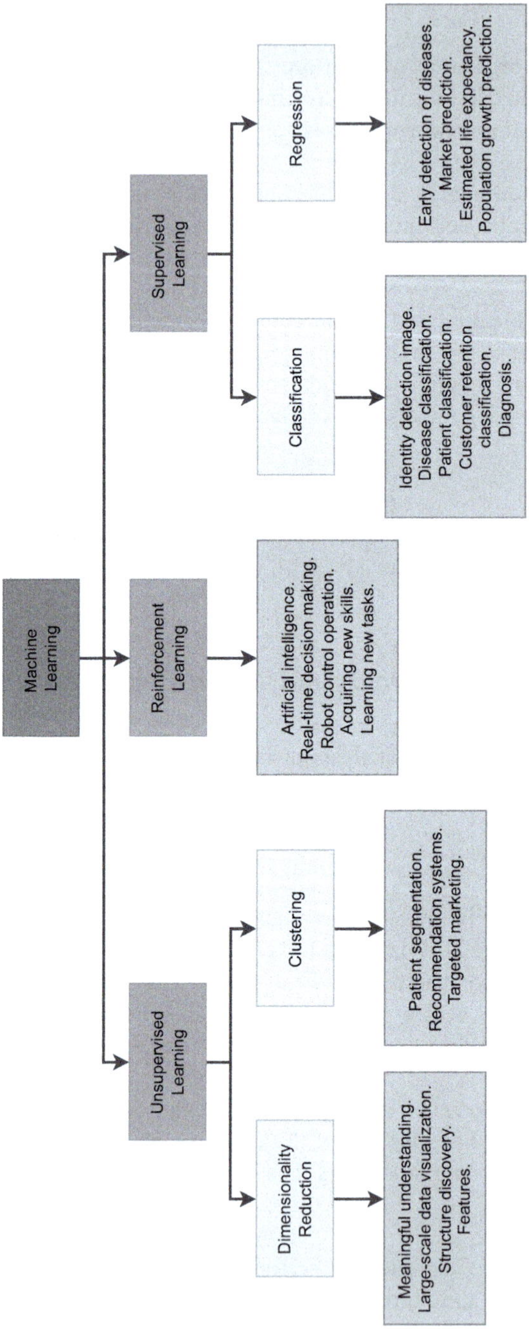


**Fig. 3** Machine learning in healthcare

diverse solutions across various healthcare domains.

**Veracity:** Accuracy is paramount in the architectures, tools, platforms, algorithms, and methodologies used for large-scale data mining and analysis in healthcare. The availability of precise information is crucial for patient safety and well-being. Given the variability and potential inaccuracies inherent in unstructured data, ensuring data quality is of utmost importance.

### *5.3 Support Vector Machine*

The Support Vector Machines (SVM) have displayed a great performance in classification problems with two or several classes [7]. The general formulation of the SVM consists in mapping the input data into a high-dimensional feature space called $\phi(x)$ and building a hyper-plane express $f(x) = w\phi + b$ to separate samples from two classes. In the case of an L1 soft-margin SVM, the hyperplane is constructed by solving the primal problem presented in Eq. 1:

$$\begin{aligned} min\frac{1}{2}\|w\|^2 + C\sum_{i=1}^{n}\xi I \\ \text{subject to } y_1(w)\cdot\phi(x_i) + b \geq 1 - \xi i, \\ \text{with } \xi i \geq 0 \text{ and } 1 \leq I \leq n \end{aligned} \tag{1}$$

where $x_i$ is a data example, and $y_i in - 1, 1$ its label of the class. This problem can be solved in its dual form with Eq. 2.

$$\begin{aligned} min\frac{1}{2}\alpha^T K\alpha - e^T\alpha \\ \text{subject to } y^T\alpha = 0 \\ \text{with } 0 \leq \alpha \leq C \text{ and } 1 \leq i \leq n \end{aligned} \tag{2}$$

where $e$ is a vector of ones and $K_{ij} = k(x_i x_j) = \phi(x_i)^T \cdot \phi(x_j)$ is a kernel function that performs the $\phi$ mapping, regarding that the kernel is the gaussian radial basis presented below:

$$k(x_i x_j) = e^{-\gamma\|x_i - x_j\|^2} \tag{3}$$

The hyper-parameters $C$ and $\gamma$ must set the values before solving the minimization presented in Eq. 3.

# 6 SVM Model for the Prediction of Complications Derived from Diabetes

This study focused on analyzing how habits influence the development of diabetes-related complications in patients, using Support Vector Machines (SVM) in machine learning. The data analyzed were extracted from the 2018 National Health and Nutrition Survey (ENSANUT) [4] conducted by INEGI, which contains information related to the health and habits of the respondents. To investigate the intricate relationship between patients' habits and the onset of diabetes-related complications, a robust experimental framework was established. Employing the Support Vector Machine (SVM) methodology, we meticulously crafted a predictive model utilizing a comprehensive set of independent variables-20 habits characteristic of diabetic patients alongside 8 dependent variables representing various complications associated with the condition.

In this study, the variable analysis reveals the complex interaction between patients' habits and the incidence of diabetes-related complications. The coefficients assigned to each habit in the importance analysis highlight its relative contribution to predicting each complication. A positive coefficient suggests a positive association between the habit and the complication, while a negative coefficient indicates an inverse association. These coefficients provide a quantitative measure of the influence of each habit on the occurrence of complications. On the other hand, the F1 score, which combines the model's precision and sensitivity, offers a comprehensive assessment of its performance in predicting the presence of complications. By considering both the model's ability to identify positive cases and its ability to avoid false positives, the F1 score provides a robust metric of the predictive model's quality. Together, these analyses allow for a deeper understanding of how patients' habits affect the course of diabetes and its complications and provide a solid foundation for developing personalized intervention strategies and improving long-term clinical outcomes.

In our experiment, each habit was meticulously scrutinized for its potential impact on the incidence of diabetes-related complications, ranging from foot or leg ulcers to diabetic encephalopathy. Through rigorous model training and validation, we aimed to unravel the intricate interplay between these habits and the occurrence of complications, with a keen focus on enhancing patient care and quality of life. The resulting insights, meticulously presented, offer a comprehensive overview of the most influential habits for each complication, thereby empowering healthcare professionals to prioritize interventions effectively. These findings not only underscore the importance of proactive habit adoption but also pave the way for personalized interventions tailored to individual patient needs, thereby heralding a new approach to diabetes care.

## 6.1 Methodology and Data Processing Tools

To perform data preprocessing and execute the model, MATLAB was employed, a widely recognized platform in numerical analysis, data processing, and machine learning. Initially, the data were curated to select only records corresponding to individuals with diabetes, resulting in a dataset of 4555 observations. Subsequently, these data were divided into 80% for model training and 20% for validation. Below are the elements for the development of the model:

**Tools and Parameters for the study**

**Tool:** Matlab
**Method:** Support Vector Machine (SVM)
**Database:** ENSANUT 2018, INEGI
**Total number of registrations:** 43,070
**Sample:** 4,555
**Training data:** (80%) = 3,644
**Test data:** (20%) = 911
**Dependent variables:** 8 Complications
**Independent variables:** 20 Habits
**Metrics for model validation:** Accuracy, Sensibility, Recall, F1 Score

Beneath is the list of variables corresponding to the habits of diabetic patients used for training the model:

**Habits in diabetic patients**

- Make a meal plan.
- Make an exercise plan.
- Consumption of alcoholic beverages.
- Daily intake of aspirin.
- Regular visits to a specialist for foot examination.
- Regular General Urine Examination.
- Regular testing for microalbuminuria and creatinine levels.
- Frequent monitoring of blood pressure levels.
- Influenza vaccination.
- Pneumococcal vaccination.
- Regular monitoring of cholesterol and triglyceride levels.
- Medication adherence for cholesterol and triglyceride control.
- Medication adherence for blood pressure control.
- History of active smoking.

- Regular visits to a specialist for a dental examination.
- Receipt of diabetes education.
- Regular electrocardiogram (ECG) screening.
- Avoidance of uncomfortable footwear.
- Participation in diabetes support groups.
- Frequent inspection of feet and legs for lesions.

The following list shows the variables corresponding to diabetes-related complications:

**Complications in diabetic patients**

- Foot or leg ulcers.
- Foot or leg amputations.
- Vision Loss
- Acute Kidney failure
- Hearth failure.
- Stroke.
- Diabetic coma.
- Diabetic encephalopathy.

## *6.2 Machine Learning Model Results*

Diabetes is a chronic disease affecting millions worldwide, with potentially devastating complications if not managed properly. Understanding how patients' habits may influence the onset of complications is pivotal to improving care and the quality of life for those living with this condition. In this study, we employ an SVM-based machine learning model to analyze the relationship between diabetes patients' habits and the incidence of associated complications.

The results are presented from Tables 1 to 8, where all analyzed habits are sorted in descending order. This allows for a clear visualization of which habits have a greater relationship with the analyzed complication, with the most influential ones placed at the top of the table. This structure facilitates the identification of the most relevant factors for each complication and provides a practical guide for the development of more effective preventive strategies.

**Table 1** Results for the foot or leg ulcers variable

| Foot or leg ulcers | | |
|---|---|---|
| Habits | Importance | Coefficient |
| Frequent inspection of feet and legs for lesions | 1.9372 | 1.9372 |
| Medication adherence for blood pressure control | 1.436 | 1.436 |
| Frequent monitoring of blood pressure levels | 1.4123 | 1.4123 |
| Make an exercise plan | 1.394 | -1.394 |
| Medication adherence for cholesterol and triglyceride control | 1.3133 | −1.3133 |
| Regular electrocardiogram (ECG) screening | 1.3046 | 1.3046 |
| Regular visits to a specialist for dental examination | 1.2908 | −1.2908 |
| Pneumococcal vaccination | 1.0803 | −1.0803 |
| Receipt of diabetes education | 1.0367 | −1.0367 |
| Regular monitoring of cholesterol and triglyceride levels | 0.8946 | −0.8946 |
| Regular testing for microalbuminuria and creatinine levels | 0.7477 | −0.7477 |
| Avoidance of uncomfortable footwear | 0.6281 | −0.6281 |
| Participation in diabetes support groups | 0.5079 | 0.5079 |
| Influenza vaccination | 0.4834 | 0.4834 |
| Make a meal plan | 0.313 | −0.313 |
| Regular visits to a specialist for foot examination | 0.2944 | 0.2944 |
| Regular general urine examination | 0.2732 | −0.2732 |
| Consumption of alcoholic beverages | 0.2635 | 0.2635 |
| Daily intake of aspirin | 0.2486 | 0.2486 |
| History of active smoking | 0.2082 | −0.2082 |
| **Accuracy** | **0.93407** | |
| **Sensitivity** | **0.93407** | |
| **Recall** | **0.93** | |
| **F1 score** | **0.93** | |

The analysis of the **foot or leg ulcers variable** presented in Table 1 revealed an accuracy of 0.93407 and an F1 Score of 0.93, which depicts significant relationships between various patient habits and the presence of this diabetic complication. The most influential feature in predicting the presence of ulcers was the frequency of inspecting feet and legs. This suggests that regular surveillance of the lower extremities is crucial for preventing ulcers in patients with diabetes. Other features positively associated with the presence of foot or leg ulcers include regular blood pressure monitoring and periodic electrocardiograms.

The analysis of the **foot or leg amputation variable** shown in Table 2, with an accuracy of 99.01% and an F1 Score of 0.99, revealed that the habit of attending support groups for diabetes patients represents one of the most significant habits in preventing this complication. Additionally, the habit of smoking tobacco showed a notable positive association with the possibility of foot or leg amputation, highlighting the importance of avoiding this harmful habit for the prevention of this complication.

The **Vision loss complication** variable with an accuracy of 98.68% and an F1 Score of 0.99. Table 3 presents the analysis of the variable related to vision loss, unveiling a series of significant factors, where the habit of receiving education about diabetes was the most influential characteristic. This underscores the critical role of education in diabetes management to mitigate the risk of complications of diabetes-related vision impairment. Additionally, maintaining up-to-date pneumococcal vaccination records and undergoing frequent ECG studies also showed considerable importance coefficients.

Table 4 presents the **acute kidney failure complication** in which the proposed model demonstrated a significant relationship with an accuracy of 98.79% and an F1 Score of 0.99. There is a notable connection between the frequency at which patients analyze their microalbuminuria and creatinine levels, as well as undergoing an electrocardiogram (ECG), and the prevention of renal failure associated with diabetes. Additionally, smoking history was found to have a detrimental impact on the development of this medical condition.

In turn, Table 5 illustrates the **heart failure complication** variable. The model demonstrated an accuracy of 97.91% and an F1 Score of 0.98. Among the most notable findings, it is observed that a history of smoking correlates with a significant negative impact on the development of heart failure or myocardial infarction. Furthermore, participation in support groups for diabetes patients, as well as frequent analysis of microalbumin, creatinine, and ECG, emerged as crucial factors in preventing this complication.

The **stroke complication** variable achieves an accuracy of 98.57% and an F1 Score of 0.99; the results are observed in Table 6. The model revealed that participation in support groups for diabetes patients and receiving education about the disease are significant influencing factors. Additionally, smoking history was identified as a detrimental characteristic and trigger for this complication.

**Table 2** Results for the Foot or leg amputations variable

| Foot or leg amputations | | |
|---|---|---|
| Habits | Importance | Coefficient |
| Participation in diabetes support groups | 3.4914 | −3.4914 |
| History of active smoking | 1.6218 | 1.6218 |
| Regular testing for microalbuminuria and creatinine levels | 1.0782 | −1.0782 |
| Regular electrocardiogram (ECG) screening | 1.0346 | 1.0346 |
| Daily intake of aspirin | 0.8934 | 0.8934 |
| Make an exercise plan | 0.653 | 0.653 |
| Make a meal plan | 0.6404 | −0.6404 |
| Consumption of alcoholic beverages | 0.6334 | 0.6334 |
| Regular general urine examination | 0.5778 | 0.5778 |
| Receipt of diabetes education | 0.4894 | −0.4894 |
| Regular monitoring of cholesterol and triglyceride levels | 0.4481 | −0.4481 |
| Pneumococcal vaccination | 0.4294 | 0.4294 |
| Frequent monitoring of blood pressure levels | 0.3388 | −0.3388 |
| Medication adherence for blood pressure control | 0.2442 | −0.2442 |
| Avoidance of uncomfortable footwear | 0.2373 | 0.2373 |
| Regular visits to a specialist for foot examination | 0.182 | 0.182 |
| Frequent inspection of feet and legs for lesions | 0.1489 | −0.1489 |
| Medication adherence for cholesterol and triglyceride control | 0.1473 | 0.1473 |
| Influenza vaccination | 0.132 | 0.132 |
| Regular visits to a specialist for dental examination | 0.1217 | −0.1217 |
| **Accuracy** | **0.99011** | |
| **Sensitivity** | **0.99011** | |
| **Recall** | **0.99** | |
| **F1 score** | **0.99** | |

**Table 3** Results for the Vision loss variable

| Vision loss | | |
|---|---|---|
| Habits | Importance | Coefficient |
| Receipt of diabetes education | 1.8186 | −1.8186 |
| Pneumococcal vaccination | 1.722 | −1.722 |
| Regular electrocardiogram (ECG) screening | 1.4844 | 1.4844 |
| Regular testing for microalbuminuria and creatinine levels | 1.2891 | 1.2891 |
| Medication adherence for cholesterol and triglyceride control | 1.2738 | 1.2738 |
| Avoidance of uncomfortable footwear | 1.2607 | −1.2607 |
| Regular visits to a specialist for foot examination | 1.1907 | −1.1907 |
| Consumption of alcoholic beverages | 1.1552 | 1.1552 |
| Medication adherence for blood pressure control | 0.9296 | −0.9296 |
| Frequent monitoring of blood pressure levels | 0.8261 | −0.8261 |
| Regular visits to a specialist for dental examination | 0.8169 | 0.8169 |
| Influenza vaccination | 0.7296 | −0.7296 |
| Make a meal plan | 0.729 | −0.729 |
| History of active smoking | 0.6731 | 0.6731 |
| Daily intake of aspirin | 0.6107 | 0.6107 |
| Frequent inspection of feet and legs for lesions | 0.4166 | 0.4166 |
| Regular monitoring of cholesterol and triglyceride levels | 0.3217 | 0.3217 |
| Make an exercise plan | 0.1658 | 0.1658 |
| Regular general urine examination | 0.1148 | 0.1148 |
| Participation in diabetes support groups | 0.05 | 0.05 |
| **Accuracy** | **0.98681** | |
| **Sensitivity** | **0.98681** | |
| **Recall** | **0.99** | |
| **F1 score** | **0.99** | |

**Table 4** Results for the Acute kidney failure variable

| **Acute kidney failure** | | |
|---|---|---|
| Habits | Importance | Coefficient |
| Regular testing for microalbuminuria and creatinine levels | 2.9845 | −2.9845 |
| Regular electrocardiogram (ECG) screening | 2.0982 | 2.0982 |
| History of active smoking | 1.5623 | 1.5623 |
| Influenza vaccination | 1.1127 | 1.1127 |
| Daily intake of aspirin | 0.7823 | −0.7823 |
| Regular visits to a specialist for dental examination | 0.6743 | 0.6743 |
| Regular visits to a specialist for foot examination | 0.6554 | −0.6554 |
| Medication adherence for cholesterol and triglyceride control | 0.634 | 0.634 |
| Consumption of alcoholic beverages | 0.5545 | −0.5545 |
| Receipt of diabetes education | 0.5372 | −0.5372 |
| Medication adherence for blood pressure control | 0.3908 | −0.3908 |
| Make an exercise plan | 0.36 | −0.36 |
| Make a meal plan | 0.3421 | −0.3421 |
| Regular general urine examination | 0.3168 | 0.3168 |
| Regular monitoring of cholesterol and triglyceride levels | 0.3006 | −0.3006 |
| Frequent inspection of feet and legs for lesions | 0.295 | 0.295 |
| Participation in diabetes support groups | 0.2253 | 0.2253 |
| Avoidance of uncomfortable footwear | 0.204 | 0.204 |
| Pneumococcal vaccination | 0.1606 | 0.1606 |
| Frequent monitoring of blood pressure levels | 0.102 | −0.102 |
| **Accuracy** | **0.98791** | |
| **Sensitivity** | **0.98791** | |
| **Recall** | **0.99** | |
| **F1 score** | **0.99** | |

Additionally, in Table 7 are presented the results for the **diabetic coma complication**. Results depict an accuracy of 98.02% and an F1 Score of 0.98; the results emphasize the importance of implementing proper dietary and exercise plans, as well as maintaining adequate control of blood pressure levels for the prevention of stroke in patients with diabetes.

Finally, in Table 8 **diabetic encephalopathy complication**, results are depicted. The analysis conducted by the model reaches an accuracy of 95.52% and an F1 Score of 0.95, showing results indicating that participation in diabetes support groups, adopting a regular exercise plan, frequent monitoring of blood pressure, daily intake of aspirin, and a history of smoking are determining factors associated with the incidence of diabetic encephalopathy.

**Table 5** Results for the Heart failure variable

| Heart failure | | |
|---|---|---|
| Habits | Importance | Coefficient |
| History of active smoking | 3.38 | -3.38 |
| Regular testing for microalbuminuria and creatinine levels | 1.6924 | -1.6924 |
| Participation in diabetes support groups | 1.1508 | 1.1508 |
| Regular electrocardiogram (ECG) screening | 1.0808 | 1.0808 |
| Influenza vaccination | 0.8011 | 0.8011 |
| Frequent inspection of feet and legs for lesions | 0.7313 | 0.7313 |
| Medication adherence for cholesterol and triglyceride control | 0.6667 | -0.6667 |
| Regular visits to a specialist for foot examination | 0.5484 | -0.5484 |
| Regular monitoring of cholesterol and triglyceride levels | 0.5102 | 0.5102 |
| Daily intake of aspirin | 0.4939 | 0.4939 |
| Regular General Urine Examination | 0.4279 | -0.4279 |
| Consumption of alcoholic beverages | 0.4054 | 0.4054 |
| Avoidance of uncomfortable footwear | 0.3875 | 0.3875 |
| Receipt of diabetes education | 0.2924 | 0.2924 |
| Make an exercise plan | 0.2833 | 0.2833 |
| Medication adherence for blood pressure control | 0.2551 | 0.2551 |
| Make a meal plan | 0.1696 | 0.1696 |
| Diabetic condition | 0.128 | 0.128 |
| Regular visits to a specialist for dental examination | 0.1042 | -0.1042 |
| Frequent monitoring of blood pressure levels | 0.0675 | 0.0675 |
| **Accuracy** | **0.97912** | |
| **Sensitivity** | **0.97912** | |
| **Recall** | **0.98** | |
| **F1 Score** | **0.98** | |

**Table 6** Results for the Stroke variable

| Stroke | | |
|---|---|---|
| Habits | Importance | Coefficient |
| Participation in diabetes support groups | 3.6873 | -3.6873 |
| Regular electrocardiogram (ECG) screening | 1.4808 | 1.4808 |
| History of active smoking | 0.9628 | 0.9628 |
| Receipt of diabetes education | 0.9473 | -0.9473 |
| Regular testing for microalbuminuria and creatinine levels | 0.9041 | 0.9041 |
| Make an exercise plan | 0.7124 | 0.7124 |
| Frequent inspection of feet and legs for lesions | 0.4299 | -0.4299 |
| Frequent monitoring of blood pressure levels | 0.421 | -0.421 |
| Regular General Urine Examination | 0.3856 | 0.3856 |
| Medication adherence for cholesterol and triglyceride control | 0.3836 | -0.3836 |
| Make a meal plan | 0.3481 | 0.3481 |
| Avoidance of uncomfortable footwear | 0.3149 | 0.3149 |
| Consumption of alcoholic beverages | 0.2746 | 0.2746 |
| Regular visits to a specialist for dental examination | 0.1877 | 0.1877 |
| Regular monitoring of cholesterol and triglyceride levels | 0.163 | 0.163 |
| Daily intake of aspirin | 0.1223 | 0.1223 |
| Medication adherence for blood pressure control | 0.0955 | -0.0955 |
| Influenza vaccination | 0.0924 | -0.0924 |
| Regular visits to a specialist for foot examination | 0.0856 | 0.0856 |
| Pneumococcal vaccination | 0.0562 | 0.0562 |
| **Accuracy** | **0.97912** | |
| **Sensitivity** | **0.97912** | |
| **Recall** | **0.98** | |
| **F1 Score** | **0.98** | |

## 7 Conclusions

The findings of this study underscore the significant role that patients' habits play in mitigating the risk of diabetes-related complications. By employing a Support Vector Machines (SVM)-based machine learning model, we have identified key associations between specific habits and the incidence of complications. These results highlight the importance of adopting healthy behaviors and lifestyle choices in managing diabetes effectively. Presenting the results in descending order within tables provides a clear hierarchy of influential habits for each complication. This approach enables healthcare professionals to prioritize interventions that target the most impactful factors, thereby optimizing preventive strategies and enhancing patient outcomes.

Moreover, the identification of detrimental habits, such as tobacco consumption, emphasizes the critical need for targeted interventions to address modifiable risk

**Table 7** Results for the Diabetic coma variable

| Diabetic coma | | |
|---|---|---|
| Habits | Importance | Coefficient |
| Make a meal plan | 2.0809 | 2.0809 |
| Make an exercise plan | 1.9407 | -1.9407 |
| Medication adherence for blood pressure control | 1.7765 | -1.7765 |
| Regular visits to a specialist for dental examination | 1.2778 | 1.2778 |
| Regular monitoring of cholesterol and triglyceride levels | 1.1976 | -1.1976 |
| Receipt of diabetes education | 1.098 | -1.098 |
| Regular visits to a specialist for foot examination | 1.0651 | 1.0651 |
| Daily intake of aspirin | 0.8724 | 0.8724 |
| Influenza vaccination | 0.8244 | 0.8244 |
| History of active smoking | 0.7343 | -0.7343 |
| Regular General Urine Examination | 0.706 | -0.706 |
| Medication adherence for cholesterol and triglyceride control | 0.6191 | 0.6191 |
| Avoidance of uncomfortable footwear | 0.3998 | 0.3998 |
| Medication adherence for blood pressure control | 0.3646 | 0.3646 |
| Regular visits to a specialist for foot examination | 0.3211 | 0.3211 |
| Participation in diabetes support groups | 0.2303 | -0.2303 |
| Consumption of alcoholic beverages | 0.121 | -0.121 |
| Frequent inspection of feet and legs for lesions | 0.1099 | -0.1099 |
| Pneumococcal vaccination | 0.0395 | 0.0395 |
| Frequent monitoring of blood pressure levels | 0.0395 | 0.0395 |
| **Accuracy** | **0.98022** | |
| **Sensitivity** | **0.98022** | |
| **Recall** | **0.98** | |
| **F1 Score** | **0.98** | |

factors. By integrating these findings into clinical practice, healthcare providers can tailor interventions to individual patient needs, promoting personalized care and improved long-term health outcomes for individuals living with diabetes.

Moreover, while this study sheds light on current associations between habits and diabetes-related complications, there remains ample room for further research. Future investigations could delve deeper into the mechanisms underlying these relationships, potentially uncovering novel insights into the pathophysiology of diabetes and its complications. Additionally, longitudinal studies tracking the evolution of habits and their impact on outcomes over time could provide invaluable data for refining predictive models and developing more targeted interventions. Furthermore, exploring the intersectionality of habits with socioeconomic factors and cultural influences may unveil nuances that are crucial for tailoring interventions to diverse patient populations effectively. By addressing these avenues for future inquiry, researchers

**Table 8** Results for the Diabetic encephalopathy variable

| Diabetic encephalopathy | | |
|---|---|---|
| Habits | Importance | Coefficient |
| Make a meal plan | 2.0809 | 2.0809 |
| Make an exercise plan | 1.9407 | -1.9407 |
| Medication adherence for blood pressure control | 1.7765 | -1.7765 |
| Regular visits to a specialist for dental examination | 1.2778 | 1.2778 |
| Regular monitoring of cholesterol and triglyceride levels | 1.1976 | -1.1976 |
| Receipt of diabetes education | 1.098 | -1.098 |
| Regular visits to a specialist for foot examination | 1.0651 | 1.0651 |
| Daily intake of aspirin | 0.8724 | 0.8724 |
| Influenza vaccination | 0.8244 | 0.8244 |
| History of active smoking | 0.7343 | -0.7343 |
| Regular General Urine Examination | 0.706 | -0.706 |
| Medication adherence for cholesterol and triglyceride control | 0.6191 | 0.6191 |
| Avoidance of uncomfortable footwear | 0.3998 | 0.3998 |
| Medication adherence for blood pressure control | 0.3646 | 0.3646 |
| Regular visits to a specialist for foot examination | 0.3211 | 0.3211 |
| Participation in diabetes support groups | 0.2303 | -0.2303 |
| Consumption of alcoholic beverages | 0.121 | -0.121 |
| Frequent inspection of feet and legs for lesions | 0.1099 | -0.1099 |
| Pneumococcal vaccination | 0.0395 | 0.0395 |
| Frequent monitoring of blood pressure levels | 0.0395 | 0.0395 |
| **Accuracy** | **0.98022** | |
| **Sensitivity** | **0.98022** | |
| **Recall** | **0.98** | |
| **F1 Score** | **0.98** | |

can continue to refine our understanding of the intricate interplay between habits, complications, and patient outcomes, ultimately paving the way for more effective strategies in diabetes management.

## References

1. World Health Organization. WHO, April 5 2023. https://www.who.int/news-room/fact-sheets/detail/diabetes. Accessed from 12 July 2024
2. Pan American Health Organization. PAHO, September 29 2022. https://www.paho.org/es/temas/diabetes. Accessed from 6 Dec 2022
3. Tattersall, R.B., Matthews, D.R.: The history of diabetes mellitus. Textbook of diabetes, pp. 1–21 (2024)

4. Romero-Martínez, M., Shamah-Levy, T., Vielma-Orozco, E., Heredia-Hernández, O., Mojica-Cuevas, J., Cuevas-Nasu, L., Rivera-Dommarco, J.: Encuesta nacional de salud y nutrición 2018-19: metodología y perspectivas. Salud Pública de México, 61(6, nov-dic):917–923, dic. (2019)
5. Boser, B.E., Guyon, I.M., Vapnik, V.N.: A training algorithm for optimal margin classifiers. In: Proceedings of the Fifth Annual Workshop on Computational Learning Theory, pp. 144–152 (1992)
6. Cortes, C., Vapnik, V.: Support-vector networks. Mach. Learn. **20**, 273–297 (1995)
7. Guido, R., Ferrisi, S., Lofaro, D., Conforti, D.: An overview on the advancements of support vector machine models in healthcare applications: a review. Information **15**(4), 235 (2024)
8. Mora, T., Roche, D., Rodríguez-Sánchez, B.: Predicting the onset of diabetes-related complications after a diabetes diagnosis with machine learning algorithms. Diabetes Res. Clin. Pract. **204**, 110910 (2023)
9. Tan, K.R., Benjamin Seng, J.J., Kwan, Y.H., Chen, Y.J., Zainudin, S.B., Hui Fang Loh, D., Liu, N., Low, L.L.: Evaluation of machine learning methods developed for prediction of diabetes complications: a systematic review. J. Diabetes Sci. Technol. **17**(2), 474–489 (2023)
10. Nicolucci, A., Romeo, L., Bernardini, M., Vespasiani, M., Chiara Rossi, M., Petrelli, M., Ceriello, A., Di Bartolo, P., Frontoni, E., Vespasiani, G.: Prediction of complications of type 2 diabetes: A machine learning approach. Diabetes Res. Clin. Pract. **190**, 110013 (2022)
11. Mosa, A.S.M., Thongmotai, C., Islam, H., Paul, T., Hossain, K.S.M.T., Mandhadi, A.V.: Evaluation of machine learning applications using real-world ehr data for predicting diabetes-related long-term complications. TAYLOR & FRANCIS LTD **5**(2), 141–151 (2022)
12. Federación a D. R. Center. CIAD, September 14 2020. https://www.ciad.mx/la-pandemia-de-la-diabetes-en-mexico/. Accessed from 2 Dec 2021
13. National Institute of Statistics and Geography. INEGI, November 12 2021. https://www.inegi.org.mx/app/saladeprensa/noticia.html?id=6923. Accessed from 16 Nov 2021
14. Mediavilla Bravo, J.J.: Complicaciones de la diabetes mellitus. diagnóstico y tratamiento. SEMERGEN - Medicina de Familia **27**(3), 132–145 (2001)
15. Vlacho, B., Rossell-Rusiñol, J., Granado-Casas, M., Mauricio, D., Julve, J.: Overview on chronic complications of diabetes mellitus. In: Chronic Complications of Diabetes Mellitus, pp. 1–10. Elsevier, (2024)
16. Ioannidis, I.: Macroangiopathy in diabetes. Diabetes in Clinical Practice: Questions and Answers from Case Studies, p. 199 (2007)
17. ElSayed, N.A., Aleppo, G., Aroda, V.R., Bannuru, R.R., Brown, F.M., Bruemmer, D., Collins, B.S., Cusi, K., Das, S.R., Gibbons, C.H. et al.: Summary of revisions: Standards of care in diabetes-2023. Diabetes Care **46**(Supplement_1), S5–S9 (2023)
18. Liang, J., Gong, X., Xuyang, H., You, C., Zhou, J., Gao, Y., Zong, J., Liu, Y.: Integrated genetic analysis of diabetic complications: Bioinformatics insights into foot ulcers, neuropathy and peripheral artery disease. Int. Wound J. **21**(2), e14748 (2024)
19. Zhang, L., Li, X., Wolfe, T.R.D., O'Connell, M.D.L., Wang, Y.: Diabetes as an independent risk factor for stroke recurrence in ischemic stroke patients: An updated meta-analysis. Neuroepidemiology **55**(6), 427–435 (2021)
20. Rossiter, A., La, A., Koyner, J.L., Forni, L.G.: New biomarkers in acute kidney injury. Crit. Rev. Clin. Laboratory Sci. **61**(1), 23–44 (2024)
21. Stitt, A.W., Curtis, T.M., Chen, M., Medina, R.J., McKay, G.J., Jenkins, A., Gardiner, T.A., Lyons, T.J., Hammes, H-P., Simo, R. et al.: The progress in understanding and treatment of diabetic retinopathy. Prog. Retinal Eye Res. **51**, 156–186 (2016)
22. Khanolkar, M.P., Bain, S.C., Stephens, J.W.: The diabetic foot. QJM: Int. J. Med. **101**(9), 685–695 (2008)
23. Raghupathi, W., Raghupathi, V.: Big data analytics in healthcare: Promise and potential. Health Inf. Sci .Syst. **2**, 3 (2014). Accessed from 31 March 2015
24. Sarker, M.: Revolutionizing healthcare: the role of machine learning in the health sector. J. Artif. Intell. General Sci. (JAIGS) **2**(1), 36–61 (2024). ISSN: 3006-4023
25. Khanra, S., Dhir, A., Najmul Islam, A.K.M., Mäntymäki, M.: Big data analytics in healthcare: a systematic literature review. Enterprise Inf. Syst. **14**(7), 878–912 (2020)

# Metaheuristic-Based Neuroevolution Framework for Improved Pneumonia Classification in X-ray Images

**Diego Campos Peña, Oscar Ramos-Soto, Javier Augusto Galvis-Chacon, Beatriz A. Rivera-Aguilar, Anuradha Thakare, and Parijata Majumdar**

**Abstract** Medical imaging, including X-ray imaging, involves capturing and interpreting visual representations of the body's internal structures, aiding in disease diagnosis and treatment planning. In this context, Artificial intelligence (AI) and image processing represent a potential tool for clinical practice and disease detection due to its ability to analyze vast amounts of medical data rapidly, aiding healthcare professionals in making accurate diagnoses and improving patient outcomes. This study explores metaheuristic optimization in neuroevolution techniques to enhance image processing tasks, focusing on X-ray image classification for pneumonia detection. The proposed framework utilizes a pre-trained Convolutional Neural Network (CNN) and different metaheuristic algorithms for hyperparameters optimization for comparison purposes. The test is conducted by using a publicly available dataset, with normal and pneumonia classes. Experimental results demonstrate improved accuracy in distinguishing pneumonia-positive and normal X-ray scans, reaching. The study

D. C. Peña · O. Ramos-Soto (✉) · J. A. Galvis-Chacon
Departamento de Ingeniería Electro-Fotónica, CUCEI, Universidad de Guadalajara, Guadalajara, Jalisco, México
e-mail: oscar.ramos9279@alumnos.udg.mx

D. C. Peña
e-mail: diego.campos7896@alumnos.udg.mx

J. A. Galvis-Chacon
e-mail: javier.galvis8583@alumnos.udg.mx

B. A. Rivera-Aguilar
División de Tecnologías para la Integración Ciber-Humana, CUCEI, Universidad de Guadalajara, Guadalajara, Jalisco, México
e-mail: beatriz.rivera@academicos.udg.mx

A. Thakare
Pimpri Chinchwad College of Engineering, Pune, India
e-mail: anuradha.thakare@pccoepune.org

P. Majumdar
Techno College of Engineering Agartala, Maheshkhola, Agartala, India



D. Oliva et al. (eds.), *Advances in Optimization Algorithms for Multidisciplinary Engineering Applications: From Classical Methods to AI-Enhanced Solutions*, Studies in Computational Intelligence 806, https://doi.org/10.1007/978-3-031-78440-8_17

showcases the potential of neuroevolution combined with metaheuristic optimization for addressing complex image processing challenges in medical diagnostics, hence advancing AI-assisted disease detection systems.

## 1 Introduction

In recent years, the fusion of artificial intelligence (AI) and medical imaging has been fruitful, as it provides innovative approaches to address complex disease diagnostic challenges [1, 2]. Among these challenges, the accurate classification of X-ray images for detecting pulmonary diseases, such as pneumonia, is a critical task that plays an important role in public health [3]. Traditional machine-learning methods have made significant improvements in this regard by completing image classification tasks with great success. However, the inherent complexity and variability of medical images keep making their classification a challenge when testing these methods in different datasets [4, 5].

Neuroevolution is a novel research area where evolutionary algorithms optimize neural network architectures and their hyperparameters [6] for fine-tuned performance. Unlike manual design or random search methods, the neuroevolution algorithms approach generates and evolves populations to optimize neural networks iteratively, allowing them to adapt and obtain better performance each iteration. This evolutionary process results in the generation of neural architectures finely tuned to the task of image classification [7, 8].

On the other hand, metaheuristic optimization represents a powerful tool to complete this fine-tuning process [9], offering enhanced exploration and exploitation capabilities within the search space to obtain the best configuration parameters for a determined problem. Metaheuristic algorithms excel in navigating complex and high-dimensional problems by leveraging heuristic search strategies inspired by natural phenomena or human behaviors [10]. When integrated into the neuroevolution approach, these metaheuristic techniques complement the evolutionary process by providing efficient mechanisms for local refinement and global exploration [9, 10], by accelerating convergence towards higher-quality solutions.

Neuroevolution and metaheuristic algorithms offer promising hybridization for improving image processing tasks. Together, neuroevolution and metaheuristic algorithms present a powerful framework for addressing the complexity and variability of image data [11, 12], offering enhanced capabilities for tasks such as object recognition, segmentation, and classification.

This chapter introduces a comprehensive study for X-ray image classification using neuroevolution augmented with metaheuristic optimization for a pre-trained Convolutional Neural Network (CNN) fine-tuning, with a particular emphasis on distinguishing between pneumonia-positive and normal cases in X-ray scans. First, in Sect. 2 the state of the art is analyzed, where important advancements in the neuroevolution and metaheuristic optimization merges are mentioned. Subsequently, in

Sect. 3 preliminary concepts are detailed, providing the reader with a general theoretical background of elements considered for this proposal, including dataset description, performance metrics, and experimental setup. Then, in Sect. 5 a quantitative analysis is done to evaluate the performance of different metaheuristic algorithms in the optimization of a CNN for X-ray image classification. Last, the conclusions of this study are included in Sect. 6.

## 2 Related Work

In image processing, the development of neural networks (NN) [13] represents a great advance since it provides high-performance results in different applications. As these networks have been studied and improved, a particular type of network has stood out due to its capacity to automatically learn hierarchical features directly from raw pixel data: the Convolutional Neural Networks (CNNs). This network type excels at automatically learning hierarchical representations from raw input data, particularly well-suited for image processing tasks [14]. Key features include convolutional layers for extracting local patterns, pooling layers for downsampling and preserving essential information, hierarchical representation learning for capturing complex features, parameter sharing for efficient learning, and local connectivity for modeling spatial dependencies [15, 16]. These characteristics make CNNs highly effective for tasks like image recognition, object detection, and segmentation.

These networks and their variants have been applied to several image applications, such as extreme low-light image enhancement [17], pedestrian and vehicle detection [18], and facial recognition systems [19], to name a few. A special research topic where CNNs have presented great interest is in the medical area; these proposals have contributed with approaches for anatomical parameter quantification [20], teleophthalmology systems [21], and animal modeling studies [22], among others. Different approaches based on CNNs have been successfully developed regarding X-ray image analysis. In 2020, Wang et al. [23] presented a custom-built deep CNN model for COVID detection in chest X-ray images, also implementing an explainability module to provide decision transparency. In the same year, Rahimzadeh and Attar [24] introduced a hybridization of the Inception and ResNet50 CNNs to provide enhanced accuracy when compared to state of the art method. Later in 2022, Dey et al. [25] applied a transfer learning using DenseNet121, VGG19, and ResNet50 and type-1 Sugeno fuzzy integral for early detection of tuberculosis. More recently, in 2023, a CNN-based proposal for Knee osteoarthritis for 5-classes grading was presented by Sajaan et al. [26] with a pre-trained model based on Xception and InceptionResNet V2 networks.

On the other hand, metaheuristic algorithms have gained special attention for their remarkable efficiency in optimizing complex and non-linear problems [27]. These algorithms, drawing inspiration from natural processes such as evolution, genetic algorithms, and particle swarm optimization, offer robust and versatile optimization techniques [28]. They serve as versatile tools across diverse fields for optimization. For instance, they optimize engineering designs, logistics, and financial portfolios, maximizing efficiency while minimizing risks [29, 30]. Moreover, metaheuristic algorithms have proven their ability to improve signal processing and robotics control strategies, while aiding in bioinformatics tasks such as gene expression analysis [31]. Additionally, energy systems benefit from their optimization in power generation and distribution [32]. Recent years have seen the effectiveness of metaheuristic algorithms in various image processing tasks, including medical image analysis [33, 34]. Also, these techniques have been implemented for the optimization of CNN hyperparameters [35]. For example, a study on the optimization using swarm intelligence-based was carried out to enhance the optimization of numerical handwriting classification [36]. Another interesting approach recently implemented was for the optimization of hyperparameters of machine learning algorithms and deep neural networks for the classification in biomedical and biological cases, such as human skin images, cancer, molecular interactions, and clinical diagnosis, among many others [37].

The neuroevolution approach emerges as an important tool for the performance improvement of AI models by using both neural networks and evolutionary algorithms [38], generating an evolutive process of the neural network structures, hence the techniques' name. The neuroevolution-based models have shown promise in medical diagnosis applications, including breast cancer classification [39, 40], magnetic resonance imaging (MRI) and computed tomography (CT) 2D and 3D segmentation [8], and blood vessels segmentation [41], to name a few. Particularly in X-ray imagery, some proposals have been presented. Most of the works are focused on COVID-19 detection [42, 43]. Other approaches, such as the one introduced by Shukla et al. [44] in 2021. By using a pre-trained GoogLeNet and a 20-fold cross-validation, a multiobjective genetic algorithm is used to optimize network hyperparameters for COVID-19 diagnosis. Gülmez [45] proposed a more recent approach in 2023, where the use of the Xception network and genetic algorithm is combined for the detection of COVID-19 in X-ray images. The parameters considered for optimization include network depth, layer count, layer type, and parameters. In the same year and with the same disease analysis, Cai et al. [46] developed a hyperparameter optimization of a Deep Convolutional Neural Network (DCNN) by using the novel metaheuristic called Chimp Optimization Algorithm (ChOA). These works denote the strengths of the mix of CNNs and optimization algorithms, allowing for efficient performance improvement of CNNs architectures and their parameters, improving the accuracy and robustness of medical image analysis systems.

## 3 Preliminary Concepts

This section explores the fundamental concepts and methodologies behind CNNs and Optimization Algorithms. First, an overview of neural networks (NNs), which are based on ANNs and find their inspiration in the biological neuron, is given. Then, the architecture of a CNN is discussed, analyzing its key components and mechanisms such as Transfer Learning and Feature Extraction.

## 3.1 Neural Networks (NNs) Overview

Neural networks (NNs) [47] are a fundamental concept in machine learning and artificial intelligence. They are computational models inspired by the structure and functioning of the human brain, consisting of interconnected nodes (neurons) organized in layers. These networks are capable of learning from data and making predictions or decisions based on that learning.

CNNs [14] are a specific type of Artificial neural networks (ANNs) specially designed for processing structured data, such as images and temporal signals. They draw inspiration from the organization and functioning of the biological visual system, where neurons in the primary visual layers respond to specific regions of the visual field. CNNs primarily use convolution filters to detect local features in the input data and, through successive layers, combine these features to recognize complex patterns.

Both ANNs and CNNs are based on simulating the human brain's behavior by interconnecting artificial nodes (neurons) that process information. This idea dates back to the 1940s, when Warren McCulloch and Walter Pitts proposed a mathematical model of neurons, laying the theoretical foundation for neural networks [48]. However, the first practical implementations of neural networks emerged in the 1950s, with Frank Rosenblatt's perceptron, a simple single-layer neural network model. Despite its initial success in applications such as simple pattern recognition, the perceptron's limitations led to a decline in neural network research in the 1960s and 1970s [49].

The resurgence of neural networks occurred in the 1980s, when more advanced learning algorithms were developed, such as the backpropagation algorithm, which allowed the training of multilayer neural networks. This led to the development of CNNs in the late 1980s and early 1990s, with the pioneering work of Yann LeCun and others in the application of convolution filters in image processing [50]. Since then, CNNs have demonstrated exceptional performance in a wide range of computer vision applications, including object recognition, image segmentation, and video processing.

### 3.1.1 Artificial Neural Networks

Biomimicry, also known as biomimetics or biomimicry, is a scientific discipline that draws inspiration from nature to solve human problems by studying natural models and processes. This methodology is based on the observation and emulation of mechanical and chemical systems present in nature to create innovative technologies [51]. One of the most outstanding applications of biomimicry is the

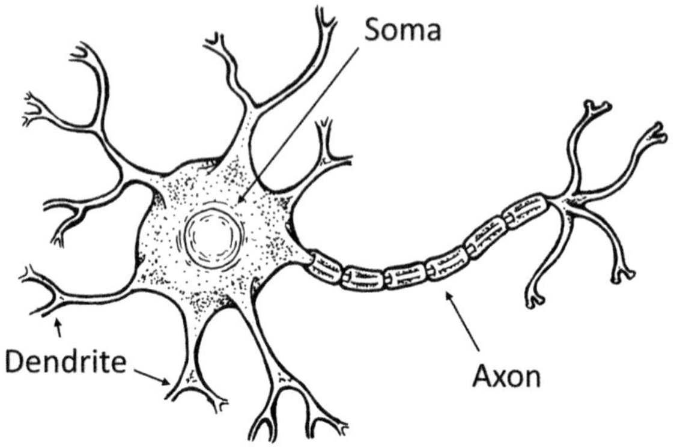


**Fig. 1** Diagram of a biological neuron based on the model proposed by Ramón y Cajal [53]

understanding and development of ANNs [52]. These networks constitute a computational model that seeks to solve problems like the human brain, although with greater abstraction.

It was Ramón y Cajal [53] who made the pioneering discovery of the cellular structure of the nervous system, identifying the fundamental unit known as the neuron. He defended the theory that neurons were interconnected in parallel with each other, as opposed to the previous belief that they formed a closed circuit similar to the blood system. A neuron consists of a cell body (also called the soma), as shown in Fig. 1. From the soma dense branching in the form of dendrites, and a tubular fiber called the axon, which can reach lengths of between 100 micrometers and one meter.

In a way, a neuron is a very simple information processor:

- Input channel: dendrites
- Processor: soma
- Output channel: axon

The connection between neurons is called a synapse, although it does not constitute a direct physical junction. These connections are unidirectional: information is transmitted electrically within the neuron itself and chemically between neurons, thanks to the action of specialized substances called neurotransmitters.

In its basic structure, an ANN is composed of units called artificial neurons, interconnected to transmit signals. Input information passes through the neural network, undergoing various operations, and produces output values. Each neuron is linked to other neurons by connections where the output value of the previous neuron is weighted by an assigned value called a weight. These weights in the connections can enhance or inhibit the activation of neighboring neurons. In addition, at the output of each neuron, there may be an activation function that modifies the result or sets a threshold before propagating to another neuron.

What is remarkable about neural networks is their ability to learn and self-train, as opposed to explicit programming. They excel in areas where solution or feature detection is difficult to express by conventional programming. To achieve this automatic learning, the minimization of a loss function that evaluates the overall performance of the network is employed. The weights of the neurons are iteratively adjusted by a process called backpropagation. This innovative approach continues to play a crucial role in the development of intelligent and efficient technologies.

### 3.1.2 Convolutional Neural Network

Convolutional Neural Networks (CNNs) have emerged as a fundamental tool in the field of computer vision and machine learning. Their ability to extract complex features from images has revolutionized numerous applications, from object recognition to medical image processing [14]. The key components of a CNN include convolutional layers, which apply filters to detect features like edges and textures, pooling layers that downsample feature maps to retain essential information, and fully connected layers that combine features for classification or regression tasks. In Fig. 2, a general scheme for the above-described CNN is presented.

CNNs operate by performing convolutions on input images using small kernels or filters. These convolutions involve taking groups of adjacent pixels from the input image and applying a scalar product with the kernel to produce a new output matrix [54]. Each kernel detects specific features such as edges, textures, or patterns in the image. These convolutional operations are mathematically represented as the dot product between the input matrix and the kernel matrix, followed by an element-wise summation and the addition of a bias term. In the context of color images, where each pixel has three color channels (red, green, blue), the kernels are three-dimensional, operating on each color channel independently and then combining the results to form a single output. This process allows CNNs to capture multi-channel features and patterns in images.

Furthermore, CNNs employ activation functions like Rectified Linear Units (ReLU) after convolutions to introduce non-linearity into the model. ReLU activation sets negative values to zero, aiding in feature extraction and model training by introducing sparsity and addressing the vanishing gradient problem. Subsampling

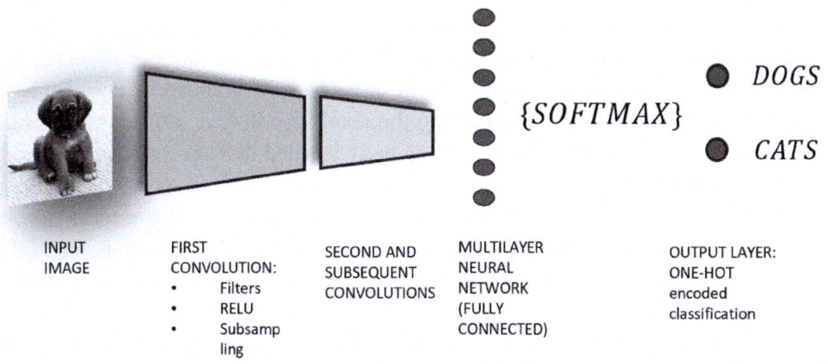


**Fig. 2** General CNN architecture

techniques such as Max-Pooling are then applied to reduce the dimensionality of feature maps while retaining important features. Max-pooling involves selecting the maximum value from a small window of pixels, effectively downsampling the feature map and preserving dominant features.

As CNNs progress through multiple convolutional and pooling layers, they become capable of detecting increasingly complex and abstract features, allowing them to learn hierarchical representations of input data. This hierarchical feature learning is crucial for CNNs to achieve high accuracy in tasks like image classification, where intricate patterns and structures need to be identified. The final layers of a CNN typically include fully connected layers followed by a Softmax activation function for classification tasks. Softmax converts the outputs of the network into probabilities, facilitating the interpretation of results as confidence scores for different classes. During training, CNNs use loss functions like categorical cross-entropy to optimize model parameters and minimize classification errors.

## 3.2 Transfer Learning

Transfer learning, a fundamental strategy in the field of machine learning, has revolutionized the approach to modeling problems in various real-world applications. This technique has become an important element in the development of new artificial intelligence approaches by allowing us to reuse the knowledge acquired in a specific task to address another related or similar task. The basic premise behind transfer learning lies in the recognition that models trained in one domain can be valuable for solving problems in another domain, even if the specific characteristics of the data differ [55]. In the context of knowledge transfer between neural networks, this practice involves taking a neural network pre-trained on one data set and adapting it for a different task. For example, one can take a CNN previously trained on a general image dataset, such as ImageNet, and adjust it to perform a specific task, such as classifying medical X-rays or detecting objects in satellite images machine learning [56].

By adjusting the weights and connections of the pre-trained neural network during a "retraining" process on data specific to the new task, the model can be adapted to capture relevant and useful features for the target task. This approach can also mitigate the risk of overfitting [57] by avoiding the need for a large amount of training data on the new task since the network has already learned useful representations of data in the source domain.

This approach offers several advantages, including a significant reduction in the time and resources required to train a model from scratch, as well as the ability to obtain highly accurate results even with small or sparse datasets. By enabling the use of pre-trained models as a starting point, the process of developing and deploying machine learning systems in a variety of practical applications [58] becomes easier.

From medical diagnostics to image recognition and natural language processing, transfer learning has proven its effectiveness in a wide range of fields and has paved the way for significant innovations in artificial intelligence and machine learning [59].

### 3.3 Feature Extraction

Feature extraction involves identifying and extracting meaningful patterns or characteristics from raw data, particularly in image processing [60]. In this context, it involves identifying and isolating distinctive patterns, textures, shapes, or other visual characteristics present within an image. When these features are correctly determined and filtered, they result in useful for subsequent analysis tasks such as classification, object detection, and segmentation [61]. There are several main approaches for image feature extraction, each with its advantages and applications. One approach is handcrafted feature extraction, where features are manually designed based on domain knowledge and expertise. Examples include Histogram of Oriented Gradients (HOG) for detecting object edges and textures, and Scale-Invariant Feature Transform (SIFT) for detecting key points and descriptors [62, 63]. However, handcrafted features may not generalize well across different datasets or applications, as they are specifically designed for a particular domain or problem.

To overcome this, the use of CNNs excels at feature extraction by automatically learning hierarchical representations directly from raw pixel data [64]. Initially, convolutional layers apply filters to detect specific features like edges or textures across the input image, generating feature maps. Activation functions introduce non-linearity, enabling the network to learn complex relationships between features. Subsequent pooling layers downsample the feature maps, retaining essential information while reducing spatial dimensions. Through stacking multiple layers, CNNs learn increasingly abstract and complex features, capturing high-level representations. Global average pooling may further reduce dimensionality before passing features to fully connected layers, which perform classification or regression tasks [65]. Utilizing this hierarchical approach, CNNs efficiently extract important features, obtaining information for further image processing tasks, such as classification.

### 3.4 Support Vector Machine (SVM)

Support Vector Machines (SVM) is a powerful approach in the field of machine learning, developed by Vladimir Vapnik and his team in the 1990s [66]. This theory has gained widespread popularity due to its ability to perform classification tasks on both linear and nonlinear data sets with high accuracy and computational efficiency [67]. SVM is based on the concept of finding the optimal hyperplane that maximizes the separation between different classes of data in a high-dimensional feature space. In other words, SVM searches for the best way to draw a decision boundary between different data sets to classify them effectively. This hyperplane is constructed in a way that maximizes the distance between the closest points in each class, known as support vectors.

The key to SVM's effectiveness lies in its ability to generalize well to new and unseen data sets. Through the optimization process, SVM finds the hyperplane that best fits the training data and minimizes the classification error. This allows SVM to perform well even on noisy or overlapping data sets between classes [68]. In addition, SVM offers different approaches to tackle multiclass classification problems, such as the One-vs-One (OvO) and One-vs-Rest (OvR) strategies, which convert the problem into a series of simpler binary classifications [69].

As mentioned, the SVM performs classification by creating a hyperplane in higher dimensions, known as a decision plane, as illustrated in Fig. 3. The center line represents the maximum margin hyperplane. In other words, the boundary line that separates the two classes at a maximum distance from the nearest data point is chosen. For two-class problems, the decision boundary or classification hyperplane is defined by Vapnik's theory [70]. If the input set is $x_i$, with dimension $mxn$, and the corresponding class is $y_i$ in $\{-1, +1\}$.

The SVM algorithm operates by creating a decision plane, or hyperplane, in higher dimensions to separate data points of different classes effectively. This decision plane is represented by a center line that maximizes the margin between classes, ensuring optimal separation. Support vectors, which are crucial data points defining the decision boundary, guarantee a wide margin between classes in the decision plane, enhancing the model's classification accuracy [69]. The concept of a "soft margin" introduced by SVM allows for some flexibility in accommodating noisy or outlier data points within the margin or on the wrong side of the hyperplane without significantly compromising classification performance [71]. To solve the optimization problem of selecting the optimal hyperplane, SVM employs the Lagrangian technique and saddle points to find the best solution efficiently.

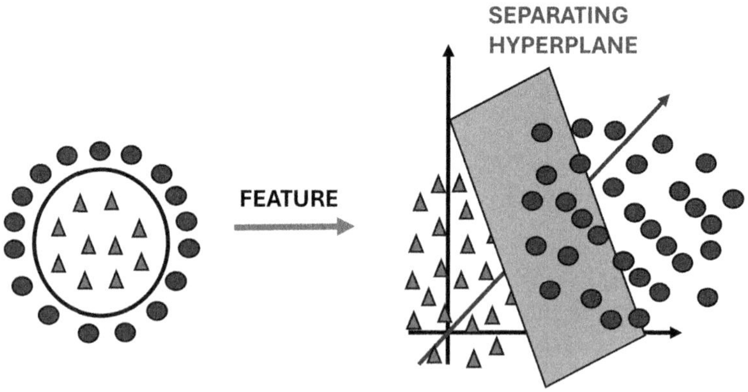


**Fig. 3** Separating Hyperplane and decision space [69]

Moreover, SVM utilizes kernel functions to handle nonlinear data by mapping it to higher-dimensional feature spaces. These functions transform the original data into spaces where linear separability is more achievable, enhancing classification accuracy in both linear and nonlinear problems. Among the most commonly used kernel functions are:

- Linear kernel function: this is a basic function that performs a dot product between input data.
- Polynomial kernel function: This function is nonlinear and directional, meaning that the output depends on the direction of the input in a low dimensional space.
- Radial basis kernel function (RBF): Widely used in SVM, this function selects smooth solutions.
- Sigmoidal function: Uses the hyperbolic tangent and is another option for SVM.

Since each kernel function has its limitations and is best suited to certain data sets, research is underway to develop new kernel functions to improve model efficiency and accuracy. The design of a universal kernel function is a constantly evolving area of research within the machine learning field.

The combination of maximum margin hyperplane selection, the concept of soft margin, and the use of kernel functions make SVM a powerful and versatile tool in the field of machine learning for data classification in both linear and nonlinear problems.

### *3.5 Optimization Algorithms*

Optimization is essential in many disciplines, always with the motivation and objective of delivering quality solutions for specific problems [72]. A good way to see it is as the search for the best possible combination of decision variables or characteristics involved for a desired result; in other words, the best possible solution among a combination of actions or situations to solve something.

A problem to optimize could be defined and formulated as follows, where the best solution is sought from a set of solutions that minimizes or maximizes an objective function $f(x)$:

$$\begin{aligned} &\text{Minimize or Maximize } f(x), \quad x = (x_1, \ldots, x_d) \in R^d \\ &\text{Subject to:} \qquad\qquad\qquad\qquad x \in X, \end{aligned} \tag{1}$$

where $x$ corresponds to the vector with the decision variables, $d$ is the problem's dimensionality, and $X$ is the set of candidate solutions.

This area of study is observed and applied to many problems in several engineering problems and everyday life, hence the importance of investigating, proposing, and developing new techniques and methods that lead to obtaining an optimal result. Always looking for the shortest path, wasting the least possible time, more profit, less

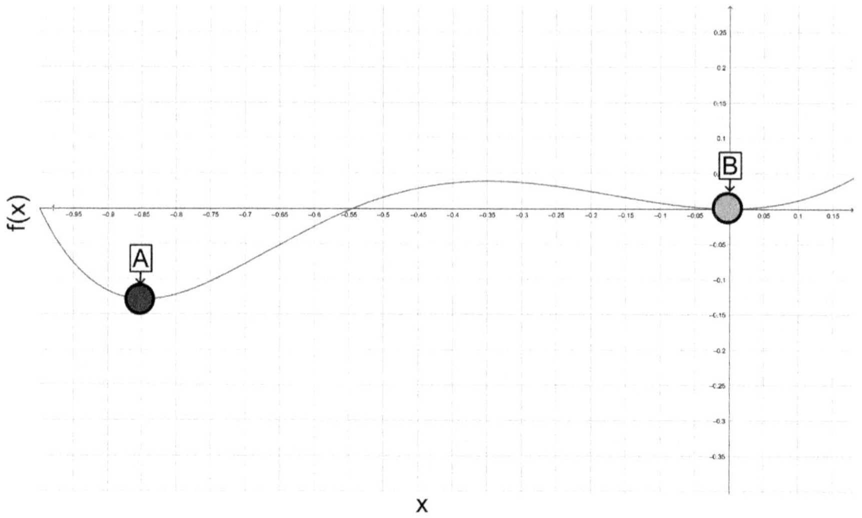


**Fig. 4** Graphical representation of a one-dimensional optimization problem

loss, any optimization has a thing in common: the best solution is sought, regardless of whether it seeks to decrease or increase the result, in other words, maximize or minimize [72].

It is important to understand these last two concepts since this is the basis for the type of search that will have to be done to find the best solution. For this purpose, the following problem is presented graphically in Fig. 4, seeking to minimize an objective function, considering a single decision variable, $d = 1$. In this figure, two solutions in different locations can be seen, called solution $A$ and solution $B$, and each of these has a fitness given by the objective function. Since it was previously defined that what is sought is to minimize, the best solution between these two candidate solutions is solution $A$, which offers a lower value than the other solution evaluated in the cost function.

Now that the foundations have been established, it is important to understand the methods used to solve optimization problems. These methods are divided into two classes for this section: classical and metaheuristic algorithms [73]. Specific metaheuristic algorithms will be explained in more detail.

Classical optimization methods include gradient descent, Newton-Raphson, secant, linear search, binary search, and random search [74]. These methods are simple to understand and easy to implement, which makes them accessible to a larger number of users. They are also very efficient in simple problems; some can be scaled to higher dimensionality problems guaranteeing convergence and low memory consumption. However, the algorithm's simplicity also means it has clear limitations, making it unsuitable for certain optimization problems [75]. For instance, when using gradient descent, the objective function must be twice derivable and sensitive to local minima. Another requirement is that the objective function is unimodal. It is important to note these limitations.

Given the complications mentioned above, it is very complicated to put these methods into practice since real optimization problems do not usually present those requirements to use classical methods; it is very common to find problems that present multimodal behavior, which leads the classical method to fall into local optima (as can be seen in point $B$ of Fig. 4).

Unlike classical methods, such as gradient descent, metaheuristic algorithms do not need information on the derivative of the objective function but use heuristic rules to construct search patterns, which often simulate natural behaviors. Metaheuristic algorithms are sometimes called population-based optimization, consisting of a set of candidate solutions that work together to optimize a problem.

Next, this section will explain some of the metaheuristic algorithms that are considered relevant and have had a significant impact on the state of the art. These algorithms were also used as a tool for hyperparameter optimization in the CNN to enable a comparison between different algorithms.

### 3.5.1 Particle Swarm Optimization (PSO)

There is much to learn from nature, where it is clear that there is collective intelligence, for example, animals living in groups. These groups are created to facilitate many activities, such as protection from predators or dangers, traveling long distances, or even finding food. Many of these things would not be possible if each particle were on its own, but are achieved through intelligence as a group when there is a goal, such as survival.

The Particle Swarm Optimization (PSO) algorithm [76] is based on the group intelligence described above, where a group collaboration is sought, and all particles have the same goal but are always looking for the best for each particle. This type of behavior is common in animals but also human behavior. For example, when humans seek to collaborate, they always have an overall goal but also seek to improve individually.

The particle swarm is given by a group $X$ of $N$ particles (solutions). Each of these particles is represented by $x_i$ so that the solutions group can be represented by $X = \{x_1, x_2, ..., x_i, ..., x_N\}$, which will have a cooperative behavior throughout the iterative process/generations. Each of these particles/solutions will move determined by the past position (local influence) and by the information of other particles (global influence).

- **Initialization**

Each of the N solutions is randomly initialized for a population $X$ such that $X = \{x_1, x_2, ..., x_N\}$. Each particle with $d$ characteristics (depends on the dimensionality of the shape problem) is given by $x_i = [x_{i,1}, x_{i,2}, ..., x_{i,d}]$. For these characteristics, it is necessary to define lower bounds ($lb$) and upper bounds ($ub$) given by the same problem; also, it has to be considered that each bound can be the same or different for each dimension $j$. The following formula describes the way to initialize:

$$x_{ij} = lb_j + rand(ub_j - lb_j) \tag{2}$$

where $x_{ij}$ corresponds to the $i$ solution in the $j$ dimension, $rand$ is a random number in a uniformly distributed distribution of 0 and 1 and $lb_j$ and $ub_j$ are the bounds in the $j$ dimension given by the problem.

- **Particle Velocity**

The particle motion for this algorithm, the search strategy, is given by a velocity $v_i$, which needs an initial value and, in its simplest form, is initialized with a value of zero so that $v_i = 0$. This velocity changes in the evolution process, and the information on the best particle globally and the best particle locally is considered. The global best refers to the best solution found through all iterations, while the local best refers to the best solution concerning the current $k$ iteration.

The speed update is determined based on the influence of two factors known as cognitive and social. The cognitive factor affects the local, while the social factor affects the global influence. This velocity for the $i$th particle is calculated as shown in Eq. 3.

$$v_i^{k+1} = v_i^k + c_1 \cdot \left(r_1^k \cdot \left(p_i^k - x_i^k\right)\right) + c_2 \cdot \left(r_2^k \cdot \left(g^k - x_i^k\right)\right) \tag{3}$$

where $p^k$ represents the local best while the global best is represented by $g^k$, the cognitive and social coefficients, also called acceleration constants, are represented by $0 < c1, c2 < 4$ and are in charge of modulating the magnitude of steps towards which the particle $x_i$ is directed. On the other hand, $r_1$ and $r_2$ are two randomly defined d-dimensional vectors of a random distribution with values in [0, 1].

- **Particle movement**

The movement of the particle $x_i$ through the search space is given by Eq. 4, where the calculated velocity $v^{k+1}$ is used in Eq. 3 and its last position in the search space. Each of the particle positions is updated by this same procedure given by the following expression:

$$\mathbf{x}_i^{k+1} = \mathbf{x}_i^k + \mathbf{v}_i^{k+1} \tag{4}$$

### 3.5.2 Differential Evolution (DE)

In 1997, Storn and Price proposed a solution to the Chebyshev polynomial fitting problem using Genetic Annealing [77], building on Price's earlier research. Although the algorithm solved the problem, Price noted that it did not perform as well as other optimization techniques. The proposal is based on a vector calculus perspective and evolutionary computation combination [78]. The generation of new solution vectors

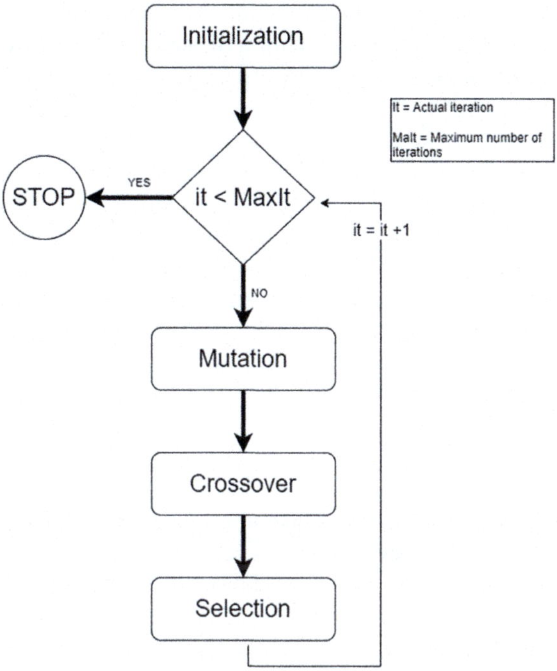


**Fig. 5** DE flowchart

is achieved by subtracting two solution vectors and adding a third one from the same population of solutions. This operation is called differential mutation (which will be explained later). This algorithm was developed following the described scheme, resulting in a proposal that outperforms other proposals in solving problems with high computational cost and ease of implementation.

Like many other algorithms, it has an iterative process where the position of the vectors changes as the optimization process progresses. The process can be summarized in four steps: Initialization, Mutation, Crossover, and Selection, as shown in Fig. 5.

- **Initialization**

Initialization is critical for any multi-population algorithm because it defines the bounds. These bounds are represented by vectors representing the lower and upper bounds of each dimension of the problem to be solved. These lower and upper bound vectors and are represented by $lb = [lb_1, lb_2, ..., lb_d]$ and $ub = [ub_1, ub_2, ..., ub_d]$. Once these bounds are defined, the first solutions are initialized. There are many initialization methods, but one of the most common and used in metaheuristic algorithms is the random initialization given by Eq. 5. This equation assigns a random value of the established bounds to each solution.

$$x_{nd} = lb_d + rand(ub_d - lb_d) \tag{5}$$

where $n$ is the number of solutions to be initialized, $d$ is the number of dimensions of the problem to be optimized, and $rand$ is a random number from a uniform distribution of values between 0 and 1.

- **Mutation**

After the initial solutions are available, the mutation operator generates new solutions by combining elements of the population to create a new candidate solution. This operator promotes the exchange of information between individuals in the population and is, therefore, also known as differential mutation. Numerous proposals have been made for modifying the operator in the literature and publications on DE. This is because the operator is both easy to understand and highly effective. The generic form of the operator is described below, along with some of the modifications found in the literature.

**DE rand/1 mutation**

$$v_k = x_{r_3,k} + F \cdot \left(x_{r_1,k} - x_{r_2,k}\right) \tag{6}$$

As mentioned, this operator aims to create new solutions by exchanging information. The new solution is referred to as the donor/mutant vector, denoted as $v$. To generate $v$, take the weighted difference of two vectors and add it to a third vector randomly selected from the same population, as shown in Eq. 6. It is important to note that $x_{r1}$, $x_{r2}$, and $x_{r3}$ are randomly selected from the population and must be distinct, considering that each of these vectors has to be different so that

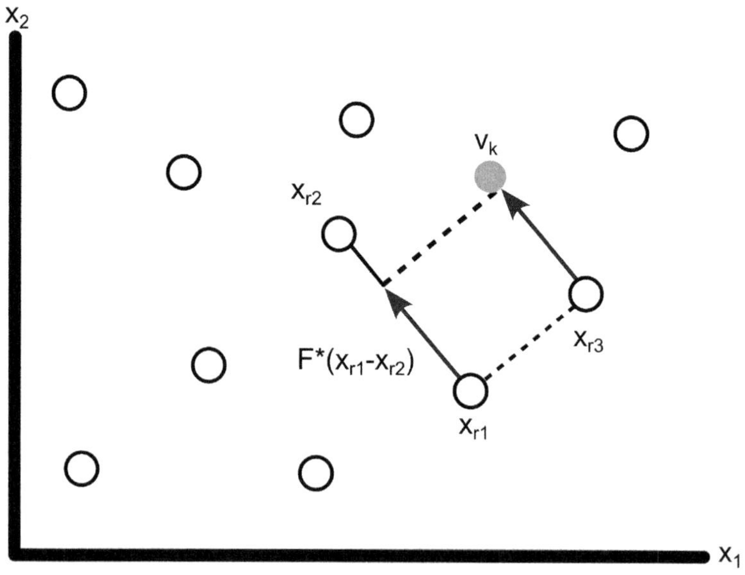


**Fig. 6** Graphical representation for generic mutation operator

$r_1, r_2, r_3 \in \{1, 2, \ldots, n\}$ and $r_1 \neq r_2 \neq r_3 \neq i$, where $n$ is the total number of solutions, $i$ represent an integer number referring to the selected solution vector of the population and $k$ the current generation. This process is represented graphically in Fig. 6.

**DE best/1 mutation**

$$v_k = x_{best,k} + F \cdot \left(x_{r_1,k} - x_{r_2,k}\right) \tag{7}$$

In this strategy, as in the previous one, random individuals are taken from the population, however, in this strategy, only two are randomly selected for the weighted subtraction $x_{r1}$ and $x_{r2}$ where $r_1, r_2 \in \{1, 2, \ldots, n\}$ and $r_1 \neq r_2 \neq i$, and instead of taking a third one to which this subtraction would be added, the global best $x_{best}$ is taken into account as shown in Eq. 7. This strategy is intended to concentrate the solutions in a promising area creating very good solutions. Given the good exploitation characteristics of this strategy, it can cause premature convergence, loss of diversity, and a higher probability of falling into a local optimum

**DE rand/2 mutation**

$$v_k = \mathbf{x}_{r_5,k} + F \cdot \left(\mathbf{x}_{r_1,k} - \mathbf{x}_{r_2,k}\right) + F \cdot \left(\mathbf{x}_{r_3,k} - \mathbf{x}_{r_4,k}\right) \tag{8}$$

The strategy described by Eq. 8 could be considered an extension of the rand/1 strategy in Eq. 6 since a new pair of randomly taken vectors ($r_4$ and $r_5$) has been added. This strategy aims to improve diversity and cause a greater perturbation since more individuals are taken into account, which means also taking into account more characteristics of the individuals, which helps in high dimensionality problems.

**DE best/2 mutation**

$$v_k = \mathbf{x}_{best,k} + F \cdot \left(\mathbf{x}_{r_1,k} - \mathbf{x}_{r_2,k}\right) + F \cdot \left(\mathbf{x}_{r_3,k} - \mathbf{x}_{r_4,k}\right) \tag{9}$$

The strategy presented in Eq. 9 combines the $rand/2$ and $best/1$ strategies to take advantage of their respective strengths. $best/1$ is excellent at exploiting promising areas, while $rand/2$ enriches diversity to aid exploration. The aim is to compensate for and strengthen the weak points of each strategy.

- **Crossover**

The purpose of this operator is to combine the mutant vector $v_k$ and the individual in question from population $x_i$ to increase diversity. This process of combining both vectors is done element by element to obtain the test vector $u$ from this combination finally. A graphical representation of this process can be seen in Fig. 7. To perform this process, two conditions are sought to be fulfilled following the scheme:

$$u_j^{(k)} = \begin{cases} v_j^{(k)} & \text{if } j = j_0 \text{ or rand}[0, 1] \leq Cr \\ x_{i,j}{}^{(k)} & \text{otherwise,} \end{cases} \tag{10}$$

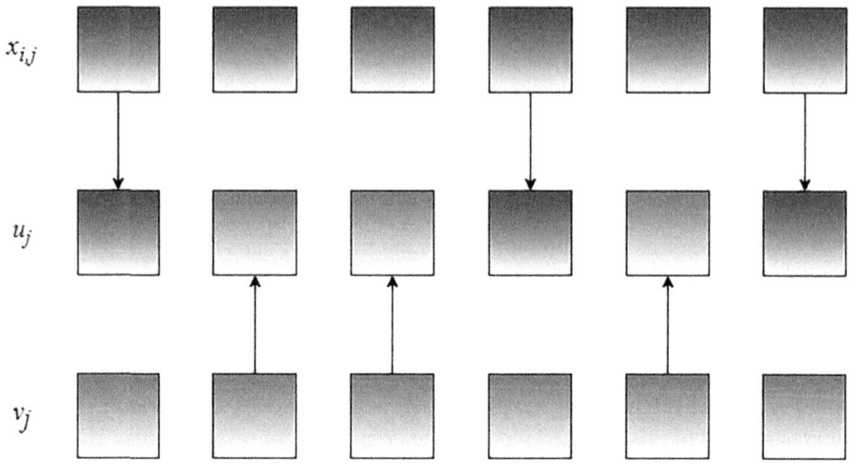


**Fig. 7** Graphical representation for crossover operator

where $j_0$ is one of the positions of the $ith$ solution vector, rand is a random value between 0 and 1 within a uniformly distributed distribution, which is compared with a $CR$ parameter. This parameter is known as a probabilistic factor that indicates when the value of the mutant vector is taken, and this is one of the parameters that must be defined at the beginning and is a value between 0 and 1; it is called the Crossover Rate.

- **Selection**

This operator seeks to maintain the best solutions within the population. This is done by subjecting the resulting test vector $u_k$, after having executed the mutation and crossover operators, to a quality criterion where it is considered to replace the current solution vector $x_i$ as long as it is a better solution.

This quality criterion considers the fitness of the test vector $u_k$ and the fitness of the current solution vector $x_i$. The rule is as follows: If the fit of the test vector $f(u_k)$ is less than the fit of the current solution vector $f(x_i)$, the test vector $u_k$ replaces the current solution vector (or individual) $x_i$; otherwise, the individual $x_i$ remains in the population. This scheme is shown below (Eq. 11):

$$x_i^{(k+1)} = \begin{cases} u_k & \text{if } f(u_k) \leq f(x_i) \\ x_i & \text{otherwise} \end{cases} \tag{11}$$

### 3.5.3 Grey Wolf Optimizer (GWO)

This algorithm, GWO [79], is inspired by the gray wolf (Canis Lupus). This species is at the top of the food chain and usually lives in groups of 5–12 wolves on average. Also, within the same group, a strict social order is maintained.

The leaders are called alphas; this Alpha is the main decision maker, from where to sleep to hunting. Although there are more hierarchies, the Alpha's decisions must be followed by the whole pack [80], even though the Alpha is not necessarily the strongest wolf in the pack. This shows that strength is not the most important thing in the group, but how the group is managed.

The second level of authority in the pack is the beta. The beta is mostly in charge of assisting the Alpha in decision-making. Also, this beta is the best candidate for the Alpha in case the Alpha needs to be replaced for some reason. The beta has to follow the Alpha's orders, but the others are under him.

Another rank among wolves is the omega. The omega is the lowest rank among wolves and is often used as a scapegoat. Although it may seem that they do not have a position or task relevant to this rank, it has been found that when this rank is missing, the group suffers differently. When a wolf does not belong to any of the ranks mentioned above (alpha, beta, omega), it is called subordinate or sometimes delta. These are below the Alpha and beta but above the omega. Scouts, hunters, elders, sentinels, and caretakers belong to this group of wolves.

- **Social hierarchy**

To mathematically model these animals' behavior and hierarchical rank, the best solution will be Alpha ($\alpha$). Consequently, the second and third solutions will be assigned as beta ($\beta$) and delta ($\delta$). The rest of the population will be referred to as omega ($\omega$). By defining this model and following the group behavior, Alpha, beta, and delta will guide the $\omega$ solutions.

- **Encircling prey**

How a group of wolves hunt their prey by rounding them up and enclosing them. To define this behavior, the following equations are defined:

$$D = \left|C \cdot X_p(k) - X(k)\right| \tag{12}$$

$$X(k+1) = X_p(k) - A \cdot D \tag{13}$$

where $k$ is the current iteration, $A$ and $C$ are coefficient vectors, $X_p$ is the position of the prey, and $X$ corresponds to the position of the wolf in question. The coefficient vectors are calculated as follows:

$$A = 2a \cdot r_1 - a \tag{14}$$

$$C = 2 \cdot r_2 \tag{15}$$

where $a$ decreases linearly from 2 to 0 throughout the iterative process, $r_1$ and $r_2$ are vectors with random values between [0, 1] in a uniform distribution.

- **Hunting**

The hunting strategy of the wolves is to use their ability to locate the prey and corner or enclose it, always guided by the $\alpha$, while the $\beta$, and $\delta$ participate in the hunt occasionally. In terms of optimization, the prey will be the optimum, so in a search space, it is unknown where this optimum is. To mathematically simulate this housekeeping behavior, it is assumed that the $\alpha$, $\beta$, and $\delta$ have better knowledge about the possible position of the prey. Therefore, the three best solutions are reserved, and the position of the other particles is updated according to the position of these three best solutions, as shown in the following formulas:

$$D_\alpha = |C_1 \cdot X_\alpha - X|, D_\beta = \left|C_2 \cdot X_\beta - X\right|, D_\delta = |C_3 \cdot X_\delta - X| \quad (16)$$

$$X_1 = X_\alpha - A_1 \cdot (D_\alpha), X_2 = X_\beta - A_2 \cdot \left(D_\beta\right), X_3 = X_\delta - A_3 \cdot (D_\delta) \quad (17)$$

$$X(t+1) = \frac{X_1 + X_2 + X_3}{3} \quad (18)$$

In Fig. 8, it is visually presented how the particles change their position considering the three best solutions. It can also be seen that the new position will be random but within the circle defined by these three best particles. In other words, the three best ones estimate the position of the prey, and the others adjust their position with that reference.

- **Attacking prey (exploitation)**

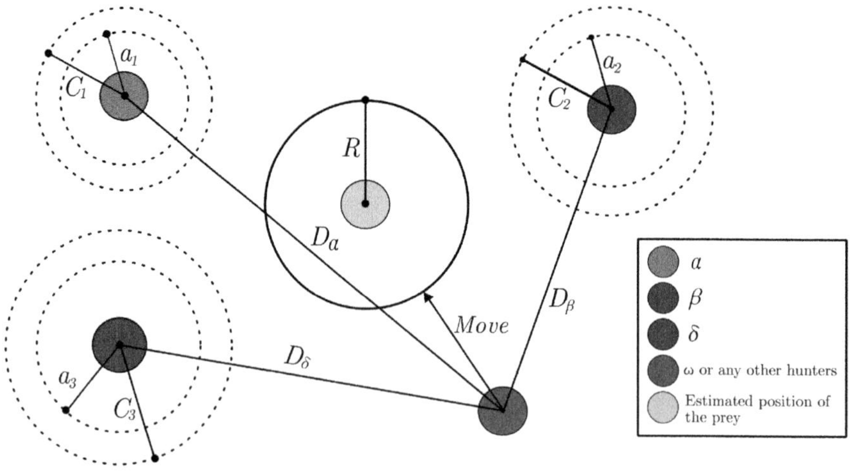


**Fig. 8** Wolf position update diagram

The way to end a hunt is to attack the prey when it is no longer moving. To model this prey approach, $a$ decreases, and consequently, $A$ decreases. Otherwise, a is $a$ random value in the interval $[-2a, 2a]$ where a decreases from 2 to 0. When there are random values in $[-1, 1]$, the next position of the particle can be any position between the current position and that of the prey so that if $|A| < 1$, the wolf is forced to attack the prey. With these operators, the algorithm can move the particles based on $\alpha$, $\beta$ and $\delta$. However, implementing another operator is necessary to avoid stalemates in local optima and do more exploration work.

- **Search for prey (exploration)**

The wolves search according to the three commands ($\alpha$, $\beta$, and $\delta$) to diverge from each other to search and converge to attack the prey. To model divergence, $A$ with random values greater than 1 and less than $-1$ is used to force search. Another component that favors search is $C$, and as can be seen in Eq. 16, this vector contains random values ranging from 0 to 2 and thus provides random weights to emphasize or de-emphasize the prey.

### 3.5.4 Whale Optimization Algorithm (WOA)

The WOA is a bio-inspired optimization algorithm [81], meaning that it uses some biological behavior as its optimization strategy, in this case, the behavior of whales. These animals have demonstrated great capabilities throughout research, such as judgment, emotions, and social behaviors. According to Hof and Van Der Gucht [82], whales have brain cells similar to humans, hence the intelligence of these animals.

This algorithm is based on humpback whales since this species has a peculiar hunting technique [83]. Their favorite prey are schools of krill and small fish near the surface, and it has been discovered that the hunting technique of these whales involves creating very distinctive bubbles in the shape of a circle when viewed from above. However, it was not until 2011, when Goldbogen [84] investigated this behavior in more detail using contact sensors, which helped to discover that these bubbles were not circular-shaped but a spiral shape, driving the fish towards the center of this spiral and to the surface.

- **Encircling prey**

Whales can know the location of their prey and enclose them with this bubble spiral mentioned above. Since the optimum is unknown in an optimization problem, this algorithm takes the best particle as the prey. After that, the other particles will adjust their position as shown below:

$$D = |C \cdot X_{best}(t) - X(t)| \tag{19}$$

$$X(t+1) = X_{best}(t) - A \cdot D \tag{20}$$

where $k$ indicates the current iteration, $A$ and $B$ are coefficient vectors, $X_{best}$ is the best solution found, and $X$ is the position vector. The coefficients $A$ and $B$ are calculated as follows:

$$A = 2a \cdot r - a \tag{21}$$

$$C = 2 \cdot r \tag{22}$$

where $a$ is linearly decreasing from 2 to 0 throughout the iterative process, and $r$ is a vector with random values between 0 and 1 in a uniform distribution.

- **Bubble-net attacking method (exploitation phase)**

To mathematically model this behavior, two approaches are given:

1. Circle shrinkage: This behavior is carried out by decreasing the value of $a$ as in Eq. 21. It is important to note that this decrease of $a$ affects $A$; therefore, a is a random value in the interval $[-a, a]$. So, if the values of $A$ are set to $[-1, 1]$, the new position of this particle will be defined anywhere between its current position and the position of the best particle found.
2. Spiral position update: This approach first calculates the distance between the whale ($X$) and the prey ($X_{best}$), then uses a spiral equation to create the helix shape based on these positions. This behavior is represented as shown below:

$$X(t+1) = D' \cdot e^{bl} \cdot \cos(2\pi l) + X_{best}(t) \tag{23}$$

where $D'$ represents the distance between the whale and the prey, $b$ is a constant to define the spiral shape, and $l$ is a random number between $[-1, 1]$ in a uniform distribution.

Given these approaches, it is noted that the whales swim around the prey in a circle and, in turn, in a spiral fashion simultaneously. To simulate this behavior, it is defined that there is a 50% probability of choosing one of these two approaches. This choice of approach is represented as shown below:

$$X(t+1) = \begin{cases} X_{best}(t) - A \cdot D & \text{if } p < 0.5 \\ D' \cdot e^{bl} \cdot \cos(2\pi l) + X_{best}(t) & \text{if } p \geq 0.5 \end{cases} \tag{24}$$

where $p$ is a random number in a uniform distribution with a value between 0 and 1, representing the probability.

- **Search for prey (exploration phase)**

The same approach of varying $A$ can be used for the prey search method. It is clear that humpback whales randomly search for prey, but this search is collaborative, communicating with each other. Random values will be assigned to $A$ greater than 1 and less than $-1$ to represent this mathematically and force the search. Unlike the exploitation phase, the positional adjustment of the whales is done by selecting an individual at random rather than the best. This representation is shown below:

$$D = |C \cdot X_{\text{rand}} - X| \tag{25}$$

$$X(t+1) = X_{\text{rand}} - A \cdot D \tag{26}$$

## 4 Proposed Framework

The proposed framework combines two powerful techniques: a pre-trained CNN by using a transfer learning technique and metaheuristic optimization to improve image classification. The CNN selected is the MobileNetV2 [85] to accurately detect pneumonia in X-ray scans, by improving its performance through the four metaheuristics described in the previous section: PSO, DE, WOA, and GWO. The selection of these four algorithms was based in their performance and relevance in the last time. While the PSO and DE are two of the classical algorithms that have had great relevance to motivating researchers to improve and develop great proposals based on these algorithms [86]. On the other hand, GWO and WOA are two algorithms very competitive and more recent on the area of bio-inspired metaheuristic algorithms.

In general, this scheme operates as follows: X-ray scans are initially processed using the pre-trained MobileNetV2 for feature extraction, which is followed by normal and pneumonia image classification based on an SVM. Subsequently, this process is iteratively repeated to optimize hyper-parameter values through metaheuristic optimizers. The stages and parameters of this process are summarized in Fig. 9, and they are detailed below.

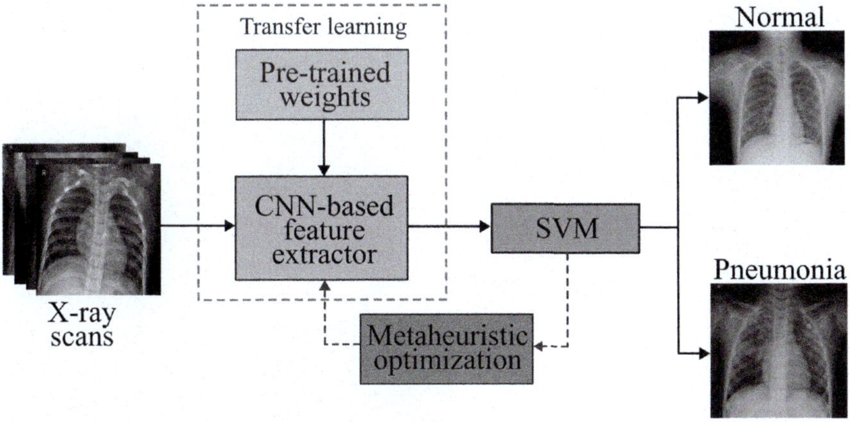


**Fig. 9** General flowchart of the proposed method

## 4.1 Dataset Description

For this proposal test, a pediatric chest X-ray dataset is used. It is composed of 5,863 JPEG format images categorized into "Pneumonia" and "Normal", distributed as shown in Fig. 10. They are sourced from the Guangzhou Women and Children's Medical Center [87, 88]. Each image, obtained from pediatric patients aged one to five years, underwent meticulous quality control to ensure readability and reliability. Expert physicians meticulously graded the diagnoses for each image, with consensus required before inclusion in the dataset. Additionally, a third-party expert evaluated the evaluation set to minimize potential grading errors. This dataset provides a comprehensive resource for training and evaluating artificial intelligence systems aimed at automating the diagnosis of pediatric chest conditions. In Fig. 11, examples of x-ray images of this set are shown.

## 4.2 CNN, SVM, and Optimizers Configuration

To evaluate the proposed framework, image partition is done allocating 70% of the dataset for training, 20% for testing, and the remaining 10% for validation samples. The PyTorch library along with Python version 3.9 was employed for the training process of the CNN model. The hardware used to run the simulations included a 12 GB RTX 3060 GPU on a Windows 10 64-bit operating system machine. In the experiments, to ensure a fair comparison, the population size of the metaheuristic algorithms used was set to 5 search agents, with 5 iterations and 5 runs. Importantly, the SVM parameters in the proposed method were tuned by a search process to find their optimal values The hyperparameters of the proposed SVM model are shown in Table 2. Additionally, three additional hyperparameters: learning rate ($Lr$),

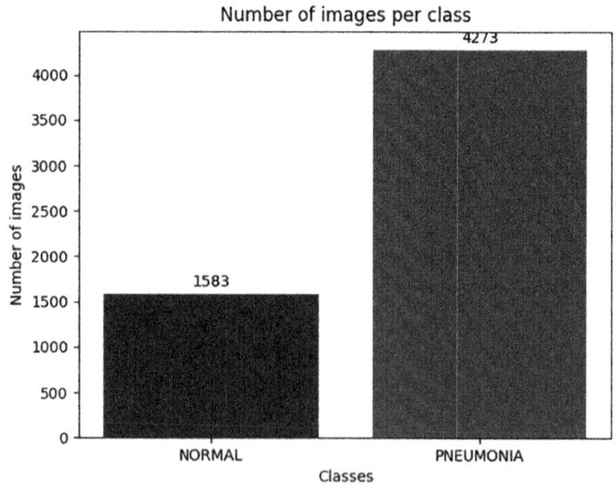


**Fig. 10** Number of images for the normal class and for the pneumonia class present in the dataset

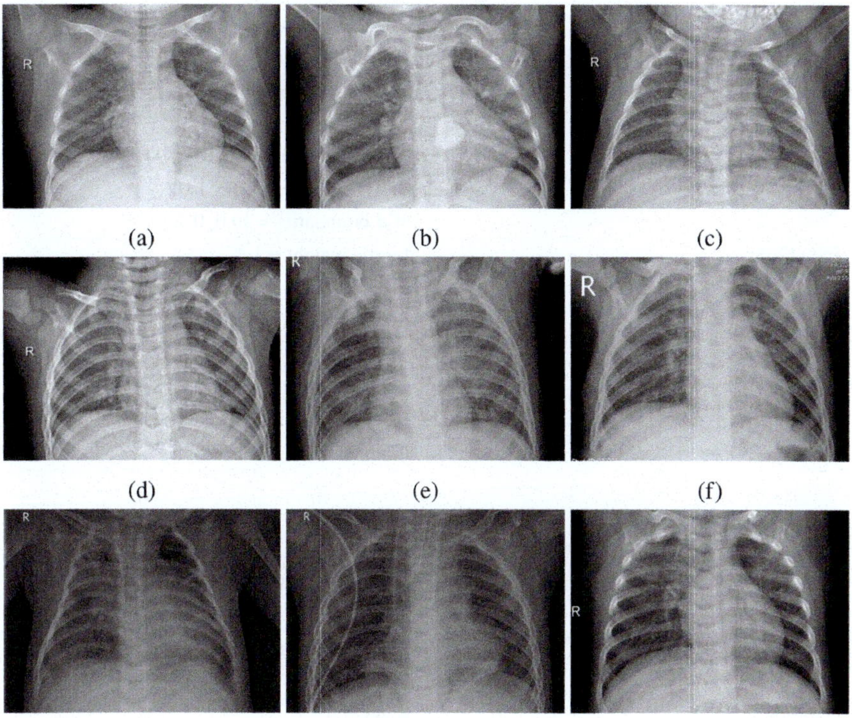

(a) (b) (c) (d) (e) (f) (g) (h) (i)

**Fig. 11** Examples of dataset images. **a–d** Normal scan **e–i** pneumonia scan

momentum rate ($Mr$), and weight decay ($Wd$) were optimized by PSO, GWO, DE, and WOA algorithms to evolve CNN structures. These parameters are detailed in Table 1. The input parameters of these optimizers were set according to the optimal values reported in their respective papers.

This study introduces the use of the MobileNetV2 [85] CNN in the medical domain, specifically for the classification of chest X-rays between normal patients and those with pneumonia. MobileNetV2 is a CNN architecture specially designed for applications in mobile devices and environments with limited computational resources. It introduces innovative design elements like depthwise separable convolutions, inverted residuals with linear bottlenecks, and efficient channel attention mechanisms. These features collectively contribute to reducing computational complexity and model size while preserving the network's ability to capture intricate features crucial for accurate classification [89]. Its optimized structure enables fast and efficient image processing, making it ideal for medical image classification tasks.

To adapt MobileNetV2 to the classification problem of chest radiographs, transfer learning is employed. This strategy utilizes the prior knowledge acquired by

**Table 1** General and particular parameter settings for the optimization process

| Algorithm | Parameter settings |
|---|---|
| General settings | Population size = 5,<br>Maximum iterations = 5<br>$Lr$ boundaries = [0.0001, 0.1]<br>$Mr$ boundaries = [0.0, 1.0]<br>$Wd$ boundaries = [0.0, 0.1] |
| PSO [76] | Inertia weight = 1.4,<br>Personal acceleration coefficient = 1.4,<br>Social acceleration coefficient = 0.4 |
| DE [77] | Scaling factor = 0.7,<br>Crossover rate = 0.9 |
| GWO [79] | Parameter set is unnecessary |
| WOA [81] | Parameter set is unnecessary |

**Table 2** Model SVM hyperparameters

| Parameter | Assigned value |
|---|---|
| C | 1.0 |
| Kernel | Radial basis function |
| Degree | 3 |
| Gamma | scale |
| Tol | 1e-3 |
| Cache size | 200 |
| Decision function shape | OvR |

MobileNetV2 during its training on a vast and diverse dataset like ImageNet, consisting of millions of images across various classes. By initializing MobileNetV2 parameters with pre-trained ImageNet weights, the model gains knowledge about general visual features, expediting the training process and enhancing its ability to generalize to new data. The transfer learning process involves fine-tuning the pre-trained MobileNetV2 architecture by adjusting the weights of its final layers to suit the task of classifying chest radiograph images into normal and pneumonia categories. This adaptation occurs through retraining the model with a labeled dataset containing chest radiographs, allowing the network to learn specific features crucial for discerning between normal and pneumonia-afflicted images.

### 4.3 Performance Metric

The metric that will be used as the objective function is the Accuracy. This metric is widely used in Machine Learning to evaluate the models. It provides the proportion of correct predictions over the total of samples, as is shown in Eq. 27.

$$Accuracy = \frac{Number\ of\ correct\ predictions}{Total\ number\ of\ samples} \tag{27}$$

The accuracy metric shows how well the model is making its predictions. A high accuracy metric means that the model is working well.

## 5 Experimental Results and Discussion

In this section, the performance of this framework is presented. Considering the above-detailed experimental setup, all described metaheuristic algorithms were tested individually for the $Lr$, $Mr$, and $Wd$ CNN hyperparameters optimization.

Table 3 presents the optimal hyperparameters obtained through exploration with various optimization algorithms for the MobileNetV2 CNN, while Table 4 shows the corresponding accuracy levels achieved. Additionally, Table 5 also presents the optimization and training/testing times.

**Table 3** Best hyperparameter global values obtained by each optimizer

| Model | Optimized hyperparameters values |
|---|---|
| MobileNetV2 | $Lr = 0.001$,<br>$Mr = 0.9$,<br>$Wd = 0.0001$ |
| MobileNetV2 + PSO | $Lr = 0.007321$,<br>$Mr = 0.014751$,<br>$Wd = 0.061103$ |
| MobileNetV2 + DE | $Lr = 0.011092$,<br>$Mr = 0.177158$,<br>$Wd = 0.026357$ |
| MobileNetV2 + GWO | $Lr = 0.025398$,<br>$Mr = 0.0$,<br>$Wd = 0.033660$ |
| MobileNetV2 + WOA | $Lr = 0.0153710$,<br>$Mr = 0.423283$,<br>$Wd = 0.012497$ |

**Table 4** Accuracy obtained with and without MobileNetV2 hyperparameters optimization ranked from lowest to highest performance

| Model | Accuracy (%) |
|---|---|
| MobileNetV2 | 95.90 |
| MobileNetV2 + PSO | 97.18 |
| MobileNetV2 + GWO | 97.20 |
| MobileNetV2 + DE | 97.35 |
| MobileNetV2 + WOA | 98.00 |

**Table 5** Training and testing times in seconds (s)

| Model | Optimization time | Train and test time |
|---|---|---|
| MobileNetV2 | – | 180 |
| MobileNetV2 + PSO | 8745 | 184 |
| MobileNetV2 + GWO | 8417 | 175 |
| MobileNetV2 + DE | 8660 | 178 |
| MobileNetV2 + WOA | 8370 | 175 |

In the absence of additional optimization algorithms, MobileNetV2 attained optimal values of a $Lr$ of 0.001, $Mr$ of 0.9, and $Wd$ of 0.0001, resulting in an accuracy of 95.9%. Applying the PSO algorithm alongside MobileNetV2 yielded different hyperparameter values: a $Lr$ of 0.007321, $Mr$ of 0.014751, and $Wd$ of 0.061103, leading to an accuracy of 97.1%. The PSO optimization time was 8745 s, with a train and test time of 184 s. Similarly, using the GWO algorithm with MobileNetV2 resulted in optimized values of a $Lr$ of 0.025398, $Mr$ of 0.0, and $Wd$ of 0.033660, achieving an accuracy of 97.2%.

The GWO optimization time was 8417 s, with a train and test time of 175 s. Employing the DE algorithm with MobileNetV2 produced optimized hyperparameters of a $Lr$ of 0.011092, $Mr$ of 0.177158, and $Wd$ of 0.026357, resulting in an accuracy of 97.4%. The DE optimization time was 8660 s, with a train and test time of 178. Lastly, integrating the WOA algorithm with MobileNetV2 resulted in optimal values of a $Lr$ of 0.015371, $Mr$ of 0.423283, and $Wd$ of 0.01249, achieving an accuracy of 98%. The WOA optimization time was 8370 s, with a train and test time of 175 s. These findings highlight the significant impact of different optimization algorithms on hyperparameter tuning and classification accuracy in CNNs.

Starting with the accuracy levels, it's evident that the MobileNetV2 model alone achieves the lowest accuracy of 95.90%. However, when combined with optimization algorithms there is a significant improvement in accuracy, reaching up to 98.00% with WOA. This demonstrates the effectiveness of these metaheuristic optimization algorithms in fine-tuning the model and improving its classification performance.

Regarding optimization times, each algorithm has varying optimization times, with PSO having the longest time of 8745 s and WOA having the shortest time of

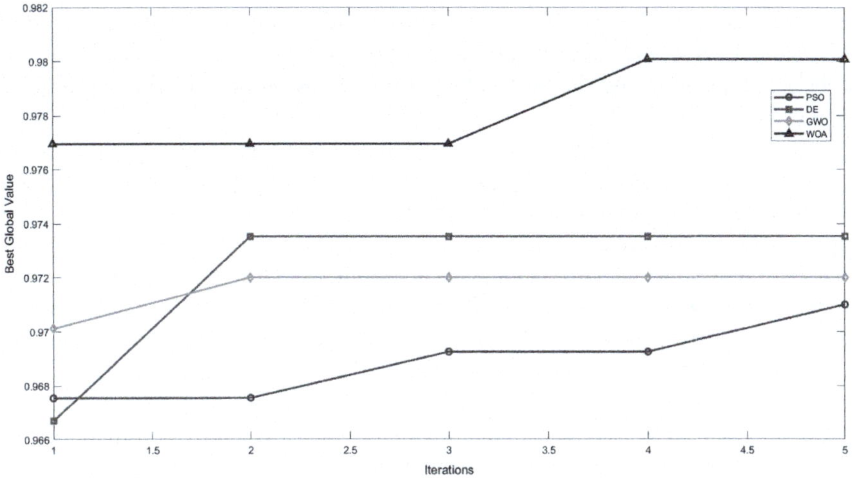


**Fig. 12** Convergence curves over 5 iterations

8370 s. Interestingly, despite the longer optimization time for PSO, the accuracy achieved is slightly lower compared to GWO and DE, which have shorter optimization times. This suggests that optimization time alone may not directly correlate with the final accuracy achieved, indicating the importance of algorithm efficiency and convergence speed. Additionally, the training and testing times for the optimized models are relatively consistent, ranging from 175 to 184 s. This indicates that the optimization process does not significantly impact the overall computational time for training and testing the models, making these optimized models feasible for practical deployment.

The behavior observed in Fig. 12, corresponds to the convergence curves plotted using the best global values, highlighting the performance and convergence trends of different optimization algorithms over iterations. Each algorithm exhibits distinct convergence patterns as it optimizes the hyperparameters of the MobileNetV2 CNN. The convergence curves provide valuable insights into the optimization process's dynamics, including the speed of convergence, stability, and overall performance of each algorithm. Such analysis aids in understanding how different metaheuristic optimization techniques influence the fine-tuning of CNN models and their impact on classification accuracy.

From the plotted curves, it is shown how the accuracy or best global values evolve over successive epochs for each algorithm. PSO demonstrates a slower increase in the best global value over epochs, indicating continuous improvement in optimization and accuracy but achieving lower performance in accuracy terms across all iterations. On the other hand, DE initially shows a slight improvement in the first two epochs, followed by stability in subsequent epochs, suggesting early convergence in the optimization process. Meanwhile, GWO exhibits consistent performance with minimal fluctuations in the best global value, indicating stable and reliable optimization across epochs. Last, the WOA showcases significant and continuous improvement in the best global value and starts its first epoch with a higher value, indicating a more robust initialization and the best potential for achieving higher accuracy levels in less epochs.

The experimental results showcase the significant impact of integrating metaheuristic optimization algorithms with MobileNetV2 for pneumonia detection in X-ray images. Each optimization algorithm, including PSO, GWO, DE, and WOA, played a crucial role in fine-tuning the hyperparameters of MobileNetV2, leading to substantial improvements in classification accuracy. The accuracy levels achieved, ranging from 97.18 to 98%, demonstrate the effectiveness of these algorithms in enhancing the performance of CNNs. It's worth noting the convergence speed and computational efficiency of each algorithm during the optimization process. Algorithms that converge faster while maintaining high accuracy levels are desirable for practical applications where computational resources are limited. Furthermore, the convergence speed and computational efficiency varied across the algorithms, with some converging faster while maintaining high accuracy levels, making them desirable for practical applications with limited computational resources. These results not only highlight the adaptability and sensitivity of metaheuristic algorithms but also emphasize their potential in optimizing deep learning models for image classification tasks in medical imaging, thereby contributing to the advancement of AI-driven disease detection systems.

## 6 Conclusions

This hybrid approach takes advantage of the evolutionary process to optimize neural network architectures with the efficient search mechanisms of metaheuristic algorithms, resulting in finely tuned models capable of enhancing the overall performance of CNN-based medical systems. In this study, four different algorithms were tested for the fine-tuning of a pre-trained MobileNetV2 for X-ray image classification: PSO, GWO, DE, and WOA, focusing on distinguishing between pneumonia-positive and normal cases.

While all algorithms contributed to overall accuracy enhancement, the WOA outperforms all others. Also, they exhibited varying performance levels in terms of optimization time. The PSO required more time to converge to optimal solutions, whereas WOA demonstrated faster optimization due to its efficient exploration and exploitation strategies. These observations suggest that the choice of optimization algorithm must be made by analyzing the specific requirements and constraints of the task at hand.

As a future research direction, it is intended to implement more robust metaheuristic approaches to obtain improved performance. Moreover, the research will focus on integrating these metaheuristic approaches with sophisticated class-balancing techniques such as oversampling the minority class, undersampling the majority class, and leveraging synthetic data generation techniques. In addition to the model performance metric, it is planned to implement precision, recall, Area Under the Curve (AUC), and F1 score, for a more thorough understanding of the class balancing behavior.

In conclusion, this analysis demonstrates the potential of neuroevolution approaches with metaheuristic optimization for improving the accuracy of X-ray image classification in the context of pulmonary disease diagnosis. Future research can explore further metaheuristic optimization techniques and different CNN architectures to improve the already high performance of imaging and AI-assisted diagnostics.

## References


1. Ting, D.S., Liu, Y., Burlina, P., Xu, X., Bressler, N.M., Wong, T.Y.: Ai for medical imaging goes deep. Nat. Med. **24**(5), 539–540 (2018)
2. Haleem, A., Javaid, M., Khan, I.H.: Current status and applications of artificial intelligence (ai) in medical field: an overview. Curr. Med. Res. Pract. **9**(6), 231–237 (2019)
3. Fusco, R., Grassi, R., Granata, V., Setola, S.V., Grassi, F., Cozzi, D., Pecori, B., Izzo, F., Petrillo, A.: Artificial intelligence and covid-19 using chest ct scan and chest x-ray images: machine learning and deep learning approaches for diagnosis and treatment. J. Pers. Med. **11**(10), 993 (2021)
4. Prevedello, L.M., Halabi, S.S., Shih, G., Wu, C.C., Kohli, M.D., Chokshi, F.H., Erickson, B.J., Kalpathy-Cramer, J., Andriole, K.P., Flanders, A.E.: Challenges related to artificial intelligence research in medical imaging and the importance of image analysis competitions. Radiol.: Artif. Intell. **1**(1), e180031 (2019)
5. Panayides, A.S., Amini, A., Filipovic, N.D., Sharma, A., Tsaftaris, S.A., Young, A., Foran, D., Do, N., Golemati, S., Kurc, T., et al.: Ai in medical imaging informatics: current challenges and future directions. IEEE J. Biomed. Health Inform. **24**(7), 1837–1857 (2020)
6. Stanley, K.O., Clune, J., Lehman, J., Miikkulainen, R.: Designing neural networks through neuroevolution. Nat. Mach. Intell. **1**(1), 24–35 (2019)
7. Real, E., Aggarwal, A., Huang, Y., Le, Q.V.: Regularized evolution for image classifier architecture search. In: Proceedings of the AAAI Conference on Artificial Intelligence, vol. 33, pp. 4780–4789 (2019)
8. Hassanzadeh, T., Essam, D., Sarker, R.: Evodcnn: an evolutionary deep convolutional neural network for image classification. Neurocomputing **488**, 271–283 (2022)
9. Abd Elaziz, M., Dahou, A., Abualigah, L., Yu, L., Alshinwan, M., Khasawneh, A.M., Lu, S.: Advanced metaheuristic optimization techniques in applications of deep neural networks: a review. Neural Comput. Appl. 1–21 (2021)
10. Oliva, D., Abd Elaziz, M., Hinojosa, S., Oliva, D., Abd Elaziz, M., Hinojosa, S.: Metaheuristic optimization. Metaheuristic algorithms for image segmentation: theory and applications 13–26 (2019)
11. Połap, D., Woźniak, M.: Meta-heuristic as manager in federated learning approaches for image processing purposes. Appl. Soft Comput. **113**, 107872 (2021)
12. Naveen, P., Diwan, B.: Meta-heuristic endured deep learning model for big data classification: image analytics. Knowl. Inf. Syst. 1–31 (2023)
13. Egmont-Petersen, M., de Ridder, D., Handels, H.: Image processing with neural networks-a review. Pattern Recogn. **35**(10), 2279–2301 (2002)
14. Albawi, S., Mohammed, T.A., Al-Zawi, S.: Understanding of a convolutional neural network. In: International Conference on Engineering and Technology (ICET), pp. 1–6. IEEE (2017)

15. O'Shea, K., Nash, R.: An introduction to convolutional neural networks (2015). arXiv:1511.08458
16. Chen, B., Li, H., Luo, W.: Image processing operations identification via convolutional neural network (2017). arXiv:1709.02908
17. Perez-Zarate, E., Ramos-Soto, O., Rodríguez-Esparza, E., Aguilar, G.: Loli-iea: low-light image enhancement algorithm. In: Applications of Machine Learning 2023, vol. 12675, pp. 230–245. SPIE (2023)
18. Xu, C., Wang, G., Yan, S., Yu, J., Zhang, B., Dai, S., Li, Y., Xu, L.: Fast vehicle and pedestrian detection using improved mask r-cnn. Math. Probl. Eng. **2020**, 1–15 (2020)
19. Said, Y., Barr, M., Ahmed, H.E.: Design of a face recognition system based on convolutional neural network (cnn). Eng. Technol. & Appl. Sci. Res. **10**(3), 5608–5612 (2020)
20. Jo, H.-C., Jeong, H., Lee, J., Na, K.-S., Kim, D.-Y.: Quantification of blood flow velocity in the human conjunctival microvessels using deep learning-based stabilization algorithm. Sensors **21**(9), 3224 (2021)
21. Calderon-Auza, G., Carrillo-Gomez, C., Nakano, M., Toscano-Medina, K., Perez-Meana, H., Gonzalez-H, A., Leon, H. Quiroz-Mercado.: A teleophthalmology support system based on the visibility of retinal elements using the cnns. Sensors **20**(10), 2838 (2020)
22. Ramos-Soto, O., Jo, H.C., Zawadzki, R.J., Kim, D.Y., Balderas-Mata, S.E.: Automated segmentation and morphometry of zebrafish anterior chamber oct scans, In: Photonics, vol. 10, p. 957. MDPI (2023)
23. Wang, L., Lin, Z.Q., Wong, A.: Covid-net: a tailored deep convolutional neural network design for detection of covid-19 cases from chest x-ray images. Sci. Rep. **10**(1), 19549 (2020)
24. Rahimzadeh, M., Attar, A.: A modified deep convolutional neural network for detecting covid-19 and pneumonia from chest x-ray images based on the concatenation of xception and resnet50v2. Inf. Med. Unlocked **19**, 100360 (2020)
25. Dey, S., Roychoudhury, R., Malakar, S., Sarkar, R.: An optimized fuzzy ensemble of convolutional neural networks for detecting tuberculosis from chest x-ray images. Appl. Soft Comput. **114**, 108094 (2022)
26. Sajaan Almansour, S.H., Singh, R., Hadrami Alyami, S.M., Sharma, N., Reshan, A., Saleh, M., Gupta, S., Mahdi Alyami, M.F., Shaikh, A.: A convolution neural network design for knee osteoarthritis diagnosis using x-ray images. In. J. Online & Biomed. Eng. **19**(7) (2023)
27. Oliva, D., Houssein, E.H., Hinojosa, S.: Metaheuristics in machine learning: theory and applications, vol. 967. Springer (2021)
28. Fausto, F., Reyna-Orta, A., Cuevas, E., Andrade, Á.G., Perez-Cisneros, M.: From ants to whales: metaheuristics for all tastes. Artif. Intell. Rev. **53**, 753–810 (2020)
29. Mehrani, K., Mirshahvalad, A., Abbasi, E.: Portfolio optimization using black hole meta heuristic algorithm. Specialty J. Account. Econ. 5 (2020)
30. Kundu, R., Chattopadhyay, S., Nag, S., Navarro, M.A., Oliva, D.: Prism refraction search: a novel physics-based metaheuristic algorithm. J. Supercomput. 1–50 (2024)
31. Banu, P.N., Andrews, S.: Gene clustering using metaheuristic optimization algorithms. Int. J. Appl. Metaheuristic Comput. (IJAMC) **6**(4), 14–38 (2015)
32. Navarro, M.A., Oliva, D., Ramos-Michel, A., Haro, E.H.: An analysis on the performance of metaheuristic algorithms for the estimation of parameters in solar cell models. Energy Convers. Manage. **276**, 116523 (2023)
33. Aranguren, I., Valdivia, A., Morales-Castañeda, B., Oliva, D., Abd Elaziz, M., Perez-Cisneros, M.: Improving the segmentation of magnetic resonance brain images using the lshade optimization algorithm. Biomed. Signal Proc. Control **64**, 102259 (2021)
34. Ramos-Soto, O., Rodríguez-Esparza, E., Balderas-Mata, S.E., Oliva, D., Hassanien, A.E., Meleppat, R.K., Zawadzki, R.J.: An efficient retinal blood vessel segmentation in eye fundus images by using optimized top-hat and homomorphic filtering. Comput. Methods Programs Biomed. **201**, 105949 (2021)
35. Gaspar, A., Oliva, D., Cuevas, E., Zaldívar, D., Pérez, M., Pajares, G.: Hyperparameter optimization in a convolutional neural network using metaheuristic algorithms. In: Metaheuristics in Machine Learning: Theory and Applications, pp. 37–59. Springer (2021)

36. Bacanin, N., Bezdan, T., Tuba, E., Strumberger, I., Tuba, M.: Optimizing convolutional neural network hyperparameters by enhanced swarm intelligence metaheuristics. Algorithms **13**(3), 67 (2020)
37. Nematzadeh, S., Kiani, F., Torkamanian-Afshar, M., Aydin, N.: Tuning hyperparameters of machine learning algorithms and deep neural networks using metaheuristics: a bioinformatics study on biomedical and biological cases. Comput. Biol. Chem. **97**, 107619 (2022)
38. Akut, R., Kulkarni, S.: Neuroevolution: using genetic algorithm for optimal design of deep learning models. In: 2019 IEEE International Conference on Electrical, Computer and Communication Technologies (ICECCT), pp. 1–6. IEEE (2019)
39. Bourouis, S., Band, S.S., Mosavi, A., Agrawal, S., Hamdi, M.: Meta-heuristic algorithm-tuned neural network for breast cancer diagnosis using ultrasound images. Front. Oncol. **12**, 834028 (2022)
40. Oyelade, O.N., Ezugwu, A.E.: Characterization of abnormalities in breast cancer images using nature-inspired metaheuristic optimized convolutional neural networks model. Concurr. Comput.: Pract. Exper. **34**(4), e6629 (2022)
41. Calimeri, F., Marzullo, A., Stamile, C., Terracina, G.: Blood vessel segmentation in retinal fundus images using hypercube neuroevolution of augmenting topologies (hyperneat). In: Italian Workshop on Neural Nets, pp. 173–183. Springer (2017)
42. Canayaz, M.: Mh-covidnet: Diagnosis of covid-19 using deep neural networks and meta-heuristic-based feature selection on x-ray images. Biomed. Signal Process. Control **64**, 102257 (2021)
43. El-Kenawy, E.-S.M., Mirjalili, S., Ibrahim, A., Alrahmawy, M., El-Said, M., Zaki, R.M., Eid, M.M.: Advanced meta-heuristics, convolutional neural networks, and feature selectors for efficient covid-19 x-ray chest image classification. IEEE Access **9**, 36019–36037 (2021)
44. Shukla, P.K., Sandhu, J.K., Ahirwar, A., Ghai, D., Maheshwary, P., Shukla, P.K.: Multiobjective genetic algorithm and convolutional neural network based covid-19 identification in chest x-ray images. Math. Probl. Eng. **2021**, 1–9 (2021)
45. Gülmez, B.: A novel deep neural network model based xception and genetic algorithm for detection of covid-19 from x-ray images. Ann. Oper. Res. **328**(1), 617–641 (2023)
46. Cai, C., Gou, B., Khishe, M., Mohammadi, M., Rashidi, S., Moradpour, R., Mirjalili, S.: Improved deep convolutional neural networks using chimp optimization algorithm for covid19 diagnosis from the x-ray images. Expert Syst. Appl. **213**, 119206 (2023)
47. Wang, S.-C., Wang, S.-C.: Artificial neural network. Interdisciplinary computing in java programming 81–100 (2003)
48. Wu, Y.-C., Feng, J.-W.: Development and application of artificial neural network. Wireless Pers. Commun. **102**, 1645–1656 (2018)
49. Dayhoff, J.E., DeLeo, J.M.: Artificial neural networks: opening the black box. Cancer: Interdiscipl. Int. J. Amer. Cancer Soc. **91**(S8), 1615–1635 (2001)
50. Macukow, B.: Neural networks–state of art, brief history, basic models and architecture. In: Computer Information Systems and Industrial Management: 15th IFIP TC8 International Conference, CISIM 2016, Vilnius, Lithuania, September 14–16, 2016, Proceedings 15, pp. 3–14. Springer (2016)
51. Dongare, A., Kharde, R., Kachare, A.D., et al.: Introduction to artificial neural network. Int. J. Eng. Innov. Technol. (IJEIT) **2**(1), 189–194 (2012)
52. Kulkarni, A., Saraf, C.: Learning from nature: applications of biomimicry in technology. In: IEEE Pune Section International Conference (PuneCon), pp. 1–6. IEEE (2019)
53. Ramón y Cajal, S.: Textura del sistema nervioso del hombre y de los vertebrados, tomo ii, primera parte. Imprenta y Libreria de Nicolas Moya, Madrid, reprinted by Graficas Vidal Leuka, Alicante **1992** 399–402 (1904)
54. Wu, J.: Introduction to convolutional neural networks. National Key Lab for Novel Software Technology. Nanjing University, China **5**(23), 495 (2017)
55. Weiss, K., Khoshgoftaar, T.M., Wang, D.: A survey of transfer learning. J. Big data **3**, 1–40 (2016)

56. Gupta, J., Pathak, S., Kumar, G.: Deep learning (cnn) and transfer learning: a review. J. Phys.: Confer. Ser. **2273**, 012029 IOP Publishing (2022)
57. Marcelino, P.: Transfer learning from pre-trained models. Towards Data Sci. **10**(330), 23 (2018)
58. Zhuang, F., Qi, Z., Duan, K., Xi, D., Zhu, Y., Zhu, H., Xiong, H., He, Q.: A comprehensive survey on transfer learning. Proc. IEEE **109**(1), 43–76 (2020)
59. Asanuma, J., Doi, S., Igarashi, H.: Transfer learning through deep learning: application to topology optimization of electric motor. IEEE Trans. Magn. **56**(3), 1–4 (2020)
60. Nixon, M., Aguado, A.: Feature Extraction and Image Processing for Computer Vision. Academic (2019)
61. Kumar, G., Bhatia, P.K.: A detailed review of feature extraction in image processing systems. In: Fourth International Conference on Advanced Computing & Communication Technologies, pp. 5–12. IEEE (2014)
62. Routray, S., Ray, A.K., Mishra, C.: Analysis of various image feature extraction methods against noisy image: sift, surf and hog. In: 2017 Second International Conference on Electrical, Computer and Communication Technologies (ICECCT), pp. 1–5. IEEE (2017)
63. Öztürk, Ş, Bayram, A.: Comparison of hog, mser, sift, fast, lbp and canny features for cell detection in histopathological images. Helix **8**(3), 3321–3325 (2018)
64. Jogin, M., Madhulika, M., Divya, G., Meghana, R., Apoorva, S.: Feature extraction using convolution neural networks (cnn) and deep learning. In: et al. (eds.) 3rd IEEE International Conference on Recent Trends in Electronics, Information & Communication Technology (RTEICT), pp. 2319–2323. IEEE (2018)
65. Liu, Y.H.: Feature extraction and image recognition with convolutional neural networks. J. Phys.: Conf. Ser. **1087**, 062032, IOP Publishing (2018)
66. Mammone, A., Turchi, M., Cristianini, N.: Support vector machines. Wiley Interdiscipl. Rev.: Comput. Stat. **1**(3), 283–289 (2009)
67. Ghosh, S., Dasgupta, A., Swetapadma, A.: A study on support vector machine based linear and non-linear pattern classification. In: 2019 International Conference on Intelligent Sustainable Systems (ICISS), pp. 24–28. IEEE (2019)
68. Pradhan, A.: Support vector machine-a survey. Int. J. Emer. Technol. Adv. Eng. **2**(8), 82–85 (2012)
69. Chandra, M.A., Bedi, S.: Survey on svm and their application in image classification. Int. J. Inf. Technol. **13**(5), 1–11 (2021)
70. Kecman, V.: Support vector machines–an introduction. In: Support Vector Machines: Theory and Applications, pp. 1–47. Springer (2005)
71. Noble, W.S.: What is a support vector machine? Nat. Biotechnol. **24**(12), 1565–1567 (2006)
72. Baldick, R.: Applied Optimization: Formulation and Algorithms for Engineering Systems. Cambridge University Press (2006)
73. Yang, X.-S.: Engineering Optimization: An Introduction with Metaheuristic Applications. Wiley (2010)
74. Mataifa, H., Krishnamurthy, S., Kriger, C.: Volt/var optimization: a survey of classical and heuristic optimization methods. IEEE Access **10**, 13379–13399 (2022)
75. Fonseca, C.M., Fleming, P.J.: Multiobjective optimization and multiple constraint handling with evolutionary algorithms. ii. application example. IEEE Trans. Syst. Man Cybern.-Part A: Syst. Hum. **28**(1), 38–47 (1998)
76. Kennedy, J., Eberhart, R.: Particle swarm optimization. In: Proceedings of ICNN'95-International Conference on Neural Networks, vol. 4, pp. 1942–1948. IEEE (1995)
77. Price, K.: Genetic annealing. Dr. Dobb's J. **19**, 127–132 (1994)
78. Storn, R., Price, K.: Differential evolution-a simple and efficient heuristic for global optimization over continuous spaces. J. Global Optim. **11**, 341–359 (1997)
79. Mirjalili, S., Mirjalili, S.M., Lewis, A.: Grey wolf optimizer. Adv. Eng. Softw. **69**, 46–61 (2014)
80. Mech, L.D.: Alpha status, dominance, and division of labor in wolf packs. Can. J. Zool. **77**(8), 1196–1203 (1999)
81. Mirjalili, S., Lewis, A.: The whale optimization algorithm. Adv. Eng. Softw. **95**, 51–67 (2016)

82. Hof, P.R., Van der Gucht, E.: Structure of the cerebral cortex of the humpback whale, megaptera novaeangliae (cetacea, mysticeti, balaenopteridae). Anat. Record: Adv. Integrat. Anat. Evol. Biol.: Adv. Integr. Anat. Evoluti. Biol. **290**(1), 1–31 (2007)
83. Watkins, W.A., Schevill, W.E.: Aerial observation of feeding behavior in four baleen whales: Eubalaena glacialis, balaenoptera borealis, megaptera novaeangliae, and balaenoptera physalus. J. Mammal. **60**(1), 155–163 (1979)
84. Goldbogen, J.A., Friedlaender, A.S., Calambokidis, J., McKenna, M.F., Simon, M., Nowacek, D.P.: Integrative approaches to the study of baleen whale diving behavior, feeding performance, and foraging ecology. Bioscience **63**(2), 90–100 (2013)
85. Sandler, M., Howard, A., Zhu, M., Zhmoginov, A., Chen, L.-C.: Mobilenetv2: inverted residuals and linear bottlenecks. In: Proceedings of the IEEE Conference on Computer Vision and Pattern Recognition, pp. 4510–4520 (2018)
86. Jain, M., Saihjpal, V., Singh, N., Singh, S.B.: An overview of variants and advancements of pso algorithm. Appl. Sci. **12**(17), 8392 (2022)
87. Kermany, D.S., Goldbaum, M., Cai, W., Valentim, C.C., Liang, H., Baxter, S.L., McKeown, A., Yang, G., Wu, X., Yan, F., et al.: Identifying medical diagnoses and treatable diseases by image-based deep learning. Cell **172**(5), 1122–1131 (2018)
88. Kermany, D., Zhang, K., Goldbaum, M.: Large dataset of labeled optical coherence tomography (oct) and chest x-ray images. Mendeley Data **3**(10), 17632 (2018)
89. Sunnetci, K.M., Kaba, E., Beyazal Çeliker, F., Alkan, A.: Comparative parotid gland segmentation by using resnet-18 and mobilenetv2 based deeplab v3+ architectures from magnetic resonance images. Concurr. Comput.: Pract. Exper. **35**(1), e7405 (2023)

# PEL: Population-Enhanced Learning Classification for ECG Signal Analysis

**Mehran Pourvahab, Seyed Jalaleddin Mousavirad, Farzaneh Lashgari, Anilson Monteiro, Kamran Shafafi, Virginie Felizardo, and Sebastião Pais**

**Abstract** In the study, a new method for analyzing Electrocardiogram (ECG) signals is suggested, which is vital for detecting and treating heart diseases. The technique focuses on improving ECG signal classification, particularly in identifying different heart conditions like arrhythmias and myocardial infarctions. An enhanced version of the differential evolution (DE) algorithm integrated with neural networks is leveraged to classify these signals effectively. The process starts with preprocessing and extracting key features from ECG signals. These features are then processed by a multi-layer perceptron (MLP), a common neural network for ECG analysis. However, traditional MLP training methods have limitations, such as getting trapped in suboptimal solutions. To overcome this, an advanced DE algorithm is used, incorporating

M. Pourvahab (✉) · F. Lashgari · A. Monteiro · V. Felizardo · S. Pais
Department of Computer Science, University of Beira Interior, Covilhã, Portugal
e-mail: mehran.pourvahab@ubi.pt

F. Lashgari
e-mail: f.lashgari@ubi.pt

A. Monteiro
e-mail: anilson.monteiro@ubi.pt

V. Felizardo
e-mail: virginie.felizardo@ubi.pt

S. Pais
e-mail: sebastiao@ubi.pt

M. Pourvahab · S. Pais
NOVA LINCS, NOVA University of Lisbon, Lisbon, Portugal

S. J. Mousavirad
Department of Computer and Electrical Engineering, Mid Sweden University, Ostersund, Sweden
e-mail: seyedjalaleddin.mousavirad@miun.se

K. Shafafi
INESC TEC and Faculty of Engineering, University of Porto, Porto, Portugal
e-mail: kamran.shafafi@inesctec.pt

S. Pais
Groupe de Recherche en Informatique, GREYC, University of Caen Normandie, Caen, France

 
D. Oliva et al. (eds.), *Advances in Optimization Algorithms for Multidisciplinary Engineering Applications: From Classical Methods to AI-Enhanced Solutions*, Studies in Computational Intelligence 806, https://doi.org/10.1007/978-3-031-78440-8_18

a partition-based strategy, opposition-based learning, and local search mechanisms. This improved DE algorithm optimizes the MLP by fine-tuning its weights and biases, using them as starting points for further refinement by the Gradient Descent with Momentum (GDM) local search algorithm. Extensive experiments demonstrate that this novel training approach yields better results than the traditional method.

## 1 Introduction

Cardiovascular diseases (CVDs) stand as the leading cause of mortality, accounting for approximately 32% of all deaths. The World Health Organization (WHO) attributes around 85% of these deaths to heart attacks and strokes [1]. CVDs, characterized by irregular heartbeats known as arrhythmias and blockages in the coronary arteries, are often linked to sudden death based on the severity of the arrhythmia. Lifestyle factors including diet, hypertension, and smoking play a significant role as primary causative factors of cardiovascular diseases.

The Electrocardiogram (ECG or EKG) has emerged as the most popular bio-signal for the early detection of the CVDs. ECG measures the electrical activity of the heart and is used to detect irregular heartbeats known as arrhythmias [2, 3]. This bio-signal is pivotal in the diagnosis and monitoring of CVDs, contributing to the early detection of cardiac events. ECGs are utilized to identify abnormalities such as irregular heart rhythms, blockages in the coronary arteries, and other indicators of cardiovascular health. This technology plays a crucial role in the proactive management of cardiovascular health, allowing for timely intervention and preventive measures.

The ECG signal analysis is a critical component in the diagnosis and monitoring of cardiac conditions. The advent of machine learning and artificial intelligence (AI) has revolutionized the field, offering new insights and improved accuracy in the detection and classification of cardiac events.

Recent advancements in the AI have significantly impacted healthcare [4], particularly in the analysis of The ECG signals for the detection and classification of cardiac arrhythmias [5]. Deep learning (DL) algorithms, such as convolutional neural networks (CNNs), long short-term memory (LSTM) networks, and hybrid models combining CNNs and LSTMs, have been extensively utilized for ECG signal analysis, achieving high accuracies in arrhythmia detection [6]. These AI models can outperform traditional methods and even experienced cardiologists, providing state-of-the-art results. This study presents a new machine learning approach for classifying ECG signals using a Multi-Layer Perceptron (MLP), optimized by an advanced Differential Evolution (DE) algorithm that integrates a region-based strategy and opposition-based learning for superior weight and bias optimization. Despite the prevalence of gradient-based methods like back-propagation in training the MLP, they have limitations, including susceptibility to local optima and initialization sensitivity. The method seeks to overcome these challenges by optimizing the MLP's architecture through meticulous experimentation to determine the optimal number of nodes in the hidden layers, thereby enhancing the prediction effectiveness for complex classification and regression tasks in scientific research.

Population-based Metaheuristic (PBMH) algorithms, such as Particle Swarm Optimization (PSO) and DE, have been recognized for their effectiveness in solving diverse optimization problems, including the training of MLPs [7–9]. These algorithms are particularly noted for their ability to improve convergence and diversity in the search for optimal solutions. Evolutionary Algorithms (EA), a subset of PBMH, have been widely utilized for MLP training and have shown superior effectiveness and efficiency compared to traditional algorithms like Back-Propagation (BP) [10]. Genetic Algorithms (GA), another form of PBMH, have been specifically tailored for rapid MLP training, demonstrating more efficiency than conventional GA-based training algorithms that combine GA and BP for MLP weight determination [11]. This specialized GA-based method has proven to outperform both GA and BP in terms of training MLPs [12].

Swarm intelligence and evolutionary algorithms are indeed prominent in the field of PBMHs due to their natural inspiration and robust optimization capabilities [13, 14]. A hybrid approach that combines accelerated Particle Swarm Optimization (PSO) with the Levenberg-Marquardt method has been shown to significantly improve PSO's convergence, particularly for medical data analysis [15]. This hybrid method is part of a broader trend of employing various algorithms, such as the artificial bee colony, grey wolf optimizer, and whale optimization algorithm, to optimize MLPs [16].

The DE is particularly adept at tackling complex optimization problems, outperforming gradient-based methods through its distinctive mutation, crossover, and selection operators [15, 16]. Recent advancements have seen DE augmented with opposition-based learning and region-based strategies, which have further enhanced its effectiveness, especially in classification tasks [13, 14]. Such innovations underscore the ongoing refinement of DE algorithms for Multi-Layer Perceptron (MLP) training within scientific research, emphasizing their growing potential to surpass traditional gradient-based techniques [17].

In the study, an innovative training model for ECG signal classification is introduced, leveraging DE enhanced with opposition-based learning, a centroid-focused approach, and a supplementary local search phase. This hybrid methodology integrates various algorithmic strengths, offering a robust solution for distinguishing ECG patterns, with significant implications for both clinical diagnostics and biomedical research.

The method focuses on classifying ECG signals into two categories: "Normal" and "Abnormal". The PTB-XL [18] dataset, an extensive collection dedicated to ECG analysis, is employed as the primary dataset. For the sake of the study's objectives, the dataset's superclass labels have been converted into a binary system: "Normal" heart rhythms are marked as "1", and all other rhythms, which might indicate any abnormality, are denoted as "0". This binary classification framework is fundamental to the approach, streamlining the ECG signal classification process and enhancing the efficiency of the model.

The methodology introduces several innovative features for ECG signal classification, detailed as follows:

1. Integration of an MLP with an advanced DE algorithm and a gradient-based local search mechanism, providing a powerful and efficient approach to classification.
2. Validation of the effectiveness and superiority of the proposed method for classifying ECG signals through comprehensive evaluations using state-of-the-art benchmarks and extensive experimentation.
3. Reinforcement of the DE algorithm with a quasi-opposition-based learning technique to elevate its exploratory potential and aid significantly in the identification of optimal solutions. The optimal set of weights generated by the advanced DE algorithm serves as the initial point for the gradient-based local search, enabling finer and more accurate model optimization.
4. Employment of a partitioning-based multi-parent crossover mechanism to improve the DE strategy's ability to examine data more thoroughly for better convergence.

## 2 Background

### *2.1 Neural Networks*

In the field of supervised pattern recognition, neural networks—ANNs—have become a key technology. Due to their broad range of applications in numerous industries, the main type of ANNs that have attracted a lot of attention are MLPs [19].

The MLPs are structured with three essential layers: an input layer, several hidden layers, and an output layer. Each node, or neuron, within these layers is equipped with an activation function. This function is critical as it dictates how the output is influenced by the weighted sum of the inputs. Among the various activation functions, the Sigmoid function is notably popular. It is mathematically represented in Eq. 1:

$$\sigma(\text{net}) = \frac{1}{1 + e^{-\text{net}}} \tag{1}$$

- $\sigma(\text{net})$ represents the Sigmoid function.
- net is the input to the neuron, calculated as the weighted sum of the inputs to that neuron.
- The Sigmoid function outputs a value between 0 and 1, making it useful for binary classification problems.

Here, "net" is the aggregated input to the neuron. The weights in an MLP represent the strength of connections between neurons in different layers. The process of training involves adjusting these weights to optimize the network's performance, often using Gradient Descent (GD) techniques.

In an MLP, the connections between layers are assigned specific weights that symbolize the strength of the connection between two nodes. These weights play a pivotal role in the network's performance. However, determining the optimal values for these weights, a process known as "training", is recognized as a complex and critical task. GD-based methods are the most commonly employed techniques for training MLPs.

The MLP comprises multiple layers of nodes that include input, hidden, and output layers. An MLP is designed to learn the mapping of input patterns to corresponding output patterns through supervised learning. This learning process involves adjusting the network's weights to minimize the discrepancy between the predicted outputs and the actual outputs.

The MLPs utilize activation functions such as the Sigmoid function to modulate the output of each node, based on the weighted sum of its inputs, as shown in Eq. 1.

Training an MLP involves fine-tuning the weights using gradient descent or other optimization algorithms. This is done to reduce the error between the network's predicted output and the actual output. The effectiveness of an MLP, or any neural network, is largely contingent on the efficiency and effectiveness of its training process. This training not only involves adjusting weights but also entails selecting the appropriate architecture, activation functions, and learning rate, among other factors.

The backpropagation algorithm is commonly used for training MLPs. It calculates the gradient of the loss function (error function) to update the weights, as shown in the following Eq. 2:

$$\Delta w_{ij} = -\eta \frac{\partial \mathcal{L}}{\partial w_{ij}} \tag{2}$$

- $\Delta w_{ij}$ represents the change that needs to be applied to the weight connecting the $i$th and $j$th neurons.
- $-\eta$ is the learning rate, which controls how much the weights are adjusted during each step of the training process.
- $\frac{\partial \mathcal{L}}{\partial w_{ij}}$ is the partial derivative of the loss function $\mathcal{L}$ with respect to the weight $w_{ij}$. This derivative indicates how much a small change in $w_{ij}$ will impact the loss.

Figure 1 illustrates a typical architecture of an MLP, a class of feedforward the ANN. This network comprises an input layer on the left, multiple hidden layers in the center, and an output layer on the right. The input layer consists of nodes (also known as neurons) labeled $X_1, X_2, \ldots, X_n$ where each node represents a feature of the input data. The hidden layers serve as computational units that transform the input data into features at different levels of abstraction. Each neuron in these layers applies an activation function to the weighted sum of its inputs and then passes this output to the subsequent layer. The output layer consists of neurons labeled $Y_1, Y_2, \ldots, Y_m$ where each neuron corresponds to a potential class or value to be predicted by the network. Connections between neurons symbolize the weights that are adjusted during the training process. The network learns to map inputs to the correct output by adjusting these weights based on a method such as backpropagation, using a loss function to

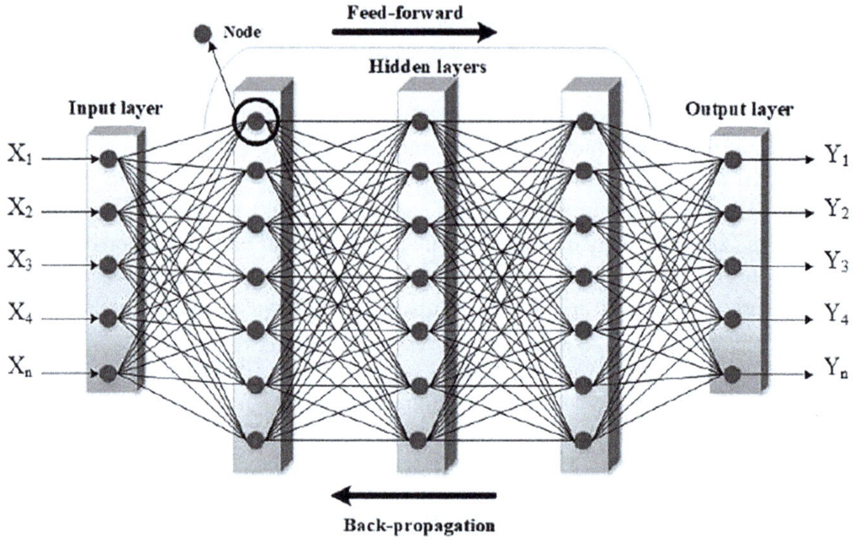


**Fig. 1** A MLP with hidden layer structure

measure the prediction error. Arrows indicate the direction of data flow from input to output. This unidirectional flow characterizes the "feedforward" nature of the network. Additionally, the back arrow at the bottom signifies the "feedback" during the training phase, where the error is propagated backward to update the weights, hence the term "backpropagation".

This MLP is capable of learning complex patterns thanks to its deep architecture, which includes multiple layers of neurons. The depth and breadth of the network are key factors in determining its capacity to model complex functions.

Overall, MLPs are powerful tools in the field of machine learning, offering significant potential for pattern recognition, data analysis, and predictive modeling across a wide array of disciplines. Their versatility and adaptability make them integral to the advancement of artificial intelligence and its applications.

## 2.2 *Gradient Descent with Momentum (GDM)*

Batch steepest descent, also known as Gradient Descent with Momentum, is a technique that improves neural network training. One particularly well-known feature of this approach is its capacity to accelerate convergence, an essential component of neural network training [20]. Fundamentally, momentum makes the network responsive to both the current trends seen on the error surface and the immediate local gradient. This feature is advantageous as it aids the network in ignoring small aberrations or variations on the error surface. Momentum essentially smoothes out noise and highlights important patterns by functioning as a low-pass filter [21].

A challenge associated with conventional techniques is the ease with which a network might become caught in shallow local minima. These are the places where a minimum is inadvertently detected by the network, but it is not the optimal or "global" minimum. In this sense, momentum is helpful as it drives the network out of these brief minima. It functions as an additional impetus for the network to keep searching for the best answer.

To implement GDM, two key training parameters are necessary: the learning rate (denoted as $\alpha$) and the momentum constant (denoted as $\gamma$). The learning rate determines the size of the steps taken during the training process, while the momentum constant helps in deciding how much of the past trends should influence current updates.

The process involves calculating the derivatives of the performance function $Q$ with respect to the network's weights and biases (collectively represented as vector $M$). This calculation is typically done using a method known as backpropagation.

Once these derivatives are obtained, GDM modifies each weight and bias vector $M_i$ according to a specific formula, as shown in Eq. 3:

$$dM = \gamma \times (dM_{\text{previous}}) + \alpha \times (1 - \gamma) \times \frac{dQ}{dM} \tag{3}$$

where:

- $dM$ represents the change made to the weight or bias in the current update.
- $\gamma$ is the momentum constant.
- $dM_{\text{previous}}$ is the change made to the weight or bias in the previous update.
- $\alpha$ is the learning rate.
- $\frac{dQ}{dM}$ is the derivative of the performance function $Q$ with respect to the weight or bias vector $M$.

In this equation, $dM_{\text{previous}}$ represents the adjustment made to the bias or weight during the previous update. This inclusion of the previous update in the current modification is what embodies the concept of Momentum in this method. In a nutshell, Gradient Descent with Momentum is a valuable technique in the field of neural networks, particularly for its ability to accelerate the convergence process and help the network navigate more effectively through the error surface. Its incorporation of both the current gradient and the trend observed in previous updates makes it a robust and efficient approach for training.

### *2.3 Pattern Partitioning*

Pattern partitioning, also known as pattern clustering, is a process fundamental to the analysis of complex datasets, where the objective is to organize patterns into distinct subsets, or partitions, that are homogenous within themselves but heterogeneous when compared to each other [22]. This strategy is pivotal in many machine learning tasks, including the classification of ECG signals, as it enhances the ability of algorithms to discern and categorize distinct patterns that correspond to different cardiac conditions. The essence of pattern partitioning lies in the manipulation of distances within a dataset. The goal is twofold: to minimize intra-cluster distances, which are the distances between samples within the same partition, and to maximize inter-cluster distances, which are the distances between samples belonging to different partitions.

Let's consider a set of $N$ patterns, denoted by $P = \{p_1, p_2, \ldots, p_N\}$, where each pattern $p_i$ is a vector in a $d$-dimensional space, $p_i = \{p_{i,1}, p_{i,2}, \ldots, p_{i,d}\}$. Here, $d$ represents the number of features or dimensions within the dataset.

The partitioning task aims to categorize these patterns into $K$ distinct partitions represented by $C = \{c_1, c_2, \ldots, c_K\}$. The conditions for an optimal partitioning are formalized as follows:

1. Each partition $c_i$ must be non-empty:

$$c_i \neq \emptyset, \quad \text{for } i = 1 \ldots K \tag{4}$$

2. All partitions together must encompass all patterns:

$$\bigcup_{i=1}^{K} c_i = P \tag{5}$$

3. Partitions must be mutually exclusive; there should be no overlap between any two partitions:

$$c_i \cap c_j = \emptyset, \quad \text{for } i, j = 1 \ldots K, \quad i \neq j \tag{6}$$

4. Euclidean distance between two patterns $p_i$ and $p_j$:

$$d(p_i, p_j) = \sqrt{\sum_{k=1}^{d} (p_{i,k} - p_{j,k})^2} \tag{7}$$

## 2.4 *Differential Evolution (DE)*

The DE is a type of evolutionary algorithm that has become famous because it is easy to use and good at solving hard problems. DE works on the same ideas as natural selection and biology, and it works especially well for problems that need to be solved over and over again [23]. DE begins with a population of potential solutions, known as candidate solutions, which are usually generated randomly within the problem's boundaries. A population-based method is used to process training, as shown in Fig. 2.

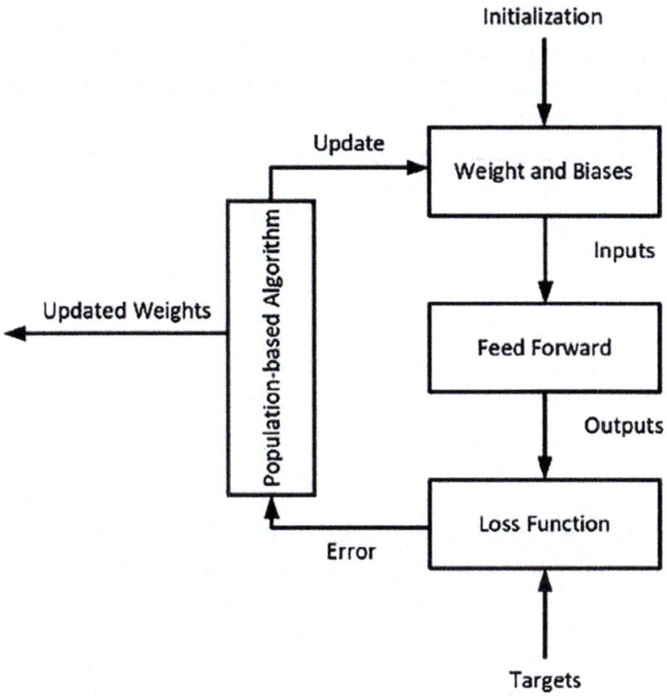


**Fig. 2** Using a population-based method to train the MLP [21]

This population undergoes a process of evolution through mutation, crossover, and selection.

**Mutation:** In DE, mutation creates a new candidate solution, referred to as a mutant vector. This process is described by the equation:

$$v_i = x_{r1} + F \cdot (x_{r2} - x_{r3}) \tag{8}$$

where $v_i$ is the mutant vector, $x_{r1}$, $x_{r2}$, and $x_{r3}$ are three distinct candidate solutions randomly chosen from the current population, and $F$ is the scale factor that controls the differential variation.

**Crossover:** The crossover operation is used to combine the mutant vector with the target vector to produce a trial vector. The binomial crossover can be defined for each component $j$ of the trial vector $u_{i,j}$ as follows:

$$u_{i,j} = \begin{cases} v_{i,j} & \text{if rand}(0, 1) \leq CR \text{ or } j = j_{\text{rand}} \\ x_{i,j} & \text{otherwise} \end{cases} \tag{9}$$

where $CR$ is the crossover rate and $j_{\text{rand}}$ is a randomly chosen index from the set $\{1, 2, ..., D\}$, ensuring that $u_{i,j}$ inherits at least one component from $v_i$.

**Selection:** After crossover, a selection step determines if the trial vector should become a member of the subsequent generation. The selection is typically based on the fitness value, and the trial vector will replace the target vector if it exhibits a better or equivalent fitness.

The DE algorithm iteratively improves the candidate solutions, which, in the context of training neural networks, correspond to the network's weights and biases. By employing mutation, crossover, and selection operations, DE navigates through the solution space to find an optimal set of parameters that minimize the loss function of the neural network.

## 2.5 Electrocardiogram (ECG)

The ECG serves as a critical tool in the non-invasive monitoring and diagnosis of heart conditions. ECG signals are the graphical representation of the electrical activity of the heart, captured via electrodes placed on the patient's body. These signals are essential for identifying various cardiac anomalies, such as arrhythmias and myocardial infarctions. Figure 3 shows the traditional ECG curve and some of its most frequent waveforms. The traditional ECG curve comprises several distinct components: the P-wave, QRS complex, ST segment, T wave, and occasionally the U wave. Each component corresponds to a specific electrical event within the heart.

- The P-wave and PR interval: The P-wave, which shows the electrical impulse propagating through the atria, is indicative of atrial depolarization. The time it takes for an electrical impulse to get from the atria to the ventricles is then measured

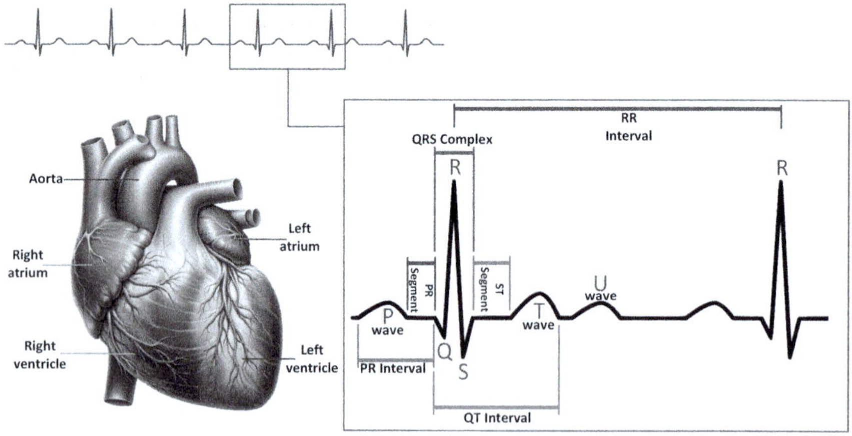


**Fig. 3** Electrocardiogram (ECG) waveforms

by the PR interval. It is essential for evaluating the function of the atrioventricular node. An extended PR interval may indicate a ventricular atrial conduction delay.
- The QRS Complex: It symbolizes ventricular depolarization and is arguably the most important element for understanding an electrocardiogram. Electrical conduction through the ventricles is revealed by the length and form of the QRS complex. Ventricular hypertrophy or problems with the ventricular conduction route may be indicated by abnormalities in the QRS complex.
- The ST Segment: The cardiac action potential's plateau phase is correlated with this area of the ECG. Careful examination of the ST segment is essential for detecting acute coronary events because alterations in it are sensitive markers of myocardial ischemia or infarction.
- The T Wave: It depicts ventricular repolarization. Abnormalities in the T wave can be indicative of electrolyte imbalances, ischemia, or other cardiac conditions. It is essential for a precise diagnosis to note the symmetry and contour of the T wave.
- The QT Interval: This measures the total time for ventricular depolarization and repolarization, essentially reflecting the electrical systole of the ventricles. The length of the QT interval is subject to change with heart rate; hence, it is adjusted for rate to assess its normality accurately.
- The RR Interval: A measure of heart rate and rhythm, the RR interval is the interval between two successive R waves. It is an essential measure for determining heart rate variability (HRV), which offers insightful data regarding the autonomic control of heart rate.

To understand the signals and make programs for automated analysis, you need to know how these ECG parts work together. The way ECGs are interpreted is changing a lot thanks to machine learning and other advanced computing methods like PEL. DE training and advanced partitioning methods are used by PEL to improve the accuracy and reliability of diagnosing cardiac disease by classifying ECG signals.

## 3 Methodology

The PEL is a training approach to classify ECG data presented in this study. The main objective of PEL is to optimize the weights and biases of a MLP. Initially, a real-valued representation is used to encode the MLP's weights and biases. An objective function based on classification error is then defined to guide the optimization process. PEL begins with a global search using an enhanced DE algorithm, which incorporates the standard DE mechanisms alongside opposition-based learning and a region-based strategy. The result of the global search gives the local search its starting points, making sure that the whole solution space is explored. The suggested PEL method involves preprocessing ECG signals to make them more accurate, feature transformation to help with classification, and optimization of neural network settings to accurately tell the difference between "normal" and "abnormal" ECG patterns. Figure 4 shows the overall architecture of the proposed method.

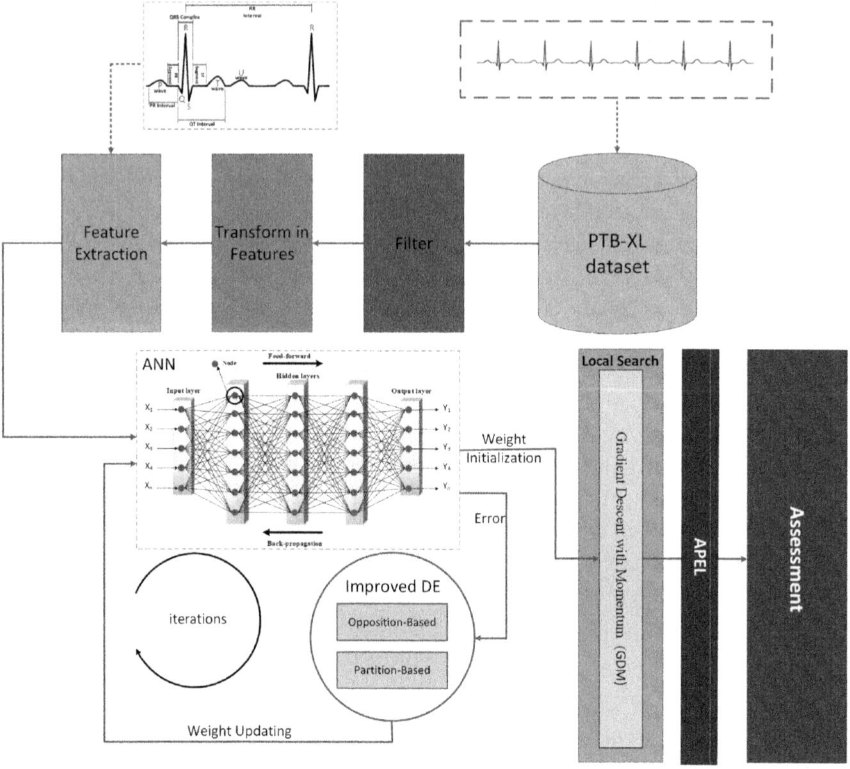


**Fig. 4** The PEL method's operating procedure

## 3.1 Dataset

The PTB-XL dataset, a substantial and well-annotated ECG collection, serves as the foundation for this research. It includes various heart conditions, offering a rich ground for developing and evaluating classification algorithms. The dataset consists of recordings from a diverse demographic, providing a balanced representation across different age groups and genders. The data used here are sampled at 100 Hz, and superclass labels are transformed into a binary format-"Normal" (1) and "Not Normal" (0)-to streamline the classification process.

## 3.2 Preprocessing and Feature Extraction

As part of preprocessing, the ECG signal is normalized and artifacts are removed using a Butterworth filter. A changed Hampel Filter [24] helps find and deal with outliers, and the Heartpy toolkit [25] has been changed to fit the needs of the dataset.

This step before training the model is very important for getting correct feature extraction and training the model well. In the PEL methodology, feature extraction plays a pivotal role. The ECG signal is transformed into a rich set of features that capture both the statistical and dynamic characteristics of the heart's activity. In order to correctly classify ECG signals, which show both normal and abnormal heart rhythms in great detail, these traits are necessary. These include both static and moving parts, such as heart rate variability (HRV) measures, and statistical values like mean and standard deviation. The extracted features and their corresponding equations are as follows:

- **BPM (Beats Per Minute)**: A measure of the heart rate, representing the average number of heartbeats per minute.
- **IBI (Interbeat Interval)**: The average time interval between successive heartbeats, reflecting the rhythm of the heart.
- **SDNN (Standard Deviation of NN intervals)**: Represents overall heart rate variability and indicates the autonomic nervous system's control over the heart.
- **SDSD (Standard Deviation of Successive Differences)**: Measures short-term variations in heart rate and is indicative of beat-to-beat variability.
- **RMSSD (Root Mean Square of Successive Differences)**: Another metric for assessing short-term variability in heart rate, particularly influenced by the parasympathetic nervous system.
- **pNN50 and pNN20**: The percentages of pairs of successive NN intervals that differ by more than 50 ms and 20 ms, respectively, providing insight into the presence of arrhythmic variations.
- **HR_MAD (Median Absolute Deviation of RR intervals)**: A robust measure of heart rate variability, less sensitive to outliers than SDNN.
- **SD1 and SD2**: Derived from a Poincaré plot, these standard deviations reflect the short-term and long-term variations in heart rate variability.
- **Ratio**: The ratio of SD1 to SD2, providing a single value that reflects the overall shape of the Poincaré plot.
- **Breathing Rate**: Extracted from the ECG signal, it provides information about the respiratory activity.

The equations of main features taken from the ECG signal are shown in Table 1. In the feature extraction formulas, the following notations are used:

- $\overline{RR}$ denotes the average of all RR intervals.
- $HR_{\text{avg}}$ is the average heart rate.
- $RR_{\text{dif}}$ represents the difference between successive RR intervals.

The signal's amplitude and distribution characteristics are some of the statistical aspects that give a broad picture of its properties. Dynamic features, on the other hand, explore the temporal patterns of the heart, such as the variability in time between consecutive heartbeats-commonly known as the RR interval. HRV, derived from the RR interval, has been demonstrated to yield vital insights into cardiac function and is widely used in diagnosing various heart-related conditions.

**Table 1** Equations related to feature extraction

| Feature | Formula |
|---|---|
| BPM | $\frac{60}{RR}$ |
| IBI | $\frac{6000}{HR_{avg}}$ |
| SDNN | $\sqrt{\frac{1}{N}\sum(RR_i - \overline{RR})^2}$ |
| SDSD | $\sqrt{\frac{1}{N-1}\sum(RR_i - RR_{i+1})^2}$ |
| RMSSD | $\sqrt{\frac{1}{N-2}\sum(RR_i - RR_{i+1})^2}$ |
| pNN50 | $\frac{NN50}{N} \times 100$ |
| pNN20 | $\frac{NN20}{N} \times 100$ |
| HR_MAD | $\frac{1}{N}\sum \lvert RR_i - \text{Median}(RR)\rvert$ |
| SD1 | $\sqrt{\text{var}(d_1)}$ |
| SD2 | $\sqrt{\text{var}(d_2)}$ |
| Ratio | $\frac{SD1}{SD2}$ |

In the PEL approach, the features are leveraged to feed into the MLP, with the aim of distinguishing "Normal" from "Not Normal" ECG signals. This binary classification approach simplifies the detection process, allowing for a clear delineation between typical heart rhythms and those that may indicate a cardiac anomaly.

## 3.3 *Enhancement of MLP Through Advanced DE Optimization*

In this research, a refined methodology for training MLPs is investigated, designed for the classification of ECG signals. The technique relies on an advanced form of the DE algorithm, carefully crafted to refine the MLP's performance. The core components of this sophisticated approach are detailed in the following.

### 3.3.1 Model Representation and Evaluation

At the heart of the proposed method lies the MLP, a neural architecture whose synaptic strengths and biases are compactly represented in a vectorial form. Each segment of this vector encapsulates the MLP's structural elements, as depicted in the model representation schematic, shown in Fig. 5.

The goal is to minimize classification errors, hence adopting a function that pivots on the discrepancies between predicted and actual classifications. The metric of focus

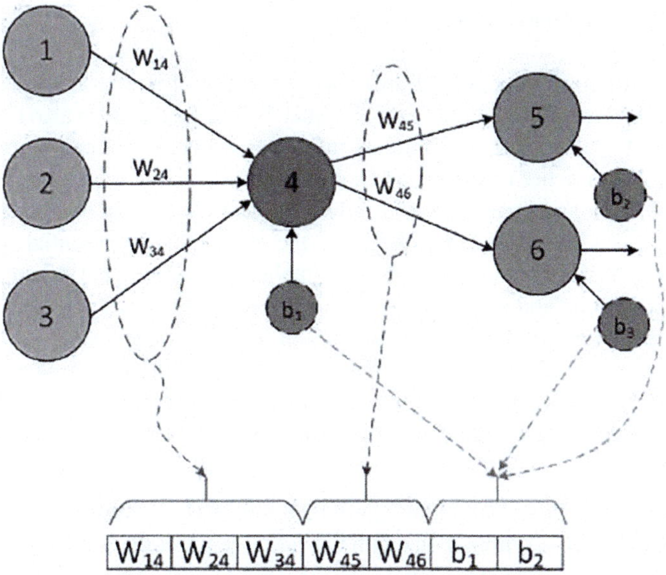


**Fig. 5** The PEL algorithm utilizes a candidate solution representation. Top: the network; Bottom: the structure resulting from the candidate solutions

is the error percentage, denoted by Eq. 10, which contrasts the forecasted output against the real-world data, granting a mirror to the model's precision.

$$E = \frac{100}{P} \sum_{p=1}^{P} \xi(x_p) \tag{10}$$

where $\xi(x_p)$ is defined as:

$$\xi(x_p) = \begin{cases} 1 & \text{if } o_p \neq d_p \\ 0 & \text{otherwise} \end{cases} \tag{11}$$

and where:

- $E$ represents the classification error.
- $P$ is the total number of test samples.
- $x_p$ is the $p$th test sample.
- $o_p$ is the predicted output.
- $d_p$ is the desired output.

### 3.3.2 Partition-Driven Strategy

To enhance the conventional DE framework, a partition-driven strategy is integrated, deploying the well-known k-means technique. This strategic move also brings about the fusion of multiple parent solutions into Partition centroids and invigorates the algorithm with a richer crossover mechanism, enhancing the search process.

The selection and generation of candidate solutions are random yet purposeful, leveraging the k-means foundation. Then, the assortment of potential solutions is sifted through, sieving out the finest to establish the subsequent population. This cycle of generation and selection, delicately paced by a defined partitioning period, ensures a rhythmic yet dynamic exploration of the solution space.

The centroid of a partition represents an amalgamation of multiple progenitors, encompassing all potential solutions within that particular partition. The update process is executed using the General Population-Based Algorithm (GPBA) as outlined below [26]:

- **Selection:** Candidate solutions should be selected randomly from the entire population. This is synonymous with the initial point of the k-means algorithm.
- **Generation:** Generate $m$ potential solutions (set $A$). PEL utilizes the k-means algorithm to generate fresh candidate solutions, with each cluster center representing a distinct candidate solution.
- **Substitution:** Choose $m$ potential solutions (set $A$) for substitution from the current population. PEL randomly selects $m$ potential solutions.
- **Update:** Choose the $m$ most optimal solutions from the set of all possible combinations $A \cup B$ and designate them as $B$. The resulting population is calculated by taking the union of the set difference between $P$ and $B$, and the set $B$.

Every instance of PEL does not utilize a partitioning algorithm. Partitioning algorithms are regularly conducted depending on a clustering period parameter, which follows a similar technique as described in [27].

### 3.3.3 Opposition, Quasi-Opposition Approaches and Local Search

In order to enhance the strength and reliability of the model, an opposition-based learning technique has been integrated, significantly expanding the model's ability to cover a wide range of solutions. Quasi-Opposition-Based Learning (QOBL) is employed as a novel approach to enhance the search process, aiming to identify the global optimum with a higher level of precision.

The process begins with the creation of a random population through the initialization process. Following this, QOBL is utilized to construct an opposing group. The combination of these two populations at the beginning and the subsequent updates significantly increases the likelihood of finding the best solutions.

Within the framework of the suggested method, the implementation of local search is the final step. In order to accomplish this goal, a method based on gradients, known as GDM, is utilized. To put it another way, the weights obtained via the application of PEL are used as the first step in the gradient-based technique.

The approach combines the DE technique with a regional focus and incorporates Quasi-Opposition-Based Learning (QOBL) to enhance the search potential in a wider and more comprehensive manner. The MLP model utilizes the GDM technique to locally modify the weight and bias modifications, adding a subtle and precise touch. The exclusive emphasis on GDM enables a thorough optimization process that is carefully calibrated to effectively navigate the complex field of ECG signal categorization with improved accuracy and flexibility.

#### 3.3.4 The proposed DE Trainer General Structure

The methodology expands upon previous methodologies and transforms the training approach for MLPs. By utilizing the complex capabilities of GDM inside a DE framework, a foundation for significant improvements in the precision of ECG signal analysis is being created. The PEL method is illustrated in Algorithm 1.

### *3.4 Evaluation Metrics*

The efficacy of the advanced strategy for ECG signal classification was meticulously gauged using a spectrum of performance metrics-accuracy, sensitivity, specificity, precision, F-score, and G-mean. These metrics, derived from the elements of the confusion matrix-true positives (TP), true negatives (TN), false positives (FP), and false negatives (FN)-furnish a multi-faceted evaluation of the classifier's ability to accurately discern positive and negative classifications within the context of ECG signal analysis.

- **True Positives (TP)**: Instances correctly identified as positive by the classifier.
- **True Negatives (TN)**: Instances correctly identified as negative by the classifier.
- **False Positives (FP)**: Instances incorrectly identified as positive by the classifier.
- **False Negatives (FN)**: Instances incorrectly identified as negative by the classifier.

The evaluation metrics are calculated as follows:

- **Accuracy(ACC):** The ratio of correctly predicted instances to the total instances.

$$\text{Accuracy} = \frac{TP + TN}{TP + TN + FP + FN} \tag{12}$$

- **Sensitivity(SEN):** Also known as recall, it measures the proportion of actual positives that are correctly identified.

$$\text{Sensitivity} = \frac{TP}{TP + FN} \tag{13}$$

**Algorithm 1** Pseudocode for PEL

**Input:** ECG data, Algorithmic Parameters
**Output:** Refined Weights
Begin the process
Establish $Size_{population}$ // Define the size of the population
    $MaxEvaluations$ // Set the maximum evaluation limit
    $Rate_{jump}$ // Determine the jumping rate
    $Period_{cluster}$ // Define the clustering interval
    $EvalCount \leftarrow 0$ // Initiate evaluation count
    $Iteration \leftarrow 1$ // Initiate iteration count
Initial population $Population$ with random distributions
Calculate fitness of $Population$ as per defined objective function
Define $Dimensionality$
Allocate $PopulationSize$
Initialize population $PopulationInitial$
Set boundaries $Lower$ and $Upper$
**for** each individual $i$ in $PopulationSize$ **do**
    **for** each dimension $j$ in $Dimensionality$ **do**
        $OppPopulation(i, j) = \text{rand}\left[\frac{Lower(i,j)+Upper(i,j)}{2}, (Lower(i, j) + Upper(i, j) - Population(i, j))\right]$
    **end for**
**end for**
Determine the fitness for each candidate solution
Update $Population$ by selecting the top-performing solutions
Increment $EvalCount$ appropriately
For each member of the population, apply:
    Mutation operation
    Crossover operation
    Objective function evaluation
    Selection operation
Increment $EvalCount$ by population size
If $Iteration$ modulus $Period_{cluster} == 0$
    Determine number of clusters
    Perform clustering on the population
    Identify new potential solutions and update population
End if
If random chance is below $Rate_{jump}$
    Update population based on advanced exploration techniques
End if
Increment $Iteration$
Utilize Gradient Descent with Momentum (GDM) for local optimization
Assign the best solution for local search initiation
Execute GDM for local refinement
Repeat mutation operation steps if $EvalCount$ is below threshold
Conclude the process

- **Specificity(SPEC):** Measures the proportion of actual negatives that are correctly identified.

$$\text{Specificity} = \frac{TN}{TN + FP} \tag{14}$$

- **Precision(PREC):** The ratio of true positive instances to the total predicted positives.

$$\text{Precision} = \frac{TP}{TP + FP} \tag{15}$$

- **F-score(F-SCR):** The harmonic mean of precision and sensitivity.

$$\text{F-score} = \frac{2 \times TP}{2 \times TP + FP + FN} \tag{16}$$

- **G-mean(G-MN):** The geometric mean of sensitivity and specificity.

$$\text{G-mean} = \sqrt{\text{Sensitivity} \times \text{Specificity}} \tag{17}$$

In addition to the traditional measurements, the win/tie/loss (w/t/l) statistic was used to compare the local-search-enhanced algorithm with the baseline strategy and determine if it performed better, equally, or worse.

A novel metric called Error Enhancement (EE) is proposed in this study. It is calculated for each method using the following formula:

$$EE = \left( \frac{\text{Error}_{\text{GDM\_alg}}(\beta) - \text{Error}_{\text{PEL\_alg}}(\beta)}{\text{Error}_{\text{GDM\_alg}}(\beta)} \right) \times 100 \tag{18}$$

where $\text{Error}_{\text{alg}}(\beta) = 100 - \beta_{\text{alg}}$ and $\beta$ represents the evaluation metric of choice.

The significance of these metrics is important in ECG classification endeavors. They not only give a detailed insight into the classifier's multiples performance but also guide in refining the method and benchmarking it against other classifiers in this domain.

## 4 Experimental Results and Discussion

In this section, the proposed model for classifying ECG signals is critically assessed, with an emphasis on its efficacy and stability. The investigation utilized the PTB-XL dataset, a comprehensive compilation of 21,799 clinical ECGs spanning 12 leads. Following an extensive preprocessing routine, 13 important features were extracted from the dataset, such as BPM, IBI, SDNN, SDSD, etc.

The next step was to utilize a MLP that had been enhanced with a differential evolution strategy and a local search phase to clearly identify the relationship between the collected features and their respective classes. Focusing exclusively on GDM for the local search, the approach was designed to optimize the MLP's performance.

Uniformity was maintained across all PBMHs used in the study, with each algorithm subjected to 25,000 function evaluations. This parameter is consistent with the iteration counts typical of conventional algorithms. The population for each PBMH was set at 50 entities. For the differential evolution parameterization, a crossover probability of 0.9, a scaling factor of 0.5, and a jumping rate of 0.3 were adopted,

**Table 2** Performance metrics for GDM and PEL algorithms

| Metric | Algorithm | Mean | Std. | Min. | Max. | Med. | Rank | W/T/L |
|---|---|---|---|---|---|---|---|---|
| Accuracy | GDM | 68.21 | 2.69 | 64.95 | 74.30 | 67.68 | 8 | W |
| | PEL | 72.11 | 3.06 | 68.46 | 77.80 | 71.45 | 6 | |
| Sensitivity | GDM | 83.50 | 4.85 | 75.68 | 89.96 | 82.63 | 2 | W |
| | PEL | 87.91 | 4.32 | 81.08 | 94.98 | 88.39 | 1 | |
| Specificity | GDM | 44.89 | 12.12 | 30.77 | 66.47 | 43.65 | 12 | W |
| | PEL | 48.03 | 13.21 | 32.94 | 69.23 | 43.53 | 11 | |
| Precision | GDM | 70.18 | 3.99 | 66.09 | 78.24 | 68.63 | 7 | W |
| | PEL | 72.53 | 4.60 | 67.89 | 80.60 | 70.65 | 5 | |
| F-Score | GDM | 76.05 | 1.47 | 72.83 | 78.85 | 75.98 | 4 | W |
| | PEL | 79.25 | 1.39 | 77.08 | 81.97 | 79.64 | 3 | |
| G-Mean | GDM | 60.46 | 6.46 | 52.04 | 72.67 | 59.37 | 10 | W |
| | PEL | 64.17 | 7.23 | 55.47 | 75.98 | 62.29 | 9 | |
| Total W/T/L | | | | | | | | 6/0/0 |

with the clustering period and regularization parameters fixed at 10 and 0.1, respectively. This rigorous experimental framework was established to ensure a robust and replicable evaluation of the proposed model's performance.

The application of advanced optimization techniques to improve the performance of machine learning models is illustrated in Table 2, which presents a comprehensive evaluation of GDM and the proposed PEL algorithm. The metrics assessed include accuracy, sensitivity, specificity, precision, F-score, and G-mean, offering a robust analysis of each algorithm's capacity to classify ECG signals.

PEL demonstrates superior accuracy over GDM, with an increase in mean accuracy from 68.21 to 72.11%. This enhancement suggests that the partitioning strategy within the evolutionary process is effective in navigating the search space more efficiently, leading to a more accurate model.

In terms of sensitivity, PEL again outshines GDM, which is indicative of PEL's improved ability to correctly classify true positive rates. Specificity, however, shows a more modest improvement, reflecting the challenge in correctly classifying true negatives, which is a common issue in imbalanced datasets.

The precision metric sees a noticeable increase from 70.18 to 72.53% when employing PEL, implying that the model is more discerning in its predictions. The F-score, which balances precision and sensitivity, also shows an increase, confirming that PEL maintains a balance between these two crucial aspects.

Finally, the G-mean, an indicator of a model's balance between sensitivity and specificity, shows an increase, indicating that PEL maintains an equilibrium between identifying positive and negative classes.

The Win/Tie/Loss (W/T/L) metric further corroborates the superiority of PEL over GDM, as demonstrated by the consistent 'W' across all metrics. This signifies that PEL not only wins in the majority of performance measures but also confirms the robustness and reliability of PEL in ECG signal classification.

The advancements in PEL, therefore, represent a significant step forward in ECG signal classification. The algorithm's fine-tuned balance between exploration and exploitation, achieved through its partitioning-based approach, translates into a model that is not only accurate but also reliable in its classifications. As such, these results support the incorporation of PEL as a potent tool in biomedical signal processing, with implications for enhancing diagnostic procedures and patient outcomes.

As determined by the EE computation in Eq. 18, the results detailed the error enhancement statistics of the PEL method in comparison to its original base algorithm in Table 3, providing a quantitative evaluation of the Enhanced Learning through Partitioning (PEL) compared to the standard Gradient Descent with Momentum (GDM) approach. The performance metrics indicate that PEL offers substantial improvements in several key areas. For instance, PEL shows a remarkable 26.73% improvement in Sensitivity, underscoring its ability to correctly identify true positive instances-an essential aspect of ECG signal classification. This enhancement could lead to more reliable detection of cardiac events, which is vital in clinical settings.

Accuracy sees a 12.24% improvement, suggesting that the overall predictive performance of PEL is more precise than that of GDM. Given the critical nature of ECG interpretation, this heightened accuracy could significantly affect patient outcomes by reducing misdiagnoses.

Moreover, the F-Score-which balances precision and recall-improved by 13.38%. This balanced improvement is critical in the medical field, where the trade-off between false positives and false negatives has to be carefully managed.

The G-Mean, which assesses the balance between sensitivity and specificity, improved by 9.38%. This is particularly encouraging as it suggests that PEL does not sacrifice specificity to improve sensitivity or vice versa.

However, the modest 5.69% increase in Specificity indicates there may be opportunities for refinement. Specificity is crucial for avoiding false alarms, which can lead to unnecessary anxiety and medical procedures.

The Table 3 also reveals an important aspect of machine learning models: the trade-off between various performance metrics. While PEL exhibits superior performance in most areas, there is a nuanced interplay between the different metrics. For example, a model might excel in sensitivity but have room for improvement in specificity. This shows how important it is to consider all aspects when evaluating and choosing models, especially in the medical field where each measure can have significant effects.

**Table 3** Error enhancement for PEL compared to GDM

| Algorithm | ACC | SEN | SPEC | PREC | F-SCR | G-MN |
|---|---|---|---|---|---|---|
| PEL versus GDM | 12.24% | 26.73% | 5.69% | 7.86% | 13.38% | 9.38% |

The results clearly show that PEL significantly improves most of the criteria. However, considering these improvements, it is important to note that no single algorithm is superior in all aspects. This means that the choice of algorithm should be measured according to the specific needs, advantages, and operational limitations.

The comprehensive gains observed across these performance metrics affirm the potential of PEL in the realm of ECG signal classification. Its ability to consistently outperform GDM suggests that PEL could be an invaluable tool for enhancing diagnostic accuracy and reliability in clinical practice.

Figure 6 presents a line chart that clearly shows the trends across multiple metrics for the PEL and GDM algorithms. The line trajectories offer a longitudinal perspective on the performance, showing how the PEL algorithm maintains a consistent lead over GDM, particularly notable in the accuracy and sensitivity metrics. The PEL's ascendency in mean ranks suggests an algorithm refined to minimize classification errors, a pivotal attribute in ECG signal analysis where misinterpretations can have serious repercussions.

The grouped bar chart in Fig. 7 corroborates these findings with a comparative clarity that only juxtaposed bars can provide. Each pair of bars serves as a stark visual reminder of the enhancements PEL brings to the table, with higher bars in key metrics like precision and F-score signaling its heightened ability to discern true positives from false alarms. Despite GDM's robustness, PEL's bars reach further into the echelons of improved performance, heralding its advantages in ECG signal classification tasks.

Figure 8 displays a box plot that illustrates the variability and distribution of the performance scores of the algorithms. The smaller interquartile range of PEL suggests a steady output, unaffected by the unpredictable fluctuations in data-a confirmation

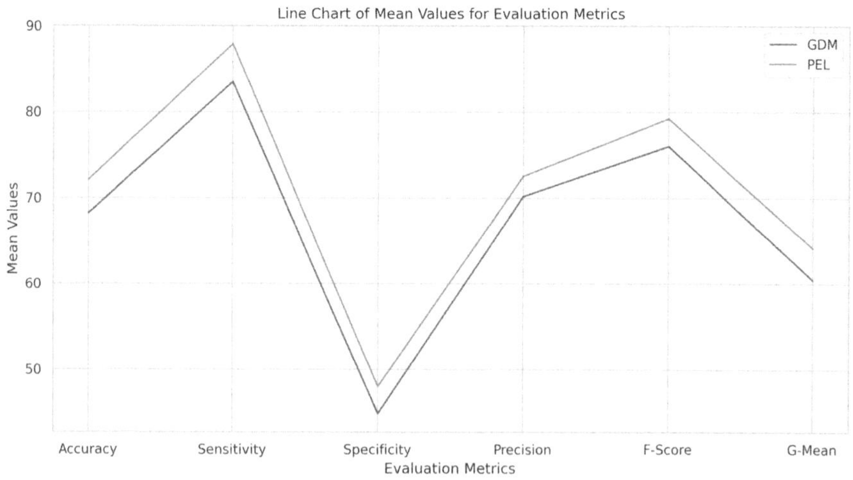


**Fig. 6** Line chart comparing mean values of evaluation metrics for GEM and PEL algorithm

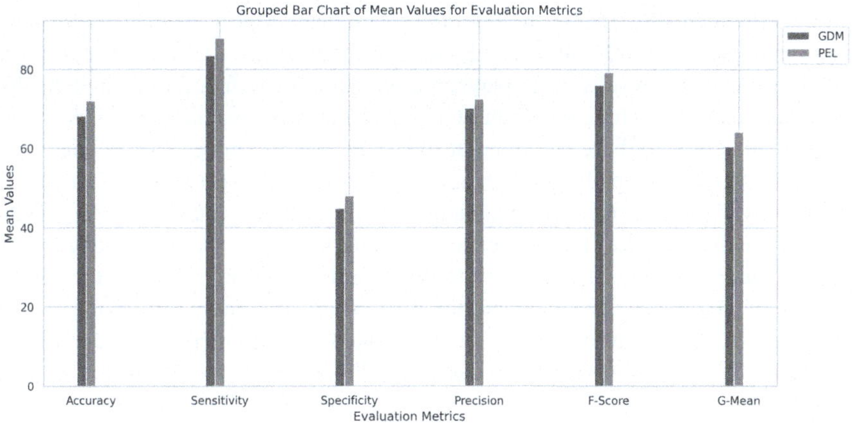


**Fig. 7** Grouped bar chart illustrating performance improvements in evaluation metrics for PEL versus GDM algorithm

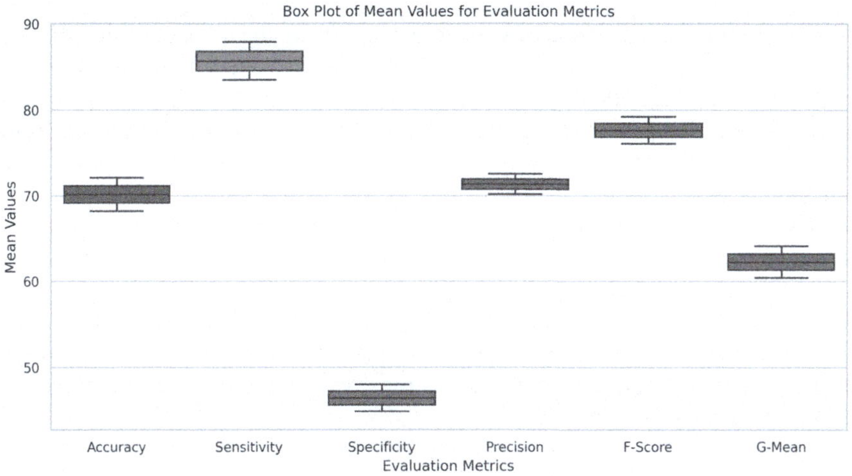


**Fig. 8** Box plot demonstrating the distribution and variability of evaluation metrics for PEL versus GDM algorithm

of its reliability. On the other hand, GDM's broader box plot indicates a significant amount of variation that could potentially disrupt the consistency needed in medical diagnostic procedures.

By considering Figs. 6, 7, and 8 together, it becomes evident that PEL has a higher capability, resulting in a cohesive narrative. Although GDM has a good starting point, the visual representation in the statistics clearly demonstrates that PEL has notable benefits in important performance areas. PEL's durability, stability, and correctness are undeniable, making it stand out in graphic depictions.

To summarize, the findings highlight the significant improvements that PEL provides compared to GDM in different situations. Although PEL is often superior to GDM in enhancing results, its implementation requires careful evaluation of the distinct characteristics inherent in each ECG signal categorization scenario. It is crucial to customize PEL to these particular attributes in order to properly exploit its capabilities. The experimental evidence, along with visual data, establishes that PEL is not just an alternative but also a high approach for furthering the study of ECG data. The thorough analysis of PEL's enhanced algorithms emphasizes the significance of using a variety of evaluation criteria to obtain a deep understanding of their capabilities. In the field of ECG data interpretation, where accurately identifying cardiac abnormalities and creating tailored treatment plans are extremely important, PEL is a valuable tool for improving the accuracy of classification algorithms.

## 5 Conclusion and Future Works

This study presents an advanced method for improving the training of the MLPs in ECG signal classification. The method combines the DE, opposition-based learning, and region-based techniques to get better results. Through the strategic implementation of these techniques, we effectively enhanced the neural network's weights, resulting in a substantial improvement in classification accuracy compared to conventional training methods.

The core of our strategy is based on the intricate integration of feature extraction and sophisticated optimization techniques. The DE method is employed as the central component of our weight optimization. It is enhanced with opposition-based learning to increase its ability to explore, and is supplemented by region-based strategies that allow for a targeted search inside the solution space. The research studies demonstrate that this combination of techniques is a strong way to improve the current level of ECG signal classification.

The precise experimentation confirms the effectiveness of the proposed method, which not only achieves its main purpose but also exhibits a high level of reliability and robustness. The MLP achieves remarkable performance in ECG classification tasks by utilizing collaborative partitioning, quasi-oppositional reasoning, and local search techniques.

While the current research has shown promising results, there are several directions in which this work can be extended and enhanced. The potential areas for future exploration include:

- **Architectural Optimization:** Investigating the optimization of the neural network architecture to improve the classification of ECG signals.
- **Multi-Objective Optimization:** Expanding the research to include multi-objective optimization methods in order to improve the model in a more comprehensive way.

- **Chaotic Dynamics:** Integrating chaotic variables into the optimization process to explore new landscapes and potentially improve convergence and solution variety.
- **Wider Application Scope:** Applying the refined DE algorithm to other complex tasks, such as image segmentation, to test its flexibility across various domains.
- **Real-time Classification:** Implementing and assessing the algorithm within a real-time classification framework to evaluate its effectiveness and adaptability in live scenarios.
- **Transferability and Adaptation:** Assessing the transferability of the proposed method to other domains and applications, broadening the impact of the research.
- **Explainable AI:** Enhancing the transparency and explainability of the algorithm to align with the increasing demand for understandable AI decision-making processes.
- **Deep Learning Synergy:** Exploring the synergies between the proposed methods and deep learning architectures, particularly in the context of sequential and time-dependent data analysis.

**Acknowledgements** The research work reported in this paper has received funding from the Pilots for Healthy and Active Ageing (Pharaon) project of the European Union's Horizon 2020 research and innovation programme under the grant agreement no. 857188. We express our sincere gratitude to the Fundação para a Ciência e a Tecnologia (FCT), Portugal, for the generous grant of computational resources that have been pivotal in conducting our research. This project, identified by reference 2023.10386.CPCA.A0, was made possible through the National Advanced Computing Network (RNCA), under the operational center HPCUÉ_Oblivion.

## References

1. Singh, G.K., Siahpush, M., Azuine, R.E., Williams, S.D.: Increasing area deprivation and socioeconomic inequalities in heart disease, stroke, and cardiovascular disease mortality among working age populations, United States, 1969–2011. Int. J. MCH AIDS **3**(2), 119 (2015)
2. Zhang, X., Gu, K., Miao, S., Zhang, X., Yin, Y., Wan, C., Yu, Y., Hu, J., Wang, Z., Shan, T.: others: Automated detection of cardiovascular disease by electrocardiogram signal analysis: a deep learning system. Cardiovascular Diagn. Therapy **10**(2), 227 (2020)
3. Abubaker, M.B., Babayiğit, B.: Detection of cardiovascular diseases in ECG images using machine learning and deep learning methods. IEEE Trans. Artif. Intell. **4**(2), 373–382 (2022)
4. Salvador, C., Felizardo, V., Zacarias, H., Souza-Pereira, L., Pourvahab, M., Pombo, N., Garcia, N.M.: Epileptic seizure prediction using EEG peripheral channels. In: 2023 IEEE 7th Portuguese Meeting on Bioengineering (ENBENG), pp. 60–63 (2023)
5. Zacarias, H., Felizardo, V., Souza-Pereira, L., Pinho, A., Ramos, S., Pourvahab, M., Garcia, N.: Gender classification using nonstandard ECG Signals-a conceptual framework of implementation. In: EAI International Conference on IoT Technologies for HealthCare, pp. 108–120 (2022)
6. Zacarias, H., Marques, J.A.L., Felizardo, V., Pourvahab, M., Garcia, N.M.: ECG forecasting system based on long short-term memory. Bioengineering **11**(1), 89 (2024)
7. Mousavirad, S.J., Schaefer, G., Jalali, S.M.J., Korovin, I.: A benchmark of recent population-based metaheuristic algorithms for multi-layer neural network training. In: Proceedings of the 2020 Genetic and Evolutionary Computation Conference Companion, pp. 1402–1408 (2020)
8. Sree Ranjini, K.S.: A study on performance of MHDA in training MLPs. Eng. Comput. **36**(6), 1820–1834 (2019)

9. Camargo, L.C., Tissot, H.C., Pozo, A.T.R.: Use of backpropagation and differential evolution algorithms to training MLPs. In: 2012 31st International Conference of the Chilean Computer Science Society, pp. 78–86 (2012)
10. Khudhur, H.M., Ibraheem, K.I.: Metaheuristic optimization algorithm based on the two-step Adams-Bashforth method in training multi-layer perceptron's. East.-Eur. J. Enter. Technol. **2**(4), 116 (2022)
11. Dweekat, O.Y., Lam, S.S.: Cervical cancer diagnosis using an integrated system of principal component analysis, genetic algorithm, and multilayer perceptron. Healthcare **10**(10), 2002 (2022)
12. Afentaki, F., Hefenbrock, M., Zervakis, G., Tahoori, M.B.: Embedding hardware approximations in discrete genetic-based training for printed MLPs (2024). arXiv:2402.02930
13. Al-Areeq, A.M., Saleh, R.A.A., Ghanim, A.A.J., Ghaleb, M., Al-Areeq, N.M., Al-Wajih, E.: Flood hazard assessment in Yemen using a novel hybrid approach of Grey Wolf and Levenberg Marquardt optimizers. Geocarto Int. **38**(1), 2243884 (2023)
14. Tunay, M., Pashaei, E., Pashaei, E.: Hybrid hypercube optimization search algorithm and multilayer perceptron neural network for medical data classification. Comput. Intell. Neurosci. **2022** (2022)
15. Abboud, A., Abouaissa, A., Shahin, A., Mazraani, R., others: A hybrid aggregation approach for federated learning to improve energy consumption in smart buildings. In: 2023 International Wireless Communications and Mobile Computing (IWCMC), pp. 854–859 (2023)
16. Islam, N., Oyekan, J.: An improved hybrid multi-objective particle swarm optimization to enhance convergence and diversity. In: Proceedings of the Companion Conference on Genetic and Evolutionary Computation, pp. 1793–1802 (2023)
17. Mousavirad, S.J., Gandomi, A.H., Homayoun, H.: A clustering-based differential evolution boosted by a regularisation-based objective function and a local refinement for neural network training. In: 2022 IEEE Congress on Evolutionary Computation (CEC), pp. 1–8 (2022)
18. Wagner, P., Strodthoff, N., Bousseljot, R., Samek, W., Schaeffter, T.: PTB-XL, a large publicly available electrocardiography dataset (version 1.0. 1). Physionet (2020)
19. Abiodun, O.I., Jantan, A., Omolara, A.E., Dada, K.V., Umar, A.M., Linus, O.U., Arshad, H., Kazaure, A.A., Gana, U., Kiru, M.U.: Comprehensive review of artificial neural network applications to pattern recognition. IEEE Access **7**, 158820–158846 (2019)
20. Hussain, Md.T., Sarwar, A., Tariq, M., Urooj, S., BaQais, A., Hossain, Md.A.: An evaluation of ANN algorithm performance for MPPT energy harvesting in solar PV systems. Sustainability **15**(14), 11144 (2023)
21. Pourvahab, M., Mousavirad, S.J., Felizardo, V., Pombo, N., Zacarias, H., Mohammadigheymasi, H., Pais, S., Jafari, S.N., Garcia, N.M.: A cluster-based opposition differential evolution algorithm boosted by a local search for ECG signal classification (2023). arXiv:2305.02731
22. Peng, D., Gui, Z., Wang, D., Ma, Y., Huang, Z., Zhou, Yu., Wu, H.: Clustering by measuring local direction centrality for data with heterogeneous density and weak connectivity. Nat. Commun. **13**(1), 5455 (2022)
23. Shilaja, C.: Implementation of differential evolution algorithm and its variants for optimal scheduling of distributed generations. Int. J. Commun. Syst. **34**(6), e4318 (2021)
24. Nagahawatte, N.D., Cheng, L.K., Avci, R., Bear, L.R., Paskaranandavadivel, N.: A generalized framework for pacing artifact removal using a Hampel filter. In: 2022 44th Annual International Conference of the IEEE Engineering in Medicine & Biology Society (EMBC), pp. 2009–2012 (2022)
25. Vićentić, T., Rašljić Rafajilović, M., Ilić, S.D., Koteska, B., Madevska Bogdanova, A., Pašti, I.A., Lehocki, F., Spasenović, M.: Laser-induced graphene for heartbeat monitoring with HeartPy analysis. Sensors **22**(17), 6326 (2022)
26. Lu, Z., Zhou, Y., Hao, J.-K.: A hybrid evolutionary algorithm for the clique partitioning problem. IEEE Trans. Cybern. **52**(9), 9391–9403 (2021)
27. Yang, Y., Liu, J., Tan, S.: A partition-based constrained multi-objective evolutionary algorithm. Swarm Evol. Comput. **66**, 100940 (2021)

# Enhancing Neural Network Generalisation with Improved Differential Evolution

**Seyed Jalaleddin Mousavirad, Seyyed Mohammad Tabatabaei, Davood Zabihzadeh, Mahshid Helali Moghadam, Mehran Pourvahab, and Diego Oliva**

**Abstract** Training feedforward neural networks (FFNN) is a complex process that requires determining the right weights and biases, crucial parameters that largely dictate the FFNN's effectiveness. Traditional methods like gradient descent are common in FFNN training but have limitations, including the risk of becoming trapped in local minima. Evolutionary algorithms present a solution to bypass these issues. This study takes into consideration the effectiveness of three differential evolution variations in FFNN training: the conventional DE, an opposition-based DE, and a novel centroid-based DE. The most successful among these, the centroid-based DE algorithm, called CQODTA, integrates a centroid-based method and dynamic opposition-based learning with differential evolution. The centroid method uses the centre point of the top-performing individuals as a reference for population updates, while the rest are adjusted using the usual crossover and mutation techniques. This method effectively affects the search space due to its reliance on the leading individuals. Simultaneously, the opposition-based learning concept, which takes the opposite

S. J. Mousavirad (✉)
Department of Computer and Electrical Engineering, Mid Sweden University, Sundsvall, Sweden
e-mail: seyedjalaleddin.mousavirad@miun.se

S. M. Tabatabaei
Department of Medical Informatics, Faculty of Medicine, Mashhad University of Medical Sciences, Mashhad, Iran

D. Zabihzadeh
Computer Engineering Department, Hakim Sabzevari University, Sabzevar, Iran

M. H. Moghadam
Mälardalens University, Västerås, Sweden

M. Pourvahab
Departamento de Informática, Universidade da Beira Interior, Covilha, Portugal

NOVA LINCS, NOVA University of Lisbon, Lisbon, Portugal

D. Oliva
Departamento de Ingeniería Electro-Fotónica, CUCEI, Universidad de Guadalajara, Guadalajara, Mexico

 
D. Oliva et al. (eds.), *Advances in Optimization Algorithms for Multidisciplinary Engineering Applications: From Classical Methods to AI-Enhanced Solutions*, Studies in Computational Intelligence 806, https://doi.org/10.1007/978-3-031-78440-8_19

of an existing solution, bolsters the algorithm's ability to search broadly. We compare our trio of DE-based training methods against 14 other variants. Our thorough testing demonstrates the DE-based trainers' adeptness at tackling various complex tasks.

## 1 Introduction

Feed-forward neural network(FFNN) is a widely adopted architecture in the domain of artificial neural networks (ANN) for addressing sophisticated classification and regression challenges [39, 45, 46]. These networks are structured around basic units known as neurons and the connections between them. In FFNNs, information flows unidirectionally from input to output through one or more hidden layers. Each link between neurons is assigned a weight that signifies its importance. The objective of training FFNNs is to adjust these weights to reduce the discrepancy between the actual outcomes and the model's predictions. Techniques grounded in gradients, notably the back-propagation algorithm, are extensively documented for their effectiveness, despite their inclination to settle at local optima [23].

Evolutionary algorithms (EA), including differential evolution (DE) [42], offer a dependable alternative for addressing the limitations of traditional algorithms. EAs can provide reasonable solutions for complex problems from lane-keeping systems in self-driving cars [18, 19] to image segmentation [26, 30]. Research comparing back-propagation (BP) and genetic algorithms (GA) in the context of FFNN training [40] highlighted GA's superiority in terms of effectiveness, and efficiency. Additionally, a study [11] introduced an enhanced GA designed for the rapid training of FFNNs, demonstrating through computational time that this revised approach surpasses traditional GA-based training methods in efficiency. Nawi et al. [33] proposed a hybrid technique that combines accelerated particle swarm optimization (PSO) with Levenberg Marquardt for enhanced convergence speed. Another study [41] introduced a variant of PSO designed for the training on medical datasets, with experimental results confirming its efficacy across various clinical datasets. Additional evolutionary strategies applied to FFNN training encompass a diverse array of algorithms such as the artificial bee colony [9], imperialist competitive algorithm [6, 22], firefly algorithm [14], grey wolf optimiser [3, 16], ant lion optimiser [44], dragonfly algorithm [10], sine cosine algorithm [17], whale optimisation algorithm [2], grasshopper optimization algorithm [8], and salp swarm algorithm [1], among others.

Differential evolution (DE) [42] is a recognised and efficacious EA known for its superior ability to tackle complex optimisation challenges, as evidenced by extensive research [4, 20, 21, 31, 32, 35] demonstrating its robust performance across various domains. Recent years have seen a surge in efforts to refine the DE algorithm [25, 27, 29], focusing on enhancing its capabilities. DE has three pivotal operations: mutation, crossover, and selection. The mutation process amalgamates information from diverse candidate solutions, whereas crossover facilitates the amalgamation of the mutant and target vectors, highlighting DE's strategic approach to solution refinement.

One of the main features of FFNNs is the ability to generalise. This means that once trained on a subset of data, FFNNs can make accurate predictions or decisions about new, unseen data that shares similar characteristics with the training set. Generalisation is a crucial aspect of neural network performance, indicating the model's capability to apply learned knowledge to novel situations, thus avoiding overfitting the training data. Achieving good generalisation requires careful design of the network architecture, including the number of layers and neurons, and the selection of appropriate training algorithms.

This paper aims to demonstrate the effectiveness of the DE algorithm in enhancing the generalisation capability of FFNNs. To achieve this, we introduce three variants of DE, including standard DE, opposition-learning-based DE, and a centroid-based DE.

The remainder of this paper is organised as follows. Section 2 briefly explains some background knowledge on neural networks, DE, and opposition-based learning. Section 3 introduces three DE-based approaches for FFNN training. Section 4 evaluates the algorithms on various benchmark problems. Finally, Sect. 5 concludes the paper.

## 2 Preliminaries

### 2.1 Differential Evolution

Differential Evolution (DE) is an exceptionally efficient evolutionary algorithm, acclaimed for its profound efficacy in diverse real world scenarios. Originating with an ensemble of $NP$ randomly generated candidates, this population-based strategy initiates a broad exploration of the search space. Central to DE's operation are three fundamental processes: mutation, crossover, and selection, each playing a crucial role in the algorithm's effectiveness.

**Mutation** The mutation is at the forefront of introducing variation, where for each candidate, a variant vector, $v_i = (v_{i,1}, v_{i,2}, \ldots, v_{i,D})$, is formulated as:

$$v_i = x_{r1} + F \cdot (x_{r2} - x_{r3}), \tag{1}$$

leveraging three distinct solutions $x_{r1}$, $x_{r2}$, and $x_{r3}$ from the pool, and a scaling factor $F$, to infuse diversity.

**Crossover** The crossover phase follows, crucial for generating a trial vector, $u_i = (u_{i,1}, u_{i,2}, \ldots, u_{i,D})$, through the integration of the mutant vector with the initial candidate. Employing a binomial crossover mechanism, it is described as:

$$u_{i,j} = \begin{cases} v_{i,j}, & \text{if } rand(0,1) \leq CR \text{ or } j = j_{rand} \\ x_{i,j}, & \text{otherwise.} \end{cases} \tag{2}$$

Here, $CR$ signifies the crossover rate, and $j_{rand}$ denotes a randomly chosen index, enriching the candidate diversity.

**Selection** Concluding the cycle, the selection process evaluates the trial vector against its progenitor, prioritising advancement based on their objective function evaluations, ensuring the perpetuation of superior solutions.

Refinement and adaptation of DE's core methodologies-mutation, crossover, and selection-herald new avenues for its application. By fine-tuning its parameters and integrating innovative selection mechanisms, DE's versatility and problem-solving prowess are significantly amplified. Such enhancements not only solidify DE's status as a premier tool in optimisation but also navigate it towards new horizons in both research and practical application realms.

Moreover, the convergence of DE with machine learning technologies marks a promising domain, potentially revolutionising our optimisation approach. This synergy between DE's evolutionary tactics and machine learning's predictive proficiency opens up unprecedented possibilities for tackling complex challenges across various fields.

## 2.2 Neural Network Training

Feed-forward neural networks (FFNN) are machine learning models, featuring neurons organised into layers. These networks typically consist of input, output, and hidden layers. Information flows in a unidirectional manner, with input values transmitted through connections to subsequent layers. The strength of each connection is governed by a weight parameter. Neurons employ activation functions, such as the commonly used Sigmoid function, to determine their output. During the training or learning phase, which involves examples, the optimisation of connection weights is crucial for enhancing the performance of FFNNs. Moreover, an activation function is employed to elicit neuron output. In this study, we utilised the Sigmoid function, recognised as one of the most prevalent activation functions, defined as:

$$\delta(net) = \frac{1}{1 + e^{-net}}. \tag{3}$$

Here, $net$ signifies the input signal. The weights play a pivotal role in FFNN performance. Consequently, determining suitable weights for connections is a primary objective in FFNNs. This task is accomplished during the learning phase of FFNNs through examples.

## 2.3 *Opposition-Based Learning*

Opposition-based Learning (OBL), as introduced by Tizhoosh [43], is a cornerstone technique designed to significantly bolster the exploratory strength of evolutionary algorithms. The OBL lies in the innovative idea of utilising the oppositional counterpart of a given point, thereby refining the solution's approximation efficiency.

In an N-dimensional space, for a point $P = (x_1, x_2, \ldots, x_N)$, the formulation of its oppositional counterpart, $\check{x}$, is articulated as:

$$\check{x} = a_i + b_i - x_i, \tag{4}$$

where $\check{x}$ embodies the opposing point of $x$, with $a$ and $b$ delineating the lower and upper boundaries of the search space, respectively, and $i$ indicating the $i$th dimension.

To date, the literature has witnessed the suggestion of various variants of this concept [13]. This study advocates for the use of quasi-opposition numbers [37], predicated on their enhanced propensity for guiding the algorithm towards optimal solutions, in comparison to traditional opposition points. A quasi-opposite number is defined by:

$$\check{x} = rand\left[\frac{(a_i + b_i)}{2}, (a_i + b_i - x)\right], \tag{5}$$

where $rand[m, n]$ signifies the generation of a random number uniformly distributed between $m$ and $n$, $a$ and $b$ are the lower and higher boundaries, and $x$ is a candidate solution (point).

The dynamic adaptation of OBL, through the incorporation of quasi-oppositional strategies, illuminates the pathway for algorithms to navigate the search space more effectively, revealing unexplored territories that might hold the key to elusive solutions.

# 3 Approches

## 3.1 *Encoding Scheme*

In this study, we introduce an approach for the representation of neural network parameters, specifically focusing on connection weights and biases through real-valued encoding. Our representation strategy dissects each candidate solution into two pivotal elements: (1) the synaptic weights that facilitate neuron-to-neuron communication, and (2) the biases that adjust the output along the activation function. The dimensionality of each candidate solution, therefore, mirrors the aggregate count of both connections and biases within the network. To elucidate, Fig. 1 depicts a standard FFNN characterised by a solitary hidden layer, serving as a model for this encoding scheme.

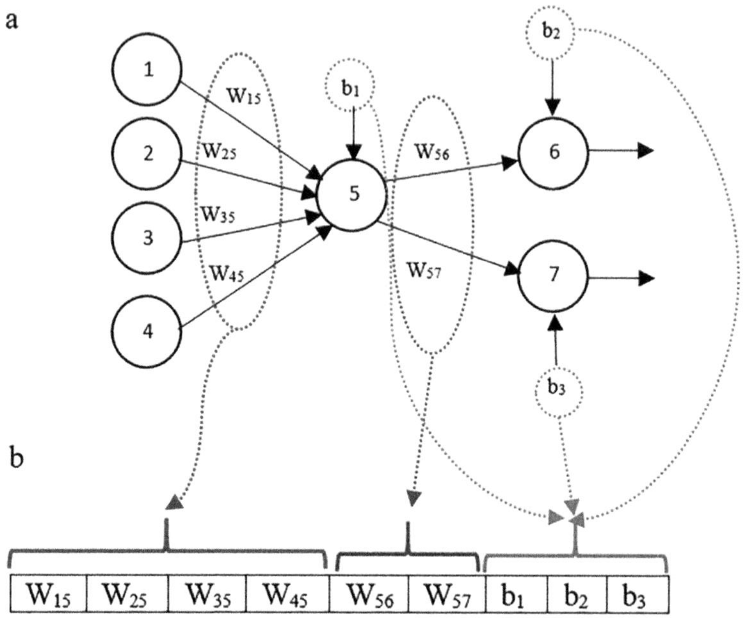


**Fig. 1** **a** A typical FFNN with 4-1-2 architecture **b** the corresponding encoding scheme

## 3.2 Cost Function

In this study, we utilised a cost function centred on classification error, as defined by:

$$\text{Classification error} = 100 \times \sum_{p=1}^{P} \xi\left(\frac{1}{P}\sum_{n=1}^{P}\vec{p}\right) \tag{6}$$

where $\xi(\vec{p})$ is defined as:

$$\xi(\vec{p}) = \begin{cases} 1 & \text{if } \vec{o_p} \neq \vec{d_p} \\ 0 & \text{otherwise} \end{cases} \tag{7}$$

Here, for each $P$-dimensional input vector in the training data, the corresponding desired output is denoted by $\vec{d_p}$, and $\vec{o_p}$ represents the predicted output. The primary objective in an FFNN learning process is to minimise the classification error by determining optimal connection weights and biases.

### 3.3 DE-Based Training Algorithm

The DE-based training algorithm (DTA) leverages the standard operations introduced in Sect. 2 for the evolutionary process. The pseudocode for DTA can be seen in Fig. 2.

### 3.4 Quasi-Opposition Infused Differential Evolution-Based Training (QODTA) Algorithm

The Quasi-Opposition Infused Differential Evolution-based training algorithm (QODTA) integrates the concept of opposition-based learning with the Differential Evolution algorithm for neural network training. The new components, compared to the DTA algorithm, are explained below.

```
Input  : D: the number of weights and biases, Max_NFC: Maximum number of function
         evaluations, N_P: Population size
Output: x*: Best solution
// Initialisation
Generate initial population Pop randomly based on Section 3.1;
Evaluate cost function for each population using Eq. 6;
while NFC < MAX_NFC do
    for i ← 1 to N_P do
        Select three parents, x_i1, x_i2, and x_i3, randomly from current population, with
          x_i1 ≠ x_i2 ≠ x_i3 ;
        v_i = x_i1 + F * (x_i2 − x_i3) ;
        for j ← 0 to D do
            if rand_j[0, 1] < C_R or j == j_rand then
                u_i,j = v_i,j
            else
                u_i,j = x_i,j
            end
        end
        Calculate cost function of u_i using Eq. 6;
        if f(u_i) < f(x_i) then
            x̄ ← u_i ;
        else
            x̄ ← x_i ;
        end
    end
    x ← x̄
end
x* ← Best solution in population;
```

**Fig. 2** Pseudo-code of DTA algorithm for FFNNs

#### 3.4.1 Initialisation Using Quasi-Opposition-Based Approach

In a manner akin to the standard DE algorithm, QODTA initially generates a population ($Pop$) uniformly at random. Subsequently, a quasi-opposition-based population ($OPop$) is created using the formula presented in Eq. 5. Finally, from the combined sets of $Pop$ and $OPop$, QODTA selects the top $N_P$ candidate solutions to form the initial population.

#### 3.4.2 Generation Jumping with Quasi-opposition-Based Approach

Upon generating new candidate solutions, QODTA proceeds to determine the quasi-opposition-based population based on a probability termed as the jumping rate ($J_r$). Typically, $J_r$ lies within the range of 0–0.4 [37]. Subsequently, the top $N_P$ candidate solutions are selected from the amalgamation of the current population and the quasi-opposition-based population. It is noteworthy that the opposition of each variable is calculated dynamically, with the maximum and minimum values of candidate solutions utilised to construct the opposite candidate solution, as depicted by the equation:

$$\check{x}i, j = minj^p + max_j^p - x_{i,j}\ i = 1, 2, ..., N_p\ j = 1, 2, ..., D \tag{8}$$

Here, $D$ represents the dimensions of the population, while $N_p$ denotes the population size. Moreover, $min_j^p$ and $max_j^p$ denote the minimum and maximum values of the population in the $j$th dimension, respectively. Figure 3 encapsulates the operation of QODTA in the form of pseudo-code.

### *3.5 Centroid Quasi-Opposition Infused Differential Evolution-Based Training Algorithm*

The centroid-centric strategy represents a pivotal advancement in evolutionary algorithms, significantly enhancing the exploratory capabilities inherent to these methods. Rahnamayan and Wang [38] reveals through Monte Carlo simulations the propensity for solutions to gravitate towards the search space's central nexus rather than dispersing randomly. This phenomenon is quantified within any given $i$th dimension spanning $[a_i, b_i]$ as the centroid:

$$c_i = \frac{a_i + b_i}{2}, \tag{9}$$

where $D$ encapsulates the search space's dimensionality, iterating over $i = 1, \ldots, D$.

```
Input  : D: Dimension count, Max_NFC: Max function evaluations, N_P: Population size,
         J_r: Jumping rate
Output: x*: Best solution
// Initialisation with Quasi-opposition
Generate initial population Pop randomly based on Section 3.1;
for i ← 0 to N_P do
|   for j ← 0 to D do
|   |   OPop(i, j) = rand[(L(i,j)+U(i,j))/2, (L_{i,j} + U_{i,j} − Pop(i, j))]
|   end
end
Evaluate cost function for each population using Eq. 6;
Select N_P best candidate solutions from (Pop, OPop) as initial population.
while NFC < MAX_NFC do
|   for i ← 1 to N_P do
|   |   Select three parents, x_i1, x_i2, and x_i3, randomly from current population, with
|   |     x_i1 ≠ x_i2 ≠ x_i3 ;
|   |   v_i = x_i1 + F * (x_i2 − x_i3) ;
|   |   for j ← 0 to D do
|   |   |   if rand_j[0, 1] < C_R or j == j_rand then
|   |   |   |   u_{i,j} = v_{i,j}
|   |   |   else
|   |   |   |   u_{i,j} = x_{i,j}
|   |   |   end
|   |   end
|   |   Calculate cost function of u_i using Eq. 6;
|   |   if f(u_i) < f(x_i) then
|   |   |   x̄ ← u_i ;
|   |   else
|   |   |   x̄ ← x_i ;
|   |   end
|   end
|   x ← x̄
|   // Opposition-Based Generation Jumping
|   if rand(0, 1) < J_r then
|   |   for i ← 1 to N_p do
|   |   |   for j ← 0 to D do
|   |   |   |   Opop(i, j) = rand[(min_j^p + max_j^p)/2, (min_j^p + max_j^p − Pop(i, j))] ;
|   |   |   end
|   |   |   Calculate cost function of Opop_i using Eq. 6;
|   |   end
|   end
|   Select N_p best candidate solutions from {Pop, Opop} as current population;
end
x* ← Best solution in population;
```

**Fig. 3** Pseudo-code of QODTA algorithm for FFNNs

Inspired by prior studies [28], our innovative Centroid-Quantified Optimisation with DE Algorithm (CQODTA) capitalises on this principle by selecting a centroid-based individual from the top $N$ performers. Unlike conventional methods, our approach updates all but one individual through standard evolutionary operations. The exception is a uniquely calculated centroid of the $N$ best performers, defined as:

$$\overrightarrow{x_{center}} = \frac{\overrightarrow{x_{b1}} + \ldots + \overrightarrow{x_{bN}}}{N}, \tag{10}$$

representing an aggregation of the top $i$th individuals' positions.

Significantly, this centroid-centric individual contributes to enhanced exploitation without necessitating additional computational overhead. It embodies the collective strength of the best performers, thereby streamlining the search process.

In addition, our CQODTA employs a *DE/local-to-best/1* approach, formulated as:

$$v_i = x_i + F \cdot (x_{best} - x_i) + F \cdot (x_{r2} - x_{r3}), \tag{11}$$

Here, $x_i$ aligns with the $i$th individual, $x_{best}$ denotes the best individual from the preceding cycle, with $x_{r1}$ and $x_{r2}$ representing two randomly selected individuals. The scaling factor $F$ modulates the mutation intensity, aiming for a harmonious blend of exploration and swift convergence. The pseudo-code of CQODTA is shown in Fig. 4.

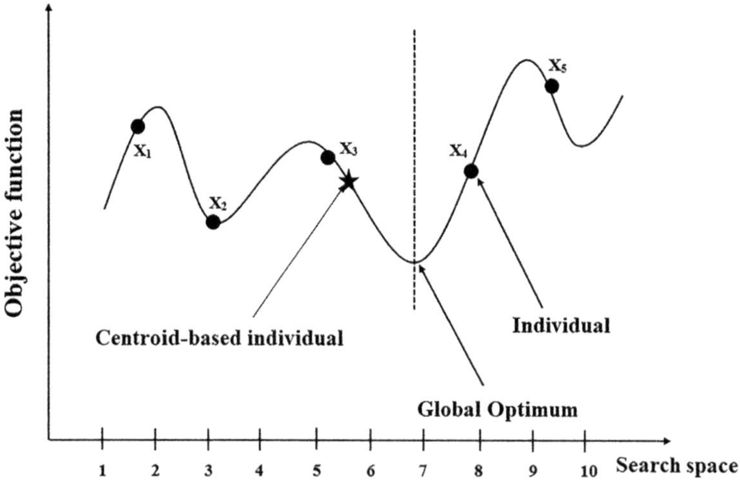


**Fig. 4** Pseudo-code of CQODTA algorithm for FFNNs

## 4 Experimental Evaluation

To ascertain the efficacy of our DE-based trainers, extensive experimental tests were established, leveraging a collection of datasets sourced from the UCI machine learning repository.[1] These datasets, covering a range of domains and present varied characteristics:

- **Iris Dataset:** Encompassing 150 samples across 3 classes with 4 attributes each, it provides a classic example of one class being linearly separable, while the other two are intertwined.
- **Breast Cancer Dataset:** Consisting of 699 samples, it features 9 attributes, including tumour size and menopause status, divided into binary classes.
- **Liver Disorder Dataset:** Sourced from BUPA Medical Research Ltd., this dataset includes 345 entries with 7 attributes, offering a binary classification challenge.
- **Pima Indians Diabetes Dataset:** A binary classification dataset with 768 samples and 8 attributes, renowned for its difficulty.
- **Seed Dataset:** Focused on the agricultural sector, this collection categorises 210 wheat kernel samples into three classes based on seven geometrical attributes.
- **Vertebral Column Dataset:** This dataset uses biomechanical attributes to classify orthopaedic conditions into three categories, providing a total of 310 instances.

For neural network architecture optimisation, we follow established methodologies [23, 24], setting the hidden layer neuron count to $2 \times D + 1$, where $D$ is the feature count. This decision yields specific connection weight counts for each dataset, respectively identified as 43 for Iris, 210 for Breast Cancer, 105 for Liver, 171 for Pima, 136 for Seed, and 105 for Vertebral. We adopt a $k$-fold cross-validation approach, with $k = 10$, for a comprehensive evaluation, ensuring each segment of the dataset is utilised once as test data.

Parameter settings, crucial for the performance of our DE algorithms, include a fixed population size of 50 and a set number of function evaluations at 25000, mirroring the iteration count of traditional approaches. Furthermore, optimisation parameters such as the crossover probability (0.9), scaling factor (0.5), selection of the top 3 best individuals, and a jumping rate of 0.3 are selected, alongside default parameters for comparative algorithms, to ensure a balanced and fair assessment of our proposed methods.

Table 1 compares the performance of three DE-based training algorithms across six different datasets. Each algorithm's performance is assessed by its average score, how much this performance varies (standard deviation), and its ranking against the others. From the table, CQODTA has the best average scores for most datasets and the best overall ranking. Its standard deviations are low, meaning it consistently performs well no matter the dataset. This consistency suggests that CQODTA could be a reliable choice for a range of different problems.

[1] https://archive.ics.uci.edu/ml/index.php.

**Table 1** The results of different DE-based training algorithms

| Algorithms | | Iris | Cancer | Liver | Pima | Seed | Vertebral | Avg. rank |
|---|---|---|---|---|---|---|---|---|
| DTA | Avg. | 92.00 | 97.36 | 67.81 | 76.94 | 70.00 | 85.16 | |
| | Std. | 5.26 | 2.06 | 8.21 | 4.97 | 11.01 | 5.31 | |
| | R | 4 | 4 | 4 | 4 | 2 | 4 | 3.67 |
| QODTA | Avg. | 95.33 | 98.10 | 76.82 | 79.55 | 67.62 | 88.39 | |
| | Std. | 6.32 | 0.99 | 9.46 | 4.95 | 3.01 | 8.76 | |
| | R | 3 | 3 | 2 | 3 | 3.5 | 1.5 | 2.58 |
| CQODTA | Avg. | 98.67 | 98.68 | 78.79 | 81.90 | 90.95 | 88.39 | |
| | Std. | 2.81 | 1.08 | 8.64 | 3.17 | 10.15 | 5.09 | |
| | R | 1 | 2 | 1 | 1 | 1 | 1.5 | 1.33 |

QODTA also performs well, particularly in the Vertebral datasets, but with more variation in its performance. It ranks second overall, indicating it is a strong algorithm but might not always perform predictably across different kinds of data. In addition, DTA ranks third.

In the next experiment, we compare the DE-based training algorithms with some other alternative training methods, including gradient descent with momentum backpropagation (GDM) [34], gradient descent with adaptive learning rate backpropagation (GDA) [5], conjugate gradient backpropagation with Fletcher-Reeves updates (CG-FR) [7], conjugate gradient backpropagation with Powell-Beale restarts (CG-PBR) [36], Levenberg-Marquardt backpropagation (LM) [12, 15], PSO, ABC, ICA, FA, GWO, ALO, DA, SCA, and WOA. The results can be seen in Table 2. From the table, the CQODTA shows the best performance. It obtains the best scores almost all the time, and secures the top rank with impressive average accuracies, dominating across all datasets. This consistent superiority in its mean ranking, which is around 1.08, underlines its robustness as a DE-based training algorithm.

Following the CQODTA, the QODTA and the ICA also deliver strong performances. The QODTA stands out with an average rank of 5.00, indicating high effectiveness across the board. However, it is worth noting that it does not completely outshine ICA, which shows a commendable performance, particularly in the Seed dataset.

GDM performs worst with lower average accuracies and higher rankings, pointing to less competitiveness compared to its counterparts. The rest of the algorithms present a mixed performance. For instance, GDA and CG-FR, while not at the bottom, do not reach the effectiveness of the top performers. The varying ranks suggest that these algorithms might have selective applicability, excelling in certain conditions and falling short in others.

**Table 2** Comparison of DE-based trainers with others (Avg. means the avefae accuracy over the folds, and R shows the rank)

| Algorithm | | Iris | Cancer | Liver | Pima | Seed | Vertebral | Avg. rank |
|---|---|---|---|---|---|---|---|---|
| GDM | Avg. | 92.00 | 92.99 | 59.47 | 67.98 | 47.62 | 76.13 | 16.00 |
| | R | 13 | 17 | 16 | 16 | 17 | 17 | |
| GDA | Avg. | 94.67 | 95.90 | 58.24 | 75.91 | 82.38 | 80.65 | 12.58 |
| | R | 7.5 | 16 | 17 | 14 | 6 | 15 | |
| CG-FR | Avg. | 95.33 | 96.19 | 60.86 | 76.69 | 86.67 | 83.87 | 9.58 |
| | R | 5.5 | 14 | 15 | 11 | 4 | 8 | |
| CG-PBR | Avg. | 94.00 | 96.49 | 67.18 | 76.04 | 88.10 | 82.90 | 9.75 |
| | R | 9 | 13 | 11 | 12.5 | 2.5 | 10.5 | |
| LM | Avg. | 96.67 | 96.04 | 65.55 | 76.04 | 88.10 | 83.55 | 8.92 |
| | R | 2.5 | 15 | 12 | 12.5 | 2.5 | 9 | |
| PSO | Avg. | 96.00 | 97.95 | 73.36 | 77.60 | 78.10 | 86.45 | 5.67 |
| | R | 4 | 5.5 | 4 | 8 | 8.5 | 4 | |
| ABC | Avg. | 84.67 | 97.95 | 70.75 | 78.26 | 72.38 | 82.90 | 9.42 |
| | R | 17 | 5.5 | 8 | 5 | 10.5 | 10.5 | |
| ICA | Avg. | 96.67 | 97.22 | 72.39 | 79.42 | 84.76 | 86.77 | 4.92 |
| | R | 2.5 | 10 | 6 | 3 | 5 | 3 | |
| FA | Avg. | 92.00 | 97.66 | 73.55 | 78.90 | 72.38 | 85.81 | 7.08 |
| | R | 13 | 7 | 3 | 4 | 10.5 | 5 | |
| GWO | Avg. | 93.33 | 98.10 | 73.01 | 67.45 | 78.10 | 81.94 | 9.50 |
| | R | 10 | 3 | 5 | 17 | 8.5 | 13.5 | |
| ALO | Avg. | 94.67 | 98.10 | 71.06 | 78.12 | 80.48 | 85.48 | 9.50 |
| | R | 10 | 3 | 5 | 17 | 8.5 | 13.5 | |
| DA | Avg. | 92.67 | 97.51 | 70.42 | 77.85 | 70.48 | 81.94 | 10.33 |
| | R | 11 | 8 | 9 | 7 | 13.5 | 13.5 | |
| SCA | Avg. | 90.67 | 97.08 | 65.50 | 74.47 | 71.43 | 82.26 | 13.00 |
| | R | 15 | 11 | 13 | 15 | 12 | 12 | |
| WOA | Avg. | 87.33 | 97.07 | 62.87 | 76.95 | 70.48 | 79.03 | 13.42 |
| | R | 16 | 12 | 14 | 9 | 13.5 | 16 | |
| DTA | Avg. | 92.00 | 97.36 | 67.81 | 76.94 | 70.00 | 85.16 | 10.67 |
| | R | 13 | 9 | 10 | 10 | 15 | 7 | |
| QODTA | Avg. | 95.33 | 98.10 | 76.82 | 79.55 | 67.62 | 88.39 | 5.00 |
| | R | 5.5 | 3 | 2 | 2 | 16 | 1.5 | |
| CQODTA | Avg. | 98.67 | 98.68 | 78.79 | 81.90 | 90.95 | 88.39 | 1.08 |
| | R | 1 | 1 | 1 | 1 | 1 | 1.5 | |

## 5 Conclusion

In this paper, we demonstrated the values of differential evolution (DE) for neural network training. Our investigation reveals that the DE offers a viable and robust framework for training feedforward neural networks (FFNNs). The study's findings confirm that the innovative centroid-based DE algorithm, CQODTA, excels in optimising FFNNs by effectively balancing the dual objectives of exploitation and exploration. By leveraging the centroid of the most promising solutions to guide the search process and employing dynamic opposition-based learning to expand the search space, CQODTA demonstrates superior performance over traditional gradient-based methods that are prone to premature convergence.

For future work, there is ample opportunity to extend the application of DE-based strategies to other types of neural networks, such as recurrent neural networks (RNNs) and convolutional neural networks (CNNs). Additionally, it would be worthwhile to explore the integration of adaptive mechanisms that can dynamically adjust DE parameters in response to the problem's landscape, potentially enhancing the training performance further.

The incorporation of parallel computing techniques to scale DE-based FFNN training for large-scale and real-time applications also presents a promising avenue for research. Furthermore, the application of DE to domain-specific FFNN tasks, such as natural language processing or image recognition, may yield insightful advancements, helping to tailor the training process to the unique challenges of these domains.

## References


1. Abusnaina, A.A., Ahmad, S., Jarrar, R., Mafarja, M.: Training neural networks using salp swarm algorithm for pattern classification. In: Proceedings of the 2nd International Conference on Future Networks and Distributed Systems, pp. 1–6 (2018)
2. Aljarah, I., Faris, H., Mirjalili, S.: Optimizing connection weights in neural networks using the whale optimization algorithm. Soft. Comput. **22**(1), 1–15 (2018)
3. Amirsadri, S., Mousavirad, S.J., Ebrahimpour-Komleh, H.: A Levy flight-based grey wolf optimizer combined with back-propagation algorithm for neural network training. Neural Comput. Appl. **30**(12), 3707–3720 (2018)
4. Ara, A., Khan, N.A., Razzaq, O.A., Hameed, T., Raja, M.A.Z.: Wavelets optimization method for evaluation of fractional partial differential equations: an application to financial modelling. Adv. Difference Equ. **2018**(1), 8 (2018)
5. Beale, H.D., Demuth, H.B., Hagan, M.: Neural Network Design. Pws, Boston (1996)
6. Duan, H., Huang, L.: Imperialist competitive algorithm optimized artificial neural networks for ucav global path planning. Neurocomputing **125**, 166–171 (2014)
7. Fletcher, R., Reeves, C.M.: Function minimization by conjugate gradients. Comput. J. **7**(2), 149–154 (1964)
8. Heidari, A.A., Faris, H., Aljarah, I., Mirjalili, S.: An efficient hybrid multilayer perceptron neural network with grasshopper optimization. Soft. Comput. **23**(17), 7941–7958 (2019)

9. Karaboga, D., Akay, B., Ozturk, C.: Artificial bee colony (ABC) optimization algorithm for training feed-forward neural networks. In: International Conference on Modeling Decisions for Artificial Intelligence, pp. 318–329 (2007)
10. Khishe, M., Safari, A.: Classification of sonar targets using an MLP neural network trained by dragonfly algorithm. Wireless Pers. Commun. **108**(4), 2241–2260 (2019)
11. Kim, D., Kim, H., Chung, D.: A modified genetic algorithm for fast training neural networks. In: International Symposium on Neural Networks, pp. 660–665. Springer (2005)
12. Levenberg, K.: A method for the solution of certain non-linear problems in least squares. Q. Appl. Math. **2**(2), 164–168 (1944)
13. Mahdavi, S., Rahnamayan, S., Deb, K.: Opposition based learning: a literature review. Swarm Evol. Comput. **39**, 1–23 (2018)
14. Mandal, S., Saha, G., Pal, R.K.: Neural network training using firefly algorithm. Global J. Adv. Eng. Sci. (GJAES) **1**(1), 7–11 (2015)
15. Marquardt, D.W.: An algorithm for least-squares estimation of nonlinear parameters. J. Soc. Ind. Appl. Math. **11**(2), 431–441 (1963)
16. Mirjalili, S.: How effective is the grey wolf optimizer in training multi-layer perceptrons. Appl. Intell. **43**(1), 150–161 (2015)
17. Mirjalili, S.: SCA: a sine cosine algorithm for solving optimization problems. Knowl.-Based Syst. **96**, 120–133 (2016)
18. Moghadam, M.H., Borg, M., Mousavirad, S.J.: Deeper at the sbst 2021 tool competition: Adas testing using multi-objective search. In: 2021 IEEE/ACM 14th International Workshop on Search-Based Software Testing (SBST), pp. 40–41. IEEE (2021)
19. Moghadam, M.H., Borg, M., Saadatmand, M., Mousavirad, S.J., Bohlin, M., Lisper, B.: Machine learning testing in an adas case study using simulation-integrated bio-inspired search-based testing. J. Softw.: Evol. Proc. e2591 (2022)
20. Moravvej, S.V., Mousavirad, S.J., Moghadam, M.H., Saadatmand, M.: An lstm-based plagiarism detection via attention mechanism and a population-based approach for pre-training parameters with imbalanced classes. In: International Conference on Neural Information Processing, pp. 690–701. Springer (2021)
21. Mousavirad, S.J., Asilian Bidgoli, A., Rahnamayan, S.: Tackling deceptive optimization problems using opposition-based DE with center-based Latin hypercube initialization. In: 14th International Conference on Computer Science and Education (2019)
22. Mousavirad, S.J., Bidgoli, A.A., Ebrahimpour-Komleh, H., Schaefer, G.: A memetic imperialist competitive algorithm with chaotic maps for multi-layer neural network training. Int. J. Bio-Inspired Comput. **14**(4), 227–236 (2019)
23. Mousavirad, S.J., Bidgoli, A.A., Ebrahimpour-Komleh, H., Schaefer, G., Korovin, I.: An effective hybrid approach for optimising the learning process of multi-layer neural networks. In: International Symposium on Neural Networks, pp. 309–317 (2019)
24. Mousavirad, S.J., Jalali, S.M.J., Sajad, A., Abbas, K., Schaefer, G., Nahavandi, S.: Neural network training using a biogeography-based learning strategy. In: International Conference on Neural Information Processing (2020)
25. Mousavirad, S.J., Moghadam, M.H., Saadatmand, M., Chakrabortty, R., Schaefer, G., Oliva, D.: RWS-L-SHADE: An effective l-shade algorithm incorporation roulette wheel selection strategy for numerical optimisation. In: International Conference on the Applications of Evolutionary Computation (Part of EvoStar), pp. 255–268. Springer (2022)
26. Mousavirad, S.J., Oliva, D., Chakrabortty, R.K., Zabihzadeh, D., Hinojosa, S.: Population-based self-adaptive generalised masi entropy for image segmentation: a novel representation. Knowl.-Based Syst. **245**, 108610 (2022)
27. Mousavirad, S.J., Oliva, D., Schaefer, G., Moghadam, M.H., El-Abd, M.: A novel two-level clustering-based differential evolution algorithm for training neural networks. In: International Conference on the Applications of Evolutionary Computation (Part of EvoStar), pp. 259–272. Springer (2024)
28. Mousavirad, S.J., Rahnamayan, S.: A novel center-based differential evolution algorithm. In: Congress on Evolutionary Computation. IEEE (2020)

29. Mousavirad, S.J., Schaefer, G., Korovin, I., Moghadam, M.H., Saadatmand, M., Pedram, M.: An enhanced differential evolution algorithm using a novel clustering-based mutation operator. In: 2021 IEEE International Conference on Systems, Man, and Cybernetics (SMC), pp. 176–181. IEEE (2021)
30. Mousavirad, S.J., Schaefer, G., Zhou, H., Moghadam, M.H.: How effective are current population-based metaheuristic algorithms for variance-based multi-level image thresholding? Knowl.-Based Syst. **272**, 110587 (2023)
31. Mousavirad, S.J., Zabihzadeh, D., Oliva, D., Perez-Cisneros, M., Schaefer, G.: A grouping differential evolution algorithm boosted by attraction and repulsion strategies for masi entropy-based multi-level image segmentation. Entropy **24**(1), 8 (2021)
32. Navarro, M.A., Ramos-Michel, A., Gaspar, A., Oliva, D., Hinojosa, S., Mousavirad, S.J., Pérez-Cisneros, M.: Improving the convergence and diversity in differential evolution through a stock market criterion. In: International Conference on the Applications of Evolutionary Computation (Part of EvoStar), pp. 157–172. Springer (2022)
33. Nawi, N.M., Rehman, M., Aziz, M.A., Herawan, T., Abawajy, J.H., et al.: An accelerated particle swarm optimization based levenberg marquardt back propagation algorithm. In: International Conference on Neural Information Processing, pp. 245–253. Springer (2014)
34. Phansalkar, V., Sastry, P.: Analysis of the back-propagation algorithm with momentum. IEEE Trans. Neural Networks **5**(3), 505–506 (1994)
35. Pourvahab, M., Mousavirad, S.J., Felizardo, V., Pombo, N., Zacarias, H., Mohammadigheymasi, H., Pais, S., Jafari, S.N., Garcia, N.M.: A cluster-based opposition differential evolution algorithm boosted by a local search for ecg signal classification (2023). arXiv:2305.02731
36. Powell, M.J.D.: Restart procedures for the conjugate gradient method. Math. Program. **12**(1), 241–254 (1977)
37. Rahnamayan, S., Tizhoosh, H.R., Salama, M.M.: Quasi-oppositional differential evolution. In: IEEE Congress on Evolutionary Computation, pp. 2229–2236 (2007)
38. Rahnamayan, S., Wang, G.G.: Center-based sampling for population-based algorithms. In: IEEE Congress on Evolutionary Computation, pp. 933–938. IEEE (2009)
39. Salvador, C., Felizardo, V., Zacarias, H., Souza-Pereira, L., Pourvahab, M., Pombo, N., Garcia, N.M.: Epileptic seizure prediction using eeg peripheral channels. In: 2023 IEEE 7th Portuguese Meeting on Bioengineering (ENBENG), pp. 60–63. IEEE (2023)
40. Sexton, R.S., Gupta, J.N.: Comparative evaluation of genetic algorithm and backpropagation for training neural networks. Inf. Sci. **129**(1–4), 45–59 (2000)
41. Si, T., Dutta, R.: Partial opposition-based particle swarm optimizer in artificial neural network training for medical data classification. Int. J. Inf. Technol. & Dec. Making **18**(5), 1717–1750 (2019)
42. Storn, R., Price, K.: Differential evolution-a simple and efficient heuristic for global optimization over continuous spaces. J. Global Optim. **11**(4), 341–359 (1997)
43. Tizhoosh, H.R.: Opposition-based learning: a new scheme for machine intelligence. In: International Conference on Computational Intelligence for Modelling, Control and Automation and International Conference on Intelligent Agents, Web Technologies and Internet Commerce, vol. 1, pp. 695–701 (2005)
44. Yamany, W., Tharwat, A., Hassanin, M.F., Gaber, T., Hassanien, A.E., Kim, T.H.: A new multilayer perceptrons trainer based on ant lion optimization algorithm. In: Fourth International Conference on Information Science and Industrial Applications, pp. 40–45. IEEE (2015)
45. Zacarias, H., Felizardo, V., Souza-Pereira, L., Pinho, A., Ramos, S., Pourvahab, M., Garcia, N., Pombo, N.: Gender classification using nonstandard ecg signals-a conceptual framework of implementation. In: EAI International Conference on IoT Technologies for HealthCare, pp. 108–120. Springer (2022)
46. Zacarias, H., Marques, J.A.L., Felizardo, V., Pourvahab, M., Garcia, N.M.: Ecg forecasting system based on long short-term memory. Bioengineering **11**(1), 89 (2024)

# Evolutionary Algorithms in Code Smell Detection: A Feature Selection Approach for Software Engineering

**Diego Avalos and Salvador Hinojosa**

**Abstract** This chapter delves into how evolutionary algorithms (EAs) can be applied to identify code smells, framing it as a problem of selecting features in software code metrics. Code smells, which signal potential flaws in software design, typically demand labor intensive techniques for their detection. We investigate the efficiency of EA driven strategies that optimize the selection of features related to code quality, ultimately improving the precision and effectiveness of detecting code smells. We cover the basics of code smells and evolutionary algorithms, then detail our methodology for using EAs to select features that significantly improve code smell classification. Through experiments on industry-relevant software datasets, our approach is shown to improve traditional rule-based methods in detection accuracy. The chapter concludes with implications for future research and the potential of AI-enhanced solutions in software engineering.

## 1 Introduction

The dynamic nature of the software development process involves multiple changes. These changes can be produced by multiple reasons, usually to address market needs, new technologies and industry trends. Although the software development process usually uses agile methodologies to fit the change, stakeholder's pressure to reduce the time-to-market can lead to poor quality products. Multiple methodologies have been designed to streamline the supporting process of software development, usually automating tasks such as software testing over iterative build increments to prevent software defects from reaching a production environment. By using automated testing frameworks, the functional evaluation of the code under test can be examined.

D. Avalos (✉) · S. Hinojosa
Tecnologico de Monterrey, Escuela de Ingeniería y Ciencias, Av. Gral Ramón Corona No. 2514, Colonia Nuevo México, 45121 Zapopan, Jalisco, Mexico
e-mail: diego.avalos@tec.mx

S. Hinojosa
e-mail: salvador.hinojosa@tec.mx

 
D. Oliva et al. (eds.), *Advances in Optimization Algorithms for Multidisciplinary Engineering Applications: From Classical Methods to AI-Enhanced Solutions*, Studies in Computational Intelligence 806, https://doi.org/10.1007/978-3-031-78440-8_20

However, there are certain bad practices that lead to the production of low-quality software artifacts which can comply with its functional description but will produce problems in the future when maintaining and modifying such code pieces. Although the code works, something is wrong and it receives the name of Code Smell [1]. A common scenario is when developers give up encapsulation principles in exchange for a shorter time-to-market; this produces the generation of a complex large class which has attributes and responsibilities that are not semantically related. Thus, this scenario would require the separation of its functionality into two or more semantically different classes with lower complexity. This bad practice is known as God Class. Code smells hinder the quality, scalability, and code comprehension [2].

Code smells can limit the ability of a team to understand and maintain code bases generated by other developers or even themselves. As the software industry grows, code smell-related problems are becoming more common [3]. This fact has motivated empirical studies to analyze the impact of this practices in software quality [4]. These studies point out that code artifacts which possess the so called code smells are prone to functional defects and have a higher chance of requiring refactoring. With all this, the code smell study has become an increasingly relevant topic for software engineering researchers over the last few years [5, 6]. As code smells cannot be detected by functional tests, for example unit tests, it is hard to provide automatized frameworks that can search for this code smells. The main challenges researchers face are the lack of relevant high-quality data sets to train automatic tools, the subjective nature of some code smells, and a high incidence of false positives in the reported literature [7, 8].

One common way to determine if a Software Under Test (SUT) presents code smells is by using rules and thresholds over multiple static analysis quality measures [9]. Rules based on quality measures are often used due to its simplicity and the use of valuable information such as the occurrence frequency of multiple static elements. By looking at such metrics, we can selectively differentiate when a SUT is likely to have a specific code smell. Although there is no standardized definition of code smells and its relationship with static analysis metrics, we take the works by Fowler as the most widely agreed definitions [1]. On his work, he established a set of twenty two code smells that cover a wide spectrum of potentially problematic bad practices on software projects, ranging from simple to complex definitions.

In this chapter, we analyze the feasibility of applying swarm-based metaheuristic algorithms to correctly classify code smells in an automatic way. To do so, the problem is translated into a classification problem aided by a feature selection scheme.

The chapter is organized as follows, Sect. 2 presents the background knowledge required to understand the problematic and a brief landscape of related approaches. Then, in Sect. 3 we discuss the overall methodology. In Sect. 4 we discuss de experimental setup details. Afterwards, on Sect. 5 we analyze the results to conclude the chapter with Sect. 6.

## 2 Background

### 2.1 Code Smells

Automatic Code Smell detections remains a challenging task no matter what approach is used to solve it. We summarize the challenges into three key aspects.

1. Lack of a formal definition. Fowler defined a total of twenty two types of code smell [2]. But this is not the only definition of code smell, since then, other researchers in the field presented different classifications and definitions of code smell [10, 11]. So far, there is not a clear consensus on what a code smell is and no official or comprehensive classification that can cover all code smells has been widely accepted. However, up to now the work of Fowler shows a more complete and widespread classification and definition.
2. Inaccurate results. Since the software project can reach to the order of millions of lines of code, the chances of code smells being present in a project can grow exponentially. As code smells usually are manually detected by experienced developers, the tools and platforms require human intervention [12]. The expert could easily be overwhelmed by the amount of information to analyze. Even more, the time and budget constraints and the pressure to minimize the time-to market many false positives could occur [13]. For this reason, proposals have already been generated in researches to make a selection process to facilitate the filtering. However, the proposed methods are limited to process a small amount of code smells or the filters applied for selection must be done through a manual adjustment by the user.
3. Not all types of code smell have been investigated and developed in the same way. For example, the God class code smell has received more attention from researchers, with many investigations and proposals [14–16]. Other code smells may be less explored due to factors such as complexity for automated detection or the infrequency with which these code smells tend to occur.

### 2.2 Feature Selection

Feature selection is a critical step in the process of developing machine learning models and data analysis. It involves selecting the most relevant and significant features from a dataset to improve model performance, reduce overfitting, and enhance interpretability [17]. By carefully choosing the right set of features, we can improve the accuracy and efficiency of the models, leading to better decision-making and a deeper understanding of the underlying data patterns.

In the context of code smell detection, feature selection plays a vital role in identifying the most discriminative code metrics and characteristics, which can help in effectively distinguishing and addressing code smells in software systems. By selecting the most pertinent features, the code smell detection process becomes more efficient and accurate, leading to improved software maintenance and reliability. Traditional feature selection methods often face challenges in handling the vast amount of data and the complex interdependencies among features.

In this chapter, we will explore how metaheuristic algorithms offer a promising solution to these challenges by leveraging their ability to efficiently search through large solution spaces and adapt to complex feature interactions. Through the integration of metaheuristic algorithms into the feature selection process, we can enhance the effectiveness of code smell detection, as several feature selection methodologies are commonly used in the field of software engineering and code smell detection [18]:

The main approaches used for feature selection are:

1. Filter Methods: These methods rank the features based on certain criteria and select the top-ranked features. Common techniques include correlation-based feature selection and information gain.
2. Wrapper Methods: Wrapper methods evaluate the performance of a specific model by considering a subset of features. They involve repeatedly training and evaluating the model on different subsets of features to find the best performing subset.
3. Embedded Methods: Embedded methods integrate feature selection as part of the model construction process. Examples include LASSO and decision tree-based feature selection.

The adaptability and flexibility of wrapper-based approaches make them an attractive option for addressing the challenges associated with feature selection in code smell detection. As we delve further into the application of metaheuristic algorithms in tandem with wrapper-based approaches, we will uncover the potential for synergistic effects in enhancing the efficiency and effectiveness of code smell detection.

## 2.3 Wrapper-Based Methods

Wrapper methods for feature selection are particularly relevant in the context of code smell detection [19]. These methods involve the evaluation of the feature subset by employing a specific model, such as a classification algorithm, and selecting the best performing subset based on the model's performance. The key advantage of wrapper methods is their ability to take into account the interaction between features and their collective impact on the model's performance, making them well-suited for code smell detection where the presence of multiple code smells may have interdependencies [18].

Wrapper-based methods typically employ a search strategy to explore different subsets of features, which can be computationally intensive, especially with a large number of features. Nonetheless, their ability to identify the optimal subset of features based on the specific model's performance makes them a valuable approach for addressing code smell detection as a feature selection problem.

The wrapper-based optimization problem in the context of feature selection for code smell detection can be formally defined as follows:

Given a set of features, or columns in the dataset, F = f1, f2,..., fn where n is the total number of features, and a classification algorithm, the goal is to find the optimal subset of features $S^* \subseteq F$ that maximizes the performance of the classification algorithm in identifying code smells within software systems.

The wrapper-based optimization problem involves the iterative evaluation of different feature subsets by training and testing the classification algorithm on each subset. This process aims to discover the subset of features that result in the highest classification performance, taking into account the interactions and dependencies among the features.

The objective function of the wrapper-based optimization problem can be defined as maximizing the performance metric of the classification algorithm, such as accuracy, precision, recall, or F1 score, while considering the computational complexity of evaluating each subset. Wrapper-based optimization algorithms employ search strategies, such as forward selection, backward elimination, metaheuristic algorithms, to explore the space of possible feature subsets and identify the most effective combination of features for code smell detection.

## 2.4 *Metaheuristic Algorithms*

Metaheuristic algorithms belong to a group of optimization techniques inspired by phenomena and problem-solving strategies found in nature. These algorithms are designed to search for optimal solutions in large and complex problem spaces using a combination of exploration and exploitation techniques. A metaheuristic algorithm may not return the true or exact solution, but rather the near-optimal solution. Generally, the optimization procedure for any metaheuristic algorithm is shown in Fig. 1 [20]. Swarm based metaheuristics, a type of metaheuristic algorithm take cues from how natural systems like bird flocks, fish schools and ant colonies operate. These algorithms leverage the concept of swarm intelligence, wherein a group of simple agents interacts locally with their environment and peers to exhibit complex, self-organized global behavior.

The most representative swarm-based metaheuristic is the Particle Swarm OptIn PSO, a group of candidate solutions called particles adjust their positions in the search space iteratively based on their best known position and the collective best-known position of neighboring particles. This collaborative exploration and exploitation process enables PSO to effectively navigate complex solution spaces and converge towards optimal solutions.

Swarm-based metaheuristics, including Particle Swarm Optimization, follow a general process for searching optimal solutions in large and complex problem spaces. The process involves the following key steps:

**Fig. 1** General flowchart for metaheuristic algorithm

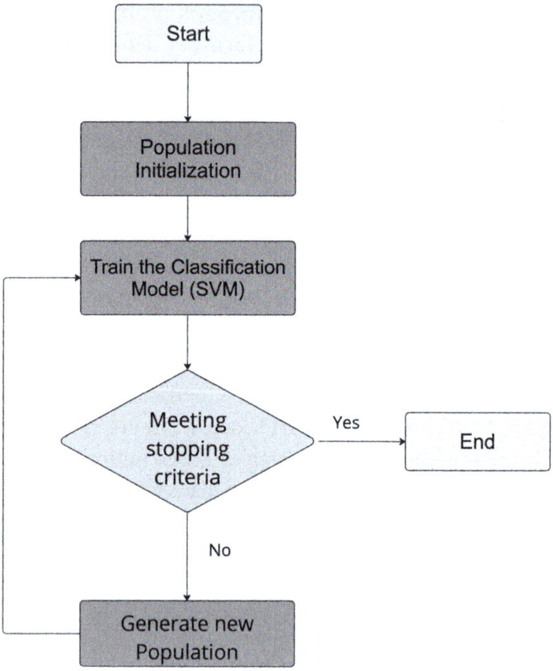


1. Initialization: The algorithm initializes a population of candidate solutions, also known as particles in the case of PSO. The initial positions of the particles are typically randomly generated within the problem space.
2. Evaluation: Each candidate solution is evaluated using the objective function, which measures the fitness or quality of the solution with respect to the problem being optimized.
3. Update: The positions of the particles are updated iteratively based on their own best-known position and the best-known position of their neighboring particles. This update mechanism involves a combination of exploration to search the solution space and exploitation to converge towards optimal solutions.
4. Determination of Stopping Criteria: The algorithm iterates through the evaluation and update steps until a stopping criterion is met. Stopping criteria can include a maximum number of iterations, reaching a satisfactory solution quality, or convergence of the particles towards a specific region of the problem space.
5. Solution Extraction: Once the algorithm terminates, the best-known solution, or the optimal subset of features in the context of wrapper-based feature selection, is extracted from the final positions of the particles.
6. Solution Utilization: The extracted solution, which represents the optimal feature subset, is utilized in the specific application. In the context of code smell detection, the feature selection problem can be formulated as finding an optimal subset of features that maximize the classification performance in identifying code.

Swarm-based metaheuristic algorithms exhibit inherent parallelism and adaptability, making them well-suited for addressing feature selection [21].

Their collaborative nature allows efficient exploration of feature subsets, while their ability to dynamically adapt to changing environments enables robust and effective search processes. Moreover, the adaptability of wrapper-based methods to different classification models and their ability to capture complex feature interactions make them a valuable asset in the pursuit of enhancing software quality and maintainability through effective code smell detection. Therefore, the utilization of swarm-based metaheuristic algorithms for feature selection in code smell detection can greatly improve the effectiveness of classifiers in identifying and addressing code.

## 3 Proposal

To carry out the code smell classification process, according to Fig. 2, as a first step a dataset is loaded from a csv file *MLCQ*. This file contains a number of records related to software features and the identification of code smells, along with their severity. By loading this data into a DataFrame structure (a two-dimensional table), for easy manipulation and further analysis. This step is crucial, as it establishes the database to be worked with.

As a next step, a specific transformation is performed on the data: the different types of "code smells" are mapped to numerical values according to a pre-established criterion. The following values were assigned to the code smells: blob = 1, data class = 2, feature envy = 3, long method = 4 and finally zero to code smells that have *none* in the value of *severity*. In addition, a function is defined and applied to each row to determine the value of a new 'Final' column based on the severity of the code smell. This step is essential to convert categorical variables (such as types of code smells and their severity) into numerical values that can be processed by machine learning algorithms.

Finally, this data set is separated into two parts: one containing the independent variables (X), and the other the target value or dependent variable (y), which in this

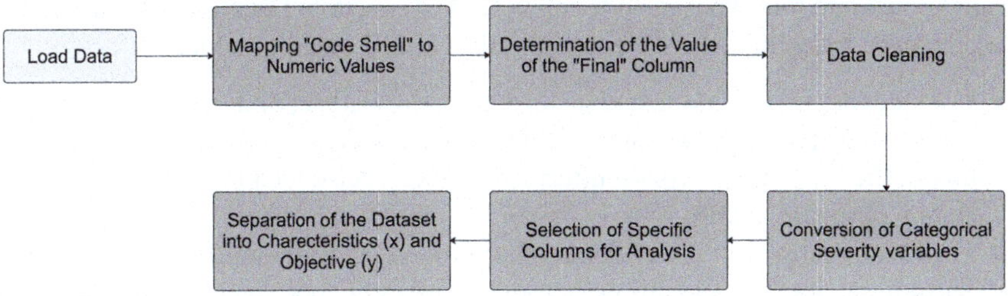


**Fig. 2** Code smell mapping and classification process

case is the value of the 'Final' column. This stage prepares the data for use in machine learning models, facilitating the identification of patterns and the prediction of new cases based on the features provided.

# 4 Experiments

## 4.1 Experimental Setup

The experiments were carried out using Google Colab, a cloud computing platform that allows the execution of Python code in a virtualized environment [22]. It consists of the following hardware resources:

- Processor: Intel(R) Xeon(R) CPU @ 2.20 GHz
- Number of logical CPU cores: 2
- Total RAM memory: 12.67 GB
- cache size: 56320 KB

All experiments were implemented using Python, a high-level programming language widely recognized for its versatility and the vast ecosystem of libraries available [23]. In particular, MAFESE (Metaheuristic Algorithms for FEature SElection), a specialized framework designed for the analysis and selection of features in complex datasets, was used to implement the feature selection process on the dataset [24]. MAFESE stands out for its ability to integrate various optimization and evaluation techniques in a cohesive manner, providing a robust platform for experimentation and analysis.

## 4.2 Dataset

The dataset we utilized for our experiments is the MLCQ dataset, an effort created by 26 software developers. They gathered code snippets from 792 source Java projects and asked developers to inspect them for code smell [25]. 15,000 code samples were categorized into four types of code smell; God Class/Blob, Data Class, Long Method and Feature Envy Table 1. The entries, in the MLCQ dataset include;

- The category of code issue and its level of impact (none, minor, major or critical).
- Details for accessing the sample code and the software project it belongs to; a link to the repository, version information and precise location of the code issue, among pertinent details.
- The identification of the developer who provided the annotation.
- The dataset notably contains numerous instances of code smells classified as God Class and Data Class.

**Table 1** MLCQ dataset severity values

| Severity | | | | |
|---|---|---|---|---|
| Smell | Critical | Major | Minor | None |
| Blob | 127 | 312 | 535 | 3045 |
| Data Class | 146 | 401 | 510 | 2964 |
| Feature envy | 24 | 142 | 288 | 2883 |
| Long method | 78 | 274 | 454 | 2556 |

- Several developers tagged some included software projects and some tags were verified. In some of the cases, they did not match the tags, although these data are negligible.

This dataset went through four stages before being implemented (Fig. 3).

1. Export of the Dataset: The 'mlcq' dataset was imported into the script so that the contained data can be used.
2. Data Exploration: The values are initially explored, in order to know the values, categories and types of data. In addition to analyzing which of these values will provide valuable information.
3. Data Mapping: The mapping process is the conversion of the categorical label to numerical data, with the purpose of giving them the indicated format for processing, for example, the code smell data were mapped as follows:

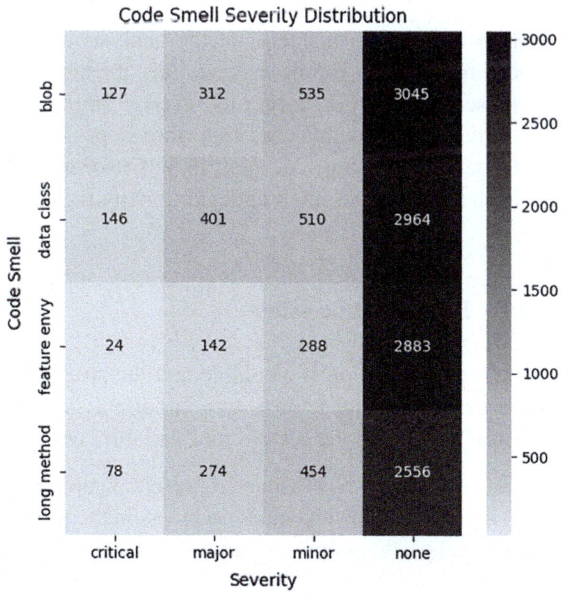


**Fig. 3** Code smell severity distribution

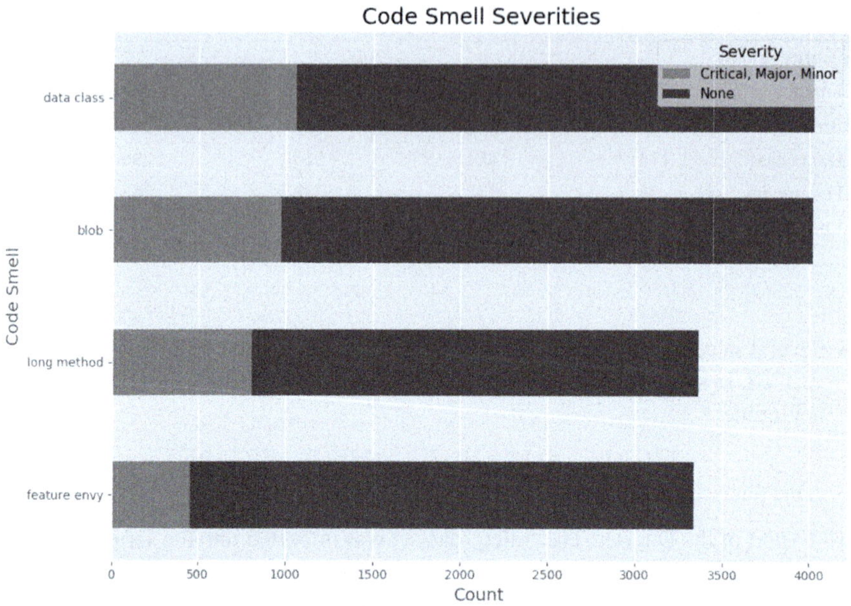


**Fig. 4** Code smell severities after processing

- 'blob': 1,
- 'data class': 2,
- 'feature envy': 3,
- 'long method': 4

4. Label Mapping: to facilitate the analysis and since the dataset does not have enough samples for all the severities, the label indicates only if the code smell is present, without regarding its severity. In Fig. 4 we can observe the distribution of the processed data used in further steps.
5. Code Smell Count: An analysis is performed to know initially the quantity of each type of code smell and additionally the severity of each code smell.

### 4.2.1 Data Set Processing

Once the information is available and the structure, types of code smell and other important details are known, two important steps are carried out, the treatment of missing values and the selection of variables or columns that provide value.

∗ *Missing Values:* NaN values represent a common problem in data analysis, since they can alter the results or cause errors in the processing of the information. In this case we chose to assign a value of zero to maintain the integrity of the data set.

$*$ *Column selection:* It was decided to exclude categorical columns and numerical columns that do not contribute to value, based on preliminary analysis that indicated a low correlation between these and the evaluation metrics of interest. This was done to simplify the model and improve performance by focusing the analysis on numerical variables with higher potential.

#### 4.2.2 Training and Testing Data Split

Once the data is formatted and selected, the data is segmented using the Scikit-learn library function *'train_test_split'*, ensuring a random distribution that maintains the original proportion of classes in both sets. This standard practice helps to fairly assess the model's ability to generalize to new data. And for our case it is 80% for training and 20% for testing.

### *4.3 Classifier*

The SVM (Support Vector Machine) classifier is a supervised learning technique used for classification. Its main goal is to find a hyperplane that best separates the different classes in the feature space. This means that the SVM identifies a line (in two dimensions) or a plane (in more dimensions) that divides the data into different categories, maximizing the distance between the classes and thus ensuring a clear and efficient separation [26]. Essentially, an SVM finds the optimal separation hyperplane by maximising the margin between classes, defined as the distance between the hyperplane and the nearest points of each class, known as the support vector.

The SVM prediction is obtained by evaluating which side of the hyperplane each data point lies. Unlike tree-based classifiers such as random forest, which build multiple decision trees and combine their predictions [27], the SVM focuses on a single optimal solution based on the support vectors. This can result in increased generalizability and robustness to noisy or complex data, although it may require careful tuning of parameters such as kernel type and regularization parameters [28].

The SVM classifier is used for the detection of code smells in software projects by minimizing the classification error in the objective function. This approach is based on identifying the most relevant features that represent code smells by optimizing the margin between different classes in the feature space [26].

#### 4.3.1 Optimization Algorithms Implementation

For the configuration of different parameters and selection of algorithms. Which were:

- *Optimizers:* Eleven swarm-based search algorithms were selected for its potential to address the specific optimization problem in code smells detection. The algo-

**Table 2** Swarm-based search algorithms used

| Acronym | Algorithm | Main features | Ref |
|---|---|---|---|
| ABC | Artificial Bee Colony | Inspired by the foraging behavior of honey bees. Used for global optimization problems. Notable for its simplicity and effectiveness | [29] |
| ACOR | Ant Colony Optimization for Continuous Domains | A version of the ant colony optimization algorithm for continuous domains. Based on ant behavior to find optimal solutions | [30] |
| PSO | Particle Swarm Optimization | Inspired by the social behavior of birds. "Particles" in the solution space move towards the best found positions | [31] |
| ALO | Ant Lion Optimizer | Simulates the interaction between ant hunters and their prey (ant lions). Used to solve optimization problems | [32] |
| GWO | Grey Wolf Optimizer | Inspired by the social hierarchy and hunting of grey wolves. Uses leadership and hunting mechanisms to guide solution space exploration | [33] |
| CA | Cellular Automata | A mathematical model for dynamic systems where space is discretized. Used in a wide range of applications from simulations to cryptography | [34] |
| SA | Simulated Annealing | Inspired by the annealing process in metallurgy. A probabilistic method to approximate the global maximum or minimum of a function | [35] |
| FA | Firefly Algorithm | Inspired by the attractive behavior of fireflies. Used to find optimal solutions based on attractiveness between agents | [36] |
| CRO | Coral Reefs Optimization | Inspired by the reproduction process and formation of coral reefs. Seeks optimal solutions by mimicking these natural processes | [37] |
| TSA | Tabu Search Algorithm | A local search method that uses a tabu list to avoid cycles and escape local optima. Versatile and applicable to various optimization problems | [38] |
| HBA | Hybrid Bat Algorithm | Combines features of the bat algorithm with other optimization methods to improve performance in finding optimal solutions | [39] |

rithms are listed in Table 2: *'ABC', 'ACOR', 'PSO', 'ALO', 'GWO', 'CA', 'SA', 'FA', 'CRO', 'TSA', 'HBA'*.

- *Parameter selection:* The choice of 20 generations and a population size of 30 was the result of a sensitivity analysis, which proved to be an optimal balance between accuracy of results and computational efficiency. These parameters allowed adequate exploration of the solution space without incurring prohibitive computational costs. The rest of the parameters required for each algorithm are set according to their standard version.

- *Statistical significance:* Repeating each experiment 30 times contributes to the statistical robustness of the results, allowing more accurate estimation of performance metrics and minimizing the impact of random variations.

### 4.3.2 Algorithm Evaluation

To evaluate the performance of the implemented optimization algorithms, three critical metrics have been meticulously selected: Accuracy, Recall and Precision. The choice of these metrics is not arbitrary; each plays a vital role in providing a comprehensive and detailed perspective on how the algorithms perform under different conditions. These selected metrics are valuable because they provide a reliable and consistent framework for comparing and evaluating the effectiveness of different models.

This methodology allows not only to identify the average capabilities of each algorithm but also to understand the variability that may arise across multiple iterations of the experiments. This provides a complete and detailed picture of the efficiency and consistency of each method evaluated.

In machine learning, in the classification process, it is essential to have metrics that allow us to evaluate the performance of the models. Among the metrics, 3 of the most important we have: Accuracy, Precision and Recall [40]. These metrics provide a quantitative way to measure the effectiveness of the models in classification. These metrics are defined by the following equations.

$$Acurracy = \frac{TP + TN}{TP + TN + FP + FN} \tag{1}$$

$$Precision = \frac{TP}{TP + FP} \tag{2}$$

$$Recall = \frac{TP}{TP + FN} \tag{3}$$

where

- *TP* are the True Positives: positive cases that the model predicted correctly.
- *TN* are the True Negatives: negative cases that the model predicted correctly.
- *FP* are False Positives: negative cases that the model incorrectly predicted as positive.
- *FN* are False Negatives: positive cases that the model incorrectly predicted as negative.

To interpret the results, we would consider the following:

- *Accuracy (AS):* It is the ratio of correct predictions to total predictions. The higher the value, the better the metaheuristic algorithm has performed the classification task.

- *Precision (PS):* Measures the proportion of positive predictions that were correct. High accuracy indicates a low number of false positives.
- *Recall (RS):* Also known as sensitivity, it measures the proportion of true positives that were correctly identified. A high recall value indicates that most of the true positives were identified.

## 4.4 Rule Based Classification

Rule-based approaches use quantitative static analysis metrics computed from the source code to identify possible occurrences of code smells [15]. The key idea is that metrics can reflect symptoms, such as the number of lines of code, complexity, cohesion, coupling could be used as markers of the presence of code smells when they exceed specified thresholds [41]. Some examples of code smells and their associated metrics are: God Class. This kind of code smell is linked to classes with a high number of methods and attributes, low cohesion among the methods and this class groups most of the functionality of the system to name a few [15, 16]. Long method. This code smell is characterized by a high number of lines of code (LOC), usually over 100. It also usually presents a high cyclomatic complexity (above 10). Additionally, long methods may have a large number of local variables, nested loops and conditions to name a few [16, 41]. Complex method. A code smell of the complex method type is usually detected or classified when the cyclomatic complexity exceeds 10, has a high value in the number of local variables, nested loops, conditionals, to name a few attributes [16, 41]. God Class and Long Method are common code smells that can be detrimental to software projects. The metrics associated with them are often used to evaluate software quality. It is important to avoid these issues to maintain high-quality software. Extensive empirical research has been conducted on the existence of code smells, revealing that the God Class and Long Method types are among the seven most frequent code smells. Eliminating these code smells provides great benefits in terms of improving maintainability and overall scalability of the code. It is recommended to eliminate these code smells to improve the propensity to change [42]. In another systematic review, the authors found that the God Class and the Long Method are highly ranked as having the most negative effect on most quality attributes. Pices of code with these smells have a higher chance of containing defects [43]. In reference, the study confirms the correlation between code smells and software defects [44]. Code smells are indicators that can accurately anticipate fault detection models. However not all types of smells are equally useful in fault detection; God Class and Long Method stand out for their effectiveness. It is also noteworthy that the interaction of code smells can escalate refactoring likelihood and impede the maintainability of source code. Reference [10, 45]. Alarmingly, code smell coexistence was found to be a pervasive factor. The same study found that God class and large class frequently coexist with other smells [46].

Following the thorough assessment of existing research on identifying code issues through code smell indicators [47]. The initial step in selecting metrics based detection methods involves reviewing the criteria for identifying complexity in code and lengthy code snippets. We excluded any rules lacking definitions, from the original source that introduced them. Subsequently, we meticulously reviewed the definitions outlined in the tool documentation used for extracting metrics and removed rules that relied on metric interpretations or metrics not supported by the tools. The finalized compilation of rules employed as benchmarks is presented in Table 3.

Table 4 the definitions of the different metrics for the God Class and Long Method code smells that will be used to compute the different heuristics. In the research, most of them focus on characterizing the God Class by implementing size (NOM, NOF, CLOC) and complexity (WMC) metrics, while some also include additional coupling (ATFD) and cohesion (LCOM) measures Table 5. Researchers characterize the Long Method mainly by the size attribute, while in other studies they add complexity as well. As the MLCQ dataset comprises a total of 792 projects, tools were implemented

**Table 3** Current formal specifications for bad smells extracted from

| Code smell | Specification |
|---|---|
| God Class/Blob | (ATFD > 2) & (WMC ≤ 47) & (TCC < 0.33) |
| | (ATFD > 5) & (WMC ≤ 47) & (TCC < 0.33) |
| | (WMC ≤ 47) & (TCC < 0.3) & (ATFD > 5) |
| | NOM > 15\| NOF > 15 |
| | CLOC > 750\|NOM + NOF > 20 |
| | NOM + NOF > 20 |
| | WMC ≤ 47 & TCC < 1/3 & ATFD > 5 |
| | CLOC ≤ CLOCV \| NOM ≤ NOMV |
| | (CLOC > 300 & NLM > 5)\|DIT > 5\|DCC > 10 |
| | LCOM3 > 0.8\|LCOM4 > 1.0\|NOF > 20\| NOM > 20 |
| | LCOM ≤ Bad & WMC ≤ Bad & NOF ≤ Bad & NOM ≤ Bad |
| | (ATFD is in top 20%) & (ATFD > 4) & (WMC > 20) & (TCC < 33%) |
| | (CLOC > 150)\|(NOM > 15)\|(NCC > 3) |
| | (NOM > 20)\|(NOF > 9)\|(CLOC > 750) |
| | CLOC > 100 \| VG > 20 |
| | ATFD > 3\|WMC > Threshold\|TCC < 0.33 |
| Data Class | (WOC < 0.33) & ((NOAP + NOAM > 4) & (WMC < 47)\|(NOAP + NOAM > 2) & (WMC < 31)) |
| | nMc (nf) ≤ nMc (nf)V\|nMc = 0 |
| | LCOM > 2\|NOAM > 10 |
| | NSC ≤ Good & DIT ≤ Good & NOF > Regular |
| | WOC < 33% & WOC < 33 & NOAP > 5)\|(NOAM > 5) |
| | (WMC < 50)\|(LCOM < 0.8) |
| | VG & LCOM is low |
| | WOC < 0.33\|NOAP > 3\| NOAM > 3 |
| | MLOC > 50 |

**Table 4** Definition of source code metrics used in the heuristic-based approach for God Class detection

| Metric | Acronym | Definition | Metric extraction tool |
|---|---|---|---|
| Number of methods | NAME | Number of methods in a class | CK tool |
| Number of fields | NOF | Number of attributes of a class | |
| Tight class cohesion | TCC | The relative number of method pairs of a class that jointly access at least one attribute of the measured class | |
| Lines of code | CLOC | Lines of code in a class | RepositorioMiner |
| Foreign data access | ATFD | The number of attributes of unrelated classes that are accessed directly or by invoking accessor methods | |
| Weighted Method Count | WMC | The sum of the static complexity of all methods in a class. McCabe complexity (cyclomatic complexity) is used to quantify the complexity of the method | |
| Lack of cohesion | LCOM5 | Provides a measure for the lack of cohesion of a class. Lack of cohesion between class methods | |

**Table 5** Metrics for code smell specification

| Metric | Acronym | Definition | Metric extraction tool |
|---|---|---|---|
| Lines of code | MLOC | Lines of code in a method | CK tool |
| The complexity of McCabe | VG | Cyclomatic complexity of the method | |
| Nested Blocks Depth | until next day | Nested Blocks Depth | |

to perform massive calculations, but taking care to have the results in analyzable formats. In addition to this, static data were obtained by means of tools, these are values that do not require code compilation.

# 5 Discussion

## *5.1 Results Rule-Based Approach*

To implement the rule-based method, we mapped the columns we have in the dataset of **MLCQ**, with the data obtained from the dataset vs. the rules of the method, we could only implement the code smell **Blob/God Class and Data Class**, while it was not possible to implement the rules for the code smell **Feature Envy and Long Method** due to the lack of values.

The rule-based method shows an efficient balance in the process of classifying positive and negative instances, being reliable and efficient for the classification

**Table 6** Rules of God Class/Blob

| Rules | AS_train | PS_train | RS_train | AS_test | PS_test | RS_test |
|---|---|---|---|---|---|---|
| gc1 | 0.750 | 0.795 | 0.225 | 0.769 | 0.729 | 0.277 |
| gc2 | 0.749 | 0.812 | 0.214 | 0.769 | 0.750 | 0.261 |
| gc3 | 0.757 | 0.644 | 0.425 | 0.769 | 0.616 | 0.462 |
| gc4 | 0.753 | 0.619 | 0.461 | 0.765 | 0.592 | 0.505 |
| gc5 | 0.753 | 0.619 | 0.461 | 0.765 | 0.592 | 0.505 |
| gc6 | 0.722 | 0.656 | 0.154 | 0.747 | 0.667 | 0.185 |
| gc7 | 0.750 | 0.623 | 0.420 | 0.760 | 0.596 | 0.440 |
| gc8 | 0.786 | 0.610 | 0.801 | 0.806 | 0.617 | 0.804 |

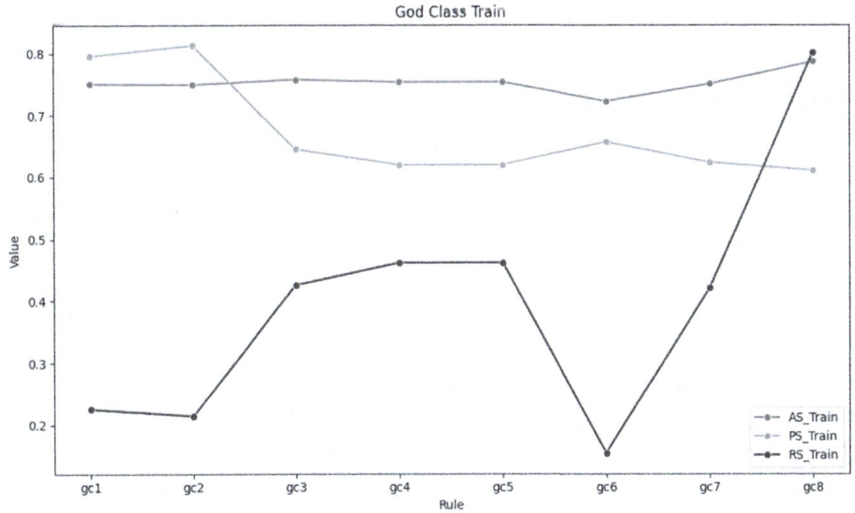


**Fig. 5** God Class metrics—train value

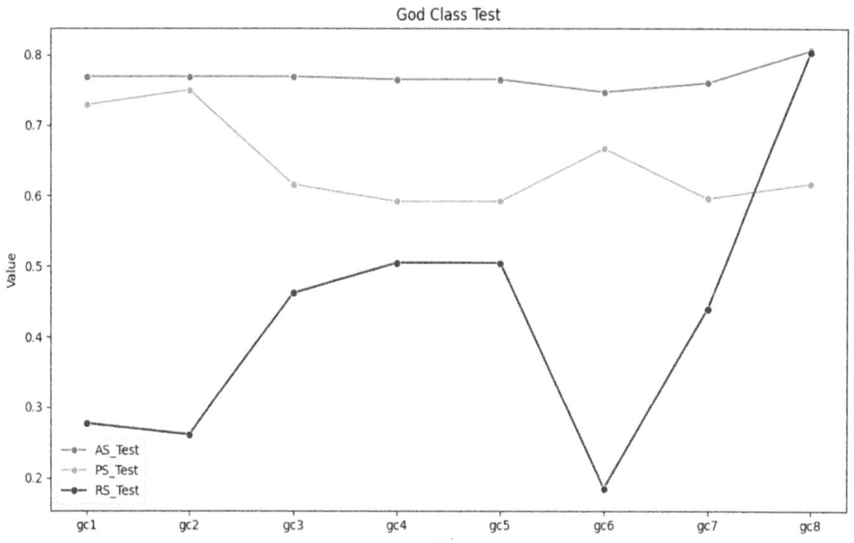


**Fig. 6** God Class metrics—test value

**Table 7** Rules of Data Class

| Rules | AS_train | PS_train | RS_train | AS_test | PS_test | RS_test |
|---|---|---|---|---|---|---|
| dc1 | 0.761 | 0.804 | 0.314 | 0.741 | 0.841 | 0.320 |
| dc2 | 0.696 | 0.509 | 0.969 | 0.712 | 0.550 | 0.970 |
| dc3 | 0.730 | 0.689 | 0.254 | 0.716 | 0.744 | 0.290 |
| dc4 | 0.752 | 0.565 | 0.901 | 0.768 | 0.617 | 0.887 |
| dc5 | 0.710 | 0.529 | 0.683 | 0.701 | 0.560 | 0.688 |

**Table 8** Average values of God Class and Data Class rules

| Rules | AS_train | PS_train | RS_train | AS_test | PS_test | RS_test |
|---|---|---|---|---|---|---|
| gc | 0.753 | 0.672 | 0.395 | 0.769 | 0.645 | 0.430 |
| dc | 0.730 | 0.619 | 0.624 | 0.728 | 0.662 | 0.631 |

process, with a good handling of false positives and negatives. However, with the values obtained, it indicates the efficiency of the model but also shows that there is room for improvement in accuracy and sensitivity.

The data obtained for these code smells for Blob are shown in the following Table 6, and Figs. 5, 6.

While the data obtained for these code smells for Data Class are shown in the following Table 7 and Figs. 7, 8 (Table 8).

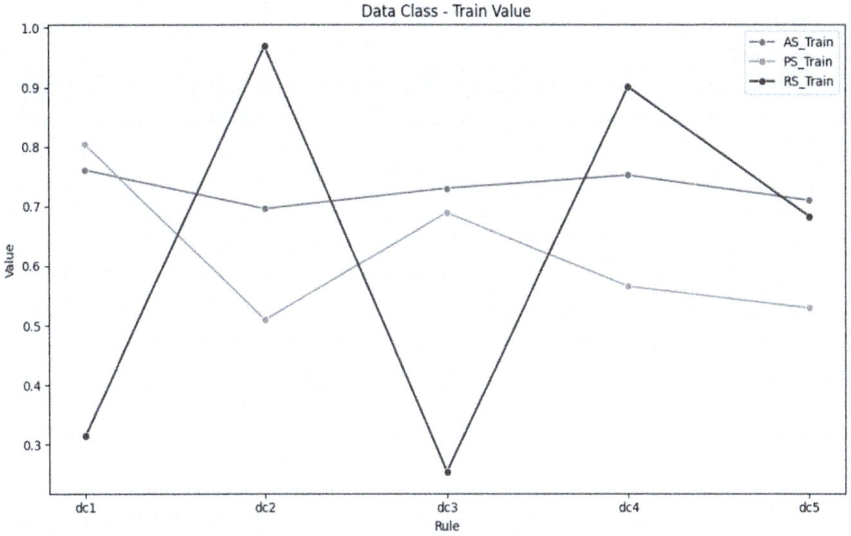


**Fig. 7** Data Class metrics—train value

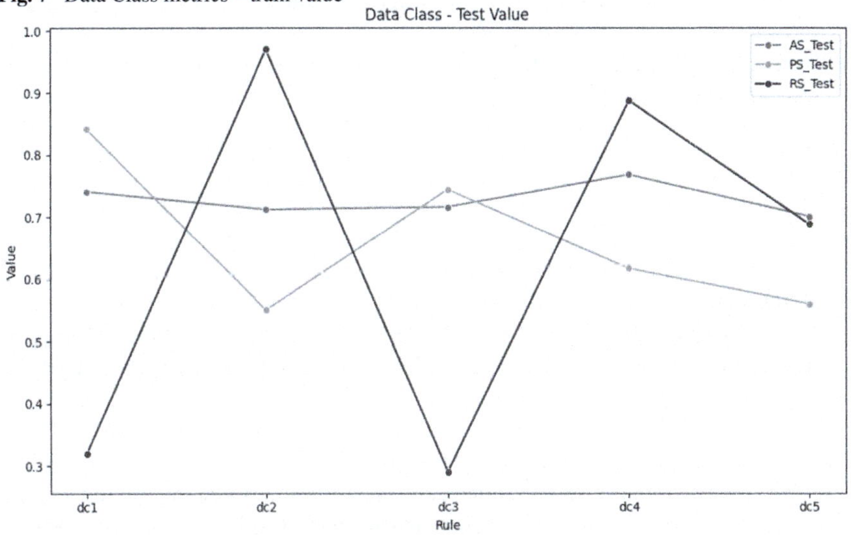


**Fig. 8** Data Class metrics—test values

## *5.2 Results on Feature Selection Using Metaheuristic Algorithms*

Although cross-validation is a commonly used technique for evaluating models, a study conducted on this topic found this technique to be one of the most biased and least reliable in the field of software defect prediction [49]. This is likely due to the randomness in the selection of the folds. This has a significant impact on performance [50].

A more reliable alternative for model evaluation is to calculate the average of the performances obtained in multiple tests. For each test, the data is divided into different training and test sets.

Considering some studies in this regard, it is advisable to adopt this strategy: The data is divided into 80 for training and 20 for testing. This division is done by random stratified sampling due to the large imbalance in class labels.

Based on the table: The "best" metaheuristic algorithms could be the one with the best balance between Accuracy, Precision and Recall in the test set, unless runtime is a primary consideration. The "worst" metaheuristic algorithm could be the one with the lowest values on these metrics in the test set and/or the longest run time.

In the Tables 7 and 8, if we only consider Accuracy, Precision and Recall for the test set: Taking into account the values of *Average values in metrics of the different optimizers* we can assume:

- The TSA metaheuristic seems to be the best in terms of Accuracy and Recall, but not in Precision.
- Metaheristico CRO has one of the shortest execution times, but its performance in terms of metrics is not the highest.
- The ABC metaheuristic has the longest execution time, which could be considered less desirable if time is a critical factor.

It is important to keep in mind that the "best" metaheuristic algorithms for optimization depend on the specific task context and performance priorities. If the trade-off between metrics is important, then it may not be sufficient to select the algorithm with the best value on a single metric. Furthermore, in rigorous research practices, these decisions are often supplemented with statistical tests to ensure that differences in performance are meaningful.

The stochastic nature of metaheuristic algorithms requires its analysis on statistically significant samples. The variability over 30 samples of each experiment can be observed in Fig. 9.

From Table 9 There is a notable difference in the runtime duration across different optimizers, with FA taking the longest time (1120.186 s) and SA being the quickest (14.638 s). This disparity suggests that certain algorithms are more efficient computationally than others. The performance metrics (AS, PS, RS) during both training and testing phases show a level of consistency within each optimizer. This consistency implies that the models are not overfitting to the training data and might have good generalization capabilities for new data. The accuracy score (AS), precision score (PS) and recall score (RS) demonstrate close values for all optimizers, both in training and testing phases. This indicates that there isn't a significant difference in how effective the optimization algorithms are based on these metrics for this specific task. FA stands out due to its notably longer runtime compared to other optimizers, without showing a proportional improvement in performance metrics. This may suggest that FA is less efficient for this particular optimization task. When comparing efficiency to effectiveness, optimization techniques such as CRO and SA strike a nice

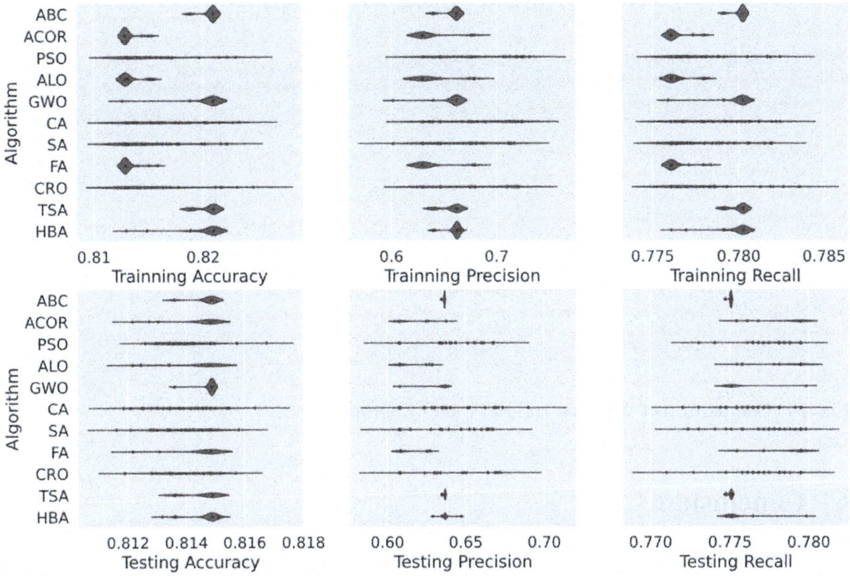


**Fig. 9** Comparison of AS_test and RS_test by optimizer

**Table 9** Average values in metrics of the different optimizers

| Optimizador | Runtime | AS_train | PS_train | RS_train | AS_test | PS_test | RS_test |
|---|---|---|---|---|---|---|---|
| ABC | 251.104 | 0.821 | 0.660 | 0.780 | 0.815 | 0.637 | 0.775 |
| ACOR | 143.488 | 0.813 | 0.635 | 0.776 | 0.815 | 0.610 | 0.779 |
| ALO | 161.776 | 0.813 | 0.639 | 0.776 | 0.814 | 0.611 | 0.779 |
| CA | 136.151 | 0.817 | 0.681 | 0.779 | 0.814 | 0.636 | 0.777 |
| CRO | 75.154 | 0.816 | 0.669 | 0.778 | 0.814 | 0.630 | 0.778 |
| FA | 1120.186 | 0.813 | 0.635 | 0.776 | 0.815 | 0.610 | 0.779 |
| GWO | 136.507 | 0.821 | 0.661 | 0.780 | 0.815 | 0.636 | 0.775 |
| HBA | 131.383 | 0.821 | 0.661 | 0.780 | 0.815 | 0.637 | 0.775 |
| PSO | 160.392 | 0.815 | 0.673 | 0.778 | 0.814 | 0.628 | 0.778 |
| SA | 14.638 | 0.815 | 0.665 | 0.778 | 0.814 | 0.632 | 0.778 |
| TSA | 122.771 | 0.820 | 0.657 | 0.780 | 0.815 | 0.636 | 0.775 |

balance between being computationally efficient (with shorter runtime) and effective (yielding competitive performance metrics) can be observate in Fig. 10. This makes them potentially better options for similar optimization tasks.

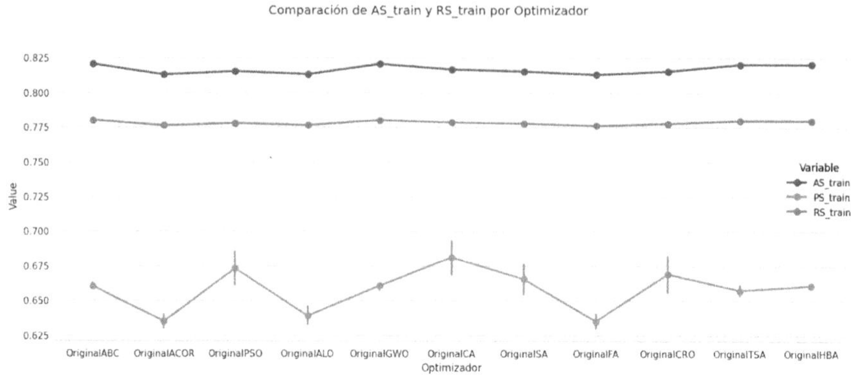


**Fig. 10** Comparison of AS_train and RS_train by optimizer

# 6 Conclusion

This chapter thoroughly examined how evolutionary algorithms (EAs) can enhance the detection of code problems by improving feature selection methods in software metrics. Code smells, which signal possible flaws in software design, typically require time consuming techniques for detection. Our investigation into using EAs has revealed a significant boost in accuracy and efficiency compared to traditional rule-based approaches for identifying these code problems. The research emphasized the ever evolving and intricate nature of software creation, pointing out how code issues arise due to factors such as rapid market changes and the drive for faster development cycles. Although these code problems may not affect functionality immediately, they can greatly impede future maintenance and scalability efforts, posing a considerable risk to software quality. Through the utilization of EAs for feature selection, our methodology has shown promise in automating and enhancing the detection of code smells more effectively. This method utilizes EAs principles of natural selection to pinpoint the most pertinent software metrics associated with code problems, ultimately simplifying the detection process. The results of our experiments, based on datasets relevant to the industry, confirm that our approach using evolutionary algorithms can outperforms traditional methods. These findings encourage further exploration of AI driven solutions to challenges in software engineering. In summary, applying evolutionary algorithms in detecting code smells marks is a growing trend in software engineering. By enhancing the precision and effectiveness of code smell detection, this research contributes to creating better quality software systems. Future research directions may involve exploring how other AI techniques can be integrated into code smell detection and expanding the use of evolutionary algorithms to other tasks in software engineering. This study sets the stage for future endeavors, particularly in improving the flexibility and scalability of these algorithms to meet the growing complexity of software systems. Furthermore, incorporating AI enhanced solutions into software engineering practices holds promising potential for automating and enhancing various aspects of software development and maintenance processes.

**Acknowledgements** I want to express my sincere gratitude to those who have been fundamental in the completion of this project. To my wife, Elizabeth, and my son, Santiago, for their love, patience and constant support, which have been my refuge and motivation. I also extend my gratitude to my advisors, whose expert guidance and valuable advice have been essential to achieving this accomplishment. Their support has been indispensable on this journey, and for that, I am deeply grateful. I would like to express my sincere gratitude to the National Council for Science and Technology (CONACYT) of Mexico for the funding provided through the Scholars Program, which allowed for the successful development and completion of this project. **Diego Avalos**.

## References

1. Fowler, M., Beck, K., Brant, J., Opdyke, W., Roberts, D.: Refactoring: Improving the Design of Existing Code. Addison-Wesley Professional (1999)
2. Khomh, F., Penta, M.D., Guéhéneuc, Y.-G., Antoniol, A.: An exploratory study of the impact of antipatterns on class change-and fault-proneness. Empir. Softw. Eng. **17**(3), 243–275 (2012)
3. Kotsiantis, S., Verykios, V.S., Tzagarakis, M.: Ai-assisted programming tasks using code embeddings and transformers. Electronics **13**(4), 767 (2024). https://www.mdpi.com/2079-9292/13/4/767
4. Hoy, Z., Xu, M.: Agile software requirements engineering challenges-solutions-a conceptual framework from systematic literature review. Information **14**(6), 322 (2023). https://www.mdpi.com/2078-2489/14/6/322
5. Wang, T., Backes, M., Chen, S.-W., Chen, Z., Gu, X., Liu, Y.: How reliable are code smell detection tools? Proc. ACM Progr. Lang. **7**(OOPSLA2), 1–29 (2023)
6. Yamashita, A., Counsell, S.: Code smells as system-level indicators of maintainability: an empirical study. J. Syst. Softw. **147**, 1–23 (2019)
7. Wang, T., Ampatzoglou, A., Avgeriou, P.: An empirical study of lazy class detection tools and techniques. Sci. Comput. Progr. **215**, 102664 (2022)
8. Mäntylä, M., Savolainen, J., Koponen, S., Mikkonen, T.: Automatically detecting concentrated technical debt: a case study in a startup company. Inf. Softw. Technol. **146**, 106951 (2022)
9. Rasool, G., Arshad, Z.: A review of code smell mining techniques. J. Softw. Evol. Process **27**(11), 867–895 (2015)
10. Yamashita, A., Moonen, L.: Code smells as system-level weaknesses. In: 2013 IEEE International Working Conference on Mining Software Repositories (MSR), pp. 380–389 (2013)
11. Abebe, S.L., Haiduc, S., Tonella, P., Marcus, A.: Lexicon bad smells in software. In: Proceedings of the 16th Working Conference on Reverse Engineering, pp. 95–99. IEEE (2009)
12. Authors: Deepscc: source code classification based on fine-tuned roberta (2021). arXiv:2110.00914
13. Souza, P., Sousa, B., Ferreira, K., Bigonha, M.: Applying software metric thresholds for detection of bad smells. In: Proceedings of the 11th Brazilian Symposium on Software Components, Architectures, and Reuse, pp. 6–16. ACM (2017)
14. Chaurasia, S.P., Shekhawat, M.: Automated approach to detect different types of code smells using software quality metrics. Int. J. Inf. Technol. **9**(3–4), 227–236 (2017)
15. Fernandes, E., Terra, R., Cirilo, E., Miranda, I., Kulesza, U.: Detecting and eliminating god classes through class cohesion metrics. In: 2018 IEEE 25th International Conference on Software Analysis, Evolution and Reengineering (SANER), pp. 517–521. IEEE (2018)
16. Rasool, G., Arshad, J., Zhang, H.: Mining and characterizing god classes in object-oriented software systems. Inf. Softw. Technol. **123**, 106279 (2020)

17. Devi, S., Sabrigiriraj, M.: Feature selection, online feature selection techniques for big data classification: - a review. In: 2018 International Conference on Current Trends towards Converging Technologies (ICCTCT) (2018). https://ieeexplore.ieee.org/document/8550928/
18. Alazba, A., Aljamaan, H.: Code smell detection using feature selection and stacking ensemble: an empirical investigation. Inf. & Softw. Technol. **138**, 106 648–106 648 (2021). https://www.sciencedirect.com/science/article/pii/S0950584921001129
19. Khoshniat, N., Jamarani, A., Ahmadzadeh, A., Kashani, M.H., Mahdipour, E.: Nature-inspired metaheuristic methods in software testing (2023). https://doi.org/10.1007/s00500-023-08382-8
20. Khajehzadeh, M., et al.: A survey on meta-heuristic global optimization algorithms. Res. J. Appl. Sci. Eng. Technol. **3**(6), 569–578 (2011)
21. Rostami, M., Berahmand, K., Nasiri, E., Forouzandeh, S.: Review of swarm intelligence-based feature selection methods. Eng. Appl. Artif. Intell. **100**, 104210 (2021)
22. Google: Google colaboratory (2023). https://colab.research.google.com/. Accedido en Marzo 2024, Jalisco, México
23. Van Rossum, G., et al.: Python (2023). https://www.python.org/. Accedido en 2024
24. MAFESE: Feature Film Selection and Extraction Analysis Framework (2023). https://mafese.readthedocs.io/en/latest/index.html. Accedido en 2024
25. Madeyski, L., Lewowski, T.: MLCQ: industry-relevant code smell data set (2020). https://zenodo.org/record/3590102
26. Cortes, C., Vapnik, V.: Support-vector networks. Mach. Learn. **20**(3), 273–297 (1995)
27. Breiman, L.: Random forests. Mach. Learn. **45**(1), 5–32 (2001)
28. Schölkopf, B., Smola, A.J.: Learning with Kernels: Support Vector Machines, Regularization, Optimization, and Beyond. MIT Press (2002)
29. Karaboga, D., Basturk, B.: A powerful and efficient algorithm for numerical function optimization: artificial bee colony (abc) algorithm. J. Global Optim. **39**(3), 459–471 (2007)
30. Socha, K., Dorigo, M.: Acor: a ant colony optimization algorithm for continuous optimization. Eur. J. Oper. Res. **185**(3), 1155–1173 (2008)
31. Eberhart, R., Kennedy, J.: Particle swarm optimization. Proc. IEEE Int. Conf. Neural Netw. **4**, 1942–1948 (1995)
32. Mirjalili, S.: The ant lion optimizer. Adv. Eng. Softw. **83**, 80–98 (2015)
33. Mirjalili, S., Mirjalili, S.M., Lewis, A.: Grey wolf optimizer. Adv. Eng. Softw. **69**, 46–61 (2014)
34. Yang, X.-S., Deb, S.: Cuckoo search via lévy flights. In: World Congress on Nature & Biologically Inspired Computing (NaBIC), pp. 210–214 (2009)
35. Kirkpatrick, S., Gelatt, C.D., Vecchi, M.P.: Optimization by simulated annealing. Science **220**(4598), 671–680 (1983)
36. Yang, X.-S.: Firefly algorithms for multimodal optimization. In: International Symposium on Stochastic Algorithms, pp. 169–178 (2009)
37. Salcedo-Sanz, S., Del Ser, J., Rahim, I., et al.: Coral reefs optimization-an experimental evaluation. In: Proceedings of the 2010 IEEE Congress on Evolutionary Computation (CEC'10), pp. 1–8 (2010)
38. Kaveh, A., Sheikholeslami, R.: Tug of war optimization: a novel meta-heuristic method. Eng. Comput. **32**(2), 215–233 (2016)
39. Anadrai, R.N., Hassan, M.K., Al-Hilali, B.T., Hussein, A.I.: Honey badger algorithm: new nature-inspired metaheuristic optimization algorithm. In: 2019 International Conference on Computational Science and Computational Intelligence (CSCI), pp. 1278–1283. IEEE (2019)
40. Gonzalez, M., Rodríguez, J., López, C.: Comprehensive analysis of classification metrics in machine learning. J. Mach. Learn. Res. **20**(1), 123–145 (2019)
41. Taibi, D., Lenarduzzi, V., Pahl, C., Janes, A.: Code smell detection: a systematic literature review. Inf. Softw. Technol. **123**, 106307 (2020)
42. Di Nucci, D., Palomba, F., Tamburri, D.A., Serebrenik, A., De Lucia, A.: Detecting code smells using machine learning techniques: are we there yet? In: 2018 IEEE 25th International Conference on Software Analysis, Evolution and Reengineering (SANER), pp. 612–621. IEEE (2018)

43. Lacerda, G., Petrillo, F., Pimenta, M., Guéhéneuc, Y.G.: Code smells and refactoring: a tertiary systematic review of challenges and observations. J. Syst. Softw. **167**, p. Article 110610 (2020)
44. Piotrowski, P., Madeyski, L.: Software defect prediction using bad code smells: a systematic literature review. Data-Centric Bus. Appl. 77–99 (2020)
45. Palomba, F., Bavota, G., Di Penta, M., Fasano, F., Oliveto, R., De Lucia, A.: On the diffuseness and the impact on maintainability of code smells: a large scale empirical investigation. Empir. Softw. Eng. **23**(3), 1188–1221 (2018)
46. Palomba, F., Bavota, G., Di Penta, M., Fasano, F., Oliveto, R., De Lucia, A.: A large scale empirical study on the lifecycle of code smell co-occurrences. Inf. Softw. Technol. **99**, 1–10 (2018)
47. Kovačević, A., Slivka, J., Vidaković, D., Grujić, K.-G., Luburić, N., Prokić, S., Sladić, G.: Automatic detection of long method and god class code smells through neural source code embeddings. Expert Syst. Appl. **204**, 117607 (2022)
48. Bafandeh Mayvan, B., Rasoolzadegan, A., Javan Jafari, A.: Bad smell detection using quality metrics and refactoring opportunities. J. Softw.: Evol. Process **32**(8), e2255 (2020)
49. Tantithamthavorn, C., McIntosh, S., Hassan, A.E., Matsumoto, K.: An empirical comparison of model validation techniques for defect prediction models. IEEE Trans. Softw. Eng. **43**(1), 1–18 (2017). http://dx.doi.org/10.1109/tse.2016.2584050
50. Hall, T., Beecham, S., Bowes, D., Gray, D., Counsell, S.: Developing fault prediction models: What the research can show industry. IEEE Softw. **28**(6), 96–99 (2011). http://dx.doi.org/10.1109/ms.2011.138

# Exploratory Analysis of Machine Learning Methodologies Optimization for Facial Expression Recognition (FER)

**Mauricio Priego, Itzel Aranguren, and Fernando Fausto**

**Abstract** Communication is one of the significant milestones of human beings; however, most of what we communicate is transmitted through facial expressions. Due to this, Facial Expression Recognition is one of the areas of most outstanding research in Artificial Intelligence for the potential benefits it can provide both economically and socially in areas such as marketing, education, security, technology, entertainment, politics, human resource management, or physical and mental health. This paper presents a review with the result of the comparative analysis of the main advances in the challenges that the identification of emotions entails, presenting the conceptual architecture model of each solution. It describes the praxis of how machine learning, artificial neural networks, deep learning, optimization algorithms, metaheuristics, and even anatomy and psychology tools—such as Action Units and the Arousal-Valence Two-Dimensional Model—drive the development of this transcendent area.

**Keywords** Facial expression recognition · Machine learning · Artificial neural network · Deap learning · Optimization algorithm

## 1 Introduction

The ability to communicate precisely, from specific basic needs such as asking for food to abstract concepts such as love or loyalty, is an inherent characteristic of human beings. The spoken language, the most recent form of communication

M. Priego (✉)
Universidad de Guadalajara, CUCEI, Av. Revolución 1500, Guadalajara, Jalisco, Mexico
e-mail: mauricio.priego2879@alumnos.udg.mx

I. Aranguren · F. Fausto
Depto. de Innovación Basada en la Información y el Conocimiento, Universidad de Guadalajara, CUCEI, Av. Revolución 1500, Guadalajara, Jalisco, Mexico
e-mail: itzel.aranguren@academicos.udg.mx

F. Fausto
e-mail: abraham.fausto@academicos.udg.mx

 
D. Oliva et al. (eds.), *Advances in Optimization Algorithms for Multidisciplinary Engineering Applications: From Classical Methods to AI-Enhanced Solutions*, Studies in Computational Intelligence 806, https://doi.org/10.1007/978-3-031-78440-8_21

acquired evolutionarily speaking, represents only 7% of the information that humans transmit, with 38% being the tone of voice and 55% being body language, also known as nonverbal language [1].

Thus, if individuals cannot comprehend nonverbal language, they will miss out on more than half of the information, which may result in misunderstandings, a lack of empathy, and weakened relationships. As a result, numerous researchers have conducted studies and research to identify and interpret facial expressions, hand gestures, and body postures.

Facial expressions are, by far, the most expressive source of nonverbal language, but they are also the most difficult for machines to identify. However, this ability is natural for most humans. Throughout this chapter we will review several proposals that seek to solve the challenges represented by the recognition of facial expressions through different approaches, methodologies and optimization algorithms, making a comparative analysis between the results and the techniques used, and ending with the conclusions from the aforementioned analysis.

## *1.1 Facial Action Coding System*

The foundations of Facial Expression Recognition (FER) -defined as the capacity of Artificial Intelligence (AI) to identify facial expressions through machine learning techniques- were established in 1970, when groups of facial muscles involved in different expressions were described by Hjortsjö, naming them "Action Units" (AU) [2]. Being discrete units of facial muscle movement, AU provides an objective and quantifiable basis for analyzing and understanding human facial expressions in different contexts and situations, such as cultural or ethnic traits. Action Units have demonstrated their potential for recognizing basic human emotions as well as micro-expressions, faint AU movements that reveal intentions, and spontaneous expressions [3]. In general, to detect AU through IA, there are mainly two ways to do it [4]:

- Geometric-based methods, where machine learning techniques seek to discover the shape of facial components (eyes, eyebrows, mouth)
- Appearance-based methods, where AI seeks to extract facial textures such as expression lines or temporal patterns within sequences.

Emotions cause a variety of facial muscle movements that humans inadvertently show. Based on facial muscles, different facial movements can be parameterized and used to represent different facial expressions. To construct the parameterization of facial movements different proposals have been presented, being Facial Action Coding Scheme (FACS) and Facial Animation Parameters (FAP) those that have obtained the most precise results in practice. FACS is a widely used AU-based facial expression coding system, which allows movements to be objectively described based on the movement of specific facial muscles by assigning a number to each movement-place for identification and analysis Fig. 1. This system provides a standardized way

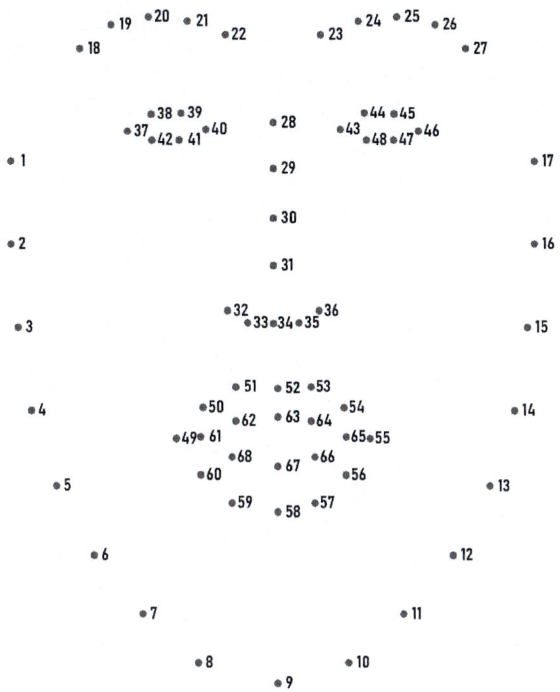


**Fig. 1** Example of action unites pointed on a face silhouette

of describing facial expressions, making it a valuable tool for psychological research, emotion detection, and facial expression recognition among other fields. On the other hand, FAP use computational models to describe and simulate facial movements [5], being used in computer animation and the creation of virtual characters to represent realistic and expressive facial expressions.

## *1.2 Arousal Valence 2D Model*

Although AU is an objective and naturally human approach to recognizing emotions, there are other common methods for modeling based on the idea that emotions are not discrete but multi-dimensional, so they should be classified into scalar dimensions. The Arousal-Valence-2D-Model (AV2D) is the commonly used approach of multi-dimensional methods, which defines two dimensions for classifying emotions: Arousal, the intensity of the emotional experience, and Valence, the positivity or negativity of the emotional experience [6]. *Sadness*, for example, is defined in AV2D as low arousal and low valence, while *happiness* is defined as high arousal and high valence Fig. 2. AV2D lands on FER, usually to evaluate the accuracy of emotion classification, commonly through Deep Learning. The 2D Arousal-Valence model is fundamental in analyzing and understanding human emotions.

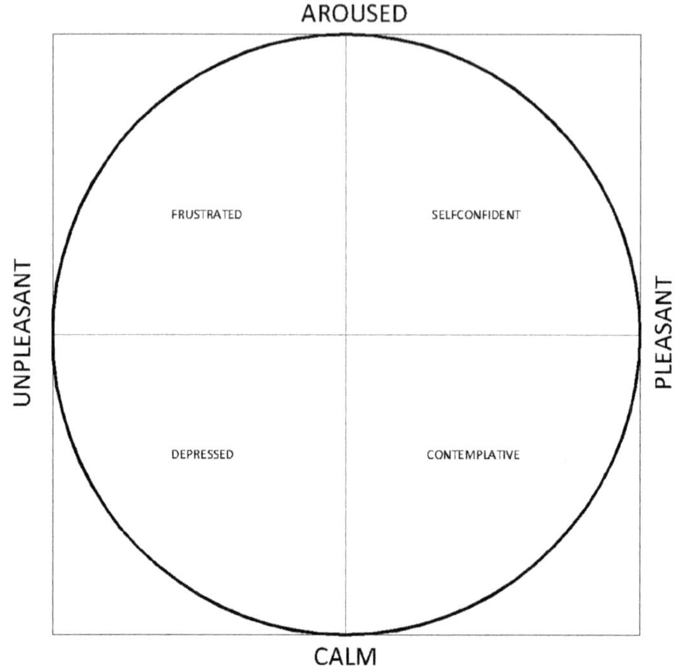


**Fig. 2** Example of arousal-valence two-dimensional model

One key advantage of the 2D Arousal-Valence model is its ability to capture the complexity and subtlety of human emotions in a relatively simple dimensional space. The model can provide information about an individual's emotional state and be used in various contexts and applications. For example, in human-computer interaction, the model can be used to develop user interfaces that intelligently respond to user emotions, adapting the user experience to meet their emotional needs. However, the AV2D also faces some challenges and limitations. For example, individual variability in emotional expression and cultural and contextual factors can affect the accuracy of emotion detection.

## *1.3 Challenges of Artificial Intelligence in Facial Expressions Recognition*

Facial expression recognition is a subfield of artificial intelligence that faces complex challenges such as facial expression variability. Expressions vary between individuals and can be influenced by aging, gender, and emotional context, making it necessary to design robust algorithms to adapt to this diversity and generalize it effectively. In addition, facial expressions can be subtle, so AI approaches must be able to detect even the slightest differences accurately. Cultural interpretation is also a critical aspect to consider. Facial expressions can vary significantly between cultures and ethnic groups, affecting the effectiveness of FER algorithms in a global context. Real-time processing is another major challenge. In applications such as human-computer interaction and security, the algorithms must be able to rapidly analyze large volumes of video data in real-time to detect and classify facial expressions efficiently and accurately.

Addressing these challenges will require labeled and representative data sets, advanced machine learning algorithms, and interdisciplinary understanding, including computer vision, psychology, and data collection among others. This combined effort has the potential to capture the complexity of facial expressions in different contexts and for different individuals.

## 2 Facial Expression Recognition with Machine Learning Techniques

In the last decade, nonverbal language recognition has experienced significant advancements because of Machine Learning (ML) tools and techniques. These algorithms offer a powerful approach to capture and process large amounts of data efficiently. These algorithms can identify subtle patterns and relevant features in the data, enabling accurate detection of emotional expressions and gestures. Moreover, machine learning algorithms are highly adaptable. They can be trained to recognize a wide variety of expressions and gestures, even under variable conditions such as changes in lighting, viewing angles, and complex backgrounds. Another essential advantage is their ability to improve over time as they are fed with more data and receive feedback, being adaptive to each application's specific needs and improving their accuracy and performance over time [7], making them ideal for security, healthcare, human-computer interaction, entertainment applications, and human emotion recognition.

### *2.1 Principal Component Analysis and K-Nearest Neighbors for FER*

The Principal Component Analysis (PCA) is an unsupervised learning technique for dimensionality reduction and exploratory data analysis. To do this, PCA takes the most representative characteristics of the object of study, feeding with them a covariance matrix that identifies those characteristics that are interdependent among themselves. After that, PCA preserves those characteristics with higher values, constructing with them a new data matrix with fewer dimensions, maintaining, on the way, the significant part of their information. Although PCA is not a supervised learning algorithm, it is typically used to pre-process data before applying supervised learning algorithms in machine learning tasks [9].

The K-Nearest Neighbors (KNN) is a supervised learning method for classification and regression. It operates as follows:

1. Training: The algorithm stores the training data during the training stage.
2. Classification/Regression: When presented with a new data point, KNN searches for the "k" nearest data points (neighbors) in the search space. The distance between points is typically calculated using Euclidean distance.
3. Classification: KNN assigns the most common class label among the "k" neighbors to the new data point based on the nearest class with the highest representation in the neighborhood. This is known as "majority voting."
4. Regression: In regression, KNN returns the average response values of the "k" nearest neighbors as the prediction value for the new data point.

The value of "k" is a parameter that must be selected before applying the algorithm. A lower value of "k" means the model is more flexible and will fit more closely to the data details, but it may also be more susceptible to noise. On the other hand, a higher value of "k" will smooth out predictions, which can help reduce the effect of noise, but it may lose detail in the data [10].

Iromini et al. proposed an approximation of FER combining the algorithms PCA and KNN, as seen in Fig. 3. In the proposal, PCA detects and extracts relevant face characteristics that allow the expression to be identified and those that distinguish the particular expression from similar expressions. It is important to remember that PCA is a nontrained machine learning technique, meaning that no humans determine the face points required for FER. Once PCA creates a features vector with FER points data, KNN is trained to classify the expression among six emotions plus neutral [8].

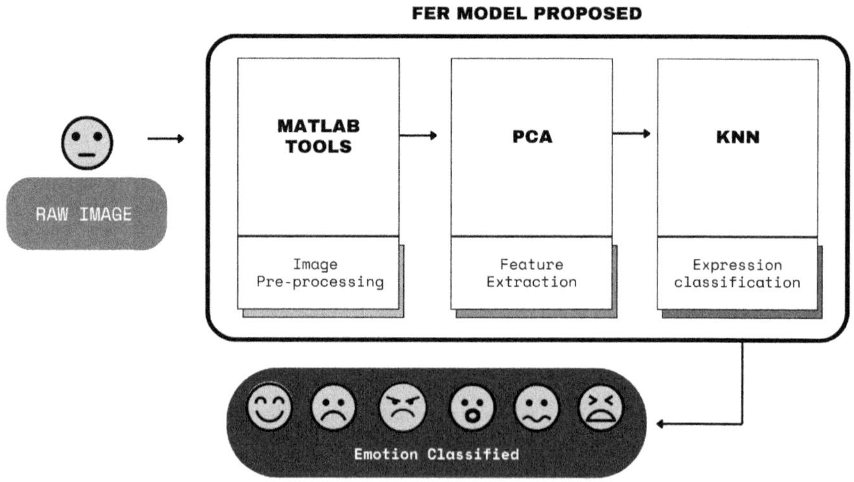


**Fig. 3** Model of the methodology proposed by Iromini et al. [8]

Japanese Female Facial Expression (JAFFE) dataset was used for training and testing, achieving an average accuracy rate of 85.71% in classifying six emotions and the neutral expression.

## 2.2 Particle Swarm Optimization and Extreme Learning Machine for FER

The Particle Swarm Optimization (PSO) is a metaheuristic algorithm; it is an optimization technique that guides the search for solutions through a search space in a heuristic manner, using general rules and principles inspired by natural phenomena or observations of human behavior. In this way, PSO is based on the behavior of swarms of particles in nature, such as flocks of birds or schools of fish. Each particle in the swarm represents a potential solution to the optimization problem. It moves through the search space, adjusting its position and velocity based on its own experience and that of neighboring particles, guiding its search toward promising regions of the search space to find high-quality solutions to complex optimization problems.

The Extreme Learning Machine (ELM) is a machine learning technique mainly used for classification and regression problems, standing out for its simplicity and computational efficiency. Its processing is based on four basic steps: Initialization of the input weights and hidden biases with random values, calculation of the hidden output through an activation function, calculation of the output weights using linear regression methods, and Prediction, which uses the output of the hidden layer along with the calculator weights to predict the output in new examples. The two main benefits of ELM lie in its computational efficiency, achieved by not requiring iterative adjustment of weights, and its ability to handle large volumes of data quickly. However, it may require more memory because all training data is stored in the memory.

Sui et al. consider facial expression feature extraction to be a critical part of the entire FER process. To achieve that, they use a hybrid feature extraction method, combining Geometric Feature Extraction and principles with Texture Feature Extraction, feeding a PCA algorithm to reduce dimensionalities, and building a feature vector (Fig. 4). In this point, PSO optimizes the initialization of ELM input weights and hidden biases, which uses the feature vector for training and testing [11]. Cohn-Kanade extended(CK+) dataset was used for training and testing, achieving an average accuracy rate of 92.8% in classifying six emotions and the neutral expression.

## 2.3 Bayesian Regression Network for FER

The Bayesian Regression Network (BLR), an extension of Bayesian networks, is a statistical model used in machine learning and probabilistic modeling that allows modeling complex relationships between observed and unobserved variables, helping

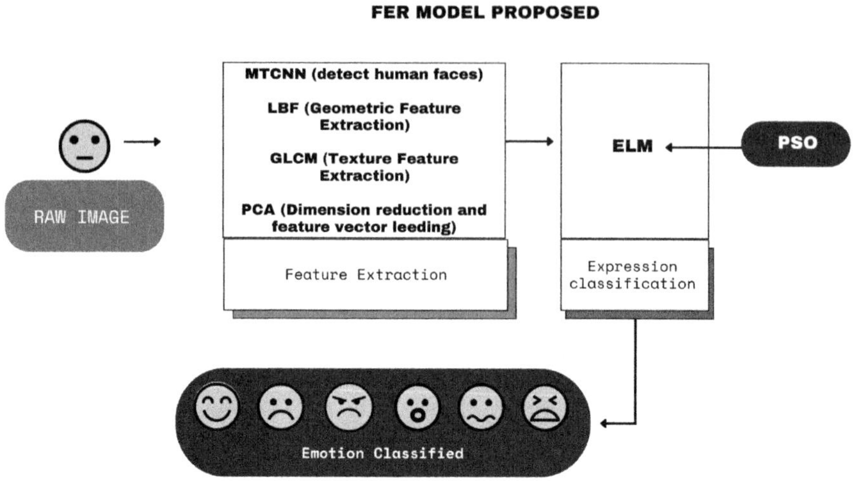


**Fig. 4** Model of the methodology proposed by Sui et al. [11]

analyze complex data and make probabilistic inferences about the data. In simple terms, a BLR combines two main components:

1. The Regression Model identifies the possible relationships between observed and unobserved variables.
2. The Bayesian Network, which models the dependencies between observed and unobserved variables. This dependencies modeling uses a Directed Acyclic Graph (DAG), where the nodes represent the variables, and the arcs represent the probabilistic relationships between them. The structure of the Bayesian network allows inferences to be made about the joint distribution of all the variables in the model.

Focusing on image recognition -including FER-, the BLR technique is used for Latent Feature Modeling, which detects visual elements relevant to image classification, such as edges, textures, or shapes. It is also used for Learning transfer and Uncertainty Modeling to determine the confidence or certainty of the model in its predictions.

Wang et al. gives an interesting point of view about FER, differentiating between posed and spontaneous expressions, as seen in Fig. 5. To reach this goal, the proposal requires several BLR models, each for expression-related factors like gender or expression categories. The BLR's latent layer, which contains unobserved variables, is fed with FSPs, which, for practical purposes, are the principal geometric feature points of the face. At the same time, the visible layer is fed with the given expression-related factors, that is to say, the observed variables [12]. For training and testing, Spontaneous vs. Posed facial expression (SPOS), USTC-NVIE (NVIE), and DISFA+ datasets were used, achieving average accuracy on the NVIE dataset of 99.32% for gender as an expression-related factor and 98.35% for expression categories without any expression-related factor, considering six emotions and the neutral expression.

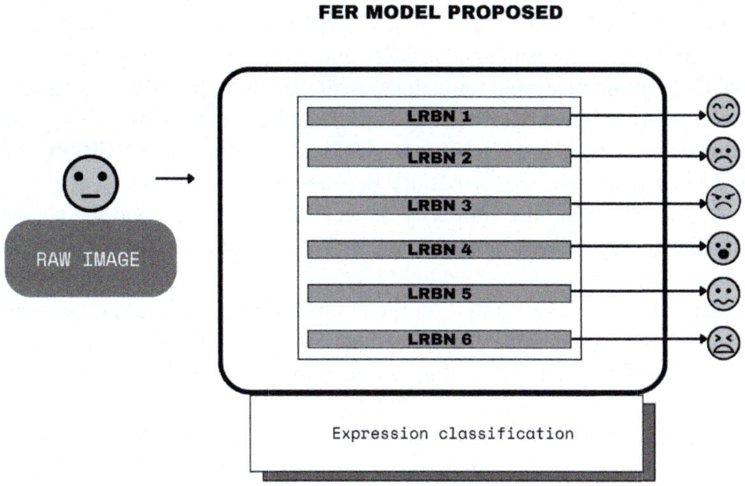


**Fig. 5** Model of the methodology proposed by Wang et al. [12]

## 2.4 *Fuzzy Rough Feature Selection Inspired by Ant Colony Optimization and Fuzzy Rough Nearest Neighbor for FER*

Fuzzy Rough Feature Selection inspired by Ant Colony Optimization (FRFS-ACO) is a feature selection technique that uses fuzzy and rough logic concepts, along with inspiration taken from the behavior of ant colonies, to identify an optimal set of features that maximizes the performance of the predictive model in terms of accuracy or computation. Its operation can be summarized in the following steps:

1. Pheromone initialization and heuristics. An initial amount of pheromones is assigned to each feature in the data set. Furthermore, heuristic measures are calculated for each characteristic, representing its initial importance in the selection process.
2. Solution construction. The algorithm uses a probabilistic approach to select features based on the number of pheromones and associated heuristics. Features are selected iteratively to form a subset of potential features.
3. Evaluation of the quality of the solutions. The quality of each subset of features is evaluated using some criterion, such as the accuracy of a predictive model trained with those features.
4. Pheromone Update. The quantity of pheromones is updated based on the quality of feature subsets. Characteristics that contribute to high-quality solutions receive an additional amount of pheromones.
5. Iteration and convergence. The process of building solutions and updating pheromones is repeated for a fixed number of iterations or until a predefined convergence criterion is reached.

The Fuzzy Rough Nearest Neighbor (FRNN) algorithm is a classification technique that combines the concepts of fuzzy and rough logic to classify instances in a data set. Combining fuzzy and rough logic with the KNN algorithm can handle uncertainty and imprecision in data, making it useful when ambiguous or noisy data. The main difference between the traditional KNN algorithm and the FRNN algorithm is that FRNN calculates the proximity between k neighbors using fuzzy proximity measures, evaluating the similarity between the data point and the nearest class with the highest representation in the neighborhood, and considering the uncertainty associated with this similarity. Additionally, for class selection, a confidence value is calculated for each class based on the contribution of the nearest neighbors and their degree of fuzzy membership. The class with the highest confidence value is assigned as the predicted class to the instance. Once the model has been built using the training set, it can classify new instances.

Gafar et al. approach the recognition of facial expressions from the perspective of the interpretation that the individual gives them, which is known as Grammatical Facial Expressions, as seen in Fig. 6. Due to the ambiguity and lack of certainty that the interpretation of a facial expression can provide, they propose a solution based on fuzzy logic algorithms to address the ambiguity of the recognition and interpretation of expressions. The proposal is based on The Feature Reduction Module and The Classifier Building Module, with which they identify emotions. In the first module, the FRFS-ACO algorithm is implemented to optimize the detection of facial landmarks, eliminate redundant attributes, and reduce the dimensions of the images. In the second module, the FRNN algorithm is implemented, which is responsible for classifying facial expressions based on the patterns of the facial landmarks extracted previously [13]. Microsoft Kinect sensor dataset was used for training and testing, achieving an average accuracy rate of 93.7% in classifying nine emotions.

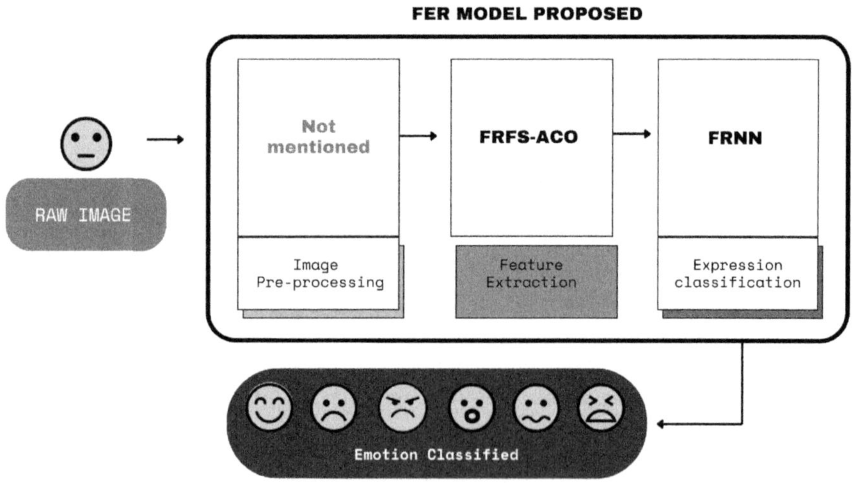


**Fig. 6** Model of the methodology proposed by Gafar et al. [13]

# 3 Nonverbal Language Identification and Interpretation with Artificial Neural Network

One of the most widely implemented machine learning classification techniques is Artificial Neural Networks (NN), and many researchers apply them to FER.

Inspired by how human neurons work together, NNs are a network of artificial neurons distributed over several layers. Each artificial neuron has one or more input values, weights that define the importance of each input, and an activation function that determines the neuron's reaction (the output value) based on each input value and weight. The typical way an NN learns is by training it with known values, comparing the output value of the entire network with the known value, and adjusting the connection weights based on differences between known value and output value through the Backpropagation method [14].

This learning process enables NNs to identify patterns and infer new knowledge, driving NNs to get them to recognize images different from the ones they were trained with, as long as the images are of the same kind. This characteristic makes Artificial Neural Networks useful for computer vision [15], self-driving vehicles [16], disease identification [17], or Facial Expression Recognition.

## *3.1 Deep Neural Network for FER*

A Deep Neural Network (DNN) has multiple hidden layers. Its depth allows the model to learn and represent increasingly abstract and complex features as it moves up the layers. This characteristic makes DNNs very effective for supervised learning tasks, where a large set of labeled data is available for training.

The dlib is a Python library that localizes important face structures such as eyes, eyebrows, nose, mouth, and jawline, mapping them like facial landmarks. To do this, dlib implements the Ensemble of Regression Trees technique and a set of manually labeled facial landmarks, their coordinates, and the probability of distance between facial landmarks for training [19].

Rao et al. expose that there is a vast bibliography that indicates upper facial features are more expressive than lower in adults, but, at the same time, there is little research on FER in children, as seen in Fig. 7. Their proposal aims to confirm if the prior hypothesis is consistent for adults and children. To reach this goal, a DNN was designed to recognize emotions using multiple hidden layers, intermixed dropout layers to avoid overfitting and a Softmax output layer for classification. In the proposed model, dlib detects and ticks principal face landmarks and saves their coordinates. The DNN is fed with the coordinates and calculates displacements from an initial position (neutral emotion), thus classifying the emotion in the analyzed image [18]. The Child Affective Facial Expression (CAFE) and the Extended Cohn-Kanade (CK+) datasets were used for training and testing, achieving an average accuracy rate of 87.5% for adults and 66.7% for children. In both cases, accuracy was approximately the same on the upper-face and lower-face landmarks.

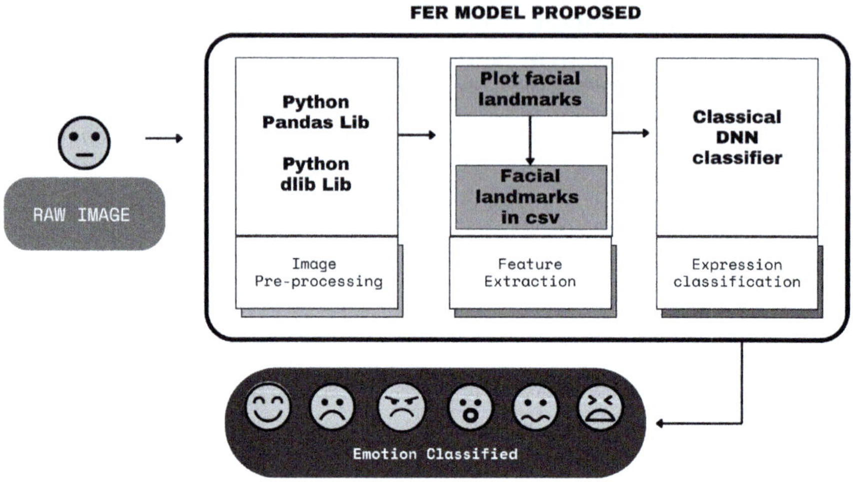


**Fig. 7** Model of the methodology proposed by Rao et al. [18]

## 3.2 *Recurrent Neural Network for FER*

Even though NNs produce excellent results with static images like photographs or drawings, they have essential limitations when processing data sequences. To resolve this challenge, Werbos improved an existing type of NN called recurrent neural network (RNN), which was used to process data sequences [20].

Unlike traditional feedforward neural networks, RNNs have connections that form directed cycles, allowing them to exhibit dynamic temporal behavior. This cyclic structure enables RNNs to maintain an internal state or memory used to capture and process information from previous time steps, which can be helpful for tasks involving sequences, such as language modeling, speech recognition, or Facial Expression Recognition on video.

A Parametric Bias Unit Context Layer (PB) is a contextual layer in an NN that introduces relevant contextual information into the learning process, which can improve model performance by allowing the NN to consider specific factors. PB is helpful in tasks where context is essential in decision-making, such as natural language processing or computer vision.

Li et al. proposed a method to detect human emotions on videos using an RNN and PB, an architecture recognized as an efficient unsupervised learning method (RNNPB). Their main contribution was adding context layers that can handle three-dimensional coordinates of 51 elements of the human upper body, provided by a 3D camera to calculate angular velocities from emotion-aroused human body behavior, eliminating the need to train the model with labeled datasets. In the proposal, PB units' values represent emotions within datasets. Similar sequences sharing the same emotion yield akin PB values using this method. Consequently, human emotions are identified unsupervised, meaning the model can generalize untrained datasets. That implies the model yields relatively consistent recognition outcomes even with limited training samples. To make it, normalization of input features is necessary before feeding fused data into the network to enhance model accuracy and convergence speed. Normalized dataset values were scaled between zero and one [21].

For experimentation, various sensor data were investigated to determine the impact on the outcomes of emotion recognition, using three datasets generated by themselves modifying the sensor's modality. Because one of the objectives of this research was that human subjectivity does not interfere with the classification process, a level of accuracy was not obtained concerning the emotion expressed by the study subjects, but rather the PB distance detected between the emotions of the same kind. The average PB distance between interclass emotions was 0.0092.

### *3.3 Multi-verse Optimizer Based on Whale Optimization Algorithm for FER*

The Multi-Verse Optimizer based on the Whale Optimization Algorithm (MV-WOA) is an optimization technique inspired by observing whale behavior and the concept of multiple universes. The WOA algorithm simulates the foraging process of whales in the ocean, where the most successful whales influence the behavior of other whales to improve foraging efficiency in the group. This process adapts to optimization problems by using a population of candidate solutions that move through a search space for the optimal solution.

VM refers to maintaining several populations of candidate solutions (universes) simultaneously while searching for the optimal solution, where each universe represents a set of candidate solutions that evolves independently but can also interact with other populations of universes to exchange information and improve the search.

The Adaptive Selection of Influential Features (ASIF) algorithm is a feature selection technique that identifies and selects the most relevant or influential features for the classification task through an iterative approach. The algorithm dynamically adapts its selection criteria during the selection process, improving its perform an accuracy.

Kumar et al. propose a four-step methodology, as seen in Fig. 8. Firstly, the images are pre-processed to identify and isolate the face in the image and reduce any noise it may have. The second step involves detecting and extracting face landmarks through the ASIFT algorithm. Subsequently, the optimal descriptors of the extracted landmarks are selected through the MV-WOA algorithm, simulating diverse affine distortions from different viewpoints. Finally, an NN optimized in its number of hidden neurons is fed with them -also with the MV-WOA algorithm- to classify expressions in emotions [22]. A free Kaggle dataset was used for training and testing, achieving an average accuracy rate of 92.95% in classifying six emotions and the neutral expression.

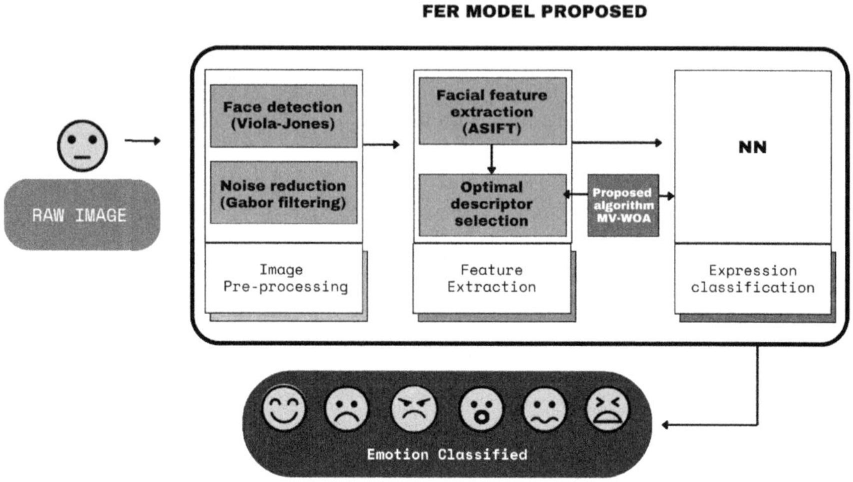


**Fig. 8** Model of the methodology proposed by Kumar et al. [22]

## 3.4 Genetic Algorithms and Fuzzy Logic for FER

Genetic Algorithms (GA) are optimization techniques inspired by natural evolution and the process of natural selection, seeking to find optimal solutions to complex problems. A GA works with an initial population of possible solutions called chromosomes. Each chromosome represents a possible solution to the problem to solve. These chromosomes have characteristics or genes that encode the solution to the problem. The genetic algorithm operates in several stages:

1. Initialization. An initial population of chromosomes is created randomly or using some criteria.
2. Aptitude evaluation. Each chromosome in the population is evaluated based on how good its solution to the problem is, using a fitness function that assigns a numerical value to each chromosome based on its performance.
3. Selection. The most suitable chromosomes are selected to reproduce and create a new generation of chromosomes. Selection operators, such as selection roulette or selection tournament, give the fittest chromosomes a greater chance of being selected.
4. Crossing. By applying crossover operators, selected chromosomes are combined to produce new offspring chromosomes. These operators simulate sexual reproduction in nature, mixing the genes of the parent chromosomes to produce offspring with characteristics of both.
5. Mutation. Variability is introduced into the population through mutation, which randomly changes some genes on descendant chromosomes. Mutation helps maintain genetic diversity and avoid premature convergence towards local optimal solutions.

6. Replacement. Offspring are used to replace less-fit chromosomes in the original population.

This process is repeated over several generations, allowing the population to evolve toward optimal solutions.

Fuzzy logic algorithms (FLA) work to handle data uncertainty and imprecision in decision-making by representing variables and rules in linguistic terms. That is, instead of classifying data into clearly defined categories such as "true" or "false," fuzzy logic allows uncertainty to be represented with terms such as "cheap," "expensive," or "very expensive." The operation of a FLA is divided into several steps:

1. Fuzzification. The input variables are converted to fuzzy values using membership functions in this stage like "cheap," "expensive," or "very expensive".
2. Fuzzy rules. Rules are expressed through "if-then" sentences establishing relationships between input and output variables and determining the system's behavior.
3. Fuzzy inference. Combines fuzzy rules using fuzzy logical operators to obtain an output. In addition, inference involves aggregating information from the different rules.
4. Defuzzification. The fuzzy output is converted into a concrete value or action.

Adaptive Neural Networks (ANNs) are a class of Artificial Neural Networks that can adapt and adjust their internal parameters when faced with new data. Like conventional NNs, they are composed of layers of interconnected neurons, weights associated with each neuron that determine the relative importance of the inputs concerning the output they generate, and a nonlinear activation function. Nevertheless, ANNs can learn incrementally, updating their weights iteratively as they progress, and can adapt to changes in the data and environment without retraining from scratch. In addition, ANNs can adapt other aspects of their architecture, such as the number of neurons in the hidden layers or the network's topology, thus improving their performance based on specific data and tasks. These features allow ANNs to adapt and improve performance in changing environments and with dynamic data.

Gupta et al. focused their proposal on facing the intrinsic problem of vagueness and lack of certainty that characterizes FER, considering fuzzy logic and soft computing over the deterministic approximations of mathematical models to be more appropriate for this objective, as seen in Fig. 9. The proposal was based on FACS methodology to overcome the vagueness challenge. The proposed architecture consists of two main modules. In the first, the Pre-processing Module, AUs and landmarks are extracted from each image's face, eliminating non-relevant features, reducing their dimensionality, and optimizing the attributes to make the classification more efficient. The second is the Prediction Module, which contains seven Fuzzy Inference Systems (FIS) to classify each emotion category. At this point, Gupta et al. decided to evaluate an ideal fuzzy model, where the Fuzzy Functions are based on linear distribution methods using Genetic Algorithms as a learning engine, and compare it against two optimized models, one with GA (GAFIS) and the other with ANN (ANFIS), in which the Fuzzy Functions are not pre-established by the researchers, but are generated based on the respective algorithms [23]. The Extended Cohn-Kanade (CK+)

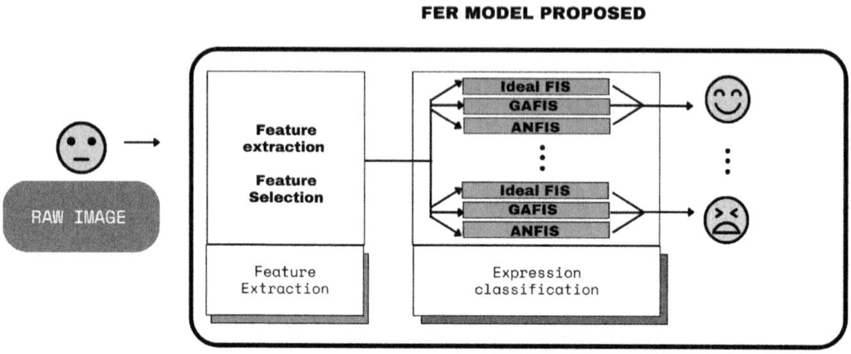


**Fig. 9** Model of the methodology proposed by Gupta et al. [23]

dataset was used for training and testing, achieving an average accuracy rate of 77% for FIS, 81% for GAFIS, and 88% for ANFIS in classifying seven emotions and the neutral expression.

# 4 Nonverbal Language Identification and Interpretation with Deep Learning

Deep learning, a branch of machine learning, focuses on developing algorithms that mimic the functioning of the human brain to process data and perform complex tasks. In the context of FER, deep learning has demonstrated an unprecedented ability to capture and analyze facial features, significantly improving the accuracy and efficiency of recognition systems. Its importance lies in its ability to learn hierarchical representations of data, allowing a deeper understanding of human emotions by interpreting subtle and complex facial expressions. This advanced analytics capability can transform diverse fields, from psychology and medicine to human-computer interaction.

## *4.1 Convolutional Neural Network for FER*

A Convolutional Neural Network (CNN) is a highly effective solution for classifying facial expressions. While traditional neural networks have proven to be robust classifiers, facial images require pre-processing to extract facial features before classification. The more features an image has, the more computing resources are needed. CNNs can overcome these challenges and streamline the process of facial expression classification.

Unlike NNs that have an input layer, an output layer, and one or more hidden layers, CNNs have deep feedforward architecture:

- A convolution layer extracts features from the image,
- A pooling layer optimizes features and reduces training parameters,
- Fully connected layers to classification, and
- Rectified Linear Unit (ReLU) as the activation function for optimizing computing resources.

Due to their architecture, CNNs intrinsically avoid overfitting, the extraction and classification stages utilize the learning process, and have strong generalization ability. Thus, they are helpful for different objectives such as speech recognition, object detection, or FER, among others [5].

Kollias et al. began their approach to FER by pointing out that existing datasets generally have two limitations: A small size in terms of the total number of frames and a discrete number of emotions for training machine learning models, while the daily behaviors of any person are a continuous space. With this idea in mind, a dataset was developed to capture the expressiveness and richness of human emotional states, as seen in Fig. 10, basing the effort on the AV2D method. In addition, six AV2D expert annotators were invited to collaborate, who labeled emotions by watching the 298 videos that comprise their dataset, called *AFF-Wild Database*. After obtaining the dataset, Kollias et al. developed an emotion recognition model based on two parts: Image pre-processing using the Menpo Software application that extracted faces and facial landmarks from each frame, and their machine learning model based on a CNN connected to two RNN layers with GRU units (a CNN-RNN network), calling it *AffWildNet* [24]. To evaluate AffWildNet's performance, the AFF-Wild Database was used for training, and the EmotiW dataset was used for testing, achieving an average accuracy rate of 45.4% in classifying six emotions and the neutral expression.

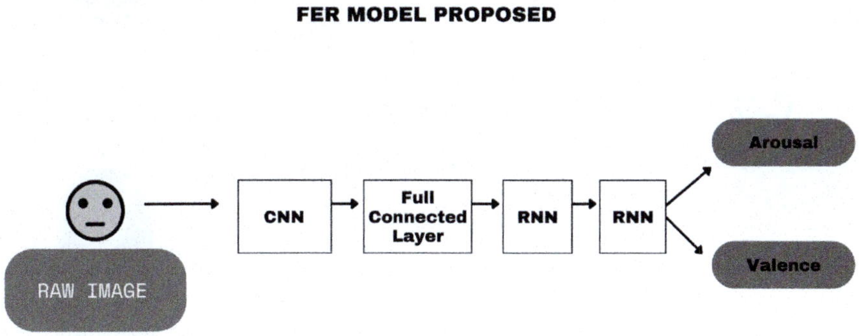


**Fig. 10** Model of the methodology proposed by Kollias et al. [24]

## 4.2 Deep Convolutional Neural Network for FER

As a CNN evolution, Deep Convolutional Neural Networks (DCNN) are CNNs with multiple convolutional layers, which implies that the more profound the layers of the network, the more it is considered a deep convolutional network. These architectures can learn increasingly abstract and complex feature representations as they go deeper into the layers of the network, making them especially suitable for more complex computer vision tasks, such as detecting objects in images or semantic segmentation. Unlike other image classification techniques that assign a single label to the entire image, semantic segmentation enables a more detailed understanding of the image structure and content by providing pixel-level label assignment [26].

In order to afford real life where there are no label data to recognize expressions, Yuanling et al. propose a Semantic Learning for Facial Expression Recognition method (SLFER), as seen in Fig. 11. This method is built for four components: ResNet18, a freely available pre-trained CNN, is used as a backbone to extract the face image's height, weight, and number of channels, getting the semantic labels for an external facial parsing model. The Segmentation Task (ST) performs the image's semantic segmentation by assigning a semantic label to each pixel. The Classification Task (CT) predicts the number of facial component classes and chooses the top dominant facial component. The Expression Classification Task (ECT) predicts the expression categories using semantic classes and a global average pooling layer for features detected by ResNet18. ECT, ST, and CT are jointly trained in a multi-task learning network [25]. The Real-world Affective Face Dataset (RAFDB) and the Static Facial Expression in the Wild

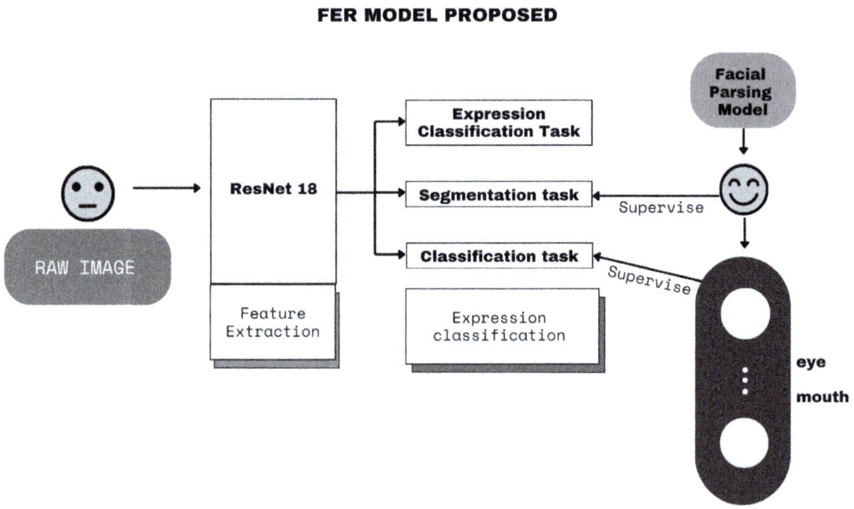


**Fig. 11** Model of the methodology proposed by Yuanling et al. [25]

(SFEW) were used for training and testing, achieving an average accuracy rate of 88.33 and 59.86%, respectively, in classifying six emotions and the neutral expression.

### 4.3 Generative Adversarial Network for FER

The Generative Adversarial Network (GAN) architecture is a classification of Deep Neural Networks that consists of two main networks: the Generator and the Discriminator. These networks are designed to work together to generate realistic images, speech synthesis, or text.

- Generator. The generating network takes as input a set of random numbers called "noise" and transforms them into data. Its objective is to learn to generate data as similar as possible to real data.
- Discriminator. The discriminative network takes real data and data generated by the Generator as input. It aims to distinguish between them and determine whether the data is real or falsified.

The training of a GAN is carried out in a competition process between the Generator and the Discriminator, where the Generator learns how to fool the Discriminator by making fake data look real, and the Discriminator learns to distinguish between real and fake data improving its ability to discriminate. Ideally, this leads to the Generator producing data that is difficult to differentiate from original data.

The Key Point Detection Network (KPDN) is a neural network used in computer vision to detect points of interest or distinctive features in an image. It consists of convolutional layers followed by essential point detection layers, which may include additional convolution layers or fully connected layers specifically designed to identify and locate the Key Points in the image.

The Lu approach to FER uses the Generative Adversarial Network (GAN) algorithm, which aims for the capacity to recognize facial expressions regardless of the angle the face presents to the observer, as seen in Fig. 12. In the architecture proposed, face images are pre-processed with the KPDN to face alignment and reduce redundant information. The Generator is composed of two CNNs. The Encoder CNN extracts face features from the image and fuses them with new angle coding information. The Decoder CNN processes the fused information and generates face images corresponding to different angles. The Discriminator ensures that the image created by the Decoder is similar to the original image. Finally, a classifier, composed of several CNNs, takes the created and original images to normalize the angles and sends them to a CNN for FER from different angles [27]. The Multi-PIE the CFP datasets were used for training and testing, achieving an average accuracy rate of 84.81% in classifying seven emotions and the neutral expression. The multi-angle face images used were 0°, 15°, 30°, 45°, 60°, 75° and 90°, being 0° and 90° the original images.

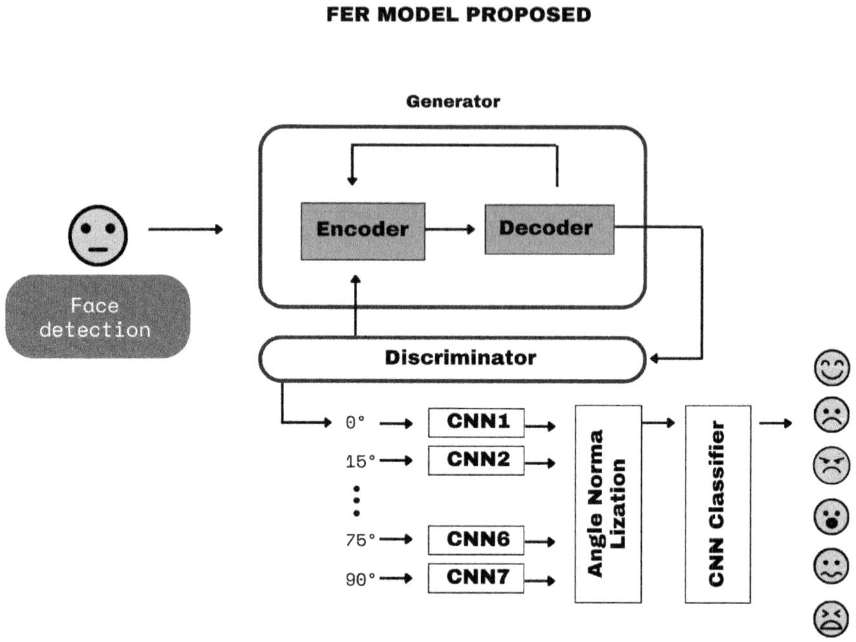


**Fig. 12** Model of the methodology proposed by Lu [27]

## 4.4 *Deep Belief Network Optimized and Spider Monkey Optimization for FER*

A Deep Belief Network (DBN) is an NN consisting of multiple layers of neural units organized in a hierarchical architecture and fed by unlabeled data. Its main differences from a traditional NN are based on the following facts:

- Connections between the layers are restricted under rules defined in its design to facilitate efficient network training.
- Uses a layer-wise training method, where each layer is trained independently before adjusting the weights across the network.
- Mainly use unsupervised learning techniques to learn representations of unlabeled data.
- After training, the network is fine-tuned using supervised learning methods to fit a specific dataset best or perform a specific task, such as classification or regression.

The Spider Monkey Optimization (SMO) is a nature-based optimization method inspired by the social behavior of spider monkeys. The SMO maintains a population of possible solutions, each represented by a set of parameters or variables. These solutions are like "monkeys" looking for the best solution to the optimization problem. Monkeys in SMO move between different positions in search space through "jumps," "climbs," and "position changes." After each move, the fitness of each solution is evaluated using an objective function that measures how good a solution is in terms of optimality. Based on the fitness assessment, the monkeys' positions in the population are updated. Better-performing monkeys can influence the movement of other monkeys in the population, encouraging exploration and exploitation of the search space. The move, evaluate, and update process is repeated for several iterations until a stopping criterion is reached, such as a maximum number of iterations or successful convergence.

Ethnicity, age, and even the appearance of sudden shadows in a video recorded outdoors establish changes in the perception of facial expressions, which can confuse people and, even more so, AI algorithms. Seeking to solve these challenges, Vedantham et al. propose a model that combines robust facial feature extraction with the optimized DBN-SMO classification algorithm, as seen in Fig. 13. The model begins with the pre-processing of the image, in which non-relevant elements are eliminated with the Adaptive Median Filtering (AMF) method, thus reducing noise and facilitating subsequent processes, to later use the Viola-Jones algorithm to detect the most relevant components of the face such as eyes, nose or mouth. The second stage of the model, feature extraction, is based on texture detection, for which it uses the operators Local Binary Pattern (LBP), Local Phase Quantization (LPQ), Weber Local Descriptor (WLD) and Discrete Cosine Transform (DCT). Once the features have been extracted, their dimensions are reduced with the PCA method, filtering only the most representative features -or face landmarks- to the third stage, which consists of two phases. In the pre-training phase, the parameterization of the DBN (bias, weight of connections) is determined through the SMO algorithm. Subsequently, in the

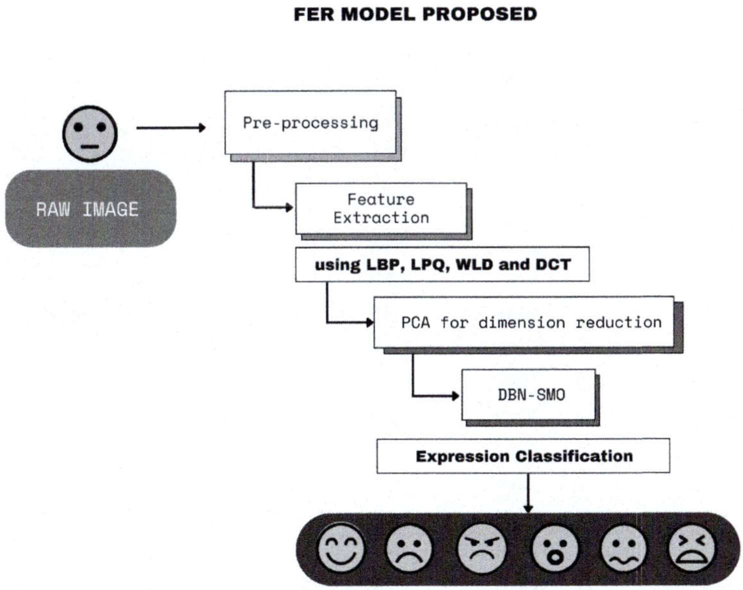


**Fig. 13** Model of the methodology proposed by Vedantham et al. [28]

fine-tuning phase, backpropagation refines the internal rules defining the network's layered learning [28]. The KDEF, CK+, MMI, Oulu CASIA, and DISFA+ datasets were used for training and testing, achieving an average accuracy rate of 97.9, 97.5, 95.4, 92.4, 95.7%, respectively, in classifying six emotions and the neutral expression.

## 5 Discussion

In most cases, the related research works were chosen for their contribution to advancing Facial expression recognition. One definition of "intelligence" is the ability to solve problems. As seen in the Challenges of AI in the FER section, there is still a long way to go. The advances achieved by the researchers mentioned above are a bridge they build for those of us who follow in their footsteps.

Although most of the investigations achieved the objectives that had been proposed, thus solving FER challenges, these objectives are dissimilar to each other, with the accuracy obtained in their testing process being their main common reference point between them. Table 1 presents a table illustrating the respective accuracy achieved by each research.

The columns provide insights into whether the proposal integrates optimization algorithms (AO), the classifier utilized for expression identification (Classification), the subfield of AI to which the proposal belongs based on integrated tools (IA Subfield), and the Facial Landmark Detection Group (FLdg) to which each proposal is categorized according to the author's methodology for identifying and extracting facial expression features: Facial Landmarks Human Defined (FLhd), Facial Landmarks Machine Detected (FLmd), Facial Landmarks Texture Driven (FLtd), Facial

**Table 1** General accuracy of the proposals' inferences

| Research | OA | Classification | IA Subfield | Accuracy FLhd (%) | FLmd (%) | FLtd (%) | Flad (%) | FLHybrid (%) |
|---|---|---|---|---|---|---|---|---|
| iromini2021 [8] | N | KNN | ML | | 85.7 | | | |
| sui2020 [11] | Y | EML | ML | | 79.5 | 88.0 | | 92.8 |
| wang2020 [12] | N | BLR | ML | | | | **99.3** | |
| gafar2019 [13] | Y | FRNN | ML | 93.7 | | | | |
| rao2020 [18] | N | DNN | NN | 87.5 | | | | |
| li2020 [21] | N | RNN | NN | | | | | |
| ashok2023 [22] | Y | NN | NN | | 92.9 | | | |
| gupta2020 [23] | Y | ANN | NN | | | | 88.0 | |
| kollias2019 [24] | N | CNN | DL | 45.4 | | | | |
| lv2022 [25] | N | DCNN | DL | | | 88.3 | | |
| lu2022 [27] | N | GAN | DL | | | 87.0 | | |
| vedantham2020 [28] | Y | DBN | DL | | | 97.9 | | |

Landmarks Action Units Defined (FLad), and the employment of hybrid methodologies (FLhybrid). The criterion for determining the accuracy ratio of the research in which the proposals were tested on multiple datasets was to reflect the highest value since the differences were just under 15% points. This leads us to another point of reflection: The methodologies are very sensitive to variations in the criteria with which the datasets are created.

The first fact that deserves to be highlighted is that the highest level of accuracy was achieved by wang2020 [12] (99.3%), using Machine Learning techniques (BLR) without any type of optimization algorithm. Unlike the previous proposal, which is a relatively simple technique with low computational cost, the second best result was obtained by vedantham2020 [28] (97.9%) using Deep Learning methodologies (DBN) optimized through metaheuristic algorithms (SMO), methodologically more complex and expensive. The third best result was obtained by gafar2019 [13] (93.7%) through Machine Learning techniques (FRFS) optimized with Metaheuristic Algorithms (ACO). In general, we can see that three of the four evaluated Machine Learning techniques obtained an accuracy greater than 90%, while only one of Artificial Neural Networks (ashok2023 [22]) reached it, and one of Deep Learning methodologies (The already mentioned vedantham2020 [28]). Another interesting fact to highlight is that Sui et al. proposal [11], in which tests were executed with the Facial Landmark Detection Groups FLmd, FLtd, and FLHybrid, training and testing all three with the same dataset, shows the synergy generated by hybridization since the FLHybrid's accuracy is greater than that of the independent groups.

Talking about the lowest results, li2020 [21] is focused on zero human participation in the classification process, which leads to no training with labeled datasets. Based on this, its accuracy indicator is not based on the correct correlation of the detected emotion versus the labeled emotion as in other investigations. It is the efficiency of clustering emotions based on the distance between them in the search area. That is why its accuracy ratio does not appear in the image. Besides, kollias2019 [24] low accuracy is because once the proposed AffWildNet network validated the unlabeled Aff-Wild Database based on the AV2D model, it was used to classify a traditional dataset without first training it on that dataset. Therefore, the accuracy ratio of this proposal will not be considered in the following analyses.

In the first instance, implementing optimization algorithms aims to reduce processing time and cost. However, heuristics are used to optimize learning in some of the research evaluated. Table 2 shows how methodologies that include optimization algorithms in their architecture have slightly higher accuracy than those that do not. It would have been enriching to compare the processing time and cost and not just its accuracy, but most of the authors did not share this information.

**Table 2** Average accuracy by optimization

| Methodology | Accuracy (%) |
|---|---|
| With optimized algorithms | 93.06 |
| Without optimized algorithms | 89.56 |

**Table 3** Average accuracy by IA subfields

| IA subfield | Accuracy (%) |
|---|---|
| Machine learning | 92.87 |
| Artificial neural network | 89.67 |
| Deep learning | 91.07 |

**Table 4** Average accuracy by face landmarks groups

| Face landmarks | Accuracy (%) |
|---|---|
| Human defined (FLhd) | 90.60 |
| Machine detected (FLmd) | 86.00 |
| Texture drived (FLtd) | 90.30 |
| Action units defined (FLad) | 93.70 |
| Hybrid methodologies (FLHybrid) | 92.80 |

Machine learning techniques are generally more efficient in computing time and cost than neural networks and deep learning. However, it is a long-held idea that NN and DL are better classifiers than ML. However, Table 3 points out that ML techniques have the best accuracy.

Two of the projects evaluated sought to dispense with dependence on humans to label the datasets. On the one hand, because of the work involved, which translates into smaller datasets than might be needed. Moreover, on the other hand, because of the reality that we do not carry labels that inform everyone about what emotion we are expressing daily. However, Table 4 indicates that developing methodologies for unsupervised learning of FER needs to catch up to the other FLgp. The most advanced, in terms of accuracy, is FLad, which could be explained by more than 40 years of research and development of the FACS framework.

## 6 Conclusions and Future Scope

As a result of the research and documentary evaluation, we found ingenious solutions to FER-related challenges. However, we also found that many of these solutions still need to land in applications that people can use daily. The high sensitivity shown by some of these tools, whose accuracy fell by more than ten percentage points just by changing the dataset in which they were initially programmed, tells us how fragile the solutions are for the chaotic world entire of unforeseen events in which we live. On the other hand, although each researcher managed to achieve the objective that was set in their hands, the reality is that none of them was comprehensive, which does not speak negatively of the researchers but instead makes clear the broad spectrum in which the Recognition of Facial Expressions means, as well as how much remains to be done. The most valuable point that this exercise leaves us with is the identification of how the learning strategy of identification and extraction of facial expressions is approached, which can lead us to more versatile tools. We propose a new evaluation model and approach based on face landmark identification groups (FLgp), Facial Landmarks Human Defined (FLhd), where people, specialists, or volunteers label the training datasets. Facial Landmarks Machine Detected (FLmd) brings together initiatives and efforts so that the algorithms' learning and training in detecting and identifying human emotions are autonomous. Facial Landmarks Texture Drived (FLtd), whose focus is the identification of landmarks based on changes in the texture of the images, and Facial Landmarks Action Units Defined (FLad), which takes advantage of the knowledge accumulated over decades of how the muscles that move our faces are what trigger, in a universal way and regardless of race or culture, human expressions. To address the above, we must add hybrid methodologies (FLhybrid), which have demonstrated their potential to improve identification accuracy through the synergy of groups when working collaboratively. This proposal will be finalized and presented later in a paper.

## References


1. Berscheid, E.S.: Review of Silent Messages: Implicit Communication of Emotions and Attitudes. American Psychological Association (1981)
2. Ekman, P., Rosenberg, E.L.: What the face reveals: Basic and applied studies of spontaneous expression using the Facial Action Coding System (FACS). Oxford University Press, USA (1997)
3. Eckman, P., Friesen, W.: Facial action coding system (facs): a technique for the measurement of facial action, **A8** **5**(3), 56–75 (1978)
4. Senechal, T., Bailly, K., Prevost, L.: Impact of action unit detection in automatic emotion recognition. Pattern Anal. Appl. **17**, 51–67 (2014)
5. Indolia, S., Goswami, A.K., Mishra, S.P., Asopa, P.: Conceptual understanding of convolutional neural network-a deep learning approach. Procedia Comput. Sci. **132**, 679–688 (2018)
6. Zhang, J., Yin, Z., Chen, P., Nichele, S.: Emotion recognition using multi-modal data and machine learning techniques: a tutorial and review. Inf. Fusion **59**, 103–126 (2020)
7. Belle, V., Papantonis, I.: Principles and practice of explainable machine learning. Front. Big Data **4**, 688969 (2021)
8. N. Iromini, BTc, A.S.I., COd, O., Gbadamosi, K.: Facial expression recognition system using KNN. Technology (ICONSEET) **6**(4), 31–38 (2021)
9. Good, R.P., Kost, D., Cherry, G.A.: Introducing a unified PCA algorithm for model size reduction. IEEE Trans. Semicond. Manuf. **23**(2), 201–209 (2010)
10. Kramer, O.: K-nearest neighbors. In: Dimensionality Reduction with Unsupervised Nearest Neighbors, pp. 13–23 (2013)
11. Sui, X., Xue, L., Li, D.: Dynamic expression recognition based on hybrid features and optimized extreme learning machine model. In: IOP Conference Series: Materials Science and Engineering, vol. 719, p. 012080. IOP Publishing (2020)
12. Wang, S., Hao, L., Ji, Q.: Posed and spontaneous expression distinction using latent regression Bayesian networks. ACM Trans. Multimed. Comput., Commun., Appl. (TOMM) **16**(3), 1–18 (2020)
13. Gafar, M.G.: Grammatical facial expression recognition basing on a hybrid of fuzzy rough ant colony optimization and nearest neighbor classifier. In: 2019 International Conference on Innovative Trends in Computer Engineering (ITCE), pp. 136–141. IEEE (2019)
14. Kohli, S., Miglani, S., Rapariya, R.: Basics of artificial neural network. Int. J. Comput. Sci. Mob. Comput. **3**(9), 745–751 (2014)

15. Zhou, Y.-T., Chellappa, R.: Artificial neural networks for computer vision, vol. 5. Springer Science & Business Media (2012)
16. Huang, Y., Deng, Y.: A hybrid model utilizing principal component analysis and artificial neural networks for driving drowsiness detection. Appl. Sci. **12**(12), 6007 (2022)
17. Gunasundari, S., Baskar, S.: Application of artificial neural network in identification of lung diseases. In: 2009 World Congress on Nature & Biologically Inspired Computing (NaBIC), pp. 1441–1444. IEEE (2009)
18. Rao, A., Ajri, S., Guragol, A., Suresh, R., Tripathi, S.: Emotion recognition from facial expressions in children and adults using deep neural network. In: Intelligent Systems, Technologies and Applications: Proceedings of Fifth ISTA 2019, India, Springer, pp. 43–51 (2020)
19. https://pyimagesearch.com/2017/04/03/facial-landmarks-dlib-opencv-python/ (2017)
20. Werbos, P.J.: Backpropagation through time: what it does and how to do it. Proc. IEEE **78**(10), 1550–1560 (1990)
21. Li, J., Zhong, J., Wang, M.: Unsupervised recurrent neural network with parametric bias framework for human emotion recognition with multimodal sensor data fusion. Sens. Mater. **32** (2020)
22. Ashok Kumar, P., Maddala, J.B., Martin Sagayam, K.: Enhanced facial emotion recognition by optimal descriptor selection with neural network. IETE J. Res. **69**(5), 2595–2614 (2023)
23. Gupta, K., Gupta, M., Christopher, J., Arunachalam, V.: Fuzzy system for facial emotion recognition. In: International Conference on Intelligent Systems Design and Applications, pp. 536–552. Springer, Berlin (2020)
24. Kollias, D., Tzirakis, P., Nicolaou, M.A., Papaioannou, A., Zhao, G., Schuller, B., Kotsia, I., Zafeiriou, S.: Deep affect prediction in-the-wild: AFF-wild database and challenge, deep architectures, and beyond. Int. J. Comput. Vis. **127**(6), 907–929 (2019)
25. Lv, Y., Huang, G., Huang, Y., Xie, Y.: Semantic learning for facial expression recognition, jasonyanglu (2022)
26. Namatēvs, I.: Deep convolutional neural networks: structure, feature extraction and training. Inf. Technol. Manag. Sci. **20**(1), 40–47 (2017)
27. Lu, L.: Multi-angle face expression recognition based on generative adversarial networks. Comput. Intell. **38**(1), 20–37 (2022)
28. Vedantham, R., Reddy, E.S.: A robust feature extraction with optimized DBN-SMO for facial expression recognition. Multimed. Tools Appl. **79**, 21487–21512 (2020)

# Prediction of Bitcoin Price Based on Optimized Support Vector Regression Using Modified Grey Wolf Optimizer

**Parijata Majumdar, Snehasish Debnath, Sanjoy Mitra, Joyjit Dhar, and Trishita Das**

**Abstract** Bitcoin has emerged as the foremost digital currency in the global financial sphere, with prognosticating its price becoming a pivotal focus in finance. This study delves into the pressing necessity for precise Bitcoin price forecasts, spurred by the notable volatility inherent in cryptocurrency markets. Such predictions play a crucial role in assisting investors to manage risks and make well-informed decisions. Moreover, accurate forecasts aid in the formulation of effective trading strategies and market analysis. Employing the Grey Wolf Optimizer (GWO), this research endeavors to forecast Ethereum's day-ahead prices. By incorporating time-varying acceleration coefficients, the GWO algorithm seeks to strike a balance between exploration and exploitation, mimicking the reflex actions of wolves in nature. Additionally, to bolster the exploration process, a modified GWO (mGWO) algorithm integrates an exponential decay function to prevent premature convergence, thereby enhancing accuracy. Subsequently, this modified algorithm is employed to optimize Support Vector Regression (SVR) parameters, ensuring the model's effectiveness in capturing intricate patterns within cryptocurrency data. Through the innovative mGWO-SVR approach, this study not only advances cryptocurrency price prediction techniques but also underscores the significance of algorithm refinement for optimal Bitcoin price forecasting outcomes. This methodology is poised to benefit intraday, interday, and short-term traders in stock markets.

P. Majumdar (✉)
Indian Institute of Information Technology Agartala, Agartala, India
e-mail: er.parijata@gmail.com

S. Debnath · J. Dhar · T. Das
Techno College of Engineering, Agartala, India

S. Mitra
Tripura Institute of Technology, Agartala, India


D. Oliva et al. (eds.), *Advances in Optimization Algorithms for Multidisciplinary Engineering Applications: From Classical Methods to AI-Enhanced Solutions*, Studies in Computational Intelligence 806, https://doi.org/10.1007/978-3-031-78440-8_22

# 1 Introduction

Despite facing concerns such as high transaction costs, regulatory uncertainties, and competition from other cryptocurrencies, Bitcoin has maintained its status as the most esteemed cryptocurrency since its inception in 2009 as the pioneering decentralized currency. With its introduction by Nakamoto [8], Bitcoin has garnered significant attention from both investors and researchers, solidifying its position as the most popular and valuable digital asset in the financial landscape. As highlighted in Yermack [9], Bitcoin exhibits notable volatility. While it offers a secure and anonymous platform for digital transactions, bypassing traditional banking systems, its substantial price fluctuations have drawn in a sizable investor community [24]. However, due to its volatile nature, Bitcoin is often viewed more as a speculative asset rather than a stable digital currency. Understanding the sources of Bitcoin's price fluctuations is imperative for grasping its distinct characteristics compared to traditional financial assets like stocks or currencies and for formulating strategies to leverage it's growth potential while mitigating risks associated with volatility [12]. A significant challenge stems from the absence of economic and accounting fundamentals for Bitcoin analysis, compounded by the multitude of interconnected factors influencing its price dynamics [10]. Accurately identifying these determinants is crucial for enhancing the accuracy of price predictions and facilitating informed market decisions. Given that all investors aspire to achieve high returns, precise prediction of Bitcoin prices holds paramount importance for them. Consequently, numerous researchers have delved into the realm of Bitcoin prediction. The subsequent section of this study reviews various approaches aimed at enhancing the accuracy of such predictions. Notably, machine learning (ML) techniques have gained widespread attention for their ability to capture non-linear features effectively. Furthermore, metaheuristic algorithms have emerged as powerful tools in optimizing complex problems, offering significant advantages in terms of computational efficiency and time effectiveness by emulating natural collective behaviors. Metaheuristic algorithms are capable of delivering superior results in shorter time frames and possess additional merits such as simplicity, derivation-free mechanisms, and adaptability [11]. These algorithms mimic natural phenomena to tackle optimization problems, treating them as black boxes where the researcher's primary focus lies in representing the problems as input to obtain desired outputs [11].

While metaheuristic algorithms offer numerous advantages, they often encounter challenges such as becoming trapped in local search space neighborhoods, lacking search diversity, and exhibiting imbalanced exploration and exploitation strategies [13]. However, in recent years, the Grey Wolf Optimizer (GWO) has emerged as a promising solution within the realm of evolutionary-based metaheuristics, showcasing remarkable optimization performance across various problem domains [14]. Its effectiveness has been demonstrated in diverse applications including energy optimization [15], computational fluid dynamics [16], frequency regulation of power generation systems [17], parameter calculation of transmission lines [20], and feature selection tasks [13, 18, 19] etc.

Despite its advantages, the Grey Wolf Optimizer (GWO) has been criticized for its slow convergence speed and limited local searching ability [22]. As no single metaheuristic is universally applicable to all optimization problems [21], ongoing research in metaheuristics aims to address these limitations by proposing novel approaches. This motivated our team to enhance the existing GWO framework to improve its accuracy and robustness in price prediction tasks. In recent years, researchers have increasingly focused on integrating chaotic theory into metaheuristic algorithms to enhance efficiency and accuracy. Chaotic mappings are known to enhance the search performance of optimization algorithms. In our modified GWO approach, we introduce chaos to govern the reflex positions of the wolves, utilizing a Support Vector Regression (SVR) objective function. Chaos theory is leveraged to introduce the butterfly effect within the GWO, achieved through time-varying acceleration coefficients that simulate the movements of leader wolves. This adjustment enhances the balance between exploration and exploitation, mirroring the reflex actions observed in natural wolf behavior. This novel approach holds promise for predicting financial securities and offers valuable assistance to intraday, interday, and short-term traders in the stock markets. Furthermore, to augment the exploration process, we incorporate an exponential decay function into the modified GWO framework as a convergence factor within the position updating mechanism [23]. The modified Grey Wolf Optimizer (mGWO) model incorporates Support Vector Regression (SVR) as its objective function, aiming to minimize (or maximize) its fitness value to optimize the cost (C) and gamma ($\gamma$) hyperparameters of SVR. Once trained by the improved GWO, SVR can be employed to forecast Ethereum prices. The hybrid methodology (mGWO-SVR) is then compared with other state-of-the-art benchmark optimization methodologies such as traditional GWO optimizing SVR (GWO-SVR), Particle Swarm Optimization optimizing SVR (PSO-SVR), and Ant Colony Optimization optimizing SVR (ACO-SVR), among others.

The results demonstrate the superior performance of this model over all comparable models, as evidenced by evaluation measures including Root Mean Squared Error (RMSE), Mean Squared Error (MSE), Mean Absolute Percentage Error (MAPE), and R-Square. The research framework aims to identify the most suitable multivariate forecasting model for Bitcoin prices, shedding light on the influence of various exogenous factors on price dynamics.

The structure of this paper is as follows:

Section 2 provides a literature review on studies of Bitcoin and the prediction of financial time series utilizing machine learning techniques. Section 3 elaborates on the modified mGWO-SVR algorithm. Section 4 outlines the dataset utilized for Bitcoin price prediction and presents the experiment results. Finally, Sect. 5 concludes the paper, highlighting potential avenues for future research.

## 2 Related Works

Numerous researchers have explored forecasting stock prices or indices using artificial neural networks (ANNs) like multi-layer perceptron (MLP), recurrent neural networks (RNNs), long short-term memory (LSTM), and others. In order to mitigate the risk associated with predicting Bitcoin prices, Chen et al. opted for the random forest regression (RF) algorithm alongside the LSTM model. Their primary focus lies in evaluating the RF regression's performance in Bitcoin price prediction when compared to LSTM's predictions [25]. In the realm of Bitcoin price prediction research, methodologies are categorized into time series and machine learning approaches. Several studies have shown that ARIMA's prediction accuracy lags behind that of machine learning methods. Livieris et al. [2] employed a hybrid CNN-LSTM model to forecast Bitcoin (BTC), Ethereum (ETH), and Ripple (XRP) prices, with BTC being the focus, given its highest market value at the time. Carbo and Gorjon [5] conducted a comparative analysis on the impact of incorporating the previous period's Bitcoin price into the explanatory variables using the LSTM algorithm. They observed a significant enhancement in the RMSE accuracy of the model when the previous Bitcoin price was included as an explanatory variable, improving from the original 21–11%. Lahmiri and Bekiros [1] were among the first to employ statistical machine learning (ML) techniques to forecast Bitcoin price. Their study utilized support vector regression (SVR), gaussian Poisson regression (GRP), regression tree (RT), k-nearest neighbors (k-NN), feedforward neural network (FFNN), bayesian regularization neural network (BRNN), and radial basis function neural network (RBFNN), highlighting BRNN's exceptional accuracy in prediction. Alonso et al. [3] predicted prices for six cryptocurrencies, including Bitcoin, utilizing four network architectures (CNN, hybrid CNN-LSTM network, MLP, and radial basis function neural network). They concluded that LSTM combined with convolutional layers yielded the most optimal results. Koo et al. [24] identified the highly concentrated nature of return data in Bitcoin prices around zero, which can lead to varying outcomes with minor changes in components. To address this, they proposed a flattening distribution strategy (FDS) based on copula theory to artificially manipulate component distributions and enhance Bitcoin price return prediction. Their assessment involved the utilization of multi-layer perceptron (MLP), recurrent neural networks (RNN), and LSTM to evaluate FDS performances. Inspired by the methodologies mentioned above, this paper also makes a significant contribution by developing a precise machine learning (ML) algorithm aimed at improving the accuracy of Bitcoin price prediction. Specifically, we utilize a modified version of the Grey Wolf Optimizer (GWO) to optimize the hyperparameters of Support Vector Regression (SVR), termed as mGWO-SVR, with the goal of achieving accurate and optimal price prediction.

# 3 Methodology

## *3.1 Grey Wolf Optimizer*

In the realm of grey wolves, teamwork is paramount, as they hunt in packs and rely on cooperation among individual members. Two notable features characterize their hunting behavior: social hierarchy and hunting tactics.

The concept of "dominance hierarchies" delineates the hierarchical structure within the wolf pack, where individuals are ranked based on their strength and power. The hierarchy consists of alphas, the leaders of the pack; betas, who follow the decisions made by alpha wolves; deltas, strong wolves lacking leadership capabilities; and omegas, the weakest members responsible for caring for the young ones.

Hunting is a collaborative effort in which wolves work together to identify prey within a herd. Once the prey is spotted, one or two wolves initiate the attack while others drive off any stray individuals. The hunting process involves approaching the prey, tracking its movements, chasing it down, and harassing it until it becomes fatigued and unable to flee. At this point, the attack is executed. During hunting, wolves encircle their prey according to the:

$$\mathbf{D} = |\mathbf{C} \cdot \mathbf{X_p}(t) - \mathbf{X}(t)| \tag{1}$$

$$\mathbf{X}(t+1) = \mathbf{X_p}(t) - \mathbf{A} \cdot \mathbf{D} \tag{2}$$

where, $t$ denotes present iteration, $\mathbf{X_p}$ is the position vector of the prey, and $\mathbf{X}$ is the current position of the wolf. Vectors $\mathbf{A}$ and $\mathbf{C}$ are coefficient vectors equal to:

$$\mathbf{A} = |2\mathbf{a} \cdot \mathbf{rand1}(t) - \mathbf{a}| \tag{3}$$

$$\mathbf{C} = 2 \cdot \mathbf{rand2} \tag{4}$$

where, **rand1** and **rand2** are random vectors with values [0, 1], computed for each wolf at each iteration, and **a** linearly decreases from 2 to 0 through iterations. Exploration and exploitation is controlled using **a**, while randomization is always added through vector **C**.

$$\mathbf{a} = 2 - t\frac{2}{iter} \tag{5}$$

In this scenario, the maximum number of iterations allowed is specified by the variable $iter$. The wolves' positions are updated in accordance with the location of the prey. By manipulating vectors **a** and **C**, it's possible to fine-tune the position of the best-performing wolf. The top three solutions, represented by the alpha, beta, and delta wolves, serve as benchmarks for the adjustment of positions by the remaining wolves.

$$\mathbf{D}_\alpha = |\mathbf{C}_1 \cdot \mathbf{X}_\alpha - \mathbf{X}| \tag{6}$$

$$\mathbf{D}_\beta = |\mathbf{C}_2 \cdot \mathbf{X}_\beta - \mathbf{X}| \tag{7}$$

$$\mathbf{D}_\delta = |\mathbf{C}_3 \cdot \mathbf{X}_\delta - \mathbf{X}| \tag{8}$$

$$\mathbf{X}_1 = \mathbf{X}_\alpha - \mathbf{A}_1 \cdot \mathbf{D}_\alpha \tag{9}$$

$$\mathbf{X}_2 = \mathbf{X}_\beta - \mathbf{A}_2 \cdot \mathbf{D}_\beta \tag{10}$$

$$\mathbf{X}_3 = \mathbf{X}_\delta - \mathbf{A}_3 \cdot \mathbf{D}_\delta \tag{11}$$

$$\mathbf{X}_{t+1} = \frac{\mathbf{X}_1 + \mathbf{X}_2 + \mathbf{X}_3}{3} \tag{12}$$

In this context, **X** represents the current location of the agent, and $\mathbf{X}_{t+1}$ denotes the updated location. It's important to note that $\mathbf{X}_{t+1}$ will not be exactly equal to the average of the positions of the three best wolves; instead, a minor random shift is introduced by **C**. This approach is implemented to ensure that the search agents are guided by the best wolves while mitigating the risk of getting stuck in local optima, as discussed by [11]. Figure 1 illustrates how the positions of the wolves are updated relative to the positions of $\alpha$, $\beta$, and $\delta$.

#### 3.1.1 Modified Grey Wolf Optimizer

Chaos theory is harnessed to introduce the butterfly effect into the GWO, enabling accurate simulation of weather prediction models and other deterministic chaotic systems. This is achieved by integrating time-varying acceleration coefficients to represent the behavior of alpha, beta, and delta search agents.

By emulating a naturally occurring reflex action sequence in wolf movement, the balance between local search capability around promising regions and exploration of the most promising areas of the search space is enhanced. This facilitates convergence towards a global optimal solution. Additionally, the number of leaders in each iteration is dynamically updated based on a chaotic sequence. Time-varying acceleration coefficient vectors are represented as:

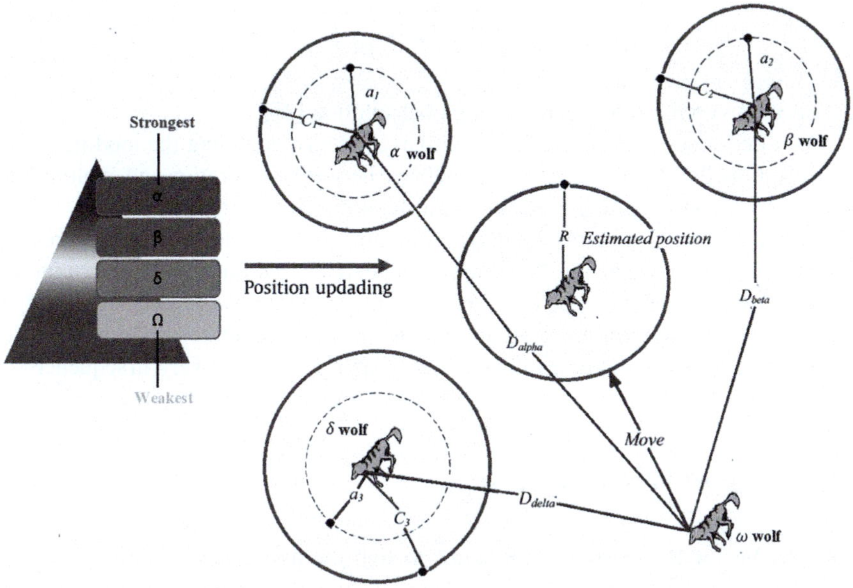


**Fig. 1** Updating position of other wolves based on the position of best three wolves [11]

$$\mathbf{C} = \frac{1}{2} + \left(\frac{1}{2 * sigmoid(-num_{wolf}/500)}\right) + \left(\frac{1.4 * (sine(num_{wolf}))}{30}\right) \quad (13)$$

$$\mathbf{C} = 1 + \left(1.4 * \left(1 - sigmoid\left(\frac{-num_{wolf}}{500}\right)\right)\right) + \left(1.4 * \left(sine\frac{num_{wolf}}{30}\right)\right) \quad (14)$$

$$\mathbf{C} = \frac{1}{\left(1 + sigmoid\left(-0.0001 * \frac{num_{wolf}}{max_{it}}\right)\right)} + \left((0.5 - 2.5) * \left(\left(\frac{num_{wolf}}{max_{it}}\right) * *2\right)\right) \quad (15)$$

where, **C** represents difference between mean based perturbation time varying parameter used during alpha wolf location update.

A sigmoid function-based acceleration coefficient is employed during the update of the beta wolf's location. In this context, $max_{it}$ denotes the maximum number of iterations, and $num_{wolf}$ ranges from 1 to n, where n represents the total number of wolves. Time-varying acceleration coefficient vectors are utilized to achieve a balance between the diversification and intensification operations of the mGWO. The investigation involves reducing the value of **a** through an exponential decay function, which is categorized within the position updating mechanism.

$$\mathbf{a} = 2\left(1 - \frac{t^2}{T^2}\right) \tag{16}$$

for speeding up the convergence to a global solution [23].

As **a** decreases, the value of **A** is similarly reduced, allowing the modified Grey Wolf Optimizer (mGWO) to achieve a balance between exploration and exploitation, ultimately facilitating the rapid identification of the global minimum.

In this context, T denotes the maximum number of iterations, while t represents the current iteration. The mGWO employs an exponential function to determine the decay of the convergence factor **a** over successive iterations. To enhance the algorithm's performance and generate promising candidate solutions, the iterations for exploration and exploitation are allocated as 70 and 30% respectively, in sequence [4].

### 3.2 Support Vector Regression

Support Vector Regression (SVR) aims to find the hyperplane within the best-fit line that encompasses the most data points. Unlike traditional regression methods, SVR focuses on fitting the best line within a specified margin rather than minimizing the difference between actual and predicted values. This margin is represented by a specific value, and the distance between the boundary line and the hyperplane is determined by this value. The data points closest to the hyperplane on either side are referred to as support vectors. The steps involved in determining the best-fitting line using SVR are as follows:

$$f(x) = (w^T \cdot \phi x_i) + b \tag{17}$$

where $b$ is the offset top, $w$ and $x$ are both n-dimensional column vectors, and $\phi x_i$ is the non-linear mapping function.

For optimizing the SVR model, the following objective function is used:

$$min \frac{1}{2}||w||^2 + c\sum_{i=1}^{l} \xi_i \tag{18}$$

where $\xi_i$ denotes the relaxation factor and $c$ denotes a constant value [7].

Selecting the parameters (C and $\gamma$) is crucial for enhancing the accuracy of the SVR model. These parameters can vary widely and have a significant impact on SVR's performance. Due to the absence of fixed rules for parameter selection, determining the optimal values for SVR poses a computational challenge and can be viewed as an optimization problem. Consequently, we employed modified Grey Wolf Optimizer (mGWO) to optimize these parameters of SVR, resulting in mGWO-SVR. The radial basis function (RBF) kernel was chosen for SVR execution. Figure 2 illustrates the flowchart depicting the optimization of SVR using mGWO-SVR.

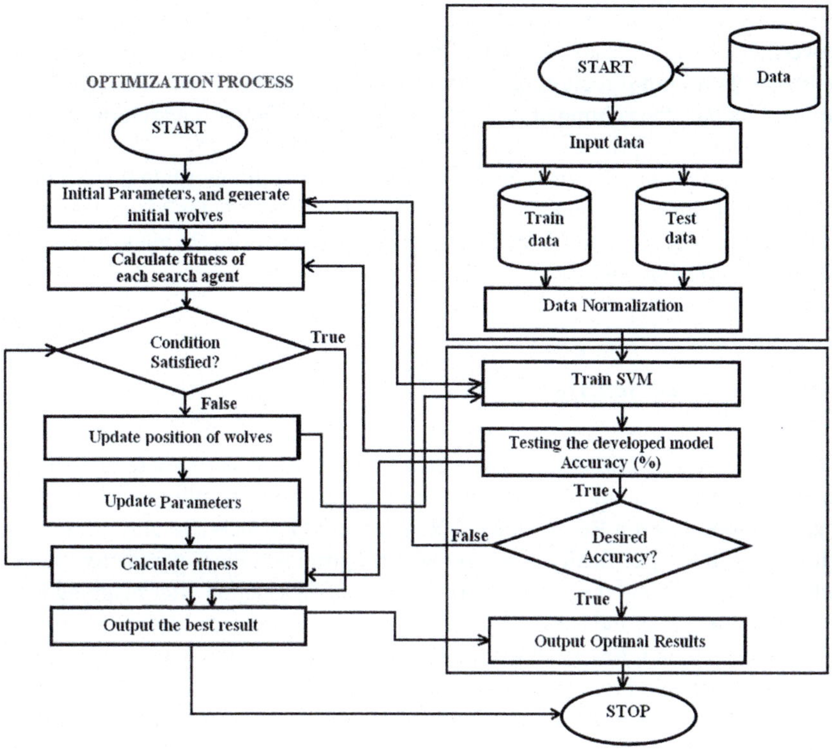


**Fig. 2** Flowchart of the support vector regression optimized using modified grey wolf optimizer (mGWO-SVR)

# 4 Experimental Results and Discussion

## *4.1 Datasets*

For our experiments, we utilize daily Bitcoin data spanning from January 28, 2013, to January 27, 2023, dividing it into training and test sets. The first 70% of the data is allocated for training purposes, while the remaining 30% reserved for testing. Each dataset comprises six types of data: volume of business, market capitalization, and prices of Bitcoin in terms of open, high, low, and close. Specifically, we consider the close price of Bitcoin to predict the direction of return. The return is calculated as $(p_t - p_{t-1})/p_{t-1}$, where $p_t$ denotes the close price of Bitcoin at time t.

Notably, there was a significant increase in values observed during the period between 2021 and 2022. We leverage all available information up to the current day to predict the return direction of the close price for the next day. Our explanatory variables include the volume of business, market capitalization, and returns for open, high, low, and close prices.

After transforming the data into return format and splitting it into a 7:3 ratio for training and testing, we apply the mGWO algorithm to each index of the data. Using the trained weights obtained from the training data, predictions are made by applying the test data to the models.

## 4.2 Parameter Settings

Certainly, ensuring fairness in performance comparison necessitates conducting experiments under identical conditions. Table 1 illustrates the initial parameters for both the Support Vector Regression (SVR) and mGWO models. Notably, the mGWO algorithm will be specifically directed towards optimizing the C parameter (penalty factor) of the SVR and the gamma parameter (data distribution after mapping to the new feature space) of the kernel. By subjecting these parameters to optimization via mGWO, we aim to refine the SVR model's performance, particularly enhancing its accuracy for Bitcoin price prediction tasks.

Here, dim refers to dimension of the problem, as we need to optimize two variables so it's value is 2, lb and ub are lower and upper bound parameters.

## 4.3 Results

The regression experiments are conducted utilizing the Mean Squared Error (MSE) loss function, aimed at assessing both the up-down prediction performance and investigating distribution changes in predictive distributions simultaneously. Although the regression approach cannot leverage the cross-entropy loss function typically used in classification tasks, it enables us to not only ascertain whether the return rises on

**Table 1** Parameter settings

| Algorithm and parameters | Values |
|---|---|
| SVR (kernel; tol; k; gamma; epsilon) | (RBF; 0.001; 0.5; 0.001; 0.1) |
| mGWO (population size; number of iterations; independent runs; $\mu$; Jr; lb; ub; dim) | (50; 500; 30; 2; 0.05; 0.01; 10; 2) |

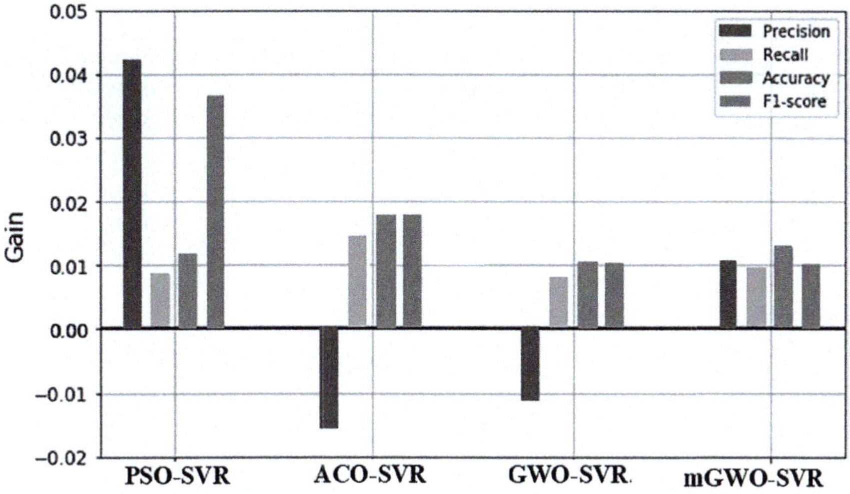


**Fig. 3** Accuracy gain (or loss) achieved by different algorithms

the next day compared to the previous day by comparing consecutive predictions but also to monitor changes in the distribution of regressive prediction. To facilitate these experiments, we artificially manipulated the provided return data to conform to a uniform distribution on the interval $[-1, 1]$. All experiments are implemented using Python 3, allowing for efficient and effective analysis of the performance of regression models in predicting Bitcoin price returns. The complete experimental results are summarized in Table 2, where each value is expressed as the mean value and standard deviation obtained from 100 experiments. Additionally, the improvement in accuracy achieved by mGWO is illustrated in Fig. 3 and quantified in Table 3. For the determined hyperparameters, the total time spent for 100 experiments of mGWO-SVR and GWO-SVR is 689 and 705 s, respectively. It is observed that the time spent in the mGWO applied case increased by 6.1% compared to the native case.

**Table 2** Experimental results

| Model | Precision | Recall | Accuracy | F1-score |
|---|---|---|---|---|
| SVR | $0.5611 \pm 0.2433$ | $0.5412 \pm 0.0281$ | $0.5098 \pm 0.0101$ | $0.4889 \pm 0.1546$ |
| PSO-SVR | $0.5902 \pm 0.2458$ | $0.5429 \pm 0.0219$ | $0.5243 \pm 0.0069$ | $0.5287 \pm 0.1246$ |
| ACO-SVR | $0.6001 \pm 0.2073$ | $0.5473 \pm 0.0213$ | $0.5277 \pm 0.0032$ | $0.5611 \pm 0.0861$ |
| GWO-SVR | $0.7312 \pm 0.0322$ | $0.5654 \pm 0.0065$ | $0.5612 \pm 0.0043$ | $0.6119 \pm 0.0122$ |
| mGWO-SVR | $0.7423 \pm 0.0349$ | $0.5747 \pm 0.0063$ | $0.5612 \pm 0.0063$ | $0.6369 \pm 0.0102$ |

**Table 3** Skill gain (or loss) achieved by the algorithms

| Model | Precision | Recall | Accuracy | F1-score |
|---|---|---|---|---|
| SVR | 0.0421 | 0.0084 | 0.0118 | 0.0365 |
| mGWO-SVR | 0.0401 | 0.0079 | 0.0121 | 0.0378 |
| GWO-SVR | 0.0389 | 0.0071 | 0.0101 | 0.0303 |
| ACO-SVR | −0.0082 | 0.0068 | 0.0003 | 0.0010 |
| PSO-SVR | 0.0001 | 0.0059 | 0.0009 | 0.0010 |

**Table 4** Model confidence set (MCS) test

| Loss function ranking | Model | RMSE | Improvement (%) | MCS |
|---|---|---|---|---|
| 1 | mGWO-SVR | 0.03319 | 13.6 | 1.000 |
| 2 | GWO-SVR | 0.03428 | 11.9 | 0.912 |
| 3 | ACO-SVR | 0.03559 | 10.9 | 0.639 |
| 4 | PSO-SVR | 0.03621 | 9.29 | 0.621 |
| 5 | SVR | 0.03742 | 1.32 | 0.009 |

To evaluate the statistical significance of the model providing the lowest Root Mean Squared Error (RMSE), we employ the Model Confidence Set (MCS) approach [6]. The MCS p-value obtained for a fixed model allows us to determine whether the model belongs to the group of the best models ($\hat{M}^*1-\alpha$, where $1-\alpha$ represents the significance level). A model with a relatively small MCS p-value is unlikely to be one of the best alternatives in $\hat{M}^*1-\alpha$. We present the MCS p-values of the models used in comparison in Table 4, considering $\alpha = 0.05$, to evaluate the quality of the models.

Initially, mGWO enhances the accuracy for each implementation model, followed by narrowing the width of error spread. As a result, the common area of distributions (probability density functions, PDFs) between the label and prediction is enlarged. The accuracy gains achieved by mGWO-SVR surpass 1% for all cases (0.0121, 0.0101, and 0.0119 for GWO-SVR, PSO-SVR, and ACO-SVR, respectively), with the largest gain observed in the mGWO-SVR case. Given that returns obtained from financial data are typically noisy, reducing the error spread of predictions for each error metric poses a challenging task. Therefore, mGWO proves to be an appropriate strategy for analyzing the noisy nature of returns. It's important to note that while the coincidence of distribution between the label and prediction doesn't guarantee the alignment of time series between them, it serves as a necessary condition for achieving alignment. Also, the mGWO-SVR transforms the distribution of Bitcoin price returns into a uniform distribution, effectively excluding outlier data. While this aspect increases the overlapping region compared to the native case, it also poses a challenge in tracking distribution changes. Furthermore, the prediction performances on both tail regions are observed to be unsatisfactory. Future work will involve addressing this issue and modifying the approach accordingly. Additionally, due to the large standard deviations of error metrics exhibited in PSO-SVR and ACO-SVR, it is reasonable to assert that the performance of mGWO-SVR emerges as the most reliable implementation model with a smaller error spread. We anticipate that mGWO-SVR can be efficiently applied in other fields where the data are concentrated near a particular point and are extremely noisy, offering potential for further exploration and application beyond Bitcoin price prediction.

## 5 Conclusion and Future Work

In recent years, predicting Bitcoin prices has become a paramount task given its emergence as one of the most sought-after investment assets in the financial market. However, forecasting Bitcoin prices is inherently challenging due to various factors including high volatility, market supply and demand dynamics, stock market conditions, US dollar exchange rates, among others. This paper focuses on leveraging ML algorithms for Bitcoin price prediction, particularly utilizing the return of Bitcoin, which is characterized by near-zero concentration and noisy movement. The SVR optimized using the mGWO algorithm is employed for prediction. The proposed model not only enhances the accuracy of simulated models but also accelerates convergence towards the global optimal solution. Comparatively, the proposed approach yields superior results in terms of performance evaluation metrics for Bitcoin price prediction when contrasted with other methodologies. Notably, transaction costs are not considered in the prediction process, although low transaction costs and high liquidity are desirable traits. Moving forward, it is imperative to evaluate the proposed approach for predicting the prices of other cryptocurrencies such as Ethereum, Bitcoin Cash, Litecoin, and Ripple, among others. Furthermore, enhancing the model by incorporating additional explanatory variables in the experiment would leverage valuable information. Consequently, future extensions of our approach should consider transaction costs, diverse types of cryptocurrencies, and an expanded set of explanatory variables to further improve prediction accuracy and robustness.

## References


1. Salim, L., Stelios, B.: Chaos, randomness and multi-fractality in Bitcoin market. Chaos Solitons Fractals 28–34 (2018)
2. Livieris, I.E., Pintelas, E., Stavroyiannis, S., Pintelas, P.: Ensemble deep learning models for forecasting cryptocurrency time-series. Algorithms 121 (2020)
3. Alonso-Monsalve, S., Suarez-Cetrulo, A.L., Cervantes, A. and Quintana, D.: Convolution on neural networks for high-frequency trend prediction of cryptocurrency exchange rates using technical indicators. Expert. Syst. Appl. 113250 (2020)
4. Seyyedabbasi, A., Kiani, F.: I-GWO and Ex-GWO: improved algorithms of the grey wolf optimizer to solve global optimization problems. Eng. Comput. 509–532 (2021)
5. Carbo, J.M., Gorjon, S.: Application of machine learning models and interpretability techniques to identify the determinants of the price of bitcoin (2022)
6. Hansen, P.R., Lunde, A., Nason, J.M.: The model confidence set. Econometrica 453–497 (2011)
7. Cortes, C., Vapnik, V.: Support-vector networks. Mach. Learn. 273–297 (1995). https://doi.org/10.1007/BF00994018
8. Nakamoto, S.: Bitcoin: a peer-to-peer electronic cash system (2008)
9. Yermack, D.: Corporate governance and blockchains. Rev. Financ. 7–31 (2017)
10. Xu, M., Chen, X., Kou, G.: A systematic review of blockchain. Financ. Innov. 1–14 (2019)
11. Mirjalili, S., Mirjalili, S.M., Lewis, A.: Grey wolf optimizer. Adv. Eng. Softw. 46–61 (2014)
12. Arora, P., Nagpal, R., de Albuquerque, V.H.C., Mirjalili, S.: Blockchain technology and its applications: a systematic review of the literature. In: Proceedings of the International Conference on Innovative Computing & Communication (ICICC) (2022)
13. Abdel-Basset, M., El-Shahat, D., El-henawy, I., de Albuquerque, V.H.C., Mirjalili, S.: A new fusion of grey wolf optimizer algorithm with a two-phase mutation for feature selection. Expert. Syst. Appl. 112824 (2020). https://doi.org/10.1016/j.eswa.2019.112824
14. Dada, E., Joseph, S., Oyewola, D., Fadele, A.A., Chiroma, H.: Application of grey wolf optimization algorithm: recent trends, issues, and possible horizons. Gazi Univ. J. Sci. 485–504 (2021). https://doi.org/10.35378/gujs.820885
15. Ghalambaz, M., Yengejeh, R.J., Davami, A.H.: Building energy optimization using grey wolf optimizer (GWO). Case Stud. Therm. Eng. 101250 (2020). https://doi.org/10.1016/j.csite.2021.101250
16. Mirjalili, S., Aljarah, I., Mafarja, M., Heidari, A.A., Faris, H.: Grey wolf optimizer: theory, literature review, and application in computational fluid dynamics problems. Nat.-Inspired Optim. 87–105 (2020). https://doi.org/10.1007/978-3-030-12127-3_6
17. Padhy, S., Panda, S.: Application of a simplified Grey Wolf optimization technique for adaptive fuzzy PID controller design for frequency regulation of a distributed power generation system. Prot. Control Mod. Power Syst. 1–16 (2021). https://doi.org/10.1186/s41601-021-00180-4
18. Hu, P., Pan, J.S., Chu, S.C.: Improved binary grey wolf optimizer and its application for feature selection. Knowl.-Based Syst. 105746 (2020). https://doi.org/10.1016/j.knosys.2020.105746
19. Kitonyi, P.M., Segera, D.R.: Hybrid gradient descent grey wolf optimizer for optimal feature selection. Biomed. Res. Int. (2021). https://doi.org/10.1155/2021/2555622
20. Shaikh, M.S., Hua, C., Jatoi, M.A., Ansari, M.M., Qader, A.A.: Application of grey wolf optimisation algorithm in parameter calculation of overhead transmission line system. IET Sci., Meas. Technol. 218–231 (2021). https://doi.org/10.1049/smt2.12023
21. Wolpert, D.H., Macready, W.G.: No free lunch theorems for optimization. IEEE Trans. Evol. Comput. 67–82 (1997). https://doi.org/10.1109/4235.585893
22. Singh, N.: A modified variant of grey wolf optimizer. Sci. Iran. 1450–1466 (2020). https://doi.org/10.24200/SCI.2018.50122.1523
23. Mittal, N., Singh, U., Sohi, B.S.: Modified grey wolf optimizer for global engineering optimization. Appl. Comput. Intell. Soft Comput. 1–16 (2016)

24. Koo, E., Kim, G.: Prediction of Bitcoin price based on manipulating distribution strategy. Appl. Soft Comput. 110 (2021)
25. Chen, K., Yi, Z., Fangyan, D.: LSTM-based method for stock returns prediction: a case study of China stock market. In: 2015 IEEE International Conference on Big Data, Big Data, pp. 2823–2824. IEEE (2015)

# Salp Swarm Algorithm Based Hyperparameter-Optimized Deep EfficientNet for COVID-19 Detection

**Parijata Majumdar and Sanjoy Mitra**

**Abstract** The outspread of COVID-19 significantly effected both the world economy and public health. Motivated by the collaborative efforts of the free resource community in compiling COVID-19 data and the successful utilization of deep learning techniques in disease identification, the aim is to create a novel COVID-19 detection system that used images from chest X-rays. We designed an advanced Convolutional Neural Network (CNN) architecture, building upon the EfficientNet-B0 model, to detect COVID-19 with high accuracy. This enhanced model incorporates a compound coefficient to precisely balance depth, breadth, and resolution. By employing predefined scaling coefficients, the network's dimensions are uniformly adjusted. However, fine-tuning the hyperparameters of EfficientNet-B0 is essential for optimizing its performance. Thus, we propose an automated hyperparameter optimization using the Salp Swarm Algorithm (SSA) to fine-tune EfficientNet-B0 for COVID-19 classification. Our SSA-EfficientNet-B0 model outperforms Particle Swarm Optimization and Genetic Algorithm. The optimized neural network achieves a remarkable accuracy rate of 98.89% in detecting COVID-19 positive cases, surpassing alternative methods.

## 1 Introduction

The COVID-19 virus can spread rapidly between individuals through small particles caused due to speaking, coughing, and sneezing when in close proximity [1]. Efficient screening to identify infected patients can help prevent the contagious COVID-19 by enabling isolation.

P. Majumdar (✉)
Indian Institute of Information Technology, Agartala, India
e-mail: er.parijata@gmail.com

S. Mitra
Tripura Institute of Technology, Agartala, India
e-mail: mail.smitra@gmail.com

 
D. Oliva et al. (eds.), *Advances in Optimization Algorithms for Multidisciplinary Engineering Applications: From Classical Methods to AI-Enhanced Solutions*, Studies in Computational Intelligence 806, https://doi.org/10.1007/978-3-031-78440-8_23

Chest X-rays (CXR) are a commonly available, easily obtainable, and more affordable clinical imaging modality that can aid in COVID-19 detection. While chest computed tomography (CT) imaging can also assist in detecting COVID-19, its significant radiation exposure limits its routine use for screening purposes.

Radiologists heavily rely on CXR images to detect chest diseases, including COVID-19, due to their simplicity and availability. CXR scans are cost-effective, and expose patients to minimal radiation. Given, the clinical importance of CXR images, researchers are investigating artificial intelligence to improve CXR interpretation [2].

Deep learning (DL) techniques are increasingly popular for medical image classification and disease detection due to their high accuracy, particularly with the presence of multiple hidden layers enabling both regression and classification tasks.

Recent research in DL technology has focused on developing intelligent diagnostic tools to improve patient care decisions. Convolutional neural networks (CNNs) have been instrumental in achieving remarkable results in COVID-19 case recognition using chest X-ray (CXR) [3]. CNNs can effectively learn and extract features [4].

When training the EfficientNet-B0 algorithm, choosing hyperparameters is essential. If learning rate increases, the algorithm may converge too quickly, potentially overlooking important data elements.

Optimizing CNN hyperparameters is essential for effective training and optimal performance. Metaheuristic algorithms, which resembles the collective behavior of species, have shown significant development in optimization problems [5]. One such metaheuristic, the Salp Swarm Algorithm (SSA), has demonstrated success in various fields [6–10]. Here, SSA is utilized in conjunction with hyperparameters of EfficientNet-B0 and compared with Particle Swarm Optimization (PSO-EfficientNet-B0) and Genetic Algorithm (GA-EfficientNet-B0) for EfficientNet-B0 hyperparameter optimization.

In Sect. 2, previous works are reviewed, outlining the existing literature in the field. Section 3 delves into the background discussing the mathematical foundations of the SSA-EfficientNet-B0 algorithm, which is a hyperparameterized form of the EfficientNet-B0 depending on SSA. The methodology and key results are elucidated in Sect. 4. Finally, Sect. 5 draws conclusions.

## 2 Literature Review

Tang et al. [11] introduced an EDL-COVID for COVID-19 identification. However, COVID-Net require retraining to enhance performance whenever new CXR images become available. Maghdid et al. [12] constructed a comprehensive database comprising CT and X-ray images from various sources to train convolutional neural networks (CNNs), including a modified version of AlexNet, for COVID-19 detection. Waisy et al. [3] used Butterworth bandpass filters to improve contrasts of CXR scans.

EfficientNetB0, VGG16, and InceptionV3, known for their efficiency in terms of parameters, were utilized for COVID-19 identification Gaur et al. [13]. Ganesh et al. [14] utilized transfer learning with AlexNet and VGG16 pretraining, to address class imbalance problem.

Singh et al. [1] distinguished individuals with COVID-19 from X-ray scans using MADE-based CNN models [15]. Kumar et al. [16] utilized ResNet152 for analyzing critical data from CXR scans. Various techniques like PSO [17, 18], and multilevel evolutionary optimization [19] have been employed to enhance CNN hyperparameters.

While the outcomes of hyperparameter optimization in these studies are promising, there is still room for improvement. Recently, Salp Swarm Algorithm (SSA) has emerged as a technique for CNN training and hyperparameter optimization in deep learning based methods, as evidenced by studies [20–22].

In none of the existing approaches utilized for COVID-19 detection, hyperparameter tuning of existing deep learning algorithms are done, though optimization techniques have witnessed resounding success in hyperparameter optimization of deep learning-based methods. Hence, this motivated us to utilize SSA to optimize the EfficientNet-B0 parameters (SSA-EfficientNet-B0 algorithm) for addressing COVID-19 classification problem from CXR images with more accuracy. Hyperparameters are used to tune the model performance. These parameters are typically set prior to the start of the training process because they cannot be learned immediately via the standard training procedure. SSA is chosen because of it's inherent advantages of simplicity, flexibility, and local optima avoidance capabilities [23].

# 3 Background

## *3.1 Salp Swarm Algorithm*

Typically, salps form salp chains in deep waters [23]. The salp at the front is the leader, whereas the other salps at the back are referred to as followers. Salp positions are specified in a search space that is dim-dimensional, where dim is the number of variables. All salp positions are stored in a 2-D matrix called "mat." It is described as follows:

$$mat_n^1 = Faim_n + r_1((u_n - l_n)r_2 + l_n), if r_3 \geq 0 \tag{1}$$

$$mat_n^1 = Faim_n - r_1((u_n - l_n)r_2 + l_n), if r_3 < 0 \tag{2}$$

Here, $mat_n^1$ represents the leader position in the $n-th$ dimension, $Faim_n$ denotes the food source position, $u_n$ and $l_n$ represents upper and lower bound of the $n-th$ dimension, $r2$, and $r3$ are uniform in range of [0, 1].

The coefficient $r_1$ balances exploration and exploitation rate.

$$r_1 = 2e^{-}(-\frac{4curr_{iter}}{max_{iter}}) \tag{3}$$

Here, $curr_{iter}$ represents the current iteration, $max_{iter}$ represents the max. no. of iterations, $mat_m^n$ represents the position of the $m-th$ follower salp in the $n-th$ dimension, e is time, and $initial_{speed}$ represents the initial speed.

The following equation is used at every iteration:

$$mat_n^m = \frac{1}{2}ce^2 + initial_{speed} \times e \tag{4}$$

where $m \geq 2$, c= $\frac{final_{speed}}{initial_{speed}}$ and speed = $\frac{mat-mat_0}{e}$. Considering $initial_{speed} = 0$ , it can be written as:

$$mat_n^m = \frac{1}{2}(mat_n^m + mat_n^{m-1}) \tag{5}$$

where $m = 1, \ldots, dim.$

Initially, N positions are generated in a random population. Each individual's fitness is calculated, and the food source is assigned to the best solution. According to predetermined formulae, the population is changed at each iteration based on updates to the positions of the leader and follower salps. The optimal solution discovered thus far is compared to the feasible solution's fitness values, and the optimal solution is modified if needed. This method is repeated until the halting requirements are met.

Figure 1 illustrates the fundamental structure of the SSA algorithm [23].

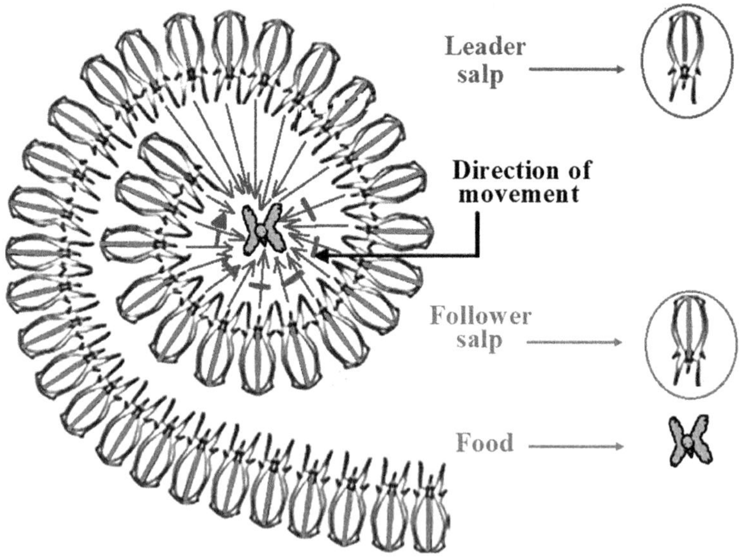


**Fig. 1** Basic structure of Salp Swarm Agorithm (SSA) [23]

## 3.2 *EfficientNet-B0*

In CNN during model training, random weights are initialized for the kernel weights which are then optimized using gradient descent training to modify CL parameters. Rectified Linear Unit (ReLU) is used to transfer features into feature space, maintaining CL values within a specific range. To mitigate overfitting, dropout layers are added. Batch normalization layers (BNL) normalize gradients. While max-pooling layers (PL) reduce feature map dimensions, they retain essential image elements. Max pooling downsamples the feature map.

The fully connected classifier layer uses a soft-max activation function in the final layer to classify the retrieved features into distinct groups. The loss function is penalized to combat overfitting. In order to prevent overfitting, CNN dropout minimizes interdependent learning.

EfficientNet's architecture aims to balance model size and performance by maximizing each parameter simultaneously. It employs a compound coefficient to precisely quantify these components. The compound scaling approach efficiently and consistently scales each dimension-depth, width, and resolution. This technique enhances network performance while maintaining manageable computational and memory requirements. Ensuring a balance across all network dimensions is crucial, achieved by scaling each dimension [24].

In the case of EfficientNet-B0, the network incorporates squeeze-and-excitation blocks. The top layer is modified to accommodate two output neuron layers, reflecting the model's classification of two distinct image classes. However, the remaining layers remain unaltered, leveraging their extensive pre-training on the ImageNet dataset.

To further enhance performance, a combination of dropout and thicker layers is employed. Subsequently, experiments are conducted using the Adam optimizer, which dictates how weight values are managed and how quickly they should be adjusted. The model is defined, and checkpoints are established during training, with precautions implemented to prevent overfitting.

The compound scaling approach is described below:

$$depth : d = \alpha^{\phi} \tag{6}$$

$$width : w = \beta^{\phi} \tag{7}$$

$$resolution : r = \gamma^{\phi} \tag{8}$$

$$s.t. \alpha \cdot \beta^2 \cdot \gamma^2 \propto 2, \alpha \geq 1, \beta \geq 1, \gamma \geq 1, \tag{9}$$

where $\alpha$, $\beta$, and $\gamma$ depends on a small grid search. $\phi$ controls the number of resources for model scaling.

The compound scaling is illustrated in Fig. 2.

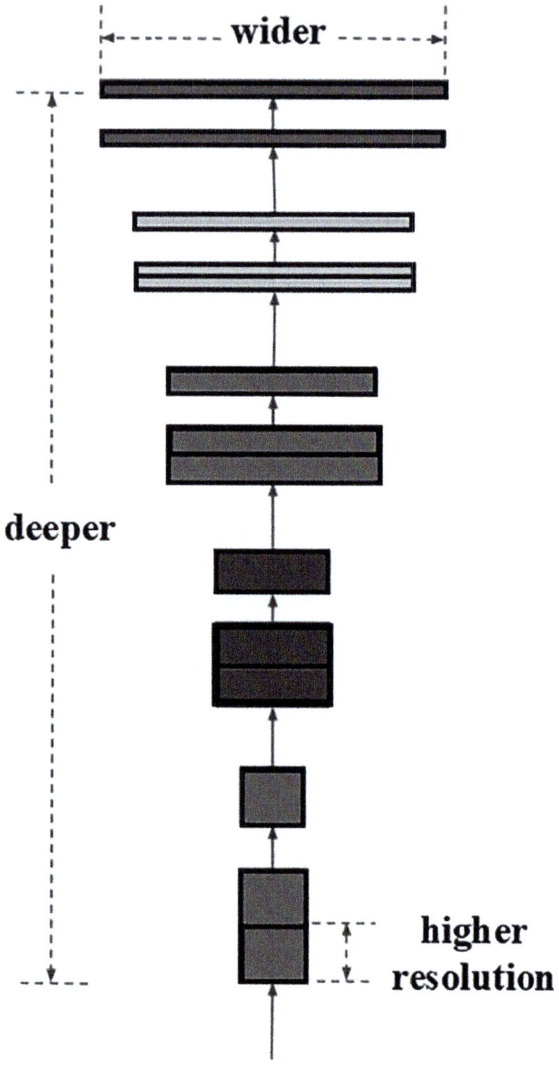


**Fig. 2** Compound scaling of EfficientNet-B0

### *3.3 Hyperparameter Optimized EfficientNet-B0 Using SSA*

Hyperparameters play a vital role in improving performance. SSA is employed to fine-tune these hyperparameters, determining the accuracy and convergence. Learning rate influences the speed of the gradient descent, while momentum influences the impact of older weights on current weights. How frequently the learning algorithm updates network parameters based on the training dataset depends on the number of epochs. Regularization prevents network overfitting. Optimizing these hyperparameters is essential for achieving accuracy, considering these factors.

The SSA is utilized for hyperparameter optimization in Stochastic Gradient Descent (SGD) training for EfficientNet-B0. Numerous hyperparameters in the model significantly impact outcomes. Here, the parameters chosen for the algorithm includes Batch Size: b = 30; number of search agents: s = 3; dimension: dim = 4; number of optimization iterations: opt = 30; hyperparameters: $H1, H2, H3, H4 = [1, 0.5, 15, 5.0000e - 04]$; hyperparameter evaluation function ($H_D$): Error rate = $(FP + FN)/(TP + TN + FP + FN)$.

Fitness can be computed according to the following equation:

$$fitness = \rho Error(D) + (1 - \rho)|s|/|d| \tag{10}$$

where $Error(D)$ is the classification error rate, $|s|$ is the selected features, $|d|$ are all the features, $\rho$ and $\phi$ are the weights to influence the classification accuracy & the feature minimization where $\rho \in [0, 1]$ [25]. We considered $\rho = 0.9$.

# 4 Results and Discussion

## 4.1 Datasets

SSA-based EfficientNetB0 was implemented in Python 3.8 and executed on a Windows 10 Pro system and Nvidia GPU. The study comprises of the CoronaHack-CXR dataset [26], the COVID CXR dataset [27], and the COVID-19 Radiography database [28].

### 4.1.1 Training and Testing

The input images were resized to dimensions of 256 × 256, ensuring consistency in the model's input. During this resizing process, random-sized data is reshaped to match the model's input format. Scaling down the images to their original size can reduce the training and testing time of the EfficientNetB0. The dataset was split randomly, allocating 30% for testing and 70% for training purposes. Hyperparameters were optimized by SSA, and parameters underwent tuning through SGD training.

The aim is to attain optimal values for hyperparameters, particularly those related to convolution, dropout, and pooling layers, following the selection of the architecture. SGD fine-tunes parameters, whereas SSA optimizes hyperparameters. Once the design of EfficientNet-B0 is established, the focus shifts to achieving values for hyperparameters related predominantly to convolution, dropout, and pooling layers that approach optimality.

Developing an appropriate fitness function capable of swiftly calculating accuracy as the fitness of EfficientNet-B0 is crucial to improving hyperparameter tuning efficiency. EfficientNet-B0 was chosen for its robust training and feature selection.

The optimized hyperparameters includes training algorithm called Stochastic Gradient Descent, $momentum = 0.6, initiallearningrate = 0.015, maximumepoch = 10, validationfrequency = 30, andL2Regularization = 1.0000e - 04$.

## 4.2 Performance Evaluation and Comparative Analysis

SSA-EfficientNet-B0 undergoes validation using various performance metrics [29]. The ROC curve visually represents the classifier's performance.

Table 1 lists the training and testing accuracies of models. In every case, SSA-EfficientNetB0 outperform others. In Table 2, the accuracy of each class was assessed. In every case, SSA-EfficientNetB0 outperform others. The computation times were calculated. For each class, individual precision, and recall were tabulated in Table 3. Table 4 shows the model's Area Under Curve (AUC) values. Table 5 shows that SSA-EfficientNetB0 outperform others in terms of accuracy. Here, a represents SSA-EfficcientNetB0, b represents PSO-EfficcientNetB0 and c represents GA-EfficcientNetB0 algorithm.

SSA obtains a minimum average error (0.3119) shown in Table 6. It has low standard deviation (0.0032) indicating it's stability.

Figure 3 illustrates Box plot's correctness.

Figure 4 illustrates the results from testing the Box plot accuracy.

# 5 Conclusion and Future Work

To save lives, it's imperative to accurately detect COVID-19. EfficientNet-B0 relies heavily on precise hyper-parameter tuning, particularly in EfficientNet-B0, for optimal performance. SSA is used to adjust the hyper-parameters of EfficientNet-B0

**Table 1** Training and testing accuracy of models

| Dataset | Number of images | Model | Training accuracy | Testing accuracy | Time elapsed (min) |
|---|---|---|---|---|---|
| COVID | | a | 89.4 | 92.2 | 133 |
| Chest | 803 | b | 85.6 | 87.9 | 132 |
| X-ray | | c | 82.8 | 84.5 | 136 |
| COVID-19 | | a | 92.9 | 95.5 | 2318 |
| Radiography | 2484 | b | 84.5 | 86.7 | 2210 |
| Radiography | | c | 81.9 | 83.9 | 1960 |
| CoronaHack | | a | 93.6 | 94.7 | 2708 |
| -Chest | 5911 | b | 89.6 | 91.9 | 132 |
| X-ray | | c | 85.8 | 87.9 | 2601 |

**Table 2** Accuracy per class

| Dataset | Model | Normal images training accuracy | Normal images testing accuracy | COVID-19 images training accuracy | COVID-19 images testing accuracy |
|---|---|---|---|---|---|
| COVID | a | 87.2 | 88.8 | 90.2 | 92.2 |
| Chest | b | 86.8 | 87.7 | 88.5 | 90.5 |
| X-ray | c | 83.8 | 84.1 | 86.6 | 88.8 |
| COVID-19 | a | 91.2 | 93.2 | 95.7 | 96.2 |
| Radiography | b | 87.2 | 88.5 | 91.2 | 93.3 |
| | c | 85.5 | 86.8 | 90.9 | 92.2 |
| CoronaHack | a | 94.6 | 95.5 | 93.5 | 92.6 |
| -Chest | b | 90.9 | 89.2 | 90.1 | 89.7 |
| X-ray | c | 88.3 | 89.9 | 88.4 | 86.6 |

**Table 3** Recall and precision per class

| Dataset | Model | Normal images recall | Normal images precision | COVID images recall | COVID images precision |
|---|---|---|---|---|---|
| COVID | a | 0.91 | 0.90 | 0.92 | 0.94 |
| Chest | b | 0.86 | 0.87 | 0.88 | 0.90 |
| X-ray | c | 0.83 | 0.84 | 0.86 | 0.88 |
| COVID-19 | a | 0.92 | 92.2 | 0.95 | 0.97 |
| Radiography | b | 0.89 | 88.5 | 0.91 | 0.93 |
| | c | 0.87 | 86.8 | 0.90 | 0.92 |
| CoronaHack | a | 0.94 | 0.96 | 0.94 | 0.93 |
| -Chest | b | 0.91 | 0.91 | 0.90 | 0.89 |
| X-ray | c | 0.88 | 0.89 | 0.88 | 0.87 |

**Table 4** Area under curve

| Dataset | a | b | c |
|---|---|---|---|
| COVID | | | |
| Chest | 0.69 | 0.68 | 0.54 |
| X-ray | | | |
| COVID-19 | 0.82 | 0.76 | 0.69 |
| Radiography | | | |
| CoronaHack | | | |
| CoronaHack | | | |
| -Chest | 0.84 | 0.77 | 0.69 |
| X-ray | | | |

**Table 5** Accuracy comparison

| Model | Accuracy (in %) |
|---|---|
| a | 74.2 |
| b | 63.4 |
| c | 54.7 |

**Table 6** Performance analysis based on different metrics

| Metric | SSA | GA | PSO |
|---|---|---|---|
| Average error | 0.3119 | 0.3215 | 0.3239 |
| Average fitness | 0.1592 | 0.1612 | 0.1549 |
| Best fitness | 1.3456 | 1.4011 | 1.4113 |
| Worst fitness | 1.5013 | 1.5102 | 1.5117 |
| Standard deviation fitness | 0.0032 | 0.0043 | 0.0036 |

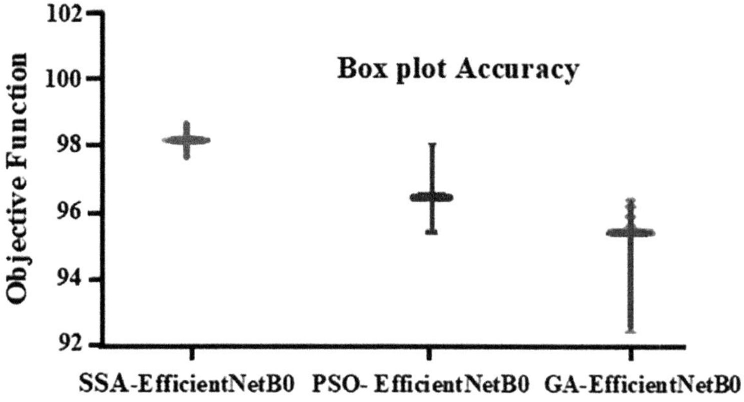


**Fig. 3** Box plot accuracy of algorithms for normal CXR images

(referred to as SSA-EfficientNet-B0). The resulting model demonstrates significant accuracy in distinguishing COVID-19 and normal CXR images compared to alternative models. Given its robust performance, SSA-EfficientNet-B0 has the potential to aid healthcare professionals in accurate COVID-19 diagnosis and prompt treatment decisions. The proposed computer-aided diagnostic approach holds promise for enhancing COVID-19 case identification significantly. Future research will focus on refining SSA-EfficientNet-B0 and incorporating recent datasets for validation. One limitation of SSA-EfficientNet-B0 is the requirement for pre-processing due to its inability to encode the input image's positional orientation. Moreover, spatial invariance in SSA-EfficientNet-B0 remains unaffected by input images. Nonetheless, the algorithm presents opportunities for addressing various real-life challenges and

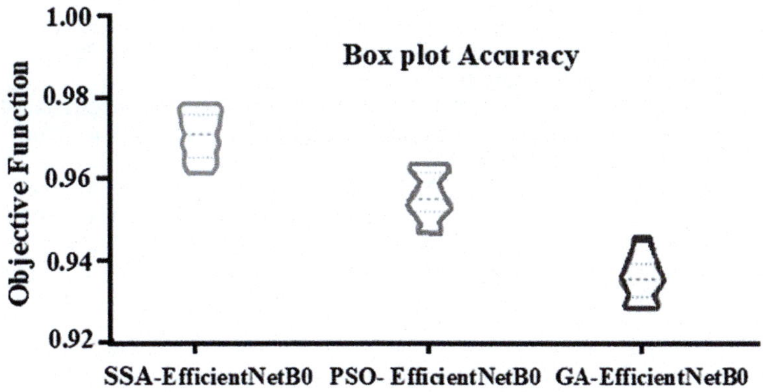


**Fig. 4** Box plot accuracy of algorithms for CXR COVID-19 images

complex optimization problems, both single and multi-objective. While the groundbreaking research in this field holds immense promise, its widespread implementation remains limited. Researchers are exploring a multifaceted approach, combining chest imaging with clinical symptoms, exposure history, and lab tests for diagnosis. Moreover, the literature highlights a deficiency in proper feature extraction techniques from image data, necessitating their inclusion in future research endeavors.

## References


1. Singh, D., Kumar, V., Yadav, V., Kaur, M.: Deep neural network-based screening model for COVID-19-infected patients using chest X-Ray images. Int. J. Pattern Recognit. Artif. Intell. **34**(12), 207–7002 (2020). https://doi.org/10.1142/S0218001421510046
2. Ginneken, B.V., Romeny, B.M.T.H., Viergever, M.A.: Computer-aided diagnosis in chest radiography: a survey. IEEE Trans. Medical Imaging **20**(12), 1228–1241 (2001). https://doi.org/10.1109/42.974918
3. Al-Waisy, A.S., Mohammed, M.A., Fahdawi, S.A., Maashi, M.S., Zapirain, B.G., Abdulkareem, K.H.: COVID-DeepNet: hybrid multimodal deep learning system for improving COVID-19 pneumonia detection in chest X-ray images. Comput. Mater. Continua **68**(3), 3733–3747 (2021). https://doi.org/10.32604/cmc.2021.012955
4. Krizhevsky, A., Sutskever, I., Hinton, G.E.: Imagenet classification with deep convolutional neural networks. Commun ACM. **60**(6), 84–90 (2017). https://doi.org/10.1145/3065386
5. Mirjalili, S., Mirjalili, S.M., Lewis, A.: Grey wolf optimizer. Adv. Eng. Softw. **69**, 46–61 (2014)
6. Heidari, A.A., Mirjalili, S., Faris, H., Aljarah, I., Mafarja, M., Chen, H.: Harris hawks optimization: algorithm and applications. Future Gener. Comput. Syst. **97**, 849–872 (2019). https://doi.org/10.1016/j.future.2019.02.028
7. Kumaran, N., Vadivel, A., Kumar, S.S.: Recognition of human actions using CNN-GWO: a novel modeling of CNN for enhancement of classification performance. Multimedia Tools Appl. **77**(18), 23115–23147 (2018). https://doi.org/10.1007/s11042-017-5591-z
8. Sanjay, R., Jayabarathi, T., Raghunathan, T., Ramesh, V., Mithulananthan, N.: Optimal allocation of distributed generation using hybrid grey wolf optimizer. IEEE Access **5**, 14807–14818 (2017)

9. Lu, C., Gao, L., Li, X., Xiao, S.: A hybrid multi-objective grey wolf optimizer for dynamic scheduling in a real-world welding industry. Eng. Appl. Artif. Intell. **57**, 61–79 (2017). https://doi.org/10.1016/j.engappai.2016.10.013
10. Venkatakrishnan, G., Rengaraj, R., Salivahanan, S.: Grey wolf optimizer to real power dispatch with non-linear constraint. CMES-Comput. Modeling Eng. Sci. **115**(1), 5–45 (2018). https://doi.org/10.3970/cmes.2018.115.025
11. Tang, S., Wang, C., Nie, J., Kumar, N., Zhang, Y., Xiong, Z., Barnawi, A.: EDL-COVID: ensemble deep learning for COVID-19 case detection from chest X-ray images. IEEE Trans. Industrial Inform. **17**(9), 6539–6549 (2021). https://doi.org/10.1109/TII.2021.3057683
12. Maghdid, H.S., Asaad, A.T., Ghafoor, K.Z., Sadiq, A.S., Khan, M.K.: Diagnosing COVID-19 Pneumonia from X-Ray and CT images using deep learning and transfer learning algorithms (2021). https://ieeexplore.ieee.org/document/9413479
13. Gaur, L., Bhatia, U., Jhanjhi, N.Z., Muhammad, G., Masud, M.: Medical image-based detection of COVID-19 using deep convolution neural networks. Multimedia Syst. **29**(3), 1729–1738 (2023). https://doi.org/10.1007/s00530-021-00794-6
14. TN, S.G., Satish, R., Sridhar, R.: Learning effective embedding for automated COVID-19 prediction from chest X-ray images. Multimedia Syst. **29**(2), 739–751 (2023). https://doi.org/10.1007/s00530-022-01015-4
15. Zhang, J., Xie, Y., Li, Y., Shen, C., Xia, Y.: COVID-19 screening on chest X-ray images using deep learning based anomaly detection (2020). https://arxiv.org/abs/2003.12338
16. Kumar, R., Arora, R., Bansal, V., Sahayasheela, V.J., Buckchash, H., Imran, J., Narayanan, N., Pandian, G.N., Raman, B.: Classification of COVID-19 from chest x-ray images using deep features and correlation coefficient (2022). https://arxiv.org/abs/2202.11243
17. Yoo, Y.: Hyperparameter optimization of deep neural network using univariate dynamic encoding algorithm for searches. Knowl.-Based Syst. **178**, 74–83 (2019). https://doi.org/10.1016/j.knosys.2019.04.019
18. Wang, Y., Zhang, H., Zhang, G.: cPSO-CNN: an efficient PSO-based algorithm for fine-tuning hyper-parameters of convolutional neural networks. Swarm Evol. Comput. **49**, 114–123 (2019). https://doi.org/10.1016/j.swevo.2019.06.002
19. Cui, H., Bai, J.: A new hyperparameters optimization method for convolutional neural networks. Pattern Recogn. Lett. **125**, 828–834 (2019). https://doi.org/10.1016/j.patrec.2019.02.009
20. Xie, H., Zhang, L., Lim, C.P.: Evolving CNN-LSTM models for time series prediction using enhanced grey wolf optimizer. IEEE Access **8**, 161519–161541 (2020). https://doi.org/10.1109/ACCESS.2020.3021527
21. Agarwal, R., Sharma, H.: A new enhanced recurrent extreme learning machine based on feature fusion with CNN deep features for breast cancer detection. Adv. Comput., Commun. Comput. Sci., AISC **1158**, 461–471 (2020)
22. Mohakud, R., Dash, R.: Designing a grey wolf optimization based hyper-parameter optimized convolutional neural network classifier for skin cancer detection. J. King Saud Univ.-Comput. Inf. Sci. **34**, 6208–6291 (2022). https://doi.org/10.1016/j.jksuci.2021.05.012
23. Mirjalili, S., Gandomi, A.H., Mirjalili, S.Z., Saremi, S., Faris, H., Mirjalili, S.M.: Salp swarm algorithm: a bio-inspired optimizer for engineering design problems. Adv. Eng. Softw. **114**, 163–191 (2017). https://doi.org/10.1016/j.advengsoft.2017.07.002
24. Tan, M., Le, Q.: EfficientNet: rethinking model scaling for convolutional neural networks. In: International Conference on Machine Learning, pp. 6105–6114 (2019). 10.48550/arXiv.1905.11946
25. Hegazy, A.E., Makhlouf, M.A., El-Tawel, G.S.: Improved Salp swarm algorithm for feature selection. J. King Saud Univ.-Comput. Inf. Sci. **32**, 335–344 (2020)
26. Praveen, O.: CoronaHack-chest X-Ray-dataset (2019). https://www.kaggle.com/datasets/praveengovi/coronahack-chest-xraydataset
27. Cohen, J.P.: A public open dataset of chest X-ray and CT images (2021). https://github.com/ieee8023/covid-chestxray-dataset
28. Rahman, T., Chowdhury, M., Khandakar, A.: COVID-19 chest X-ray images and lung masks database (2021). https://www.kaggle.com/tawsifurrahman/covid19-radiography-database

29. El-Kenawy, E.S., Eid, M.M., Saber, M., Ibrahim, A.: MbGWOSFS: modified binary grey wolf optimizer based on stochastic fractal search for feature selection. IEEE Access **8**, 107635–107649 (2020). https://doi.org/10.1109/ACCESS.2020.3001151

# Engineering Applications

# Intelligent Model to Build an Smart Grid for Electrical Vehicles

**Shaban Mousavil, Alberto Ochoa-Zezzatti, Roberto Contreras-Masse, Vianey Torres, and Saúl Gonzalez**

**Abstract** Rapidly growing smart cities escalate energy demands straining traditional grids, often leading to increased fossil fuel reliance and emissions. This chapter proposes an adaptive Smart Grid model to promote sustainable urban development. The model prioritizes ecological responsibility by integrating renewables (solar, wind) with traditional sources, reducing greenhouse gas emissions. It also fosters electric vehicle (EV) adoption through strategic charging station placement and incentives, lowering urban transportation emissions. For efficient deployment and user needs, the model utilizes adaptable regulations that seamlessly integrate evolving technologies while maintaining grid stability. Additionally, advanced data analytics optimize energy forecasting, enabling proactive management and minimizing waste. In essence, this model tackles rising energy demands through ecological responsibility, efficient deployment, optimized forecasting, and EV promotion, paving the way for a sustainable future for smart cities.

S. Mousavil
Universidad Autónoma de Ciudad Juárez (UACJ), Ciudad Juárez, Chih MX, Mexico
e-mail: al237855@alumnos.uacj.mx

A. Ochoa-Zezzatti · V. Torres · S. Gonzalez
UACJ, Ciudad Juárez, Chih, Japan
e-mail: alberto.ochoa@uacj.mx

V. Torres
e-mail: vianey.torres@uacj.mx

R. Contreras-Masse (✉)
TecNM Campus Ciudad Juárez (ITCJ), Ciudad Juárez, Mexico
e-mail: roberto.cm@cdjuarez.tecnm.mx

 
D. Oliva et al. (eds.), *Advances in Optimization Algorithms for Multidisciplinary Engineering Applications: From Classical Methods to AI-Enhanced Solutions*, Studies in Computational Intelligence 806, https://doi.org/10.1007/978-3-031-78440-8_24

# 1 Introduction

A Smart Grid is an intelligent electrical network that uses advanced technology to control and coordinate electrical energy generation, distribution and consumption. The main components of a Smart Grid are:

- Renewable energy generation: Power generation from renewable sources, such as solar, wind, hydroelectric, and geothermal energy, is an integral part of a Smart Grid.
- Smart Distribution Grid: The distribution network of a Smart Grid uses advanced technology to monitor and control the distribution of electric power in real-time.
- Smart meters: Smart meters measure electricity consumption in real-time and provide detailed information about electricity consumption in a home or building.
- Energy storage systems: Energy storage systems, such as batteries, can store electrical energy generated by renewable sources and distribute it when needed.
- Charge management systems: Charge management systems enable controlling and coordinating the charging of electric vehicles and other devices that consume electrical energy.
- Control and automation systems: Control and automation systems allow real-time monitoring and controlling of electric power generation, distribution and consumption.
- Communications networks: Communication networks allow transmitting information in real time between the components of a Smart Grid.

The main components of a Smart Grid are shown in Fig. 1.

Here are some key points about the Smart Grid required for electric vehicles in a Smart City:

- Bi-directional communication: the Smart Grid must allow bi-directional communication between electric vehicles and the electric grid. This allows electric vehicles to charge their battery and, at the same time, can send power back to the grid at times of highdemand or when the vehicle battery is full.
- Smart charging systems: Smart charging systems are necessary for charging electric vehicles in a Smart City. These systems can control the charging speed, the amount of energy supplied, and the timing of charging to optimize energy use and avoid overloading the grid.
- Energy storage: Energy storage is essential for the integration of electric vehicles in a Smart City. This allows energy generated by renewable sources, such as solar and wind power, to be stored and used to charge electric vehicles when needed.
- Demand management: Demand management is key to ensuring that the grid can handle the charging of electric vehicles at times of high demand. This includes managing charging based on power availability and charging at off-peak times [7, 8].
- Remote monitoring and control: Remote monitoring and control systems are necessary to monitor the status of the power grid, electric vehicles and charging systems. This allows operators to control power flow and make adjustments in real time [7].

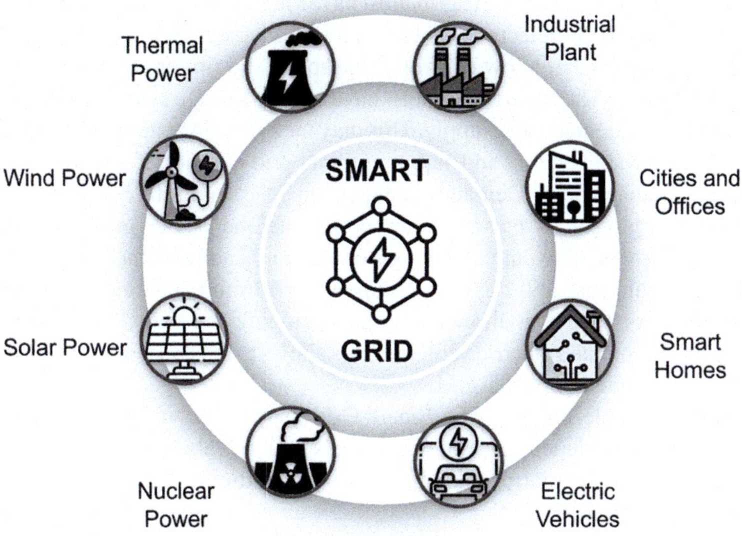


**Fig. 1** Components of a smart grid

In summary, the Smart Grid required for electric vehicles in a Smart City must enable two-way communication, smart charging systems, energy storage, demand management, and remote monitoring and control to ensure efficient and safe charging of electric vehicles, while maintaining grid stability [1, 5, 12].

This chapter is organized in sections. Section 2 discusses the requirements of a smart grid for an electric vehicle fleet. The use of electric vehicles in a smart city is discussed in Sect. 3. The connection and relation between smart grid and electric vehicles is presented in Sect. 4. Then, Sect. 5 shows the benefits of integrating all those elements by mathematical relations, challenges, obstacles, and case studies of successful implementations. Finally, Sect. 6 presents conclusions, recommendations and future research.

## 2 Requirements of a Smart Grid for a Electric Vehicle Fleet

The size of the Smart Grid required to support 10,000 electric vehicles in a Smart City will depend on several factors, such as population density, number of charging stations, energy storage capacity and the use of renewable energy in the grid. Generally, a Smart Grid with the capacity to handle a load of at least 10 MW will be needed to charge 10,000 electric vehicles. However, this figure may vary depending on the specific characteristics of the Smart City. The main characteristics that a Smart Grid must have to support 10,000 electric vehicles include the following:

- Charging capacity: the electric grid must have the capacity to supply enough energy to charge the 10,000 electric vehicles. This can be achieved by using innovative charging systems that allow control of the speed and timing of charging.
- Energy storage: The Smart Grid must have energy storage systems to handle fluctuating energy demand and supply. This may include using energy storage batteries to store renewable energy and supply it to the grid at times of high demand.
- Demand management: Demand management is essential to ensure that electric vehicle charging does not cause disruptions to the power grid. This includes controlling power flow and load based on demand and power availability.
- Remote monitoring and control: The power grid must have remote monitoring and control systems that allow operators to monitor the status of the power grid, charging stations and electric vehicles. This allows a quick response to any problem or interruption in the power grid.
- Use of renewable energy: The Smart Grid should encourage using renewable energy to reduce dependence on fossil fuels and greenhouse gas emissions. This may include using solar panels, wind turbines, and other renewable resources to generate power.

In conclusion, to support 10,000 electric vehicles in a Smart City, the Smart Grid must have the capacity to charge enough electric vehicles, store energy, manage demand, remotely monitor, and control the grid, and encourage the use of renewable energy. The size and specific features will depend on the needs and particularities of the Smart City in question [6].

## 3 Use of Electric Vehicles in a Smart City

The operation of an electric vehicle is relatively simple. The energy stored in the battery is used to power the electric motor, which drives the vehicle's wheels. When the accelerator is pressed, the controller sends power from the battery to the motor, spilling the wheels and moving the vehicle forward. The kinetic energy generated by the vehicle's motion can be recovered by regenerative braking technology, which converts the kinetic energy into electrical energy and stores it in the battery.

Figure 2 shows the electric motor at the bottom of the vehicle and the lithium-ion battery at the rear. In addition, the electric charger can be seen at the front of the vehicle, which is used to charge the battery.

The advantages of electric vehicles are as follows:

- Zero emissions: Electric vehicles do not emit carbon dioxide or other greenhouse gases, making them a more environmentally friendly option.
- Lower operating cost: Electric vehicles have a lower operating cost than gasoline vehicles since electricity is cheaper than fossil fuel.
- Higher efficiency: Electric vehicles are more energy efficient than gasoline vehicles, which means they can travel more distances with the same amount of energy.

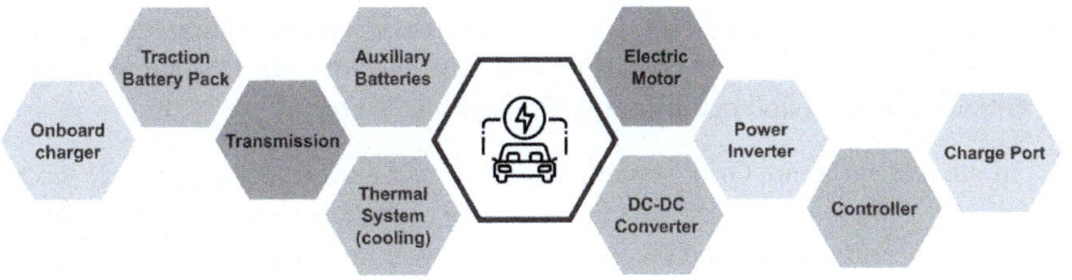


**Fig. 2** The main components of electric vehicles

- Higher performance: Electric vehicles have superior performance in terms of acceleration and speed compared to gasoline vehicles.
- Less noise: Electric vehicles produce less noise than gasoline vehicles, which can benefit urban areas.

And the disadvantages are:

- Limited range: Electric vehicles are still limited compared to gasoline vehicles, requiring more frequent recharging.
- Charging time: Electric vehicles' charging time is longer than gasoline vehicles, which can be a disadvantage on long trips.
- Higher initial cost: Electric vehicles have a higher initial cost than gasoline vehicles, although this cost decreases as technology advances.
- Limited infrastructure: The charging infrastructure for electric vehicles is less widespread than gasoline stations, making charging difficult in certain areas.
- Battery wear and tear: The battery life of electric vehicles is limited and can be affected by extreme environmental conditions or how the battery is charged.
- Electric vehicles have essential advantages such as zero greenhouse gas emissions, lower operating costs and higher efficiency. Still, they also have disadvantages such as limited range, higher initial cost and the need for a more extensive charging infrastructure.

The advantages of using a Smart Grid in a Smart City are diverse and significant. Some of the main advantages are mentioned below:

- Greater energy efficiency: Smart Grid allows more efficient and coordinated electrical energy generation, distribution, and consumption management. This translates into a reduction of energy waste and an improvement in the city's energy efficiency.
- Integration of renewable energy sources: Smart Grids can effectively integrate renewable energy sources, such as solar and wind, into the city's power grid. This allows for a reduction in greenhouse gas emissions and greater energy autonomy.
- Reduced energy costs: Implementing a Smart Grid allows utilities and end users to optimize energy consumption, which reduces energy costs for all stakeholders.
- Increased reliability and resilience: Smart Grids enable early detection and rapid resolution of power grid problems, increasing system reliability and resilience. This translates into reduced power outages and greater energy security for the city.

- Integration of electric mobility: Smart Grids are a crucial requirement for effectively integrating electric vehicles into the city's power grid. This allows for more efficient charging and better energy demand management.

In summary, Smart Grid implementation in a Smart City can provide several significant benefits, including increased energy efficiency, integration of renewable energy sources, reduced energy costs, improved reliability and resilience, and effective integration of electric mobility [4].

## 4 Interconnection Between the Smart Grid and Electric Vehicles

Electric vehicles can interact with the Smart Grid through several technologies and systems that enable efficient charging and better energy demand management. Some of how electric vehicles interact with the Smart Grid are described below:

- Smart charging: Electric vehicles can be charged through intelligent charging stations connected to the Smart Grid. These charging stations can adjust electric vehicle charging based on the grid's power demand.
- Demand management: Electric vehicles can participate in demand management programs that allow vehicle charging to be automatically adjusted based on energy supply and demand on the grid.
- Energy storage: Electric vehicles in the Smart Grid can be used as energy storage systems. This allows energy generated by renewable sources, such as solar and wind power, to be stored in the batteries of electric vehicles and used to supply the grid at times of high demand.
- Communication with the Smart Grid: Electric vehicles can communicate with the Smart Grid through two-way communication systems. This allows the grid to send signals to electric vehicles to adjust their load or release energy stored in their batteries.

In short, electric vehicles can interact with the Smart Grid through various technologies and systems that enable efficient charging and better energy demand management. This contributes to building a more efficient and sustainable electric grid and improving electric vehicle integration in a Smart City [12].

The Smart Grid can leverage the battery of electric vehicles in various ways as an energy storage system for the power grid to manage energy demand and reduce energy costs at peak times. Some of these ways are described below:

- Energy storage system: the battery in electric vehicles can be used as an energy storage system for the power grid. Electric vehicles can be charged through the grid during low energy demand and store the energy in their batteries. At times of high energy demand, the energy stored in the batteries of electric vehicles can be released and used to supply the grid.

- Demand management: Electric vehicles can be programmed to charge at times of low energy demand, such as in the evening hours when energy costs are lower. This helps reduce energy demand at peak times and can help reduce overall energy costs.
- Cost reduction: The battery in electric vehicles can reduce energy costs during peak hours. During high energy demand, when energy costs are highest, electric vehicles can be programmed to discharge the energy stored in their batteries to the grid, thereby reducing energy demand and costs.

In summary, the battery of electric vehicles can be utilized by the Smart Grid in a variety of ways to improve grid efficiency and reduce energy costs. Using the battery of electric vehicles as an energy storage system and to manage energy demand can help create a more efficient and sustainable electric grid in a Smart City [3].

Integrating electric vehicles in a Smart City with Smart Grid can offer many benefits, both for the users and the city in general. Some of these benefits are detailed below:

- Reduced pollution and improved air quality: electric vehicles do not emit exhaust gases that harm the environment and human health. Therefore, their use can help reduce air pollution in the city and improve air quality.
- Reduced dependence on oil and energy diversification: Electric vehicles run on electricity, reducing dependence on oil and diversifying energy sources in the city.
- Reduced energy costs: Using electric vehicles with a Smart Grid can enable more efficient management of energy demand and renewable energy, reducing energy costs in the city.
- Reduced traffic congestion: Electric vehicles can be part of a more efficient and sustainable public and private transportation system, which can help reduce traffic congestion in the city.
- Improved quality of life for citizens: Reduced air pollution, energy diversification, reduced energy costs, and improved public and private transportation can contribute to improving the quality of life for Smart City citizens.

In summary, integrating electric vehicles in a Smart City with a Smart Grid can offer several significant benefits to the city and its inhabitants, including reduced pollution, energy diversification, energy costs, reduced traffic congestion, and improved quality of life. Electric vehicle charging infrastructure is an integral part of a Smart Grid. Proper implementation of charging infrastructure can ensure availability and access to charging at all times, which is crucial to encourage the adoption of electric vehicles in the Smart City. The following are some of the characteristics that a Smart Grid's electric vehicle charging infrastructure should have:

- Public charging stations: the Smart Grid must have enough public charging stations strategically distributed throughout the city to ensure charging availability for all EV users.
- Private charging stations: In addition to public charging stations, the Smart Grid should also have private charging stations for businesses and households using

electric vehicles. These private charging stations can be installed in company parking lots, residential buildings and garages.

- Advanced technology: The charging infrastructure must have advanced technology that enables fast and efficient charging of electric vehicles. This includes the use of high-capacity plugs and charging sockets and fast charging technology.
- Accessibility: Charging stations must be accessible to all users, including those with physical disabilities. This means charging stations must have accessibility features like ramps and unique parking spaces.
- Integration with the Smart Grid: Charging infrastructure must be integrated with the Smart Grid to enable efficient load management and the use of renewable energy. This may include power demand management and electric vehicle charging management during periods of low power demand.

In summary, the charging infrastructure for electric vehicles by a Smart Grid must be accessible, have advanced technology, be strategically distributed, have accessibility features, and be integrated with the Smart Grid. With these features, availability and access to charging can be guaranteed at all times, which is essential to encourage the adoption of electric vehicles in the Smart City. To match the power grid and energy demand in a Smart Grid for electric vehicles in a Smart City, the following aspects must be taken into account:

- Proper planning: planning must be detailed and long-term to foresee future energy demand and ensure that the power grid can support the additional load of electric vehicles.
- Use of renewable energy: Using renewable energy to charge electric vehicles should be encouraged, which will help reduce dependence on fossil fuels and greenhouse gas emissions.
- Demand management: The Smart Grid must be able to manage the energy demand of electric vehicles and ensure that they are charged at times of lower energy demand. This avoids overloading the grid and optimizes its use.
- Intelligent charging technology: The charging infrastructure must be able to manage the charging of electric vehicles intelligently; this allows charging to be scheduled based on energy availability and the time of day they are charged.
- Flexible charging rates: Charging rates must be flexible and adjust to the charging schedules of electric vehicles. In this way, users are incentivized to charge their vehicles at times of lower energy demand and avoid overloading the electric grid.

In general, matching the power grid and energy demand in a Smart Grid for electric vehicles in a Smart City involves intelligent management of energy and charging infrastructure to optimize its use and reduce its environmental impact [2].

## 5 Benefits of Integrating Electric Vehicles in a Smart City with Smart Grid

The costs of implementing and adopting the grid and energy demand of a Smart Grid for a large group of electric vehicles in a Smart City can be high. Still, they can vary depending on several factors, such as:

- Size of the Smart City: the size of the Smart City will influence the expected number of electric vehicles and the amount of charging infrastructure that will be needed. A larger Smart City will require a more significant investment in the electric grid and charging infrastructure.
- Existing infrastructure: The existing electrical infrastructure in the city can affect the costs of implementing and adopting the Smart Grid's electrical grid and energy demand. If a robust grid and adequate charging infrastructure exist, costs will be lower.
- Charging technology: Charging technology costs vary depending on the type and quality of equipment. For example, fast charging may be more expensive than slow charging.
- Use of renewable energy: Investing in renewable energy, such as solar and wind, may increase the initial costs of implementing a Smart Grid, but in the long run, it can be a profitable investment.
- Government policies and regulations: Government policies and regulations can affect the costs of implementing and adopting a Smart Grid. For example, tax incentives and subsidies can reduce costs, while regulatory restrictions can increase costs.

It is important to note that although upfront costs can be significant, implementing a Smart Grid for electric vehicles in a Smart City can have long-term benefits, such as reducing dependence on fossil fuels and greenhouse gas emissions, which can contribute to sustainability and air quality in the city. In addition, using electric vehicles can have economic benefits, such as increased energy efficiency and decreased maintenance and operating costs in the long term. The cost of implementation and adoption of the electric grid and the energy demand of a Smart Grid for a group of 10,000 electric vehicles in a Smart City can vary depending on several factors, such as the size of the city, the number of charging stations needed, the current electric infrastructure, and others. The investment required to establish an electric grid suitable for electric vehicles can be high, but the long-term benefits can be significant. Below are some of the main costs to consider [11]:

- Charging infrastructure: an adequate network of charging stations is essential for the widespread use of electric vehicles. The cost of installing these stations varies depending on the type of charger and its capacity and can range from $5,000 to $100,000 per unit. Generally, one charging station for every 10-20 electric vehicles is recommended.
- Upgrading the electrical grid: To meet the power demand of electric vehicles, the existing electrical infrastructure may need to be upgraded. This may include

the installation of additional transformers, power lines, and other electrical equipment. The cost of these upgrades varies depending on the specific situation but is estimated to range from $1,000 to $10,000 per unit.
- Maintenance: Regular maintenance of the electrical infrastructure and charging stations is essential to ensure the proper operation of the grid. This includes regular inspection of equipment, repair of breakdowns, and cleaning of charging stations. The maintenance cost varies depending on the type of equipment but is estimated to range from $100 to $500 per unit.
- Training and personnel: Trained personnel will be needed to install and maintain the electrical infrastructure and charging stations. In addition, additional employees will be needed to manage and monitor the electrical grid and its integration with electric vehicles. The cost of training and staffing can vary depending on the size of the city and the number of charging stations but is estimated to range from $500 to $2,000 per employee.

## *5.1 Objective Function Proposal*

The objective function that could be considered to evaluate the Smart Grid model suitable with the ecological footprint, agile to implement, optimal for predictions of energy demand and ideal for electric vehicles in a Smart City, considering a population of 1.57 million inhabitants and 10 thousand electric vehicles, could be the following.

Minimize:

- The total cost of implementation and operation of the Smart Grid (including the installation of infrastructures and the acquisition of equipment and technologies necessary for the implementation of the model), considering the objectives of environmental sustainability, energy efficiency and optimal energy demand.
- The greenhouse gas emissions and ecological footprint of the Smart Grid, including the use of renewable energy sources and energy efficiency practices.
- The response time to energy demands and the capacity to supply energy to users through the prediction of demand and optimization of the generation and distribution of electrical energy in the Smart Grid.
- The Smart Grid's users' satisfaction, considering the availability and accessibility of electrical energy, the quality of service, and the stability and security of the network.
- The promotion of the use of electric vehicles in the Smart City through the installation of charging stations, incentives for their use and the integration of electric vehicles into the Smart Grid network.

In general, the target function should balance reducing costs and promoting environmental sustainability, energy efficiency and satisfaction of Smart Grid users while promoting the use of electric vehicles in the Smart City. The target function for the

Smart Grid model is suitable with the ecological footprint, agile to implement, optimal for energy demand predictions and ideal for electric vehicles in a Smart City; considering a population of 1.57 million inhabitants and ten thousand electric vehicles where:

C Total cost of implementation and operation of the Smart Grid
E Greenhouse gas emissions and ecological footprint
T Response time to energy demands and power supply capacity to users
S Smart Grid user satisfaction
V Promoting the use of electric vehicles in the Smart City

The objective function can be expressed mathematically in Eq. 1:

$$F(C, E, T, S, V) = a_1C + a_2E + a_3T + a_4S + a_5V \tag{1}$$

where $a_1, a_2, a_3, a_4, and a_5$ are the weights associated with each variable, which can be determined according to the priorities and specific objectives of the Smart Grid model. The value of the objective function F represents the overall balance of the objectives and priorities of the Smart Grid model. It seeks to minimize this value to achieve an optimal solution. In summary, implementing and adopting an electric grid suitable for 10,000 electric vehicles in a Smart City may require a significant upfront investment. However, the long-term benefits, such as reduced greenhouse gas emissions, improved air quality, and fuel cost savings, can far outweigh the initial costs. In Fig. 3, you can see the seven segmentation radii of the location of the connectors for electric vehicles considering the year of 2037 and 10,000 Electrical Vehicles [2].

To program the target function in Python, you can define the smart grid objective function with the following parameters: C: total cost of implementation and operation of the Smart Grid; E: greenhouse gas emissions and ecological footprint; T: response time to energy demands and power supply capacity to users; S: satisfaction of Smart Grid users; V: promotion of the use of electric vehicles in the Smart City.

To calculate V, there are three variables needed, Population $(P)$: Number of inhabitants of the city; Electric vehicles $(E_V)$: number of electric vehicles in the city; and Average Use $(\mu)$: average daily use of electric vehicles in hours.

$$V = \frac{\mu E_V}{P} \tag{2}$$

The result of our objective function is shown in Fig. 4; the red bar is represented in Megawatt hours, the green bar line is the increase in electric vehicles in a Smart City for a time horizon and finally, the blue bar is represented in Mexican pesos considering the fluctuating price of electric vehicles for the said period.

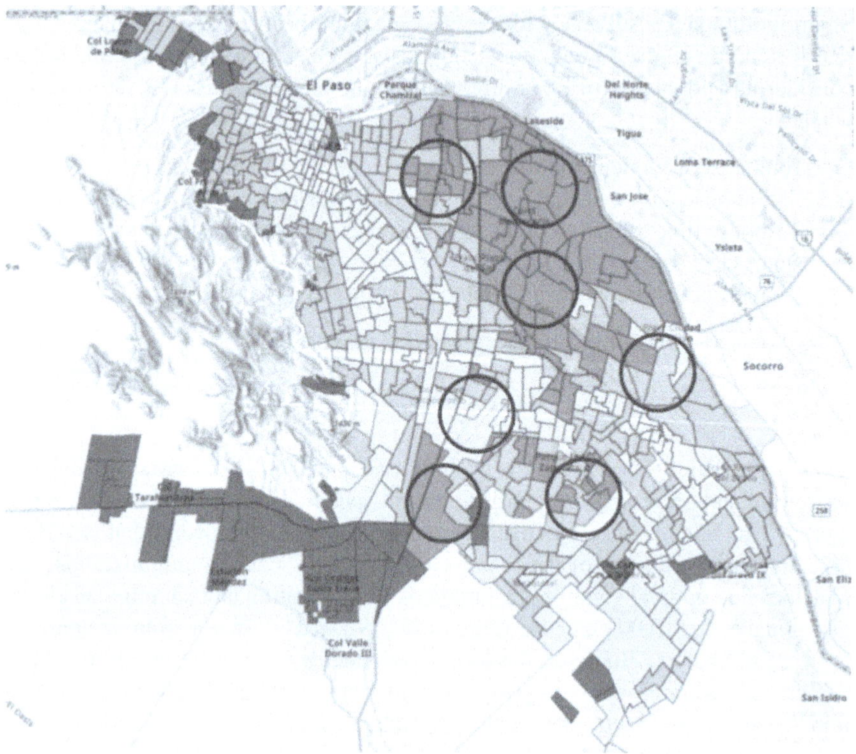


**Fig. 3** Segmentation by distribution pole of the location of 10,000 electric vehicles associated with a smart city and their need for electrical energy through a smart grid

## 5.2 *Simulate Consumption Within the City*

It is important to analyze how EVs will be growing within the population. Ciudad Juárez, Chihuahua, México (CJS) has a population of 1,512,450 inhabitants. This represents the 40.42% of Chihuahua State population. Also, the number of vehicles (particular) in CJS is 46,808 (as of October 2023) and there are almost 5,366 EVs starting January 2024. In addition, based on INEGI, the national car growth rate per month is between 0.12 and 0.14%. This is the growth rate used for the simulation, which for EVs will be 150 cars per month.

The period of time studies is 2024-01-01 through 2037-12-31, that is 156 months or 5114 days. The simulation will consider growing car fleet month by month. In other words, each month has a constant number of cars. Although the growth rate is 150, it is not feasible to have a constant number of cars. Therefore, for the simulation, the growth rate will be the number of cars in the prvious month multiplied by growth factor and add a random number of extra cars from zero to 40.

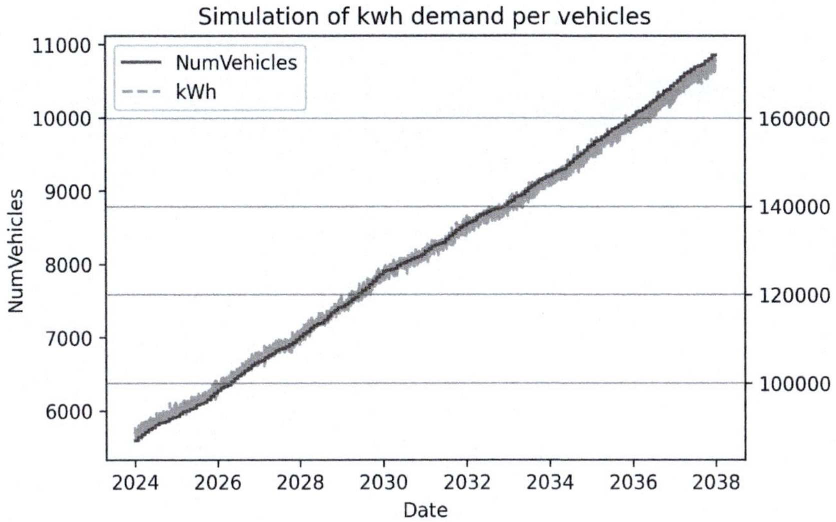


**Fig. 4** Results of simulation about energy demand per vehicle

Then, for each day, each car will drive diferent distances. The car can drive between zero to 150 kilometers. More than 150km is not a realistic distance, as the city is not that big. For instance, Fig. 5 shows a very long drive of 81.6km, thus 150km is a safe value as the maximum kilometers drove by a car per day.

With this simulation, it is then easy to calculate the number of kWh demanded by day, only for EVs. To have autonomy of 160.9 km, it needs 34 kWh. Figure 4 shows the results of energy required in adition to growing line of EVs. It is normal to have a direct relationship between these two variables.

Even further, the $CO_2$ emisions saved has a inverse relationship, saving 0.192 kg of $CO_2$ saved per km [10].

### *5.3 Challenges and Obstacles for the Integration of Electric Vehicles in a Smart City with Smart Grid*

Integrating electric vehicles in a Smart City with Smart Grid presents several challenges and obstacles. Here are some of the most important:

- Charging infrastructure: One of the biggest challenges for integrating electric vehicles in a Smart City is the charging infrastructure. Many charging points need to be installed to meet the demand for electric vehicles, requiring significant investment in charging infrastructure. Additionally, charging infrastructure must be located strategically, which can be challenging due to the need for more space and competition for land use.

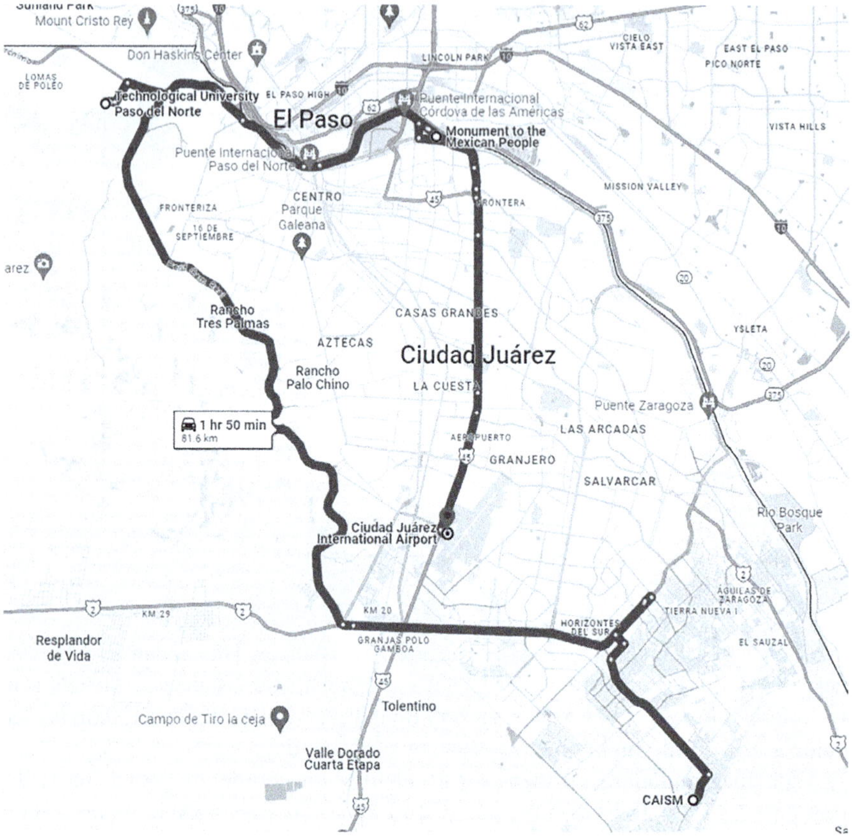


**Fig. 5** Example of one of largest driving routes in Ciudad Juarez

- Interoperability: Another major challenge is interoperability between charging systems and electric vehicles. It is necessary to establish common standards and ensure that charging systems are compatible with all models of electric vehicles. In addition, it is essential to ensure that charging systems are safe and reliable.
- Load demand management: Load demand management is a significant challenge in a Smart City with Smart Grid. It is necessary to balance the supply and demand of electricity to ensure that the electrical grid is not overloaded and remains stable. In addition, it is necessary to develop intelligent load demand management systems to optimize energy use and minimize costs.
- Costs: Electric vehicles are still more expensive than internal combustion engines, which may discourage consumers from buying them. Additionally, charging infrastructure can be expensive to install and maintain, and costs can significantly hinder the adoption of electric vehicles in a Smart City.

- Regulation and Politics: Regulation and politics can also be obstacles to integrating electric vehicles in a Smart City. It is necessary to establish policies and regulations that encourage the adoption of electric vehicles, such as fiscal and financial incentives for users and the elimination of access barriers to the electrical network. In addition, it is essential to establish security and privacy regulations to ensure that charging systems are safe and reliable.

In summary, integrating electric vehicles in a Smart City with a Smart Grid presents significant challenges regarding charging infrastructure, interoperability, charging demand management, costs, regulation, and policy. It is essential to address these challenges to ensure a successful and sustainable integration of electric vehicles in a Smart City.

### *5.4 Case Studies and Examples of Successful Implementation*

Here are some examples of successful Smart Grid implementations for electric vehicles in different societies and from different parts of the world:

- Amsterdam, The Netherlands: Amsterdam has one of the largest EV fleets in Europe and has implemented a smart grid for EV charging. The city has installed intelligent charging stations that automatically adjust charging based on power demand and weather conditions.
- California, United States: California has launched an initiative called ChargePoint America, which has deployed more than 4,600 electric vehicle charging stations in the region. The initiative has achieved significant adoption of electric vehicles in the area.
- Tokyo, Japan: Tokyo has implemented a smart grid for electric vehicle charging. The grid uses renewable energy, such as solar panels, to power charging stations and reduce reliance on fossil fuels.
- Malmö, Sweden: The Swedish city of Malmö has implemented an intelligent electric vehicle charging grid in collaboration with the energy company E.ON. The network uses fast charging technology and includes renewable energy storage systems.
- Dubai, United Arab Emirates: The city of Dubai has launched an initiative called "Green Charger" to deploy electric vehicle charging stations throughout the city. The initiative has been a success and has helped foster the adoption of electric vehicles in the city.

In short, these are just a few examples of cities that have successfully implemented an innovative EV charging grid. These cases show how Smart Grid technology can help drive electric vehicle adoption worldwide while reducing reliance on fossil fuels and promoting renewable energy.

In addition, several cities around the world have successfully implemented electric vehicles and Smart Grid; here are some examples:

- Amsterdam, The Netherlands: Amsterdam has been recognized for its sustainability and electric mobility approach. The city has more than 3,000 electric vehicle charging stations. It has implemented a Smart Grid system that allows drivers to charge their vehicles using renewable energy, such as solar and wind power.
- San Diego, United States: San Diego has been a pioneer in the implementation of electric vehicles and Smart grids. The city has installed more than 300 electric vehicle charging stations and has implemented a Smart Grid system that allows drivers to charge their vehicles at times of low energy consumption.
- Stockholm, Sweden: Stockholm has implemented a Smart Grid system that allows electric vehicle drivers to charge their vehicles using renewable energy generated by the city. In addition, the city has established a fleet of electric taxis and has developed a public transport plan that encourages using electric vehicles.
- Barcelona, Spain: Barcelona has implemented a network of electric vehicle charging stations throughout the city and has established incentives to encourage using electric vehicles. The city has also developed a Smart Grid system that allows drivers to charge their vehicles using locally generated renewable energy.

These examples demonstrate that implementing electric vehicles and Smart Grid in a Smart City can be successful if planned and implemented effectively, considering factors such as infrastructure, energy demand, costs, and adoption. by the users.

## 6 Conclusions, Recommendations, and Future Research

Implementing electric vehicles and Smart Grid in a Smart City can be a complex challenge requiring careful planning and coordination between stakeholders. Below are some recommendations to facilitate the successful implementation of electric vehicles and Smart Grid in a Smart City: Establish a comprehensive strategy: A comprehensive strategy should be established that involves all stakeholders, including local authorities, energy companies, electric vehicle manufacturers, charging infrastructure providers, and citizens. This strategy must consider charging infrastructure, system interoperability, security and privacy. Develop a detailed action plan: A detailed action plan should include the objectives, deadlines, necessary resources and milestones for implementing electric vehicles and Smart Grid in a Smart City. Identify and address regulatory and market barriers: It is essential to identify and address regulatory and market barriers that may hinder the adoption of electric vehicles and Smart grids. This may include creating fiscal and financial incentives for users, the removal of access barriers to the electricity grid, and promoting technical and safety standards.

Develop adequate charging infrastructure: An adequate and accessible charging infrastructure for electric vehicles must be developed throughout the Smart City. This can include the installation of charging points in public and private places, the implementation of payment systems and the use of intelligent technologies to managing to charge demand. Promote education and public awareness: Education and public awareness about the benefits of electric vehicles and Smart grids should be promoted. This can include public awareness campaigns, the organization of events and the promotion of incentives for adopting electric vehicles. In summary, successfully implementing electric vehicles and Smart Grid in a Smart City requires careful planning, coordination between stakeholders, and addressing regulatory and market barriers. Creating adequate and accessible charging infrastructure and promoting education and public awareness are also critical to facilitating the adoption of this technology.

In future research, the integration of technology with a focus on the human factor in Industry 4.0 necessitates a comprehensive approach to virtual training. This is crucial in order to effectively combine a range of capabilities and determine an optimal learning curve to supplement the tacit knowledge involved in product development [9]. This will be essential for meeting the needs of Generation Z and resulting in a paradigm shift in operations. To ensure success, it will be imperative to identify the necessary resources while also considering future sustainability and minimizing any negative impact on the environment.

## References


1. Alomar, M.A.: An IoT based smart grid system for advanced cooperative transmission and communication. Phys. Commun. **58**, 102069 (2023)
2. Couillet, R., Perlaza, S.M., Tembine, H., Debbah, M.: Electrical vehicles in the smart grid: a mean field game analysis. IEEE J. Selected Areas Commun. **30**(6), 1086–1096 (2012)
3. De Ridder, F., D'Hulst, R., Knapen, L., Janssens, D.: Applying an activity based model to explore the potential of electrical vehicles in the smart grid. Procedia Comput. Sci. **19**, 847–853 (2013)
4. Hadizade, A.: An agent-based model for optimal voltage control and power quality by electrical vehicles in smart grids. In: Distributed Computing and Artificial Intelligence, Special Sessions, 15th International Conference 15, pp. 388–394. Springer (2019)
5. Hasan, M.K., Habib, A.A., Shukur, Z., Ibrahim, F., Islam, S., Razzaque, M.A.: Review on cyber-physical and cyber-security system in smart grid: standards, protocols, constraints, and recommendations. J. Netw. Comput. Appl. **209**, 103540 (2023)
6. Hou, K., Xu, X., Jia, H., Yu, X., Jiang, T., Zhang, K., Shu, B.: A reliability assessment approach for integrated transportation and electrical power systems incorporating electric vehicles. IEEE Trans. Smart Grid **9**(1), 88–100 (2016)
7. Kumar, R.S., Saravanan, S., Pandiyan, P., Tiwari, R.: Impact of artificial intelligence techniques in distributed smart grid monitoring system. In: Smart Energy and Electric Power Systems, pp. 79–103. Elsevier (2023)
8. Mythreyee, M., Nalini, A.: Genetic algorithm based smart grid system for distributed renewable energy sources. Comput. Syst. Sci. Eng. **45**(1) (2023)
9. Noruzi Azghandi, M., Shojaei, A.A., Toosi, S., Lotfi, H.: Optimal reconfiguration of distribution network feeders considering electrical vehicles and distributed generators. Evol. Intell. **16**(1), 49–66 (2023)
10. Ritchie, H., Roser, M.: Cars, planes, trains: where do co2 emissions from transport come from? Our world in data (2023)

11. Sanduleac, M., Eremia, M., Toma, L., Borza, P.: Integrating the electrical vehicles in the smart grid through unbundled smart metering and multi-objective virtual power plants. In: 2011 2nd IEEE PES International Conference and Exhibition on Innovative Smart Grid Technologies, pp. 1–8. IEEE (2011)
12. Shi, Y., Tuan, H.D., Savkin, A.V.: Mixed integer nonlinear programming for joint coordination of plug-in electrical vehicles charging and smart grid operations. In: 2019 IEEE International Conference on Communications, Control, and Computing Technologies for Smart Grids (SmartGridComm), pp. 1–6. IEEE (2019)

# Integrated Distribution and Warehousing Optimization for Motorcycle Parts Using Dijkstra Algorithm, Container Packaging and Order Picking

**Daniel Moreno-Galíndez, Miguel Ángel Ruíz-Jaimes, Irma Yazmín Hernández-Báez, Samantha Abigail Muñiz-Domínguez, and Alberto Ochoa-Zezzatti**

**Abstract** Optimizing spare parts management in motorcycle stores is paramount for maintaining operational fluidity and guaranteeing customer satisfaction. This research delves into the application of Dijkstra's algorithm for route planning, complemented by the container packing distribution model and the order picking concept within parts warehousing. This synergistic integration not only ensures a comprehensive optimization but also holds the transformative potential to revolutionize both delivery and warehousing operations within the motorcycle industry. By leveraging Dijkstra's algorithm, we aim to enhance the efficiency of route planning, facilitating streamlined transportation processes. Simultaneously, the container packing distribution model contributes to spatial efficiency, minimizing wastage and maximizing storage capacity. Furthermore, the incorporation of the order picking concept refines the retrieval process, reducing lead times and increasing overall operational agility. This holistic approach, combining route planning, container packing, and order picking, presents a paradigm shift in spare parts management, promising unparalleled efficiency gains. As industries evolve, embracing such optimization strategies becomes imperative for staying competitive and meeting the growing demands of a dynamic market.

D. Moreno-Galíndez · S. A. Muñiz-Domínguez
Instituto Tecnológico de Tláhuac, Mexico City, Mexico
e-mail: samntha.md@tlahuac.com.tecnm.mx

M. Á. Ruíz-Jaimes · I. Y. Hernández-Báez (✉)
Universidad Politécnica del Estado de Morelos, Jiutepec, Mexico
e-mail: ihernandez@upemor.edu.mx

M. Á. Ruíz-Jaimes
e-mail: mruiz@upemor.edu.mx

A. Ochoa-Zezzatti
Universidad Autónoma de Ciudad Juárez, Ciudad Juarez, Mexico
e-mail: alberto.ochoa@uacj.mx



D. Oliva et al. (eds.), *Advances in Optimization Algorithms for Multidisciplinary Engineering Applications: From Classical Methods to AI-Enhanced Solutions*, Studies in Computational Intelligence 806, https://doi.org/10.1007/978-3-031-78440-8_25

**Keywords** Optimization · Dijkstra algorithm · Order picking · Bin packing

## 1 Introduction

In the motorcycle industry, there is a constant loss of competitiveness on a yearly basis. This challenge is mainly attributed to the lack of systems to optimize the distribution of motorcycle parts during critical periods. The complications of moving within a busy city, the need to consider the demand by prioritizing the most popular points of sale for specific spare parts, and the efficient arrangement of delivery units highlight the urgency of an innovative solution in this area. Consequently, the proposal to employ a logistics optimization system opens the door to a significant transformation in the routing and distribution process. The inclusion of established logistics approaches such as Order Picking and Bin Packing, supported by the Dijkstra Algorithm approach, promises an effective optimization of the distribution of products within time windows, thus improving the efficiency of warehousing and delivery operations.

In the fiercely competitive motorcycle industry, businesses face a constant struggle to maintain their market share. A critical obstacle hindering their success lies in the inefficient distribution of spare parts, particularly during peak demand periods. Traditional methods often fall short, resulting in sluggish delivery times and stock-outs at crucial moments. Navigating congested urban environments further complicates the issue. Additionally, prioritizing deliveries to high-volume points of sale for specific parts requires a nuanced approach that considers both location and demand. Efficient packing of delivery vehicles to maximize space utilization presents another challenge [5, 6].

Current approaches to motorcycle parts distribution often rely on manual route planning and rudimentary storage practices. While these methods may have sufficed in the past, they struggle to keep pace with the demands of a dynamic market. Manual route planning is time-consuming and prone to errors, leading to inefficient delivery times. Traditional storage methods may not optimize space utilization, resulting in wasted storage capacity and difficulty locating parts quickly.

This research proposes a novel solution that integrates three key components: Dijkstra's algorithm for optimized route planning, container packing distribution models for efficient space utilization, and the order picking concept for streamlined retrieval of parts. By leveraging these elements, we aim to achieve significant improvements in the efficiency of spare parts distribution within the motorcycle industry.

The anticipated results of this research include: (a) Reduced delivery times: Through optimized route planning using Dijkstra's algorithm, parts will reach their destinations faster, enhancing customer satisfaction, (b) Increased storage efficiency: Container packing models will minimize wasted space in delivery vehicles, allowing for more parts to be transported per trip and (c) Improved order fulfillment: The order picking concept will streamline parts retrieval from the warehouse, reducing lead times and improving overall operational agility.

This article is organized as follows: It begins with an introduction that presents the context and the problem being addressed. Then, a review of Dijkstra's algorithm is presented, the main section of the article describes the proposed methodology and the results obtained. Finally, conclusions are presented and future directions for research are suggested.

## 2 Dijkstra's Algorithm for Efficient Route Planning

Dijkstra's algorithm is a widely used pathfinding algorithm for finding the shortest path between two nodes in a graph. It was developed by Dutch computer scientist Edsger Dijkstra in 1956 [4]. Dijkstra's algorithm is a popular choice for solving routing and optimization problems due to its efficiency and simplicity.

Dijkstra's algorithm works by maintaining a set of tentative distances from the source node to each other node in the graph. It starts by marking all nodes as unvisited and assigning them an infinite distance. The algorithm then selects the unvisited node with the smallest tentative distance, marks it as visited, and updates the tentative distances of its neighbors. This process is repeated until the destination node is reached, at which point the shortest path from the source node to the destination node has been found.

Dijkstra's algorithm has several applications in spare parts management, particularly in optimizing warehouse operations and delivery routes. It can be used to:

**Plan efficient routes for picking and packing spare parts**: Dijkstra's algorithm can be used to determine the most efficient route for warehouse workers to pick and pack spare parts, minimizing travel time and maximizing productivity.

**Optimize delivery routes for spare parts distribution**: Dijkstra's algorithm can be used to plan the most efficient routes for delivery trucks to distribute spare parts to dealerships or customers, reducing transportation costs and improving delivery times.

**Minimize travel distance and time in warehouse operations**: By using Dijkstra's algorithm to optimize routes within the warehouse, businesses can reduce the overall travel distance and time spent by employees, leading to increased efficiency and productivity.

Dijkstra's algorithm can be integrated with other optimization techniques, such as the container packing distribution model and the order picking concept, to create a comprehensive spare parts management system. For instance, Dijkstra's algorithm can be used to determine the most efficient route for picking spare parts, while the container packing distribution model can be used to optimize the allocation of spare parts to containers, and the order picking concept can be used to streamline the retrieval process.

## 3 Methodology

In the motorcycle parts manufacturing and distribution industry in the Tláhuac delegation, a problem arises that affects the efficiency and competitiveness of the sector. The lack of systems to optimize the distribution process of spare parts during crucial time intervals represents a significant obstacle. In addition, the road complexities of Tláhuac, efficient packaging, compliance with time windows and demand by concessionaires, aggravate the difficulty of improving the logistics operation.

By examining the characteristics of the motorcycle spare parts distribution environment in the Tláhuac delegation, considering temporal and spatial limitations together with a data collection on the distribution units we will be able to use specialized mathematical models that allow the realization of Bin Packing and Dijkstra Algorithms validating that their characteristics are adequate for the resolution of our problem.

Evaluate the effectiveness of the system by comparing the results with conventional distribution methods to quantify improvements in terms of operational efficiency and time window compliance.

The Fig. 1 shows the expected operational diagram of the concluded optimization system.

This research implements a synergistic integration of three methods to optimize spare parts management in motorcycle stores: order picking concept, the container packing distribution model and Dijkstra's algorithm.

**Data Acquisition**. The first stage involved data acquisition. We assumed that the orders are already placed in the warehouse. Information regarding the location of each spare part within the warehouse was essential for this process.

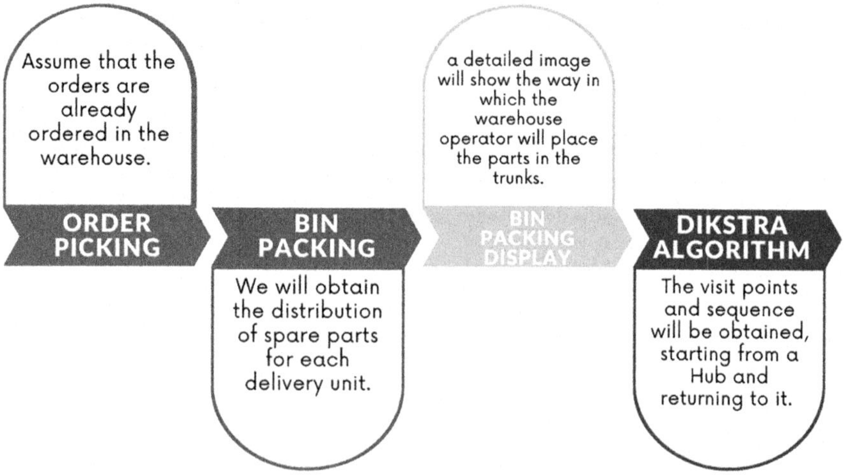


**Fig. 1** System operating diagram

**Container Packing Distribution Model**. Following the route planning stage, the container packing distribution model was implemented to assign spare parts to bins. This ensured that each bin was not overloaded and that parts were efficiently distributed throughout the bins to minimize wasted space.

**Dijkstra's Algorithm for Route Planning**. Dijkstra's algorithm was then employed to determine the most efficient route for the warehouse operator to pick the required spare parts. The algorithm identified the visit points and sequence, starting from a designated hub within the warehouse and returning to it after all parts are collected.

**Integration and Evaluation**. The proposed methodology integrates these three components to achieve an optimized spare parts management system. The effectiveness of this methodology will be evaluated through metrics such as picking time, travel distance within the warehouse, and storage space utilization.

## 4 Data Collection

Using the Botsol Crawler software (a program used for scraping or extracting data from Google Maps), we were able to obtain a list of 19 different auto parts stores in Tláhuac Borough along with their respective geographical locations. These were taken as the starting point for the beginning of our methodology (Fig. 2).

After collecting data with the help of the Botsol Crawler, we used Google Earth to create a map showing the concessionaires, we made the decision to divide the Tláhuac area into two sectors. We assigned a Hub to each of them in strategic locations to reduce the distances and, consequently, the time needed to move around the two sectors (Fig. 3).

The Tables 1 and 3 is a list of the distribution of the refactoring depots with matrix tables of distances in kilometers, the distances were calculated as Euclidean

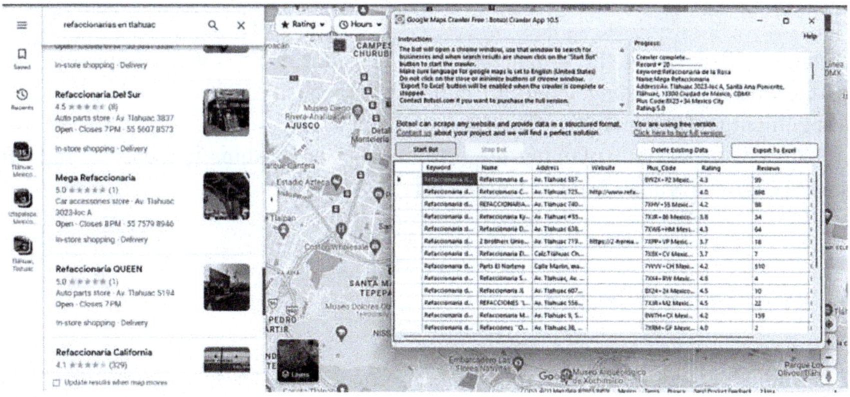


**Fig. 2** Data collection by Botsol Crawler

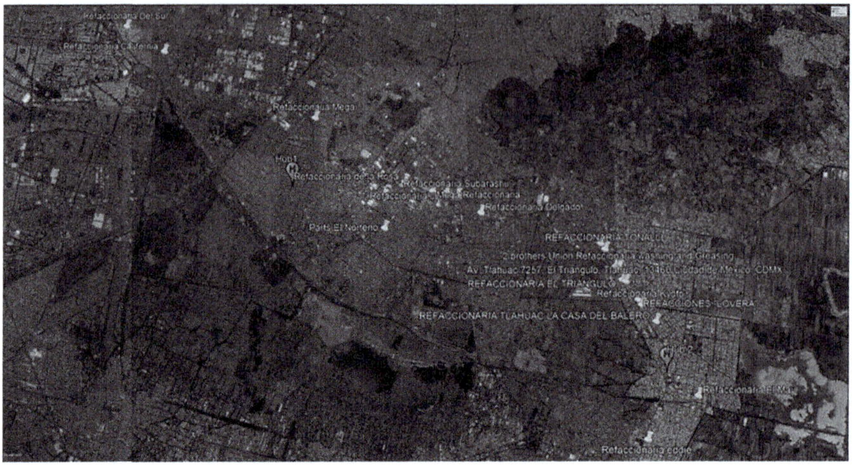


**Fig. 3** Distribution map of dealerships in the Tláhuac delegation

**Table 1** Distances between Hub 1 and concessionaires in Zone 1

| Units [Km] | Hub1 | A | B | C | D | E | F | G | H |
|---|---|---|---|---|---|---|---|---|---|
| Hub1 | 0 | 0,6 | 0,15 | 1,09 | 0,16 | 1,27 | 0,59 | 2,51 | 1,29 |
| A | 0,06 | 0 | 0,21 | 1,15 | 0,22 | 1,33 | 0,55 | 2,57 | 1,23 |
| B | 0,15 | 0,21 | 0 | 0,94 | 0,37 | 1,13 | 0,8 | 2,68 | 1,09 |
| C | 1,09 | 1,15 | 0,94 | 0 | 0,77 | 1,4 | 0,88 | 3,29 | 0,84 |
| D | 0,16 | 0,22 | 0,37 | 0,77 | 0 | 1,5 | 0,85 | 2,76 | 1,02 |
| E | 1,27 | 1,33 | 1,13 | 1,4 | 1,5 | 0 | 1,49 | 1,39 | 2,09 |
| F | 0,59 | 0,55 | 0,8 | 0,88 | 0,85 | 1,49 | 0 | 2,12 | 0,71 |
| G | 2,51 | 2,57 | 2,68 | 3,29 | 2,76 | 1,39 | 2,12 | 0 | 3,8 |
| H | 1,29 | 1,23 | 1,09 | 0,84 | 1,02 | 2,09 | 0,71 | 3,8 | 0 |

distances. To improve the visualization of the tables, a letter was assigned to each of the dealerships, using letters to indicate their position in the table. The letter assignment is show in Tables 2 and 4.

We will focus on explaining the resolution by algorithms in Zone 1, since the same programs will be used in both areas.

## 5 Hub's Order Picking

The implementation of the Order Picking concept in the spare parts warehouse is essential to speed up the distribution process. By strategically arranging parts according to their demand and accessibility, the time required to pick parts and

**Table 2** Letter assignment for Hub 1

| Refactoring depots | Letter |
|---|---|
| Refaccionaria de la Rosa | A |
| Refaccionaria California | B |
| Refaccionaria Tláhuac La Casa del Balero | C |
| Refaccionaria Kyoto | D |
| Refaccionaria Delgado | E |
| 2 brothers Union Refaccionaria washing and Greasing | F |
| Refaccionaria El May | G |
| Parts El Norteño | H |

**Table 3** Distances between Hub 2 and concessionaires in Zone 2

| Units [Km] | Hub2 | A | B | C | D | E | F | G | H | I | J |
|---|---|---|---|---|---|---|---|---|---|---|---|
| Hub2 | 0 | 1,15 | 0,09 | 0,22 | 0,14 | 0,09 | 0,11 | 0,9 | 0,12 | 0,13 | 0,29 |
| A | 0,15 | 0 | 0,24 | 0,37 | 0,26 | 0,15 | 0,11 | 0,25 | 0,03 | 0,2 | 0,22 |
| B | 0,09 | 0,24 | 0 | 0,13 | 0,11 | 0,12 | 0,22 | 0,09 | 0,24 | 0,18 | 0,07 |
| C | 0,22 | 0,37 | 0,13 | 0 | 0,14 | 0,24 | 0,28 | 0,32 | 0,41 | 0,55 | 0,07 |
| D | 0,14 | 0,26 | 0,11 | 0,14 | 0 | 0,21 | 0,16 | 0,22 | 0,37 | 0,41 | 0,15 |
| E | 0,09 | 0,15 | 0,12 | 0,24 | 0,21 | 0 | 0,17 | 0,15 | 0,23 | 0,12 | 0,29 |
| F | 0,11 | 0,11 | 0,22 | 0,28 | 0,16 | 0,17 | 0 | 0,15 | 0,16 | 0,05 | 0,34 |
| G | 0,09 | 0,25 | 0,09 | 0,32 | 0,22 | 0,15 | 0,15 | 0 | 0,2 | 0,26 | 0,36 |
| H | 0,12 | 0,03 | 0,24 | 0,41 | 0,37 | 0,23 | 0,16 | 0,2 | 0 | 0,21 | 0,37 |
| I | 0,13 | 0,2 | 0,18 | 0,55 | 0,41 | 0,12 | 0,05 | 0,26 | 0,21 | 0 | 0,48 |
| J | 0,29 | 0,22 | 0,07 | 0,07 | 0,15 | 0,29 | 0,34 | 0,36 | 0,37 | 0,48 | 0 |

**Table 4** Letter assignment for Hub 2

| Refactoring depots | Letter |
|---|---|
| Refaccionaria JL | A |
| Refacciones Lovera | B |
| Refaccionaria Oyuky | C |
| Refaccionaria El Triángulo | D |
| Refaccionaria Tonalli | E |
| Refaccionaria Eddie | F |
| Refaccionaria El Jefe de Jefes | G |
| Refaccionaria Del Sur | H |
| Mega Refaccionaria | I |

prepare shipments is minimized. Order picking principles also influence the design and layout of the warehouse to optimize workflow. The strategy impacts the efficiency of the warehouse and its layout, optimizing workflow in an increasingly competitive marketplace, this strategy becomes critical to effectively meet customer demands and expectations. In our case we will assume that Order Picking is already implemented in the processes performed in the Hubs.

## 6 Distribution by Bin Packing for Spare Parts in Delivery Units

The container packaging distribution model is a smart strategy to optimize the loading of motorcycle parts by grouping similar parts in trunks (or "containers"). This maximizes space efficiency, reduces operating costs and ensures the integrity of parts during delivery. This technique is essential to maintaining an agile and cost-effective supply chain in the industry, addressing the challenge of optimizing the loading of spare parts into delivery motorcycles, reducing the number of trips required and operating costs while ensuring that parts arrive in optimal condition.

Regarding the problem of container packaging, a field investigation was carried out in collaboration with a spare parts warehouse. In this research, measurements were taken of the parts that would represent our orders, a total of 45 parts, including electrical and mechanical components, as well as the containers that would be used for their transportation (containers with internal dimensions of 400 mm * 450 mm * 360 mm). The results of these measurements are expressed in terms of volume in cubic millimeters in the Table 5.

To carry out the application of our optimization methods, an order simulation will be carried out, where a maximum quantity of one unit per item is established for each of the dealerships. In the following, we present the order corresponding to Zone 1 through a table linking the dealership with its assigned letter in Table 1 and the item with its assigned ID in Table 6.

## 7 Solving the Bin Packing

In the framework of solving the Bin Packing problem, we are faced with the following question: with a finite set of garbage cans, all with equal capacity, our goal is to determine the minimum number of garbage cans needed to hold all the supplied items. To address this task, we will make use of the Python pywraplp library and rely on an example provided by Google OR-Tools, which will be detailed in the following section.

We generate the data from the dimensions we have obtained from the motorcycle parts and the capacities of our compartments. For this, we use a vector called

**Table 5** Spare parts dimensions

| Refurbishment | ID | Averages in millimeters | | | |
|---|---|---|---|---|---|
| | | x | y | z | Volume |
| Cylinder | 0 | 110 | 156 | 162 | 277,992 |
| Spark plug box | 1 | 85 | 110 | 41 | 383,350 |
| Spark plug 1 | 2 | 80 | 20 | 21 | 33,600 |
| Spark plug 2 | 3 | 80 | 22 | 21 | 36,960 |
| Bearings 1 | 4 | 179 | 124 | 11 | 244,156 |
| Bearings 2 | 5 | 170 | 123 | 11 | 230,010 |
| Bearings 3 | 6 | 138 | 135 | 40 | 745,200 |
| Spark plug nipple 1 | 7 | 260 | 140 | 40 | 145,600 |
| Spark plug nipple 2 | 8 | 80 | 90 | 40 | 288,000 |
| Harness 1 | 9 | 60 | 30 | 40 | 72,000 |
| Fuse Starter | 10 | 50 | 100 | 35 | 175,000 |
| Directional | 11 | 190 | 90 | 60 | 102,600 |
| Skull | 12 | 100 | 140 | 50 | 700,000 |
| Gas cap | 13 | 60 | 60 | 60 | 216,000 |
| Rearview Mirror | 14 | 300 | 100 | 45 | 135,000 |
| Footrest | 15 | 125 | 50 | 50 | 312,500 |
| Starter Lever | 16 | 190 | 70 | 30 | 399,000 |
| Electric Lever | 17 | 80 | 70 | 130 | 728,000 |
| Double Harness | 18 | 55 | 40 | 100 | 220,000 |
| Auxiliary Light | 19 | 70 | 135 | 65 | 614,250 |
| Brake Lever | 20 | 130 | 220 | 100 | 286,000 |
| Clutch Cover | 21 | 400 | 200 | 65 | 520,000 |
| Exhaust | 22 | 450 | 130 | 130 | 760,500 |
| Led Lights | 23 | 77 | 115 | 25 | 221,375 |
| Valve Cap | 24 | 100 | 135 | 77 | 103,950 |
| Valve Plug 2 | 25 | 60 | 95 | 20 | 114,000 |
| Spark Plug Wire 1 | 26 | 133 | 245 | 42 | 136,857 |
| Spark Plug Wire 2 | 27 | 130 | 210 | 38 | 103,740 |
| Spark Plug Wire 3 | 28 | 120 | 240 | 45 | 129,600 |
| Spark Plug Wire 4 | 29 | 140 | 200 | 42 | 117,600 |
| Accelerator | 30 | 135 | 50 | 50 | 337,500 |
| Chicote 1 | 31 | 100 | 110 | 11 | 121,000 |
| Chicote2 | 32 | 110 | 120 | 11 | 145,200 |
| Chicote 3 | 33 | 130 | 100 | 12 | 156,000 |
| Bolt for Piston 1 | 34 | 15 | 45 | 15 | 10,125 |
| Bolt for Piston 2 | 35 | 16 | 50 | 16 | 12,800 |

(continued)

**Table 5** (continued)

| Refurbishment | ID | Averages in millimeters | | | |
|---|---|---|---|---|---|
| | | x | y | z | Volume |
| Piston 1 | 36 | 70 | 50 | 70 | 245,000 |
| Piston 2 | 37 | 65 | 70 | 65 | 295,750 |
| Piston 3 | 38 | 55 | 40 | 55 | 121,000 |
| Connecting Rod 1 | 39 | 143 | 46 | 20 | 131,560 |
| Connecting Rod 2 | 40 | 140 | 41 | 18 | 103,320 |
| Crankshaft C/B | 41 | 231 | 143 | 118 | 389,789 |
| Crankshaft S/B | 42 | 236 | 118 | 118 | 328,606 |
| Cylinder 2 | 43 | 90 | 126 | 132 | 149,688 |
| Spark plug case 2 | 44 | 85 | 110 | 41 | 383,350 |

**Table 6** Zone 1 order

| Dealership | Part ID |
|---|---|
| A | 1, 2, 4, 5, 6, 7, 8, 9, 10, 13, 16, 17, 18, 20, 21, 24, 29, 30, 31, 34, 35, 37, 38, 40 |
| B | 1, 3, 5, 10, 12, 14, 18, 21, 22, 23, 24, 32, 33, 36, 37, 38, 39, 44 |
| C | 0, 1, 3, 5, 6, 7, 8, 9, 11, 12, 13, 14, 16, 19, 24, 26, 27, 29, 30, 31, 33, 37, 40, 41, 43 |
| D | 1, 2, 3, 4, 6, 7, 8, 9, 14, 16, 18, 19, 21, 23, 28, 29, 30, 32, 33, 43, 44 |
| E | 0, 1, 2, 3, 5, 6, 8, 10, 12, 14, 18, 19, 22, 24, 26, 28, 30, 31, 34, 36, 38, 39, 41 |
| F | 1, 2, 3, 5, 7, 8, 10, 11, 12, 15, 16, 17, 20, 21, 23, 25, 27, 29, 30, 32, 35, 37, 40, 42, 44 |
| G | 0, 1, 2, 4, 6, 8, 9, 11, 12, 14, 15, 16, 17, 18, 21, 23, 24, 28, 30, 33, 34, 36, 38, 39, 43 |
| H | 0, 1, 2, 3, 6, 7, 8, 9, 10, 13, 15, 17, 19, 21, 26, 29, 32, 34, 35, 42 |

"weights" which contains the weights of the elements, and "bin_capacity" which represents the indicator of the capacity of the containers (Fig. 4).

We continue by declaring the solver and creating the variables. We define an array of variables, named x[(i, j)], whose value is equal to 1 if element i is housed in container j, and 0 otherwise. We also set an array of variables, y [j], where its value is equal to 1 if discretization j is used, that is, if any element is packed in that bin, and 0 otherwise (Fig. 5).

```
Function CreateDataModel()
    Initialize variables `data`, `weights` and `items`.
    Assign values to variables `weights`,`items` and _
    `capacity`
    Add data to dictionary `data`.
    Return the data dictionary `data`.
```

**Fig. 4** Function create data model

```
Function Main()
  data <- CreateDataModel(weights, capacity)
  solver <- Create solver with SCIP algorithm
  if the solver is not created correctly
     Terminate the execution.
  #Decision variable: x[i,j]
  for i in items
    for j in bins
       x[i,j]<- 1 if item i is packed in bin j
  #Decision variable: y[i]
    for j in bins
       y[j]<- 1 if bin j is used
```

**Fig. 5** Function main

We set the constraints as follows:

1. Each item must be placed in exactly one container. To meet this constraint, we require that the sum of x[i][j] for all j discretization's equals 1.
2. The total weight packed in each container cannot exceed its capacity.

Figure 6 shows the pseudocode to handle the constraints.

Since our objective is to minimize the sum, where y[j] is equal to 1 if discretization j is used and 0 otherwise, the sum of y[j] represents the number of discretization's used. Therefore, we define our objective as shown in Fig. 7.

When running our solver, we obtained the following results that allow us to conclude that the program works correctly and maximizes the use of the space in the containers, using as an example the results for bin 1.

The distribution of the spare parts within the program would be as shown in Table 7.

```
Constraints
 # 1. Each item must be in exactly one bin
 for i in data["items"]:
   solver.Add(sum(x[(i, j)] for j in data["bins"])_
   == 1)

# 2. The amount packed in each bin cannot exceed its_
      capacity
 for j in data["bins"]:
   solver.Add(sum(data["weights"][i]*x[(i, j)]_
   for i in data["items"]) <= data["bin_capacity "])
```

**Fig. 6** Constraints

```
Objective
# Minimize the number of bins used
solver.Minimize(solver.Sum([y[j] for j_
in data["bins"]]))
```

**Fig. 7** Minimize the objective function

**Table 7** Assignment of numbering for the items

| ID | Number of spare parts | Item number within programming | ID | Number of spare parts | Item number within programing |
|---|---|---|---|---|---|
| 0 | 4 | 0 —> 3 | 22 | 2 | 105 —> 106 |
| 1 | 8 | 4 —> 11 | 23 | 4 | 107 —> 110 |
| 2 | 6 | 12 —> 17 | 24 | 5 | 111 —> 115 |
| 3 | 6 | 18 —> 23 | 25 | 1 | 116 |
| 4 | 3 | 24 —> 26 | 26 | 3 | 117 —> 119 |
| 5 | 5 | 27 —> 31 | 27 | 2 | 120 —> 121 |
| 6 | 6 | 32 —> 37 | 28 | 3 | 122 —> 124 |
| 7 | 5 | 38 —> 42 | 29 | 5 | 125 —> 129 |
| 8 | 7 | 43 —> 49 | 30 | 5 | 130 —> 134 |
| 9 | 5 | 50 —> 54 | 31 | 3 | 135 —> 137 |
| 10 | 5 | 55 —> 59 | 32 | 5 | 138 —> 142 |
| 11 | 3 | 60 —> 62 | 33 | 4 | 143 —> 146 |
| 12 | 5 | 63 —> 67 | 34 | 4 | 147 —> 150 |
| 13 | 3 | 68 —> 70 | 35 | 3 | 151 —> 153 |
| 14 | 5 | 71 —> 75 | 36 | 3 | 154 —> 156 |
| 15 | 3 | 76 —> 78 | 37 | 4 | 157 —> 160 |
| 16 | 5 | 79 —> 83 | 38 | 4 | 161 —> 164 |
| 17 | 4 | 84 —> 87 | 39 | 3 | 165 —> 167 |
| 18 | 5 | 88 —> 92 | 40 | 3 | 168 —> 170 |
| 19 | 4 | 93 —> 96 | 41 | 2 | 171 —> 172 |
| 20 | 2 | 97 —> 98 | 42 | 2 | 173 —> 174 |
| 21 | 6 | 99 —> 104 | 43 | 3 | 175 —> 177 |
| | | | 44 | 3 | 178 —> 180 |

Using the Painter library, we can create a three-dimensional representation of our packaging, providing the dimensions on the x, y, and z axes of each motorcycle part. As a result of the first bin, we will have the image shown in Fig. 8.

For bins 2 and 3, we will obtain similar results, following the same resolution approach used for bin 1. Three-dimensional representation of these results are shown in Figs. 9 and 10, respectively.

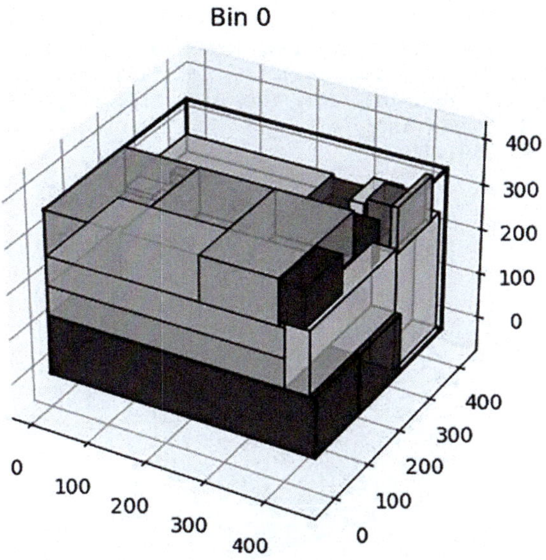


Fig. 8 Three-dimensional representation for first bin packing

```
Bin number 1
  Items packed: [99, 100, 101, 102, 103, 104, 105, 106, 111, 112, 113,
114, 115, 117, 118, 119, 120, 121, 122, 123, 124, 137, 161]
  Total weight: 61918010
```

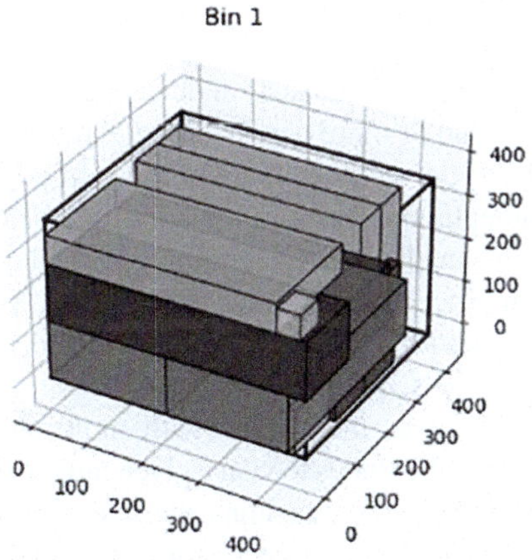


Fig. 9 Three-dimensional representation for second bin packing

```
Bin number 2
  Items packed: [125, 126, 127, 128, 129, 130, 131, 132, 133, 134, 138,
139, 140, 141, 142, 143, 144, 145, 146, 154, 155, 156, 157, 158, 159, 1
60, 162, 163, 164, 165, 166, 167, 168, 169, 170, 171, 172, 173, 174, 17
5, 176, 177, 178, 179, 180]
  Total weight: 31911746
```

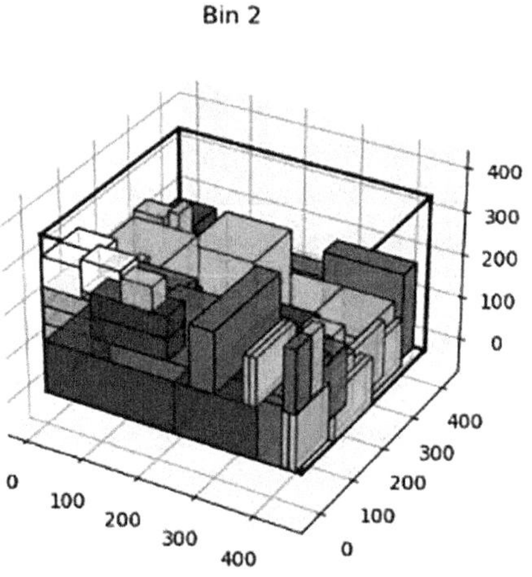


**Fig. 10** Three-dimensional representation for third bin packing

## 8 Dijkstra Algorithm

The Vehicle Routing Problem (VRP) consists of determining a set of routes for a fleet of vehicles departing from one or more depots or warehouses to satisfy the demand of several geographically dispersed customers proposed by Hillier and Lieberman en 2006 [1]. It is a fundamental challenge in logistics optimization and vehicle fleet management, this problem can also address minimizing total costs, such as distance traveled or time spent, and meeting certain constraints [7]. These constraints may include vehicle capacity limitations, time windows in which deliveries must be made, and possibly restrictions on the sequence of visits [8].

It is proposed to use a solver that works with multi-objective integer linear programming that aims to find the routes to visit a given number of customers seeking to satisfy their demand and having as decision criteria the minimization of the total distance traveled and the balance in the occupation of vehicles represented in the number of units transported by each one [2].

To start the process, we constructed a graph representing the various possible routes for the delivery of motorcycle parts to different dealerships (Fig. 11).

During our research, we noticed similarities between our problem and the example proposed by Dantzig and Ramser in 1959 [3], which describes the search for the most efficient routes between points in the system, considering the demand for products

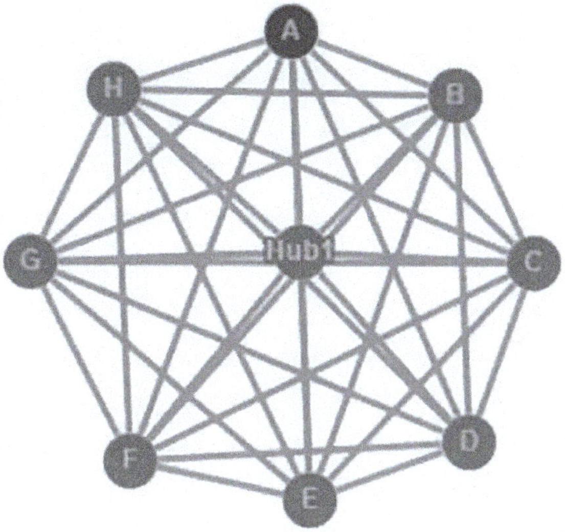


**Fig. 11** Path network for zone 1

at various distribution stations. The objective is to allocate our delivery units to the stations in a way that satisfies the demands and minimizes the total fleet mileage.

The tool we use for the solution is based on providing a list of addresses which, in turn, is used to search for points on Google Maps and calculate the most efficient route (Fig. 12). In our case, we start from "Hub 1", visit all the other delivery points and then return to the starting point, again "Hub 1".

With the information of the points and the order of visit, we generate a.json file that we introduce in the GeoJSON mapping platform (Fig. 13).

The solution for the distribution of spare parts in the Tláhuac Delegation is presented in the Fig. 14. It is important to note that you can start from any of the dealerships (represented in gray) that are connected to our distribution center "Hub1" (represented in green).

| A | B | C |
|---|---|---|
| **Destinos** | | **Google** |
| | {"type":"FeatureCollection", | {"type":"FeatureCollection","features":[{"type":"Feature","properties": |
| Jacobo de Lieja 16, Agrícola Metropolitana, Tláhuac, 13280 Ciudad de M | 19.2905659, -99.0533025 | |
| Av. Tlahuac 5579, Santa Ana Poniente, Tláhuac, 13300 Ciudad de México | 19.3018643, -99.0524719 | |
| Av. Tlahuac 7257, El Triangulo, Tlahuac, 13460 Ciudad de México, CDMX | 19.2848054, -99.0120184 | |
| Av. Tlahuac 6382, Santa Ana Centro, Tláhuac, 13300 Ciudad de México, ( | 19.2963992, -99.03826699999999 | |
| Calle Marlin, mar, Tláhuac, 13270 Ciudad de México, CDMX | 19.2897522, -99.0593065 | |
| Av. Tlahuac 7404, Santa Cecilia, Tlahuac, 13010 Ciudad de México, CDM) | 19.2781029, -99.0070004 | |
| Av. Tlahuac #5560 Col. Santa Cecilia esquina Enrique Lombardo contra e: | 19.2811024, -99.010115 | |
| Calz.Tláhuac Chalco 167-Local 3, San Andrés, Tláhuac, 13099 Ciudad de I | 19.26604, -99.000317 | |
| Av. Tlahuac 7190, San Francisco Tlaltenco, Tláhuac, 13450 Ciudad de Mé | 19.2871713, -99.0131512 | |

**Fig. 12** List of addresses of spare parts stores

```
    {
      "type": "Feature",
      "properties": {
        "marker-color": "#7e7e7e"
      },
      "geometry": {
        "type": "Point",
        "coordinates": [
          -99.0533025,
          19.2905659
        ]
      }
    },
    {
      "type": "Feature",
      "properties": {
        "marker-color": "#7e7e7e"
      },
      "geometry": {
        "type": "Point",
        "coordinates": [
          -99.0120184,
          19.2848054
        ]
```


**Fig. 13** JSON File for the GeoJSON mapping platform

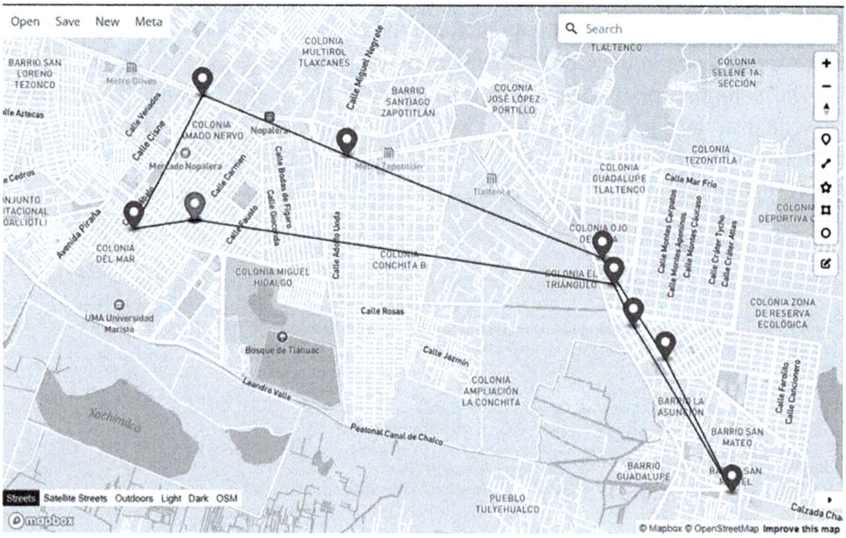


**Fig. 14** Solution for the distribution of spare parts in the Tláhuac Delegation

## 9 Conclusion and Future Research Directions

The presented project represents an innovative approach to spare parts management for motorcycle dealerships by integrating Dijkstra's algorithm, Bin packing distribution model and order picking optimization. The benefits of this implementation are remarkable for both the company and its customers. Not only has an improvement in customer service been achieved, but also considerable savings in terms of fuel and delivery times would be generated.

In terms of future research directions, opportunities for optimization in inventory management and delivery logistics can be further explored. In addition, the incorporation of emerging technologies, such as computer vision to facilitate the ordering process by dealership owners, could elevate this solution to a higher level.

This initiative demonstrates that innovation in motorcycle parts management is not only desirable, but also essential to remain competitive in an ever-evolving marketplace.

## References

1. Trabajo de Grado Presentado Como Requisito Parcial Para Obtener, E. T. de. (s/f).: Estudio del problema de ruteo de vehículos con balance de carga: Aplicación de la meta-heurística Búsqueda Tabú. Edu.co. Recuperado el 1 de septiembre de (2023)
2. Joaquín, P.-O., Hilda, C.-Z., Darnes, V.-A., Adriana, M.-S., -Díaz, Z., Crispín, J., Alicia, M.-R., Hugo, E.-E. (s/f).: Una nueva estrategia heurística para el problema de Bin packing a new heuristic strategy for the bin packing problem. Org.mx. Recuperado el 1 de septiembre de (2023)
3. Dantzig, G.B., Ramser, J.H.: The truck dispatching problem. J. Article (1959)
4. Dijkstra, E.W.: A note on two problems in connection with graphs. Numer. Math. **1**, 269–271 (1959)
5. Mockova, D., Rybickova, A.: Application of genetic algorithms to vehicle routing problem. Neural Netw. World **24**, 57–78 (2014). https://doi.org/10.14311/NNW.2014.24.003
6. Ostermeier, M., Holzapfel, A., Kuhn, H., Schubert, D.: Integrated zone picking and vehicle routing operations with restricted intermediate storage. OR Spect. **44** (2022). https://doi.org/10.1007/s00291-021-00664-7
7. Ho, W., Ho, G.T.S., Ji, P., Lau, H.C.W.: A hybrid genetic algorithm for the multi-depot vehicle routing problem. Eng. Appl. Artif. Intell. **21**(4) (2008)
8. Gendreau, M., Laporte, G., Potvin, J.-Y.: The vehicle routing problem. Soc. Indus. Appl. Math. 129–154 (2001)

# Numerical Approximation of the Maximum Absorption Capacity of the MEA-$H_2O$ Solution for a Post-combustion $CO_2$ Capture Process

**Ochoa-Zezzatti Alberto, Navarrete Procopio Miriam, Zezatti Victor, Castañeda Erick, Guzmán Irahan, and López Rodriguez Julián**

**Abstract** This study focuses on a numerical approximation to determine the maximum absorption capacity of the MEA-$H_2O$ solution in a post-combustion $CO_2$ capture process. Through simulation methods and computational modeling, the efficiency of this chemical solution in carbon dioxide absorption is thoroughly assessed, providing a detailed understanding of the factors that influence its capture capacity. The results contribute to the optimization of $CO_2$ capture processes, providing valuable insights for the efficient design and implementation of emission reduction systems in industrial applications. This numerical approach not only deepens the understanding of the MEA-$H_2O$ absorption capacity, but also suggests promising

O.-Z. Alberto
CADIT, Universidad Autónoma de Ciudad Juarez, Universidad Anáhuac Campus Norte, Av. Plutarco Elías Calles No. 1210, Fovisste Chamizal, 32310 Ciudad Juárez, ChihuahuaNaucalpan, Mexico
e-mail: laalberto.ochoa@uac.mx

N. P. Miriam · Z. Victor (✉) · C. Erick
Universidad Autónoma del Estado de Morelos, Av. Universidad No. 1001, Chamilpa, 62209 Cuernavaca, Mor., Mexico
e-mail: zezatti@uaem.mx

N. P. Miriam
e-mail: miriam.navarrete@uaem.mx

C. Erick
e-mail: erick.castanedamag@uaem.edu.mx

G. Irahan
Instituto Tecnológico Superior de Misantla, Km. 1.8 Carretera a Loma del Cojolite, 93850 Misantla, Veracruz, Mexico
e-mail: iojoseg@itsm.edu.mx

L. R. Julián
Universidad Autónoma de Sinaloa, Ciudad Universitaria SN Ciudad Universitaria, 80040 Culiacán Rosales, Sinaloa, Mexico

 
D. Oliva et al. (eds.), *Advances in Optimization Algorithms for Multidisciplinary Engineering Applications: From Classical Methods to AI-Enhanced Solutions*, Studies in Computational Intelligence 806, https://doi.org/10.1007/978-3-031-78440-8_26

paths for the continuous improvement of technologies related to the capture and mitigation of greenhouse gases.

**Keywords** $CO_2$ capture · Post-combustion process · MEA solution · Numerical approximation · Computer simulation · Artificial intelligence · Carbon dioxide absorption

# 1 Introduction

Climate change, marked by several impacts, such as rising temperatures, water shortages and the frequency of natural disasters, as well as the increased risk of waterborne diseases, has generated concern in society as it is a threat to present and future generations [1, 2]. In addition, displacement of species and human beings, rising sea levels, altered ocean acidity, species disappearance and melting of glaciers, as well as decreased oxygen levels in the sea have been observed [3, 4]. This phenomenon is attributed to global warming, which is the result of the greenhouse effect generated by the solar radiation that hits the Earth. The Earth's surface absorbs solar radiation and emits infrared radiation, but due to the greenhouse gases present in the atmosphere, this infrared Radiation cannot escape, causing the planet's average temperature to rise [5]. In a 2012 study, it was estimated that 48% of solar radiation reaches the Earth's surface, 31% being reflected and succeeding in transmitting greenhouse gases, while the remaining 21% is absorbed. Greenhouse gases include methane ($CH_4$), ozone ($O_3$) and carbon dioxide ($CO_2$) [6]. Carbon dioxide emissions are worrying, as they are considered the main driver of global climate change. Therefore, it is widely recognized that there is an urgent need to reduce emissions to avoid the worst effects of climate change [7]. In response to this concern, the United Nations (UN) has established the Paris Agreement, an international treaty aimed at limiting global warming to well below 2 °C. As part of the agreement, countries submit their climate action plans, known as Contributions, in which they communicate the measures they will take to reduce greenhouse gas emissions and the targets set [8]. The best technology to capture $CO_2$ emissions from thermal power plants is a crucial aspect of mitigating the effects of greenhouse gases. Various technologies have been developed for this purpose. Post-combustion $CO_2$ capture processes, such as aqueous post-combustion $CO_2$ capture using water as a solvent, have shown promising results in achieving high $CO_2$ capture ratios [9–11]. Aqueous solution of monoethanolamine (MEA) has been used as an absorbent for $CO_2$ capture [12–14]. The absorption capacity of monoethanolamine (MEA) for $CO_2$ capture varies based on different solvent formulations and conditions. Studies have shown that MEA can achieve an optimal absorption capacity of 0.83 mol $CO_2$/mol amine at specific temperatures and pressures [15]. The maximum absorption capacity occurs when the absorbent solution is saturated. The experimental saturation technique obtains the saturation time when the maximum temperature increase in the MEA solution is reached [16]. In this work, a mathematical model was developed using neural networks to achieve a numerical approximation to the maximum absorption capacity of the MEA solution.

The objective function is expressed as:

$$f\left(C_{MEA}, m_{MEA}, C_{CO_2}, F_{gas}\right) = \Delta T_{\max -gas} \quad (1)$$

where:

- $C_{MEA}$ represents the percentage of monoethanolamine in the solution (wt.%).
- $m_{MEA}$ indicates the amount of solution in the absorption column (mL).
- $C_{CO2}$ refers to the percentage of carbon dioxide present in the gas (vol.%).
- $F_{gas}$ denotes the gas flow through MEA in the absorption column (L/min)
- ΔT indicates the greatest temperature difference observed when comparing max—gas the gas before entering the MEA and the gas once it crossed it.

Through this analysis, a tool is sought to predict the maximum temperature reached in the $CO_2$ capture process, which will be of great use for the development and optimization of future systems of absorption and reduction of emissions of polluting gases.

# 2 Proposed Methodology

## *2.1 Description of the System*

The experimental data for the development of the numerical approximation model were obtained in the laboratory-scale post-combustion prototype installed in the Unitary Operations Laboratory (LOU) of the Universidad Autónoma del Estado de Morelos (UAEM). The Fig. 1 indicates the piping and instrumentation diagram of the absorption column where the MEA solutions were saturated to ensure its maximum absorption capacity. The gas mixture to be purified enters the packed absorption column (C.O.-101) from the bottom (stream 1). The gas mixture is air and $CO_2$. The air is supplied by a compressor (C-101) and the $CO_2$ comes from a storage tank (T.A-101). The absorbent solution is supplied to the absorption column from the top (stream 2) through a peristaltic pump (P-101). The solution is brought into contact with the gas flow that is fed continuously through the random packings of the column. The $CO_2$-free gas leaves the top of the absorption column (STREAM 4) through a gas analyzer (T.T.-101). In the experimental tests, the temperature of the absorption process was monitored with thermocouples located inside the packed column. Once each experimental test is finished, the flow valve 3 is opened to collect the saturated solution in the storage tank.

For this experiment, an absorbent solution of water-dissolved monoethanolamine (MEA) was used. The MEA used was of high quality, with an analytical degree and a purity equal to or greater than 99%, provided by the supplier Sigma-Aldrich. The

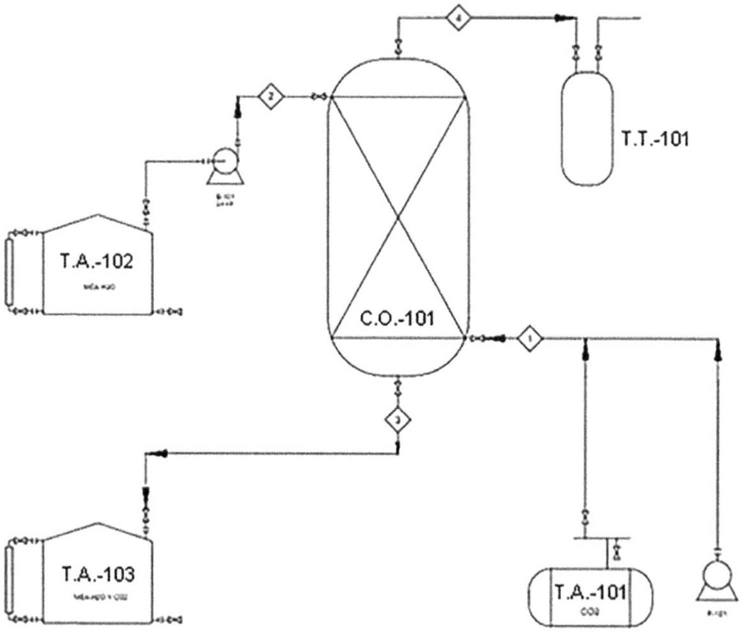


**Fig. 1** Piping and instrumentation diagram (P&ID) of the absorption column

gas mixture used in the process consisted of commercial grade $CO_2$ with a purity equal to or greater than 99%, supplied by Infra, and compressed air obtained through the piston compressor. To determine the saturation time of the MEA solution, an experimental technique was developed that consisted of loading the column with known amounts of the solution of MEA (0.175, 0.65, 1.3 and 1.7 kg). A continuous flow of air and an indifferent $CO_2$ flow at known concentrations of $CO_2$ (40, 30, 20 and 12.5% in volume) was then passed through the column while monitoring the temperature in the MEA solution. The saturation time was defined as the time at which the maximum temperature increase in the solution was achieved, indicating that the maximum $CO_2$ load according to the stechiometric reaction (0.5 mol-$CO_2$/mol-MEA) had been achieved. After that, tests were carried out to capture $CO_2$ by saturation of MEA. Table 1 indicates the operating conditions such as the MEA solution amount, $CO_2$ concentration and MEA concentration were varied.

## 2.2 *Regression Techniques*

The model for the numerical approximation of this work was developed in the Python programming language, using the seaborn library to perform a correlation analysis between the variables that were being searched. Once the correlation analysis is

**Table 1** Operating conditions for experimental tests

| Parameters | Data |
|---|---|
| Packing type | 1/2″ ceramic Berl saddle |
| Absorber diameter (m) | 0.08 |
| Absorber packed height (m) | 0.70 |
| Gas flow rate (L/min) | 10 and 16 |
| $CO_2$ concentration (vol.%) | 40, 30, 20, and 12.5 |
| MEA solution amount (mL) | 175, 650 1300, and 1700 |
| MEA concentration (wt.%) | 10, 15, 20, 25, and 30 |
| Pressure (kPa) | 80 |

completed, the Scikit-learn (sklearn) library is used to perform a linear regression and a second-degree polynomial regression. These methods fit the data using least squares. Subsequently, Keras from TensorFlow in Python is used to create 36 neural networks with different topologies chosen at random. Of the 36 neural networks, 18 networks used the ReLU activation function which is widely used due to its low computational cost and demonstrated superiority in many tasks [17, 18] and the rest of the topologies used the Tanh activation function, which has shown superior results in regression work compared to the ReLU function [19]. The 18 topologies with ReLU were identical to the 18 with Tanh, with the only difference being the activation function used. In both cases, 20,000 iterations were performed and the "adam" optimizer was used. Each of the topologies was trained 4 times, this is because the initial weights and biases were selected at random, this strategy was implemented to prevent any bias from arising due to the randomness in the initialization of the parameters. The squared error of the 4 runs was averaged to select the best topology. After training the topologies with the ReLU activation function, the topologies that had a lower mean squared error were selected. From these results, 7 topologies were created with the same architecture, but with the Leaky ReLU activation function. This variant has been shown to be more effective than the standard ReLU in almost all tasks, as it solves the "dead neurons" and gradient fading problems that are sometimes associated with the ReLU function [20]. This completed the process of building and selecting the neural networks for the study in question. Figure 2 shows the schematic diagram of the described methodology.

### 2.3 *Topologies of the 18 Neural Networks*

The 18 implemented neural networks had different topologies, which are described in Table 2. In the development of neural networks, it was considered that all topologies had in the last layer a single neuron with a linear activation function.

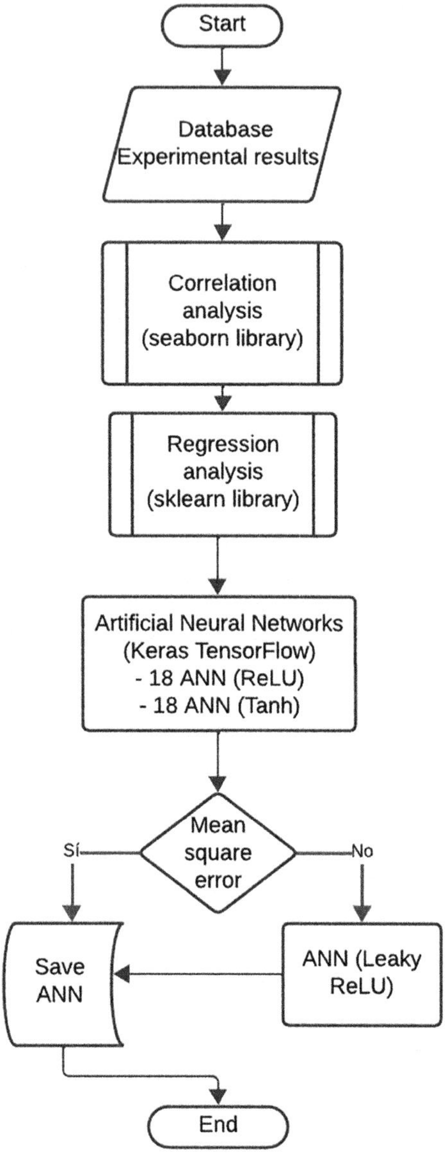


**Fig. 2** Schematic diagram of methodology

## 3 Experimentation and Results

The database for training the neural networks was obtained with the methodology described in Sect. 2.1 of this work and with the operating conditions of Table 1. 24 experimental tests were developed, varying the amount of absorbent solution ($m_{MEA}$), the concentration of the absorbent solution ($C_{MEA}$), the concentration of $CO_2$ ($C_{CO_2}$)

**Table 2** Topology of neural networks used

| Model number | Number of neurons |
|---|---|
| 0 | 10, 7, 4, 1 |
| 1 | 10, 5, 1 |
| 2 | 8, 8, 1 |
| 3 | 6, 4, 1 |
| 4 | 5, 5, 1 |
| 5 | 10, 5, 3, 1 |
| 6 | 8, 8, 8, 8, 1 |
| 7 | 6, 4, 2, 1 |
| 8 | 5, 3, 2, 1 |
| 9 | 10, 5, 2, 1 |
| 10 | 8, 4, 1 |
| 11 | 8, 8, 8, 1 |
| 12 | 10, 8, 8, 4, 1 |
| 13 | 10, 10, 10, 10, 1 |
| 14 | 10, 5, 5, 5, 1 |
| 15 | 10, 8, 5, 3, 1 |
| 16 | 12, 6, 3, 1 |
| 17 | 15, 13, 10, 7, 5, 2, 1 |

and the flow of the gaseous stream that enters the absorption process ($F_{\text{gas}}$). Table 3 shows the experimental test setups and shows the maximum temperature difference recorded indicating the saturation of the absorbent solution ($\Delta T_{max-gas}$).

The result of the correlation map is presented in Fig. 3 and clearly reveals the relationships between the variables analyzed. A strong correlation is highlighted between the MEA concentration in the solution, the MEA solution amount in the absorption column, and the $CO_2$ concentration in the gas flow with the maximum temperature difference. In addition, an inverse correlation with the gas flow with the maximum temperature difference is observed, which is also beneficial for this study.

## 4 Conclusions and Future Research

This study has addressed the pressing issue of carbon dioxide ($CO_2$) emissions and their impact on climate change. Through the numerical approximation of the maximum absorption capacity of the MEA solution in a post-combustion $CO_2$ capture process, significant advances have been made in understanding the efficiency of this chemical solution. A detailed understanding of the factors affecting $CO_2$ capture capacity has been established using the MEA solution. The results provided crucial information for the optimization of $CO_2$ capture processes in industrial applications,

**Table 3** Experimental tests setups

| Max ($\Delta T$) Temperature increase (°C) | MEA concentration (%w/w) | MEA solution amount (mL) | $CO_2$ concentration (%v/v) | Gas flow (mL/min) |
|---|---|---|---|---|
| 4.7027 | 10 | 175 | 12.5 | 16 |
| 7.1313 | 10 | 1300 | 12.5 | 16 |
| 7.7947 | 15 | 175 | 12.5 | 16 |
| 7.8684 | 10 | 650 | 12.5 | 16 |
| 8.4518 | 20 | 175 | 12.5 | 16 |
| 8.6892 | 10 | 1700 | 20 | 10 |
| 9.2302 | 15 | 1300 | 12.5 | 16 |
| 9.2549 | 25 | 175 | 12.5 | 16 |
| 9.3762 | 10 | 1700 | 30 | 10 |
| 9.8508 | 15 | 650 | 12.5 | 16 |
| 10.6131 | 30 | 175 | 12.5 | 16 |
| 11.3232 | 10 | 1700 | 40 | 10 |
| 11.9713 | 20 | 1300 | 12.5 | 16 |
| 12.1769 | 25 | 1300 | 12.5 | 16 |
| 12.2859 | 20 | 650 | 12.5 | 16 |
| 12.7191 | 25 | 650 | 12.5 | 16 |
| 14.5198 | 20 | 1700 | 20 | 10 |
| 14.9057 | 25 | 1700 | 20 | 10 |
| 14.9703 | 30 | 650 | 12.5 | 16 |
| 16.7629 | 25 | 1700 | 40 | 10 |
| 17.4497 | 30 | 1300 | 12.5 | 16 |
| 18.1408 | 30 | 1700 | 20 | 10 |
| 18.7586 | 20 | 1700 | 40 | 10 |
| 24.5402 | 30 | 1700 | 40 | 10 |

contributing to the reduction of emissions. By considering the Mahalanobis distance we can determine in a simpler way the correct design of experiments associated with the precedence values of each selected sample, as shown in Fig. 4.

Environmental Relevance: The implementation of the post-combustion capture technique has significant potential to mitigate $CO_2$ emissions in power plants, contributing to the greenhouse gas reduction targets set out in international agreements such as the Paris Agreement.

Technological Approach: Numerical analysis has proven to be a valuable tool for evaluating the efficiency of the MEA solution, providing a solid basis for future research and technological developments.

Identified Challenges: The existence of disadvantages in the post-combustion capture process, such as associated costs and solvent degradation, has been.

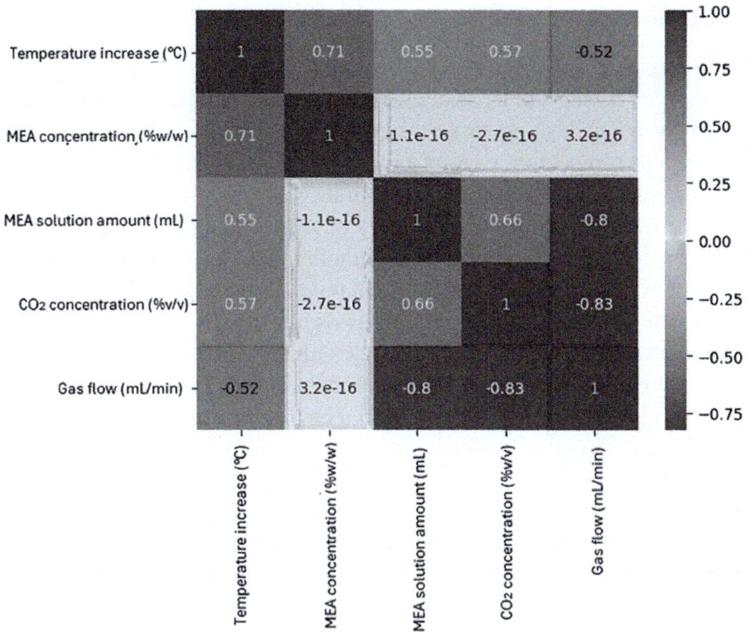


**Fig. 3** Correlation map of the absorption process variables

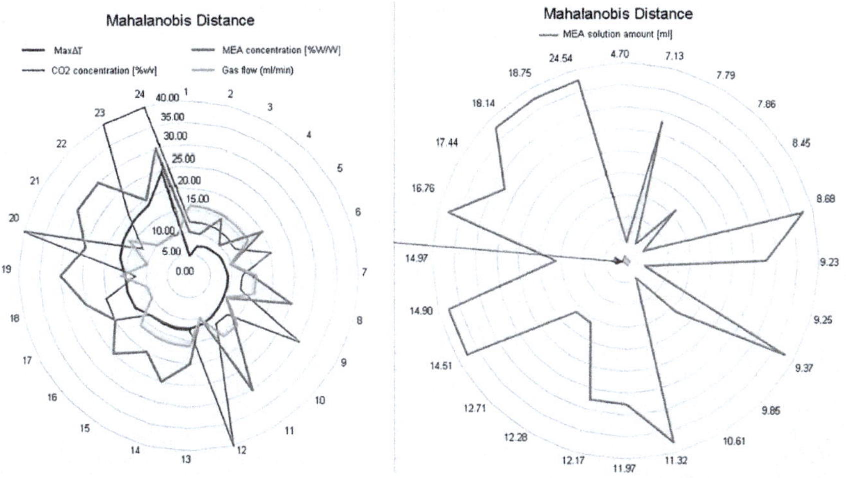


**Fig. 4** Mahalanobis distance, in which issues 12, 23 and 24 are identified as having the highest correlation in the behavior of the group associated with the design of experiments

## Future Research

Solvent Improvement using continuous research is suggested in the search for alternative or improved solvents that can address current limitations, such as thermal degradation and corrosion. Using a Process Optimization in Future studies could focus on optimizing post-combustion capture processes to minimize negative impacts, such as loss of efficiency in power plants. Exploration of Emerging Technologies is very important to improve this research what could be expanded to explore emerging technologies, such as the use of artificial intelligence in computational simulations to predict the behavior of the MEA solution more accurately. In Economic Analysis, is very important a detailed economic analysis is recommended to assess the financial viability of the large-scale implementation of post-combustion catch, considering the long-term costs and benefits. In Real Industrial Applications is very important the Experimental validation in real industrial environments could provide additional data and confirm the effectiveness of the MEA solution in practical conditions. In short, this study lays the foundations for future research and development in $CO_2$ capture, highlighting the importance of addressing environmental challenges through innovative and sustainable technological solutions. Finally, it is proposed to use an innovative Metaheuristic based on Belugas [21] that can correctly optimize the search for the estimation values of the absorption tower, as shown in Fig. 5:

1. Parameter Optimization: The $CO_2$ capture process involves several parameters, such as MEA concentration, amount of MEA solution, $CO_2$ concentration in the gas, and gas flow rate. BOA can be used to find the optimal combination of these parameters that maximizes the $CO_2$ absorption capacity.

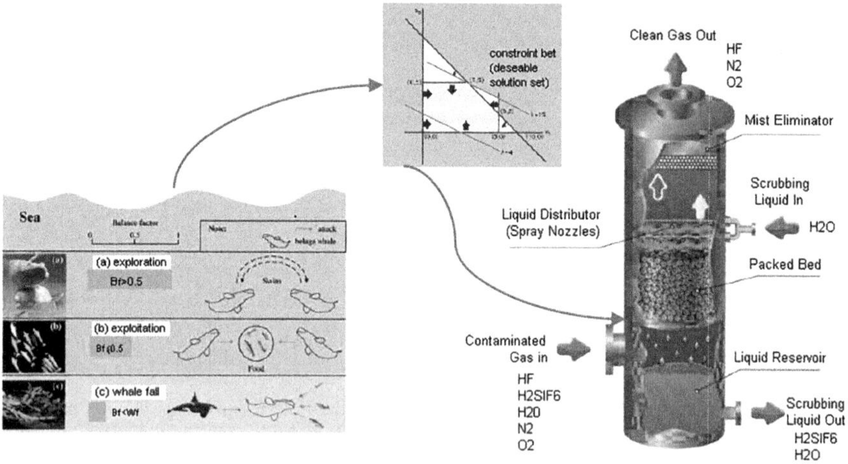


**Fig. 5** Representative model of the selection of a Metaheuristic in the process of optimization of an absorption column model

2. Energy Efficiency Optimization: BOA can be applied to optimize the energy efficiency of the process. It can look for parameter configurations that reduce energy consumption without compromising $CO_2$ absorption capacity.
3. Sensitivity Analysis: BOA can help perform a sensitivity analysis to identify which parameters are most critical in terms of impact on absorption capacity. This can guide attention to key areas for improvement or adjustment.
4. Solution Space Exploration: BOA is known for its ability to efficiently explore the solution space. This can be beneficial for finding solutions that lie in regions not explored by other optimization methods [20, 22].
5. Continuous Optimization: The $CO_2$ capture process can be a dynamic system, and BOA can be applied to continuously optimize parameters in response to changes in operating conditions or process objectives.
6. Constraint Management: If there are constraints on the process, such as limits on the amount of MEA that can be used or restrictions on gas flow, BOA can be adapted to handle these constraints and find optimal solutions within the defined limits.
7. Dynamic Process Optimization: The $CO_2$ capture process can have complex dynamics that vary with time. BOA can be employed for dynamic process optimization, adapting to changes in operating conditions or $CO_2$ capture requirements over time. BOA's ability to handle dynamic problems can be valuable for adjusting process parameters in real time, thereby improving system efficiency and responsiveness. This is especially beneficial in industrial environments where conditions may change over time due to variations in energy demand or $CO_2$ emission patterns.

It is very important to note that the successful application of BOA will depend on the specific nature of this type of problem and the quality of the knowledge representation associated with the problem in the objective function. Experiments and adjustments are recommended as necessary to adapt BOA to the specific context of the $CO_2$ capture process.

## References

1. Zilin, S.: World $CO_2$ emissions: simple analysis and its relationship with global temperature change. Highlights Sci. Eng. Technol. (2022). https://doi.org/10.54097/hset.v25i.3415
2. Skvortsov, V.A., Chudnenko, K.V.: Thermodynamic model of the carbon emission in the atmosphere and climate change. Atmos. Oceanic Opt. (2013). https://doi.org/10.1134/S1024856013010144
3. An, S.I., Franck, H., Bengkel, M.M.: Irreversible changes in surface temperature and precipitation to $CO_2$ forcing (2023). https://doi.org/10.5194/egusphere-egu23-1664
4. Fedorov, V.M., Ivan, A., Denis, F.: Influence of anthropogenic carbon dioxide on the thermal regime of the atmosphere and its changes. Žizn′ zemli (2022). https://doi.org/10.29003/m3115.0514-7468.2022_44_4/402-414
5. Zickfeld, K., Arora, V.K., Gillett, N.P.: Is the climate response to $CO_2$ emissions path dependent. Geophys. Res. Lett. (2012). https://doi.org/10.1029/2011GL050205

6. Elias, S.A.: Greatly Increased $CO_2$ (2012). https://doi.org/10.1016/B978-0-12-409548-9.09927-9
7. $CO_2$ Emissions equivalences (2023). https://doi.org/10.18356/9789210023047c010
8. Delbeke, J., Runge-Metzger, A., Slingenberg, Y., Werksman, J.: The paris agreement. In: Towards a Climate-Neutral Europe, pp. 24–45. Routledge (2019)
9. Vaz, S., de Souza, A.P.R., Baeta, B.E.L.: Technologies for carbon dioxide capture: a review applied to energy sectors. Clean. Eng. Technol. (2022). https://doi.org/10.1016/j.clet.2022.100456
10. Cristian, D.: Technical, environmental, and economic assessment of $CO_2$ absorption chemical process integration in the power plant technologies (2012)
11. Chavez, R.H., Guadarrama, J.J.: Optimized process from post-combustion $CO_2$ capture in thermoelectric power plant, using structured packing. Chem. Eng. Trans. (2011)
12. Yun, S.H., Kim, Y.E., Seong, Y.H., Nam, S.C., Chu, S., Yoon, Y.: $CO_2$ absorption of chemical phase transitional absorbents: absorption capacity and reaction mechanism (2014). https://doi.org/10.14257/ASTL.2014.65.02.
13. Aghel, B., Heidaryan, E., Sahraie, S., Nazari, M.: Optimization of monoethanolamine for $CO_2$ absorption in a microchannel reactor. J. $CO_2$ Utilization (2018). https://doi.org/10.1016/J.JCOU.2018.10.005
14. Naruphol, J., Thananat, L., Atipoang, N., Nakorn, T.: Performance analysis of a laboratory monoethanolamine-based carbon absorption and stripping setup. Nucl. Atmos. Aerosols **10**(1063/5), 0115844 (2021)
15. Ulus, N., Syed Ali, S.A., Khalifa, O., Orhan, O.Y., Elkamel, A.: Optimization of novel nonaqueous hexanol-based monoethanolamine/methyl diethanolamine solvent for $CO_2$ absorption. Int. J. Energy Res. (2022). https://doi.org/10.1002/er.7779
16. Procopio, M.N., Urquiza, G., Castro, L., Zezatti, V.: Saturation of the MEA solution with $CO_2$: absorption prototype and experimental technique. Results Eng. **19**, 101286 (2023). https://doi.org/10.1016/j.rineng.2023.101286
17. Rathinasamy, A., Mayavel, P.: Strong convergence and almost sure exponential stability of balanced numerical approximations to stochastic delay Hopfield neural networks. Appl. Math. Comput. **438**, 127573 (2023)
18. Karafyllis, I., Vokos, F., Krstic, M.: Output-feedback control of viscous liquid–tank system and its numerical approximation. Automatica **149**, 110827 (2023)
19. Bose, S., Mukherjee, K.: A fast uniformly accurate global numerical approximation to solution and scaled derivative of system of singularly perturbed problems with multiple diffusion parameters on generalized adaptive mesh. Comput. Appl. Math. **42**(4), 180 (2023)
20. Sforza, D., Pellacini, F.: Numerical approximations for energy preserving microfacet models. Comput. Graph. **114**, 36–44 (2023)
21. Zhong, C., Li, G., Meng, Z.: Beluga whale optimization: A novel nature-inspired metaheuristic algorithm. Knowl.-Based Syst. **251**, 109215 (2022). https://doi.org/10.1016/j.knosys.2022.109215
22. Li, L.L., Fan, X.D., Wu, K.J., Sethanan, K., Tseng, M.L.: Multi-objective distributed generation hierarchical optimal planning in distribution network: improved beluga whale optimization algorithm. Expert Syst. Appl. **237**, 121406 (2024)

# Optimizing Smart Home Energy Analysis with Sailfish and Random Forest Algorithms

G. Shanmugasundar

**Abstract** Smart homes incorporate numerous devices that automate processes and simplify lives, assisting in managing temperature, lighting and security access. The capacity of smart homes to learn and adapt to human owners' routines is powered by the Internet of Things (IoT) network. Due to the multitude of devices and the necessity for efficiency and cost savings, scrutinizing power consumption for each device is crucial. Furthermore, higher energy consumption contributes to a larger carbon footprint, greater climate risk and higher supply demands, necessitating monitoring of energy usage. We employ sophisticated machine learning algorithms for an exhaustive study of energy consumption in smart homes, introducing a novel sailfish-optimized random forest algorithm (SORFA) to predict energy usage. SORFA optimizes the hyperparameters of the random forest algorithm to enhance its predictive accuracy and efficiency. Our proposed method, when compared with baseline methods, demonstrates superior performance in estimating energy consumption by reducing computational time and improving prediction accuracy. These improvements suggest significant potential for energy savings and cost reduction in smart home environments.

**Keywords** Internet of Things (IoT) · Smart homes · Energy consumption · Random forest algorithm · Optimization

## 1 Introduction

The worldwide expansion of smart homes and microgrids, driven by the necessity for green energy systems and grid-independence to counter escalating environmental damage, is evident. In contemporary smart cities, all interconnected households are collectively managed to optimize electricity consumption [1]. This is facilitated by a

G. Shanmugasundar (✉)
Department of Mechanical Engineering, Sri Sai Ram Institute of Technology, Chennai 600044, India
e-mail: shanmugasundar.mech@sairamit.edu.in



D. Oliva et al. (eds.), *Advances in Optimization Algorithms for Multidisciplinary Engineering Applications: From Classical Methods to AI-Enhanced Solutions*, Studies in Computational Intelligence 806, https://doi.org/10.1007/978-3-031-78440-8_27

smart grid-based network control system that interlinks the house and the meters of the Home Energy Management System (HEMS), leading to a centralized system that controls generation, use and storage of electricity. HEMS can assist consumers in decreasing their energy costs by enhancing the efficiency of their in-home renewable energy sources [2].

The integration of smart meters into the power distribution network enables the recording of energy consumption in smart homes at an exceptionally detailed level. Power grids are buttressed by the data and communication technology sector, which produces datasets with heterogeneity. With the rising demand in this field, new standards and techniques are being established. The grid now has an additional information technology (IT) layer over the traditional electrical layer. This IT layer, developed from the synergy of electrical, communication and computer technologies, collects data from networks of smart sensors and meters [3]. However, gathering high-quality data is a challenging endeavor.

Smart homes generate vast amounts of energy usage data continuously, typically in the form of timestamps and measurements. This data is crucial for understanding energy consumption patterns, billing generation, load profile, forecasting, contingency study, device health studies, etc. [4].

Past studies have employed machine learning algorithms for comprehensive investigation of energy use in smart homes [5]. These investigations have used methods like Decision Trees (DT), Random Forest (RF), Extreme Gradient Boosting (XGBoost) and k-Nearest Neighbor (KNN) for estimating power consumption [5]. Further, they proposed a DT-RF-XGBoost-based Ensemble method to analyze consumption and contrast it with standard methods. Other studies offered classifications of smart homes based on several interest areas such as energy, data and communication, privacy, health, environment, home entertainment and household appliances [6]. Some identified current trends and challenges for potential future improvements in smart home energy management systems, highlighting quality issues such as security, privacy, scalability and interoperability [7]. Some efforts have been made to summarize previously suggested energy conservation methods and recommendations for energy-saving measures [8]. Efforts like the IntelliHome aim to reduce household consumption of electrical energy [9], while the Home Energy Management System (HEMS) targets an energy-optimized ecosystem controlling IoT devices [10]. Another study explored the relationship between sensor performance and energy consumption forecasting [11]. Each of these studies contributes to the growing field of smart home energy management.

Based on the extensive literature study the following literature gaps are noticed:

- While existing research has made strides in studying energy consumption in smart homes, there is a lack of comprehensive studies that simultaneously consider energy consumption, cost savings, security, and time delay.
- Existing studies mainly use conventional machine learning algorithms like Decision Trees (DT), Random Forest (RF), and Extreme Gradient Boosting (XGBoost) for estimating power consumption. There is a need to explore new, potentially more efficient methods. Efficiency in this context refers to accomplishing the

task of energy consumption prediction using the least amount of computational resources, such as time and energy. Effectiveness, on the other hand, pertains to how well the method achieves accurate and reliable predictions.
- Though quality issues such as security, privacy, scalability, and interoperability have been highlighted, comprehensive and comparative studies addressing these concerns are insufficient.
- Real-world implementation and scalability of new algorithms are not extensively explored, leaving a gap in the application of theoretical models in real-life scenarios.

In light of the above literature gaps, this paper makes the following contributions:

- A comprehensive study that addresses energy consumption, cost savings, security, and time delay in smart homes.
- This paper introduces the Selfish Optimization Random Forest Algorithm (SORFA) as a novel technique for optimizing smart home energy management.
- The study addresses quality issues by providing a comparative analysis of the SORFA approach with traditional methods in terms of security and time delay, thus providing a more detailed understanding of these parameters.
- This study does not just propose a theoretical model, but also demonstrates the superiority of the SORFA approach in comparative analyses, thus contributing to the knowledge on real-world implementation of new algorithms in smart homes.

The rest of the paper is arranged as follows: Sect. 2 contains the methodology used in this research. In this section, the dataset, the optimization and the ML methods are introduced. In Sect. 3, the results obtained are visually presented and discussed critically. Section 4 summarizes the paper by proving a crisp conclusion. The limitations and the future scope of this research work is also detailed in the final part of Sect. 4.

## 2 Methodology

Analyzing energy usage data is vital to identifying trends and opportunities for energy conservation in a smart home. In this section, we begin by collecting and preprocessing data through normalization. Following this, we implement our innovative method, SORFA. The technique is visually represented in Fig. 1.

### *2.1 Dataset*

A power study using G * power software suggested a sample size of approximately 1,400, based on a moderate effect size and an alpha error probability of 0.05. After eliminating respondents with attention check scores less than 50%, our final sample

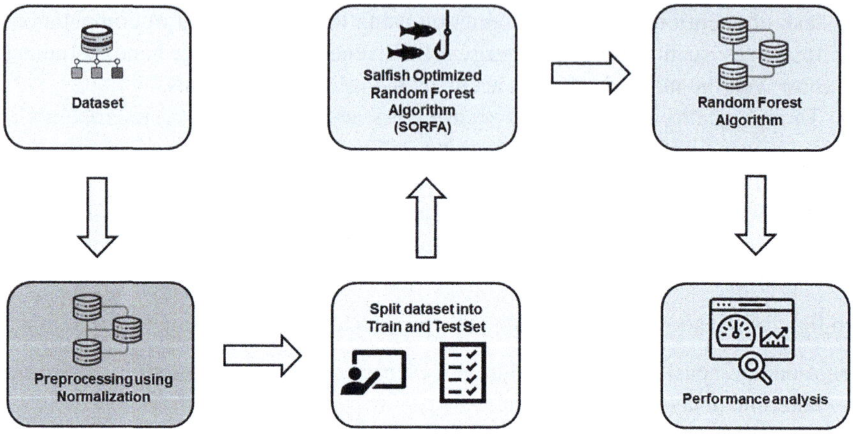


**Fig. 1** Methodology used in this research

size amounted to 1,457. We received nearly 1,600 responses. The participants, averaging 35.74 years with a standard deviation of 11.3, consisted of 52.3% males and 47.7% females. The distribution of ethnicities was as follows: 75.4% white, 13.5% Asian, 8.5% black, 1.4% American Indian or Alaskan and 1.1% others. Education-wise, 10% had completed high school, 34% had attended some college and 56% had a bachelor's degree or higher. The employment status was divided as follows: 65% were paid employees, 17% self-employed, 6% job-seeking, 8% not in the job market and 4% were students [12]. The data that support the findings of this study are openly available in https://github.com/khamesi/Perceived-Value-User-Utility.git.

## 2.2 *Preprocessing Using Normalization*

Normalization, a common preprocessing method in machine learning, is used to scale numerical data into a standard range without distorting differences in value ranges or losing information. It ensures each variable contributes proportionally to the study and enhances the effectiveness of machine learning algorithms. Normalization is particularly crucial when features have varied scales or units, as some algorithms might be sensitive to these variations. One popular normalization technique is Min–Max scaling, among others. By subtracting the minimum value of the feature and dividing it by its range, Min–Max scaling scales a feature's values between 0 and 1 is done using Eq. (1):

$$Y_{normalized} = \frac{Y - Y_{\min}}{Y_{\max} - Y_{\min}} \tag{1}$$

$Y$ denotes the original parameter of the feature, $Y_{\min}$ denotes the minimum value of the feature, $Y_{\max}$ denotes the maximum value of the feature and $Y_{normalized}$ represents the normalized value of the feature.

Sailfish optimized algorithm

Inspired by the hunting behaviors of sailfish, known for their remarkable speed and agility, the Sailfish Optimization Algorithm (SOA) is a meta-heuristic optimization technique [13]. It mathematically simulates sailfish hunting strategies, using a population-based approach where a group of sailfish is initialized randomly and updated iteratively following guidelines that mirror the sailfish behavior. SOA has been successful in solving various optimization problems, including multi-objective, continuous and discrete issues. It has demonstrated competitiveness with other meta-heuristic methods like particle swarm optimization, genetic algorithms and ant colony optimization [14].

Initializing the location of the sardine and sailfish populations

In the SFO method, the population of sailfish represents a candidate solution, while the sardine population signifies multiple solutions. Locating the sailfish and sardines in the search space aids in solving the HDEED problem. Hence, we use random generation to create the population in the solution space. The location matrices of the initialized sailfish can be seen in Eq. (2):

$$Y_{SF}^{(d)} = \left[Y_{j,i}^{(d)}\right]_{n\times t} = \begin{bmatrix} Y_{1,1}^{(d)} & Y_{1,2}^{(d)} & Y_{1,t}^{(d)} \\ Y_{2,1}^{(d)} & Y_{2,2}^{(d)} & Y_{2,t}^{(d)} \\ \vdots & \vdots & \vdots \\ Y_{n,1}^{(d)} & Y_{n,2}^{(d)} & Y_{n,t}^{(d)} \end{bmatrix} \tag{2}$$

Here, $n$ denotes the amount of sailfish in the population of sailfish, $d$ denotes the dimension of the parameter $Y_{j,i}^{(d)}$ denotes the values of the $j$ th dimension of the $i$ th sailfish and $d$ is the number of recent iterations. The matrices are displayed upon initialization of the sardine location can be seen in Eq. (3):

$$Y_{S}^{(d)} = \left[Y_{j,i}^{(d)}\right]_{n\times t} = \begin{bmatrix} Y_{1,1}^{(d)} & Y_{1,2}^{(d)} & Y_{1,t}^{(d)} \\ Y_{2,1}^{(d)} & Y_{2,2}^{(d)} & Y_{2,t}^{(d)} \\ \vdots & \vdots & \vdots \\ Y_{n,1}^{(d)} & Y_{n,2}^{(d)} & Y_{n,t}^{(d)} \end{bmatrix} \tag{3}$$

$m$ denotes the set of sardines in the population, $Y_{j,i}^{(d)}$ denotes the values of the $i$th dimension of the $j$th sardine.

The fitness function of the approach utilized to assess the fitness of the sailfish and sardine solution is the target role of the HDEED issue. The fitness function is $L$ and is saved as a matrix. Here $Y_{S}^{(d)}$ is a sailfish fitness matrix as shown in Eq. (4):

$$Y_{SF}^{(d)} = \begin{bmatrix} L_1^{(d)} \\ L_2^{(d)} \\ \vdots \\ L_n^{(d)} \end{bmatrix} = \begin{bmatrix} L\left(Y_{1,1}^{(d)} Y_{1,2}^{(d)} Y_{1,t}^{(d)}\right) \\ L\left(Y_{2,1}^{(d)} Y_{2,2}^{(d)} Y_{2,t}^{(d)}\right) \\ \vdots \\ L\left(Y_{n,1}^{(d)} Y_{n,2}^{(d)} Y_{n,t}^{(d)}\right) \end{bmatrix} \tag{4}$$

Here, $L_{GL}^{(t)}$ denotes the fitness matrices of the sailfish, $m$ denotes the number of sailfishes and $L_j^{(t)}$ is the fitness value of the $i$th sailfish. The sardine fitness matrix is displayed as shown in Eq. (5):

$$Y_S^{(d)} = \begin{bmatrix} L_1'^{(d)} \\ L_2'^{(d)} \\ \vdots \\ L_n'^{(d)} \end{bmatrix} = \begin{bmatrix} L'\left(Y_{1,1}'^{(d)} Y_{1,2}'^{(d)} Y_{1,t}'^{(d)}\right) \\ L'\left(Y_{2,1}'^{(d)} Y_{2,2}'^{(d)} Y_{2,t}'^{(d)}\right) \\ \vdots \\ L'\left(Y_{n,1}'^{(d)} Y_{n,2}'^{(d)} Y_{n,t}'^{(d)}\right) \end{bmatrix} \tag{5}$$

Here, $L_G^{(d)}$ denotes the sardine fitness matrix, $m$ denotes the quantity of sardines and $L_j^{(d)}$ denotes the fitness value of the $j$th sardine.

### 2.2.1 Selection of Elite Method

The elite sailfish is determined by the SFO method's iterative process and its position is logged as $Y_{elite}^{(d)}$. Sardines suffer harm from sailfish assaults as well. Seriously injured sardines are those in the sardine population with the lowest fitness and their position is noted as $Y_{injured}^{(d)}$.

### 2.2.2 Hunting Approach

When a captured sardine's fitness is lower than that of a sailfish, the sailfish's location is used instead of the sardine, as seen in Eq. (6):

$$Y_i^{(d)} = Y_j'^{(d)} \; if \; L\left(Y_i^{(d)}\right) > L'\left(Y_j'^{(d)}\right) \tag{6}$$

The sailfish's location during iteration $T$ is shown here and $Y_j^{(d)}$ is the sardine's location in the $d$ iteration.

## 2.3 Random Forest Algorithm

The training of each tree exhibits the "randomness" of the random forest. To learn the method, a selection of datasets with an assortment of the same number of elements is made at random from all training examples [15]. Bootstrap sampling is the term used to describe this data-gathering technique [16]. The optimum segmentation method of subset feature partitioning by purity calculations is obtained by randomly selecting many subsets of all features in every branch node variable when building the tree [17]. The number of trees impacts the ultimate choice value since the "forest" is represented by each tree growing to its full potential without being pruned [18]. The CART method and bootstrap sampling form the foundation of the random forest technique. The subsequent stages make up the random forest algorithm's procedure:

- **Produce training sets**: A training set is associated with each tree. N training sets must be produced to build N decision trees. To create a subset of training sets, every training set is collected using bootstrap samples. Using bootstrap aggregation, also referred to as bagging technology, all training set subsets are combined to predict the outcomes. High diversity and reliable outcomes are guaranteed by the substantial variances among the sub-training sets.
- **Create every decision tree**: Depending on the node splitting rule, which improves the gain of information and data gain rate while minimizing the Gini coefficient, node splitting is typically performed from the root node via the intermediate node and to the leaf node. For node splitting, the CART method typically uses the least Gini coefficient. The procedure of node splitting is repeated. According to a specified probability distribution, this particular technique enables the random selection of many attributes from among all attributes to take part in the node-splitting process.
- **Develop the forest**: The random forest is created by repeatedly doing the first two processes to create a vast number of decision trees. The samples in the OOB data are categorized using each tree in the forest. The votes for each category are tallied and the category with the most votes is regarded as the sample category. One instance of the categories is indicated as one vote appropriately. The error rate for the OOB data is the percentage of samples that were incorrectly categorized in the OOB data.

In the series of decision trees, M gets bigger $r(y_1), r(y_2) \ldots r(y_M)$ denotes the random forest for the majority of random variable patterns $\theta_1, \theta_2, \ldots .\theta_r$. Equation (7) are the convergences of the generalization error GE*:

$$\lim_{M\to\infty} GE^* = B_{yx}\left\{B_\theta\left[r(y,\theta) = X\right] - \max_{i\neq X} B_\theta(r(y,\theta) = i) < 0\right\} \tag{7}$$

This demonstrates that the probability upper limit is within the range of the generalization error GE*. As the number of decision trees rises, overfitting does not, which is a sign of strong robustness.

### 2.4 Integration of Sailfish Optimization Algorithm with Random Forest

SORFA enhances the random forest by optimizing its key hyperparameters, including the number of trees, maximum depth, and minimum samples split. The integration process involves the following steps:

- The sailfish and sardine populations are initialized with random values representing the hyperparameters of the random forest.
- The fitness of each sailfish and sardine is evaluated based on the predictive accuracy of the random forest on a validation dataset.
- The sailfish optimization algorithm iteratively updates the hyperparameters by simulating the hunting behavior of sailfish, where better-performing configurations are propagated and poor-performing ones are discarded.
- The algorithm converges to an optimal set of hyperparameters that maximize the accuracy of the random forest while minimizing computational time.

The primary variables optimized in the SORFA process is Number of Trees (*n_ estimators*) i.e., the total number of decision trees in the random forest.

The objective function used to evaluate the fitness of each individual is the mean squared error (MSE) of the random forest predictions on a validation dataset. Lower MSE values indicate better predictive performance.

## 3 Results and Discussion

The efficiency of energy usage in a smart home is contingent upon the strategies and techniques employed to maximize energy savings. This can yield both positive and negative outcomes. By implementing effective energy efficiency measures and addressing potential negative implications, smart homes can significantly influence the environment and their inhabitants. This section analyzes several traditional systems including Home Energy Management Systems (HEMS) [19], Linear Regression Model (LRM) [20], Perceived-Value Driven (PVD) [12], as well as our proposed Selfish Optimization Random Forest Algorithm (SORFA).

### 3.1 Energy Consumption

Energy consumption refers to the aggregate amount of energy utilized by a machine, system, or home over a given period. Typically, energy consumption is quantified in units of energy such as kilowatt-hours (kWh), joules (J), or British thermal units (BTUs). Figure 2 presents a comparison of energy consumption across traditional and proposed techniques. In terms of energy consumption, the figure demonstrates that

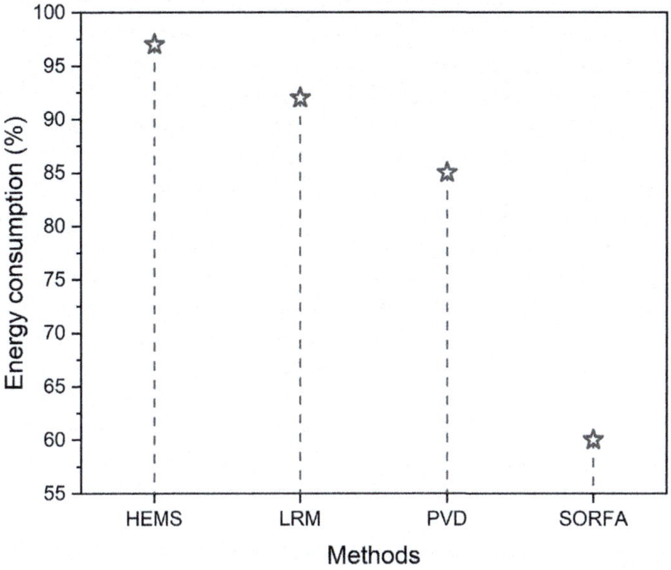


**Fig. 2** Comparative analysis of energy consumption by SORFA, HEMS [19], LRM [20], PVD [12]

SORFA performs significantly better than all the other methods considered. HEMS, LRM and PVD exhibit energy consumption of 97, 92 and 85% respectively, while SORFA manages to reduce this to 60%. This implies that SORFA is more efficient in energy usage, offering improvements of approximately 38.14, 34.78 and 29.41% over HEMS, LRM and PVD respectively. Notably, the proposed SORFA method demonstrates a decrease in energy consumption relative to traditional methods.

## *3.2 Security*

Security is the practice of protecting systems, assets and data from unauthorized access, theft, damage, or disruption. Given the ubiquity of digital technology in our personal and professional lives, security is integral to modern living. The high potential for hacker attacks and physical intrusions makes security a paramount concern in smart homes, aiming to prevent unauthorized access and damage to devices, data and networks.

Figure 3 denotes the comparison of security with traditional and suggested techniques. For Dataset 1, SORFA attains a security level of 94%, considerably outperforming HEMS (80%), LRM (65%) and PVD (70%). In Dataset 2, while PVD offers improved security at 81%, it is still surpassed by SORFA's security level of 92%. HEMS and LRM exhibit lower security percentages at 65 and 70% respectively. Dataset 3 follows a similar pattern with SORFA reaching 91% in security, outpacing

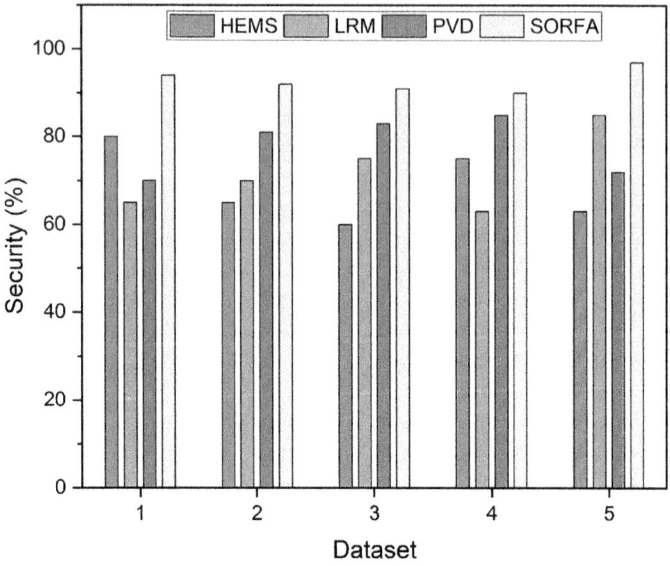


**Fig. 3** Comparative analysis of security measures with SORFA, HEMS [19], LRM [20], PVD [12]

HEMS at 60%, LRM at 75% and PVD at 83%. For Dataset 4, the security measures of all the techniques are relatively closer, but SORFA still leads with a security level of 90% against HEMS (75%), LRM (63%) and PVD (85%). Lastly, Dataset 5 shows a slight variation with LRM providing a high security measure of 85%, but SORFA still surpasses all with a security level of 97%. HEMS and PVD offer security percentages of 63 and 72% respectively. Thus, this analysis shows that SORFA consistently achieves the highest security levels across different datasets, making it a superior choice in terms of security measures. However, it's important to note that this is a singular facet of system performance. Other factors like energy consumption, cost savings and time delay also need to be taken into account while considering the overall performance of these techniques.

### *3.3 Cost Savings*

Cost savings represent a reduction in expenses over time. There are various strategies for cost savings including decreasing raw material costs, enhancing operational efficiency and reducing waste. Smart home solutions offer remote monitoring and control of appliances and systems, potentially saving on maintenance and repair costs by eliminating the need for onsite visits.

Figure 4 highlights the comparative analysis of cost savings across different methods. In terms of cost savings, SORFA again outperforms the traditional methods. While HEMS, LRM and PVD are associated with cost savings of 97, 83 and 72%

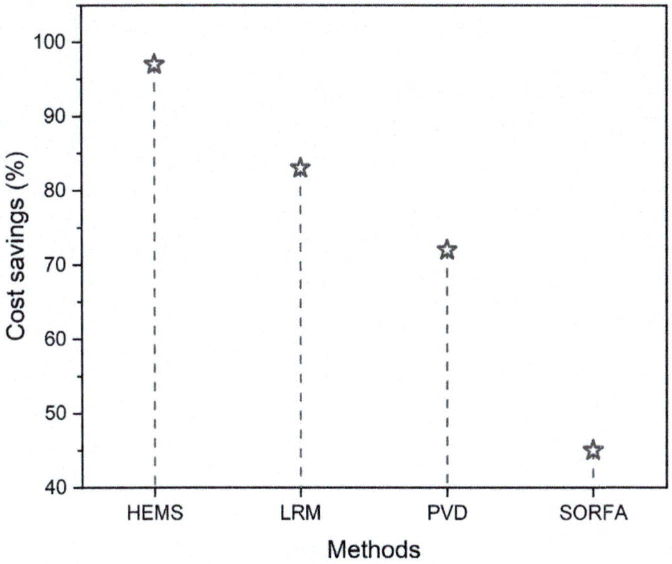


**Fig. 4** Comparative analysis of cost savings by SORFA, HEMS [19], LRM [20], PVD [12]

respectively, SORFA boosts this figure to 45%. This suggests that SORFA is more efficient in achieving cost savings, realizing improvements of approximately 53.61, 45.78 and 37.50% over HEMS, LRM and PVD respectively. In contrast to traditional methods, the proposed SORFA method achieves higher cost savings.

### *3.4 Time Delay*

Time delay refers to the duration of a signal or data journey from one point to another. Systems such as communication, control and transportation can experience time delays, leading to instability or oscillations. This is particularly problematic for robotic and aviation control systems that require swift responses.

Figure 5 presents a comparison of time delay across various methods. In each of the five datasets, SORFA consistently demonstrates the shortest time delay. For Dataset 1, the time delays for HEMS, LRM and PVD are notably longer at 96, 91 and 85 s respectively, while SORFA records a significantly shorter delay of 70 s. In Dataset 2, the trend continues with SORFA having the shortest delay (60 s), outperforming HEMS (82 s), LRM (73 s) and PVD (85 s). For Dataset 3, SORFA maintains its lead, achieving a time delay of 65 s. This is considerably shorter than the delays observed with HEMS (90 s), LRM (83 s) and PVD (76 s). Dataset 4 exhibits similar results. Here, SORFA records a time delay of 50 s, which is shorter than the delays of HEMS (82 s), LRM (70 s) and PVD (73 s). For Dataset 5, SORFA once again has the shortest time delay (45 s) when compared to HEMS (72 s), LRM (65 s) and

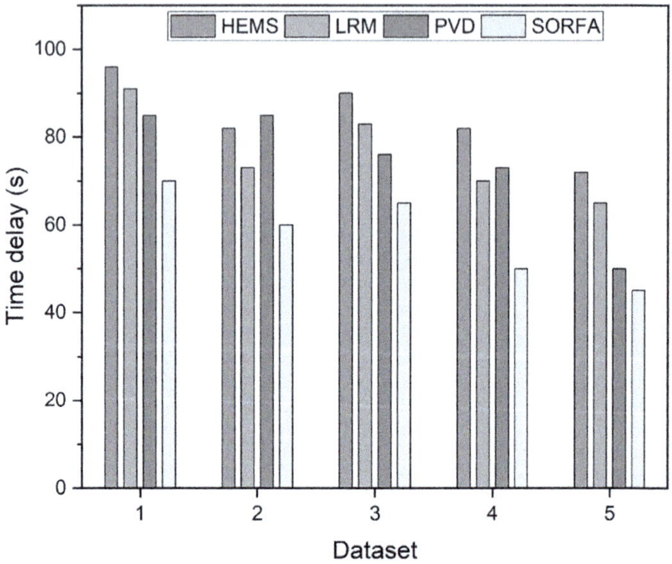


**Fig. 5** Comparative analysis of time delay with SORFA, HEMS [19], LRM [20], PVD [12]

PVD (50 s). This analysis indicates that, in terms of time delay, SORFA consistently outperforms HEMS, LRM and PVD across the datasets, making it the most efficient method among those tested.

# 4 Conclusion

Effective analysis of energy consumption is vital for optimizing energy use and cost savings in a smart home. By collecting and scrutinizing energy use data, homeowners can identify patterns and areas where energy efficiency can be improved. This study aids homeowners in identifying energy-saving strategies to reduce energy usage and costs. Our proposed SORFA method provides 60% energy consumption, 45%-time delay, 50% cost savings and 97% security, demonstrating superior efficiency to other traditional methods. Future research will focus on leveraging machine learning algorithms and artificial intelligence to enable smart homes to evaluate energy usage data in real-time, identify patterns and improve energy efficiency.

Despite the promising results, this study is not without its limitations. Firstly, the analysis was limited to specific datasets and methods such as HEMS, LRM, PVD and SORFA. Therefore, the findings may not generalize to other methods or different data configurations. Secondly, while the study examined several key aspects including energy consumption, cost savings, security, and time delay, other significant aspects like user comfort, ease of installation, or compatibility with different smart devices were not explored. Thirdly, the SORFA approach was used for the first time in this context, and its real-world application, feasibility, and scalability are yet to be fully explored.

Despite these limitations, the study opens up several avenues for future research. Given the rapidly evolving nature of smart home technologies, further investigation can be conducted with a larger variety of methods, more comprehensive datasets, and additional performance measures. In the future, real-world testing of the SORFA approach in various types of smart homes, across diverse geographical locations and different user profiles, could provide a more comprehensive understanding of its effectiveness. This could include tests in different residential settings, from individual houses to large apartment complexes. Moreover, incorporating machine learning and artificial intelligence more extensively into smart home energy management could be a promising direction for future work. These advanced technologies could facilitate real-time analysis and optimization of energy use, enhancing efficiency, security, and cost-effectiveness. Future research could also explore user interface designs and automated recommendation systems to make these technologies more accessible and user-friendly. Lastly, with the rise of the Internet of Things (IoT), the integration of SORFA with other IoT devices could provide an exciting direction for future research, potentially enhancing the synchronization and efficiency of smart homes.

## References


1. Kasaraneni, P.P., Kumar, Y.V.P., Moganti, G.L.K., Kannan, R.: Machine learning-based ensemble classifiers for anomaly handling in smart home energy consumption data. Sensors (Basel) **22**, 9323 (2022). https://doi.org/10.3390/s22239323
2. Alzoubi, A.: Machine learning for intelligent energy consumption in smart homes. Int. J. CIM **2** (2022). https://doi.org/10.54489/ijcim.v2i1.75
3. Prakash, P., Kumar, P.: Simple and effective descriptive analysis of missing data anomalies in smart home energy consumption readings. J. Energy Syst. **5**(3), 199–220. https://doi.org/10.30521/jes.878318
4. Priyadarshini, I., Sahu, S., Kumar, R., Taniar, D.: A machine-learning ensemble model for predicting energy consumption in smart homes. Internet Things **20**, 100636 (2022). https://doi.org/10.1016/j.iot.2022.100636
5. Huang, Y., Wang, K., Gao, K., Qu, T., Liu, H.: Jointly optimizing microgrid configuration and energy consumption scheduling of smart homes. Swarm Evol. Comput. **48**, 251–261 (2019). https://doi.org/10.1016/j.swevo.2019.04.007
6. Kamel, E., Memari, A.M.: State-of-the-art review of energy smart homes. J. Archit. Eng. **25**, 03118001 (2019). https://doi.org/10.1061/(ASCE)AE.1943-5568.0000337
7. Aliero, M.S., Qureshi, K.N., Pasha, M.F., Jeon, G.: Smart home energy management systems in Internet of Things networks for green cities demands and services. Environ. Technol. Innov. **22**, 101443 (2021). https://doi.org/10.1016/j.eti.2021.101443
8. Fakhar, M.Z., Yalcin, E., Bilge, A.: A survey of smart home energy conservation techniques. Expert Syst. Appl. **213**, 118974 (2023). https://doi.org/10.1016/j.eswa.2022.118974
9. Paredes-Valverde, M.A., Alor-Hernández, G., García-Alcaráz, J.L., Salas-Zárate, M.D.P., Colombo-Mendoza, L.O., Sánchez-Cervantes, J.L.: IntelliHome: an Internet of Things-based system for electrical energy saving in smart home environment. Comput. Intell. **36**, 203–224 (2020). https://doi.org/10.1111/coin.12252

10. Caldera, M., Hussain, A., Romano, S., Re, V.: Energy-consumption pattern-detecting technique for household appliances for smart home platform. Energies **16**, 824 (2023). https://doi.org/10.3390/en16020824
11. Adams, S., Greenspan, S., Velez-Rojas, M., Mankovski, S., Beling, P.A.: Data-driven simulation for energy consumption estimation in a smart home. Environ. Syst. Decisions **39**, 281–294 (2019). https://doi.org/10.1007/s10669-019-09727-1
12. Khamesi, A.R., Silvestri, S., Baker, D.A., Di Paola, A.: Perceived-value-driven optimization of energy consumption in smart homes. ACM Trans. Internet Things **1**, 1–26 (2020). https://doi.org/10.1145/3375801
13. Shadravan, S., Naji, H.R., Bardsiri, V.K.: The Sailfish optimizer: a novel nature-inspired metaheuristic algorithm for solving constrained engineering optimization problems. Eng. Appl. Artif. Intell. **80**, 20–34 (2019). https://doi.org/10.1016/j.engappai.2019.01.001
14. Kalita, K., Ramesh, J.V.N., Cepova, L., Pandya, S.B., Jangir, P., Abualigah, L.: Multi-objective exponential distribution optimizer (MOEDO): a novel math-inspired multi-objective algorithm for global optimization and real-world engineering design problems. Sci. Rep. **14** (2024). https://doi.org/10.1038/s41598-024-52083-7
15. Gupta, K.K., Kalita, K., Ghadai, R.K., Ramachandran, M., Gao, X.-Z.: Machine learning-based predictive modelling of biodiesel production-a comparative perspective. Energies **14**, 1122 (2021). https://doi.org/10.3390/en14041122
16. Gayathri, R., Rani, S.U., Čepová, L., Rajesh, M., Kalita, K.: A comparative analysis of machine learning models in prediction of mortar compressive strength. Processes **10** (2022). https://doi.org/10.3390/pr10071387
17. Ganesh, P., Jain, P., Choudhury, A., Dutta, P., Kalita, K., Barsocchi, P.: Random forest regression-based machine learning model for accurate estimation of fluid flow in curved pipes. Processes (Basel) **9**, 2095 (2021). https://doi.org/10.3390/pr9112095
18. Burande, D.V., Kalita, K., Chohan, J.S.: Machine learning-based predictive modelling of spot-welding process parameters. In: Lecture Notes in Networks and Systems, vol. 729 LNNS, pp. 337–346 (2023). https://doi.org/10.1007/978-3-031-36246-0_32
19. Nilsson, A., Wester, M., Lazarevic, D., Brandt, N.: Smart homes, home energy management systems and real-time feedback: lessons for influencing household energy consumption from a Swedish field study. Energy Build. **179**, 15–25 (2018). https://doi.org/10.1016/j.enbuild.2018.08.026
20. Moletsane, P.P., Motlhamme, T.J., Malekian, R., Bogatmoska, D.C.: Linear regression analysis of energy consumption data for smart homes. In: 2018 41st International Convention on Information and Communication Technology, Electronics and Microelectronics (MIPRO). Opatija (2018). https://doi.org/10.23919/MIPRO.2018.8400075

# Energy Consumption Optimization in Thread Machining by Various Hybrid Dragonfly Algorithms

**Hasan Öktem, Burak Öztürk, and Kanak Kalita**

**Abstract** Energy consumption plays a major role in evaluating the thread machining of cast iron materials. In particular, heat-treated ductile cast iron produced as austempered ductile cast iron (ADI) has mechanical properties such as high strength, ductility and toughness. However, when fitting parts which have thin walls are produced with ADI, the high cooling rates with this method result in loss of strength and fracturing in the walls. This work focuses on a novel study on the effect of response variables in determining energy-power consumption in the thread machining of fitting samples produced with ADI materials under different conditions. The experiments are designed using an L8 orthogonal array. Linear regression is used for developing predictive models and nature-inspired metaheuristic algorithms are employed for optimization. Dragonfly algorithm (DA), a recently developed and popular metaheuristic is considered the primary optimization algorithm. Seven advanced variants of the DA namely, biogeography-based Mexican hat wavelet dragonfly algorithm (BMDA), chaotic dragonfly algorithm (CDA), hybrid memory-based dragonfly algorithm with differential evolution (DADE), hybridization of dragonfly algorithm and artificial bee colony (HDA), hybrid Nelder-Mead algorithm and dragonfly algorithm (INMDA), memory-based hybrid dragonfly algorithm (MHDA) and quantum-behaved and Gaussian mutational dragonfly algorithm (QGDA) are also deployed to investigate the comparative advantage of each in such industrial optimization problem. The results of this study showed that heat treatment and

H. Öktem · B. Öztürk
Department of Machine and Metal Technology, Hereke Asim Kocabıyık Vocational School, Kocaeli University, İzmit, Turkey
e-mail: burak.ozturk@bilecik.edu.tr

K. Kalita (✉)
Department of Mechanical Engineering, Vel Tech Rangarajan Dr. Sagunthala R&D Institute of Science and Technology, Avadi 600 062, India
e-mail: drkanakkalita@veltech.edu.in

Department of Machining, Assembly and Engineering Metrology, Faculty of Mechanical Engineering, VSB-Technical University of Ostrava, Ostrava 70800, Czechia

 
D. Oliva et al. (eds.), *Advances in Optimization Algorithms for Multidisciplinary Engineering Applications: From Classical Methods to AI-Enhanced Solutions*, Studies in Computational Intelligence 806, https://doi.org/10.1007/978-3-031-78440-8_28

the types of machines are two important response variables in determining energy-power consumption. Among the algorithms, DA outperforms the other algorithms in locating the optima.

**Keywords** Ductile cast iron · Heat treatment · Thread machining · Energy-power consumption · Dragonfly algorithm · Interactive design

## 1 Introduction

Spheroidal graphite cast iron products are widely used in various fields such as the aerospace, automotive and fittings industries [1, 2]. It has certain advantages such as rapid production machinability, heat treatment ability and low cost [3, 4]. Due to these use advantages, ductile iron casting of cast iron installation pipe fittings according to TS 11 and EN 10,242 standards is preferred in the industry [5]. Spheroidal graphite cast iron consists of pearlite and spheroidal graphite dispersed in ferrite. Spheroidal graphite is often formed in place of the lamellar gray cast iron graphite during most of the carbonation. This matrix is called "nodular graphite" in the literature (Fig. 1). Ferrite and pearlite volume ratios formed in the matrix structure may vary depending on the cooling rate, the contents of the alloying elements, and the amount of graphite. Ductile iron with a pearlitic matrix generally consists of graphite spheres surrounded by a ferritic microstructure produced by adding 0.1% Mg to a solution containing 3–4% C, and 1.8–2.8% Si [6]. The ductile iron material is produced from gray cast iron spheroized by magnesium. In addition, since casting conditions are not ideal, there is always segregation of the casting materials, depending on the cooling rate of the casting molds and the purity of the materials used. To keep the segregation and carbides to a minimum, the charge entrances should be optimally controlled. This ensures that the product has the desired mechanical properties and that the undesired compounds remain at the lowest level [7].

Austempered ductile iron (ADI) is solidified as spheroidal graphite cast iron and then subjected to heat treatment by "austempering". In the austempering process, the workpiece is first subjected to a pre-treatment at temperatures of 871–927 °C to form austenite and spheroidal graphite structure. It is cooled in a salt bath to a temperature above the temperature at which the martensite structure begins to form. For the isothermal treatment, it is usually kept in a salt bath at a temperature of 250–370 °C for a specified time and then cooled to room temperature in Fig. 2. In the austempering heat treatment, if the maximum toughness and ductility is to be achieved, quenching from the austenite temperature to the isothermal conversion temperature must be fast enough for the pearlitic transformation. Spheroidal graphite cast iron is different from isothermal carbon steel as it is austempered. In the cast iron austempering heat treatment, firstly, austenite and ferrite (ausferrite) are formed instead of a bainitic structure (ferrite and carbide) [8].

Since fittings are thin-walled products, they have a fast cooling rate. This cooling rate is the result of the isothermal transformation of the bainitic microstructure,

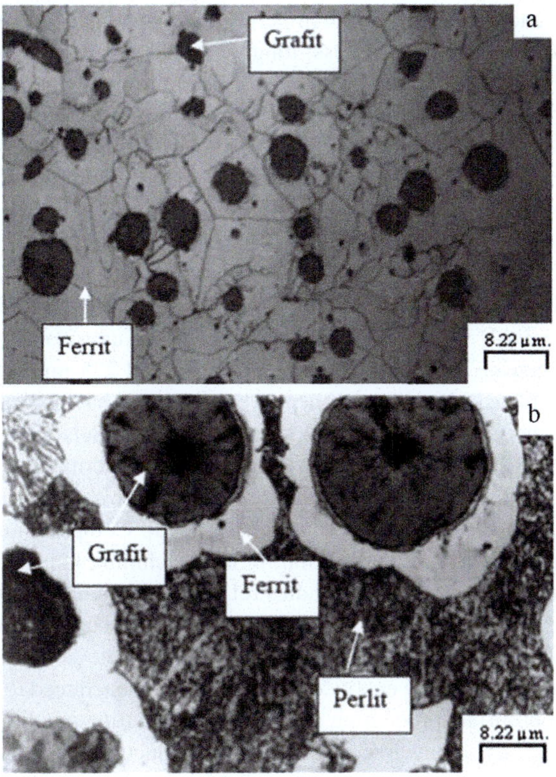


**Fig. 1** Optical micrographs of ADI materials: **a** Ferrite **b** Pearlite + ferrite + graphite

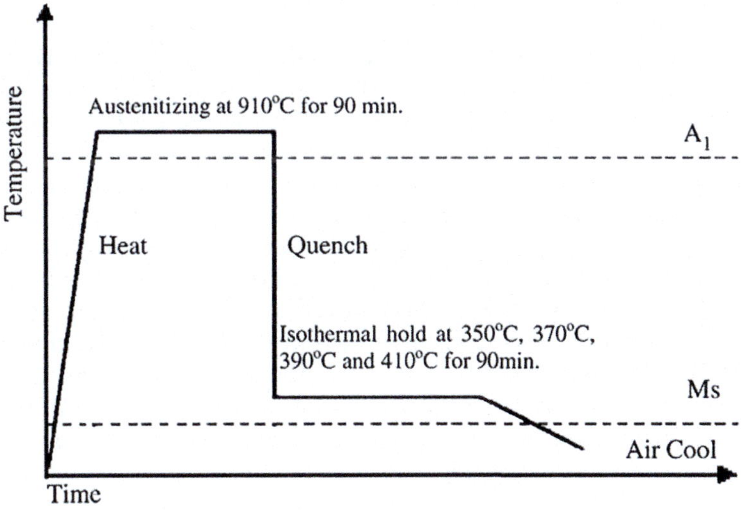


**Fig. 2** Schematic description of heat treatment of ductile cast iron with salt bath

which is hard and fragile. The austempering heat treatment reduces the machinability while increasing the hardness, toughness and mechanical properties of the fittings. There are studies related to different austempering temperatures and cooling rates of cast iron depending on the thickness of the walls. One study on ductile cast iron investigated the effects on the microstructure of austempering at temperatures of 350, 370, 390 and 410 °C in different amounts of saline solution. In addition, impact strengths were examined [9]. Cooling rates of 1.9–17.67 °C/s were detected for material designs produced in different dimensions from gray cast iron cast at 1420 °C. As the cooling rate increased, the dendrite arm spacing (DAS) characteristics increased and thin dendrites formed during rapid and slow cooling [10]. The effect of the cooling rate on DAS was investigated in austempered cast iron [11] and the chill formation and tensile strength at 25, 75, 125, 175 and 225 mm were determined. Twelve different tensile model specimens were produced having wall thicknesses of 1–20 mm and the microstructure of 2, 6 and 10 mm models were examined after casting [12]. Pipe fittings are subjected to internal and external threading according to BSP standards. A review of the literature revealed several studies investigating the gradual pipe threading process on a lathe [13–16]. Computer numerical control (CNC) is commonly used for processing in the machining industry. In the context of this machining process, various studies related to the optimization of the cutting parameters have been published. Most of them have used the surface roughness, shear force, cutting force, tool life and material removal rate as optimization criteria. In manufacturing operations, energy consumption can be reduced to optimum levels by choosing appropriate processing parameters and cutting conditions [17–24]. To achieve better energy saving associated with the production process, the relationship between the electrical energy consumption and machine tools and cutting operations must be investigated.

Mori et al. drilled of S45C carbon steel using a 10 mm-diameter drill, end milling with a 10 mm-diameter cutter and face milling with an 80 mm-diameter cutter in a vertical machining center. To determine the effect of cutting conditions on power consumption, the total machine power consumption and shaft power consumption of the vertical machining center were measured by connecting a clamp-type ammeter to the input power supply cable. Reduced power consumption could be achieved for drilling and face/end milling by setting the cutting conditions high, yet within a value range which does not compromise tool life or surface finish, and thereby shortening the machining time [17]. Liu et al. performed slot milling of 273 × 50 × 90 mm ASSAB 760 steel samples under dry cutting conditions on a Makino-V55 3-axis CNC vertical machining center using an end mill with a radius of 5 mm. The forces were measured with a 3-channel dynamometer (Kistler 9265B). The power consumption of the CNC vertical machining center was measured by the Panasonic Eco-Power Meter KW9M [18]. In a turning operation by Camposeco-Negrete on AISI 6061 T6 aluminium material, Taguchi methodology was employed for optimizing a roughing turning process involving aluminium AISI 6061 T6 as the material to be machined and a silicon carbide-reinforced ceramic grade cutting tool. Optimum values for the cutting parameters were found to minimize cutting power, cutting energy, energy-power consumption and surface roughness ($R_a$) during the machining process [19].

Oda et al. used die steel 40CrMoV5 to manufacture 5-axis machine tools for various inclined angles. An optimized inclined angle was determined and cutting experiments were conducted to reduce energy consumption. Power consumption was varied by using different inclined angles when the contour machining was carried out on an inclined surface. In this experiment, the ball end milling was optimal at an inclined angle of 158°, with the lowest power consumption per material removal unit and the longest tool life [20]. Shokoohi et al. conducted turning tests on AISI 1045 steel using a universal lathe. Surface roughness was measured at three different points to increase accuracy. A HIOKI ammeter was used to record the electric current and measure the processing power indirectly. The study was aimed to investigate the role of cooling and lubrication effects on machining performance by implementing different procedures in the turning operation [21]. Yurdakul investigated the effect of cutting depth and feed rate of the saw on the power level consumed during the cutting of granite using circular saw blades. In this context, two-column multi-blade block cutters were used to evaluate twelve different cutting conditions. The energy consumption in the implementation of these test designs was investigated [22]. Neugebauer et al. machined EN-GJL-250 Gy cast iron Ø7 mm in diameter for a drilling depth five times that of the drill bit. Energy consumption changes were investigated by processing under different parameters such as cutting speed, feed rate, tool type and dry and cooled machining conditions. Despite variations in product objectives, the analysis of the two machining methods demonstrated general trends related to an increase in energy efficiency. The design of an efficient machining process must extend beyond the study of the process level alone since this can provide only a partial picture of the energy consumption [23]. Escalona et al. used a 14 mm diameter finger milling cutter to remove chips from rectangular 303 stainless specimens to investigate energy consumption. After analysing the results of the 303 stainless steel machined under different cutting environments, it was concluded that under the specified cutting parameters the dry environment was more suitable for the machining process as it achieved the lowest results for surface roughness, power consumption and material micro-hardness [24]. The power index (PI) was determined by an ammeter assembly during chip removal from a spheroidal workpiece and the machining process for two different cutting tools was optimized using Taguchi methodology in the study of Nas and Öztürk [25].

In this study, optimum power and chip removal conditions were investigated by determining the energy-power consumption in the machining of many different types and sizes of material. In the present study, the threading process was applied to samples of austempered spheroidal graphite cast iron fittings based on the $L_8$ orthogonal array Taguchi design using two different threading machines and four different heat treatment parameters. Based on the analytical formula, the energy-power consumption was calculated according to the values obtained and the optimum machining parameters were determined via nature-inspired optimization algorithms. The focus of this work was energy-power consumption calculated from the values measured during processing with a CNC mill and a manual threading machine in serial production of fittings, a subject that has not been extensively investigated in the literature. For this process, the ¾ inch size most commonly used for plumbing systems was chosen and the external threading was compared with the internal threading. Here, spheroidal graphite cast iron subjected to three heat treatment conditions (30, 60 and 90 min at 900 °C) together with non-heat-treated material and two different types of threading machines were selected as the experimental parameters. The obtained results were analysed and the optimum response variables were determined.

## 2 Methodology

### 2.1 Dragonfly Algorithm (DA)

The classical DA [26] is a simple yet powerful metaheuristic algorithm mimicking the swarm behaviour of dragonflies. The social behaviour of dragonflies exhibited during searching and gathering of food as well as during foe avoidance forms the basis of the equation-based rules used to simulate and actuate the DA. The DA algorithm is realized by using the following five parameters namely, separation, alignment, cohesion, attraction and distraction. Collision avoidance with neighbouring dragonflies is governed by separation. Velocity matching to the neighbouring individuals is done by alignment. Attraction towards the center of mass of the neighbourhood and towards a food source are governed by cohesion and attraction respectively. Away movement from the enemy is governed by distraction.

### 2.2 Biogeography-Based Mexican Hat Wavelet Dragonfly Algorithm (BMDA)

An enhanced DA called BMDA (biogeography-based algorithm, Mexican hat wavelet and dragonfly algorithm) was proposed by Shirani and Faramarz [27] to rectify premature convergence in high workloads while introducing a mutation phase based on a combination of biogeography-based optimization (BBO) technique's migration process and the Mexican hat wavelet's transformation process in DA.

### 2.3 Chaotic Dragonfly Algorithm (CDA)

Sayed et al. [28] employed ten chaotic maps to fine-tune the weights involved in the separation, alignment, cohesion, attraction and distraction parameters of classical DA. The authors argue that as compared to DA, CDA has improved convergence rate, with the algorithmic complexity being at par with DA. The overall complexity of CDA is $O(dM + MC)$, where $d$, $M$ and $C$ are the dimensions of the problem, the number of dragonflies and objective function complexity respectively.

## 2.4 *Hybrid Dragonfly Algorithm with Differential Evolution (DADE)*

Differential evolution (DE), in general, has high computational ability and a fast convergence rate. Akin to genetic algorithm (GA), DE explores the search space by using crossover and mutation. At the end of each cycle, DADE [29] stores the best solution in its memory and continues the search with DE which promotes population diversity by employing mutation.

## 2.5 *Hybridization of Dragonfly Algorithm and Artificial Bee Colony (HDA)*

Ghanem and Jantan [30] argue that the presence of Levy Flight in the position updation phase of DA makes it unable to effectively carry out local search. To rectify this, the authors suggest the hybridization of DA and artificial bee colony (ABC) algorithm to make use of the exploitation and exploration ability of DA along with the exploration ability of ABC.

## 2.6 *Hybrid Nelder-Mead Algorithm and Dragonfly Algorithm (INMDA)*

Xu and Yan [31] argue that too many social interactions in DA are responsible for reduced solution accuracy and premature convergence to local optima. These are caused due to the improper balance between diversification and intensification. Xu and Yan [31] suggest hybridizing DA with an improved Nelder-Mead algorithm to improve its local search capacity.

## 2.7 *Memory-Based Hybrid Dragonfly Algorithm (MHDA)*

The lack of internal memory in the classical DA can cause premature convergence to local optima. To overcome this, Sree Ranjini and Murugan [32] introduced certain features of PSO into the DA and called the hybrid algorithm as MHDA. By endowing the DA with internal memory, MHDA allows each dragonfly to keep track of its DA-*pbest* solution i.e., the coordinates of the best solution obtained by it so far. MHDA also has access to DA-*gbest* i.e., the coordinates of the overall best solution obtained by the algorithm so far. After the initial exploration of the search space by DA, exploitation of the promising search space zones is initialized by PSO by using the DA-*pbest* and DA-*gbest* solutions.

### 2.8 *Quantum-Behaved and Gaussian Mutational Dragonfly Algorithm (QGDA)*

By implementing the concept of quantum rotation gate, Yu et al. [33] tried to strike a better balance between the exploration and exploitation traits of DA. Gaussian mutation was also incorporated in the algorithm to help generate diverse solutions.

## 3 Experimental Methods

The response variables for the threading operation such as heat treatment time and machine type was initially considered. The experimentation was designed using $L_8$ orthogonal array of Taguchi method. Experimentation was carried out for the experimental design. Nature-inspired optimization is implemented to determine the optimal response variables. In addition, the microstructural analysis of the samples was conducted to study the heat treatment effect, energy-power consumption and hardness. These steps were further elaborated in this section.

### 3.1 *Calculation of Energy-Power Consumption*

Using current transformers, the CNC vertical machining center was connected to the drive input phases for measuring energy consumption during machining in Fig. 3. The power index (PI) with $\pm 1\%$ accuracy was measured using an ammeter connection compatible with the current transformers. In the manual threading machine, the clamp-on ammeter was connected to the input phase of the asynchronous motor and the instantaneous PI was measured as an ampere value (A).

The power index value (PI) of the spindle servo motor drive measured via ammeter during the chip removal process was converted from the kW-type power consumption of the 3-phase motor using Eq. 1 for energy-power conversion [21].

$$P\,\text{Total} = \sqrt{3}.V.I.\text{Cos}\sigma \tag{1}$$

Following a review of the literature, significant cutting parameters such as material removal rate (MRR), the total amount of energy consumed (SEC) and material removal volume can be calculated using the energy power conversion equations and machining operations. The MRR is defined as the amount of chip ($mm^3$) removed from the workpiece in one second and Q is the total amount of chip processed during manufacturing. The MRR is calculated using Eq. (2) given below. The SEC is the total amount of energy consumed to remove 1 $mm^3$ of the chip and is calculated by the formula in Eq. (3) [17, 25].

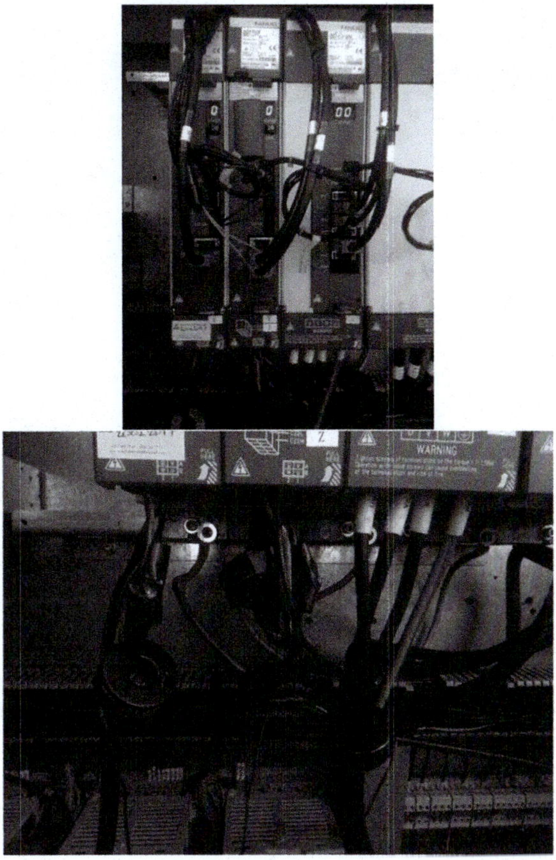


**Fig. 3** Connection of current transformers to drive inputs

$$MRR = (ap \times ae \times F)/1000 \tag{2}$$

$$SEC\left(\frac{\mathrm{J}}{\mathrm{mm}^3}\right) = \frac{P\,total(\mathrm{W})}{MRR\left(\frac{\mathrm{mm}^3}{\mathrm{s}}\right)} \tag{3}$$

The maximum torque required for the threading process in the manual threading machine (using Wagner head and chasers) was calculated using the classical torque equation (Eq. 4). For the CNC mill, this was calculated by taking into account the maximum amount of torque required by the spindle servo (Eq. 5).

$$Torque\ TM\ (\mathrm{Nm}) = \frac{9550 \times P(\mathrm{kW})}{N(dev/dak)} \tag{4}$$

$$Torque\,CNC\ (\mathrm{Nm}) = CNC\ Max\ Torque\ (\mathrm{Nm}) \times Spindle\ Power\ Strain\ Rate\ (\%) \tag{5}$$

**Fig. 4** Experimental specimens: (left) 3/4″ fitting, (right) 1V4–3/4″ reduction

**Table 1** Chemical compositions (wt. %)

| Process | Fe | C | Si | Mn | P | S | Cr | Mo | Ni |
|---|---|---|---|---|---|---|---|---|---|
| Before spheroidization | 94.4 | 3.64 | 1.63 | 0.14 | 0.02 | 0.013 | 0.02 | 0.0036 | 0.017 |
| After spheroidization | 93.3 | 3.58 | 2.64 | 0.14 | 0.03 | 0.015 | 0.02 | 0.0050 | 0.018 |

## *3.2 Austempering Heat Treatment of Cast Iron Fittings*

In this study, 1¾ inch joint fittings with 1¼–¾ inch reductions were produced according to TS 11 and EN 10,242 standards. Casting was carried out in a mass production industrial enterprise after the form was moulded by mechanical pressing in Fig. 4. Magnesium and silicon were added to carry out the spheroidizing process, thus increasing the toughness of the material. Table 1 shows the results of the spectral analysis before and after the spheroidizing heat treatment of the material being cast. Test samples austempered for 30, 60 and 90 min at a temperature of 900 °C were immediately cooled in a salt bath and austempering heat treatment was applied at a temperature of 280 °C for 30, 60 and 90 min. Thus, by reducing the fragility of the material and improving its mechanical properties, the workability was reduced.

## *3.3 Thread Machining of Cast Iron Fitting Samples*

The heat-treated 11/4-3/4″ reduction pipe fittings were threaded in the CNC milling machine shown in Fig. 5a at a speed of 170 RPM using an HSS cutting tool. However, the process was carried out for the ¾″ fitting samples on the manual threading machine with the Wagner head and chasers shown in Fig. 5b at a speed of 170 RPM.

According to the literature, the specific energy consumption changes during external threading using Wagner head and chasers has been investigated for the

(a) (b)

**Fig. 5** Threading process: **a** CNC mill **b** Manual threading machine

**Table 2** $L_8$ orthogonal design of experiments

| Experiment no. | Input parameters (actual) | |
|---|---|---|
| | Heat treatment (min) | Machine type |
| 1 | 0 | CNC mill |
| 2 | 0 | Threading machine |
| 3 | 30 | CNC mill |
| 4 | 30 | Threading machine |
| 5 | 60 | CNC mill |
| 6 | 60 | Threading machine |
| 7 | 90 | CNC mill |
| 8 | 90 | Threading machine |

first time in this study. The Joshua Heap mass production threading machine with Wagner head and chasers performs threading with a 3 kW asynchronous motor. Internal threading was carried out using a Microcut Mold milling machine using a BT 40 tool holder and having a maximum spindle speed of 15,000 RPM and a spindle motor cos $\alpha$ value of 0.6. In addition, the spindle motor can generate torque of 103–110 Nm at 15–18.5 kW power. The design of the experiments is show in Table 2.

# 4 Results and Discussion

## *4.1 Energy-Power Consumption*

For selected test designs and levels, thread machining was performed twice for the test run. When making internal threads for the test specimens in the CNC mill, in particular for those which had been subjected to the austempering heat treatment and had a high amount of hardness, the machine was locked as the guide progressed because the maximum torque values were exceeded. However, for the specimens with a high hardness value subjected to manual threading using the Wagner heads, a decrease in the number of revolutions was observed as the torque values reached the maximum values during the opening process. The calculations were made taking into account the highest value shown on the ammeter. During the threading process, ammeters were used to calculate the current indices according to the design of the experiment, by performing the experiment twice and taking the averages. Power indices (PI) were measured according to the experimental design. Cutting power (P), torque (T) and specific energy consumption (SEC) were calculated with the equations used in the literature. The energy-power consumption results and measured hardness values obtained from the test results are given in Table 3.

When the austempering heat treatment was applied at different times, the hardness values were changed in the range of 25–30 HRc. It was observed that these hardness values made the threading difficult because hard samples require high torque values. Thus, it was determined that threading on a manual threading machine might be more economical. External threading on the manual threading machine required higher torque values. The main reason for this was thought to be that a higher amount of processing is required in the casting of these materials than is required for those with internal teeth.

In addition, it was thought that this was a reason for power losses during asynchronous transfer. As a single threading operation was applied to the pipe fittings in series production, up to a rate of 50% of torque changes occurred in both operations.

**Table 3** Experimental results for threading process according to Taguchi design

| Experiment no. | Hardness (HRc) | PI (A) | $P_{total}$ (kW) | T (Nm) | SEC (J/mm$^3$) |
|---|---|---|---|---|---|
| 1 | 9 | 41.8 | 16.5 | 96.82 | 108.55 |
| 2 | 9 | 5 | 2.6 | 147 | 17.11 |
| 3 | 28 | 45.2 | 17.8 | 104.03 | 117.11 |
| 4 | 28 | 5.2 | 2.7 | 153 | 17.76 |
| 5 | 30 | 46 | 18.1 | 106.09 | 119.08 |
| 6 | 30 | 5.5 | 2.9 | 162.5 | 19.08 |
| 7 | 25 | 42.3 | 16.7 | 97.85 | 109.87 |
| 8 | 25 | 5.1 | 2.6 | 150 | 17.11 |

When the current indices were compared, there was a very great difference between the two operations. The manual threading machine is designed to be more efficient for mass production of fittings. For very high torque values, more powerful asynchronous motors may be needed, and the number of revolutions can be provided by a transmission system change. Thus, higher torque values can be obtained for a slightly lower number of revolutions. The CNC vertical machining center with a power of 15 kW can achieve high torque values in a short time by increasing the power to 18.5 kW. However, this series is not suitable for threading production applications.

## 4.2 Regression Analysis

The current study focuses on the specific energy-power consumption (SEC) minimization. To achieve this a statistical model of SEC is developed. A simple linear model of the following form is developed as shown in Eq. 6,

$$SEC\left(\mathrm{J/mm^2}\right) = 208.63 + 0.36 \times \textit{Heat Treatment Time} - 95.89 \times \textit{Machine Type} \tag{6}$$

The analysis of variance (ANOVA) for the linear model is presented in Table 4 and it is seen that all the terms have very low p-value. The $R^2$, adjusted $R^2$ and predicted $R^2$ of the model is 99.55, 99.37 and 98.76% respectively.

The analysis of the normal probability of the residuals shows that the model is appropriate. All the residuals are observed to be within the 95% confidence interval in Fig. 6. The mean and the standard deviation are observed to be close to 0 and 1 respectively.

Figure 7 shows the comparison of the actual and the predicted SEC for various experiments. The deviation in the experimental and the predicted data is seen to be minimal. Thus, the model is appropriate and can be used in optimization tasks. Based on the developed empirical relation the effect of heat treatment time and machine type is studied in Fig. 8. It is observed that the CNC mill in general will have more SEC as compared to the threading machine.

**Table 4** Analysis of variance for SEC (J/mm$^3$)

| Source | DF | Adj SS | Adj MS | F | P |
|---|---|---|---|---|---|
| Regression | 2 | 18,390.1 | 9195.1 | 553.73 | 0.000 |
| A | 1 | 111.3 | 111.3 | 88.26 | 0.0090 |
| B | 1 | 18,388.8 | 18,388.8 | 1107.39 | 0.000 |
| Error | 5 | 83.0 | 16.6 | | |
| Total | 7 | 18,473.2 | | | |

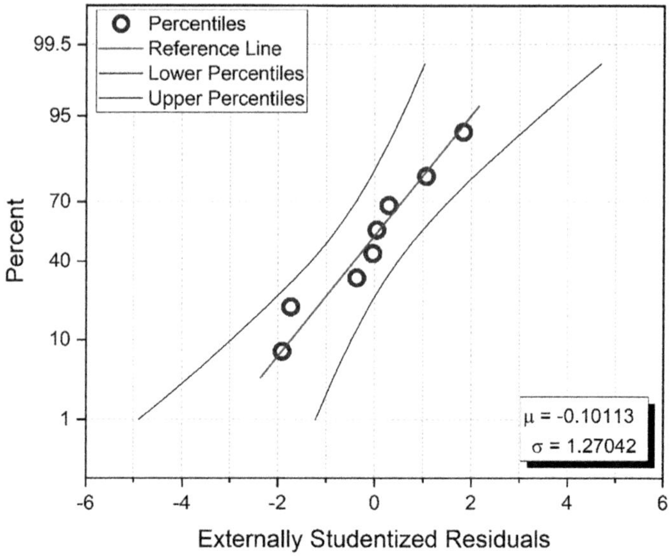


**Fig. 6** Normal probability plot of the residuals

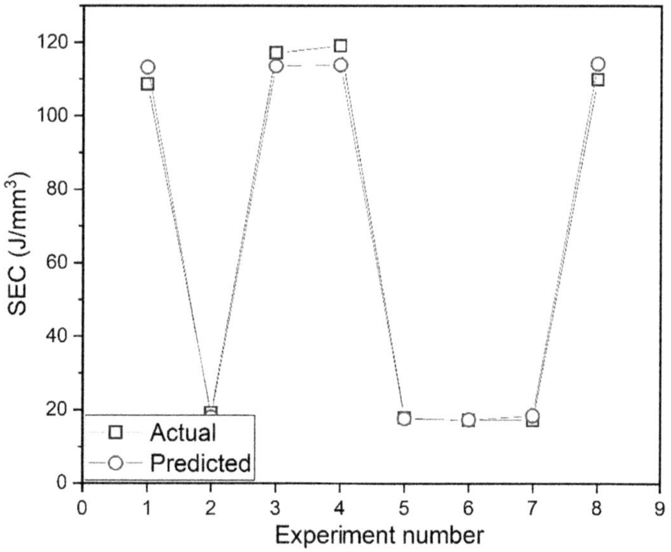


**Fig. 7** Experimental versus the predicted SEC for different experiments

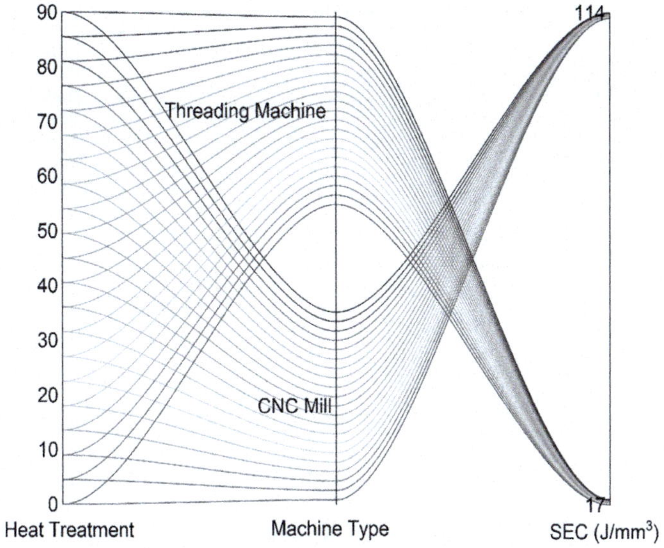


**Fig. 8** Effect of heat treatment time and machine type on the SEC

## 4.3 *Optimization Using Hybrid Dragonfly Algorithms*

The minimization of SEC was the objective for the present optimization problem (Eq. 7) which was defined considering the empirical relation (Eq. 6) generated by the predictive model to get the best response variables.

$$\begin{aligned}
&\text{Find Z } = \text{ (Heat treatment time \& Machine type);}\\
&\text{To minimize f(Z)} = \text{d(Specific energy consumption);}\\
&\text{Subjected to,}\\
&0 \le \text{Heat treatment time (min.)} \le 90\\
&\text{Machining type} \le 0 \text{ or Machining type} \ge 0
\end{aligned} \tag{7}$$

Various hybrid dragonfly algorithms are utilized to solve this optimization problem. For sake of unbiased comparison, in all the algorithms the search agents and maximum iteration limit is considered as 30 and 100 respectively. Figure 9 shows the convergence of the various dragonfly algorithms. It is evident that BMDA and DA converge to the same point i.e., 17.226 J/mm$^3$. QGDA converges to 17.248 J/mm$^3$ whereas CDA, DADE, MHDA, INMDA and HDA converges to 17.257, 17.305, 17.362, 17.425, and 17.442 J/mm$^3$ respectively. Thus, in terms of the optimum achieved, the tested algorithms can be ranked as BMDA ~ DA > QGDA > CDA > DADE > MHDA > INMDA > HAD.

The convergence of BMDA is very rapid. BMDA reaches the best fitness value by the 3rd iteration whereas for HDA, CDA, DA and INMDA this occurs in the 4, 8, 12

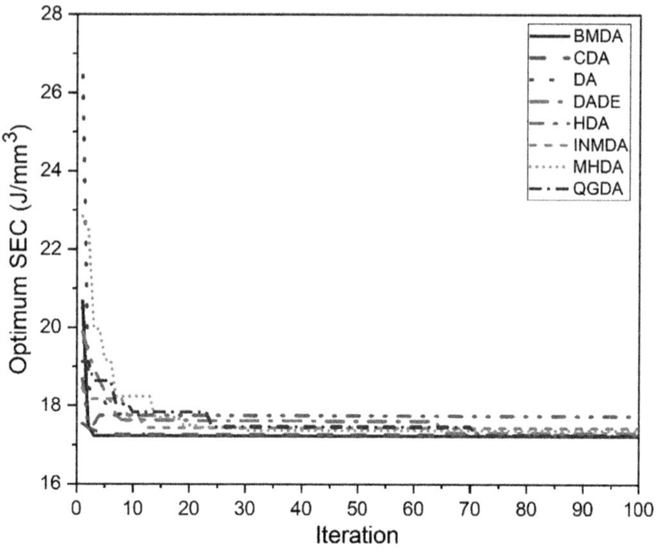


**Fig. 9** Convergence of the various hybrid dragonfly algorithms

and 12th iteration respectively. On the other hand, for MHDA, DADE and QGDA, the convergence to their respective best fitness values occur in 30th, 65th and 71st iteration. Thus, in terms of convergence rate the tested algorithms can be ranked as BMDA > HDA > CDA > DA ~ INMDA > MHDA > DADE > QGDA.

Table 5 shows the performance of the various hybrid dragonfly algorithms. Based on 10 independent trials of each algorithm the best fitness achieved is reported. Since dragonfly algorithms are stochastic it is important to look at the consistency of the algorithms as well. Thus, the mean and median of 10 trials are also reported in Table 5. The standard deviation which is an indication of the fluctuations in results in successive independent trials is also reported. The success rate i.e., the number of times the best optimum value (i.e. 17.226) is achieved out of 10 trials is also studied. It is observed that for DA the success rate is 100% whereas for BMDA it is 20%. However, for all the other algorithms, the success rate is 0% since they all were unable to converge to the best optimum value (i.e. 17.226). In terms of the optimum achieved and the success rate, the algorithms are ranked as DA > BMDA > QGDA > CDA > DADE > MHDA > INMDA > HDA. Figure 10 shows the spread of the evaluated functions for various algorithms in a typical trial. For DADE and QGDA it is seen that the diversity in the evaluated functions is very less as compared to the other algorithms.

**Table 5** Performance of the various hybrid dragonfly algorithms

| Algorithm | Best | Mean | Median | Standard deviation |
|---|---|---|---|---|
| BMDA | 17.226 | 18.174 | 17.685 | 1.902 |
| CDA | 17.257 | 17.852 | 17.928 | 0.336 |
| DA | 17.226 | 17.226 | 17.226 | 0 |
| DADE | 17.305 | 17.494 | 17.475 | 0.147 |
| HDA | 17.742 | 20.305 | 18.313 | 3.72 |
| INMDA | 17.425 | 18.363 | 17.958 | 0.992 |
| MHDA | 17.362 | 18.388 | 18.032 | 1.028 |
| QGDA | 17.248 | 17.27 | 17.263 | 0.053 |

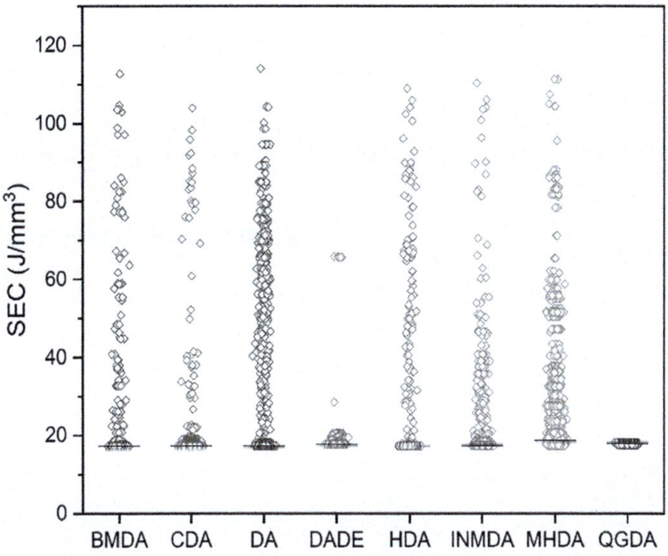


**Fig. 10** Scatter of evaluated functions for various algorithm in a typical trial

## 4.4 Microstructural Evaluations

The microstructure of the cast iron (ADI) produced via the austempering process under various heat treatments conditions was examined with optical and scanning electron microscopy (SEM) methods as shown in Fig. 11a–h. From these micrographs, it can be seen that the amount of nodular graphite was approximately 150/$mm^2$, the percentage of spheroidization was approximately 88% and the carbide content was approximately 10%. Furthermore, the graphite nodules are embedded in a fine pearlite matrix.

After austempering heat treatment, the matrix of ADI was changed from a fine pearlite matrix to a typical ausferritic matrix and the microstructure consisted of

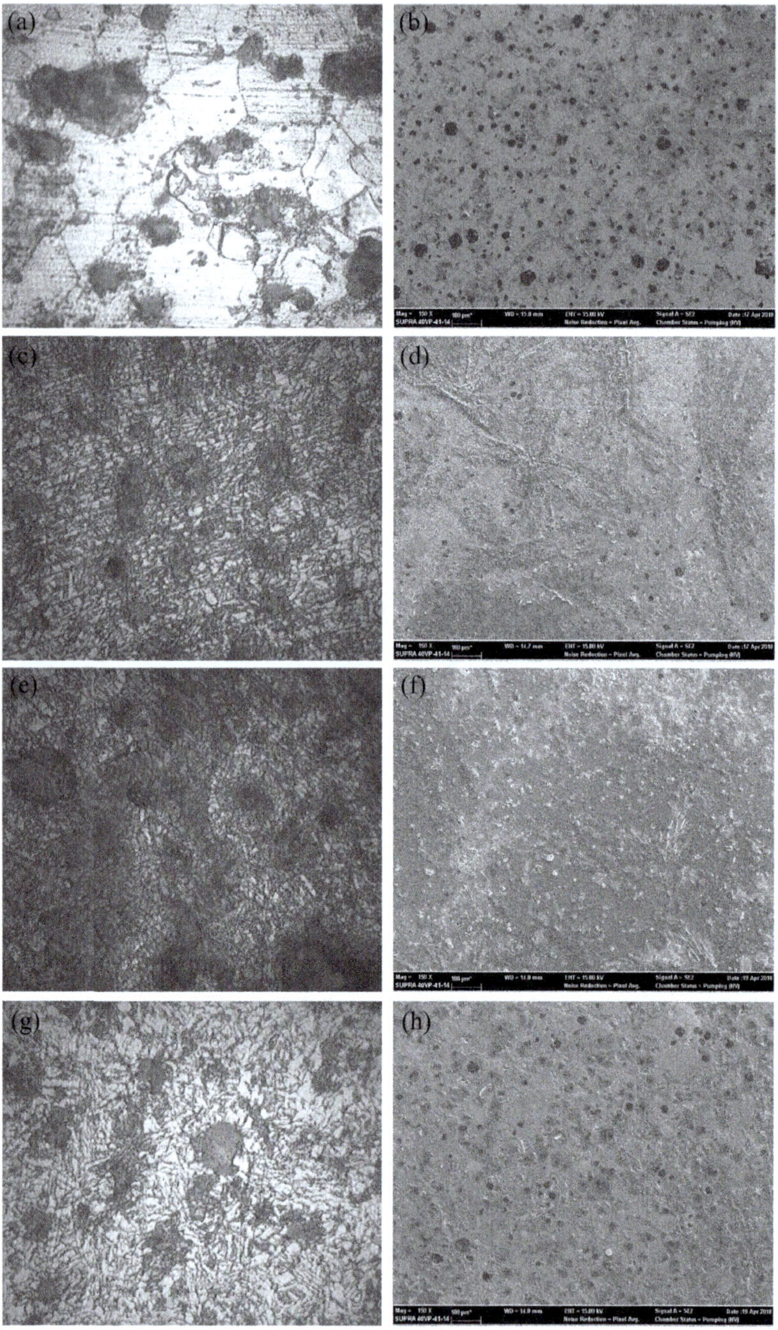


**Fig. 11** Optic (**a**, **c**, **e**, **g**) and SEM (**b**, **d**, **f**, **h**) images based on austempering heat treatment times: (1) without heat treatment, (2) 30 min, (3) 60 min, (4) 90 min

graphite nodules and a certain volume of carbides within an ausferrite matrix. Figure 11a–h shows that the acicular ferrite matrix content continuously decreased as the water content increased.

This could be related to the diffusion rate of carbon from ferrite to austenite. From these figures, it can also be observed that the retained austenite content played an effective role in the microstructure and properties of the ADI. The lower carbon-retained austenite was transformed into the brittle martensite phase easily during the cooling process from the austempering temperature to room temperature. This may be the reason for the high hardness and low impact toughness at the higher cooling rate because the microstructural examination revealed that the retained austenite content and the white-bright zones in the ADI had increased.

The main purpose of the austempering heat treatment in spheroidal graphite cast iron is to form an acicular bainitic matrix structure with high carbon austenite and carbide ausferrite. Ausferrite consists of ferrite + carbon-rich austenite. The carbon-rich austenite in ausferrite is thermodynamically stable and can be converted to martensite at high strain rates. Figure 11a, b show that a proper spheroidization process was applied after casting process. The SEM and optical photographs show that the graphite displays a uniform distribution in the matrix in Fig. 11c, d. The ausferrite volume ratio was determined by the critical austenite temperature. In addition, the greatest ausferrite volume was obtained at 60 min in Fig. 11e, f, while the least ausferrite volume was observed at 90 min in Fig. 11g, h. Experimental results showed that the pearlitic structure of the spheroidal graphite cast iron had been converted to ausferrite by heat treatment conditions. These microstructure results determined that the amount of hardness increased with the cooling rate and growth of ausferrite volume, which is the bainitic structure in the matrix, and with the increase of hardness and ausferrite there was an increase in the amount of energy-power consumption in the machining process. Both the CNC mill and the manual threading machine exhibited parallel changes in the results.

## 5 Conclusions

Austempered ductile cast iron fitting parts were produced using a CNC milling machine and a threading machine to compare the internal and external threading processes. Furthermore, the optimum response variables for energy-power consumption related to the microstructure and hardness of the austempered ADI were determined by using the Taguchi method and nature-inspired optimization algorithms. Based on the experimental and the optimization study, it can be concluded that.

- For minimum cost and maximum cycle time, the threading process in serial production must be performed in a single operation. The increasing of the hardness and wear resistance of ADI fitting samples reduced its machinability and increased the rates of electrical resistance. As a result, the threading operation was finished with an increasing electrical current.

- After the austempering heat treatment, the hardness value of the fitting samples increased with the increasing of the ausferrite volume in the ADI matrix and thus, energy-power consumption displayed a similar trend with the threading process.
- The importance of ensuring optimum energy consumption conditions in threading is different from other machining processes. Because it is determined in this article that the consumption of energy consumed during chip removal is much higher than the other types of operations.
- In regard to the variations of torque between the two different machines, the manual threading machine displayed more efficiency than the CNC threading machine by consuming less energy and by producing greater torque during the machining process.
- Significant savings in specific energy consumption is achieved by the various dragonfly algorithms. In terms of the specific energy consumption savings the various dragonfly algorithms can be ranked as DA > BMDA > QGDA > CDA > DADE > MHDA > INMDA > HDA. The outperformance of sophisticated algorithms like QGDA, CDA by a much simpler algorithm like DA proves that for such low dimension industrial problems it is seen that complex algorithms may not be best.
- Moreover, one of the most important findings of this study was that the manual threading machine had consumed less energy-power by the use of the asynchronous motor at high torque values and a low spindle speed.
- From microstructure examinations, it can be determined that the amount of hardness increased with the cooling rate and growth of ausferrite volume and with the increase of hardness and ausferrite there was an increase in the amount of energy-power consumption in the machining process.

This research opens several avenues for future work. Firstly, further studies can explore the application of hybrid dragonfly algorithms to other machining processes and materials to validate their robustness and generalizability. Secondly, incorporating real-time data and adaptive algorithms could enhance the practical implementation of these optimization techniques in dynamic industrial environments. Additionally, future research could investigate the integration of these optimization methods with advanced manufacturing technologies such as additive manufacturing and smart factories.

## References

1. Şirin, E., Şirin, Ş, Turgut, Y., Korkut, İ: Optimization of surface roughness using the Taguchi method in milling of AISI D2 cold work tool steel. Düzce Univ. J. Sci. Technol. **3**, 132–144 (2015)
2. Yazman, Ş.: The effects of cutting parameters on machining for austempered ferritic ductile iron. Master's thesis, Selcuk University, Graduate School of Natural and Applied Sciences, Konya (2006)

3. Hasırcı, H.: Effects of alloying elements (Cu and Ni) and austempering time on microstructure and mechanical properties in austempered spheroidal graphite cast irons. Master's thesis, Gazi University, Graduate School of Natural and Applied Sciences, Ankara (2000)
4. Zhang, J.Z., Chen, J.C., Kirby, E.D.: Surface roughness optimization in an end-milling operation using the Taguchi design method. J. Mater. Process. Technol. **184**, 233–239 (2007). https://doi.org/10.1016/j.jmatprotec.2006.11.029
5. Öztürk, B.: Investigation of the changes in microstructure and hardness of nodular cast irons used in fitting manufacturing depending on wall thickness with the effect of cooling rate. Master's thesis, Bartın University, Institute of Science, Bartın (2013)
6. Kuş, H.: The wear behaviour of austempered ductile iron. Graduate thesis, Selçuk University, Institute of Science and Technology, Konya (2007)
7. Çetin, M.: Investigation of spherical graphite cast iron production from $H_2$ peak. Graduate thesis, Gazi University Graduate School of Natural and Applied Sciences, Ankara (1998)
8. Korkmaz, Y.: Investigation of mechanical properties and wear behaviour of borontempered ductile iron. Undergraduate thesis, Karabük University, Technical Education Faculty, Karabük (2012)
9. Kim, Y.J., Shin, H., Park, H., Lim, J.D.: Investigation into mechanical properties of austempered ductile iron (ADI) in accordance with austempering temperature. Mater. Lett. **62**, 357–360 (2007). https://doi.org/10.1016/j.matlet.2007.05.028
10. Behnam, M.M.J., Davami, P., Varahram, N.: Effect of cooling rate on microstructure and mechanical properties of gray cast iron. Mater. Sci. Eng. A **528**, 583–588 (2010). https://doi.org/10.1016/j.msea.2010.09.087
11. Hemenanth, J.: Effect of cooling rate on dendrite arm spacing (DAS), eutectic cell count (ECC) and ultimate tensile strength (UTS) of austempered chilled ductile iron. Mater. Des. **21**, 1–8 (2000). https://doi.org/10.1016/S0261-3069(99)00052-7
12. Alabbasian, F., Boutorabi, S.M.A., Kheirandish, S.: Effect of inoculation and casting modulus on the microstructure and mechanical properties of ductile Ni-resist cast iron. Mater. Sci. Eng. A **651**, 467–473 (2015). https://doi.org/10.1016/j.msea.2015.09.024
13. Akyildiz, H.: Evaluating of cutting forces in thread machining. Int. J. Adv. Manuf. Technol. **68**, 1601–1612 (2013). https://doi.org/10.1007/s00170-013-4957-2
14. Khoshdarregi, M., Altintas, Y.: Generalized modeling of chip geometry and cutting forces in multi-point thread turning. Int. J. Mach. Tools Manuf **98**, 21–32 (2015). https://doi.org/10.1016/j.ijmachtools.2015.08.005
15. Lee, S.W., Kasten, A., Nestler, A.: Analytic mechanistic cutting force model for thread milling operations. In: 14th CIRP Conference on Modeling of Machining Operations, pp. 546–551 (2013). https://doi.org/10.1016/j.procir.2013.06.148
16. Bairamov, G., Abdulaliev, Z.: Optimization of cutting conditions and tool geometry in multiple threading of tool joint threads. Candidate Eng. Sci. S 35–37 (1965)
17. Mori, M., Fujishima, M., Inamasu, Y., Oda, Y.: A study on energy efficiency improvement for machine tools. CIRP Ann. Manuf. Technol. **60**, 145–148 (2011). https://doi.org/10.1016/j.cirp.2011.03.099
18. Liu, N., Zhang, Y.F., Lu, W.F.: A hybrid approach to energy consumption modelling based on cutting power: a milling case. J. Clean. Prod. **104**, 264–272 (2015). https://doi.org/10.1016/j.jclepro.2015.05.049
19. Camposeco-Negrete, C.: Optimization of cutting parameters for minimizing energy consumption in turning of AISI 6061 T6 using Taguchi methodology and ANOVA. J. Clean. Prod. **53**, 195–203 (2013). https://doi.org/10.1016/j.jclepro.2013.03.049
20. Oda, Y., Mori, M., Ogawa, K., Nishida, S., Fujishima, M., Kawamura, T.: Study of optimal cutting condition for energy efficiency improvement in ball end milling with tool-workpiece inclination. CIRP Ann. Manuf. Technol. **61**, 119–122 (2012). https://doi.org/10.1016/j.cirp.2012.03.034
21. Shokoohi, Y., Khosrojerdi, E., Rassolian Shiadhi, B.H.: Machining and ecological effects of a new developed cutting fluid in combination with different cooling techniques on turning operation. J. Clean. Prod. **94**, 330–339 (2015). https://doi.org/10.1016/j.jclepro.2015.01.055

22. Yurdakul, M.: Effect of cutting parameters on consumed power in industrial granite cutting processes performed with the multi-disc block cutter. Int. J. Rock Mech. Min. Sci. **76**, 104–111 (2015). https://doi.org/10.1016/j.ijrmms.2015.03.008
23. Neugebauer, R., Schubert, A., Reichmann, B., Dix, M.: Influence exerted by tool properties on the energy efficiency during drilling and turning operations. CIRP J. Manuf. Sci. Technol. **4**, 161–169 (2011). https://doi.org/10.1016/j.cirpj.2011.06.011
24. Muñoz-Escalona, P., Shokrani, A., Newman, S.T.: Influence of cutting environments on surface integrity and power consumption of austenitic stainless steel. Robot. Comput. Integr. Manuf. **36**, 60–69 (2015). https://doi.org/10.1016/j.rcim.2014.12.013
25. Nas, E., Öztürk, B.: Optimization of surface roughness via the Taguchi method and investigation of energy consumption when milling spheroidal graphite cast iron materials. Mater. Test. **60**(5), 519–524 (2018). https://doi.org/10.3139/120.111181
26. Mirjalili, S.: Dragonfly algorithm: a new meta-heuristic optimization technique for solving single-objective, discrete, and multi-objective problems. Neural Comput. Appl. **27**, 1053–1073 (2016). https://doi.org/10.1007/s00521-015-1920-1
27. Shirani, M.R., Safi-Esfahani, F.: BMDA: applying biogeography-based optimization algorithm and Mexican hat wavelet to improve dragonfly algorithm. Soft. Comput. **24**, 15979–16004 (2020). https://doi.org/10.1007/s00500-020-05340-6
28. Sayed, G.I., Tharwat, A., Hassanien, A.E.: Chaotic dragonfly algorithm: an improved meta-heuristic algorithm for feature selection. Appl. Intell. **49**, 188–205 (2019). https://doi.org/10.1007/s10489-018-1261-8
29. Debnath, S., Baishya, S., Sen, D., Arif, W.: A hybrid memory-based dragonfly algorithm with differential evolution for engineering application. Eng. Comput. **37**, 2775–2802 (2021). https://doi.org/10.1007/s00366-020-00958-4
30. Ghanem, W.A.H.M., Jantan, A.: A cognitively inspired hybridization of artificial bee colony and dragonfly algorithms for training multi-layer perceptrons. Cogn. Comput. **10**, 1096–1134 (2018). https://doi.org/10.1007/s12559-018-9588-3
31. Xu, J., Yan, F.: Hybrid Nelder-Mead algorithm and dragonfly algorithm for function optimization and the training of a multilayer perceptron. Arab. J. Sci. Eng. **44**, 3473–3487 (2019). https://doi.org/10.1007/s13369-018-3536-0
32. Sree Ranjini, K.S., Murugan, S.: Memory based hybrid dragonfly algorithm for numerical optimization problems. Expert Syst. Appl. **83**, 63–78 (2017). https://doi.org/10.1016/j.eswa.2017.04.033
33. Yu, C., Cai, Z., Ye, X., Wang, M., Zhao, X., Liang, G., Chen, H., Li, C.: Quantum-like mutation-induced dragonfly-inspired optimization approach. Math. Comput. Simul **178**, 259–289 (2020). https://doi.org/10.1016/j.matcom.2020.06.012

# Neural Network Optimization of a Current Flow Meter with Applications in Hydroelectric Power Plants

**Zezatti Victor, Camarena López Ana, Márquez Soto Cristian, Castro Laura, Navarrete Miriam, and Ochoa-Zezzatti Alberto**

**Abstract** Hydroelectric power plants necessitate precise measurement of water flow velocity in their inlet pipes to optimize water resource management. While various devices such as flow sensors have traditionally been used for this purpose, our research delves into the operational dynamics of water turbines. By deriving characteristic equations of turbine blades using a design constant 'K'—a coefficient relating blade angular velocity to flow velocity—we aim to enhance flow velocity estimation accuracy. Our methodology involves conducting rigorous experimental tests within a wind tunnel setup. Subsequently, we leverage advanced computational techniques, including the training of neural networks, to develop a robust mathematical model. This model utilizes input parameters such as blade angular velocity and fluid temperature to predict flow velocity. Through rigorous validation against Computational Fluid Dynamics (CFD) simulations, our approach ensures

Z. Victor (✉) · M. S. Cristian · C. Laura · N. Miriam
Universidad Autónoma del Estado de Morelos, Av. Universidad No. 1001, 62209 ChamilpaCuernavaca, Mor., Mexico
e-mail: zezatti@uaem.mx

M. S. Cristian
e-mail: cristian.marquezst@uaem.edu.mx

C. Laura
e-mail: lauracg@uaem.edu.mx

N. Miriam
e-mail: miriam.navarrete@uaem.mx

C. L. Ana
Universidad de Guadalajara, Blvd. Gral. Marcelino García Barragán 1421, 44430 OlímpicaGuadalajara, Jal., Mexico
e-mail: ana.camarena4700@alumnos.udg.mx

O.-Z. Alberto
Universidad Autónoma de Ciudad Juarez, CADIT, Universidad Anáhuac Campus Norte, Av. Plutarco Elías Calles No. 1210, Fovisste Chamizal, 32310 Ciudad Juárez, ChihuahuaNaucalpan, Mexico
e-mail: laalberto.ochoa@uac.mx

 
D. Oliva et al. (eds.), *Advances in Optimization Algorithms for Multidisciplinary Engineering Applications: From Classical Methods to AI-Enhanced Solutions*, Studies in Computational Intelligence 806, https://doi.org/10.1007/978-3-031-78440-8_29

high-fidelity flow velocity estimation for improved hydroelectric plant efficiency and resource optimization.

**Keywords** Current flowmeter · Stream flowmeter · Helix-flowmeter

## 1 Introduction

Society has found itself in the need to study hydraulic technologies, the force that moves each tributary, quantify consumption by assessing resource availability, and plan the respective management of closed flows [1]. These latter are pipes or conduits that are completely in contact with rigid boundaries, all caused by climate change, rising temperatures, water scarcity, the risk of waterborne diseases, among other factors, which have raised concern as they pose a threat to current and future generations [7, 8]. This concern requires a more precise measurement of turbine efficiency in hydraulic power plants, necessitating a comprehensive understanding of water flow measurement in large channels. Flow meters are one of the most reliable instruments for measuring the discharge rate of natural and artificial channels [3].

In Fig. 1, the shape of a water turbine is depicted, which is an electromechanical instrument used for flow measurement. This device operates by the movement of its blades driven by the flow, providing several revolutions proportional to the flow rate. The propeller-type flow meter is the most reliable method among practical methods for measuring flow rates in hydroelectric power plants. Another advantage of using this instrument is the opportunity to determine the flow velocity profile at any arbitrary cross-section [3]. The water turbine integrates the measurement of a velocity profile using blades placed at precise points along a pressure pipe, arranged in a matrix pattern and secured to the pipe via a rack [4].

There are different types of flow meters, such as the horizontal-axis propeller-type flow meters shown in Fig. 1, which have a rotating element. The number of rotations is mechanically transmitted to a register that displays the flow velocity [5].

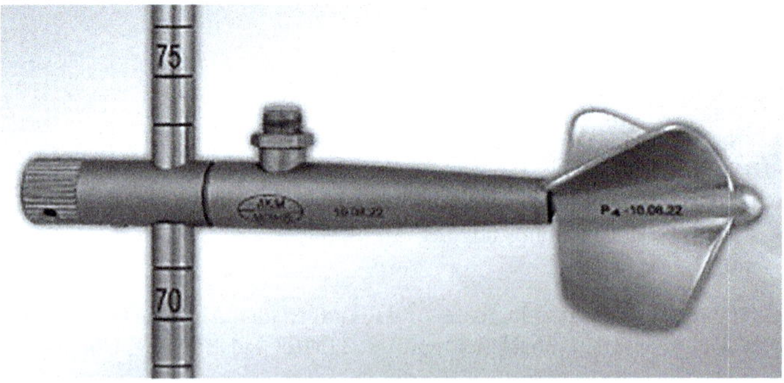


**Fig. 1** Helix current flow meter

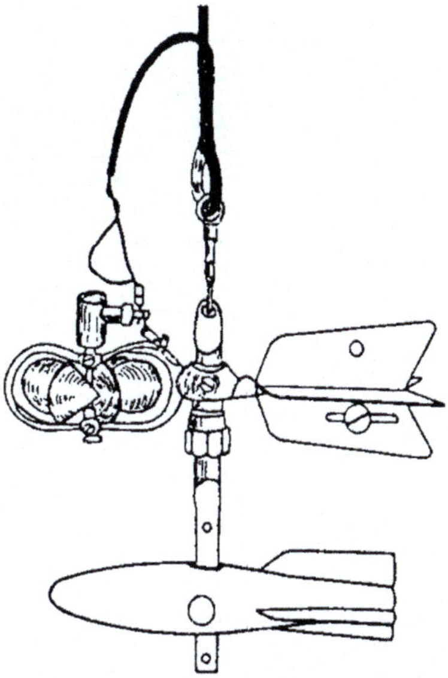

**Fig. 2** Vertical axis current flow meter [1]

Vertical-axis propeller-type flow meters (Fig. 2) are a variant that operates on the same principle, but with a vertical turbine, allowing their use in open channels [3].

Flow meters can be divided into three groups: low, medium, and high axial pitch. This study presents a specific design of the high axial pitch propeller-type flow meter, where axial pitch refers to the advance of the propeller for each revolution it completes [3].

## 1.1 Characteristic Equations of the Propellers

The constant 'K' is a factor that relates the angular velocity of the propeller to the wind speed. There are diverse ways to express the constant 'K' in the literature; K = ω/V or K = V/ω. In this research work, the constant 'K' in its design stage is denoted as follows:

$$K = \frac{\omega}{V} \tag{1}$$

$$k = \frac{n}{V} \tag{2}$$

where $\omega$ is the angular velocity in (rad/s), $n$ is the rotational speed in (rpm) and $V$ flow velocity [3]. From Eqs. 1 and 2, it can be inferred that flow meters with a larger constant have a greater sensitivity to changes in fluid velocity, as the angular velocity $\omega$ will change with slight changes in flow velocity [3].

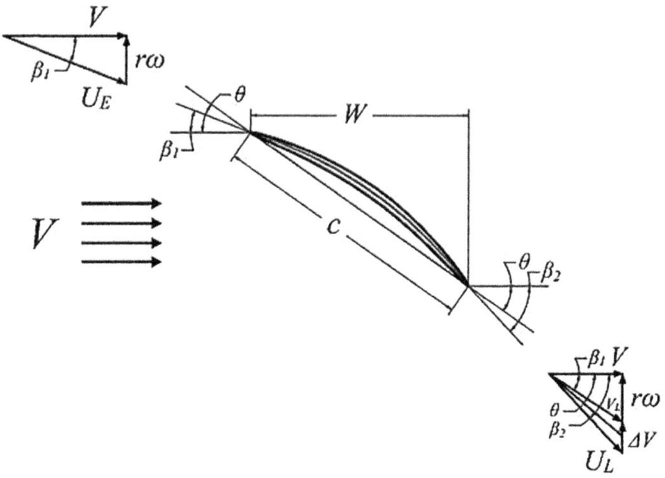


**Fig. 3** Velocity triangles inside and outside of the blades [3]

Figure 3 shows the velocity triangles of the propeller blades.

Where $\theta$ is defined as camber angle, α as attack angle, $\beta_1$ and $\beta_2$ are blade angles at inlet and outlet respectively, defined as follows [3]:

$$\theta = \frac{\beta_2 + \beta_1}{2} \tag{3}$$

$$\alpha = \frac{\beta_2 - \beta_1}{2} \tag{4}$$

The relationship between $c$ defined as chord length and $W$ defined as blade width is given by:

$$c = \frac{W}{\cos\theta} \tag{5}$$

The Ott theory from 1925 [6] was used to establish fundamental equations for the design of propeller-type flow meters [4]. There are 5 parameters for designing the flow meter; $\sigma$, $N$, $W^*$, $L^*$, and $K$. General recommendations were investigated for choosing $\sigma$ and $N$, as well as for calculating the values of the blade angles $(\theta, \beta_1, \beta_2)$ [6]:

$$\theta = \tan^{-1}\left(\frac{2\pi r}{L}\right) \tag{6}$$

$$\beta_1 = \tan^{-1}(rK) \tag{7}$$

**Table 1** Geometric criteria

| Angles |
|---|
| $0 < \theta < \pi/2$ |
| $\beta 1 < \theta$ |
| $\theta < \beta 2$ |
| $0 < \beta 1$ |
| $0 < \beta 2$ |
| $\beta 2 < \pi/2$ |

$$\beta_2 = 2\tan^{-1}\left(\frac{2\pi r}{L}\right) - \tan^{-1}(rK) \tag{8}$$

It is important to consider the geometric criteria as shown in Table 1. By substituting Eqs. 6, 7, and 8 into the Equations from the table, it is fulfilled that:

$$K = \frac{2\pi}{L} \tag{9}$$

The constraint for the medium axial pitch constant is expressed as follows [6]:

$$K > \frac{1}{\pi\sigma_r^2 D_t L^*}\left[(\pi\sigma_r)^2 - L^{*2}\right] \tag{10}$$

$L$ and $\sigma$ are design geometrical parameters. For $L^* < \sigma_r\pi$, the range for the medium axial pitch must satisfy the following range $L^* < \sigma\pi$ [3].

## 2 Methodology

The process begins with a literature review on the flow meter, its design, and the equations governing it. A circuit is fabricated for revolution measurement, and experimental data is acquired through a Labview graphical interface. The flow meter is designed in CAD, its behavior is simulated to manufacture the parts using conventional machining and 3D printing. It is assembled and installed in a wind tunnel, experimental tests are conducted with data acquisition, the neural network is trained, a comparison of simulated results against numerical simulation results is performed, and finally, a data analysis is carried out. In Fig. 4, the methodology to be followed is illustrated in a graphical organizer.

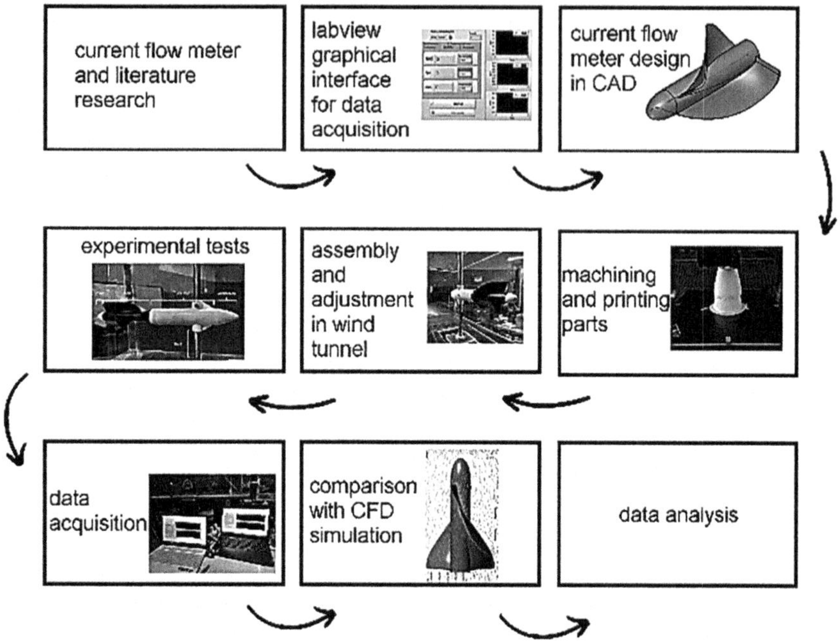


**Fig. 4** Methodology graphic organizer

## 2.1 *Instrumentation and Acquisition of the Angular Frequency Variable in (rpm)*

Currently, techniques have been developed for monitoring and maintaining machinery quickly and reliably, placing everything in the hands of instrumentation to avoid large costs and to enable predictive maintenance. Reliable systems for machinery monitoring and diagnosis, equipped with correct measurement input signals, have been developed. Various equipment, such as the tachometer, is used to measure speed. A tachometer is a mechanical or electronic device that quantifies the rotation frequency of an object over a certain period or under certain stop conditions. A tachometer is defined as an electronic device that obtains the pulses generated by an encoder and an optical sensor. An encoder is a detection device that converts motion into an electrical signal [5], which are then processed by a development board. The pulses generated are proportional to the rotational speed of the object.

There are various methods for counting pulses generated by the encoder. In this work, direct pulse counting (DPC) was used, with a relative error of 0.21–0.02. This method involves the direct counting of pulses generated by the encoder. The number of counted pulses is described as $C_m$, where the subscript m represents the number of steps (marks) of the encoder. The speed is given by the following Equation:

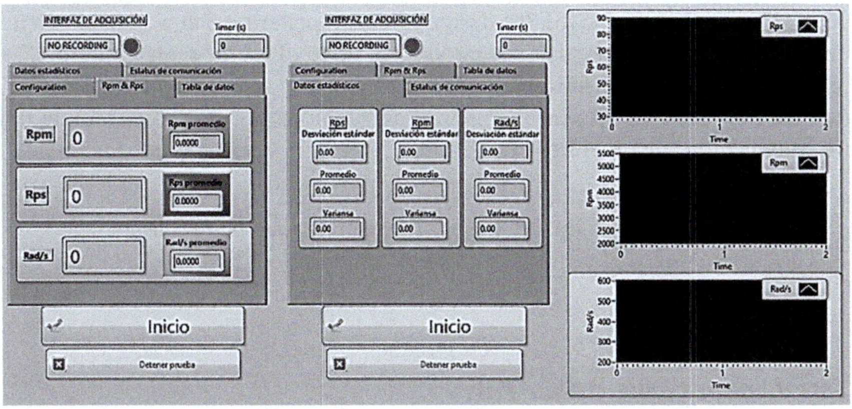


**Fig. 5** Lab view graphical interface

$$\omega = \frac{2\pi}{m} * C_m \frac{1}{T_c} \tag{11}$$

## 2.2 *Signal Acquisition*

For data processing and acquisition, LabView 2020 software with serial communication was used. This software provides a powerful graphical development environment for designing data acquisition, processing, and analysis applications, thanks to a block-based programming language, without the complexity of other development tools.

Figure 5 shows the LabView graphical interface developed for data acquisition and processing. The interface collects data from the sensor through an output cable connected from the computer to the development board, which records the pulses from the tachometer internally in the flow meter. It records and saves this data obtained in the form of frequency (rps). The interface converts it to (rpm) and then to (rad/s). In addition to this, it provides statistical data such as standard deviation in (rps), (rpm), and (rad/s), as well as real-time graphical representation. The information is saved in Excel format (.csv).

## 2.3 *CAD Design*

The objective of the research is to illustrate a practical model for designing propeller-type flow meters, seeking optimization of geometric parameters to achieve the highest performance under different flow conditions. Considering the concepts previously defined by the authors [3, 13], the design of the propeller-type flow meter is limited due to the established standards regarding the body of the flow meter design. The propeller method establishes the technical parameters and physical characteristics that the flow meter must possess, as well as the mounting rack, for accurate flow measurement in a hydroelectric power plant [9–11].

- IEC 60,041 [1991] standard: Field acceptance tests to determine the hydraulic performance of hydraulic turbines, storage pumps, and pump-turbines.
- The current flow meters must meet the requirements of ISO 2537 and ISO 3455, where applicable.

## 2.4 Current Flow Meter Parts

The components of the flow meter were designed using Computer-Aided Design (CAD) software SolidWorks. The flow meter was divided into four assemblies: propeller, transmission, counting, and mounting.

The propeller design was determined to have a high axial pitch to reduce device sensitivity, thus achieving better measurement precision and accuracy [3].

For the calculation of the design constant, Eq. 10 was used and plotted against $\sigma$.

Figure 6 shows the appropriate range for selecting the design constant 'K', which ranges from 9 to a number less than 10.5 [9].

The transmission consists of an encoder, a stainless-steel shaft, and 2 bearings at each end. Electrical signals are emitted by the optical sensor and generated by the encoder, which are read by an internal tachometer attached to the body of the flow meter. The assembled device is fixed in the wind tunnel by calculating the center of mass to balance the device with a counterweight, neutralizing the moment generated (lever arm) by the weight of the device, providing stability and reducing any potential stress.

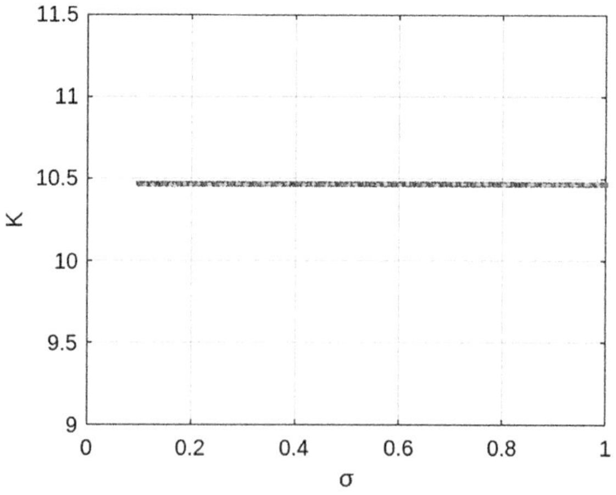


**Fig. 6** Appropriate area for K

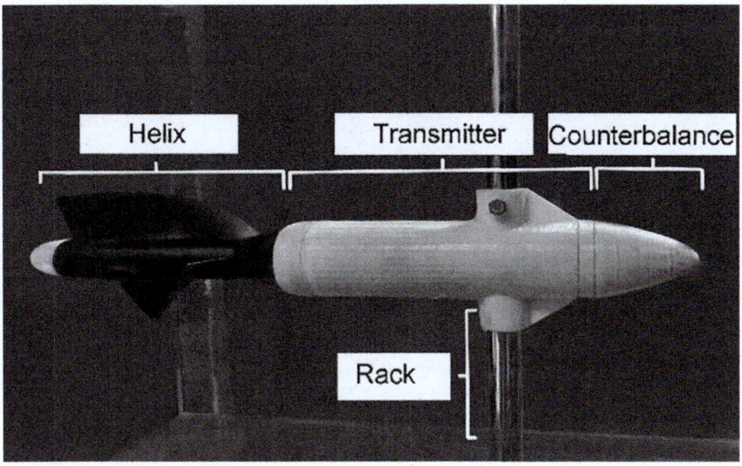


**Fig. 7** Current flow meter parts

## 2.5 *Manufacturing of Parts*

The components were designed in SolidWorks, processed using Ultimaker Cura software, and subsequently 3D printed with PLA material using an ANET 3D printer. The configuration parameters were based on fine print density and gyroid infill type, with 3 layers in the wall, 20% infill density for non-structural parts, and 80% for parts with mechanical functions, as shown in Fig. 7.

## 2.6 *System Description*

The device meets the working fluid (air) inside the wind tunnel at a specified distance from the entrance. The rotor begins to rotate from a zero angular velocity when the starting torque exceeds the retarding torque. The propeller accelerates to reach its steady-state speed, where the driving torque equals the retarding torque, causing the flow meter to neither accelerate nor decelerate. This means that the number of rotations is proportional to the fluid velocity. Accessories enable the placement of the rotor at the desired distance, its alignment with the flow, and the ability to count rotations through an electrical circuit [2, 3].

During the experimental stage, no significant variations in humidity or temperature of the working fluid (air) were observed. The wind tunnel is of the aspiration type, as shown in Fig. 8.

**Fig. 8** Wind tunnel fans

## 2.7 *Assembly and Adjustment in Wind Tunnel*

With the manufactured parts ready, the assembly and adjustment of the different pieces and internal components of the flow meter and tachometer were conducted. The rack was adjusted from the top to the bottom face of the wind tunnel, ensuring complete stability and balance.

The flow velocity at the wind tunnel inlet is controlled by a pair of fans arranged in series, as shown in Fig. 8. These fans can manually decrease or increase their rotational speed using a manual control, thereby adjusting the velocity at the tunnel inlet.

An anemometer was installed inside the tunnel at the axis height of the impeller as a reference instrument, which has a measurement error percentage of 2%. The anemometer used is a Climomaster (Fig. 9), which measures the velocity, temperature, and humidity of the fluid at the inlet, as well as atmospheric pressure. The results are exported in *.txt* and *.csv* formats.

## 2.8 *Experimental Test*

For the experimental tests, care was taken to ensure that they were conducted under controlled conditions, avoiding disturbances in the environment due to the high sensitivity of the device. Any movement inside or outside the fluid mechanics laboratory (CIICAp-UAEM) directly affected the behavior of the device's impeller, generating noise in the collected data.

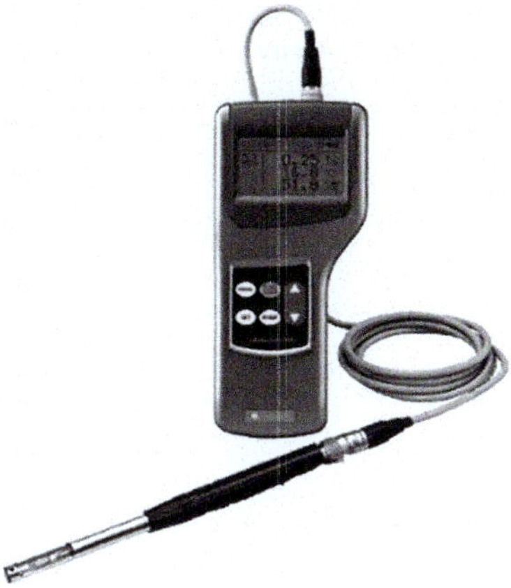

**Fig. 9** Climomaster Anemometer Mod. AK-6501

Six experimental tests were conducted, comprising three upward tests and three downward tests. In the upward tests, the wind tunnel was configured at 10 Hz frequency, increasing to 56 Hz in 2 Hz frequency intervals between each step. Simultaneously, data acquisition of wind flow velocity was performed using the anemometer. For each data acquisition, data were recorded every second for two minutes. For the downward tests, the frequency was set to 56 Hz, decreasing to 10 Hz in the same 2 Hz intervals between each step, while also collecting flow velocity data.

## 3 Results

Using the experimental data obtained from temperature ($T$), flow velocity ($V$), and angular velocity ($\omega$) during the conducted tests, an analysis was performed to understand their trend and verify if they adhere to the theory developed in this study.

Figure 10 depicts the experimental data of the angular velocity of the propeller ($\omega$) in radians per second (rad/s) plotted against the flow velocity in meters per second (m/s). A curve fitting was calculated for the data, revealing a linear trend with a variance percentage of 99.9%. The graph illustrates a direct proportionality between angular velocity and flow velocity ($\omega \propto v$).

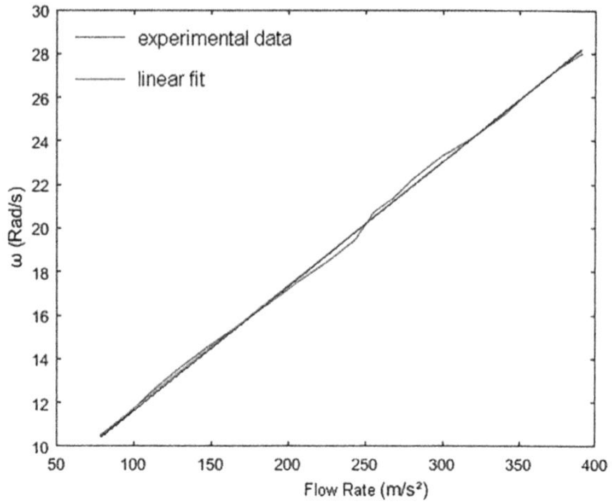


**Fig. 10** Angular velocity versus flow velocity

## 3.1 *Obtaining the Constant 'K'*

The arithmetic mean of the obtained experimental results is calculated, and the constant 'K' is determined for each data point of the helix angular velocity using Eq. 1, as shown in Table 2.

The constant 'K' was obtained by averaging the data ($\omega$ and V*)* for each test interval, where these are defined by the control frequency of the fans (f). However, the frequency ($\omega$) does not change linearly with the constant 'K'. Therefore, using an average constant [3] for this application is unfeasible, as it would significantly reduce the precision and accuracy of the developed device. Additionally, the fluid used for the experimental tests is air, so factors such as humidity, atmospheric pressure, and temperature change the physical characteristics of the fluid, factors that Eq. 1 does not involve in its input variables.

## 3.2 *Data Analysis*

The objective of using a model based on artificial neural networks (ANN) instead of the statistical model based on an average constant [3] is to obtain a model independent of such constant, thus achieving a model that calculates the flow velocity for each input value ($\omega$ and $T$) and thereby increasing the precision and accuracy of the flow meter. The goal is to estimate the flow velocity, with the following input values: the angular velocity of the meter, provided by the tachometer and processed by the LabView interface, and the ambient temperature, recorded by the anemometer. A total of 16,000 data points were processed. A neural network was developed using Python. The angular velocity of the propeller and the ambient temperature were used as input variables, and the flow velocity was used as the output variable, which is

**Table 2** Constant 'K'

| f (Hz) | T (°C) | v (m/s$^2$) | ω (Rad/s) | K |
|---|---|---|---|---|
| 18 | 9.359 | 25.297 | 63.255 | 6.765 |
| 20 | 10.461 | 25.336 | 78.487 | 7.505 |
| 22 | 11.465 | 25.249 | 96.714 | 8.536 |
| 24 | 12.444 | 25.236 | 110.729 | 8.899 |
| 26 | 13.486 | 25.215 | 128.268 | 9.512 |
| 28 | 14.524 | 25.178 | 147.400 | 10.148 |
| 30 | 15.476 | 25.151 | 166.243 | 10.740 |
| 32 | 16.556 | 25.164 | 187.924 | 11.355 |
| 34 | 17.552 | 25.143 | 206.549 | 11.773 |
| 36 | 18.512 | 25.120 | 226.376 | 12.232 |
| 38 | 19.463 | 25.126 | 243.472 | 12.500 |
| 40 | 20.725 | 25.095 | 255.465 | 12.359 |
| 42 | 21.378 | 25.077 | 268.209 | 12.548 |
| 44 | 22.209 | 25.079 | 280.148 | 12.613 |
| 46 | 23.303 | 25.058 | 299.480 | 12.848 |
| 48 | 24.277 | 25.043 | 322.438 | 13.280 |
| 50 | 25.251 | 25.016 | 341.278 | 13.515 |
| 52 | 26.126 | 25.007 | 354.560 | 13.572 |
| 54 | 27.163 | 25.060 | 372.419 | 13.710 |
| 56 | 28.016 | 24.989 | 391.500 | 13.975 |

the variable to be predicted. Several configurations of different activation functions were evaluated to obtain a model with the best statistical parameters. Additionally, the aim was to use the minimum number of layers and neurons to obtain a simple and dependable model.

Figure 11 presents the diagram of the applied neural network:

It has a topology of (2:1:1), where the input variables are $\omega$ and $T$, and the output variable is $V$. Additionally, the weights, biases, intermediate values z, and output values have a superscript indicating the layer to which they belong: 2 for the hidden layer and 3 for the output layer. The weights, except for the mentioned superscript, all receive similar names: $wij$, where $i$ is the number of the neuron to which the link arrives, and $j$ is the number of the neuron from which the link comes. The biases also now have a subscript indicating the neuron with which they are associated. The set of values as from the output layer forms the prediction $V$.

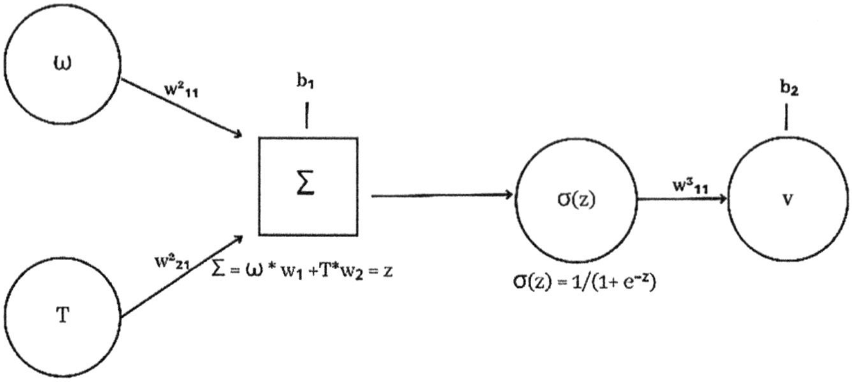


**Fig. 11** Neural network diagram

The data was initially normalized using the Min-MaxScaler tool. For data splitting, 80% was taken as 'train', which are the data used for training, and the remaining 20% were marked as 'test', which are the data used for model evaluation. For the applied model, a layer of one neuron with different activation functions was programmed, where the Sigmoid function, given by Eq. 12, yielded an $R^2$ value close to 1 (Table. 3).

$$\sigma(z) = \frac{1}{1 + e^{-z}} \tag{12}$$

where z is the argument of the function (Fig. 11).

The weights and bias obtained in the neural network allow us to find the function of the model.

- layer 0, bia: [1.205012], pesos: $\begin{bmatrix} -1.5502840 \\ -3.0936885 \end{bmatrix}$
- layer 1, bia: [1.1422722], pesos: [0.13057527]

Through Fig. 12, the trend of the applied model to the real values can be estimated. These results provide an initial insight into the performance of the non-linear regression.

**Table 3** Neural network topology

| Metrics | Value |
|---|---|
| Mean square error | 1.0679543018341064 |
| Mean absolute percentage error | 5.000591711945336% |
| $R^2$ | 0.9721593988695414 |
| Tight $R^2$ | 0.9721428122736936 |

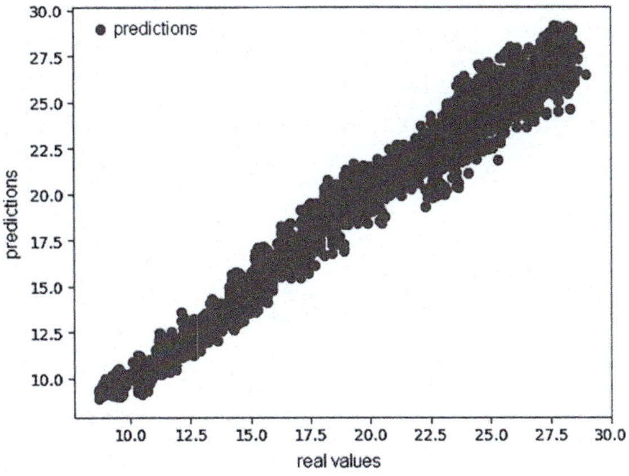


**Fig. 12** Predicted regression values

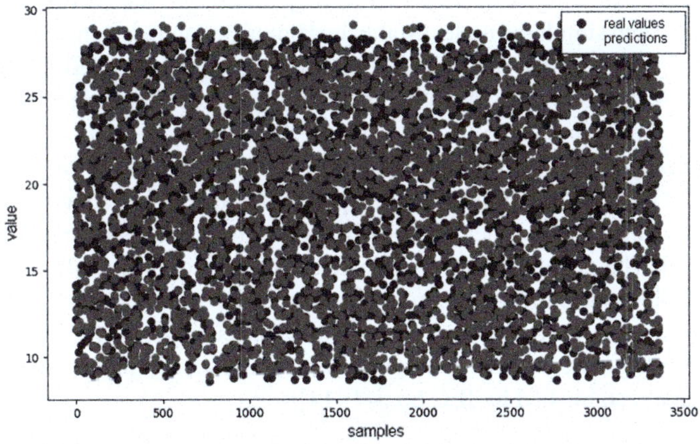


**Fig. 13** Behavior of our model and the behavior of the real data

The estimations provided by the ANN model are shown in Fig. 13 with respect to the real values (Target). The isolated data point appearing in Figs. 12 and 13 corresponds to a false positive in the database used.

## *3.3 Comparison of Results Between Models*

To perform validation and compare the accuracy of the artificial neural network (ANN) based model, Table 4 compares the different results obtained with different prediction models for estimating velocity. Three different models are compared: the first model uses the average constant 'kprom', based on Khozaei's proposal [3], the

**Table 4** Comparison of models for different flow rates

| V real | Kprom | CFD | RNA |
|---|---|---|---|
| 28.01 | 34.28 | 27.50 | 27.60 |
| 22.20 | 19.44 | 22.00 | 22.55 |
| 9.36 | 5.539 | 9.00 | 9.77 |

**Table 5** Comparison of error percentages for different models

| V real | Kprom | CFD | RNA |
|---|---|---|---|
| 28.01 | 22.385 | 1.821 | 1.442 |
| 22.20 | 12.432 | 0.901 | 1.607 |
| 9.36 | 40.823 | 3.846 | 4.361 |
| Prom | 25.213 | 2.189 | 2.470 |

second model uses data obtained for specific velocities using the Computational Fluid Dynamics (CFD) software ANSYS CFX [15], where the rotation of the propeller was modeled and simulated under established boundary and working conditions. The third model applied is the one obtained from the previously defined ANN.

In Table 4, the error percentages were calculated for each model and its respective velocity, as shown in Table 5.

Characterizing the ANN-based model statistically with only three measurement points is not feasible, as a good statistical population is required for characterization. Therefore, using the experimentally obtained data, this model was compared with the ANN model. The percentage error and standard deviation were calculated for each measured velocity, resulting in a percentage error of 3.76% and a standard deviation of 0.467. These statistical parameters, along with the accuracy and precision parameters of the tachometer, are of vital importance for characterizing the anemometer within the guidelines established by IEC 60,041 [14].

# 4 Conclusions

Through the experimental results, it was confirmed that high axial pitch helical flow meters are sensitive devices to changes in velocity. The variables of temperature or wind speed fluctuate in decimal orders, affecting the behavior and dynamic response of the anemometer. Therefore, the experimental tests conducted in the laboratory had to be under highly controlled conditions to reduce the noise in our data and analyze them effectively.

The neural network model with the Sigmoid activation function provides a good approximation to the experimental data with a model accuracy of 96.23719%, an average percentage error of 3.76281005%, and a precision of ±0.497544 m/s. This model has only one neuron layer, which makes its computation time low, and the model optimized. Additionally, the weights and biases of the model were obtained, allowing the mathematical function of the model to be determined. Applying this model to experimental tests within the ranges it was trained on.

The experimental results were compared with the data obtained from the neural network with the simulation performed in CFD, where it is observed that the precision of both the neural network and the data obtained from the simulation are close to the real values.

## References


1. Pozo, L.: Construcción de un prototipo del molinete universal para la medición de la velocidad del agua en ríos, basado en el molinete universal utilizado en el INAMHI [Tesis, Escuela Politécnica Nacional] (2011)
2. Pedroza, E.: El molinete, importancia, fundamentos y buenas prácticas. Instituto Mexicano de Tecnología del Agua (2017)
3. Khozaei, M.H., Nourbakhsh, S.A.: Analytical and numerical study of fluid flow in propeller-type current-meters. Int. J. Fluid Mach. Syst. (2020). https://doi.org/10.5293/ijfms.2020.13.2.437
4. Khozaei, M.H., Ahmadnourbakhsh.: The influence of the geometrical parameters of propeller-type current-meter on its performance (2016). https://doi.org/10.13140/RG.2.1.1528.9363
5. Ferrada, S.: Diseño y construcción de un medidor portátil de caudal para canales de tierra [Tesis, UDP Facultad de ingeniería] (2014)
6. Ott, L.A.: Theorie und Konstantenbestimmung des hydrometrischen Flügels. Springer (1925)
7. Mondragón-Suarez, J.H., Sandoval-Villalvazo, A., Brena-Ramos, F.: Calentamiento global: una secuencia didáctica. Revista Mexicana de Física E **65**(2019), 52–57 (2019)
8. Roca Villanueva, B., Beltrán Salvador, M., Gómez Huelgas, R.: Cambio climático y salud. Rev. Clin. Esp. **219**(5), 260–265 (2019)
9. International Organization of Standards (ISO).: Hydrometry—Measurement of liquid flow in open channels using current-meters or floats. ISO 748-2007
10. International Organization of Standards (ISO).: Measurement of clean water flow in closed conduits—Velocity-area method using current-meters in full conduits and under regular flow conditions. ISO 3354-2008
11. International Organization of Standards (ISO).: Measurement of fluid flow in closed conduits—Velocity-area methods of flow measurement in swirling or asymmetric flow conditions in circular ducts by means of current-meters or pitot static tubes. ISO 7194-2008.
12. International Organization of Standards (ISO). Hydrometry—Calibration of current-me-ters in straight open tanks. ISO 3455-2007
13. Khozaei, M.H.: Numerical and experimental study of effects of geometrical parameters on propeller-type current-meters [M.Sc. thesis]. University of Tehran, Tehran (2016)
14. International Standard. Hydraulic turbinesrss—Hydraulic energy engineering. IEC 60041
15. ANSYS Inc. ANSYS CFX—Solver Theory Guide (2013)

# Analysis on the Performance of Evolutionary Strategies for Solving the Wind Farm Layout Optimization Problem

**Eduardo H. Haro, Luis A. Beltran, Eric L. Marin, Omar Avalos, and Octavio Camarena**

**Abstract** The energy-saving optimization problems have gained popularity in recent decades due to their implications for the benefit of the world. In that sense, the Wind Farm Layout Optimization (WFLO) problem has maintained its relevance in evolutionary computation since its inception in the 90s. Essentially, this problem involves the task of allocating the minimum feasible quantity of wind turbines in a constrained wind farm. However, the optimization problem deals with the fact of placing the precise amount of turbines in the correct distribution in order to generate the greatest possible energy at the lowest installation cost. It is well-known that this problem has been addressed by many computational schemes, nonetheless, the evolutionary computation has probed its efficiency over other methodologies in terms of computational resources and robustness. In this study, the Differential Evolution (DE) algorithm and some of its ancient and contemporary variants are analyzed to solve the WFLO problem from the perspective of two different practical cases. The obtained results demonstrate the advantages of evolutionary strategies for solving these types of energy-saving problems such as the capabilities of each DE variant.

E. H. Haro (✉) · L. A. Beltran · E. L. Marin · O. Avalos · O. Camarena
Centro Universitario de Ciencias Exactas e Ingenierias CUCEI, Av. Revolucion 100, Guadalajara, Mexico
e-mail: eduardo.hernandezh@academicos.udg.mx

L. A. Beltran
e-mail: luis.beltran7882@alumnos.udg.mx

E. L. Marin
e-mail: eric.lopez7892@alumnos.udg.mx

O. Avalos
e-mail: omar.aalvarez@academicos.udg.mx

O. Camarena
e-mail: octavio.camarena@academicos.udg.mx

 
D. Oliva et al. (eds.), *Advances in Optimization Algorithms for Multidisciplinary Engineering Applications: From Classical Methods to AI-Enhanced Solutions*, Studies in Computational Intelligence 806, https://doi.org/10.1007/978-3-031-78440-8_30

## 1 Introduction

The climate change has modified the people's perspective about the use of renewable energy resources, understanding it not as an option but as the only way to protect the environment and ourselves [1]. This has yielded the search of different sustainable power systems and the optimization of energy-saving problems [2]. Precisely, one of the most studied energy sources is wind power due to its simple principle of operation, the large amounts of energy it generates and its design feasibility almost anywhere in the world [3]. In general terms, the installation of a wind farm consists of the allocation of a certain number of wind turbines in a limited space naturally characterized by a constant flow of wind with enough strength to rotate the turbine blades. These turbine blades are connected to an electric generator that produces electricity through the kinetic energy of the wind. Finally, this produced energy is sent by wires to the sub-station of the wind farm where it is distributed for domestic and industrial consumption [4]. Although the principle of operation is simple, there must be considered two main aspects for the installation of a wind farm, its installation cost and the wind wake effect. A wind turbine is quite expensive in terms of manufacturing, installation and maintenance. Moreover, a profitable wind farm must be composed by tens of wind turbines in order to generate the enough energy to meet the demand, which makes a wind farm too big an investment. Hence, from an economic perspective an optimal wind farm must balance its number of turbines in order to generate the maximum power at the lowest installation cost [5]. On the other hand, when two wind turbines are placed one followed by the other one and both in the direction of the wind, the second turbine receives less wind power than the first turbine. This phenomenon is known as wind wake effect and it affects to the entire wind farm. Therefore, the allocation of the wind turbines must be well-planned in order to reduce the wind wake effect as much as possible [6]. These two aspects can be modeled simultaneously from an optimization perspective called Wind Farm Layout Optimization (WFLO) problem and it has been gained relevance over the years.

The WFLO problem was originally presented by Mosetti et al. [7] using the Jensen model [8, 9] under an offshore wind farm design, and addressing the problem with a Genetic Algorithm (GA) [10]. The impact of the Mosetti work set a precedent in the area of Evolutionary Computing (EC), and a race for solving the WFLO started. Some examples include the improvement of the Mosetti work under new parameters to the GA by the Monte Carlo method [11]. Furthermore, the analysis of the GA for the WFLO problems includes several related works that have significantly impacted in the enhanced of the problem itself [12–14]. In addition, some wind farm systems are based on real offshore wind farms and addressed by EC strategies [15]. The Particle Swarm Optimization (PSO) [16] has also reported remarkable results through the Jensen, Katic and Larsen models, developed by Pookpunt [17], Hou [18] and Pillai [19], respectively. Finally, some other meta-heuristic approaches utilized for solving the WFLO problem include the Whale Optimization Algorithm (WOA) [20], the Ant Colony Optimization (ACO) [21], the Crow Search Algorithm (CSA) [22], the Grey Wolf Optimizer [23], and in more recent years the Teaching-Learning-Based Optimization (TLBO) [24], just for mention a few.

All the aforementioned strategies have demonstrated their capabilities for these types of energy-saving optimization systems. Nevertheless, the WFLO problem involves a complex programming structure that is reflected in the computational cost which in turn is increased by the complexity of some of these approaches. In that sense, the Differential Evolution (DE) originally proposed by Storn and Price in 1997 [25] has been widely studied and the state-of-the-art confirms its efficiency in terms of simplicity of low computational resources [26]. This simple structure of the DE has motivated the improvement of the algorithm through several variations on its stages, its hybridization with other computational method or its enhance by a novel principle [27]. Moreover, many of the capabilities of the original DE tend to be part of its variants like its stability, processing time or easy structure. Therefore, the performance of these evolutionary strategies under energy-saving optimization problems has been widely studied in the literature achieving interesting results. Some examples of this include the Ela's work where different EC schemes were compared to reach the best parameter settings of the Optimal Power Flow (OPF) problem [28]. On its part, Shilaja compared different evolutionary methods with a proposed DE variant for optimizing the scheduling of distributed generations [29]. Also, a modified hybrid DE was presented by Lakshminarasimman to solve the optimization of hydrothermal power systems with cascaded reservoirs, and its results were compared with different DE variants [30], just for mention a few related works.

As can be noted, the DE and its several modifications have demonstrated to be highly competitive for solving different energy-saving optimization problems. Furthermore, many of the qualities of the DE are beneficial for these problems due to the simple adaptation that characterizes the DE. In addition, it must be considered that the majority of energy-saving problems tend to be discrete systems given its own nature. In that sense, most of the classical meta-heuristic approaches work with continuous models and its adaptation for solving discrete systems is a completely different and hard task that makes more complex its treatment [31]. In counterpart, the DE was originally developed for solving continuous problems but its adaptation to discrete models is quite simple due to its evolutionary structure like GA. Hence, these and many other features become the DE and its variants in excellent methods for solving energy-saving optimization problems. Regarding the WFLO problem, it has been already addressed by the DE like in the Wang's work where a new encoding mechanism was introduced [32] or the Liu's work where the DE was modified with a binary structure[33]. Nonetheless, an analysis on the performance of the original DE and some of its more classical and contemporary variants has not been reported in the literature to the best author's knowledge.

To solve this gap of information, in this work a study on the performance of the original DE for solving the WFLO problem is presented. In addition, the DE is compared against other two evolutionary strategies that have been widely studied in the state-of-the-art, the Covariance Matrix Adaptation Evolutionary Strategy (C-MAES) [34] and the JADE [35]. Moreover, two contemporary DE variants that have gained popularity in recent years due to their optimization capabilities have been included in the study, the Bernstein-Levy Differential Evolution algorithm (BDE) [36] and the Bezier Search Differential Evolution algorithm (BESD) [37]. About the

WFLO design, it has been decided to test the algorithm in the Jensen model recently adapted by Kunakote et al. [38], where two cases are considered based on the wind wake effect eclipsing totally the turbine and a partial eclipse. The obtained results probe the benefits of using evolutionary strategies for solving the WFLO problem by achieving remarkable results in terms of metric values by the JADE. Additionally, different statistical and non-parametric tests demonstrate the effectiveness of JADE and it seems positioned to be the best option for future improvements for solving the WFLO problem in more complex conditions. However, the rest of algorithms probe efficiency in terms of stability, robustness and accuracy in other parameters, demonstrating their reliability for other future research.

The remainder of the chapter is structured as follows: Sect. 2 presents in detail the WFLO problem and its mathematical formulation, the DE and the applied variants in this study are described in Sect. 3, the obtained computational results are reported and explained in Sect. 4, the discussions of the work are shown in Sect. 5 while Sect. 6 concludes the chapter and presents the future research.

## 2 The Wind Farm Layout Optimization Problem

In this section, the formulation of a WFLO problem is described based on the wind turbine wake model and the objective function evaluated for this work. It is worth mentioning that as mentioned in the Introduction, different wind wake models have been proposed in the literature. However, in this work the Jensen model is applied for its resolution. In the same way, several objective functions have been modeled to optimize the WFLO problem. Nonetheless, since this work is focused on the sustainability of the wind farm, then an objective function that implies the installation cost and the generated power is considered.

### *2.1 Wind Turbine Wake Model*

As explained before, the wind wake effect consists in the fact that two wind turbines allocated in a horizontal-axis and both of them in the direction of the wind yield that the second wind turbine receive less wind power than the first one. This model can also consider other designs like adding noise, cost, among many other perturbations and constraints [39]. However, the power energy generated by the wind turbine is not only dependent of its allocation in the wind farm but also of its mechanical parameters. Therefore, the wind speed, wind direction, wind turbine efficiency and turbine thrust coefficient must be considered to estimate the total power produced by the wind turbine. Another remarkable consideration is the conservation of energy due to the physical phenomenon that the velocity in the turbine wake is increased as it moves away from the turbine. Based on the above, the mathematical formulation for the wind turbine has always been attractive for researchers around the world.

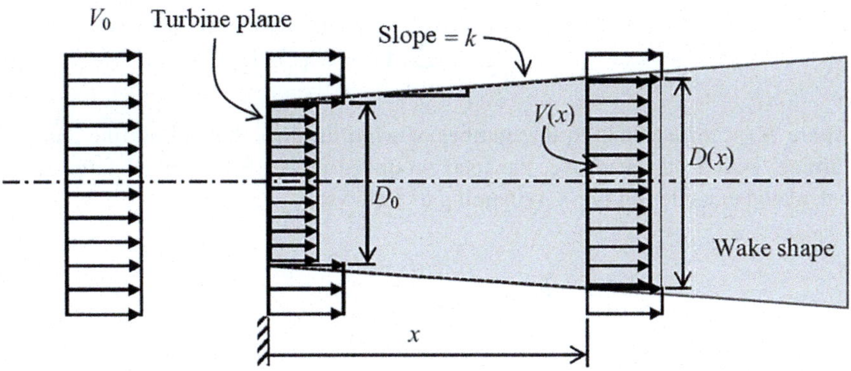


**Fig. 1** Wind wake decay model [38]

Nonetheless, the Jensen model focused on the wake decay effect has been the most popular model applied in several related work due to its simplicity and efficiency in terms of accuracy with a real-world wind turbine [40]. In addition, the Jensen's model has been the inspiration for many improvements like the Larsen or the Frandsen models [41, 42], respectively.

The Jensen's model is graphically shown in Fig. 1 based on the Kunakote work. As can be seen, this model assumes the initial wake with the same diameter as the turbine diameter, such as the wind velocity is considered uniform. When the wind affects the turbine it rotates and the wake is generated at the same time that it is linearly expanded as the wake moves away from the turbine. This wind speed tends to be described as follows:

$$1 - \frac{V(x)}{V_0} = \alpha \left( \frac{D_0^2}{D(x)^2} \right) \tag{1}$$

where $V_0$ and $V(x)$ are the full wind speed and the wind wake speed, respective. As can be seen in Fig. 1, $V(x)$ is away from the turbine at a distance $x$. Also, $D_0$ is a turbine diameter and $D(x)$ is a wake diameter at a distance $x$ from the respective turbine and it can be described by Eq. 2, where $k$ is a constant that defines the hub height and terrain roughness of the wind farm. On the other hand, $\alpha$ is known as the speed deficit in the wake area and it can be modeled as shown in Eq. 3, where $C_T$ is the turbine thrust coefficient.

$$D(x) = D_0 + 2kx \tag{2}$$

$$\alpha = 1 - \sqrt{1 - C_T} \tag{3}$$

It must be also considered the scenario where the wind turbine is affected by many wakes from the upstream turbine. This kinetic energy effect can be computed as follows:

$$\left(1-\frac{V}{V_0}\right)^2=\sum_{i=1}^{N_{EC}}\left(1-\frac{V(x_i)}{V_0}\right)^2 \tag{4}$$

where $N_{EC}$ corresponds to the number of wind turbines that eclipse the analyzed turbine. Based on the above, the total produced power by the wind farm can be calculated considering Eq. 4, remaining as follows:

$$P_{total}=\sum_{i=1}^{N_{WT}}\frac{1}{2}\eta_i\rho_{air}A_iV_i^3 \tag{5}$$

where $N_{WT}$ is the total number of wind turbines in the wind farm, while $\eta_i$ and $\rho_{air}$ are the turbine efficiency of the respective turbine and the air density, respectively. On the other hand, the wind speed $V_i^3$ is given in $m/s$ while the area of the disk of the turbine $A_i$ and its intersection with other turbine is given as follows:

$$A_{i,j}=r_i^2 cos^{-1}\left(\frac{d_{i,j}^2+r_i^2-r_j^2}{2d_{ij}r_i}\right)+r_j^2 cos^{-1}\left(\frac{d_{i,j}^2+r_j^2-r_i^2}{2d_{ij}r_j}\right)$$
$$-\frac{1}{2}\sqrt{\left(-d_{i,j}+r_i+r_j\right)\left(d_{i,j}-r_i+r_j\right)\left(-d_{i,j}+r_i-r_j\right)\left(d_{i,j}+r_i+r_j\right)} \tag{6}$$

where $d_{i,j}$ corresponds to the distance between the centers of both wind turbines, $r_i$ and $r_j$ are the wake radius of the turbines $i$ and $j$ at the sample plane turbine $j$, respectively. In order to clarify the interpretation of Eq. 6, Fig. 2 shows its graphical representation. Essentially, the turbine $j$ is being partially eclipsed by the turbine $i$

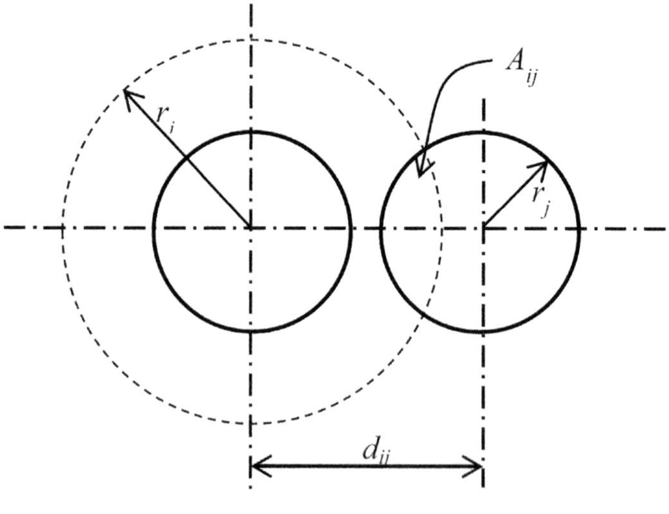


**Fig. 2** Intersection of two turbines [40]

through the wind wake effect, and the intersection of these two turbines is given by $A_{ij}$ correspondent to the diameter of the eclipse.

## 2.2 Optimization Formulation

The optimization of the WFLO problem include different perturbations or constraints. However, in this case the most popular design is use as optimization model of an offshore wind farm which involves the minimization of its installation cost and the maximization of its generated power. Usually, this model is designed as a grid-based approach where the center of each cell of the grid represents a feasible position where a wind farm can be allocated [32]. This design has been attractive for researchers over the years given the fact that it allows to implement wind turbines with different hub heights and different turbines. In addition, constraints and uncertainties can be added to the design to make it more complex and robust [43, 44]. Therefore, the objective function of this problem is to minimize the following expression:

$$f(x) = \frac{Cost}{P_{total}} \tag{7}$$

where $x$ is the respective individual in the population, $P_{total}$ corresponds to the total generated power of the wind farm which in turn is the model described in Eq. 5, while $Cost$ is the total installation cost of the wind farm. This total cost only depends on the number of turbines ($N_T$), and it can be expressed as follows:

$$Cost = N_T\left(\frac{2}{3} + \frac{1}{3}e^{-0.00174 \cdot N_T^2}\right) \tag{8}$$

Traditionally, a real-world offshore wind farm requires a design of $2000m^2$ grid area. However, in this work said surface is normalized to $10 \times 10$ grids, which corresponds to a surface where 100 wind turbines can be allocated. In the same way, it is assumed that the center of two consecutive grids have a distance of $5D_0$. This allows to make more realistic the wind farm since the aforementioned distance is considered based on the length of the turbine blades. Moreover, the internal parameters of the turbine such as the natural characteristics are described based on real parameters measured for other purposes. These features are reported in Table 1.

About the encoding/decoding of this WFLO problem involves the creation of the $x$ vector correspondent to each individual in the population. Each individual is designed as a vector of 100 dimensions where each dimension is interpreted as a grid of the wind farm. In addition, each dimension stores a continuous value distributed between 0 and 1. It is worth mentioning that a same random population is generated as initial population for each meta-heuristic approach in order that all the scheme compete on equal opportunities. Once the respective algorithm has finished its optimization process, the best result is based on the fitness value of the individuals. Subsequently,

**Table 1** Characteristics of the wind turbine [38]

| Feature | Value |
|---|---|
| Thrust coefficient $(C_T)$ | 0.88 |
| Ground surface roughness $(h_0)$ | 0.3 |
| Hub height $(h)$ | 60 m |
| Rotor diameter $(D_0)$ | 40 m |
| Turbine efficiency $(\eta)$ | 40% |
| Air density $(\rho_{air})$ | 1.225 $Kg/m^3$ |
| Wind speed $(V_0)$ | 12 m/s |
| Wind directions $(degree)$ | 10°, 20°, 30°, ..., 360° |
| | Wind direction angle $w.r.t. +x$ |
| | (East) |

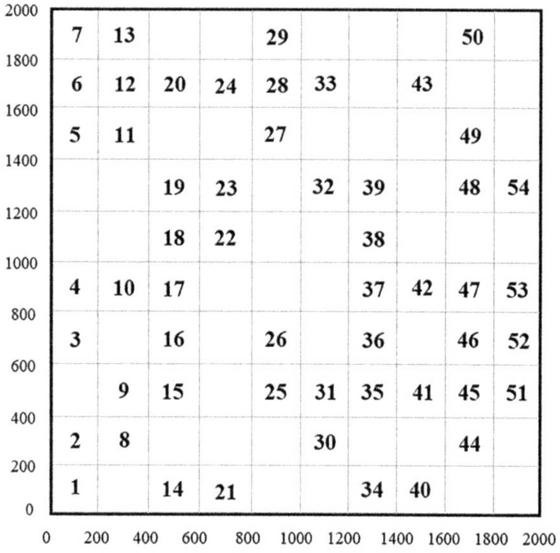


**Fig. 3** Example of turbine placement in the 2000 $m^2$ grid area

the optimal solution is rounded in the sense that each dimension value less than or equal to 0.5 becomes 0, and each dimension value greater than 0.5 becomes 1. This discretization process allows to determine the proposed allocation of turbines in the wind farm, since a value 1 corresponds to a grid that must contain a wind turbine, while a value 0 means exactly the opposite. Finally, it is worth mentioning that the graphical representation of the vector $x$ must be read as shown in Fig. 3. As can be seen, the $x$ vector must be distributed in ten sub-vector of ten grids, then each sub-vector must be transposed followed by the next sub-vector. In addition, only the grids with a value 1 must be considered for the placement of a wind turbine, while the rest of grids are empty. In the case of Fig. 3 an $x$ vector of 54 wind turbines are allocated in the wind farm of 2000 $m^2$.

# 3 The Evolutionary Computation Strategies

In this section, the DE algorithm such as the rest of evolutionary strategies (C-MAES, JADE, BDE and BESD) are summarized. It is worth remembering that the selection of this approaches is based on the analysis of each one for the solution of related energy-saving optimization problems.

## *3.1 Differential Evolution*

The differential evolution algorithm (DE), proposed in 1997 by Storn and Kennet, is a popular optimization algorithm because it is a member of the evolutionary algorithm family (EA), used to solve complex optimization problems. DE algorithm is a population-based method that is commonly used to solve various types of optimization problems [45].

The differential evolution is a multi-population meta heuristic algorithm where the population members are considered vectors of parameters, defined on a multidimensional space, and the population is modified through the optimization process by iterations or access functions. The DE is divided into two phases: Initialization and the evolutionary process. In the first phase, the population is stochastically established over the limits of the problem's search space to be optimized. Each solution, represented as a vector $X_i^g$, is defined in Eq. 9.

$$X_i^g = X_{low} + (X_{upp} - X_{low}) * rand(0, 1) \tag{9}$$

where $X_{low}, X_{upp}$ are lower and upper bounds of search space and $rand(0, 1)$ denotes a vector of random numbers uniformly distributed between 0 and 1. In the second phase, the evolutionary process uses mutation, crossover and selection operators. This evolutionary process is repeated until a satisfactory solution is reached or a predefined limit of iterations is exceeded.

### 3.1.1 Mutation

Mutation is the most important operator in the DE since mutation produces a new individual in the population. The operator is based on obtaining a mutant vector through the differences between the different vectors that make up the population to promote the exchange of information between the elements of the population to guide the search strategy to the most promising area or the optimal global solution of the optimization problem. The literature presents many mutation strategies, but Rand/1 is the most widely used in applications because it maintains a high degree of diversity in the population and has low computational costs [46]. The vector mutant is represented as $v_i^{g+1}$, is defined in Eq. 10.

$$v_i^{g+1} = x_{r3}^g + F(x_{r1}^g - x_{r2}^g) \tag{10}$$

where $F$ is the scaling factor and value vary from 1 to 0. Furthermore, the values of the vectors are randomly selected $x_{r1}^g, x_{r2}^g, x_{r3}^g \in \{1, 2, \ldots, N_P\}$, considering that they will be different $x_{r1}^g \neq x_{r2}^g \neq x_{r3}^g$.

### 3.1.2 Crossover

After the mutation process, the crossover operator is performed to increase the diversity of the vectors. This indicates that the crossover operation complements the mutation strategy; the crossover is used occasionally, increasing the diversity of the population. Crossover increases population diversity, controlled by the constant $Cr$ with a value in the range [0,1]. A trial vector $u_i^{g+1}$ in Eq. 11 is generated by combining the target vector $x_{i,d}^g$ and the mutant vector $v_{i,d}^g$.

$$u_i^{g+1} = u_d^g \begin{cases} v_{i,d}^g & \text{if } d = d_{rand} \text{ or } rand(0, 1) \leq Cr \\ x_{i,d}^g & \text{otherwise} \end{cases} \tag{11}$$

The crossover operation randomly switches positions between the individual $X_i^g$ of the population and the mutant vector $v_i^g$. If the output of rand(0,1) is less than or equal to the value of $CR$, the parameter for the test vector is inherited from the mutant vector; otherwise, the parameter for the test vector is inherited from the individual vector. The crossover mechanism in DE plays a crucial role in exploring the search space and promoting diversity within the population, aiding in the convergence towards optimal or solutions [25].

### 3.1.3 Selection

The selection operator aims to maintain the best solutions in the population throughout the optimization process, so after the mutation and crossover operations have been performed, it allows for determining whether the trial vector $u_i^g$ will replace the corresponding individual vector $X_i^g$, following greedy criteria, as shown in the Eq. 12.

$$x_i^g = \begin{cases} u_{i,j}^g & \text{if } f(u_i^{g+1}) \leq f(x_i^g), \\ x_{i,j}^g & \text{otherwise,} \end{cases} \tag{12}$$

The greedy criteria consider the fitness of the trial vector $u_i^g$ and the fitness value of the corresponding individual vector $X_i^g$. If the fitness of the trial vector is less than that of the individual vector, the trial vector will replace the individual vector so that this vector will be part of the population for the next iteration. The process is repeated until the stop condition is met, usually when the maximum number of iterations $K_{Max}$ is reached, defined by the principle of the iterative process.

## 3.2 Covariance Matrix Adaptation Evolution Strategy

The CMA-ES (Covariance Matrix Adaptation Evolution Strategy) algorithm is a stochastic optimization method used to solve optimization problems [47]. The CMA-ES algorithm is a classical and advanced black-box optimization technique. It is a member of the evolutionary strategy family. It is especially used in numerical optimization problems where we do not have information about the objective function, such as optimization problems of cost functions or optimization problems of functions with multiple local minimums.

A black box search scenario is considered when we need to optimize a problem where we do not have information about the objective function. In the context of minimizing a function $f : R^n \rightarrow R$, expressed as $x \mapsto f(x)$, where the objective is to discover one or more search points (candidate solutions), $x \in R^n$, with the smallest possible function value $f(x)$. CMA-ES has a population of new search points (individuals, offspring) generated by a multivariate normal distribution sampling. The equation for sampling the search points is defined as follows:

$$\mathbf{x}_k^{(\tau+1)} \sim \mathbf{m}^\tau + \sigma^\tau \times \mathcal{N}(\mathbf{0}, \mathbf{C}^\tau) \ for \ k = 1, \ldots, \lambda \tag{13}$$

where:
$\mathbf{x}_k^{(\tau+1)}$ is $k$-th search point (individual/offspring) from generation $\tau + 1$
$\sim$ denotes the same distribution on the left and right side
$\mathbf{m}^\tau$ is the mean value of the search distribution
$\sigma^\tau$ is the "overall" standard deviation or step-size, at generation $\tau$.
$\mathcal{N}(\mathbf{0}, \mathbf{C}^\tau)$ is a multivariate normal distribution with zero mean and covariance matrix $\mathbf{C}^\tau$, at generation $\tau$.

For an iteration to be completed in the iterative process, $\mathbf{m}^{\tau+1}$, $\sigma^{\tau+1}$, and $\mathbf{C}^{\tau+1}$ must be calculated for the next generation.

The new mean value $m^{\tau+1}$ of the search distribution is determined through weighted recombination of the $\mu$ best candidate solutions from the sample $x_1^{(\tau+1)}, \ldots, x_\lambda^{(\tau+1)}$

$$\mathbf{m}^{\tau+1} = \sum_{i=1}^{\mu} \omega_i \mathbf{x}_{i:\lambda}^{\tau+1} \tag{14}$$

where $\mu \leq \lambda$ is the parent population size, i.e. the number of selected points $x_{i:\lambda}^{\tau+1}$ is the best individual out of $x_1^{(\tau+1)}, \ldots, x_\lambda^{(\tau+1)}$ and the positive weight coefficients for recombination $\sum_{i=1}^{\mu} \omega_i = 1$ with $\omega_i > 0$, are typically chosen linearly on the log scale with $\mu \leq \lambda/2$
Therefore, the Eq. 14 can be rewritten as follows by updating $m$

$$\mathbf{m}^{\tau+1} = \mathbf{m}^\tau + c_m \sum_{i=1}^{\mu} \omega_i (\mathbf{x}_{i:\lambda}^{\tau+1} - \mathbf{m}^\tau) \tag{15}$$

where $c_m$ is a learning rate and usually set to 1.

To construct the evolution path $\mathbf{p}_c \in \mathbf{R}^n$, starting with $\mathbf{p}_c = 0$, the step size $\sigma$ is disregarded:

$$\mathbf{p}_c^{\tau+1} = (1 - c_c)\mathbf{p}_c^{\tau} + \sqrt{c_c(2 - c_c)\mu_{\text{eff}}}\frac{\mathbf{m}^{\tau+1} - \mathbf{m}^{\tau}}{\sigma_{\tau}} \tag{16}$$

$c_c \leq 1$, and $1/c_c$ is the backward time horizon of the evolution path $\mathbf{p}_c$. The factor $\sqrt{c_c(2 - c_c)\mu_{\text{eff}}}$ is a normalization constant for $\mathbf{p}_c$. Because, in general, the expected length of the evolution path $\mathbf{p}_c^{\tau+1}$ from Eq. 16 depends on its direction, a conjugate evolution path is constructed, and to control the step-size $\sigma^{\tau}$, it utilized a sum of successive steps. The method is applied independently of the covariance matrix update and is also called commutative path length control or cumulative step length adaptation (CSA). The evolution path is updated as follows:

$$\mathbf{p}_{\sigma}^{\tau+1} = (1 - c_{\sigma})\mathbf{p}_{\sigma}^{\tau} + \sqrt{c_{\sigma}(2 - c_{\sigma})\mu_{\text{eff}}}\mathbf{C}^{\tau^{-1/2}}\frac{\mathbf{m}^{\tau+1} - \mathbf{m}^{\tau}}{\sigma_{\tau}} \tag{17}$$

For Default Parameters, see Table 1 in Appendix A of the document [48].

### 3.3 *JADE*

The JADE algorithm is an evolutionary strategy originally proposed by Zhang and Sanderson in 2009 [35]. Essentially, the JADE introduces a new mutation method called "DE/current-to-*p*best" that aims to improve to fast convergence skills of the DE/current-to-best method. This proposed mutation generates a trial vector $V_{i,g}$ described as follows:

$$V_{i,g} = X_{i,g} + F_i \cdot (X_{best,g}^{p} - X_{i,g}) + F_i \cdot (X_{r1,g} - X_{r2,g}) \tag{18}$$

where $X_{best,g}^{p}$ and $X_{i,g}$ correspond to the best and the current individuals of the current population, respectively. $X_{r1,g}$ and $X_{r2,g}$ are two individuals randomly selected from the current population with the only condition of being different between each other such as being different to $X_{best,g}^{p}$ and $X_{i,g}$. Meanwhile, $F_i$ is the mutation factor associated to $X_i$. It is worth mentioning that $F_i$ changes every generation following the next instructions:

$$F_i = randc_i(\mu F, 0.1) \tag{19}$$

$$\mu F = (1 - c) \cdot \mu F + c \cdot mean_L(S_F) \tag{20}$$

where $c$ is a positive constant defined between [0,1], the location parameter $\mu F$ is initially set as 0.5 and then updated at the end of each generation with Eq. 20, while $mean_L(S_F)$ is the Lehmer mean [49]. It is worth mentioning that the trial vectors generated by Eq. 18 are stored in an archive known as population $A$, then when $A$ can show the evolution of the process, the mutation method is modified to be applied as follows:

$$V_{i,g} = X_{i,g} + F_i \cdot (X^p_{best,g} - X_{i,g}) + F_i \cdot (X_{r1,g} - \tilde{X}_{r2,g}) \tag{21}$$

where $\tilde{X}_{r2,g}$ is an individual randomly chosen from the union between $A$ and the current population $P$. As can be noted, the principle of operation of the JADE algorithm consists of the creation of a second population based on the individuals of the trial vectors. Then the next trial vectors are formed by the these two populations. The aim of this process is to enhance the exploration phase based on the evolution process of the algorithm. This skill allows the JADE to be highly competitive in many optimization tasks, which in turn has become the JADE one of most popular and well-known DE variants in the literature.

## 3.4 Bernstein-Levy Differential Evolution Algorithm

The Bernstein-Levy Differential Evolution Algorithm (BDE) is a population-based optimization algorithm designed to solve complex optimization problems [36]. It combines the concepts of differential evolution and Bernstein polynomials to efficiently explore the search space and find optimal solutions. This algorithm has been successfully applied to various problems, such as function optimization, parameter estimation, and feature selection. The BDE is a powerful optimization technique that leverages the strengths of both differential evolution and Bernstein polynomials to navigate complex search spaces and find optimal solutions efficiently. The algorithm is particularly effective for problems where the search space is large or multidimensional, as it can efficiently explore and exploit different regions of the space. Moreover, the BDE offers computational efficiency and transparency in finding optimal solutions.

The algorithm effectively solves non-differentiable, complex, and discontinuous numerical problems. It achieves this by using a unique elitist-mutation operator and a stochastic parameter-free crossover operator based on Bernstein polynomials. The elitist-mutation operator preserves the best solutions in each iteration, ensuring that valuable information is not lost during the optimization process. The stochastic parameter-free crossover operator based on Bernstein polynomials allows for efficient exploration of the search space, increasing the algorithm's ability to find optimal solutions. In addition to its effectiveness in numerical problem-solving, the BDE has also demonstrated success in real-world optimization problems. It utilizes three direction vectors, dv1, dv2, and dv3, to determine the evolutionary direction and step size values. The algorithm generates these direction vectors using permutation functions and leverages the Levy distribution to generate the evolutionary step size value. The crossover process is controlled by a matrix and is updated iteratively. The objective function values of mutation patterns are used to update corresponding patterns and their objective function values. The global sub-optimum solution pattern and its related objective function value are decided based on the best solutions acquired. The BDE, is not just a theoretical concept; it has proven to be successful in real-world optimization problems. The algorithm uses direction vectors, permutation functions, and the Levy distribution to determine evolutionary directions and step sizes, which helps it adapt and navigate through difficult optimization landscapes. Overall, the BDE is a reliable and transparent optimization technique that combines the strengths of its individual components to efficiently address a wide range of optimization challenges.

### 3.5 Bezier Search Differential Evolution Algorithm

Evolutionary computing algorithms, such as the differential evolution algorithm, are widely employed for solving complex numerical optimization problems. The Bezier Search Differential Evolution Algorithm (BESD) is a variation of the differential evolution algorithm that integrates properties of Bezier curves and utilizes a unique elitist-mutation operator and a Bernstein polynomials-based stochastic parameter-free crossover operator [37]. The BESD is designed to enhance the problem-solving ability of the differential evolution algorithm by introducing the concept of Bezier curves. The algorithm starts by describing a set of candidate solutions, called patterns, in a pattern matrix $S$. This is defined by using this equation:

$$S_{q,j} \sim U(\text{low}_j\ ,\ \text{up}_j\ ) \mid q\ =\ [1 : K \cdot N], j = [1 : Dim] \tag{22}$$

where $q, J \in Z^+$. At the start of each iteration, the pattern matrix is regenerated by randomly selecting $N$ patterns from the pre-pattern matrix. The size of the pre-pattern matrix is $K$ times the size of the pattern matrix, $P$. The algorithm uses two sets of patterns, dv1, and dv2, to define the evolutionary direction. The patterns of dv1 are bijective, evolving into different patterns, while dv2 is calculated using cubic Bernstein polynomials. BESD uses an efficient strategy to generate the evolutionary scaling factor and control the evolutionary direction through a binary crossover operator. BESD maintains a partially elitist approach, updating the best global solution in each iteration. The performance of the algorithm is influenced by the efficiency of the generation of evolutionary directions and the size of the pattern matrix. BESD is an efficient and simple evolutionary search algorithm designed to solve global numerical problems. It uses concepts from linear algebra and pattern recognition without relying on complex bio/geographic models and has proven to be competitive compared to other evolutionary algorithms in global optimization.

## 4 Computational Experiments

In this section, the obtained comparative results of the five algorithms under the objective function of Eq. 7 for both cases is presented. For both experiments the wind turbine considered such as the environmental conditions have the features shown in Table 1. It is worth mentioning that each approach has been computed with an initial population of 100 individuals and 250 iterations for 30 independent runs. In all cases, the internal parameters of each algorithm are based on the data proposed in their respective original articles. Finally, all the computational experiments have been carried out in the MATLAB programming environment with a Notebook (16 GB RAM and 3.60 GHz AMD 8-Core processor) in Windows 10 OS.

## *4.1 Full Eclipse Results*

For this first experiment, it is considered a 36 wind directions with a constant wind speed of 12 m/s. Under these conditions the best, average and worst fitness values reached by each algorithm are exposed in Table 2. As can be noted, the JADE approach outperforms the rest of approach in these three metrics, being followed by the DE in terms of best and average fitness, and by the BDE in terms of worst fitness. Additionally, Table 2 also presents the standard deviation and the average computational time expended by each scheme for the 30 independent runs. In this case, the BDE presents a better performance for these two remain metrics compared with the other algorithms. However, it is worth noting that the JADE reports the second best average computational time.

In order to analyze the data of Table 2 from other perspectives, the convergence rate and the computational times of each algorithm are shown in Figs. 4 and 5, respectively. It is worth mentioning that all the convergence rate of Fig. 4 correspond to the average value of each one of the 250 iterations for the 30 independent runs. Based on the above it can be seen that the DE presents the best convergence in terms of speed and homogeneity, while the BESD reports totally the opposite result. However, when the computational times are analyzed through the box plots of Fig. 5 it is clear that the DE has the second worst performance, only surpassed by the CMAES. In counterpart, the JADE and BDE present the best performances and actually they are very similar in terms of median, best and average computational time.

On the other hand, Table 3 shows the numerical characteristics of the wind farms generated by each algorithm. These wind farms are based on the best fitness values previously reported in Table 2. In this case, it is quite complicated to define the best approach since each parameters set different optimal specifications. In other words, WFLO problem specifies the minimization of the number of wind turbines such as

**Table 2** Metric results of the algorithms for Experiment 1

| Method | Best | Average | Worst | STD | Time (s) |
|---|---|---|---|---|---|
| DE | 0.001454 | 0.001458 | 0.001467 | 2.61E-06 | 714.35 |
| CMAES | 0.001463 | 0.001468 | 0.001471 | 2.10E-06 | 971.42 |
| JADE | **0.001451** | **0.001456** | **0.001460** | 2.76E-06 | 528.07 |
| BDE | 0.001455 | 0.001459 | 0.001461 | **1.87E-06** | **526.81** |
| BESD | 0.001463 | 0.001468 | 0.001472 | 2.06E-06 | 556.09 |

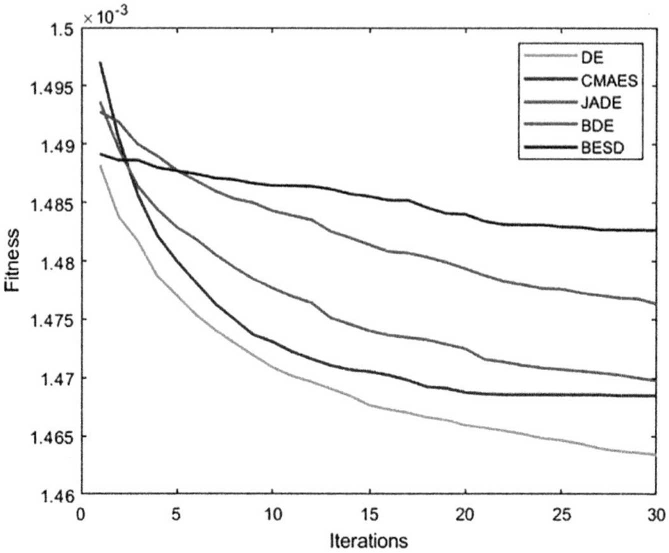


**Fig. 4** Convergences of the algorithms for Experiment 1

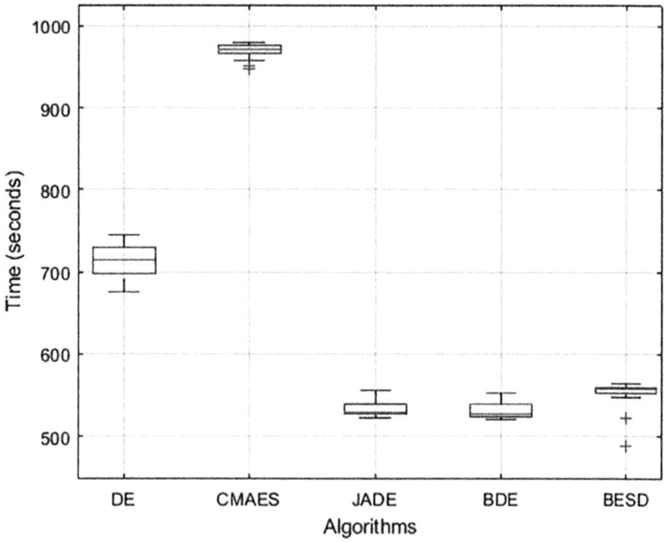


**Fig. 5** Computational times for Experiment 1

**Table 3** Obtained parameters of the wind farms for Experiment 1

| Method | Wind Turbines | Installation Cost | Supplied Power (kW) |
|---|---|---|---|
| DE | 43 | 29.24093 | 20116.75 |
| CMAES | 39 | 26.92165 | 18404.80 |
| JADE | 43 | 29.24093 | 20150.88 |
| BDE | 44 | 29.83840 | 20512.19 |
| BESD | 42 | 28.65031 | 19577.22 |

the installation cost. In that sense the CMAES offers the best result with a wind farm of only 39 wind turbines and an installation cost 26.92165 units. However, the WFLO also implies the maximization of the power supplied by the wind farm. Under this condition, the CMAES offers the worst wind farm with an only 18,404.80 kW power supply. In counterpart, the best wind farm in terms of power supply is the BDE wind farm with 20,512.19 kW. Nonetheless, the BDE also offers the worst wind farm in terms of installation cost with an investment of 29.83840 units due to its 44 turbines. Therefore, the JADE wind farm offers the best solution since it is positioned in a middle point between the lowest installation cost and the highest power supply. The JADE wind farm is composed of 43 turbines with a cost of 29.24093 units and 20,150.88 kW power supply. Moreover, the DE and JADE reports the same wind farm, but Table 2 showed that the JADE outperforms the DE in almost all metrics.

Finally, an optimal wind farm not only considers its number of turbines but also their distribution throughout the entire search space. That is the reason why in Table 3 the DE and JADE report the same number of turbines at the same installation cost but they offer different power supply. This is also well-described previously in Sect. 2.1. Based on the above, Fig. 6 show graphically the distributions of the wind farm of each algorithm according to the data of Table 3. It can be appreciated a constant in all wind farms which is that the majority of turbines are allocated along the perimeter of the wind farm. This is due to the fact that this distribution minimize the wind wake effect described in Sect. 2.1. However, the internal part must be used correctly to minimize said effect. It can be seen a completely different distribution between the DE and JADE wind farms (Fig. 6a and Fig. 6c, respectively). The JADE utilizes almost all the perimeter of the search space and the rest of wind turbines are allocated in the middle of the wind farm with a considerable separation between them and the perimeter. This allows to minimize enough the wind wake effect which in turn allows that all the turbines offer the maximum feasible power. In counterpart, the DE allocates the same number of turbines than the JADE wind farm but with a counterproductive distribution since the perimeter is wasted and the turbines in the middle of the space are positioned close to the perimeter, which affects considerably their optimal performance.

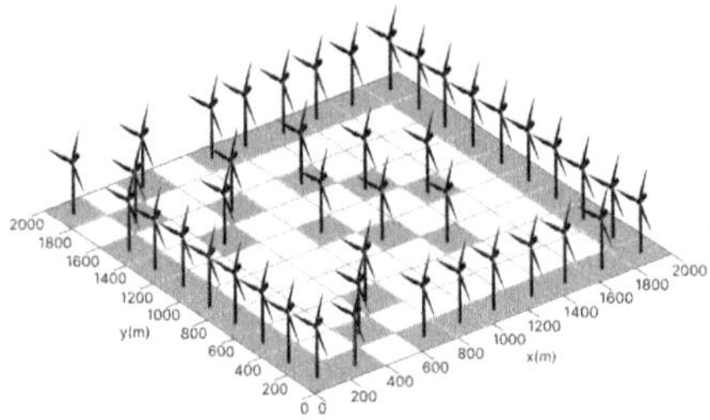


(a) DE: Cost 29.24093, Power 20116.75 *k*W.

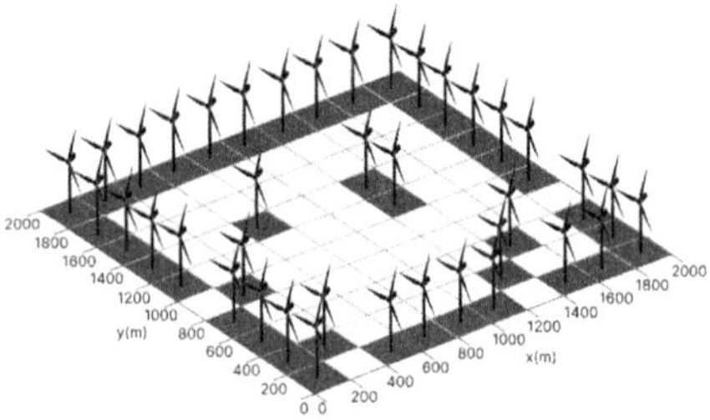


(b) CMAES: Cost 26.92165, Power 18404.80 *k*W.

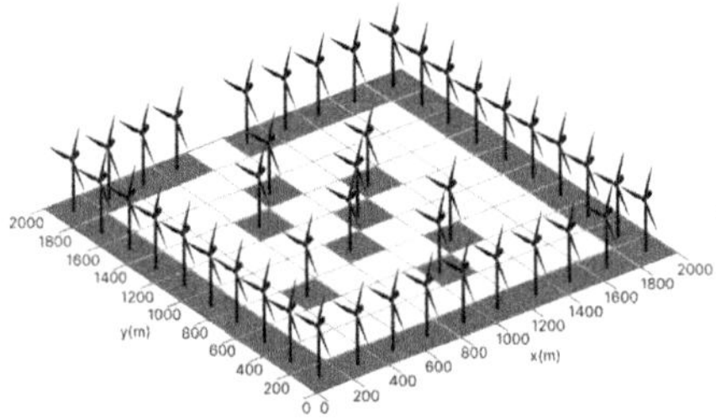


(c) JADE: Cost 29.24093, Power 20150.88 *k*W.

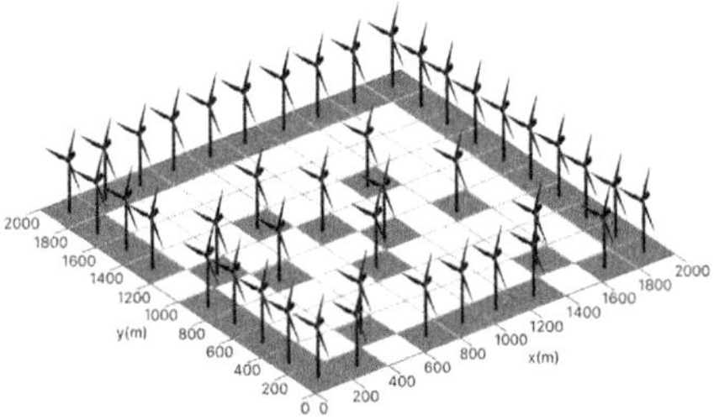


(d) BDE: Cost 29.83840, Power 20512.19 *k*W.

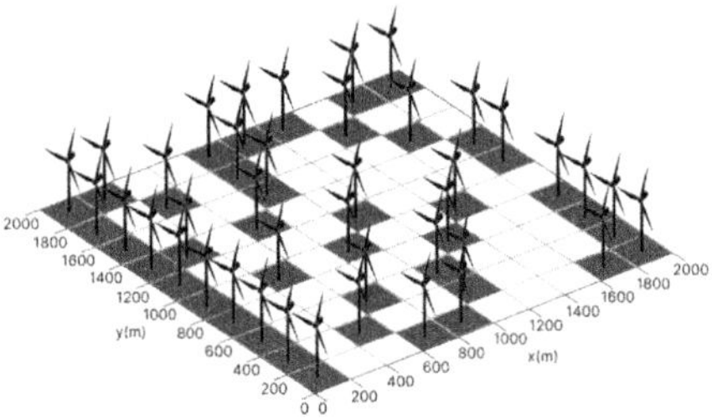


(e) BESD: Cost 28.65031, Power 19577.22 *k*W.

**Fig. 6** Wind farms generated by the algorithms for Experiment 1

About the rest of approaches, the CMAES (Fig. 6b) allocates the lowest number of turbines but wasting the perimeter of the search space, while the BDE (Fig. 6d) places the highest number of turbines which increases the installation cost of the wind farm. However, it is quite remarkable that the proposed distribution of the BDE is highly competitive. Finally, the BESD (Fig. 6e) presents the worst performance in terms of allocating a high number of turbines under a dreadful distribution that affects the power supply of the wind farm.

## 4.2 Partial Eclipse Results

The second computational experiment considers that the analyzed wind turbine is partially eclipsed by other turbines, which involves that the total power is 7.10 MW. It is worth mentioning that the same five optimization approaches are compared under the aforementioned conditions of 100 individuals, 250 iterations for 30 independent runs. Following the same structure of the last experiment, in Table 4 the metric results of the algorithms are presented. As can be noted, the JADE reports the best performance for the best, average and worst fitness value from the 30 independent results. Actually, this same performance was reported in the previous experiment. This can be attributed to the precision and robustness of the scheme. On the other hand, the standard deviation and the computational time are lead by the BDE algorithm. Furthermore, the BDE presented the same performance for the last experiment, which confirms the stability of the algorithm.

To have a better comprehension of the metric results, the convergence rates of the five approaches are shown in Fig. 7. It is worth remembering that these convergences correspond to the average convergence of each one of the 30 independent rates. As can be seen, the DE presents the best performance closely followed by the CMAES. On the other hand, the BDE and BESD have the worst rates in terms of speed and accuracy. This is quite relevant since it must not be confused the convergences of Fig. 7 with the data of Table 4, where the DE or the BDE presented competitive performances. In counterpart, the JADE shows a competitive convergence positioned as the best third result. When Figs. 4 and 7 are compared it is quite easy to note that the algorithms follow the same trend, which demonstrates the homogeneity of the results. On the other hand, the computational times expended by each algorithm for this experiment are shown in Fig. 8. It is clear that the JADE and BDE correspond to the best techniques in terms of computational cost while the CMAES is totally the opposite. These information was already presented in Table 4, however it is remarkable the visual difference between the computational costs of the approaches. Finally, the BESD presents the worst stability with scattered results.

In Table 5 the engineering parameter values are reported for each algorithm. As can be seen, the best scheme in terms of the lower number of wind turbines is the CMAES with 42 turbines, while the worst approach in the same terms are the BDE and BESD with 45 turbines. In the same way, the number of turbines is directly related to the

**Table 4** Metric results of the algorithms for Experiment 2

| Method | Best | Average | Worst | STD | Time (s) |
|---|---|---|---|---|---|
| DE | 0.001411 | 0.001415 | 0.001420 | 2.40E-06 | 656.55 |
| CMAES | 0.001416 | 0.001421 | 0.001426 | 2.34E-06 | 950.37 |
| JADE | **0.001408** | **0.001411** | **0.001415** | 2.06E-06 | 551.64 |
| BDE | 0.001412 | 0.001415 | 0.001418 | **1.48E-06** | **549.72** |
| BESD | 0.001419 | 0.001423 | 0.001429 | 1.92E-06 | 611.04 |

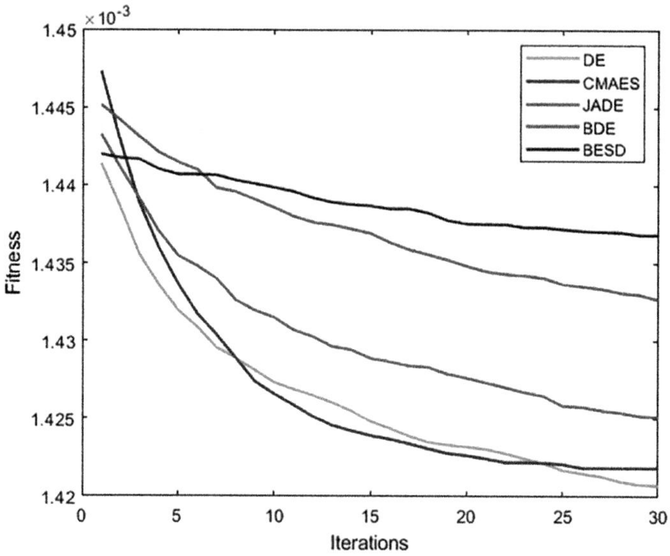


**Fig. 7** Convergences of the algorithms for Experiment 2

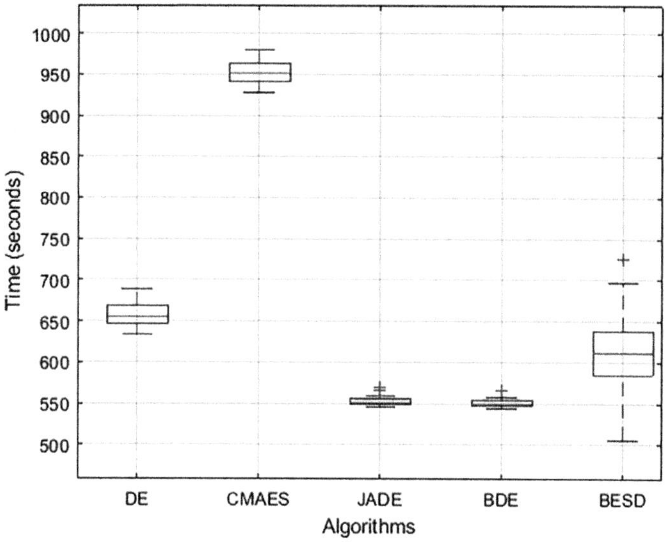


**Fig. 8** Computational times for Experiment 2

**Table 5** Obtained parameters of the wind farms for Experiment 2

| Method | Wind Turbines | Installation Cost | Supplied Power (kW) |
|---|---|---|---|
| DE | 44 | 29.83840 | 21153.81 |
| CMAES | 42 | 28.65031 | 20235.65 |
| JADE | 44 | 29.83840 | 21190.56 |
| BDE | 45 | 30.44244 | 21566.14 |
| BESD | 45 | 30.44244 | 21449.72 |

installation cost of the wind farm as explained in Sect. 2. Therefore, the wind farm proposed by the CMAES corresponds to the cheapest solution with 28.65031 units. Meanwhile, the BDE and BESD report the most expensive solutions with 30.44244 units. Nonetheless, the number of turbines is not directly related to the generated power by the wind farm since said power supply also involves the distribution of the turbines. This is confirm by the fact that the BDE and the BESD present the same installation cost but different power supply. Hence, the maximum power supply is given by the BDE with 21,566.14 kW, while the worst generation is given by the CMAES with only 20,235.65 kW. Actually, a similar performance was reported in Table 3, which probes the homogeneity of the results by the five approaches.

As previously mentioned, the power supply of the wind farm is related not only to the number of wind turbines but also to their distribution in the search space. Therefore, in Fig. 9 the reached distributions by each scheme are shown. In the same sense of Sect. 4.1, the perimeter of the search space is the best option to place a wind turbine due to the fact that these positions are not quite affected by the wind wake effect. In counterpart, the middle part of the wind farm must be well-distributed to counteract the aforementioned effect and to generate the highest possible energy. As can be see, the DE distribution (Fig. 9a) has almost all its perimeter covered by turbines while the rest of space has a competitive distribution. However, it is worth mentioning that the DE and the JADE have the same number of turbines, 44 wind turbines specifically. Nonetheless, the JADE distribution (Fig. 9c) presents the entire perimeter covered and the rest of the turbines well-distributed in the middle of the farm. On the other hand, the BESD (Fig. 9e) presents the worst distribution with 45 turbines allocated in parallel groups in the middle of the wind farm, which is counterproductive due to the wind wake effect, which in turn demonstrates the low viability of the algorithm for solving this experiment. Moreover, the BDE (Fig. 9d) allocates the same 45 turbines but with a completely different distribution which is actually much more competitive based on the aforementioned effect. Finally, the CMAES distributes its turbines in a highly competitive way as can be seen in Fig. 9b. Unfortunately the CMAES only allocates 42 turbines which is reflected on its low power supply.

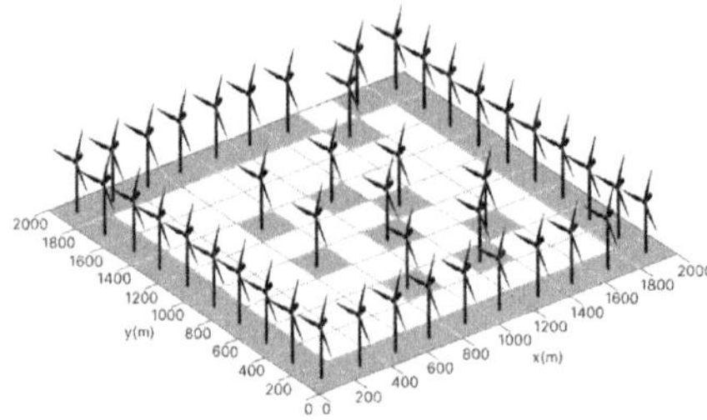


(a) DE: Cost 29.83840, Power 211533.81 *k***W**.

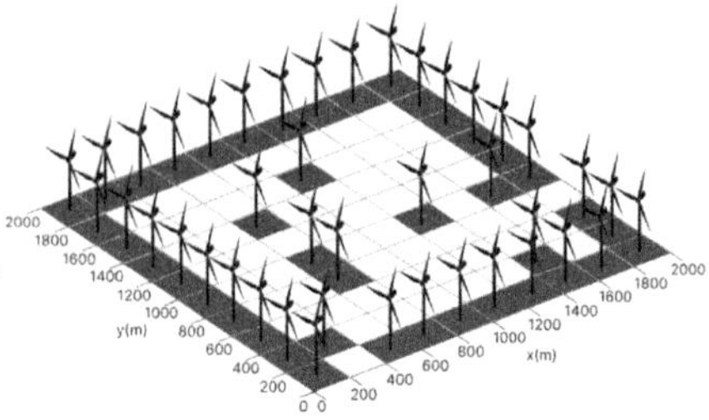


(b) CMAES: Cost 28.65031, Power 20235.65 *k***W**.

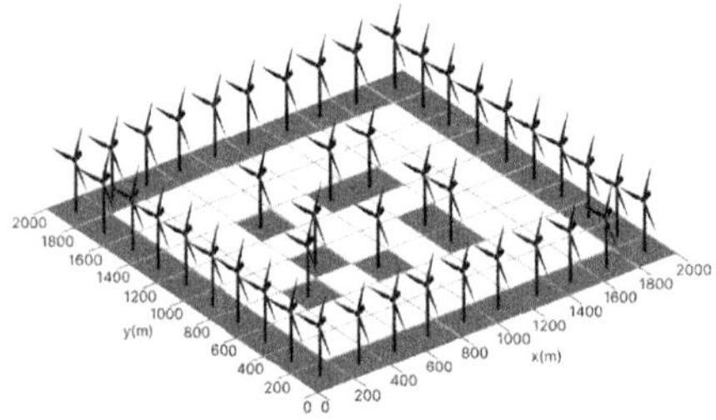


(c) JADE: Cost 29.83840, Power 21190.56 *k***W**.

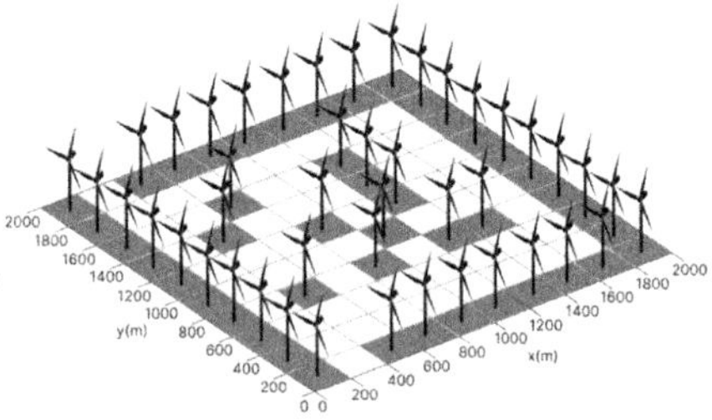


(d) BDE: Cost 30.44244, Power 21566.14 *k***W**.

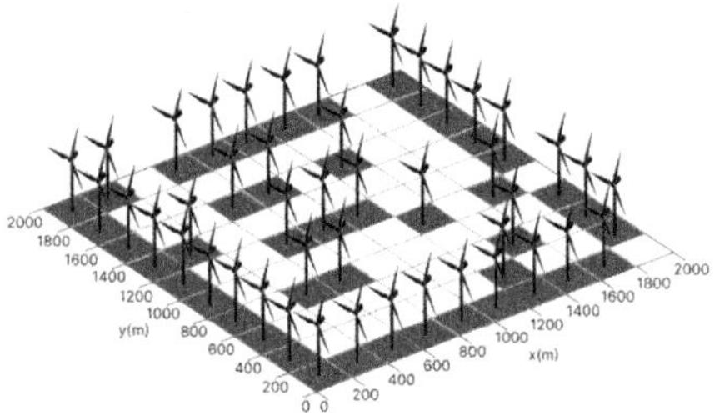


(e) BESD: Cost 30.44244, Power 21449.72 *k***W**.

**Fig. 9** Wind farms generated by the algorithms for Experiment 2

## *4.3 Statistical Analysis*

The computational experiments of Sects. 4.1 and 4.2 have demonstrated the capabilities of the algorithms for solving the WFLO problem in terms of metric and engineering results, convergence rate, computational cost such as wind turbines distribution. All these given information is enough to determine the best optimization approach for solving this energy-saving problem. However, it is quite important to understand the performances of the schemes under more sophisticated statistical methods. Therefore, the first method corresponds to the Friedman's test, which determines the competitiveness of a certain approach through the rejection or approval of a null-hypothesis [50]. Based on the above, Table 6 presents the Friedman ranking of

**Table 6** Friedman rank test

| Friedman rank for FullRotor | | | Friedman rank for PartialRotor | | |
|---|---|---|---|---|---|
| Algorithm | Friedman rank | Final ranking | Algorithm | Friedman rank | Final ranking |
| DE | 2.28E+00 | 3 | DE | 3.03E+00 | 3 |
| CMAES | 4.27E+00 | 4 | CMAES | 3.27E+00 | 4 |
| JADE | **1.80E+00** | **1** | JADE | **1.58E+00** | **1** |
| BDE | 2.22E+00 | 2 | BDE | 2.87E+00 | 2 |
| BESD | 4.43E+00 | 5 | BESD | 4.25E+00 | 5 |
| | p-value | 4.67E-20 | | p-value | 9.58E-13 |

**Table 7** Big O notation for the compared algorithms

| Algorithm | Notation |
|---|---|
| DE | $7x^3 + 23x^2 + 16x + 36$ |
| CMAES | $6x^3 + 23x^2 + 28x + 49$ |
| JADE | $4x^3 + 26^2 + 28x + 23$ |
| BDE | $3x^3 + 20x^2 + 28x + 23$ |
| BESD | $10x^3 + 19x^2 + 36x + 32$ |

the algorithms. It is seen that the JADE outperforms the rest of approaches for both computational experiments. This is quite relevant since the JADE demonstrated to be highly competitive in statistical terms through Sects. 4.1 and 4.2.

On the other hand, it is well-known that the computational cost of an algorithm is not enough to understand their computational requirements. This is due to the fact that every computer equipment involves different features that can affect or benefit the performance of an optimization approach [51]. Therefore, the Big O analysis has covered the attention of many researchers in the EC field in the last decade. At first, the Big O analysis was developed for data mining, but its methodology become useful for EC. Essentially, the Big O analysis uses mathematical structures to define the complexity of an algorithm. This is the difference between a classical boxplot graphic an a Big O analysis, since in this case, it does not matter the computer equipment to carry out the approach but the computational structure of itself. For more information about the Big O analysis, the reader is referred to the Bae's work [52]. Based on the above, Table 7 presents the Big O notation for the five compared algorithms in this study under their respective adaptation for solving the WFLO problem. The first thing to note is that all the approaches have a polynomial structure of third grade which corresponds to a quadratic complexity according to the Big O's rules. This is actually something normal in EC since the majority of algorithms are composed of similar stages like the initialization and exploration and exploitation phases. However, it can be noted that the BDE reports the lowest complexity compared with the rest of schemes followed by the JADE, while the BESD present the highest complexity. These mathematical models can also be graphically seen in Fig. 10. It is clear that the

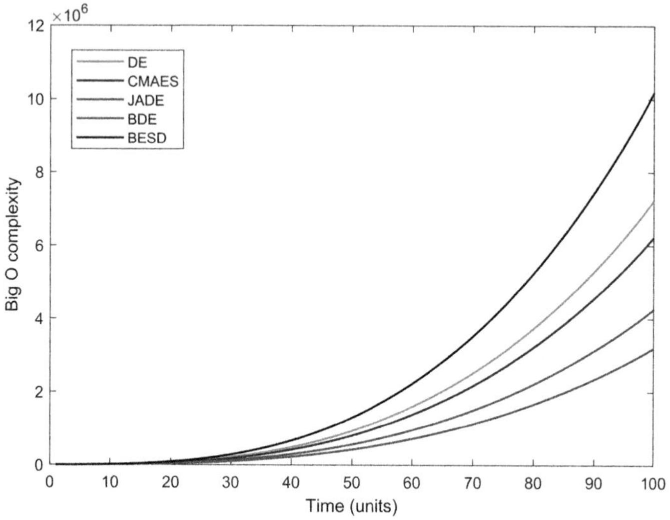


**Fig. 10** Big O graphic representation

algorithms can be divided into three main groups according to their complexity, the BDE and JADE being the less complex schemes, followed by the CMAES and DE with a medium complexity, and finally the BESD with the highest complex structure.

## 5 Discussion

Based on the computational and statistical results of Sect. 4 it is relatively simple to understand the performance of each algorithm for solving the WFLO problem. First of all, the DE in its base structure demonstrates to be highly competitive in terms of convergence rates for both experiments. In the same way, the DE probes efficiency on the distribution of the wind turbines. Nonetheless, the metric results of both experiments confirm that it is a non-viable scheme due to its low stability and its high computational cost. In addition, the Friedman test and the Big O analysis position the DE as one of the worst approaches to solve this energy-saving optimization problem. On the other hand, the CMAES is not precisely a DE variant, but its structure is based in the basic concepts of evolutionary strategies like the DE. Nevertheless, the metric results of both experiments define the CMAES as the worst strategy in terms of computational cost and it reports a non-competitive performance in the rest of metric values. The distributions proposed by the CMAES are interesting due to the fact that it proposed the cheapest wind farms in both experiments. However, at the same time the CMAES offers the worst farms in terms of power supply, which is crucial in this study. The statistical tests also position the CMAES as a non-viable approach for solving the WFLO problem. About the BDE algorithm, the computational experiments demonstrate its high competitiveness in terms of standard deviation and computational cost. Moreover, both experiments have been lead by the BDE. However, its convergence rates for both cases tends to be poor being the second worst method. In addition, the distributions proposed by the BDE are not quite relevant since it proposes the worst installation cost such as the BESD. In the same way, the Friedman test shows the BDE as some of the worst strategies. Nonetheless, the Big O analysis probes that the BDE corresponds to the lowest complex method. In counterpart, the BESD presents a relatively competitive computational cost and general metrics. Additionally, the stability of the method is the second best result for both experiments, only surpassed by the BDE. Unfortunately, the rest of results for both experiments position the BESD as the worst algorithm for solving the WFLO problem. The convergence rates of the BESD are the worst compared with the rest of schemes, the proposed distributions present the worst allocation of turbines which in turn involves the increment of the wind wake effect. Moreover, the Friedman test demonstrates that the BESD has the worst ranking for both cases while the Big O notation also positions the BESD as the highest complexity compared with the rest of methods. Finally, the JADE probes to be the best approach for both experiments in terms of general metrics (best, average and worst results). Also, the method is the second best approach in terms of computational cost being quite close to the BDE's results. It is worth mentioning that the stability of the JADE is not precisely the best since it presents the worst performance. For both cases, the convergence rates of the method are positioned as the third best rates in terms of speed and accuracy. However, it is remarkable that the JADE proposes the best distribution of turbines in the search space in general terms, sine it allocates 43 and 44 turbines with competitive installation costs and power supplies, respectively. In addition, the Friedman test demonstrates the high competitiveness of the scheme since it leads the analysis for both cases while its computational complexity corresponds to the second best structure.

Based on the above analysis, it is easy to note the advantages of the JADE over the rest of algorithms for solving the WFLO problem. In general terms, it is clear that the evolutionary strategy of the DE is quite competitive for solving energy-saving optimization problems, such as the state-of-the-art demonstrates. However, every application problem involves certain specifications that make its optimization much more complex for some approaches and vice-versa. Under that sense, the JADE probes to outperform the rest of methods in terms of accuracy, distribution and robustness. Nonetheless, it is worth mentioning the BDE's performance, which with the correct improvements may be a highly competitive scheme for solving this energy-saving optimization problem.

## 6 Conclusions

In this work, an analysis of different evolutionary strategies for solving the WFLO problem has been presented. To test the approaches, two cases of study have been carried out under an objective function related to the installation cost and power supply of the wind farm. In addition, the non-parametric Friedman test and a Big O analysis have been considered to enhance the comparison of the algorithms. The computational results of both cases such as the statistical analysis demonstrate that the JADE is the most competitive approach in general terms compared with the rest of methods. The capabilities of the JADE for proposing efficient distributions of the wind turbines in the farm is relevant. In addition, its computational complexity is low enough to be considered as a promising strategy for future research in the field. Moreover, the study confirms the reliability of using DE variants for solving energy-saving optimization problems through the obtained results of methods like the DE or the BDE. These two aforementioned strategies probed to have great potential for future improvements that allow them to compete under these types optimization systems. Finally, it is worth mentioning that an interesting future study is the analysis of similar methodologies for solving the WFLO problem under a multi-objective perspective.

**Acknowledgements** If you want to include acknowledgments of assistance and the like at the end of an individual chapter please use the `acknowledgement` environment – it will automatically be rendered in line with the preferred layout.

**Competing Interests**
The authors have no conflicts of interest to declare that are relevant to the content of this chapter.

## References

1. Reddy, S.R.: Wind farm layout optimization (WindFLO): an advanced framework for fast wind farm analysis and optimization. Appl. Energy **269**, 115090–115111 (2020)
2. Khan, M.J., Mathew, L.: Comparative study of optimization techniques for renewable energy system. Arch. Comput. Methods Eng. **27**, 351–360 (2020)
3. Porté-Agel, F., Bastankhah, M., Shamsoddin, S.: Wind-turbine and wind-farm flows: a review. Bound.-Layer Meteorol. **174**, 1–59 (2020)
4. Pao, L.Y., Johnson, K.E.: A tutorial on the dynamics and control of wind turbines and wind farms. In: American Control Conference, vol. 2009, pp. 2076–2089 (2009)
5. Rippel, D., Jathe, N., Becker, M., Lütjen, M., Szczerbicka, H., Freitag, M.: A review on the planning problem for the installation of offshore wind farms. IFAC-PapersOnLine **52**, 1337–1342 (2019)
6. Shakoor, R., Hassan, M.Y., Raheem, A., Wu, Y.-K.: Wake effect modeling: a review of wind farm layout optimization using Jensen's model. Renew. Sustain. Energy Rev. **58**, 1048–1059 (2016)
7. Mosetti, G., Poloni, C., Diviacco, B.: Optimization of wind turbine positioning in large wind-farms by means of a genetic algorithm. J. Wind Eng. Ind. Aerodyn. **51**, 105–116 (1994)

8. Katic, I., Hojstrup, J., Jensen, N.O.: A simple model for cluster efficiency. In: European Wind Energy Association Conference and Exhibition, vol. 1, pp. 407–410 (1986)
9. Frandsen, S.: On the wind speed reduction in the center of large clusters of wind turbines. J. Wind Eng. Ind. Aerodyn. **39**, 251–265 (1992)
10. Holland, J.H.: Genetic algorithms and adaptation. In: Adaptive Control of Ill-Defined Systems, vol. 16, pp. 317–333 (1984)
11. Grady, S.A., Hussaini, M.Y., Abdullah, M.M.: Placement of wind turbines using genetic algorithms. Renew. Energy **30**, 259–270 (2005)
12. Serrano, G.J., Gonzalez, R.A.G., Castro, M.J., Riquelme, S.J., Burgos, P.M.: Optimization of wind farm turbines layout using an evolutive algorithm. Renew. Energy **35**, 1671–1681 (2010)
13. Saavedra, M.B., Salcedo, S.S., Paniagua, T.A., Prieto, L., Portilla, F.A.: Seeding evolutionary algorithms with heuristics for optimal wind turbines positioning in wind farms. Renew. Energy **36**, 2838–2844 (2011)
14. Elkinton, C.N., Manwell, J.F., McGowan., J.G.: Algorithms for offshore wind farm layout optimization. Wind Energy **32**, 67–84 (2008)
15. Salcedo, S.S., Gallo, M.D., Pastor, S.A., Carro, C.L., Portilla, F.A., Prieto, L.: Evolutionary computation approaches for real offshore wind farm layout: a case of study in northern Europe. Expert Syst. Appl. **40**, 6292–6297 (2013)
16. Kennedy, J., Eberhart, R.: Particle Swarm Optimization. In: Proceedings of ICNN'95-International Conference on Neural Networks, vol. 4, pp. 1948–1948 (1995)
17. Pookpunt, S., Ongsakul, W.: Optimal placement of wind turbines within wind farm using binary particle swarm optimization with time-varying acceleration coefficients. Renew. Energy **55**, 266–276 (2013)
18. Hou, P., Hu, W., Chen, C., Soltani, M., Chen, Z.: Optimization of offshore wind farm layout in restricted zones. Energy **113**, 487–496 (2016)
19. Pillai, A.C., Chick, J., Johanning, L., Khorasanchi, M.: Offshore wind farm layout optimization using particle swarm optimization. J. Ocean. Eng. Mar. Energy. **4**, 73–88 (2018)
20. Le, T.T., Vo, D.N.: Optimal layout for offshore wind farms using metaheuristic search algorithms. GMSARN Int. J. **11**, 1–15 (2017)
21. Aggarwal, S.K., Saini, L.M., Sood, V.: Large wind farm layout optimization using nature inspired meta-heuristic algorithms. IETE J. Res. **69**, 2683–2700 (2023)
22. Rehman, S., Ali, S.S., Khan, S.A.: Wind farm layout design using cuckoo search algorithm. Appl. Artif. Intell. **30**, 899–922 (2016)
23. Kiji, S., Ahmad, M.A.: A modified grey wolf optimizer for improving wind plant energy production. Indones. J. Electr. Eng. Comput. Sci. **18**, 1123–1129 (2020)
24. Yu, X., Zhang, W.: A teaching-learning-based optimization algorithm with reinforcement learning to address wind farm layout optimization problem. Appl. Soft Comput. **151**, 1111135 (2024)
25. Storn, R., Price, K.: Differential evolution—a simple and efficient heuristic for global optimization over continuous spaces. J. Global Optim. **11**, 341–359 (1997)
26. Thamarai, I., Murugavalli, S.: Competitive advantage of using differential evolution algorithm for software effort. Commun. Power Eng. **62** (2017)
27. Wu, G., Shen, X., Li, H., Chen, H., Lin, A., Suganthan, P.N.: Ensemble of differential evolution variants. Inf. Sci. **423**, 172–186 (2018)
28. Ela, A.A. Abou, E., Abido, M.A., Spea, S.R.: Optimal power flow using differential evolution algorithm. Electr. Power Syst. Res. **80**, 878–885 (2010)
29. Shilaja, C.: Implementation of differential evolution algorithm and its variants for optimal scheduling of distributed generations. Int. J. Commun. Syst. **34**, e4318 (2021)
30. Lakshminarasimman, L., Subramanian, S.: A modified hybrid differential evolution for short-term scheduling of hydrothermal power systems with cascaded reservoirs. Energy Convers. Manage. **49**, 2513–2521 (2008)
31. Aghaee, Z., Ghasemi, M.M., Beni, H.A., Bouyer, A., Fatemi, A.: A survey on meta-heuristic algorithms for the influence maximization problem in the social networks. Computing **103**, 2437–2477 (2021)

32. Wang, Y., Liu, H., Long, H., Zhang, Z., Yang, S.: Differential evolution with a new encoding mechanism for optimizing wind farm layout. IEEE Trans. Industr. Inf. **14**, 1040–1054 (2018)
33. Liu, Y., Xu, Z., Yu, Y., Chang, X.: A novel binary genetic differential evolution optimization algorithm for wind layout problems. AIMS Energy **12**, 321–349 (2024)
34. Hansen, N., Ostermeier, A.: Completely derandomized self-adaptation in evolution strategies. Evol. Comput. **9**, 159–195 (2001)
35. Zhang, J., Sanderson, A.C.: JADE: adaptive differential evolution with optional external archive. IEEE Trans. Evol. Comput. **13**, 945–958 (2009)
36. Civicioglu, P., Besdok, E.: Bernstein-Levy differential evolution algorithm for numerical function optimization. Neural Comput. Appl. **35**, 6603–6621 (2023)
37. Civicioglu, P., Besdok, E.: Bezier search differential evolution algorithm for numerical function optimization: a comparative study with CRMLSP, MVO, WA, SHADE and LSHADE. Expert Syst. Appl. **165**, 113875 (2021)
38. Kunakote, T., Sabangban, N., Kumar, S., Tejani, G.G., Panagant, N., Pholdee, N., Bureerat, S., Yildiz, A.R.: Comparative performance of twelve metaheuristics for wind farm layout optimisation. Arch. Comput. Methods Eng. **29**, 717–730 (2022)
39. Wu, X., Hu, W., Huang, Q., Chen, C., Jacobson, M.Z., Chen, Z.: Optimizing the layout of onshore wind farms to minimize noise. Appl. Energy **267**, 114896 (2020)
40. Jensen, N.O.: A note on wind generator interaction. Roskilde, Denmark: Riso National Laboratory, vol. 2411 (1983)
41. Larsen, G.C.: A simple stationary semi-analytical wake model. Riso National Laboratory for Sustainable Energy, vol. 510 (2009)
42. Frandsen, S., Barthelmie, R., Pryor, S., Rathmann, O., Larsen, S., Hojstrup, J., Thøgersen, M.: Analytical modelling of wind speed deficit in large offshore wind farms. Wind Energy **9**, 39–53 (2006)
43. Abdulrahman, M., Wood, D.: Investigating the Power-COE trade-off for wind farm layout optimization considering commercial turbine selection and hub height variation. Renew. Energy **102**, 267–278 (2017)
44. Chen, K., Song, M.X., Zhang, X., Wang, S.F.: Wind turbine layout optimization with multiple hub height wind turbines using greedy algorithm. Renew. Energy **96**, 676–686 (2016)
45. Bilal, Pant, M., Zaheer, H., Hernandez, L. G., Abraham, A.: Differential evolution: a review of more than two decades of research. Eng. Appl. Artif. Intell. **90**, 103479 (2020)
46. Jamil, M., Yang, X.-S.: A literature survey of benchmark functions for global optimisation problems. Int. J. Math. Model. Numer. Optim. **4**, 150–194 (2013)
47. Liang, Y., Wang, X., Zhao, H., Han, T., Wei, Z., Li, Y.: A covariance matrix adaptation evolution strategy variant and its engineering application. Appl. Soft Comput. **83**, 105680 (2019)
48. Hansen, N.: The CMA evolution strategy: a tutorial (2016). arXiv:1604.00772
49. Chu, Y.-M., Wang, M.-K.: Optimal Lehmer mean bounds for the Toader mean. RM **61**, 223–229 (2012)
50. Sheldon, M.R., Fillyaw, M.J., Thompson, W.D.: The use and interpretation of the Friedman test in the analysis of ordinal-scale data in repeated measures designs. Physiother. Res. Int. **1**, 221–228 (1996)
51. Fan, J., Han, F., Liu, H.: Challenges of big data analysis. Natl. Sci. Rev. **1**, 293–314 (2014)
52. Bae, S.: Big-O Notation. JavaScript Data Structures and Algorithms, pp. 1–11 (2019)

# Optimization of Radial Distribution Networks Through an Improved African Vulture Optimization Algorithm

**Eduardo H. Haro, Mario A. Navarro, Primitivo Diaz, Omar Avalos, and Eric L. Marin**

**Abstract** One of the most important energy-saving optimization problems is the Optimal Capacitor Placement (OCP). Essentially, this problem involves the correct allocation of capacitor banks in a bounded Radial Distribution Network (RDN). This is done in order to reduce the power loss of the system and to improve its feeder's voltage profile. This problem has been addressed by different computational perspectives but the meta-heuristic methods have demonstrated their advantages over other schemes. In that sense, in this work the African Vulture Optimization Algorithm (AVOA) is used to solve the OCP problem due to its efficiency solving other energy-saving tasks. The AVOA is a relatively recent approach that has gained popularity over the years due to its robustness and stability in terms of accuracy and computational resources. In addition, the AVOA is modified with a chaotic Opposition-Based-Learning (OBL) initialization phase to increase the population diversity. The obtained results are compared against other nine classical and contemporary meta-heuristic algorithms for solving two IEEE RDN' tests. The experimentation shows that the AVOA outperforms the rest of schemes in terms of accuracy and stability. Moreover, statistical and non-parametric tests probes the efficiency of the algorithm and its implication for future research in the field.

E. H. Haro · M. A. Navarro (✉) · P. Diaz · O. Avalos · E. L. Marin
Centro Universitario de Ciencias Exactas e Ingenierias CUCEI, Av. Revolucion 100, Guadalajara, Mexico
e-mail: mario.navarro@alumnos.udg.mx

E. H. Haro
e-mail: eduardo.hernandezh@academicos.udg.mx

P. Diaz
e-mail: primitivo.diaz@academicos.udg.mx

O. Avalos
e-mail: omar.aalvarezo@academicos.udg.mx

E. L. Marin
e-mail: eric.lopez7892@alumnos.udg.mx



D. Oliva et al. (eds.), *Advances in Optimization Algorithms for Multidisciplinary Engineering Applications: From Classical Methods to AI-Enhanced Solutions*, Studies in Computational Intelligence 806, https://doi.org/10.1007/978-3-031-78440-8_31

# 1 Introduction

The global warming has yielded that computational scientific communities be more attracted to energy-saving problems due to their sustainability implications [1, 2]. This is why the Optimal Capacitor Placement (OCP) is positioned as one of the most relevant optimization problems in electrical engineering [3, 4]. This non-polynomial problem treats with the progressive decline of voltage in the buses of a Radial Distribution Network (RDN). This feature corresponds to the disability of power supplies to meet the increasing energy demands by different social sectors [5]. Thus, to counteract this effect, capacitor banks tend to be allocated on the RDN to furnish it with reactive power compensation. This process allows to improve the feeder's voltage profile, such as reducing installation costs and minimize power loss [6]. Nonetheless, this reactive compensation depends on the types of capacitors and their positions on the network. Therefore, the OCP in RND is considered a constrained optimization problem due to the fact that it requires the precise number and size of the capacitors such as their optimal bus allocation on the system [7, 8].

The OCP problem was originally addressed by different statistical and mathematical methodologies [9, 10]. These models are well-known by its simple structure and low computational resources. However, these characteristics usually imply low robustness and high trend to be trapped into local optima. In counterpart, meta-heuristic algorithms have probed a competitive performance on the solution of this types of energy-saving optimization problem. Some of the first related works were proposed by Salam and Chis who presented diverse reduction approaches for the minimization of reactive and power losses, respectively [11, 12]. On its part, Boone published the first application of a Genetic Algorithm (GA) to the OCP problem [13], while Yu did his own with the Particle Swarm Optimization (PSO) [14]. The advantages of these meta-heuristic schemes for solving this problem were probed through the aforementioned works. In the same way, the most novel algorithms of the current century have been implemented to the OCP problem given the complex challenge it represents for any approach. Some examples include Mendes's study where different Differential Evolution (DE) variants were compared on a RDN system [15], while the Teaching Learning Based Optimization (TLBO), Crow Search Algorithm (CSA) and the Whale Optimization Algorithm (WOA) were implemented by Sultana, El-Fergany and Prakash, respectively [16–18]. Moreover, in recent years novel mathematical models of the OCP problem have been exposed and optimized by techniques like the GA [19], or the Cuckoo Search (CS) [20], just for mention a few.

However, the use of meta-heuristic approaches for solving the OCP problem has involved different troubles, most of them linked to complications balancing the diversity and swiftness in the population. At the same time, this problems yield a divergence in the solutions that is almost uncontrollable. Therefore, the method tends to be competitive in different perspectives but its accuracy can be misinterpreted [21, 22]. In addition, the majority of related works present RDN models with simplified objective functions or even worst, theoretical systems far away from reality in engineering terms [23]. On the other hand, different studies have probed the capabilities of enhanced meta-heuristic algorithms through the hybridization of different schemes over the original methods [24]. Essentially, this modifications aim to overcome the limitations of the original algorithms for general optimization problems or specific applications [25]. Nevertheless, the use of these hybrid schemes for solving the OCP problem has not been analyzed as much as the original methods. This can be attributed to the fact that these hybridization tend be computationally more expensive and with a more complex structure [25].

Hence, in this work, the African Vulture Optimization Algorithm (AVOA) originally proposed by Abdollahzadeh et al. [26] is applied for solving the OCP problem. This relatively recent algorithm is characterized by the creation of two sub-populations based on its initialization phase that increases its population diversity. Subsequently, these two sub-populations are treated through exploration and exploitation phases based on the natural vultures hunting method. The OCP corresponds to a non-polynomial hard (NP-Hard) problem and it is well-founded that this types of tasks are well optimized by techniques focused on the population diversity and exploration due to the complex combinatorial model of the system [27]. In that sense, this work proposes a modification in the AVOA's initialization phase based on the use of ten chaotic maps to generate chaotically the initial population. Next, this chaotic population is compared against its opposite population through the Opposition-Based-Learning (OBL) technique [28]. This modification that the AVOA be enhanced on its population diversity, which in turn is benefit for solving this energy-saving optimization problem. In addition, the AVOA is compared against other nine classical and contemporary meta-heuristic approaches for solving two different IEEE's RDN system, defined as electrical networks of 33 and 69 buses, respectively. The obtained results show the efficiency of increasing the population diversity in the AVOA since it can outperforms the rest of schemes in terms of accuracy, robustness and stability. Furthermore, statistical and non-parametric test demonstrate that this modified AVOA is a promising alternative for the resolution of future energy-saving optimization problems and NP-hard systems.

The rest of this chapter is organized as follows: Sect. 2 presents the mathematical model of the OCP problem and the specifications of the RDN models; the proposed approach and its integration for solving this energy system problem is exposed in detail in Sect. 3; on the other hand, Sect. 4 explains and presents the obtained results of all computational tests, while the conclusions of the work are discussed in Sect. 5.

## 2 Mathematical Model for the OCP Problem

The RDN can be addressed from different perspectives based on the application study. Some examples include solving the power flow problem [29] or its voltage stability analyzed by statistical methods [30]. However, the most common way to address the RDN is optimizing it through different computational approaches to improve its voltage profile and reduce its power losses. This resolution scheme has been attractive for researchers for many years due to its impact on energy-saving. Moreover, this problem takes more relevance over the years since the RDN systems are getting more complex according the energy demand increases [31, 32]. Therefore, in this section the RDN model and its mathematical formulation for its optimization is presented focusing the analysis in terms of installation cost.

## 2.1 Objective Functions

Different objective functions have been proposed in the state-of-the-art to solve the OCP problem. Some of these functions are based on the optimization of the reactive power loss, the characteristics of the wires or the voltage profile performance of the network, just for mention a few. These perspectives are quite relevant since they enhance the system. However, in terms of energy-saving optimization the most important objective is to minimize the installation cost of the entire grid [33]. An accurate installation of the OCP on a RDN implies the correct allocation of specific capacitor banks on ideal buses of the network to reduce the power losses and improving the voltage profile of the system. In addition, the idea is to allocate the minimum number of shunt capacitor in the system to reduce its installation cost at the same time that the aforementioned electrical benefits are reached. It must be understood, that from an industrial perspective it is not profitable to obtain these energy-saving improvements if it means an exaggerated investment that may not be recovered in the short time. Thus, in this work two objective functions are presented to optimize the OCP problem. Both functions are related to the installation cost of the system, but the different between them lies in their accuracy to a real-world application. The first objective function belongs to the Su's work [34]:

$$C = K_p \times P_{LOSS} + \sum_{j=1}^{n} k_j^c \times Q_j^c \tag{1}$$

where $C$ corresponds to the total installation cost (\$) of a determined RDN with a certain number of $n$ buses. Added to this, $K_p$ is the cost of power loss per hour of every unit expressed in $\$/kVAR$ while $P_{LOSS}$ represents the total power loss of the RDN ($kW$). On the other hand, $k_j^c$ and $Q_j^c$ define the cost per unit of reactive power expressed in $\$/kVAR$ and the size of the shunt capacitor expressed in $kVAR$, respectively. It is quite easy to note that Eq. 1 is composed by two main parts which are the installation cost according to the power loss and cost related to the total reactive power. The state-of-the-art confirms the reliability of this function and the advantages that it involve [34, 35]. Nonetheless, Eq. 1 does not considers the installation costs related to the calculation of the total cost of the system. This is quite important since the installation cost of the system involves the man power cost which tends to be ignored by many researchers. In addition, this allows to optimize the RDN from a more accurate and realistic perspective, which in turn adds relevance to this work. In that sense, El-Fergany et al. [36] proposed the following modification to counteract the aforementioned problem:

$$C = K_p \times t \times P_{LOSS} + K_{ic} \times N_q + \sum_{j=1}^{n} k_j^c \times Q_j^c \tag{2}$$

where $t$ is the period of time expressed in hours, while $K_{ic}$ corresponds to the installation cost of the system expressed in \$ and $N_q$ is the number of capacitors to place in the network. It is worth mentioning that both Eqs. 1 and 2 must take in consideration the Kirchhoff's laws in order to maintain the maximum reactive power compensation delivered from the system as the same or lower than the applied power to its input [41]. The Kirchhoff's laws for the OCP problem can be expressed as follows:

$$\sum_{j=1}^{n_c} Q_j^c \leq \sum_{i=1}^{n} Q_i^c \tag{3}$$

where $n_c$ denotes the buses compensated by a shunt capacitor while $Q_i^c$ corresponds to the reactive power measured at bus $i$.

## 2.2 *Voltage Constraints*

In engineering terms, almost all elements from Eqs. 1 and 2 can be calculated through statistical analysis and electrical experimentation. Nevertheless, the power loss ($P_{LOSS}$) corresponds to an estimation that must be determined by polynomial iterative methods like Gauss-Seidel or Newton-Raphson schemes [37, 38]. In essence, the power loss is an electrical characteristic that tends to decrease with the next bus in the network. Moreover, this decrement is unstable and it does not follow a determined mathematical model. Actually, the power loss decreases based on the physical features of the RDN. It is worth mentioning that the calculation of this value is done based on the load flow calculation presented by Das et al. [39], where precisely the Gauss-Seidel model is implemented. Under these schemes, the RDN can be considered as a three-phase balanced distribution grid like the diagram of Fig. 1. Therefore, the network's total power loss is estimated through the flow equations Eqs. 4 and 5.

$$P_{LOSS}^{i,i+1} = R_{i,i+1} \times \left( \frac{P_i^2 + Q_i^2}{|V_i|^2} \right) \tag{4}$$

$$P_{LOSS} = \sum_{i=0}^{n} P_{LOSS}^{i,i+1} \tag{5}$$

where $R_{i,i+1}$ corresponds to the line resistance between buses $i$ and $i+1$, $P_i$ and its quadratic value is the real line power out bus $i$, while $Q_i$ and its quadratic value

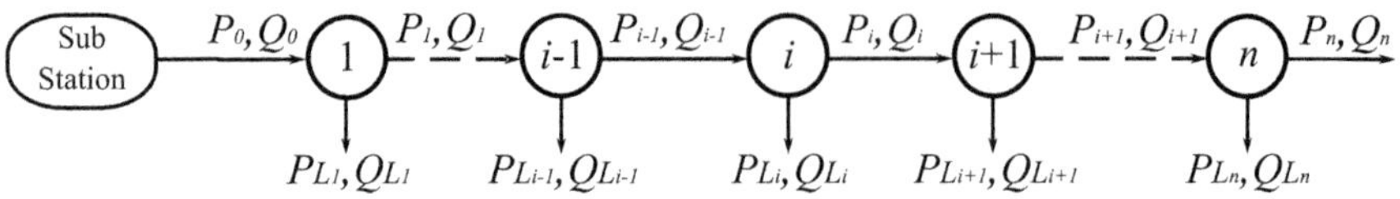


**Fig. 1** RND composed of $N$ buses

corresponds to the reactive line power from bus $i$. On the other hand, $P_{L_i}$ and $Q_{L_i}$ are the real load power and the reactive load power at bus $i$ from all the feasible buses. Based on the above, it is quite easy to note that the total power loss of the system is equivalent to the summation of the power loss for every single line. Finally, the voltage at bus $i$ is denoted by $|V_i|$ and it is constrained to certain energy limits due to safety reasons [40], these limits are described as follows:

$$V_{min} =\leq |V_i| \leq V_{max} \tag{6}$$

where the minimum and maximum bus voltage constraints are given by $V_{min}$ and $V_{max}$, respectively. In addition, $i$ corresponds to the bus number set from 1 to the maximum number of buses $n$.

# 3 The Optimization Process for the OCP

## *3.1 The African Vulture Optimization Algorithm*

Lately, bio-inspired approaches have given a lot to talk about, on the one hand their lack of novelty is criticized, while other researchers take advantage of their approaches and operators to solve application problems in different areas, such is the case of this chapter, which addresses modification of a relatively new algorithm called AVOA published in 2021 by Abdollazadeh et al. [42]. This algorithm was analyzed and studied in its structure in order to modify it and apply it to a real problem. In this section the details of the original AVOA are described to fully understand how it was modified.

AVOA is an algorithm inspired by the feeding habits of African vultures. The authors divided the vulture population (candidate solutions) of the algorithm into two different groups; one group had the best vulture based on its fitness, and another group had the second-best vulture with the best fitness. By creating two distinct groups, the authors provided more diversity to the possible solutions. Different operators and search strategies that benefited the exploration and exploitation of the solution space by emulating the typical behavior of vultures in the wild were applied.

The authors divide the AVOA algorithm method into 4 different phases, each synthetically described below.

**First phase: Determining the best vulture of each group.**

1. Initial evaluation: The fitness (quality) of all the initial solutions (vultures) is calculated.
   The two best solutions are selected as the best vultures forming two groups:
   The best solution as the best vulture of the first group, and the second best is the best vulture of the second group.
2. Attraction to the best vultures:
   Equation 7 is used to move the other solutions toward the best vultures, and assigns the probability of moving towards one of the best vultures based on two parameters ($L_1$ and $L_2$):

$$R(i) = \begin{cases} BestVulture_1 \text{ if } p_i = L_1 \\ BestVulture_2 \text{ if } p_i = L_2 \end{cases} \tag{7}$$

   Where, $L_1$ and $L_1$ are values between 0 and 1, and their sum is 1.
   A value of $L_1$ close to 1 and $L_2$ close to 0 favors intensification (convergence to the best solution found).
   A value of $L_1$ close to 0 and $L_2$ close to 1 favors diversification (exploration of the search space).
   The selection of the best vulture towards which a solution will move is performed using the roulette of wheel (Eq. 8).

$$p_i = \frac{F_i}{\sum_{i=1}^{n} F_i} \tag{8}$$

3. Recalculation:
   The fitnesses of all solutions are recalculated at each iteration.

**Second phase: Vulture starvation rate.**

Hungry vultures have less energy to explore and move closer to stronger vultures. Satiated vultures have more energy and can explore greater distances. This behavior is modeled mathematically as follows:

Equation 9 represents the satiety/hunger rate of vultures.

$$t = h \times \left( \sin^w \left( \frac{\pi}{2} \times \frac{iteration_i}{maxiterations} \right) + \cos \left( \frac{\pi}{2} \times \frac{iteration_i}{maxiterations} \right) - 1 \right) \tag{9}$$

Equation 10 determines whether a vulture enters the exploration or exploitation phase.

$$F = (2 \times rand_1 + 1) \times z \times \left( 1 - \frac{iteration_i}{maxiterations} \right) + t \tag{10}$$

The variables represented in the previous equations are described in detail in the following list:

- $F$: Represents the satiation of the vulture.
- $t$: Random value that influences satiation.
- $iteration_i$: Current iteration.
- $maxiterations$: Total number of iterations.
- $z$: Random value that changes in each iteration.
- $h$: Random value that affects the general satiation behavior.
- $rand_1$: Random value between 0 and 1.

The criteria and influence of these variables are as follows:

A low value of $|F|$ indicates that the vulture is hungry and will enter the exploitation phase (search near good solutions).

A high value of $|F|$ indicates that the vulture is satiated and will enter the exploration phase (search in new areas).

The value of $w$ controls the frequency of switching between exploration and exploitation. The high values of $w$ increase the probability of late-stage exploration, and low values of $w$ favor exploitation.

**Third phase: Exploration**.

In the wild, vultures have excellent eyesight and can find dying animals at great distances.

Mathematically represented vultures explore different areas based on two strategies:

A parameter denominated $P_1$, between 0 and 1, determines how often each strategy is used. For this purpose, a random number ($rand_{P1}$) between 0 and 1 is generated.

$$P(i+1) = \begin{cases} Equation(12) \text{ if } P_1 >= rand_{P_1} \\ Equation(14) \text{ if } P_1 < rand_{P_1} \end{cases} \tag{11}$$

If $rand_{P1}$ is greater than or equal to $P_1$, the vultures search randomly (as real vultures).

If $rand_{P1}$ is less than $P_1$, each vulture searches according to its satiation level (a measure of its performance).

The main exploration equations are: Eq. 12 that describes the random search based on a random direction and the current position and the Eq. 14 that describes exploration based on the best vulture's position adjusted for satiation and randomness. This equation uses the upper($ub$) and lower($lb$) bounds of the search space to constrain the movement.

$$P(i+1) = R(i) - D(i) \times F \tag{12}$$

$$D(i) = |X \times R(i) - \times P(i)| \tag{13}$$

In Eq. 13, $R(i)$ is one of the best vultures, which is selected by the use of Eq. 7 in the current iteration. Moreover, $X$ is where the vultures move randomly to protect food from other vultures. $X$ is used as a coefficient vector that increases the random motion, which changes in each iteration, and is obtained using the formula $X = 2 \times rand$, where rand is a random number between 0 and 1. $P(i)$ is the current vector position of the vulture.

$$P(i+1) = R(i) - F + rand_2 \times ((ub - lb) \times rand_3 + lb) \tag{14}$$

A key factor ($rand_3$) controls randomness, where values close to 1 create similar search patterns with some random movements that trigger the exploration of different areas of the search space.

**Fourth phase: Exploitation**.

The exploitation phase in AVOA is conducted in two phases depending on the value of $|F|$:

- **Phase 1**: Rotational flight and siege fighting ($|F|$ between 1 and 0.5).
  A random number ($rand_{P2}$) determines the strategy:
  If $rand_{P2}$ is greater than or equal to $P_2$, a "siege fight" strategy (slower approach) is used.
  $$P(i+1) = D(i) \times (F + rand_4) - d(t) \tag{15}$$
  On the other hand, if $rand_{P2}$ is less than $P2$, a "rotating flight" strategy (spiral search) is used.
  $$P(i+1) = R(i) - (S_1 + S_2) \tag{16}$$
- **Phase 2**: Accumulation and aggressive fighting ($|F|$ less than 0.5).
  A random number ($rand_{P3}$) determines the strategy:
  If $rand_P3$ is greater than or equal to $P3$, the vultures accumulate around the food source.
  $$P(i+1) = \frac{A_1 + A_2}{2} \tag{17}$$
  On the other hand, if $rand_{P3}$ is less than $P3$, an "aggressive siege fighting" (chaotic foraging) strategy is used.
  $$P(i+1) = R(i) - |d(t)| \times F \times Levy(d) \tag{18}$$

## 3.2 *Improved Initialization Process*

As already explained, the AVOA comprises two well-defined main stages that establish the exploration and exploitation phases of the algorithm. However, the initialization of the AVOA is based on the classical random structure for the generation of the initial population. It is at this stage that it is envisioned to implement an improvement to the original AVOA to integrate a different initialization scheme that promotes a healthier diversity and at the same time increases the performance of the algorithm by solving the OCP optimization problem.

The improved initialization scheme implemented in AVOA is reported in a paper from 2022, where Ahmad et al. [43] proposes to change the classical initialization scheme of the DE for a proposal using a chaotic generation of solutions integrated with a learning scheme based on the opposition-based-learning (OBL) strategy [28]. The results they obtained in their study are very promising, reason why we integrate the same initialization scheme this time on the AVOA. The original proposal involves a defined selection of ten chaotic maps to produce each individual of the initial population and then this population is treated by the OBL approach. The selection of the chaotic map is given by the following formulation:

$$\psi_{z+1} = \begin{cases} \frac{\psi_z}{0.7} & \psi_z < 0.7 \\ \frac{10}{3}(1-\psi_z) & \psi_z \geq 0.7 \end{cases} \tag{19}$$

where $\psi_z$ corresponds to the previous map defined between 0 and 1 with $z = 0$. This formulation allows to select all the maps with a balanced trend. However, this method is counterproductive in the sense of limiting the diversity in the selection process. Therefore, in this study the Eq. 19 is omitted and replaced by a simple random process to increase the population diversity. Based on the above, the differences between the distributions of the random initial population and the chaotic OBL process is graphically seen through a circle map in Fig. 2.

We can summarize in a list the initialization process used for the improvement, while later, we graphically represent the AVOA algorithm with the modification made in a flowchart on Fig. 3. The modification scheme (chaotic map + OBL) for solution initialization used in CODE is as follows:

1. Generate chaotic population using selected chaotic map.
2. Compute opposite solutions using OBL.

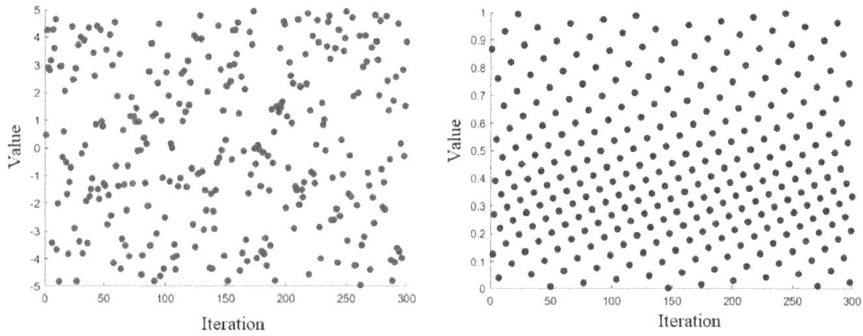


(a) Classical random search. (b) Circle chaotic distribution.

**Fig. 2** Produced distribution by the classical initialization and the CODE method [43]

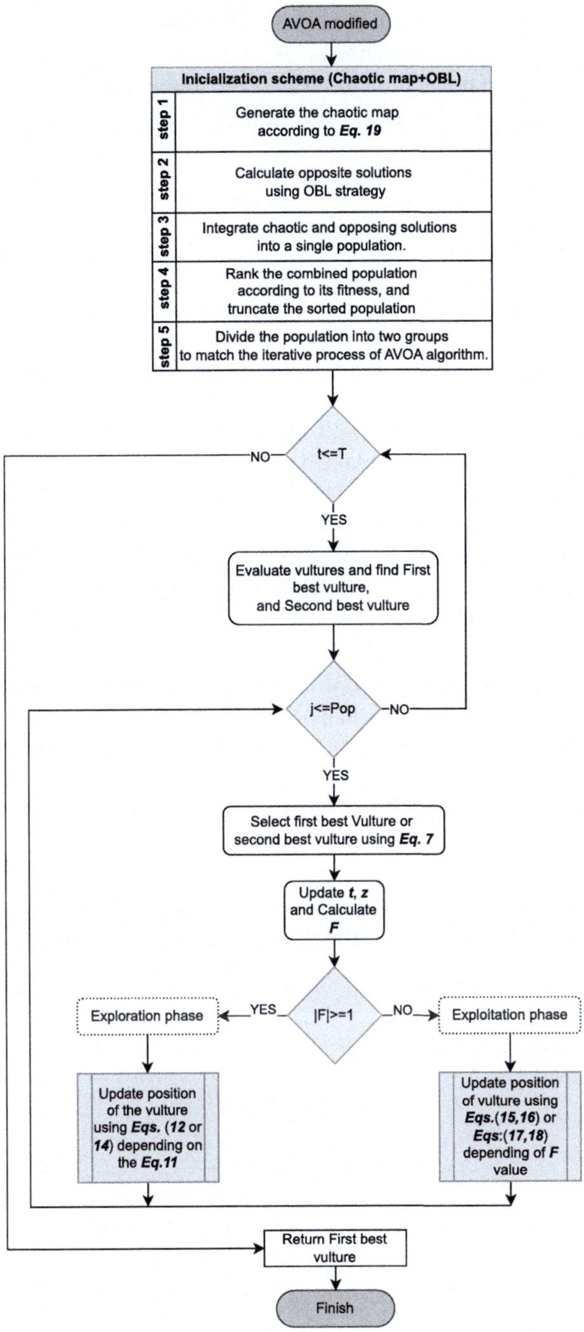


**Fig. 3** General structure of the algorithm AVOA modified

3. Merge chaotic population and opposite population.
4. Classify the combined population according to its fitness.
5. Truncate the sorted population to form a new initial population.
6. Establish the current best solution and fitness.

## 4 Computational Experimentation

In this section, the proposed improved AVOA is tested on two IEEE's RDN systems of 33-bus and 69-bus, respectively. These two models have been selected due to their resolution complexity and their implications in the robustness for any optimization approach. In addition, the obtained results of the improved AVOA are compared against the performances of other well-known optimization schemes that have been applied to the OCP problem as explained in Sect. 1, which are the GA [44], PSO [45], DE [46], WOA [47] and CSA [48]. In addition, other four classical and contemporary optimization schemes are used for comparison in these computational experiments which are the Ant Colony Optimization (ACO) [49], the Rat Swarm Optimizer (RSO) [50], the Battle Royale Optimization (BRO) [51] and the Social Spider Optimization (SSO) [52]. It is worth mentioning that all the algorithms have been configured in order to reach their best possible performance for solving the OCP problem. In that sense, the internal parameters of the PSO, RSO, BRO and AVOA are the set parameters in their respective original articles, while the internal parameters of the rest of approaches are exposed in Table 1. Also, for both experiments each algorithm has been executed 30 independent times with 100 individuals as initial population and 50,000 function accesses. The experimentation has been carried out in the MATLAB programming environment on a Notebook (8-BG RAM and 2.7 GHz AMD processor) with Windows 10 OS.

It is worth mentioning that a same set of shunt capacitors are considered for both computational experiments. The features of these capacitors are presented in

**Table 1** Parameter values of the compared approaches

| Method | Parameter | Value |
|---|---|---|
| GA | Crossover rate | 0.8 |
| | Mutation rate | 0.1 |
| DE | Scaling factor | 0.8 |
| | Crossover rate | 0.2 |
| ACO | Pheromone decay coefficient | 0.5 |
| CSA | Awareness probability | 0.1 |
| | Flight length | 2 |
| BRO | Constant | 0.7 |
| SSO | Rate of vibration atenuation | 1/3 |
| | Change probability | 0.4 |
| | Probability of ONE | 0.2 |

**Table 2** Parameters of the commercial capacitors used for the OCP problem

| Capacitor index | Capacitor size | Cost of reactive power |
|---|---|---|
| (No) | (kVAR) | ($/kVAR) |
| $q_j$ | $Q_j = Q(q_j)$ | $k(q_j)$ |
| 0 | 0 | 0 |
| 1 | 150 | 0.500 |
| 2 | 350 | 0.350 |
| 3 | 450 | 0.253 |
| 4 | 600 | 0.220 |
| 5 | 750 | 0.276 |
| 6 | 900 | 0.183 |
| 7 | 1050 | 0.228 |
| 8 | 1200 | 0.170 |
| 9 | 1350 | 0.207 |
| 10 | 1500 | 0.201 |
| 11 | 1650 | 0.193 |
| 12 | 1800 | 0.187 |
| 13 | 1950 | 0.211 |
| 14 | 2100 | 0.176 |
| 15 | 2250 | 0.197 |
| 16 | 2400 | 0.170 |
| 17 | 2550 | 0.189 |
| 18 | 2700 | 0.187 |
| 19 | 2850 | 0.183 |
| 20 | 3000 | 0.180 |
| 21 | 3150 | 0.195 |
| 22 | 3300 | 0.174 |
| 23 | 3450 | 0.188 |
| 24 | 3600 | 0.170 |
| 25 | 3750 | 0.183 |
| 26 | 3900 | 0.182 |
| 27 | 4050 | 0.179 |

**Table 3** Parameters for the objective functions

| Parameter | Value | Objective function |
|---|---|---|
| $K_p$ | $168 | F1 |
| | $0.06 | |
| $t$ | 8760 h | F2 |
| $K_{ic}$ | $1000 | |
| $K_c$ | $3000 | |

Table 2. As can be noted, 27 capacitors are considered with sizes between 150 and 4050 kVAR, and a cost of reactive power that oscillates between 0.170 $/kVAR and 0.5 $/kVAR. In the same way, the reminder of the variables for the objective functions of Eqs. 1 and 2 are exposed in Table 3.

### 4.1 IEEE's 33-Bus RDN System

As explained before, the first experiment involves the optimization of the IEEE's 33-bus RDN. Figure 4 shows a graphical representation of the distribution of this electrical model. As can be seen, it is composed of a sub-station that supplies 12.66 kV to a main branch and other three sub-branches that start at the second, third and sixth buses, respectively. In addition, the total active power and reactive power by the remaining 32 buses correspond to 3,715 kW and 2,300 kVAR, respectively. This is quite important since each sub-branch represents in electrical terms a new and independent branch with its own voltage decay, which makes much more complex the optimization of the network [53].

Based on the above, the statistical results of the compared algorithms for both objective functions are reported in Tables 4 and 5, respectively. The metrics used in this experiment include the Total Power Losses (TPL), the Power Losses Total Cost (PLTC), the Total Reactive Power (TRP), the Capacitors Cost (CC), the Total Annual Cost (TAC), the Network Savings (NS) in percentage and economic terms, and the minimum and maximum voltage in the voltage profile, represented as mV and MV, respectively. In addition, the RDN without any optimization technique is reported as the base case in order to compare the improvements by each approach. In the first objective function, the BRO obtains the best result in terms of the TPL, which in turn is reflected in metric values like the PLTC, TRP and CC. However, as explained before, the reduction of the total power losses is important for the optimization of the RDN. Nonetheless, this reduction must be related with the minimization of the installation cost of the system such as specify both objective functions. Furthermore, the improvement of the voltage profile is a relevant task that must be considered. These two aforementioned enhancements are much more relevant than the total

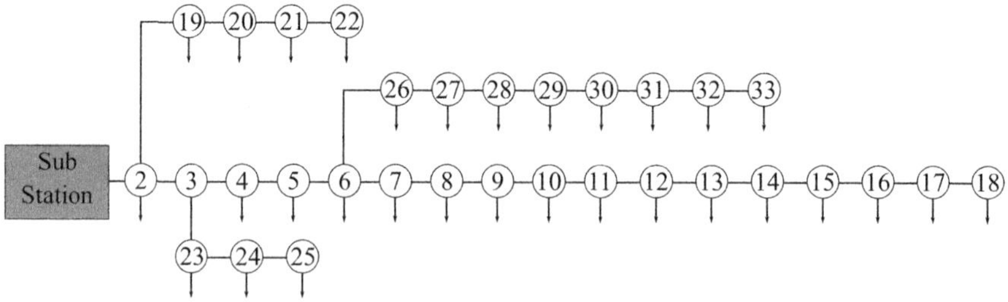


**Fig. 4** IEEE's 33-bus RDN

**Table 4** Parameter results for the IEEE's 33-bus RDN under F1

| Feature | Base | GA | PSO | DE | ACO | RSO | CSA | WOA | BRO | SSO | AVOA |
|---|---|---|---|---|---|---|---|---|---|---|---|
| TPL (kW) | 210.99 | 135.79 | 148.61 | 135.61 | 139.46 | 135.78 | 135.98 | 135.68 | 135.58 | 140.27 | 135.76 |
| PLTC (%) | 35,445.91 | 22,812.74 | 24,967.29 | 22,782.55 | 23,428.49 | 22,810.99 | 22,844.96 | 22,794.04 | 22,777.84 | 23,565.18 | 22,806.86 |
| TRP (kVAR) | 0 | 2,400 | 3,950 | 2,400 | 2,100 | 2,300 | 2,600 | 2,400 | 2,400 | 2,850 | 2,400 |
| CC ($) | 0 | 1,032 | 904.75 | 1,032 | 1,050 | 930 | 1,247.50 | 1,088.85 | 1,032 | 972 | 930 |
| TAC ($) | 35,445.91 | 23,844.74 | 25,872.04 | 23,814.55 | 24,478.49 | 23,740.49 | 24,092.46 | 23,882.89 | 23,809.84 | 24,536.88 | **23,736.36** |
| NS ($) | 0 | 11,601.17 | 9,573.87 | 11,631.36 | 10,967.42 | 11,705.42 | 11,353.45 | 11,563.02 | 11,636.06 | 10,909.03 | **11,709.55** |
| NS (%) | 0 | 32.73 | 27.01 | 32.81 | 30.94 | 33.02 | 32.03 | 32.62 | 32.83 | 30.78 | **33.03** |
| mV (p.u) | 0.9037 | 0.9317 | 0.9261 | 0.9330 | 0.9335 | 0.9346 | 0.9363 | 0.9327 | 0.9337 | 0.9322 | **0.9477** |
| MV (p.u) | 0.9970 | 0.9983 | 0.9991 | 0.9992 | 0.9997 | 0.9985 | 1.0001 | 1.0005 | 0.9999 | 1 | **1.0008** |

**Table 5** Parameter results for the IEEE's 33-bus RDN under F2

| Feature | Base | GA | PSO | DE | ACO | RSO | CSA | WOA | BRO | SSO | AVOA |
|---|---|---|---|---|---|---|---|---|---|---|---|
| TPL (kW) | 210.99 | 135.79 | 148.61 | 135.61 | 139.46 | 135.78 | 135.98 | 135.68 | 135.58 | 140.27 | 135.76 |
| PLTC (%) | 110,895.06 | 71,371.28 | 78,111.95 | 71,276.82 | 73,297.71 | 71,365.80 | 71,472.09 | 71,312.79 | 71,262.11 | 73,725.34 | 71,352.88 |
| TRP (kVAR) | 0 | 2,400 | 3,950 | 2,400 | 2,100 | 2,300 | 2,600 | 2,400 | 2,400 | 2,850 | 2,400 |
| CC ($) | 0 | 1,032 | 904.75 | 1,032 | 1,050 | 929.50 | 1,247.50 | 1,088.85 | 1,032 | 971.70 | 929.50 |
| TAC ($) | 110,895.06 | 72,403.28 | 79,016.70 | 72,308.82 | 74,347.71 | 72,295.30 | 72,719.59 | 72,401.64 | 72,294.11 | 74,697.04 | **72,282.38** |
| NS ($) | 0 | 38,491.77 | 31,878.36 | 38,586.24 | 36,547.35 | 38,599.75 | 38,175.47 | 38,493.42 | 38,600.94 | 36,198.02 | **38,612.67** |
| NS (%) | 0 | 34.71 | 28.75 | 34.80 | 32.96 | 34.81 | 34.42 | 34.71 | 34.81 | 32.64 | **34.82** |
| mV (p.u) | 0.9037 | 0.9319 | 0.9218 | 0.9330 | 0.9178 | 0.9346 | 0.9380 | 0.9327 | 0.9337 | 0.9089 | **0.9475** |
| MV (p.u) | 0.9970 | 0.9988 | 0.9993 | 0.9995 | 1.0004 | 0.9990 | 0.9992 | 1.0001 | 0.9992 | 1.0006 | **1.0007** |

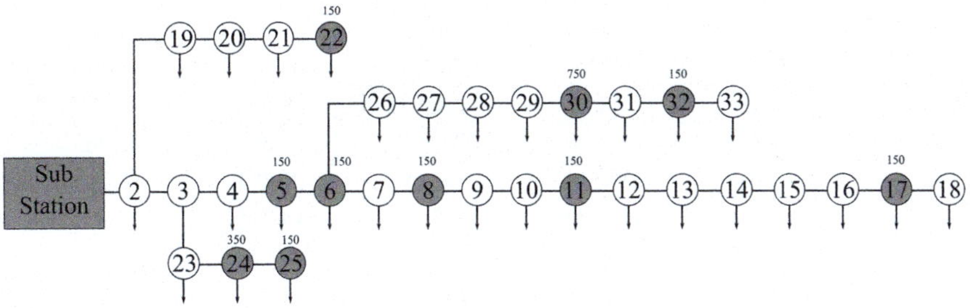


**Fig. 5** Optimal distributions of shunt capacitors for IEEE's 33-bus RDN

power losses since it is much more close to the optimization of a real-world RDN. Hence, the AVOA presents the best results for the first objective function with a TAC of $23,736.36, which represents a net saving of 33.03% respect to the original system. Moreover, the voltage profile generated by the AVOA reaches a minimum and maximum voltages of 0.9477 and 1.0008 V, respectively.

About the second objective function (Table 5), the lower TPL is given newly by the BRO algorithm with 135.58 kW. However, as explained before the most important aspect in this study are the metrics related to the installation cost of the network. Under these conditions, the improved AVOA reports the best results with a network saving of 34.82% respect to the base case. Moreover, the AVOA's voltage profile has a minimum and maximum voltages of 0.9475 and 1.0007 V, respectively. Because of this, the AVOA reports the optimal results for both objective functions in economical terms and related to the enhanced power loss.

The obtained results of Tables 4 and 5 cannot be seen graphically through convergence rates due to the own nature of the system. Nonetheless, it is possible to visualize the optimal capacitors allocation in the network through a similar diagram of Fig. 4. Therefore, in Fig. 5 the optimal shunt capacitor, their sizes and allocation are shown based on the results of the AVOA. It is worth mentioning that the AVOA proposes the same capacitors distribution for both objective functions, that is why Fig. 5 represents the optimal distribution for both cases. As can be see, the proposed distribution involves using only ten capacitors, eight capacitors of 150 kVAR, and two more of 350 kVAR and 750 kVAR, respectively. This confirms the reliability of the result since its installation cost is quite lower than the rest of proposals.

There has been already explained the relevance of the Total Power Losses in this study. In spite that, it is quite important to analyze this metric and its best, mean and worst result for each approach since it is the only way the determine their statistical performance. Hence, Tables 6 and 7 present the metric results of the algorithms for objective functions F1 and F2, respectively. As can be noted the BRO reports the best result for both tables. In addition, this algorithm reaches also the best average result for F1, which can be attributed to the stability of the method for the first objective function. On the other hand, the AVOA presents the best results in terms of worst values which in turn is reflected on its standard deviation which is better than the BRO one. Finally, it is also worth mentioning that the AVOA reaches the best computational time (measured in minutes) for both Tables.

**Table 6** Metric results for IEEE's 33-bus RDN under F1

| Approach | Best | Mean | Worst | STD | Time |
|---|---|---|---|---|---|
| GA | 135.73 | 136.94 | 136.99 | 0.513 | 18.35 |
| PSO | 148.61 | 151.35 | 156.44 | 1.357 | 19.06 |
| DE | 135.61 | 137.71 | 138.09 | 1.284 | 18.73 |
| ACO | 139.46 | 175.33 | 175.02 | 14.315 | 16.94 |
| RSO | 135.78 | 136.12 | 138.34 | 0.745 | 21.78 |
| CSA | 135.98 | 135.99 | 136.05 | 0.341 | 15.66 |
| WOA | 135.68 | 135.89 | 135.98 | **0.046** | 20.89 |
| BRO | **135.58** | **135.66** | 137.27 | 2.348 | 85.76 |
| SSO | 140.27 | 150.55 | 157.43 | 35.073 | 21.05 |
| AVOA | 135.76 | 135.90 | **136.01** | 0.156 | **14.95** |

**Table 7** Metric results for IEEE's 33-bus RDN under F2

| Approach | Best | Mean | Worst | STD | Time |
|---|---|---|---|---|---|
| GA | 135.79 | 136.98 | 136.99 | 0.618 | 18.91 |
| PSO | 148.61 | 150.27 | 156.44 | 1.466 | 19.35 |
| DE | 135.61 | 137.72 | 139.16 | 1.290 | 19.06 |
| ACO | 139.46 | 177.09 | 179.68 | 15.004 | 17.33 |
| RSO | 135.78 | 136.88 | 138.34 | 0.755 | 21.99 |
| CSA | 135.98 | 136.03 | 136.06 | 0.381 | 15.98 |
| WOA | 135.68 | 135.92 | 136.72 | **0.070** | 22.67 |
| BRO | **135.58** | 136.01 | 137.27 | 3.417 | 85.88 |
| SSO | 140.27 | 150.59 | 157.43 | 37.407 | 23.91 |
| AVOA | 135.76 | **135.87** | **135.90** | 0.081 | **15.72** |

To finalize this experiment, it is quite important to analyze the voltage profiles reached by the proposed method compared with the original RDN. This information is graphically seen in Fig. 6 for both objectives. The reading of these figures consists of understand that each node represents literally a node in the RDN of Fig. 4. These figures show perfectly the aforementioned voltage decay effect since the ideal performance of the network would mean that each node be loaded by the initial power supply of the sub-station. In both figures is seen that the three sub-branches of the grid are interpreted as a new voltage decay performance visualized at bus 19. Based on the above, the best voltage profile from the algorithms is the AVOA's profile. In addition, the rest of voltage profiles are distributed within the base case and the AVOA's profiles as presented in Tables 4 and 5 for F1 and F2, respectively.

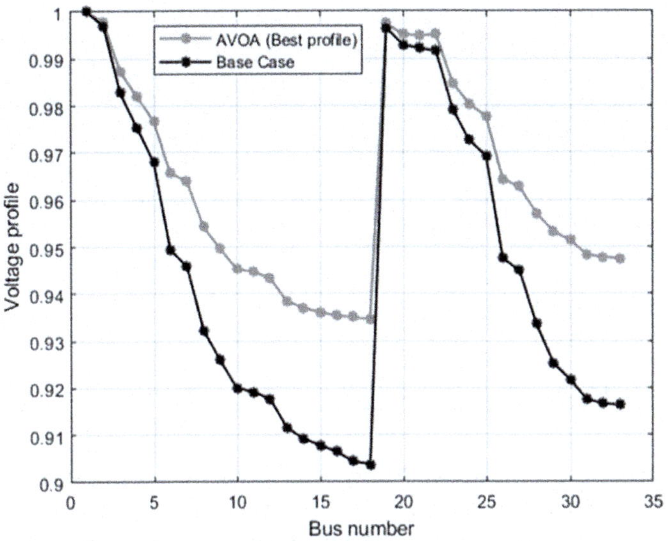


(a) AVOA profile for objective F1.

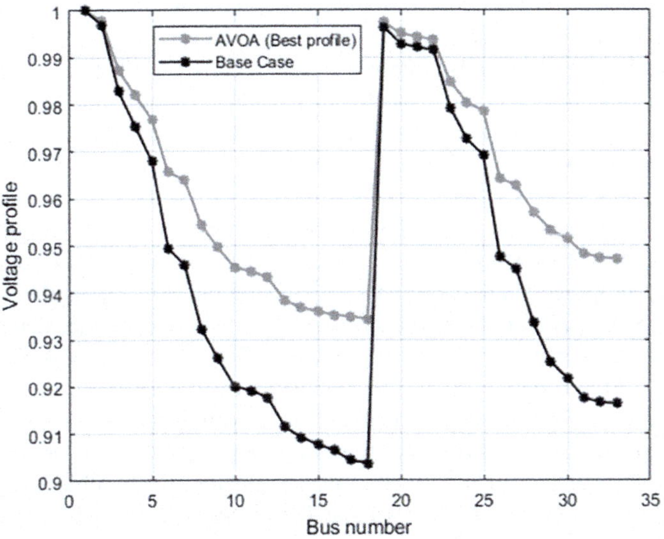


(b) AVOA profile for objective F1.

**Fig. 6** AVOA's voltage profiles for the IEEE's 33-bus RDN

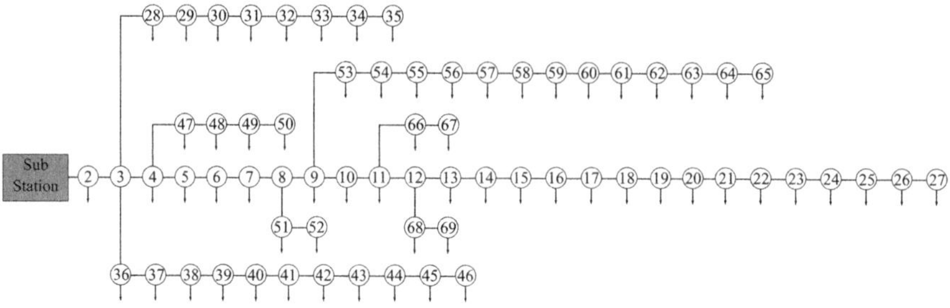


**Fig. 7** IEEE's 69-bus RDN

## 4.2 *IEEE's 69-Bus RDN System*

For this second experiment, the IEEE's 69-bus system has been selected due to its complex distribution through different sub-branches, as can be seen in Fig. 7. This RDN is characterized by a main branch of 26 buses that generate a total active and reactive power of 3,801.89 kW and 2,693.60 kVAR, respectively. These buses are supplied by a sub-station that provides 12.66 kW. In the same sense, it is seen that seven sub-branches compose the rest of the network, which in turn makes much more complex its optimization process.

The obtained results by all the algorithms and reported in Tables 8 and 9 for the first and second objective function, respectively. The utilized metrics are exactly the same as the Experiment 1 while the distribution of the tables have the same distribution like Tables 4 and 9. As can be noted for the first objective function, the best result in terms of Total Power Losses is reached by the BRO approach with 143.71 kW while the worst results is given by the PSO with 508.45 kW. This last data is quite relevant since it means that the PSO is not even able to reduce the power losses of the system through its proposed shunt capacitor's distribution but it increases the inefficiency of the network respect to the base case. Actually, this a performance that is repeated for the second objective since the PSO reports the worst result with a TPL of the same value. At first view, it seems that both tables expose the same results but that is incorrect since the rest of metrics demonstrate the differences. However, it has been already explained that in this study the most important parameters are those related to the installation cost of the system given the nature of the objective functions. Under this condition, it is clearly seen that the AVOA presents the best results for the Total Annual Cost, the Network Savings and the voltage profiles for both objectives. Furthermore, the AVOA reports a greater saving for the second function than for the first one, with a saving of 33.38 and 35.16% for F1 and F2, respectively. This same performance was reported for the first Experiment which confirms the robustness of the proposed AVOA for solving different models of this problem. Returning to Tables 8 and 9, the voltage profiles of the AVOA for both functions outperform the rest of approaches with 1.0514 and 1.0471 V as minimum and maximum voltages for the first objective, respectively. Meanwhile, the AVOA's profile reaches 0.9716 and 1.0417 V as minimum and maximum voltages for the second function, respectively.

**Table 8** Parameter results for the IEEE's 69-bus RDN under F1

| Feature | Base | GA | PSO | DE | ACO | RSO | CSA | WOA | BRO | SSO | AVOA |
|---|---|---|---|---|---|---|---|---|---|---|---|
| TPL (kW) | 224.89 | 144.13 | 508.45 | 143.90 | 163.27 | 155.51 | 150.37 | 143.83 | 143.71 | 144.53 | 143.91 |
| PLTC (%) | 37,782.35 | 24,213.88 | 85,418.79 | 24,175.66 | 27,429.05 | 26,126.24 | 25,262.64 | 24,163.16 | 24,143 | 24,281.19 | 24,177.05 |
| TRP (kVAR) | 0 | 3,150 | 30,600 | 5,550 | 5,200 | 4,200 | 12,050 | 2,900 | 3,200 | 3,350 | 2,900 |
| CC ($) | 0 | 1,352.70 | 5,783.40 | 1,591.95 | 2,495 | 1,211 | 3,593.35 | 1,229.50 | 1,379.50 | 1,001 | 1,001 |
| TAC ($) | 37,782.35 | 25,566.58 | 91,202.19 | 25,767.51 | 29,924.05 | 27,336.74 | 28,855.99 | 25,392.66 | 25,522.16 | 25,282.39 | **25,177.80** |
| NS ($) | 0 | 12,215.76 | -53,419.85 | 12,014.73 | 7,858.30 | 10,445.61 | 8,926.35 | 12,389.68 | 12,260 | 12,499.95 | **12,604.54** |
| NS (%) | 0 | 32.33 | -141 | 31.80 | 20.80 | 27.65 | 23.63 | 32.79 | 32.45 | 33.08 | **33.38** |
| mV (p.u) | 0.90901 | 0.9491 | 0.9486 | 0.9513 | 0.9501 | 0.9487 | 0.9814 | 0.9856 | 0.9919 | 0.9891 | **1.0514** |
| MV (p.u) | 0.99999 | 0.9935 | 1.0171 | 1 | 1.0001 | 1 | 1.0007 | 1.0035 | 1.0002 | 1 | **1.0471** |

**Table 9** Parameter results for the IEEE's 69-bus RDN under F2

| Feature | Base | GA | PSO | DE | ACO | RSO | CSA | WOA | BRO | SSO | AVOA |
|---|---|---|---|---|---|---|---|---|---|---|---|
| TPL (kW) | 224.89 | 144.13 | 508.45 | 143.90 | 163.27 | 155.51 | 150.37 | 143.83 | 143.71 | 144.53 | 143.91 |
| PLTC (%) | 118,204.77 | 75,754.86 | 267,238.79 | 75,635.29 | 85,813.74 | 81,737.80 | 79,035.98 | 75,596.19 | 75,532.05 | 75,965.44 | 75,639.64 |
| TRP (kVAR) | 0 | 3,150 | 30,600 | 5,550 | 5,200 | 4,200 | 12,050 | 2,900 | 3,200 | 3,350 | 3100 |
| CC ($) | 0 | 1,352.70 | 5,783.40 | 1,591.95 | 2,495 | 1,210.50 | 3,593.35 | 1,229.50 | 1,379.50 | 1,001.20 | 1,000.75 |
| TAC ($) | 118,204.77 | 77,107.56 | 273,022.19 | 77,227.24 | 88,308.74 | 82,948.30 | 82,629.33 | 76,825.69 | 76,911.55 | 76,966.64 | **76,640.39** |
| NS ($) | 0 | 41,097.20 | -154,817.42 | 40,977.53 | 29,896.03 | 35,256.47 | 35,575.44 | 41,379.08 | 41,293.22 | 41,238.12 | **41,564.38** |
| NS (%) | 0 | 34.77 | -130.97 | 34.67 | 25.29 | 29.83 | 30.10 | 35.01 | 34.93 | 34.89 | **35.16** |
| mV (p.u) | 0.90901 | 0.9328 | 0.9321 | 0.9315 | 0.9325 | 0.9565 | 0.9332 | 0.9365 | 0.9319 | 0.9323 | **0.9716** |
| MV (p.u) | 0.9992 | 0.9495 | 1 | 1 | 0.9999 | 1 | 0.9990 | 1 | 0.9791 | 1 | **1.0417** |

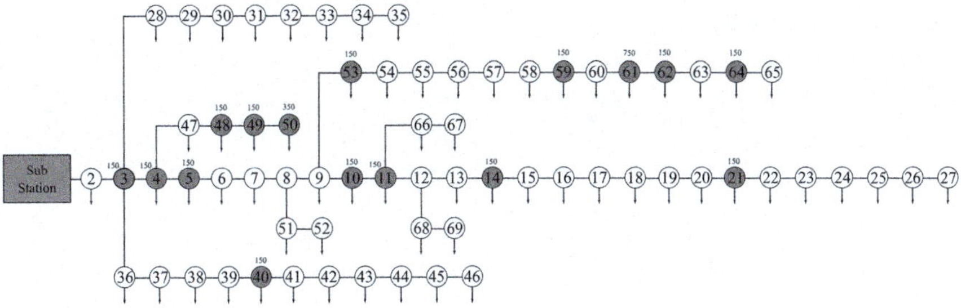

**Fig. 8** Optimal distributions of shunt capacitors for IEEE's 69-bus RDN

**Table 10** Metric results for IEEE's 69-bus RDN under F1

| Approach | Best | Mean | Worst | STD | Time |
|---|---|---|---|---|---|
| GA | 144.13 | 146.91 | 151.92 | 0.934 | 71.35 |
| PSO | 508.45 | 588.01 | 615.07 | 3.671 | 72.13 |
| DE | 143.90 | 166.29 | 199.22 | 2.852 | 70.19 |
| ACO | 163.27 | 198.28 | 285.66 | 11.015 | 84.65 |
| RSO | 155.51 | 157.93 | 161.34 | 0.682 | 116.91 |
| CSA | 150.37 | 151.32 | 152.37 | 0.284 | 79.06 |
| WOA | 143.83 | **145.04** | 147.89 | 0.217 | 73.14 |
| BRO | **143.71** | 143.28 | 146.25 | 0.561 | 97.95 |
| SSO | 144.53 | 246.75 | 268.05 | 27.172 | 74.81 |
| AVOA | 143.91 | 145.19 | **147.41** | **0.205** | **70.01** |

The optimal results presented by the AVOA in Tables 8 and 9 correspond to a single optimal distribution of shunt capacitors through the network. This optimal distribution is graphically shown in Fig. 8 under the same nomenclature of Fig. 5. As can be seen, this proposed solution involves the installation of only sixteen reactive loads for the 68 remaining buses of the system, which confirms the reliability of the proposal as previously exposed. Furthermore, 14 of the 16 shunt capacitors have a size of 150 kVAR, while the two remaining loads have sizes of 350 kVAR and 750 kVAR, respectively. It is worth mentioning that the sub-branch of buses 28–35 do not include any capacitor, such as the sub-branches of buses 51–52, 66–67 and 68–69. All this allows to minimize the installation cost of the network at the same time that the general voltage profile is kept as high as possible.

Following the same structure of Sect. 4.1, Tables 10 and 11 present the metric results of the algorithms for objective functions F1 and F2, respectively. It is worth remembering that these metric values are based on the total power losses of each algorithm since it allows to determine the general capabilities of each approach. As can be noted, the BRO reaches the best result for both functions such as the best average and best worst results for the second function. In counterpart, the proposed AVOA reaches the best result in terms of worst value for the second objective. This comparison is remarkable since the rest of approaches are not capable to compete

**Table 11** Metric results for IEEE's 69-bus RDN under F2

| Approach | Best | Mean | Worst | STD | Time |
|---|---|---|---|---|---|
| GA | 144.13 | 148.12 | 164.17 | 1.074 | 68.41 |
| PSO | 508.45 | 576.02 | 612.09 | 3.178 | 72.55 |
| DE | 143.90 | 161.83 | 195.22 | 2.828 | 70.98 |
| ACO | 163.27 | 195.33 | 277.05 | 11.167 | 84.66 |
| RSO | 155.51 | 157.84 | 160.99 | 0.591 | 117.24 |
| CSA | 150.37 | 152.26 | 152.44 | 0.277 | 70.85 |
| WOA | 143.83 | 144.47 | 147.91 | 0.285 | 73.51 |
| BRO | **143.71** | **143.74** | **146.95** | 0.513 | 97.67 |
| SSO | 144.53 | 238.01 | 258.46 | 26.171 | 75.07 |
| AVOA | 143.91 | 144.44 | 147.99 | **0.188** | **70.16** |

against these two techniques. Moreover, a similar performance was previously presented for the first experiment, which confirms the robustness of these two strategies for solving this optimization problem. Nonetheless, the improved AVOA outperforms the rest of methods in terms of standar deviation and computational time for both function, which in turn confirms the stability and reliability of the proposal. In addition, when Experiment 1 (Tables 6 and 7) and this Experiment 2 (Tables 10 and 11) are compared, it is easy to note that the complexity of this second system is quite superior. This argue is appreciated specially in the computational time required by the approaches. In the first experiment, the average required computational time is 15 min to complete the 30 independent runs, while in this second case that average expended time is increased to 75 min. Under this consideration, it is remarkable that the AVOA is still the most reliable approach given the enormous change of complexity from a RDN to another.

To finalize the analysis of this second experiment the obtained voltage profiles by the AVOA and their comparison with the original RDN are visualized in Fig. 9. It is worth mentioning that the AVOA's profile is the only profile compared with the base case due to it corresponds to the best given solution from the algorithms for both objective functions. Therefore, it must be understood that the remainder of voltage profiles are distributed between the AVOA's profile and the base case. Additionally, the Kirchhoff's law defined in Eq. 3 establishes that the generated voltage cannot surpass the power supplied by the sub-station. However, the capacitor performance precisely allows to keep energy in the respective bus which in turn increases the generated voltage of the bus by simple concepts of added energy. That is the reason why Fig. 9 apparently presents much more voltage than the offered by the sub-station. In addition, it must be seen that the Y-axe confirms the actual dimension of this voltage, which is increased only a few thousandth. Based on the above, it is seen that the AVOA's profile significantly outperforms the base case for both functions in basically all buses. Respect the original voltage profile, it can be seen that it is composed of three main phases which include the buses 2–27, the second phase is

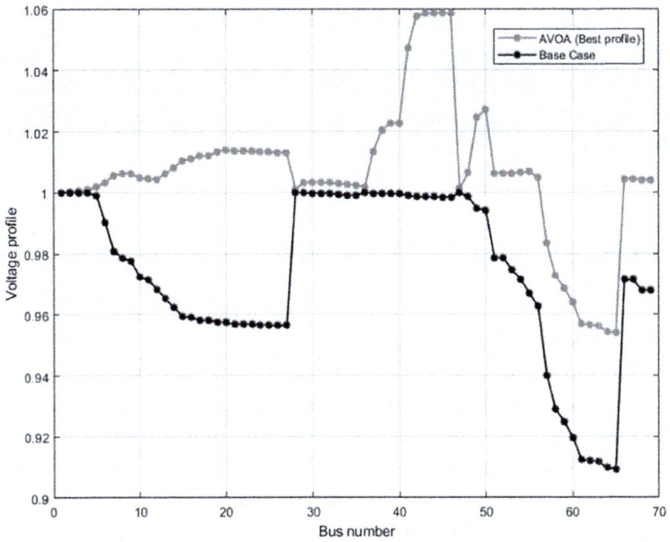


(a) AVOA profile for objective F1.

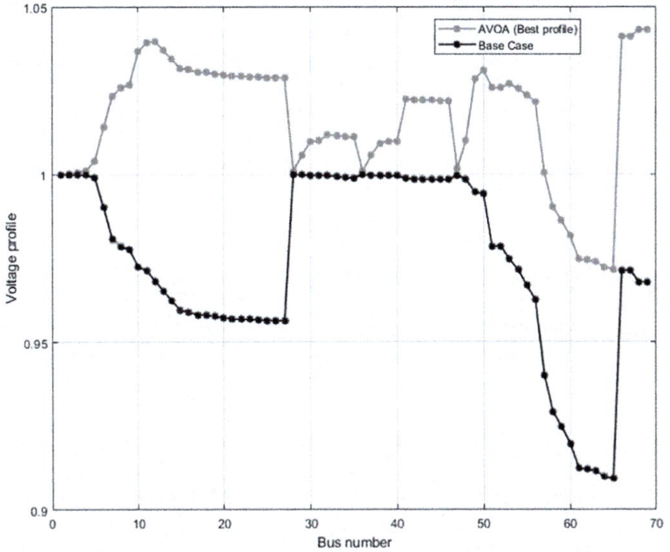


(b) AVOA profile for objective F1.

**Fig. 9** AVOA's voltage profiles for the IEEE's 69-bus RDN composed from buses 28–65, while the remaining four buses define the last phase. These phases have agreement with the RDN system of Fig. 8, where the first phase corresponds to the main branch of the network, the second phase is related to the sub-branches of buses 3, 4 and 9, while the third phase is based on the final sub-branches of buses 11 and 12.

## 4.3 Statistical Analysis

Finally, the Friedman's test is utilize to determine the complete performance of the algorithms. The Friedman's test is a non-parametric analysis that is based on the approval or rejection of a null-hypothesis according to the comparison of a certain number of distributed data [54]. This is quite important since this method allows to determine the relation between the obtained results of a methodology under a critical value. In this case the critical value is defined as the 5% of the established distribution and the ranking results for both objective functions are reported in Table 12. As can be seen, the $p$-value obtained by the AVOA for each experiment is lower than the established critical value which confirms the rejection of the null-hypothesis. In addition, the ranking test positions to the AVOA as the best non-parametric performance compared with the rest of approaches. This result confirms the robustness of the proposal and its superiority over the rest of methods.

**Table 12** Friedman ranking for the computational experiments

| Method | Objective F1 | | Objective F2 | |
|---|---|---|---|---|
| | Test | Rank | Test | Rank |
| GA | 5.23E+00 | 5 | 6.03E+00 | 5 |
| PSO | 9.23E+00 | 8 | 9.91E+00 | 8 |
| DE | 6.85E+00 | 7 | 7.01E+00 | 7 |
| ACO | 1.08E+01 | 10 | 1.13E+01 | 10 |
| RSO | 6.02E+00 | 6 | 6.11E+00 | 6 |
| CSA | 4.90E+00 | 4 | 5.38E+00 | 4 |
| WOA | 3.05E+00 | 2 | 4.01E+00 | 2 |
| BRO | 3.38E+00 | 3 | 4.52E+00 | 3 |
| SSO | 9.71E+00 | 9 | 10.41E+00 | 9 |
| **AVOA** | **1.88E+00** | **1** | **2.79E+00** | **1** |
| $p$-value | 4.4504E-10 | | 5.1284E-10 | |

## 5 Discussion

The AVOA is a relative recent meta-heuristic algorithm and it has attracted the attention of many researchers around the world due to its novel design and competitive capabilities. Hence, it is quite important to understand the limitations of this approach and the best way to determine it is testing it on different optimization scenarios. Precisely, one of the optimization problems where the AVOA has not been analyzed yet is the OCP problem, which is one of the most relevant tasks on energy-saving optimization. As explained before, the OCP problem lies in the fact of determining the optimal number and position of capacitor banks in a radial distribution network composed by a certain number of buses. The optimal allocation of these shunt capacitors will allow to keep the voltage profile of the system as higher as possible at the same time that the installation and maintenance costs of the RDN are as lowest as possible. This energy problem is characterized by its constraints which in turn represent a challenge for any optimization scheme. Thus, in this work the AVOA has been improved with a new initialization phase based on chaotic sequences enhanced by the well-known OBL technique. This procedure increases the population diversity of the algorithm and this also generates a greater exploration of the search space, which is quite important for these types of optimization tasks. The experimentation consisted of testing the improved AVOA on two RDN models (33-bus and 69-bus) and evaluated on two objective functions related to the installation costs. Furthermore, the AVOA has been compared against other nine classical and contemporary optimization algorithms that have probed their performances on the same problem in the state-of-the-art.

The obtained results demonstrate that the AVOA outperforms the rest of approaches in terms of voltage profile and reaching the best solutions with the lowest installation costs. At the same time, a Friedman ranking positions the AVOA as the best scheme compared with the other methodologies. Nonetheless, it is important to mention that some of the limitation of the proposal include a low stability since the standard deviation metric shows a non-competitive performance on that sense. It is also worth mentioning that even though the improved AVOA reports a great computational time for both experiments, its codification is complex given the own nature of the AVOA. In spite of that, the AVOA is presented as a promising algorithm capable to compete against the classical meta-heuristic approaches on energy-saving optimization problems. It is quite important to keep studying the algorithm on different tasks and analyzing its limitation and benefits in order to know its actual place on the evolutionary computation literature. Summarizing, this work belongs to the applied engineering optimization field, and the obtained results probe the effectiveness of a promising improved algorithm that can be studied in other different energy-saving optimization tasks.

## 6 Conclusions and Future Work

The optimal capacitor placement has been positioned as one of the most relevant topics in energy-saving optimization due to its implications for the improvement of our electrical networks and their reliability. Under that context, in this work an improved African Vulture Optimization Algorithm has been proposed to solve the OCP problem. This improved AVOA has been enhanced through the incorporation of a chaotic opposition-based-learning phase on its initialization stage in order to increase its population diversity. This simple but effective change yield that the AVOA be highly competitive against other nine classical and contemporary evolutionary strategies that have demonstrated their capabilities for solving this same problem in the state-of-the-art. All the approaches have been tested on two complex IEEE radial distribution networks of 33 and 69 buses, respectively. Moreover, the analyzed RDNs have been optimized in terms of installation cost through two different objective functions that allow to understand the performance of the algorithms considering different variables. The obtained results have demonstrated the robustness, stability and reliability of the proposed AVOA over the rest of methods in terms of accuracy, computational requirements and voltage profiles. Furthermore, a non-parametric test have positioned the improved AVOA as the best scheme for both objectives, which confirms the competitiveness of the proposal. Based on the above, the AVOA has probed to be a robust strategy for solving this energy-saving optimization problem, which in turn allow to explore much more capabilities of the AVOA on this topic. Some examples of future research include the testing the AVOA under other more complex and real-world adapted RDNs. Also, the analysis to reduce the computational complexity of the AVOA for solving the OCP is an interesting way to continue this work. Finally, the treatment of the topic under a multi-objective perspective is a remarkable way to improve this work.

**Competing Interests** The authors have no conflicts of interest to declare that are relevant to the content of this chapter.

## References

1. Taslimov, A.D., Rakhmonov, I.U.: Optimization of complex parameters of urban distribution electric networks. J. Phys: Conf. Ser. **1399**, 055046 (2019)
2. Mostafa, Md.G., Shingh, J.G., Haque, H.M.E.: An extensive literature review and new proposal on optimal capacitor placement in distribution systems. J. Eng. Adv. **1**, 150–169 (2020)
3. Sharma, P., Tandon, A.: Techniques for optimal placement of DG in radial distribution system: a review. Commun., Control Intell. Syst. (CCIS). **2015**, 453–458 (2015)
4. Abass, A.Z., Pavlyuchenko, D.A., Hussain, Z.S.: Methods comparison for optimal capacitor placement in distribution system. In: 2020 International Multi-conference on Industrial Engineering and Modern Technologies (FarEastCon). **5**, 1–6 (2020)
5. McGrath, B.P., Holmes, D.G.: Analytical determination of the capacitor voltage balancing dynamics for three-phase flying capacitor converters. IEEE Trans. Ind. Appl. **45**, 1425–1433 (2009)

6. Delfanti, M., Granelli, G. P., Marannino, P., Montagna, M.: Optimal capacitor placement using deterministic and genetic algorithms. In: Proceedings of the 21st International Conference on Power Industry Computer Applications. Connecting Utilities, vol. 99, pp. 331–336 (1999)
7. El-Ela, A.A.A., El-Sehiemy, R.A., Kinawy, A.-M., Mouwafi, M.T.: Optimal capacitor placement in distribution systems for power loss reduction and voltage profile improvement. IET Gener. Transm. Distrib. **10**, 1209–1221 (2016)
8. Augusto, A.A., de Souza, J.C.S., Do Coutto Filho, M.B., de Oliveira Rocha, H.R., Tafur, J.E.V.: Optimized capacitor placement considering load and network variability. J. Control, Autom. Electr. Syst. **31**, 1489–1498 (2020)
9. Schmill, J.V.: Optimum size and location of shunt capacitors on distribution feeders. IEEE Trans. Power Appar. Syst. **84**, 825–832 (1965)
10. Dura, H.: Optimum number, location, and size of shunt capacitors in radial distribution feeders a dynamic programming approach. IEEE Trans. Power Appar. Syst. **PAS-87**, 1769–1774 (1968)
11. Abdel-Salam, T.S., Chikhani, A.Y., Hackam, R.: A new technique for loss reduction using compensating capacitors applied to distribution systems with varying load condition. IEEE Trans. Power Deliv. **9**, 819–827 (1994)
12. Chis, M., Salama, M.M.A., Jayaram, S.: Capacitor placement in distribution systems using heuristic search strategies. IEEE Proc.-Gener., Transm. Distrib. **144**, 225–230 (1997)
13. Boone, G., Chiang, H.-D.: Optimal capacitor placement in distribution systems by genetic algorithm. Int. J. Electr. Power Energy Syst. **15**, 155–161 (1993)
14. Yu, X.M., Xiong, X.Y., Wu, Y.W.: A PSO-based approach to optimal capacitor placement with harmonic distortion consideration. Electr. Power Syst. Res. **71**, 27–33 (2004)
15. Carpinelli, G., Lauria, D., Varilone, P.: Voltage stability analysis in unbalanced power systems by optimal power flow. IEEE Proc.-Gener., Transm. Distrib. **153**, 261–268 (2006)
16. Sultana, S., Roy, P.K.: Optimal capacitor placement in radial distribution systems using teaching learning based optimization. Int. J. Electr. Power Energy Syst. **54**, 387–398 (2014)
17. El-Fergany, A.A., Abdelaziz, A.Y.: Capacitor allocations in radial distribution networks using cuckoo search algorithm. IET Gener. Transm. Distrib. **8**, 223–232 (2014)
18. Prakash, D.B., Lakshminarayana, C.: Optimal siting of capacitors in radial distribution network using whale optimization algorithm. Alex. Eng. J. **56**, 499–509 (2017)
19. El-Saeed, M.A.E.-S.M., Abdel-Gwaad, A.F., Farahat, M.A.-F.: Solving the capacitor placement problem in radial distribution networks. Results Eng. **17**, 100870 (2023)
20. Devabalaji, K.R., Yuvaraj, T., Ravi, K.: An efficient method for solving the optimal sitting and sizing problem of capacitor banks based on cuckoo search algorithm. Ain Shams Eng. J. **9**, 589–597 (2018)
21. Singh, S.P., Rao, A.R.: Optimal allocation of capacitors in distribution systems using particle swarm optimization. Int. J. Electr. Power Energy Syst. **43**, 1267–1275 (2012)
22. Parouha, R.P., Verma, P.: Design and applications of an advanced hybrid meta-heuristic algorithm for optimization problems. Artif. Intell. Rev. **54**, 5931–6010 (2021)
23. Díaz, P., Cisneros, M.P., Cuevas, E., Camarena, O., Martinez, F.F.A., González, A.: A swarm approach for improving voltage profiles and reduce power loss on electrical distribution networks. IEEE Access **6**, 49498–49512 (2018)
24. Blum, C., Puchinger, J., Raidl, G.R., Roli, A.: Hybrid metaheuristics in combinatorial optimization: a survey. Appl. Soft Comput. **11**, 4135–4151 (2011)
25. Jordehi, A.R.: Optimisation of electric distribution systems: a review. Renew. Sustain. Energy Rev. **51**, 1088–1100 (2015)
26. Abdollahzadeh, B., Gharehchopogh, F.S., Mirjalili, S.: African vultures optimization algorithm: a new nature-inspired metaheuristic algorithm for global optimization problems. Comput. Ind. Eng. **158**, 107408 (2021)
27. Paschos, V.T.: An overview on polynomial approximation of NP-hard problems. Yugosl. J. Oper. Res. **19**, (2016)
28. Mahdavi, S., Rahnamayan, S., Deb, K.: Opposition based learning: a literature review. Swarm Evol. Comput. **39**, 1–23 (2018)

29. Exposito, A.G., Ramos, E.R.: Reliable load flow technique for radial distribution networks. IEEE Trans. Power Syst. **14**, 1063–1069 (1999)
30. Chakravorty, M., Das, D.: Voltage stability analysis of radial distribution networks. Int. J. Electr. Power Energy Syst. **23**, 129–135 (2001)
31. Paramati, S.R., Shahzad, U., Doğan, B.: The role of environmental technology for energy demand and energy efficiency: evidence from OECD countries. Renew. Sustain. Energy Rev. **153**, 111735 (2022)
32. Sedghi, M., Ahmadian, A., Aliakbar-Golkar, M.: Assessment of optimization algorithms capability in distribution network planning: review, comparison and modification techniques. Renew. Sustain. Energy Rev. **66**, 415–434 (2016)
33. Aman, M.M., Jasmon, G.B., Bakar, A.H.A., Mokhlis, H., Karimi, M.: Optimum shunt capacitor placement in distribution system-a review and comparative study. Renew. Sustain. Energy Rev. **30**, 429–439 (2014)
34. Su, C.T., Tsai, C.C.: A new fuzzy-reasoning approach to optimum capacitor allocation for primary distribution systems. In: Proceedings of the IEEE International Conference on Industrial Technology (ICIT'96), pp. 237–241 (1996)
35. Balamurugan, K., Srinivasan, D.: Review of power flow studies on distribution network with distributed generation. In: IEEE Ninth International Conference on Power Electronics and Drive Systems, pp. 411–417 (2011)
36. El-Fergany, A.A.: Optimal capacitor allocations using evolutionary algorithms. IET Gener. Transm. Distrib. **7**, 593–601 (2007)
37. Li, W., Pardalos, P.M., Han, C.G.: Gauss-seidel method for least-distance problems. J. Optim. Theory Appl. **75**, 487–500 (1992)
38. Akram, S., Ann, Q.: Newton Raphson method. Int. J. Sci. Eng. Res. **6**, 1748–1752 (2015)
39. Das, D., Kothari, D.P., Kalam, A.: Simple and efficient method for load flow solution of radial distribution networks. Int. J. Electr. Power Energy Syst. **17**, 335–346 (1995)
40. Cook, R.F.: Analysis of capacitor application as affected by load cycle. Trans. Am. Inst. Electr. Eng. Part III: Power Appar. Syst. **78**, 950–956 (1959)
41. Priyandoko, G.: Bouc-Wen hysteresis parameter optimization for magnetorheological damper using cuckoo search algorithm. AIP Conf. Proc. **2284**, 020016 (2020)
42. Abdollahzadeh, B., Gharehchopogh, F., Mirjalili, S.: African vultures optimization algorithm: a new nature-inspired metaheuristic algorithm for global optimization problems. Comput. Ind. Eng. **158**, 107408 (2021)
43. Ahmad, M.F., Isa, N.A.M., Lim, W.H., Ang, K.M.: Differential evolution with modified initialization scheme using chaotic oppositional based learning strategy. Alex. Eng. J. **61**, 11835–11858 (2022)
44. Holland, J.H.: Genetic algorithms and adaptation. Adapt. Control Ill-Defin. Syst. **16**, 317–333 (1984)
45. Kennedy, J., Eberhart, R.: Particle swarm optimization. In: Proceedings of ICNN'95 - International Conference on Neural Networks, vol. 4, pp. 1942–1948 (1995)
46. Storn, R., Price, K.: Differential evolution—a simple and efficient heuristic for global optimization over continuous spaces. J. Global Optim. **11**, 341–359 (1997)
47. Mirjalili, S., Lewis, A.: The whale optimization algorithm. Adv. Eng. Softw. **95**, 51–67 (2016)
48. Askarzadeh, A.: A novel metaheuristic method for solving constrained engineering optimization problems: crow search algorithm. Comput. Struct. **169**, 1–12 (2016)
49. Dorigo, M., Birattari, M., Stutzle, T.: Ant colony optimization. IEEE Comput. Intell. Mag. **1**, 28–39 (2006)
50. Dhiman, G., Garg, M., Nagar, A., Kumar, V., Dehghani, M.: A novel algorithm for global optimization: rat swarm optimizer. J. Ambient. Intell. Humaniz. Comput. **12**, 8457–8482 (2021)
51. Farshi, T.R.: Battle Royale optimization algorithm. Neural Comput. Appl. **33**, 1139–1157 (2021)
52. Yu, J.J.Q., Li, V.O.K.: A social spider algorithm for global optimization. Appl. Soft Comput. **30**, 614–627 (2015)

53. Rana, A.D., Darji, J.B., Pandya, M.: Backward/Forward sweep load flow algorithm for radial distribution system. Int. J. Sci. Res. Dev. **2**, 398–400 (2014)
54. Sheldon, M.R., Fillyaw, M.J., Thompson, W.D.: The use and interpretation of the Friedman test in the analysis of ordinal-scale data in repeated measures designs. Physiother. Res. Int. **1**, 221–228 (1996)

# Smart Vehicle Charging with Variable Step Crow Search

**Ignacio Vicente Peña Fregoso, Arturo Valdivia Gonzalez, A. Fausto M. Fernando, and Gonzalez B. Adrian**

**Abstract** The charging of electric vehicles in present times is a challenge to the current installed electrical infrastructure since it is not prepared to face the demand for the future scenario where the entire vehicle fleet is electric. It is essential to find feasible alternatives to meet this growing energy demand for this increasingly numerous sector. The present will raise the possibility of coordinating and balancing the supply of electrical energy for this purpose using metaheuristic algorithms to maximize the delivery of energy, in this coordinated and balanced delivery.

## 1 Introduction

In present times, major cities face significant challenges related to pollution and energy generation. In this context, There is an urgent need in current times to generate energy from renewable sources and thus be able to reduce greenhouse gas emissions generated mainly by the use of fossil fuels [3]. As a result, there has been a notable increase in the promotion of alternative transportation means, particularly partially and fully electric vehicles, to mitigate pollution generated by transportation [15].

Among these alternatives, plug-in electric vehicles (with electric charging capability), have gained widespread adoption. This is due to their ability to be charged at home using the existing electrical grid. The use of the electrical grid is solely

I. V. P. Fregoso (✉) · A. V. Gonzalez · A. F. M. Fernando · G. B. Adrian
Universidad de Guadalajara, Blvd. Gral. Marcelino García Barragán 1421, Guadalajara Jalisco, Mexico
e-mail: ignacio.pena0184@alumnos.udg.mx

A. V. Gonzalez
e-mail: arturo.valdivia@academicos.udg.mx

A. F. M. Fernando
e-mail: abraham.fausto@academicos.udg.mx

G. B. Adrian
e-mail: adrian.gonzalezb@academicos.udg.mx



D. Oliva et al. (eds.), *Advances in Optimization Algorithms for Multidisciplinary Engineering Applications: From Classical Methods to AI-Enhanced Solutions*, Studies in Computational Intelligence 806, https://doi.org/10.1007/978-3-031-78440-8_32

for charging their batteries, aiming to replace the use of fossil fuels in urban transportation. However, this poses a challenge to the existing electrical grid, especially during peak demand hours, where electric vehicles could potentially serve not only as consumers but also as suppliers in case the grid requires it. This resurgence in interest has led to significant research and development in the electric industry.

Electric vehicles, in comparison to conventional combustion vehicles, come with numerous advantages and challenges. Advantages include zero emissions, contributing to pollution reduction, and simpler mechanics resulting in fewer failures and less costly repairs. However, challenges arise primarily from the battery, which limits the vehicle's range. The time required for charging is longer than filling a gas tank, with fast charging stations taking around 1 h for a full charge scenario. Additionally, the batteries are heavy, adding a challenge to the design of the vehicle structure. Factors such as the useful life with a limited number of charges before they need to be replaced and the cost of replacing them. Considering an increasing number of electric vehicles needing to connect to the urban electrical grid, it poses a huge challenge to the existing infrastructure [8]. These charging stations are highly energy-demanding and can potentially overload and collapse the grid. Hence, the use of efficient algorithms and mechanisms to coordinate energy supply for vehicle charging, considering certain criteria and objectives, is crucial (Fig. 1). This involves meeting charging demand,

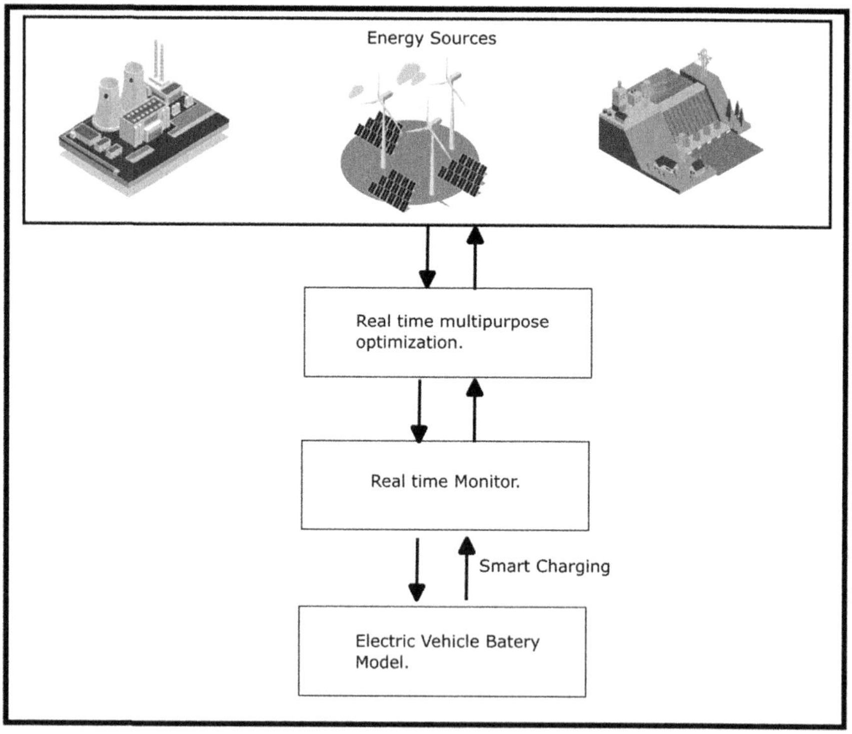


**Fig. 1** High scale PHEV charging infrastructure in a smart grid environment

satisfying customers, and balancing the load to protect the electrical grid and prevent its collapse [16]. Therefore, implementing a load control system governing these new charging centers becomes necessary.

The model of a smart charging network comprises three main components:

1. Pre-existing electrical grid, conventional and renewable energy-generating plants.
2. Smart charging ports.
3. Electric vehicles.

Employing an optimization methodology in this model allows for the efficient distribution of energy to vehicles. Factors considered for smart charging include hourly energy price variations, energy demand, charging time, battery capacity, and power grid capacity. The Crow search algorithm with variable pitch variation (CROW-VS [2]) will be used, which will aim to magnify the state of charge and distribute energy brilliantly and successfully. Limitations to consider include the state of charge of the vehicle and the power transport rate of the station. With a variable number of vehicles, a global agreement on vehicle charging demand is necessary for this experiment to minimize grid stress and improve the energy supply for this purpose.

Electric vehicle charging is classified into two types: home charging by the owner and charging at public charging centers. This research will focus on charging at public charging centers. The most important thing about this research is to find the balance between the additional demand for charging electric vehicles and maintaining the stability of the network.

According to the kind of optimization problem determines the algorithm to use for obtaining the best result. Optimization problems are classified based on requirements, function type, variable types, function scope, and the number of objectives, among other factors. There are three types of optimization algorithms: deterministic, stochastic, and metaheuristic.

A metaheuristic optimization algorithm was used to obtain the best result which is to efficiently distribute the energy supply for charging electric vehicles [1]. These algorithms are characterized by having stochastic and deterministic approaches. The operation of these algorithms in general terms consists of two main stages: first, in a first exploration stage where a possible best result is found, which tends to converge at the end of all iterations, and second, adding a randomness factor to the solution to ensure diversity in different scenarios and not have a result in a local optimal value.

The search processes employed by metaheuristic optimization methods incorporate randomized components, leading to distinct solution paths being traversed during each execution due to their non-deterministic, probabilistic characteristics. Therefore, comparing two meta heuristic algorithms applied to the same problem is challenging, as one cannot conclusively determine which is better based solely on apparently superior results. To accurately evaluate which algorithm is better, a variance analysis must be conducted. Variance analysis is considered simple to implement, using basic algebra for calculations, and allows comparisons of observations from more than two groups.

The Crow variable step search (CROW-VS) algorithm was used. Given the high exploration and search characteristics of this algorithm as well as the rapid convergence in finding the best value, it was chosen to carry out this investigation, which must find an optimal result in the distribution of the available power of the electrical network with the purpose of charging electric vehicles efficiently. The data used for this research includes information from 50 and 500 vehicles connected for charging.

The Crow algorithm with variable step will be compared with other renowned metaehuristic algorithms. Among which are Accelerated Particle Swarm Optimization Algorithm (APSO), Invasive Weed Optimization Algorithm (IWO), Sine Cosine Algorithm (SCA), Genetic Algorithm (GA) and Radial Movement Optimization Algorithm (RMO). CROW-VS is expected to obtain the best results with respect to the algorithms selected in this experiment. The objective of this comparison is to demonstrate that Crow is capable of finding the best value and that in addition to doing so more efficiently compared to the algorithms selected for this comparison.

## 2 Problem Statement

The purpose of this research is to adequately and wisely distribute the energy obtainable at the moment in the electrical grid to charge the batteries of electric vehicles. Let P be the total energy of the network, and N be the total number of vehicles to be charged. 24 h will be considered to contemplate scenarios of lower and higher available energy. In this exercise, it is considered that each vehicle must be removed from the station upon completion of its loading and before the deadline for its removal from the station. Each vehicle can only be charged once per day. The state of charge is the principal parameter to maximize to brilliantly satisfy energy demand without compromising the electrical grid (Fig. 3).

The fitness function is given in:

$$MaxH(k) = \sum_{i}(SC_i(k+1) \times Pt_i(k)) \tag{1}$$

$$Pt_i(k) = f(Bt_{r,i}(k), Zt_{r,i}(k), M_{dl,i}(k)) \tag{2}$$

$$Bt_{r,i}(k) = (1 - SC_i(k)) \times C_p \tag{3}$$

where $Bt_{r,i}$ is the battery is the rest of the battery capacity required to be filled for every iteration in a given step k. The Figure. 2 depicts this feature.

$Bt_{r,i}(k)$ is the time remaining to complete charging the a given vehicle(i) at the time step $k$, $Zt_{r,i}(k)$ is the disparity between the actual-time electricity price and the price that the ith vehicle's owner accept to remunerate.

$Bt_i$ is the assessed battery capacity of the ith vehicle, $SC_i(k)$ and $SC_i(k+1)$ are the states of charge of the ith car at time steps $k$ and $k+1$, and $Pt_i(k)$ is the coefficient weight of the ith vehicle, which is directly linked with the current SoC, energy price, and charging time (Fig. 2).

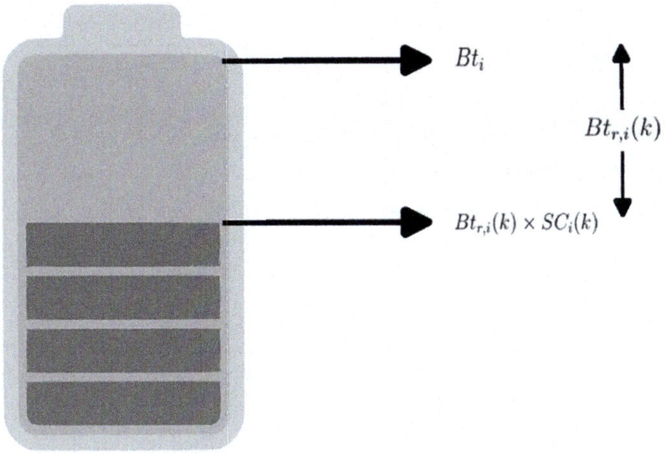


**Fig. 2** Illustration of battery charging of electric vehicle

If a particular car has a significantly lower SoC and less time to charge, but the owner is willing to pay an extra cost, the charging system will allocate more power to that vehicle. In this research, the weighting term is randomly assigned to the vehicles, ensuring that the focus remains on distributing the available grid power to the maximum number of vehicles. This example indicates the relation between $Pt_i(k)$,$M_{dl,i}(k)$,$Zt_{r,i}(k)$ and $Bt_{r,i}$. That is to say:

$$Pt_i(k)\alpha\left(Bt_{r,i} + M_{dl,i}(k) + \frac{1}{Zt_{r,i}(k)}\right) \tag{4}$$

The terms above described Normalized:

$$bt_i(k) = \frac{Bt_{r,i}(k) - min(Bt_{r,i}(k))}{max(Bt_{r,i}(k)) - min(Bt_{r,i}(k))} \tag{5}$$

$$zt_i(k) = \frac{Zt_{r,i}(k) - min(Zt_{r,i}(k))}{max(Zt_{r,i}(k)) - min(Zt_{r,i}(k))} \tag{6}$$

$$m_{dli}(k) = \frac{M_{dl,i}(k) - min(M_{dl,i}(k))}{max(M_{dl,i}(k)) - min(M_{dl,i}(k))} \tag{7}$$

Depending on the preferences for these parameters, each factor holds different levels of importance, as illustrated in (9). In this study, the charging current is assumed to remain constant over the time $\Delta t$.

$$Pt_i = \omega_1 \cdot bt_i(k) + \omega_2 \cdot zt_i(k) + \omega_3 \cdot mdl_i(k) \tag{8}$$

$$Cp_i \times SC_i(k+1) - SC_i(k) = Q = I_i(k) \times \Delta t \tag{9}$$

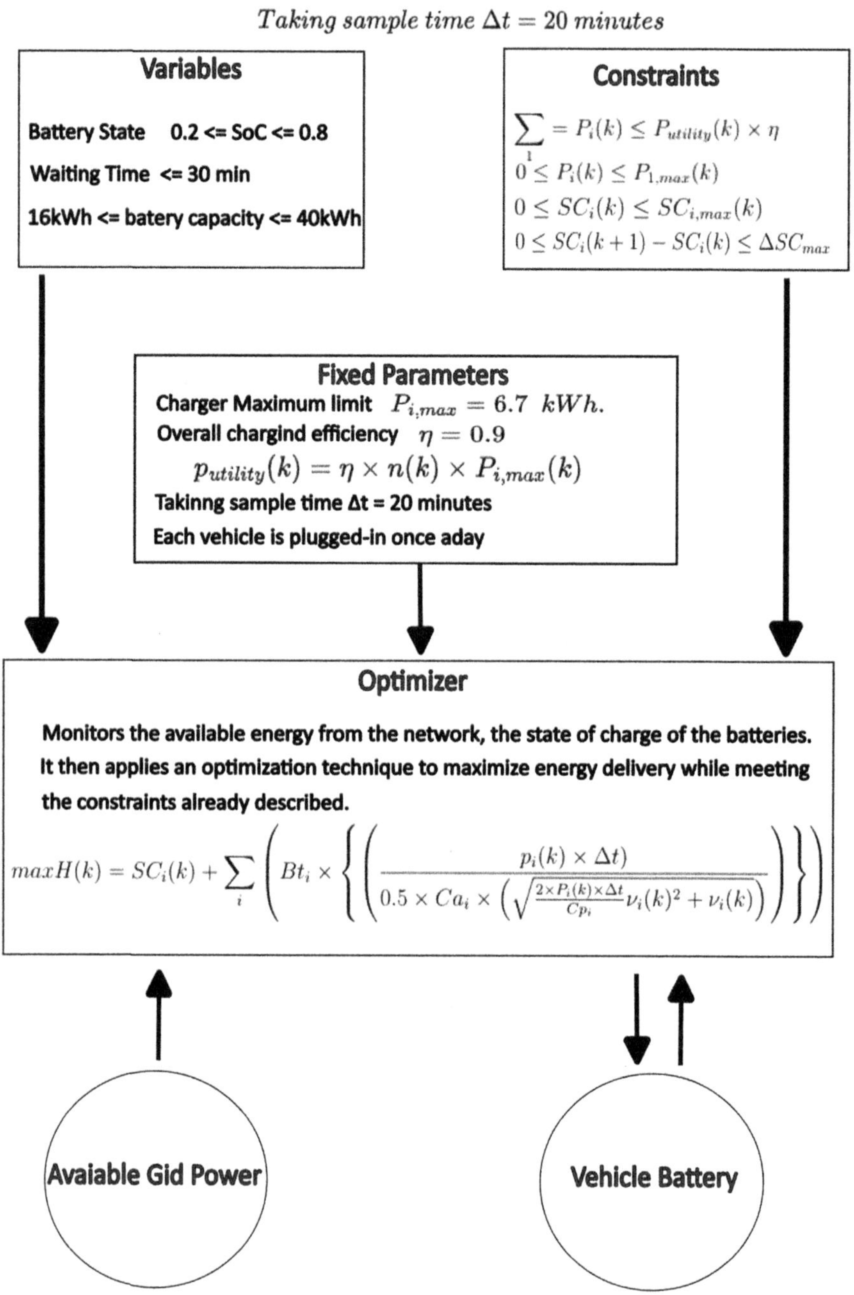


**Fig. 3** Problem statement description flow diagram

$$SC_i(k+1) = SC_i(k) + \left\{ \tfrac{I_i(k)}{Cp_i} \times \Delta t \right\} \tag{10}$$

During the $\Delta t$ a charging current $I_i(k)$ is apply and a $Q$ is the charge of the of the battery of the ith Electric Vehicle. With the approach of capacitor circuit with $Cp_i$ as Battery model and denoting the battery capacitance in Farads, the charging equations for the battery is

$$Cp_i \times \frac{dV_i}{dt} = I_i \tag{11}$$

Regarding a succinct time term, the voltage shift can be regarded as linear.

$$\left[ \tfrac{V_i(k+1) - V_i(k)}{\Delta t} \right] \times Cp_i = I_i(k) \tag{12}$$

$$V_i(k+1) - V_i(k) = \frac{I_i(k)\Delta t}{Cp_i} \tag{13}$$

substituting $I_i(k)$ with $P_i(k)$ Since power needs to be allocated intelligently, this will serve as the prime decision parameter.

$$\frac{2 \times P_i(k)}{V_i(k+1) + V_i(k)} \times Cp_i = I_i(k) \tag{14}$$

Put $I_i(k)$ from (14) into (12):

$$V_i = \frac{2 \times P_i}{I_i} - \frac{I_i \times \Delta t}{Cp_i(k)} \tag{15}$$

Putting (13) into (15) results in a formula for $V_i(k)$, which will be utilized in the algorithm coding to determine the battery's voltage at time step $k$.

$$V_i(k+1) = \sqrt{\frac{P_i \times \Delta t}{0.5 \times Cp_i} + V_i^2(k)} \tag{16}$$

Putting (12) into (11) gives $SC_i(k+1)$, which, along with (15), is substituted into (1) to provide the final form of the objective function. This is the cost function that needs to be maximized in this proposal CROW-VS is employed to fulfill this task along other well known algorithms:

$$SC_i(k+1) = SC_i(k) + \left\{ \tfrac{P_i(k) \times \Delta t}{0.5 \times Cp_i \times (V_i(k+1) + V_i(k))} \right\} \tag{17}$$

$$max\, J(k) = SoC_i(k) + \sum_i \left( wt_i \times \left\{ \left( \frac{P_i(k) \times \Delta t)}{0.5 \times Ca_i \times \left( \sqrt{\frac{2 \times P_i(k) \times \Delta t}{Ca_i}} + v_i(k)^2 + v_i(k) \right)} \right) \right\} \right) \tag{18}$$

## 2.1 Constraints

The time the electric vehicle is connected to the charger, the agreed-upon price of electric energy, the charging speed, and the condition of the battery are some practical constraints that must be considered. Additionally, monitoring the status of the load center relies on an internet connection.

The primary energy constraints include the energy available in the electrical grid and the maximum load required by the electric vehicle. Charging efficiency will be presumed to remain constant at 0.90 throughout the simulation, denoted as $\mu$.

In the reminder that the maximum battery SC limit for the ith PHEV is defined by the user and is the term $SC_{i,\max}(k)$. Regarding the once $SC_i$ of the ith vehicle and it achieves its limit $SC_{i,\max}(k)$, the charger will stop demanding more energy until another vehicle is connected.

$\Delta SC_{\max}$ limits the ramp rate of the SoC. The information is refreshed to the Regulator system when:

- There is a shift in the Utility information
- A recent Electric Vehicle is plugged-in
- $\Delta t$ has been fulfilling

System constraints are defined as:

$$\sum_{1} = P_i(k) \leq P_{utility}(k) \times \eta \tag{19}$$

$$0 \leq P_i(k) \leq P_{1,\max}(k) \tag{20}$$

$$0 \leq SC_i(k) \leq SC_{i,\max}(k) \tag{21}$$

$$0 \leq SC_i(k+1) - SC_i(k) \leq \Delta SC_{\max} \tag{22}$$

Here's my attempt to rephrase the given text to avoid plagiarism detection:

The inequality constraint expressed in Eq. 19 serves to prevent overloading the electrical grid by mandating that the cumulative energy supplied for charging all connected electric vehicles does not exceed the available grid capacity. Equations 20 and 21 restrict the state of charge for each individual electric vehicle within prescribed minimum and maximum bounds. Furthermore, Eq. 22 introduces a constraint that regulates the rate of change in the charging states, thereby averting abrupt surges in charge delivery that could potentially damage the vehicle batteries.

### 2.2 Variables

The following system variables are determined for this work as:

$$0.2 \leq SC \leq 0.8 \tag{23}$$

$$Waiting\ Time \leq 30\,\text{min} \tag{24}$$

$$16\,\text{KwH} \leq Battery\ Total\ Capacity \leq 40\,\text{KwH} \tag{25}$$

### 2.3 Fixed Parameters

The fixed parameters remain constant during the charging park scenario simulation in synergy with a specific algorithm.

- The charger's maximum restriction is $P_{i,\max} = 6.7$ KwH.
- Charging grid efficacy is taken as $\mu = 0.9$.
- $P_{utility}(k) = \mu \times n(k) \times P_{i,\max}(k)$, regard $P_{i,\max} = 6.7$ KwH at any one time step k and n is the total of electric vehicles connected in a time step k of the simulation scenario.
- The time lapsed in each $\Delta t = 30$ mins.
- can not charge multiple times a vehicle per day.

## 3 Meta Heuristic Algorithms

The algorithms used in this experiment will be described with special attention to the CROW-VS algorithm. They will be briefly described and in pseudo code the specific operation of the same will be explained in detail.

### 3.1 Variable Step Crow Search Algorithm (CROW-VS)

The crow search algorithm (CROW-VS) contemplated by Askarzadeh [2] is a meta-heuristic optimization algorithm based on the crows hiding food behavior. These birds have a great intelligence, are capable of hide its own food, and keep track of other birds' food. Crows uses this knowledge and experience to find better place to hide its own food.

The CROW-VS simulates the behavior to find the optimal value (hidden food) in a search area by implementing a crow population with their corresponding hidden food place (variable initialization) and updating these locations using a searching function (25). In each iteration each crow randomly choses another crow to try to find its food (personal best). Once selected a crow, a random value $r_{j,iter}$ determine how the new position will be updating accordingly (25)

$$X_{i,iter+1}^{new} = \begin{cases} r_{j,iter} \geq AP: & x_{i,iter}^{new} + r_i \times fl^{i,iter} \times \left(m^{j,iter} - x_{new}^{j,iter}\right) \\ r_{j,iter} < AP: a\,radom\ position \end{cases} \tag{26}$$

where $r_{[}j, iter]$, represent a random number between [0,1], $m^{j,iter}$ represent the personal best position of the other crow, $x_{new}^{i,iter}$ the current position, $fl^{i,iter}$ is the step flight, and AP is the awareness probability. The new location is compared with its personal best solution, and if the new location presents a better solution, the memory is updated to allocate the best solution of each crow.

The Variable Step Crow Search Algorithm (CROW-VS) proposed an improvement to the original CSA method where instead of a fixed $fl$ value, a cosine function allows to make a search step in any direction, providing a better exploration capacity to the algorithm. The position updating equation is describe as:

$$X_{i,iter+1}^{new} = \begin{cases} r_{j,iter} \geq AP: & x_{new}^{i,iter} + |cos(r_i)| \times \left(m^{j,iter} - x_{new}^{j,iter}\right) \\ r_{j,iter} < AP: 0.5 \times (ub_i \times rs_i + m^{j,iter}) \end{cases} \tag{27}$$

where $r_i$ represent a number between $[-1, 1]$, and $ub_i$is the searching upper bound. The pseudo code is expressed in more detail in Fig. 4.

---

Crow Search with Variable Step(CROW-VS) Algorithm

---

Initialize all parameters. Input function f(.). Searching upper bound and lower bound. Set $itermax$. Set $iter = 1$. Population size N. Evaluate the position of the crows. Initialize the memory of each crow.

```
while iter : itermax do
    for j = 1 : N do
        Randomly choose one of the crows to follow (for example, j).
        Define an awareness probability.
        if r_{j,iter+1} ≥ AP^{i,iter} then
            x_{new}^{i,iter+1} = x_{new}^{i,iter} + |cos(r_i)| × f_l^{i,iter} × (m^{j,iter} − x_{new}^{i,iter})
        else
            x_{new}^{i,iter+1} = a random position
        end if
    end for
    Check the feasibility of new positions.
    Evaluate the new position of the crows.
    Update the memory of crows:
    if f(x_{new}^{i,iter+1}) is better than f(m^{i,iter}) then
        m^{i,iter+1} = x_{new}^{i,iter+1}
    else
        m^{i,iter+1} = m^{i,iter}
    end ifiter = iter + 1
end while
End program
```

---

**Fig. 4** CROW-VS algorithm

Accelerated Particle Swarm Optimization (APSO)

Initialize particles' positions $\{X_i\}$ randomly within the search space
Initialize particles' velocities $\{V_i\}$ randomly
Initialize personal best positions $\{P_{best_i}\}$ to the initial positions
Initialize global best position $G_{best}$ to the position of the particle with the best fitness
**while** stopping criterion not met **do**
  **for** each particle $i$ **do**
    Update velocity: $V_i \leftarrow \omega \cdot V_i + c_1 \cdot rand() \cdot (P_{best_i} - X_i) + c_2 \cdot rand() \cdot (G_{best} - X_i)$
    Update position: $X_i \leftarrow X_i + V_i$
    **if** $f(X_i) < f(P_{best_i})$ **then**
      Update personal best: $P_{best_i} \leftarrow X_i$
      **if** $f(X_i) < f(G_{best})$ **then**
        Update global best: $G_{best} \leftarrow X_i$
      **end if**
    **end if**
  **end for**
  Adapt parameters $\omega$, $c_1$, and $c_2$ if necessary
**end while**

**Fig. 5** APSO algorithm

## 3.2 *Accelerated Particle Swarm Optimization Algorithm (APSO)*

The APSO algorithm is an evolution of the PSO algorithm, and is a well-known and efficient metaheuristic algorithm originally proposed in the late 90 s, it is an algorithm based on replicating mainly the behavior of flocks of birds for instance.

The APSO algorithm uses a combination of exploration and exploitation to move particles toward promising solutions, with an acceleration strategy that improves the speed of convergence toward the optimal solution. This makes it more efficient and effective in finding high-quality solutions to optimization problems. Below is a brief description of how it works Fig. 5.

## 3.3 *Invasive Weed Optimization Algorithm (IWO)*

The Invasive Weed Optimization (IWO) algorithm is a computational optimization methology based on by the behavior of invasive weeds in nature. It was first suggest by A. Mehrabian and C. Lucas in 2006 is capable of solving continuous optimization problems.

The concept behind the algorithm is inspired by the way invasive weeds spread and colonize an area. In this algorithm, "invasive weeds" are candidate solutions that represent possible solutions to the optimization problem, and the propagation and colonization process refers to how these solutions move and spread in the search

```
Invasive Weed Optimization (IWO) Algorithm
Initialize a population of weeds X randomly
while termination criterion is not met do
    for each weed x_i in X do
        Calculate the fitness value of x_i
    end for
    Sort the weeds in X based on their fitness
    Select a subset of weeds X' based on their fitness
    for each weed x_i in X' do
        Perform germination operation to generate new solutions around x_i
        Perform reproduction operation to create new solutions from x_i and other weeds
        Perform competition operation to select the best solutions among the new and existing ones
    end for
end while
return Best solution found in X
```

**Fig. 6** IWO algorithm

space. Below, and for your better understanding, the pseudo-code of the algorithm is presented in Fig. 6.

### 3.4 *Sine Cosine Algorithm (SCA)*

The Sine Cosine Algorithm (SCA) is a metaheuristic optimization technique that employs the combination of the operations of sine and cosine. The main idea behind the SCA algorithm is to perform a search strategy using sine and cosine values over the n-dimensional vector space, with the purpose of exploring and exploiting the solution search space efficiently. Below and for your better understanding, the pseudo-code of the algorithm is shown in Fig. 7.

### 3.5 *Genetic Algorithm (GA)*

A Genetic Algorithm (GA) is a type of evolutionary algorithm inspired by the principles of natural selection and genetics. Developed by John Holland in the 1970s, genetic algorithms are used to find solutions to maximization or minimization problems by imitating natural selection principles. Below, and for your better understanding, the pseudo-code of the algorithm is shown in Fig. 8.

```
Sine Cosine Algorithm (SCA)
Initialize population of solutions X randomly within the search space
Initialize amplitude factor A and scale factor r
while termination criterion is not met do
    for each solution x_i in X do
        Calculate fitness value of x_i
    end for
    Sort solutions in X based on their fitness
    for each solution x_i in X do
        Update position of x_i using sine and cosine functions
    end for
    Update amplitude factor A and scale factor r
end while
return Best solution found in X
```

**Fig. 7** SinCosine algorithm

```
Genetic Algorithm (GA)
Initialize population P with random solutions
while termination criterion not met do
    Evaluate fitness of each individual in P using objective function
    Select individuals for reproduction based on fitness (e.g., roulette wheel
selection, tournament selection)
    Apply genetic operators (e.g., crossover, mutation) to create offspring
    Evaluate fitness of offspring
    Select individuals for survival (e.g., elitism, replacement strategies)
    Replace old population with new population
end while
return Best solution found in P according to fitness
```

**Fig. 8** Genetic algorithm

## *3.6 Radial Movement Optimization Algorithm (RMO)*

The Radial Movement Optimization Algorithm (RMOA) is an metaheuristic technique inspired by radial motion observed in nature. It was first proposed by Dingwei Wang et al. in 2019. The algorithm is based on radial motion observed in phenomena such as wave propagation, particle scattering, and animal migration. Below and for your better understanding, the pseudo-code of the algorithm is shown in Fig. 9.

```
Radial Movement Optimization Algorithm (RMOA)
Initialize population P with random solutions
Initialize parameters such as maximum iterations, movement step size, etc.
while termination criterion not met do
    for each solution s_i in P do
        Evaluate fitness of s_i using objective function
    end for
    Select a reference solution s_ref from P
    for each solution s_i in P do
        Calculate radial direction vector d_i from s_i to s_ref
        Normalize d_i
        Update solution s_i by moving in radial direction with step size
    end for
    Apply constraints and bounds to the updated solutions
end while
return Best solution found in P
```

**Fig. 9** RMOA algorithm

# 4 Methodology

The previously defined objective function (1) will be evaluated by the algorithms already mentioned in the prior section. The sequence of steps to follow for this experiment is described below.

1. Primary is the creation of a group of candidate solutions according to the meta-heuristics metaphor, basically a group of random solutions according to its initialization procedure. Regarding this problem, each element is a group of powers assigned to the PHEVs Park throughout the simulation, and these powers must fulfill the limits given in the constraint, it is mandatory to satisfy the constraint, considering that the sum of every power must be equal to or less than a production of $P_{utility}$ the efficiency value.
2. Look for the power values of each PHEV encoded in each individual solution
3. Search for the best fitness value by placing the power and voltage values in (17) and considering the pondering terms, battery capabilities, and charging park scenario variables.
4. Apply the algorithm's movement operators and stages corresponding to each meta-heuristic methodology with the maximization approach in this problem.
5. Apply and fulfill every constraint in each iteration.
6. Update power values in the new population.
7. Apply step 2 until the iteration-stopping criteria are met.
8. Present the best solution.

# 5 Experiments

This section describes the general setup of the experiment, CROW-VS, and five previously mentioned metaheuristic algorithms to contrast the result of the cost function of the equation. The results of 50 and 500 PHEV will be presented. All simulations were conducted on a computer with the following hardware and software configuration:

**System type**: Ubuntu 18.04, ×64 architecture processor

**Processor**: Intel(R) Core(TM) i5-10400 CPU @ 2.90GHz 2.90 GHz

**RAM**: 16.00 GB

**Software**: MATLAB R2023b Update 7

The algorithm is applied to the scenario of a 50 PHEV set, the outcome of 100 iterations employing a population of 100 individuals or search agents. All the algorithms are performed under the same fair conditions, and the experiments are performed with the same data. The parameters of the algorithm are taken from the authors and are shown in the following Table 1.

## 5.1 *The Result Set for 50 PHEVs*

The result set of the 35 test with 50 PHEVs

As can be seen, the results (Table 2) in each of the 35 tests carried out with the 50 vehicles, the best performing algorithm by a notable difference is CROW-VS.

Analyzing the best record for each algorithm (Table 3), we can conclude that in terms of a relationship of best value found versus execution time invested, it seems to

**Table 1** Algorithms parameters in the experimentation

| Algorithms | Parameters | Reference |
|---|---|---|
| SCA | The constant value used is set to a = 2. | [12] |
| CSA-VS | Crow Search Algorithm. $AP = 0.1$ y $fl = 2$ | [5] |
| RPO | $W_{\max} = 1$ $W_{\min} = 0$ The coefficients of RPO are $C_1 = 0.6$ and $C_2 = 0.7$ | [14] |
| APSO | $\alpha = 0.2$ and $\beta = 0.45$ | [4] |
| GA | The mutation probability is set to $pm = 0.05$ and the crossover probability to $pc = 0.7$ | [10] |
| IWO | Peak number of population $m = 50$, Initial standard deviation $\sigma_{initial}1.2$, Final standard deviation$\sigma_{final}0.01$, Maximum number of seeds $s^{\max} = 5$, Smallest number of seeds $s^{\mathrm{Min}} = 0$ and Variance Reduction Exponent $\gamma = 2$ | [11] |

**Table 2** Result set of the 35 test with 50 PHEVs

| APSO | CROW-VS | GA | RMO | SinCosine | IWO | Test |
|---|---|---|---|---|---|---|
| 14.0715 | 17.2021 | 12.8690 | 5.7986 | 10.3542 | 10.4763 | 1 |
| 13.4627 | 16.1298 | 9.52703 | 6.1439 | 11.3227 | 7.8753 | 2 |
| 13.4314 | 16.2049 | 13.2396 | 12.5357 | 9.8578 | 7.5259 | 3 |
| 14.9201 | 17.2168 | 14.5438 | 7.9825 | 11.9707 | 11.6360 | 4 |
| 14.7154 | 17.2228 | 13.4963 | 8.4344 | 10.1565 | 11.3822 | 5 |
| 15.4604 | 17.8960 | 12.4358 | 5.0699 | 13.0131 | 11.6835 | 6 |
| 14.3994 | 16.6559 | 7.5336 | 8.5415 | 10.6913 | 13.2180 | 7 |
| 14.2709 | 16.7229 | 13.4435 | 13.0828 | 10.7880 | 13.2757 | 8 |
| 13.5862 | 16.3077 | 12.6336 | 11.5836 | 8.3321 | 9.5441 | 9 |
| 14.0788 | 16.9740 | 7.63116 | 13.5694 | 12.7170 | 7.2157 | 10 |
| 15.1429 | 17.6725 | 14.9853 | 14.3771 | 13.5222 | 13.0010 | 11 |
| 15.6673 | 17.6036 | 14.6138 | 14.8631 | 13.5493 | 14.5074 | 12 |
| 14.5728 | 17.3872 | 10.3027 | 13.9857 | 13.0026 | 12.9500 | 13 |
| 14.3569 | 16.9499 | 8.37133 | 13.2945 | 10.4819 | 9.3330 | 14 |
| 14.8370 | 17.6552 | 14.6538 | 6.2120 | 10.8607 | 9.3030 | 15 |
| 16.0313 | 18.2851 | 10.8975 | 15.0983 | 13.3883 | 15.3449 | 16 |
| 13.4055 | 16.0703 | 6.52635 | 12.8933 | 9.7564 | 8.2047 | 17 |
| 14.5716 | 17.3267 | 8.768763 | 13.2423 | 10.0370 | 8.2328 | 18 |
| 13.5508 | 16.4100 | 10.98963 | 13.0444 | 10.7778 | 10.6703 | 19 |
| 15.1127 | 17.9447 | 14.40795 | 8.9354 | 13.7540 | 13.6803 | 20 |
| 14.7795 | 17.0242 | 6.0228 | 13.5274 | 10.7173 | 6.2656 | 21 |
| 13.9553 | 16.6952 | 10.8747 | 10.0011 | 10.4573 | 11.4224 | 22 |
| 14.8098 | 17.2080 | 14.3818 | 13.1873 | 11.7218 | 13.8391 | 23 |
| 14.3860 | 17.8876 | 14.6526 | 14.6075 | 11.5876 | 7.1664 | 24 |
| 14.3892 | 17.0470 | 6.06290 | 13.2577 | 10.0615 | 9.9464 | 25 |
| 14.3300 | 17.4377 | 8.48231 | 7.2927 | 10.7966 | 8.5365 | 26 |
| 15.3644 | 17.8492 | 14.8145 | 14.0851 | 14.1460 | 14.4360 | 27 |
| 14.2389 | 16.8803 | 13.6721 | 6.9270 | 10.1291 | 10.4989 | 28 |
| 14.2308 | 16.4383 | 13.7387 | 9.5267 | 10.1373 | 7.0861 | 29 |
| 14.0172 | 16.9396 | 13.4641 | 10.6095 | 11.0969 | 10.0574 | 30 |
| 13.5257 | 15.9454 | 8.0585 | 10.8103 | 10.1934 | 10.7349 | 31 |
| 15.3151 | 17.8782 | 6.5006 | 14.3481 | 11.8763 | 11.5079 | 32 |
| 13.9171 | 16.5663 | 10.6973 | 4.1966 | 9.3746 | 6.3306 | 33 |
| 14.3188 | 17.0255 | 11.2188 | 13.7195 | 11.4340 | 10.2867 | 34 |
| 14.6761 | 17.3503 | 14.0871 | 13.8849 | 11.3919 | 13.3688 | 35 |
| 14.4543 | **17.0860** | 11.3884 | 11.1049 | 11.2416 | 10.5870 | Average |
| 0.66314 | **0.6033** | 2.9638 | 3.2709 | 1.4000 | 2.5445 | STD DEV |

**Table 3** Resume of the 35 test with 50 PHEVs

| Algorithm | BestFitness value | Computational time(s) |
|---|---|---|
| APSO | 16.031317 | 15.4223 |
| CROW-VS | 18.2858613 | 15.5177 |
| GA | 14.9853819 | 69.9572 |
| IWO | 15.0983336 | 23.2424 |
| RMO | 14.1460525 | 54.9137 |
| SinCosine | 15.3449216 | 14.1601 |

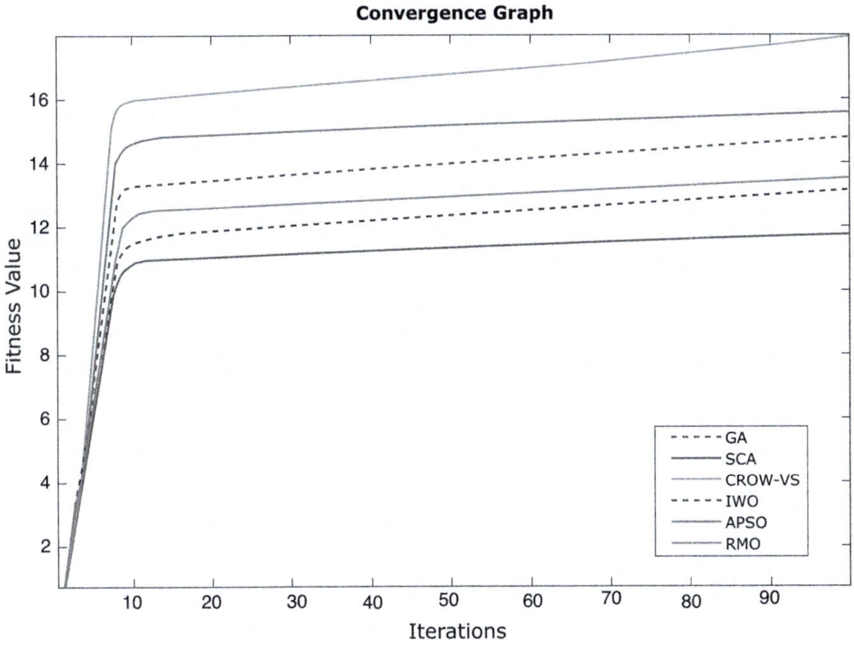


**Fig. 10** Convergence curve for the dataset of 50 PHEV

us that it is the best option in the comparison. Since the difference in time is minimal compared to the difference of the best value obtained where it is significant.

And in the same way, if we analyze the time or number of iterations (Fig. 10) to find the best result in the search space, the result becomes the same, obtaining the best performance CROW-VS.

Table 4 shows a direct comparison between CROW-VS and each algorithm in this comparison. In all cases, the CROW-VS algorithm is better. However, the difference in percentage perhaps for this exercise with few vehicles means that it is not significant, but despite this there is still a tendency to obtain better results with CROW-VS.

**Table 4** Algorithms performance comparison in 50 PHEVS scenario

| Comparison | p-value |
|---|---|
| CROW-VS versus APSO | 5.3257E-34 |
| CROW-VS versus GA | 1.2005E-13 |
| CROW-VS versus IWO | 9.6414E-13 |
| CROW-VS versus RMO | 2.6297E-27 |
| CROW-VS versus SinCosine | 7.1654E-18 |

### 5.2 *The Result Set for 500 PHEVs*

The Result set of 35 test of 500 PHEV....

With similar results to the test with 50 vehicles, it can be seen that the results (Table 5) in each of the 35 tests carried out with the 500 vehicles, continues to be the best performing CROW-VS algorithm.

In the same way that with the test of 35 vehicles, if the best record for each algorithm is analyzed (Table 6), we can conclude that in terms of the relationship between the best value found versus execution time invested, it seems to us that it is the best option in the comparison. Since the difference in time is minimal compared to the difference in the best value obtained where it is significant. And the goal is to find the best possible value.

Unlike the test with 50 vehicles, the convergence curve that projects the speed at which the best value is found (Fig. 11). In the case of the test of 500 vehicles, the difference compared to the other test algorithms is more significant.

Finally, Table 7 shows a direct comparison between CROW-VS and each algorithm in this comparison. In all cases, the CROW-VS algorithm is better. The difference in this table is in percentage difference which demonstrates the significantly better results obtained.

## 6 Results and Discussion

In this section, a comparison among well-known optimization methods is conducted. These methods include APSO (Accelerated Particle Swarm Optimization) [4], GA (Genetic Algorithm) [10], IWO (Improved Invasive Weed Optimization Algorithm) [11], RMO (Radial Movement Optimization) [14], and SCA (Sine Cosine Algorithm). Each of them was tested under the same conditions, and the results obtained by each method were evaluated. Likewise, an analysis of the resources consumed and the time taken to find the best solution was conducted.

The experimental evaluation consisted of 35 test scenarios, considering problem instances with 50 (Table 5) and 500 plug-in hybrid electric vehicles (PHEVs) (Table 2). The performance of the algorithms was assessed based on two key metrics: the best average solution quality achieved up to the current iteration (AB), and

**Table 5** Result set of the 35 runs with 500 PHEVs

| APSO | CROW-VS | GA | RMO | SinCosine | IWO | Test |
|---|---|---|---|---|---|---|
| 158.2372 | 243.1027 | 217.9994 | 209.7274 | 107.2417 | 209.7876 | 1 |
| 160.2960 | 246.5784 | 218.5761 | 206.5655 | 105.4519 | 198.5833 | 2 |
| 168.4100 | 253.5984 | 221.5431 | 211.9550 | 106.0701 | 188.6898 | 3 |
| 164.0294 | 244.4803 | 216.5800 | 148.0223 | 107.5254 | 205.2397 | 4 |
| 169.2199 | 251.6980 | 127.0451 | 121.8505 | 102.2247 | 220.7177 | 5 |
| 166.0960 | 248.7449 | 221.0294 | 206.6170 | 103.8686 | 99.5435 | 6 |
| 165.9588 | 249.5228 | 189.5309 | 214.0073 | 101.1586 | 103.3776 | 7 |
| 165.0109 | 256.0007 | 188.7302 | 141.9333 | 110.8914 | 212.8469 | 8 |
| 159.8005 | 243.4723 | 211.8945 | 201.1776 | 101.3310 | 150.6305 | 9 |
| 157.2543 | 234.4441 | 171.0349 | 98.9566 | 97.5627 | 197.7931 | 10 |
| 161.1935 | 241.9180 | 114.8941 | 197.4924 | 94.5978 | 139.6933 | 11 |
| 157.7422 | 242.8736 | 215.7355 | 58.9967 | 97.1757 | 202.5237 | 12 |
| 162.4722 | 251.0174 | 126.2600 | 137.0020 | 109.6054 | 210.6832 | 13 |
| 167.2313 | 246.3882 | 113.4573 | 204.8810 | 95.3393 | 143.6282 | 14 |
| 155.6955 | 239.5562 | 200.9688 | 194.7943 | 103.1858 | 112.1525 | 15 |
| 161.6033 | 243.4585 | 95.2538 | 112.5377 | 99.3921 | 206.4727 | 16 |
| 164.5421 | 249.5548 | 124.8277 | 211.2609 | 103.3739 | 211.4775 | 17 |
| 161.8582 | 244.8161 | 217.8940 | 135.9688 | 106.7837 | 198.8692 | 18 |
| 160.5577 | 246.3643 | 217.2588 | 211.4960 | 96.7510 | 107.9083 | 19 |
| 170.5777 | 248.9479 | 166.9861 | 214.4347 | 105.6770 | 118.9190 | 20 |
| 162.7616 | 252.0972 | 223.7391 | 87.5378 | 104.4335 | 85.1886 | 21 |
| 160.2488 | 245.6149 | 217.3892 | 190.4419 | 103.8283 | 204.1244 | 22 |
| 163.7928 | 253.6895 | 228.8038 | 209.1915 | 108.3560 | 212.1760 | 23 |
| 156.4914 | 232.1662 | 133.1549 | 154.7120 | 95.3154 | 192.7266 | 24 |
| 158.5707 | 245.9708 | 216.1701 | 140.6446 | 99.6863 | 205.3477 | 25 |
| 161.0496 | 245.7855 | 98.0170 | 205.2589 | 105.2096 | 213.9953 | 26 |
| 165.3480 | 248.5665 | 102.4225 | 209.5049 | 97.8182 | 205.8836 | 27 |
| 160.2031 | 241.6391 | 217.6803 | 64.0081 | 105.4256 | 208.1345 | 28 |
| 167.2053 | 256.3138 | 215.9434 | 213.5124 | 112.7246 | 117.6492 | 29 |
| 158.9249 | 248.7391 | 222.7422 | 212.6323 | 100.8902 | 106.0989 | 30 |
| 161.0820 | 246.7163 | 187.8238 | 121.2112 | 116.5034 | 110.4808 | 31 |
| 161.8115 | 252.5692 | 223.4114 | 211.1842 | 100.0708 | 206.1105 | 32 |
| 155.3708 | 242.1813 | 211.3569 | 171.7187 | 95.6825 | 202.6087 | 33 |
| 161.6540 | 252.1204 | 228.2018 | 212.7552 | 106.3257 | 208.6099 | 34 |
| 165.5403 | 245.5258 | 155.3669 | 207.7976 | 99.0892 | 203.0756 | 35 |
| 162.22404 | **246.749517** | 184.563512 | 172.908238 | 110.405122 | 174.907077 | Avg |
| **3.86637805** | 5.38887873 | 45.323413 | 47.9142684 | 5.26556718 | 44.7302444 | STD DEV |

the standard deviation (DE) of the best solutions found so far. The averaged results across the 35 runs are tabulated and presented, highlighting the best average performance and standard deviation for each algorithm, presented in the Tables 3 and 6. An analysis of the tabulated results reveals that, in a majority of the test cases, the

**Table 6** Resume of the 35 test with 500 PHEVs

| Algorithm | BestFitness value | Computational time(s) |
|---|---|---|
| APSO | 116.503389 | 49.0941 |
| CROW-VS | 256.313849 | 63.6541 |
| GA | 228.80375 | 93.1109 |
| IWO | 214.434725 | 88.9925 |
| RMO | 170.577716 | 65.4879 |
| SinCosine | 220.717729 | 46.2013 |

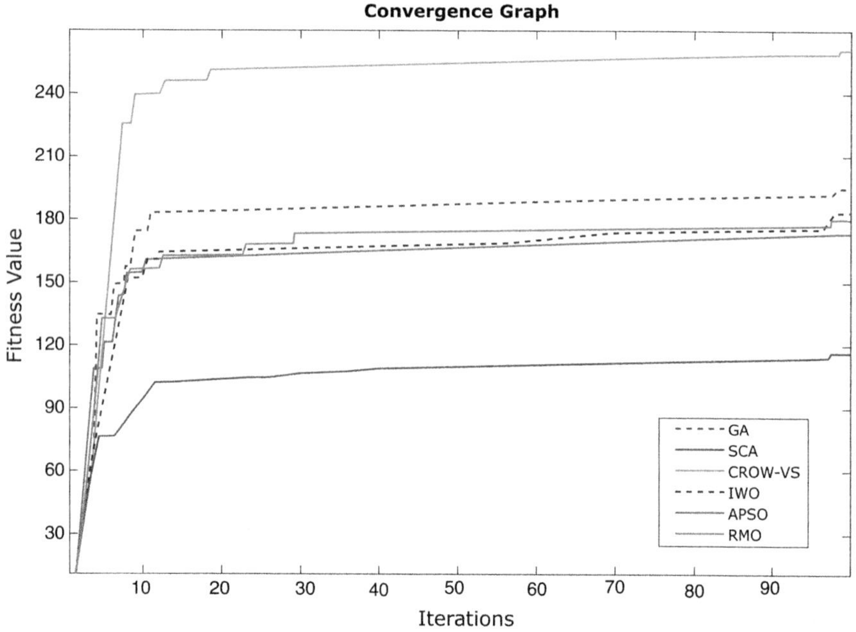


**Fig. 11** Convergence curve for the dataset of 500 PHEV

**Table 7** Algorithms performance comparison in 500 PHEVS scenario

| Comparison | p-value |
|---|---|
| CROW-VS versus APSO | 5.50E-47 |
| CROW-VS versus GA | 7.14E-10 |
| CROW-VS versus IWO | 6.55E-11 |
| CROW-VS versus RMO | 4.01E-11 |
| CROW-VS versus SinCosine | 2.64E-49 |

CROW-VS algorithm exhibited superior performance compared to the other methods under consideration. This notable performance advantage can be attributed to the algorithm's effective balance between exploiting promising regions of the search space and exploring new areas to escape local optima.

To substantiate the observed performance differences, statistical significance testing was conducted using the non-parametric Wilcoxon rank-sum test for independent samples, as described in the cited reference. This statistical validation was applied to the results obtained from the 35 independent runs for both the 50 PHEV and 500 PHEV problem instances. Tables 4 and 7 show the results of these tests. The differences shown in these tables correspond to a percentage of improvement with respect to the algorithm in question.

The Wilcoxon test consists of comparing 2 independent aptitude samples and the null hypothesis is established that the difference must be greater than 5 %. This null hypothesis assumes that the difference shown by each comparison is significant if it exceeds or is equal to this percentage.

Contrarily, the alternative hypothesis posits that if the observed difference in performance falls below a predetermined threshold percentage, then the disparity between the results is deemed statistically insignificant.

The values shown in the Tables 4 and 7 demonstrate that in all the comparisons the p-values do not meet the criteria of the null hypothesis and therefore this hypothesis is rejected. This proves that the results obtained are statistically significant and that they were not obtained by coincidence.

## 7 Conclusion

In this paper, a Crow Search algorithm with variable step (CROW-VS) is proposed to optimize the charging of PHEVs without compromising the power grid. In summary, CROW-VS takes its metaphor from the social dynamics of crows. This algorithm is based on observing the conduct of crows in the wild, where they share information and coordinate to find food sources. The key difference in our proposal to solve the problem lies in applying the variable step value at the different phases of the algorithm. This is essentially impacting in the exploration phase, where a short step is established before exploring areas with promising results, and otherwise, a long step is established to expand the search area. This is combined with controls within the algorithm where the size of the steps is also conditioned to see that the convergence occurs too quickly and prevent it from stagnating at a better local value.When tested with well-known methods in the literature, such as accelerated particle swarm optimization (APSO), genetic algorithms (GA), sine-cosine algorithm (SCA), RMO (radial motion optimization), and IWO (improved elimination of invasive weeds). Experimental results show that CROW-VS has high functionality in terms of solution quality. The performance is influenced by two main reasons: (1) Step size. 1. Step size adaptation, 2. The way crows behave in the exploration versus exploitation phase, allowing better results to be found even when a good fitness value is found.

This allows me to conclude that CROW-VS is an effective algorithm, with low computational cost and fast in finding the prime values at the domain of the problem. And for the problem posed, it proved to be the ideal solution.

## References


1. Abdel-Basset, M., Abdel-Fatah, L., Sangaiah, A.K.: Chapter 10 - metaheuristic algorithms: a comprehensive review. In: Sangaiah, A.K., Sheng, M., Zhang, Z. (eds.) Computational Intelligence for Multimedia Big Data on the Cloud with Engineering Applications, pp. 185–231. Academic Press (2018). https://doi.org/10.1016/B978-0-12-813314-9.00010-4
2. Askarzadeh, A.: A novel metaheuristic method for solving constrained engineering optimization problems: crow search algorithm. Comput. Struct. **169**, 1–12 (2016). https://doi.org/10.1016/j.compstruc.2016.03.001
3. Environmental Protection Agency (EPA).: Sources of greenhouse gas emissions (2023). Retrieved from https://www.epa.gov/ghgemissions
4. Fakhar, M.S., Kashif, S.A.R., Liaquat, S., Rasool, A., Padmanaban, S., Iqbal, M.A., Baig, M.A., Khan, B.: Implementation of APSO and improved APSO on non-cascaded and cascaded short term hydrothermal scheduling. IEEE Access **9**, 77784–77797 (2021). https://doi.org/10.1109/ACCESS.2021.3083528
5. Fan, Y., Yang, H., Wang, Y., Xu, Z., Lu, D.: A variable step crow search algorithm and its application in function problems. Biomimetics **8**(5), 395 (2023). https://doi.org/10.3390/biomimetics8050395
6. Gandomi, A.H., Yun, G.J., Yang, X.-S., Talatahari, S.: Chaos-enhanced accelerated particle swarm optimization. Commun. Nonlinear Sci. Numer. Simul. **18**(2), 327–340 (2013). https://doi.org/10.1016/j.cnsns.2012.07.017
7. García, S., Molina, D., Lozano, M.: A study on the use of non-parametric tests for analyzing the evolutionary algorithms' behaviour: a case study on the CEC'2005 special session on real parameter optimization. J. Heuristics **15**(6), 617–644 (2009). https://doi.org/10.1007/s10732-008-9080-4
8. Haidar, A.M.A., Muttaqi, K.M., Sutanto, D.: Technical challenges for electric power industries due to grid-integrated electric vehicles in low voltage distributions: a review. Energy Convers. Manag. **86**, 689–700 (2014). https://doi.org/10.1016/j.enconman.2014.06.025
9. Hedar, A.R., (Consultado en enero de).: Test functions for optimization. Recuperado de (2022). http://www-optima.amp.i.kyoto-u.ac.jp/member/student/hedar/Hedar-files/TestGO-files/Page364.htm
10. Michalewicz, Z.: Evolution strategies and other methods. In: Genetic Algorithms + Data Structures = Evolution Programs, pp. 159–177. Springer Berlin Heidelberg (1996). https://doi.org/10.1007/978-3-662-03315-9-9
11. Mehrabian, A.R., Lucas, C.: A novel numerical optimization algorithm inspired from weed colonization. Ecol. Inform. **1**(4), 355–366 (2006). https://doi.org/10.1016/j.ecoinf.2006.07.003
12. Mirjalili, S.: SCA: a sine cosine algorithm for solving optimization problems. Knowl.-Based Syst. **96**, 120–133 (2016). https://doi.org/10.1016/j.knosys.2015.12.022
13. Rahmani, R., Yusof, R.: A new simple, fast and efficient algorithm for global optimization over continuous search-space problems: radial movement optimization. Appl. Math. Comput. **248**, 287–300 (2014). https://doi.org/10.1016/j.amc.2014.09.102
14. Rahmani, R., Yusof, R.: A new simple, fast and efficient algorithm for global optimization over continuous search-space problems: radial movement optimization. Appl. Math. Comput. **248**, 287–300 (2014). https://doi.org/10.1016/j.amc.2014.09.102

15. Zhao, J., Xi, X., Na, Q., Wang, S., Kadry, S.N., Kumar, P.M.: The technological innovation of hybrid and plug-in electric vehicles for environment carbon pollution control. Environ. Impact Assess. Rev. **86**, 106506 (2021). https://doi.org/10.1016/j.eiar.2020.106506
16. Zweistra, M., Janssen, S., Geerts, F.: Large scale smart charging of electric vehicles in practice. Energies **13**(2), 298 (2020). https://doi.org/10.3390/en13020298

# A Meta-Heuristic Approach to Improving Compressor Scheduling in Refrigerated Warehouses

**Arturo Valdivia-G, Daniel Abou-Karoub-O, Itzel Aranguren-N, Fernando AFausto, and Rosaura Hernández M**

**Abstract** The work presented here details how several meta-heuristic methods may be used to control refrigeration systems by optimizing compressor schedules. Generally, compressors operate in an on-off state which sequence can be optimized for minimizing energy usage and therefore operational costs. This proposal considers certain limitations and restrictions cold rooms usually have, such as keeping food fresh in a food warehouse, for example. The framing of the problem is of great importance for it to be solvable by meta-heuristic algorithms as it will be further explained. This work explores binary and continuous optimization of this problem. In a Binary Search Space, the algorithms used are Binary IWO, classic GA, BPSO and Binary Grey Wolf, whereas in Continuous Search Spaces as well as in an Indirect Continuous Formulation; IWO, DE, PSO and Grey Wolf.

## 1 Introduction

Refrigeration systems are arguably some of the most influential technological developments in our era. They are critical for the safe storage of food and medicine, as well as for the advancement of engineering processes and scientific innovation. It is safe to say the world would not be the same without them. However, they are nonetheless riddled with challenges and improvement potential. The thermodynamic principles involved in these processes make them energy intensive and therefore costly to operate and acquire, since they also require advanced and delicate engineering [1]. These points by themselves make the case for further efforts in their optimization and betterment of existing refrigerating methods.

Refrigeration technology usually functions by making use of compressed gases and their repetitive decompression. This decompression, when done in a controlled

A. Valdivia-G · D. Abou-Karoub-O (✉) · I. Aranguren-N · F. AFausto · R. Hernández M
Universidad de Guadalajara, Centro Universitario de Ciencias Exactas e Ingenierías, Guadalajara, Mexico
e-mail: daniel.abou-karoub7551@alumnos.udg.mx



D. Oliva et al. (eds.), *Advances in Optimization Algorithms for Multidisciplinary Engineering Applications: From Classical Methods to AI-Enhanced Solutions*, Studies in Computational Intelligence 806, https://doi.org/10.1007/978-3-031-78440-8_33

manner, absorbs thermal energy from its surroundings, allowing the cooling effect to take place. This gas is then compressed again in an isolated environment and the process repeats itself [2]. The constant use of these compressors is the principal reason why this process is energetically expensive, since it requires the device to be turned on and off several times to maintain a desired temperature [1]. The optimizing of this on-off sequence through meta-heuristic algorithms is the primary focus of this work.

Meta-heuristic algorithms or methods have grown in popularity in recent years as they prove to be a very time and computation efficient way to solve complex optimization problems [3]. The way they achieve this is by formulating the problem in a discrete manner and employing stochastic methods to approach a solution [3]. In contrast to this new approach, optimization problems had been previously tackled from a continuous and analytical perspective, since these were the perspectives most of these problems were studied from, like fluid dynamics for example where the physics approach calls for analytical solutions. As other optimization problems arise with different dynamics between variables, the advent of meta-heuristic methods demonstrates other solution approach that might prove superior. These methods rely much more in randomized changes and progress evaluation in some cases, which then lends itself to a simpler solution to discrete problems with deterministic factors and controls, and more easily tested with logical tools and computation in a binary space.

Several meta-heuristic methods or algorithms have been studied and developed, some of which will be used and explained in this work to address a specific optimization problem and employed to produce experimental results for comparison.

This work proposes a meta-heuristic approach to optimize energy demand in large refrigerated systems. The methods employed consider the temporary cooling momentum and insulation within the infrastructure of refrigerated warehouses, the heat transfer dynamics of certain facilities, the temperature changes of the surrounding environment and the cooling power of these compressors. The different cost energy has during certain periods during the day are also considered since the main objective of this study is to prove a monetary cost reduction can be achieved when a compressor schedule is obtained through meta-heuristic optimization, as well as comparing several different results.

These proposed schedules could then represent methodologies that could then be applied to real refrigerated facilities independent of their conditions. The problem formulation will make the methods applicable for similar systems, even if they will be obtained through computer simulation. The results may prove useful for the future approach to compressor programming, as this would be the only aspect that could be modified and result in significant cost savings. Refrigerated systems may in the future utilize meta-heuristics for their operational efficiency and may prove to be a powerful technology taken directly to consumers soon.

This work presents three different formulations to solve the optimization problem. A natural discrete formulation and a direct and indirect continuous formulation. This way, discrete and continuous meta-heuristic search methods are used with the purpose of creating a greater set for comparison between them. Objective functions are evaluated using a realistic refrigerated warehouse simulator. Through this, the experiment can perform with constraints that would exist in a real setting, aiding the stochastic process. The results are then compared between each other to discuss insights into their performance and suitability for similar optimization problems.

The effectiveness of employing meta-heuristic algorithms in continuous problems is a growing field of study, since they are generally used in naturally discrete formulations [3]. This work will provide new insight and testing into the discretization of continuous variables. The application of these methods in a whole set of engineering problems may prove to be a very efficient and novel approach to optimization.

## 2 Search Spaces

When solving a problem one expects to eventually arrive at a particular solution. This solution could take many forms depending on the restrictions and context of the issue, but it will generally represent an option or value that exists around a set of other alternatives [4]. In a computing or optimization context the space in which the solution and its alternatives exist is called a "search space" or "state space" [5]. What this means is that a particular value or result can be found in a group of information through several different computation methods, by which this information will be analyzed under certain criteria to return an answer. In the case of a Genetic Algorithm for example, the search space is not well defined generally, and consists of a few points. As the algorithm progresses several other points, or possible solutions, are generated and then evaluated to then be selected depending on their "fitness".

Sets of data could exist in different forms, as discrete or continuous [6]. The difference between both data types comes from their nature and distribution. Discrete data generally represents information as a defined option with a distinct nature, so to speak, where its points represent a fixed value, and no points can exist between them. Examples of discrete data could be colors, like red, blue and green, which are qualitative, but also natural numbers like 1, 2 and 3. In a binary search space, the options are only represented by two variables 1 or 0, or TRUE or FALSE, which makes the search space discrete in nature. Continuous search spaces on the other hand represent measurable data over a certain domain where points can exist between each other. The clearest example of this is a physical dimension in a certain unit, like length in meters, where variables can take a value anywhere between 0 and infinity, and points can exist between others, the way 1.5 exists between 1 and 2.

The meta-heuristic methods employed in this work consider the nature of the search space as different formulations of the problem must be made to facilitate optimization [3]. More details on this will be explored further on.

# 3 Meta-Heuristic Algorithms

In this work, several different meta-heuristic algorithms are employed to optimize the compressor schedule of a refrigeration system with the objective of comparing their performance and effectiveness. Some of these methods can be employed in different problem formulations with certain modifications, hoping observations and conclusions could be made as to also promote the benefit of different approaches to the same problem.

## 3.1 IWO Invasive Weed Optimization

The Invasive Weed Optimization algorithm is an evolutionary method which emulates the expansion of weed in a field. As the population develops, each new individual's position is determined from the weed that came before it in a Gaussian distribution, but more interestingly is the number of new individuals a previous weed produces [7]. This number depends on a directly proportional relationship to its fitness value. The strength of this method relies on its exploratory capacity over the entire search space [7].

The Invasive Weed Optimization (IWO) meta-heuristic method, proposed by Mehrabian and Lucas in 2006, is an algorithm based on population dynamics inspired by the expansion of weeds in a field. The algorithm comprises four main steps: Initialization, Reproduction, Spatial Localization, and Competitive Exclusion [7].

### 3.1.1 Initialization

- A population $P$ of $m$ weeds is distributed in a random order over the search space of $n$-dimensions.
- Each weed's position $\mathbf{p}_i$ in the population is represented as a vector $(p_{1i}, p_{2i}, ..., p_{ni})$.

### 3.1.2 Reproduction

- Each weed $\mathbf{p}_i$ produces a population of $E_i$ individuals inside a specific region around its position.
- The total number of individuals $E_i$ generated by $\mathbf{p}_i$ is analyzed relative to its fitness compared to the best and worst individuals in the population.
- Parameters $E_{\text{worst}}$ and $E_{\text{best}}$ correspond to the number of secondary individual solutions generated by the initial weeds with the worst and best fitness values, respectively.

- The number of seeds produced by any other individual $\mathbf{p}_i$ linearly varies between $E_{\text{worst}}$ and $E_{\text{best}}$ based on its fitness.

#### 3.1.3 Spatial Localization

- The position of every element in $E_i$ is calculated randomly around $\mathbf{p}_i$ using a Gaussian distribution with zero mean and standard deviation $r$.
- The standard deviation $r$ is adjusted during the evolution process, gradually reducing the further expansion of newly produced elements into unknown spaces, as the number of reproductions, and therefore the total population, increases.
- The adjustment is controlled by a model involving the generation number $k$, $r_{\min}$, $r_{\max}$, $k_{\max}$, and a calibration parameter $c$.

#### 3.1.4 Competitive Exclusion

- During the process of optimization, new solutions or individuals are created through reproduction, and the total population size increases.
- An elimination mechanism is applied when the population size exceeds a limit $m_{\max}$.
- The excess individuals are ranked based on their fitness value, and those that have a better fitness value are retained for reproduction, and the rest are ignored.

### *3.2 GA Genetic Algorithm*

Genetic algorithms are some of the oldest meta-heuristic methods explored in computer science since the concept they rely on is based on basic biological evolutionary processes. Changes are made randomly to a certain population as it reproduces, and the remaining population is selected depending on certain fitness criteria. This simple yet effective method is used as a tried-and-true procedure for optimizing a problem without the need of dynamically modeling it or relying on differentiable conditions. The modification of genes is done by probability with a certain genetic weight taken from the strongest reproducers and the new population replaces the least strong of them, converging into a solution [8].

The process reproduces the classifiers which turn out to be the most adapted, favoring those with higher strength, and produces "offspring" as new, slightly different solutions. These "offspring" replace the less favorable solutions or classifiers in the population and undergo fitness testing similar to the classifiers from which they deviated. The GA involves selecting pairs of classifiers with maximum strength, reproducing them, modifying the copies through genetic operators, selecting classifiers with minimum strength for replacement, and then replacing them with the new ones [9].

Genetic operators, such as mutation and crossover, are employed [9]. Mutation involves replacing one character in the classifier's rule with another character randomly chosen from a specified set. Crossover combines rules of two "parents" by swapping randomly selected sets of characters between them. The application of genetic algorithms mimics the process of evolution, enabling the generation of new and potentially improved classifiers. This approach allows a classifier system that has reached a dead end or sub-optimal goal to explore new solutions and potentially find a better solution with the newly generated solutions [9].

### *3.3 PSO Particle Swarm Optimization*

To understand Particle Swarm Optimization, one must first understand the concept of Swarm Intelligence (SI). SI is a branch of meta-heuristic optimization techniques which rely on the collective search of a certain number of agents over a search space of solutions. They intend to replicate the behavior of self-organizing groups of natural agents, like fish schooling or bird flocking [10]. Particle Swarm Optimization is one of the most popular methods that rely on SI. PSO is a swarm-based stochastic algorithm initially developed by Kennedy and Eberhart which intends to exploit or emulate social behavior of certain animals like birds in a flock [10]. Each particle or "agent" is equal and represents a possible solution and "moves" in a search space over a certain number of iterations. The solution is evaluated and shared with other agents so as to influence the direction of their search over the next iteration, ultimately achieving convergence at an optimal solution. PSO has proven to be extremely useful and versatile, as well as simple to configure thanks to a lesser amount of tuning needed, which has resulted in its popularity [11].

A binary variant of this method exists since continuous meta-heuristic methods cannot handle certain optimization problems [10]. The variation consists basically on having the distinct decision variables 1 and 0 of a binary search space, and thus having them as potential values for each decision variable [10].

### *3.4 GWO Grey Wolf Optimizer*

This algorithm, first proposed by Mirjalili et al., attempt to mimic the leadership hierarchy and hunting mechanism of grey wolves. This meta-heuristic functions with a certain population of "wolves" with a distinct hierarchy, also utilizing the strengths of SI previously discussed. Three solutions are evaluated at a time, taking hierarchy depending on the fittest of solutions and assigning alpha, beta and delta. The rest of candidate solutions take an omega or last hierarchy in the system [12].

The GWO algorithm's optimization is guided by alpha, beta, and delta.

For encircling prey, equations are proposed to calculate position vectors based on the current iteration, coefficient vectors, and the position vector of the prey. These

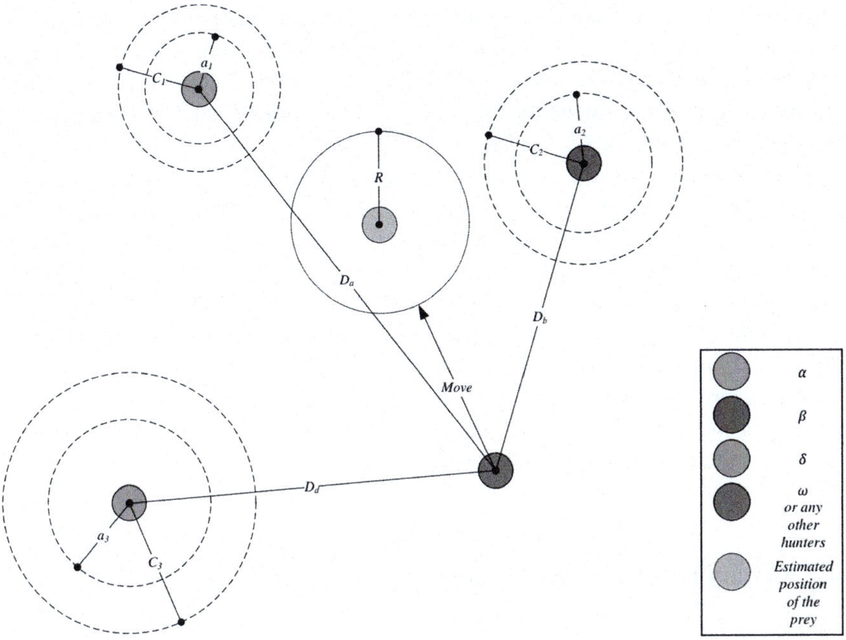


**Fig. 1** Position updating in GWO [12]

equations allow grey wolves to update their positions around the prey in a controlled manner [12]. A visual representation is seen in Fig. 1.

The hunting behavior is modeled by considering alpha, beta, and delta as wolves with better knowledge about the prey's potential location. Other search agents update their positions based on the positions of these top three solutions [12].

Attacking prey involves decreasing a random value (A) in the interval [−a, a] over iterations. The GWO algorithm finishes the hunt by having wolves attack the prey when it stops moving [12].

To encourage exploration, grey wolves diverge from the prey based on random values greater than 1 or less than −1. Another component, C, introduces random weights for prey, emphasizing exploration and avoiding local optima. The search process begins with a random population of grey wolves, and over iterations, alpha, beta, and delta estimate the prey's probable position. The algorithm terminates when a specified end criterion is met [12].

### *3.5 DE Differential Evolution*

The Differential Evolution (DE) method is a parallel direct search technique utilizing a fixed-size population ($NP$ parameter vectors) for each generation $G$. The population size $NP$ remains constant during the minimization process. The initial population is randomly chosen if system information is unknown, with a uniform probability distribution for random decisions. If a preliminary solution is available, the initial population is generated by adding normally distributed random deviations to the nominal solution $x_{\text{nom},0}$ [13].

DE represents a consolidated method for generating a set of trial parameter vectors by the addition of the weighted difference vector between two population members to a third member. If the resulting vector improves the objective function value compared to a predetermined population member, it replaces that member. The comparison vector may or may not be part of the current generation. Additionally, the best parameter vector $x_{\text{best},G}$ is evaluated for every generation to monitor progress during minimization [13].

An adaptive solution with satisfactory convergence properties is achieved by extracting the information about the distance and direction from the population to then generate random deviations [13].

## 4 Refrigerated Room Model

Refrigerated rooms and warehouses must consider several variables that will prove directly relevant to the operation. Some of these variables are the desired temperature of the products that are wished to be stored, humidity, insulation, packaging, energy demands, etc. Such complex systems are notoriously difficult to model, being one of the reasons meta-heuristic methods are worth exploring for optimizing them. Even if the approach of this study is the application of these novel methods, a model is still discussed as it provides the information needed to simulate the temperature profile and the effects of the refrigeration system.

The refrigerated warehouse modeling in this work, visualized in 2 and 3 dimensions in Fig. 2, employs a zero-dimensional approach. This approach utilizes a simple energy balance and heat transfer analysis, which doesn't consider the exchange of heat between the uniform temperature of the air circulating within the refrigerated room (denoted as $T_s$) and the outer layers of the products stored within the warehouse, which also have a uniform temperature ($T_p$), but only finds relevant the exchange between the inner air of the refrigerated room and the ambient temperature outside of it. The reasoning behind this being that this is the ultimate boundary of the problem where the heat exchange the compressor will be compensating for will be the one the heat is lost through. One can then assume the temperature restrictions had for the product must be applied to the temperature of the room, in a way considering them as the same object. This is also a practical assumption since the restrictions usually refer to the temperature of the room, and measured in this way, rather than the temperature of the object, since one controls the other.

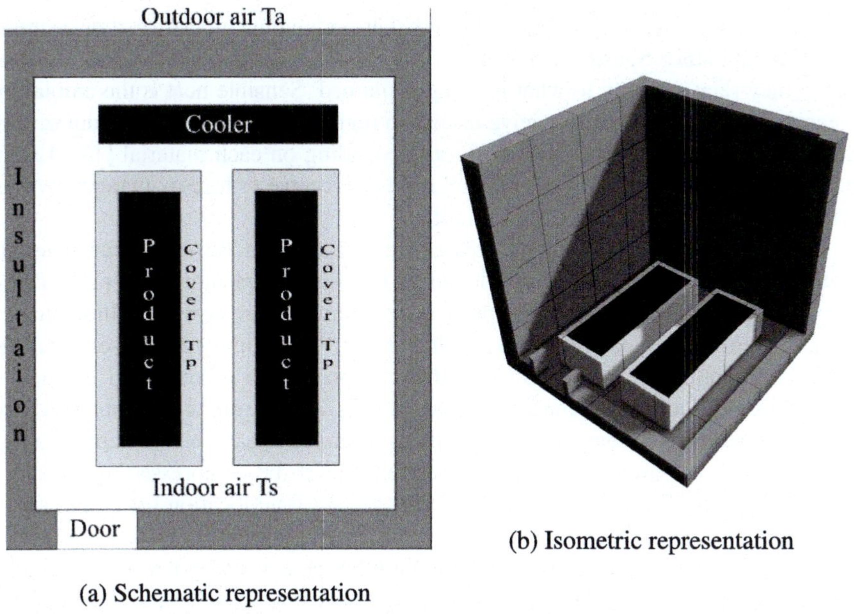


**Fig. 2** Model used to describe a refrigerated warehouse

## 4.1 Heat Transfer and Thermal Energy

Heat transfer, as a transport phenomenon, occurs primarily through three methods: conduction, which happens through a solid interface; convection, through a fluid-like interface, such as a gas or liquid; and radiation, which happens electromagnetically and is generally not relevant enough in macroscopic systems that take place in non-vacuum conditions [14]. Each of these mechanisms is modeled differently since they consider the energy dynamics of each material phase and how it interacts with the energy source. In the case of a refrigerated warehouse, both conduction and convection must be considered.

Understanding thermal energy is crucial to comprehend refrigeration. Temperature represents the kinetic energy of a material at a molecular level and can be understood as a sort of vibration, generally measured in degrees [14]. Thermal energy, or heat, is measured in energy units or work and is different than temperature. The hotter a surface or material is, referring to its temperature, is also related to the energy state in the form of thermal energy stored in the medium [14]. Two objects with the same temperature can have very different amounts of thermal energy, since different materials can "hold" a different quantity of energy yet have the same kinetic energy at a molecular scale. An analogy could be made for two objects with the same acceleration, yet having a different amount of force because their mass is different. Refrigeration can then be understood as a method through which heat can be stripped from an object or medium through thermodynamic processes, thus decreasing its temperature. Sensible and latent heat are also important concepts to understand since they relate exactly to what is being explained. Sensible heat is the amount of energy that a material must receive or lose to register a change in temperature, and, as presented before, this can be different depending on each material [14]. Latent heat is the amount of energy a material needs to receive or lose to change physical phase, from a liquid to a gas, for example.

As latent heat is generally much greater than sensible heat, certain special materials called refrigerants have been developed to harness this phenomenon [15]. These special materials have a desirable pressure and temperature at which phase change occurs, allowing the latent heat to be harnessed by manipulating these variables. They are continuously compressed and decompressed in a controlled manner to elicit a phase change between gas and liquid [15]. As a liquid is decompressed and then evaporates into a gas, it absorbs a great amount of heat from its surroundings. This gas is taken away from the refrigerated area to be compressed again into a liquid, to repeat the process. The reason for the separate location for compression and decompression is that the material will become hotter as it is compressed again. This additional heat must be transferred out of the system so as not to return it to the refrigerated environment. Some of these refrigerants don't even have to change phase as the simple compression and decompression of the gas absorbs enough energy by itself [16].

## 4.2 Dynamic Model

The fundamental premise of the model assumes uniform temperatures within the inner product layer ($T_p$) and the air contained within the chamber ($T_s$), with variations occurring solely over time. This simplification allows for the development of straightforward energy conservation equations governing $T_p$ and $T_s$. In essence, any dissipation of energy to the external environment with temperature $T_a$ must be counterbalanced by the fluctuating cooling capacity of the internal cooling system $Q_{\mathrm{r}}(\tau)$, typically subject to temporal changes.

We create a dynamic model and energy balance for a refrigerated room first considering heat loss through the walls and the energy removed by a refrigeration system without assuming steady-state conditions, while taking into account the transient behavior of the system. The approach taken in this work makes the following assumptions:

1. One-dimensional heat transfer through the walls.
2. Uniform properties across the walls.
3. Convective heat transfer dominates over radiative heat transfer.
4. Heat transfer through the walls can be modeled using Fourier's Law of heat conduction.

5. The refrigeration system operates at a constant power. (Only for modelling purposes)

After this, the model takes into account a set of components that first must be described by themselves. This develops a context for a dynamic energy balance equation.

The rate of heat transfer through the walls of the refrigerated room can be described using Fourier's Law of heat conduction (Eq. 1):

$$q_{\text{wall}} = -k \cdot A \cdot \frac{dT}{dx} \tag{1}$$

where:

- $q_{\text{wall}}$ is the heat flux through the walls (W/mÂ²),
- $k$ is the thermal conductivity of the wall material (W/(mÂ·K)),
- $A$ is the surface area of the walls (mÂ²),
- $\frac{dT}{dx}$ is the temperature gradient across the walls (K/m), which in this case would represent the difference between $T_a$ and $T_s$ at the edges of the wall.

The energy then removed by the refrigeration system is the cooling capacity of the system, typically specified in watts (W). According to our assumptions about the problem and it's system we can then proceed to construct an energy balance that equals these two components.

The energy balance equation for the refrigerated room (Eq. 2), would be:

$$\frac{dU}{dt} = q_{\text{wall}} - Q_{\text{refrigeration}} \tag{2}$$

where:

- $U$ is the internal energy of the room (J),
- $t$ is time (s).

This energy balance represents accurately the nature of the problem addressed in this work. This process can't be considered to operate in a steady-state since the compressors operation is not continuous and the temperatures aren't constant. This inevitably leads to accumulation of energy denoted by the internal energy of the room. The temperature profile of the system, as well as the specific properties of the insulation of the room and the efficiency of the compressor to refrigerate must be known previously to accurately solve the theoretical energy use and energy balance of the system. Furthermore, real operation will always deviate from models since there are additional factors that affect the process, everything from the composition of the air to the pressure of the room can have an effect in real processes.

This is why the initial estimation of $Q_r(\tau)$ usually relies on technical documentation and later undergoes adjustments to align with historical data, encompassing temperature readings and compressor operation schedules. This work introduces a more intricate numerical model, as described in [17], which computes $Q_c(\tau)$ based on either provided or optimized compressor schedules. This model encompasses the entire refrigeration system, including compressor power and efficiency, the impact of phase-change material in auxiliary tanks, and the performance of coolers within the storage chamber. The Python implementation of the model, along with comprehensive documentation, is accessible via [18]. The simulator retrieves external temperature data from weather forecasts.

## 5 The Problem

This work addresses the optimization of a refrigerated warehouse's compressor schedule using meta-heuristic methods to minimize energy costs while maintaining temperature constraints.

As previously mentioned, refrigeration compressors have a particular operational nature. They can only operate in an on-off state; at least for the purpose of this work, this assumption is made, which renders the problem discrete in nature. The schedule of this on-off sequence is to be optimized, and this is only possible when the outside temperature and energy cost profile during the time horizon of the experiment are known in advance.

Energy cost and temperature constraints are the principal limitations of the system's operation and are the dimensions that are desired to be optimized. Various meta-heuristic algorithms are attractive applications to be evaluated based on their ability to optimize energy costs effectively.

### *5.1 Problem Formulation*

In order to determine if different meta-heuristic optimization methods are suitable for optimizing a compressor schedule for a refrigerated facility, two approaches to solving this problem are explored: binary and continuous multidimensional search spaces, while also exploring a third solution as an indirect continuous formulation. The necessary background information of some meta-heuristics has already been reviewed, as well as the nature of search spaces. Now the algorithmic variants used in each case are briefly discussed as well as the general details of the study.

### *5.2 Discrete Formulation*

The optimization problem of compressor scheduling can be naturally formulated as a discrete optimization problem, where a binary vector represents each candidate solution indicating the compressor's on or off state. However, discrete optimization problems are challenging to solve as it is not possible to evaluate all possibilities to identify the optimum solution [3]. Therefore, meta-heuristic algorithms that use trial and error methods are commonly used. Continuous optimization problems, on the other hand, the solution becomes simpler as some kind of gradient approximation is a posibility, which can guide the search [3]. Moreover, continuous meta-heuristics are rapidly developing, resulting in some successful methods that require no parameters. In addition to the natural discrete problem formulation, this work also considers two alternative continuous formulations, enabling the application of continuous search methods, with the hope of finding a better solution without extensive tuning.

#### 5.2.1 Binary Search Spaces with Meta-Heuristics: (GA, BPSO, BIWO, BGWO)

The objective of the problem is to determine an on-off sequence for the compressor over the next 24 h, while meeting cooling chamber temperature requirements and minimizing energy costs. Continuous control of the compressor is not feasible; it can only be switched on or off. The refrigerated room has a high level of inertia, and changing the compressor state frequently is not practical [19]. Therefore, the problem is formulated as a discrete-time problem with a time step of 15 min, resulting in a cooling schedule represented by a binary vector of length value equal to 96. After formulating the search space in a binary fashion, discrete optimization methods that function on binary strings can be applied to solve for the optimal solution. In the case of Genetic Algorithms, GA, these are explored as the classic version, as well as binary versions of the continuous meta-heuristics PSO, IWO, and GWO [20–22].

In the case of BGWO [21], the algorithm involves adapting the principles of the Continuous Gray Wolf Optimization (CGWO) to handle binary 0, 1 values. The algorithm iteratively updates the positions of wolves in the search space to find an optimal feature subset. For feature selection, the algorithm utilizes a fitness function that combines classification quality (or error rate) with the ratio of selected features to total features. This function aims to maximize classification quality while minimizing the number of selected features.

The algorithm is applied to feature selection problems in classification tasks, where the objective is to find the best combination of features for optimal classification performance. The K-nearest neighbor (KNN) classifier is often used alongside the binary gray wolf optimization as the search method.

### *5.3 Direct Continuous Formulation*

One approach to the compressor schedule optimization problem in a continuous fashion is to proportionally transform the discrete problem into a continuous one, achieved through different methods. A popular technique for using continuous meta-heuristics for binary search spaces is two-step binarization [20]. In this approach, the range of a continuous variable is first reduced to the interval [0, 1] using an S-shaped

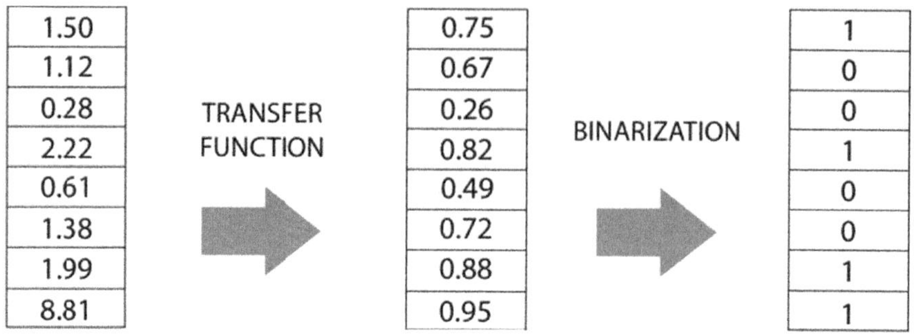


**Fig. 3** Two-step binarization method [20]

or V-shaped transfer function. The second step involves binarization, which converts the continuous result into a binary space, as the example in Fig. 3 shows. In the standard binarization method for converting into a binary search space, the result of the transfer function is considered as the probability of the truth value, and a binary sequence or string is created by sampling from the 1 or 0 value space according to these probabilities. However, in this work, the step involving the transfer function is skipped, and the optimization algorithm will work on continuous variables restricted to the [0, 1] range.

#### 5.3.1 Continuous Search Spaces with Meta-Heuristics: (DE, IWO, GWO, PSO)

The development of continuous meta-heuristics is generally directed currently towards methods not requiring parameters and which utilize automatic adaptation, which makes their application simple even for inexperienced users. This approach also will be analyzed in both direct and indirect continuous formulations of the problem.

### *5.4 Indirect Continuous Formulation*

Another approach to the problem formulation in a continuous manner is to indirectly optimize. This can be achieved by setting a sequence of target temperatures along the compressor's schedule time frame (24 hrs.) and optimizing these temperatures instead. The algorithm's objective is to enhance the target temperature ($T_s$) settings of the warehouse hourly. Consequently, the objective function utilized for this optimization task consists of 24 continuous values, each denoting the target temperature for a specific hour. Through the optimization of these temperature values, the algorithm aims to reduce the energy consumption necessary to uphold the desired temperature thresholds within the refrigerated warehouse.

These target temperatures would then be converted into an on-off, or binary schedule by a deterministic controller or switch. Specifically, when the warehouse temperature exceeds the target temperature, the compressor is activated. Conversely, the compressor is deactivated when the temperature falls below the target level. Utilizing this on/off controller to govern the compressor operation enables the refrigerated warehouse to effectively sustain the desired temperature settings while minimizing energy usage.

In this work, the compressor is limited to operating in discrete time, meaning it can only take an on or off state for a fixed time interval, in this case being set arbitrarily at 15 min.

## 6 Experimental Study

The experimental results presented in this work are composed of three parts. The first presents the results obtained from four discrete optimization methods which were applied with a binary search space formulation of the problem. The second one uses four continuous optimization meta-heuristics with the direct continuous formulation of the problem, based preemptively on the standard binarization method, and the third part compares these same continuous optimization methods but applied to the indirect continuous formulation of the problem, that is to optimize a set of target temperatures instead of the compressor's schedule, and then establishing the schedule based on a deterministic control of the compressor based on the comparison between the target temperature and the actual temperature.

The fitness function, or criteria for optimization, is composed of the weighted sum of energy cost (Eq. 3), as well as a penalty for violating the temperature constraints of the problem (Eq. 4).

$$p(x) = q(x) + \alpha f(x), \tag{3}$$

$$f(x) = \sum_{u:j<t_u} (t_u - j)^2 + \sum_{u:z>t_u} (z - t_u)^2 \tag{4}$$

where $\alpha = 10^9$, $u$ represents a time step index, $t$ represents the temperature inside the refrigerated room, $z$ is the temperature lower limit which is equal to 1Â° C, and $j$ is the temperature upper limit which is equal to 3Â° C. The problem considers two possible energy tariffs as well as an outer temperature profile presented in Fig. 2a. The methods start from a random point or population, depending on the method. They are subsequently run 35 times independently and presented in box-plots.

## 6.1 Common-Sense Schedule

A schedule is created by the analysis of the problem from an expert's perspective as well as a common-sense use of the energy, with regards to the energy cost at a certain time of day, as this is a variable condition in the problem. This provides a point of comparison to evaluate the meta-heuristic optimization effectiveness as well as an understanding as to how a meta-heuristic method might provide a solution which was not obvious initially. This common-sense solution takes strongly into account the cost of the electricity during the schedule, optimizing for cost savings.

Most interestingly, as can be seen in Fig. 4, these meta-heuristic algorithms show a counter intuitive result, the compressor is used somewhat intermittently during the high cost of energy period demonstrating that it is more efficient to energize it when the difference between the outside and inside is not as large, even if it means the use of more expensive energy. Additionally, it shows it is also more effective to trigger the compressor periodically than continuously, even at the low-cost time period, therefore using it just enough to reach the desired temperature constantly.

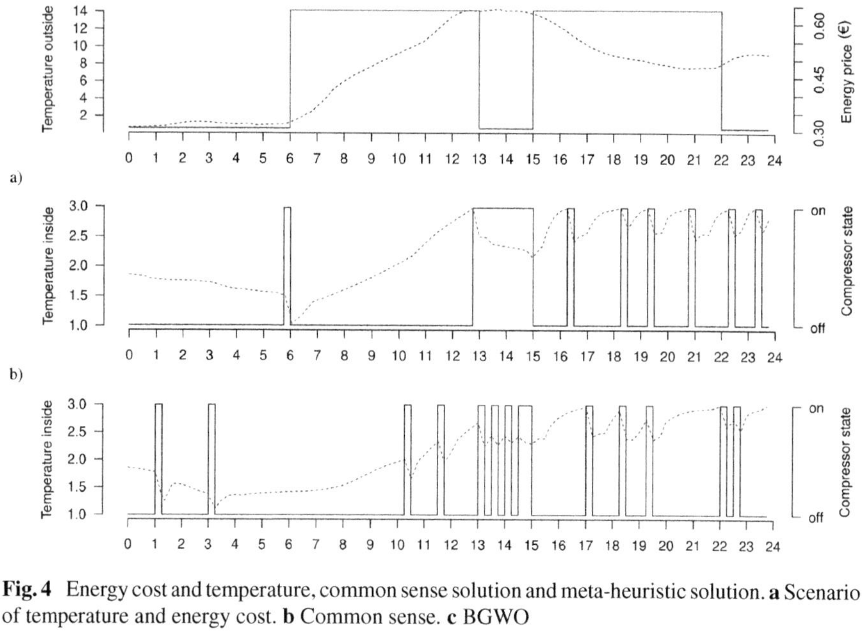


**Fig. 4** Energy cost and temperature, common sense solution and meta-heuristic solution. **a** Scenario of temperature and energy cost. **b** Common sense. **c** BGWO

## 6.2 Binary Search Space Methods Discrete Formulation

A discrete formulation of the problem is developed to allow the application of four discrete or binary optimization meta-heuristic methods as an objective of this work. These are BIWO, GA, BPSO, and BGWO, some of which are discrete or binary versions of continuous meta-heuristic algorithms.

As presented in Fig. 5, the best results were achieved with BGWO (Binary Grey Wolf Optimization) [21], adjusting the Exploration-Exploitation parameter as $a = 2$.

The next best solution proved to be the GA (Genetic Algorithm) [23], adjusting the mutation probability set to $p_m = 0.05$ and the crossover probability to $p_c = 0.7$. BIWO (Binary Invasive Weed Optimization) [20], had an intermediate performance adjusted with the following parameters: Number of generations = 30, $N_{\text{iter}} = 400$, $P_{\text{init}} = 100$, $P_{\max} = 20$, $S_{\min} = 20$, $S_{\max} = 80$, $\theta_{\text{init}} =$ Problem Dimension, $\theta_{\text{final}} = 1$, and $\theta_{\text{mod}} = 3$. Finally, the worst performance by a larger margin proved to be BPSO (Binary Particle Swarm Optimization) [22], which was adjusted with $C1$, $C2 = 2.5$, $v_{\max} = 6$, sigmoid transfer function.

Regarding the amount of iterations needed to reach convergence, a clear advantage can be observed by the Genetic Algorithm as seen in Fig. 6. The GA achieved convergence in less than 10 iterations, a close second was the BGWO algorithm after less than 20 iterations, which also reached the best solution, and further away around 80 iterations are both the BPSO and BIWO algorithms.

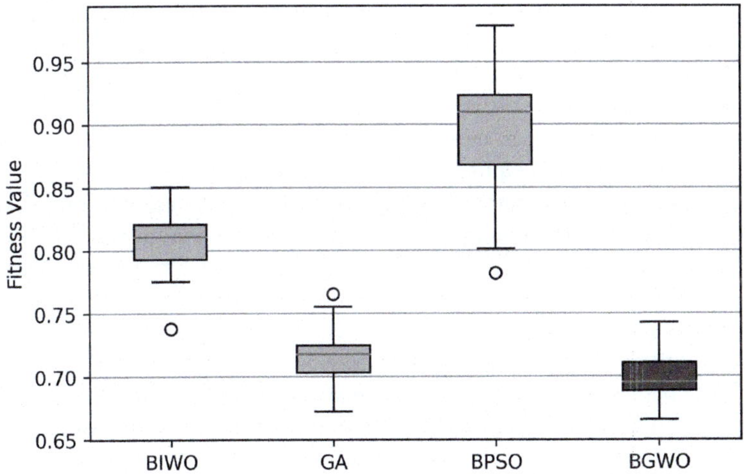


**Fig. 5** Performance of meta-heuristic methods that operate on a bit sequence (Discrete formulation)

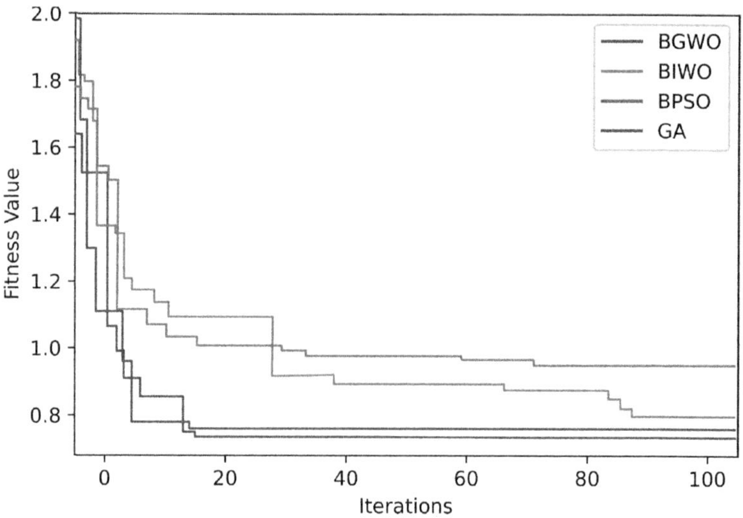


**Fig. 6** Comparison of methods that operate in binary Space 1

## *6.3 Continuous Search Space Methods with Standard Binarization Direct Continuous Formulation*

The first continuous approach to solving the problem is applied through the following four optimization meta-heuristic methods that operate in a continuous search space: IWO, DE, PSO, and GWO, coupled with the standard binarization method to adequate the continuous search space to the discrete nature of the problem, an on-off compressor sequence or schedule. As seen in Fig. 7, the best results were achieved with Grey Wolf Optimization, using a parameter $a = 2$ for Exploration-Exploitation [12], with a minor advantage over DE with the crossover rate value set to $CR = 0.5$, and the differential weight value established as $F = 0.2$ [13]. The IWO algorithm is slightly inferior in performance in this problem formulation compared to DE and GWO, when configured with the following parameters: $m = 50$, $\sigma_{\max} = 1.2$, $\sigma_{\min} = 0.01$, $\gamma = 0.5$, $E^{\text{best}} = 3$, $E^{\text{worst}} = 0$, and $m_{\max} = 60$ [20]. PSO proves to be the worst optimizer according to our objective or fitness value with the cognitive and social coefficients used in the algorithm at $c1 = 2.0$ and $c2 = 2.0$, respectively, while the inertia factor decreases in a linear fashion from 0.9 to 0.2 while the search process continues [24].

It took a similar amount of iterations for three of the meta-heuristic algorithms to reach their convergence value, as seen in Fig. 8, where DE was the longest to reach it's final solution.

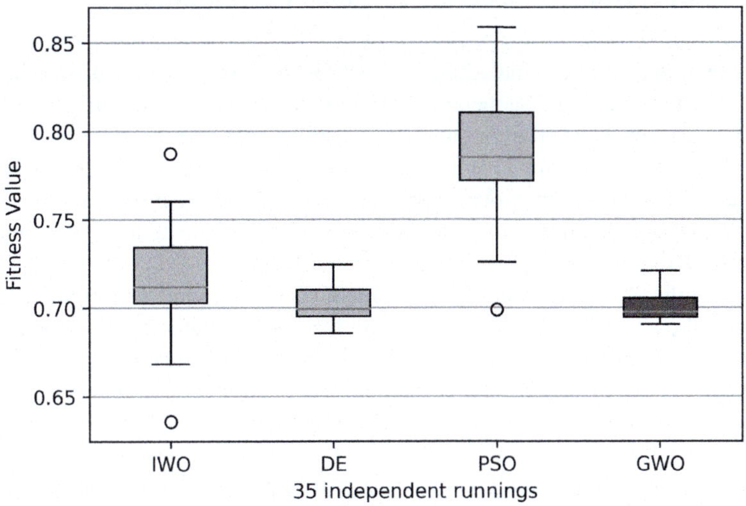


**Fig. 7** Performance of selected meta-heuristics which work with the probabilities of the true value (Standard Binarization)

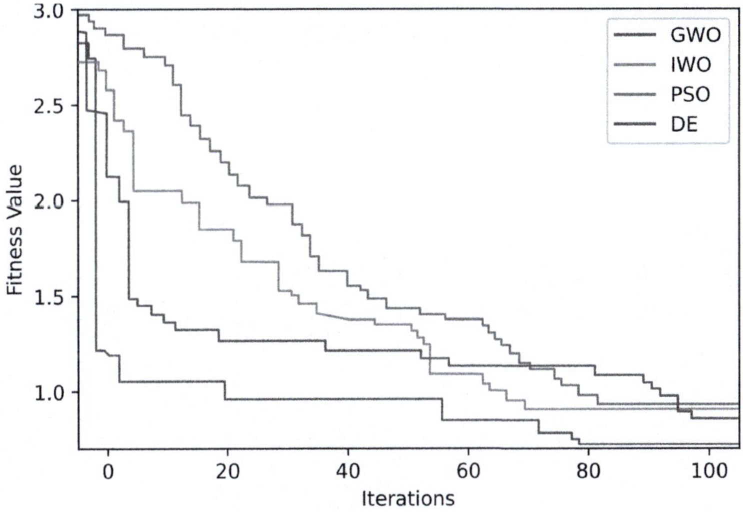


**Fig. 8** Comparison of methods that operate in continuous Space 2

## 6.4 *Continuous Methods for the Optimization of the Target Temperature Indirect Continuous Formulation*

In this section continuous meta-heuristics IWO, DE, PSO, and GWO are again compared when optimizing a sequence of target temperatures of the refrigerated warehouse which then dictate the compressor's schedule in a deterministic control, relating to the Indirect Continuous Formulation of the problem. The parameters employed are the same as in the previous implementation for the algorithms tuning, simply then changing the nature of the search space to a continuous temperature range in a given time frame.

The results of the four meta-heuristic methods in the Indirect Continuous formulation as shown in Fig. 9, prove to be quite like those obtained with standard binarization as the superior meta-heuristic method proves to be GWO, having a slightly better fitness value over DE in this problem formulation as well and demonstrating the fastest convergence as seen in Fig. 10. IWO performed slightly better than in the direct continuous formulation while PSO demonstrated the largest difference of all 4, performing significantly better. It is also important to note that even if IWO achieved a better fitness value than PSO, the latter converged sooner in this problem formulation.

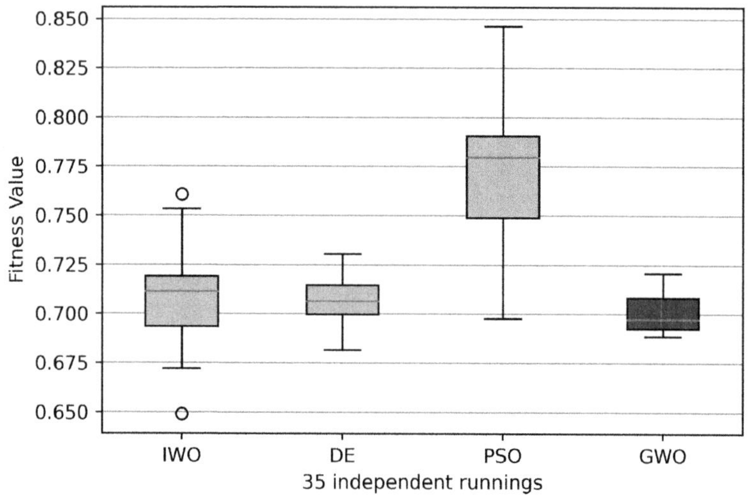


**Fig. 9** Comparison of the selected algorithm's performance in the optimization of the target temperature sequence and the best algorithm from the indirect continuous formulation

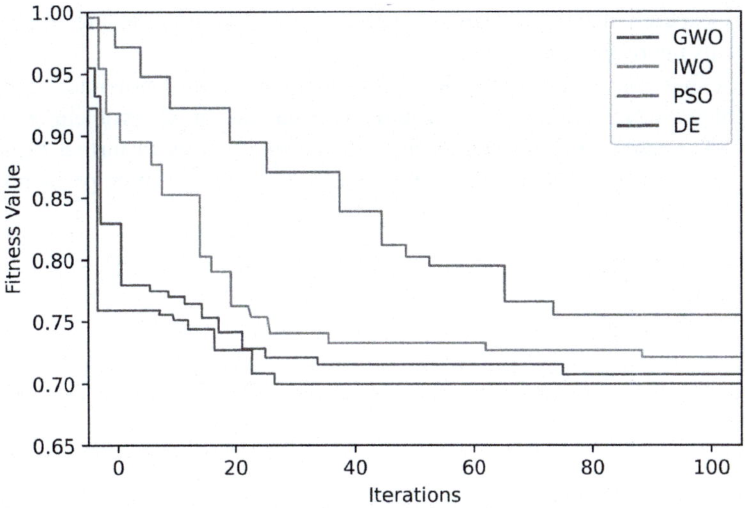


**Fig. 10** Comparison of methods that operate in continuous Space 3

## 7 Conclusions

In this chapter an analysis of several meta-heuristic methods is made. Through a general introduction to what meta-heuristic algorithms are, as well as a brief description of each of the methods of interest in this work, which includes a combination of modern and classic meta-heuristic methods, a comprehensive understanding of the significance and general characteristics of these problem solving tools is reached. Further on, by conceptually modeling a real-world application of these methods, the optimization of a refrigeration system, we associate the meta-heuristic approach to problems of an engineering perspective, where a delicate dimensional analysis and modeling of the dynamic of the system is usually needed to apply classical optimization methods, hereby exposing the value of this work.

The following concrete conclusions can be ascertained:

- Meta-heuristic methods can provide attractive solutions to optimization problems in engineering fields without the need of complex dynamic modeling.
- The formulation of the problem is a very significant factor in the performance of meta-heuristic optimization methods.
- Current meta-heuristic algorithms provide fitter solutions when working in a continuous formulation.

Even though binary modifications to current meta-heuristic algorithms have been developed, when testing and comparing them, the strength of these optimization methods in a continuous search space cannot be ignored. All four of the methods explored performed significantly better in optimizing the problem in both the direct and indirect continuous formulations of the problem over the binary search space.

Further experimentation with the objective of identifying the nature of this issue represents interesting future studies.

The expansion and innovation in other techniques for the continuous formulation of problems which are discrete in nature, as is the case in the problem chosen for this work, could confidently represent valuable future work to further exploit the advantages of meta-heuristic algorithms in all sorts of optimization problems.

## References

1. Çengel, Y.A., Boles, M.A.: Thermodynamics: An Engineering Approach. McGraw-Hill (2015)
2. Stoecker, W.F., Jones, J.W.: Refrigeration and Air Conditioning. McGraw-Hill Education (2016)
3. Glover, F., Kochenberger, G.A.: Handbook of Metaheuristics. Springer Science & Business Media (2003)
4. Goldberg, D.E.: Genetic Algorithms in Search, Optimization, and Machine Learning. Addison-Wesley (1989)
5. Russell, S., Norvig, P.: Artificial Intelligence: A Modern Approach. Pearson (2009)
6. Wasserman, L.: All of Statistics: A Concise Course in Statistical Inference. Springer Science & Business Media (2013)
7. Cuevas, E., Rodríguez, A., Valdivia, A., Zaldívar, D., Pérez, M.: A hybrid evolutionary approach based on the invasive weed optimization and estimation distribution algorithms. Methodologies and Applications. Published online: 20 March 2019. Springer-Verlag GmbH Germany, part of Springer Nature (2019)
8. Coello Coello, C.A. (ed.): Computación Evolutiva. ACADEMIA MEXICANA DE COMPUTACIÓN, A, C. Segunda Edición 2019. ISBN: 978-607-97357-9-1. Autores: Iván Mauricio Amaya Contreras, Carlos A. Brizuela, Joel Chacón Castillo, Carlos Artemio Coello Coello, Santiago Enrique Conant Pablos, Mario Graff, Julio Juárez, Efrén Mezura Montes, Sabino Miranda Jiménez, Daniela Moctezuma, Alicia Morales Reyes, José Carlos Ortiz Bayliss, Rafael Rivera López, Carlos Segura, Eric S. Tellez, Hugo Terashima Marín
9. Alfonseca, M.: Genetic Algorithms. Affiliation: IBM SofTec Lab. Paseo de la Castellana, 4, 28046 Madrid, Spain
10. Gad, A.G.: Particle swarm optimization algorithm and its applications: a systematic review. Arch Comput Methods Eng **29**(2022), 2531–2561 (2022)
11. Yazdi, A.K., Kaviani, M.A., Emrouznejad, A., Sahebi, H.: A binary particle swarm optimization algorithm for ship routing and scheduling of liquefied natural gas transportation. Transp. Lett. (2019)
12. Mirjalili, S., Mirjalili, S.M., Lewis, A.: Grey Wolf Optimizer (2014). *Authors: Seyedali Mirjalili, Seyed Mohammad Mirjalili, Andrew Lewis. Affiliations: 1 School of Information and Communication Technology, Griffith University, Nathan, Brisbane, QLD 4111, Australia. 2 Department of Electrical Engineering, Faculty of Electrical and Computer Engineering, Shahid Beheshti University, G. C. 1983963113, Tehran, Iran. Emails:* seyedali.mirjalili@griffithuni.edu.au, mohammad.smm@gmail.com, a.lewis@griffith.edu.au
13. Storn, R., Price, K.: Differential evolution—a simple and efficient adaptive scheme for global optimization over continuous spaces. (No. TR-95-012). International Computer Science Institute (1995)
14. Çengel, Y.A., Afshin, J.: Fundamentals and Applications. McGraw-Hill, Ghajar, Heat and Mass Transfer (2015)
15. Domanski, P.A., Peterson, P.F.: Refrigeration Systems and Applications, 2nd edn. Wiley (2000)
16. American Society of Heating, Refrigerating and Air-Conditioning Engineers. (2010). *Refrigeration Handbook*

17. Kwidziński, R.: In Polish: model dynamiki temperatur w komorze przechowalniczej z materiałem zmiennofazowym w pośrednim obiegu chłodniczym. Tech. Rep, Institute of Fluid-Flow Machinery Polish Academy of Sciences (2015)
18. (2021) Implementation of the model of refrigerated warehouse and objective function. https://staff.elka.pw.edu.pl/~rbiedrzy/publ/rwm.zip
19. Zong, Y., Bindner, H.W., Gehrke, O.: Different optimal control strategies for exploitation of demand response in the smart grid. In: IFAC Proceedings Volumes 15th IFAC Workshop on Control Applications of Optimization, vol. 45, no. 25, pp. 188–193 (2012)
20. Crawford, B., Soto, R., Legüe, I.F., Olguín, E.: A binary invasive weed optimization algorithm for the set covering problem. Adv. Intell. Syst. Comput. **464**, 459–468 (2016)
21. Emary, E., Zawbaa, H.M., Hassanien, A.E.: Binary Gray Wolf Optimization Approaches for Feature Selection. Scientific Research Group in Egypt (SRGE) (2021). http://www.egyptscience.net
22. Liu, J., Fan, X.: The analysis and improvement of binary particle swarm optimization. In: CIS 2009—2009 International Conference on Computational Intelligence and Security, vol. 1, pp. 254–258 (2009). https://doi.org/10.1109/CIS.2009.261
23. Miller, B.L., Goldberg, D.E.: Genetic algorithms, selection schemes, and the varying effects of noise. Evol. Comput. **4**(2), 113–131 (1996). https://doi.org/10.1162/evco.1996.4.2.113
24. Kennedy, J., Eberhart, R.: Particle swarm optimization. In: Proceedings of ICNN'95—International Conference on Neural Networks, vol. 4, pp. 1942–1948 (1995). https://doi.org/10.1109/ICNN.1995.488968

# A Metaheuristic Task Scheduling of FOG Servers Using a Hybridization of Crow Search Algorithm with Non-Monopolize Search

**Itzel Aranguren, Fernando Fausto, Adrián González, and Axel L-Aguiñaga**

**Abstract** Contemporary technologies, such as the Internet of Things (IoT), facilitate real-time access to information, influencing decision-making processes in smart cities and industries. Data is transmitted to cloud storage systems daily, undergoing analysis and processing within the network. Implementing Fog Computing (FC) technology introduces inherent challenges, particularly in managing energy consumption and enhancing response speed. This paper proposes a novel hybridization of Crow Search and Non-Monopolize Search (CSA-NMS) to address the task scheduling problem in FC, aiming to enhance the quality of services supplied by IoT devices. Recent metaheuristic algorithms were examined to identify potential solutions to the energy consumption problem. A comparison of various algorithms, such as the Genetic Algorithm (GA), Flower Pollination Algorithm (FPA), Particle Swarm Optimization Algorithm (PSO), Bat Algorithm (BAT), Crow Search Algorithm (CSA), and the amended version proposed in this study, the Crow Search en synergy Non-Monopolize Search (CSA-NMS), is presented in this work. The study's findings indicate the superiority of the (CSA-NMS) compared to other analyzed methods. This discovery suggests a more efficient alternative for addressing the energy challenges associated with fog computing.

## 1 Introduction

The Internet of Things (IoT for short) is one paradigm that has transformed how we interact with our environment, contributing to developing an interconnected world. In this concept, different sensors and actuators have been embedded into

I. Aranguren (✉) · F. Fausto · A. González · A. L-Aguiñaga
Depto. de Innovación Basada en la Información y el Conocimiento, Universidad de Guadalajara, CUCEI, Av. Revolución 1500, Guadalajara, Jalisco, Mexico
e-mail: itzel.aranguren@academicos.udg.mx

F. Fausto
e-mail: abraham.fausto@academicos.udg.mx

A. González
e-mail: adrian.gonzalezb@academicos.udg.mx

 
D. Oliva et al. (eds.), *Advances in Optimization Algorithms for Multidisciplinary Engineering Applications: From Classical Methods to AI-Enhanced Solutions*, Studies in Computational Intelligence 806, https://doi.org/10.1007/978-3-031-78440-8_34

many real-world objects to enable them to exchange information, make decisions, invoke actions, and offer different services.

This technology has witnessed rapid growth in recent years due to the advancements in sensor miniaturization, the decreasing cost of IoT devices, and the improvement of wireless networks. Therefore, the adoption of this technology can be found in more and more fields like energy management, healthcare, agriculture, home automation, environmental monitoring, surveillance, and vehicle systems, among many others [1, 2].

Cloud data centers are centralized facilities that house computing resources, including servers, storage, networking equipment, and other infrastructure components. They are the main tools for processing and storing all the data these devices generate. These data centers form the backbone of cloud computing services, providing on-demand internet access to computing resources and services.

However, due to their centralized processing characteristics, cloud servers could face a complex task processing all the data generated by an enormous number of IoT devices. In addition, some applications, especially those dealing with sensitive data, may have privacy and security concerns when the information travels through the internet. Moreover, cloud computing may only sometimes be suitable/adequate for latency-sensitive applications due to the tendency of cloud servers to be located at a considerable distance from end devices.

To overcome these problems, in 2012, Cisco proposed a new computing concept known as Fog Computing [3, 4], which seeks to manage a portion of the workloads generated by end devices near the users' network rather than sending all processing operations to the cloud. This results in some advantages, such as the capacity of devices to make real-time decisions due to no longer needing to send every piece of information to the cloud, helping indirectly to optimize bandwidth usage and minimizing congestion in the network. Furthermore, it enhances privacy and security by minimizing the exposure of sensitive information to the cloud.

Although fog computing can meet the needs of many IoT applications, it still faces many challenges inherent in its structure and capabilities. For example, because the fog nodes have different capabilities, scheduling and resource allocation could become complex tasks in the fog environment [5]. Applications in fog devices must be assigned according to their needs. Certain applications require more speed, while others may have other priorities, such as data privacy, memory for storing data, etc.

Consequently, an effective method for resource management in cloud and fog environments and task scheduling becomes imperative. This work utilizes a metaheuristic technique to achieve close to optimal solutions. The benefit of employing a metaheuristic, as opposed to an exact method, lies in the simplicity and reliability of the obtained results [6, 7].

In this proposal, the popular metaheuristic algorithm, the Crow search algorithm (CSA), is hybridized with the Non-Monopolize Search method. This new hybrid algorithm, CSA-NMS, is also applied to task scheduling in fog computing and yields good results. The remainder of this work is composed of the following sections:

Section 2: Provides an overview of the current literature.

Section 3: Describes our innovative proposal, a hybridization between Crow Search

Algorithm with non-monopolized search (CSA-NMS).
Section 4: Outlines the application of our proposed optimization approach to solving the task scheduling problem in FOG computing.
Section 5: Presents our experiments and obtained results.
Section 6: Conclusions and potential directions for future research are outlined.

## 2 State of the Art

A considerable body of literature has delved into improving overall scheduling costs for tasks with diverse constraints. Arshed et al. [8] introduced a scheduling technique based on Genetic Algorithms (or GA), encoding a chromosome-based representation that considers attributes of both the application and fog devices. The approach evolves solutions through GA's crossover and mutation operators, and its performance is evaluated against existing methods regarding runtime, bandwidth, and financial cost.

In [9], a task-scheduling method for cloud systems (CS) utilizing a grey wolf optimizer (GWO) was presented to map workloads onto available resources. The objective is to minimize execution costs, power consumption, and make-up time. Tasks are dispatched to the queue CS's job, and the VM controller assigns tasks to existing VMs or generates new VMs based on resource evaluation.

Subramoney and Nyirenda [10] developed a multi-swarm particle swarm optimization (MS-PSO) method to enhance scientific activity scheduling in cloud-fog frameworks. MS-PSO addresses the early convergence issue of classic PSO by dividing particles into swarms with unique factors. The approach introduces an optimal task-scheduling solution with multiple criteria that focus on balancing the load in cloud/fog layers, power, price, and computing time. Evaluation against DE, GA, PSO, and a hybrid between the latter two demonstrate its effectiveness.

Electric fish optimization (EFO), inspired by electric fish's electrolocation and electrocommunication abilities [11], has also been employed to solve TSFC problems by yielding good results thanks to its interesting balance between local and global search.

In the current landscape, various techniques employing metaheuristics and evolutionary algorithms have been proposed to address the challenges of Task Scheduling in FOG and Cloud Servers.

## 3 Crow Search Algorithm Based in Non-monopolize Search

### *3.1 Crow Search Algorithm (CSA)*

The Crow Search Algorithm (CSA), introduced in [12], is an optimization approach that takes inspiration from the foraging conduct of crows. Crows, known for their high intelligence, can hide food and recall its location over extended periods. Simultaneously, they monitor the food-hiding behavior of other crows, leveraging this knowledge to enhance their food-hiding strategies.

The CSA comprises an iterative process that emulates finding the optimal value (hidden food) within a search area by modeling a population of crows, each associated with a specific hidden food location (variable initialization). At each iteration of the search process, CSA updates these locations by applying the following equation:

$$x_{new}^{i,iter+1} = \begin{cases} r_{j,iter} \geq AP & x_{new}^{i,iter} + r_i \times f_l^{i,iter} \times (m^{j,iter} - x_{new}^{i,iter}) \\ r_{j,iter} < AP & \text{a random position} \end{cases} \tag{1}$$

Essentially, at each iteration, each crow randomly selects another crow to explore and improve its food-hiding strategy (personal best). The new position is determined by a random value $r_{i,iter}$, influencing the update of the position accordingly. This process is depicted in Fig. 1.

In the context of this algorithm, $r_{j,iter}$ signifies a randomly generated number within the range [0,1]. The variable $m^{j,iter}$ corresponds to the optimal position of another crow, while $x_{new}^{i,iter}$ indicates the current position. Additionally, $f_l^{i,iter}$ represents the step flight, and $AP$ represents the awareness probability. The algorithm assesses the new position about the crow's personal best solution. If the new location yields an improved solution, the memory is revised to store the optimal solution for each crow. The pseudocode of CSA is presented in the Algorithm 1.

## 3.2 Non-monopolize Search Algorithm

Non-Monopolize Search (NO) [13] was designed as a local search approach that can be applied as a stand-alone method or in tandem with other sophisticated techniques. This scheme belongs to the single-based optimization algorithms class since it employs a single particle (or solution), which two operators modify over the course of iterations.

Analogous to other metaheuristics, this algorithm encompasses a transition between exploration/exploitation phases. Throughout these stages, NO brings about alterations to candidate solutions across each of the decision variables. The design of NO's operators is deliberately kept straightforward to ensure ease of implementation. For instance, the exploration phase unfolds in the initial half of the iterations, employing a specified equation:

$$x_{new(j)} = rand \times x(rp) \tag{2}$$

In Eq. 2, $x_{new}$ represents the $j_{th}$ decision variable within the new solution, while $rand$ implies a randomly created value within the range [0, 1]. The index $j$ pertains to a specific dimension, ranging from 1 to the total number of decision variables

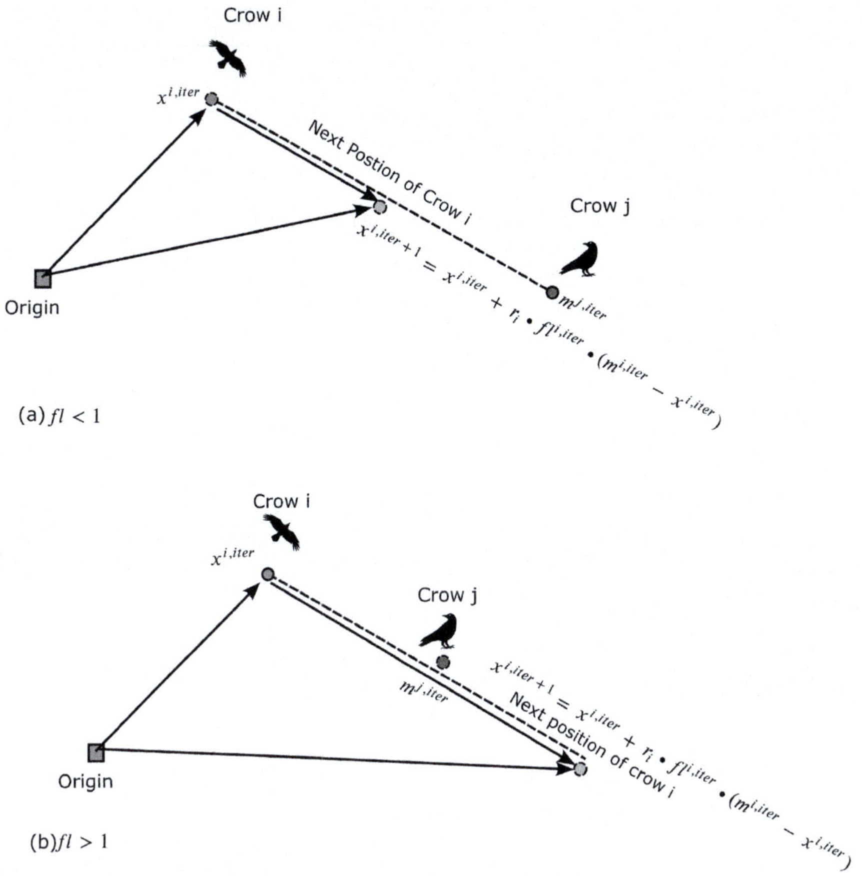


**Fig. 1** Crow search strategies **a** $fl < 1$ and **b** $fl > 1$. The j crow can go in any possible position marked by the dotted line

$(D)$. Here, $x$ refers to the current solution, and $x(rp)$ designates a randomly chosen dimension from the current solution, determined by the random number $rp$, which spans from 1 to D. It's important to note that the application of Eq. (2) is contingent upon the condition that the iteration counting variable (t) is less than or equal to half the total number of iterations (T). This equation is employed specifically during this phase to introduce a random search space exploration.

The exploitation phase is strategically designed to capitalize on the proximity of the current solution. It comes into play exclusively when the iteration count surpasses half of the total iterations (T). This phase is not just a random perturbation but

**Algorithm 1** General structure of the Crow Search Algorithm

Initialize all parameters.
Input function f(.).
Searching upper bound and lower bound. Set $itermax$. Set $iter = 1$. Population size N. Evaluate the position of the crows. Initialize the memory of each crow.
**while** $iter : itermax$ **do**
  **for** $j = 1 : N$ **do**
    Randomly choose one of the crows to follow (for example, j).
    Define an awareness probability.
    **if** $r_{j,iter+1} \geq AP^{i,iter}$ **then**
      $x_{new}^{i,iter+1} = x_{new}^{i,iter} + r_i \times f_l^{i,iter} \times (m^{j,iter} - x_{new}^{i,iter})$
    **else**
      $x_{new}^{i,iter+1}$ = a random position
    **end if**
  **end for**
  Check the feasibility of new positions.
  Evaluate the new position of the crows.
  Update the memory of crows:
  **if** $f(x_{new}^{i,iter+1})$ is better than $f(m^{i,iter})$ **then**
    $m^{i,iter+1} = x_{new}^{i,iter+1}$
  **else**
    $m^{i,iter+1} = m^{i,iter}$
  **end if**$iter = iter + 1$
**end while**
**End program**

a strategic move to explore potential improvements in the solution's vicinity. The subsequent equation delineates the mechanics of this process.

$$x_{new(j)} = x(j) - (x(sr) * rand) \times \tau - (x(j) - N_0) \tag{3}$$

Nevertheless, to broaden the exploration range and prevent a reduction in population diversity.

## 3.3 Crow Search Algorithm with Non-monopolize Search (CSA-NMS)

This proposal combines the Crow Search Algorithm (CSA) and Non-Monopolize Search (NO). The primary idea is to use the NO algorithm to improve the memory of the CSA in the latest iterations of CSA. The improvement of NO over the memory of the CSA star was at 68% of the total iterations. The parameters $N_0$, $\tau$, $AP$, and $fl$ of each algorithm were regarded and defined in the parameters Table 1. The pseudocode of this proposal is presented in Algorithm 2.

**Algorithm 2** General structure of the proposed CSA-NMS

Initialize all parameters. Input function f(.). Searching upper bound and lower bound. Set $itermax$. Set $iter = 1$. Population size N. Evaluate the position of the crows. Initialize the memory of each crow.
**while** $iter : itermax$ **do**
  **for** $j = 1 : N$ **do**
    Randomly choose one of the crows to follow (for example, j).
    Define an awareness probability.
    **if** $r_{j,iter+1} \geq AP^{i,iter}$ **then**
      $x_{new}^{i,iter+1} = x_{new}^{i,iter} + r_i \times f_l^{i,iter} \times (m^{j,iter} - x_{new}^{i,iter})$
    **else**
      $x_{new}^{i,iter+1}$ = a random position
    **end if**
  **end for**
  **if** $iter \geq itermax \times 0.68$ **then**
    **for** $j = 1 : N$ **do**
      **if** $rand > 0.71$ **then**
        $m_{Bst}^{j,iter} = rand \times m^{j,iter}(rp)$
      **else**
        $m_{Bst}^{j,iter} = m^{j,iter} \times (m^{j,iter}(rp) \times rand) \times \tau \times (m^{j,iter} - N_0)$
      **end if**
    **end for**
    If $f(m_{Bst}^{j,iter}) < f(m^{j,iter})$: $m^{j,iter} \leftarrow m_{Bst}^{j,iter}$
  **end if**
  Check the feasibility of new positions.
  Evaluate the new position of the crows.
  Update the memory of crows:
  **if** $f(x_{new}^{i,iter+1})$ is better than $f(m^{i,iter})$ **then**
    $m^{i,iter+1} = x_{new}^{i,iter+1}$
  **else**
    $m^{i,iter+1} = m^{i,iter}$
  **end if**$iter = iter + 1$
**end while**
**End program**

## 4 CSA-NMS Applied to Tasks Scheduling Problems in Fog Computing

The structure of fog computing is distinguished by its impressive scalability. It functions as an abstract entity that delivers varied service levels to cloud users. Through virtualization, fog computing enables economies of scale by providing on-demand and dynamic content and services. Fog computing extends cloud resources over the Internet, including servers, networks, applications, storage, and services. These resources can be rapidly allocated and released with minimal managerial intervention or engagement with service providers. In Fig. 2 is presented the architecture of the FOG server with a single Host.

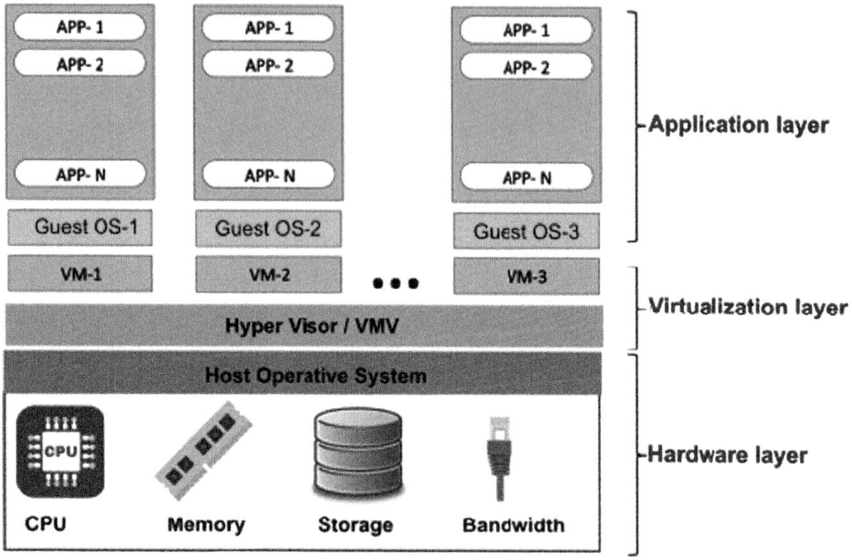


**Fig. 2** Architecture of a unique host of a FOG server

## *4.1 Initialization*

Since **CSA-NMS** was indicated to address the continuous optimization issue and **TSFC** is a discrete problem, where every crow is a vector of size $m$ (where $m$ represents the number of tasks requested by the users to be scheduled by the fog node). Each vector is initialized with a random number within the 1 to $n$ range. $n$ represents the number of available virtual machines represented by $VM$. Figure 3 represents the assigned tasks to the virtualization machines.

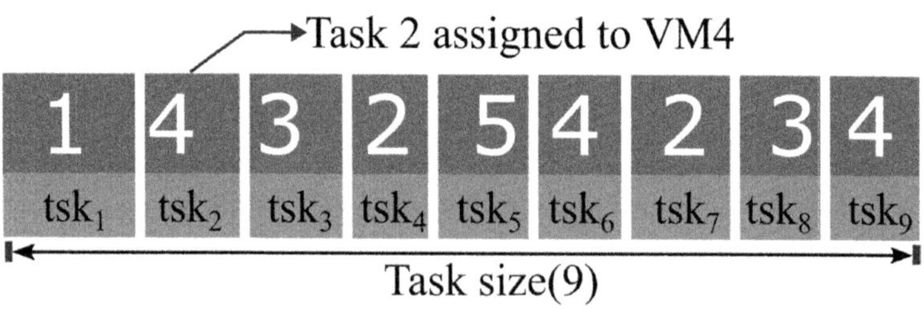


**Fig. 3** Representation of the task and solutions

## 4.2 Evaluation

During the assessment phase, four performance metrics, makespan, energy consumption, flow time, and the rate of carbon dioxide emissions, were employed to evaluate the quality of the acquired solutions. Each metric serves a specific purpose, which will be elaborated on in the following sections.

## 4.3 MakeSpan

It is regarded that each task is performed at one virtual machine $VM_j$ and continues until finish all the assigned tasks to the $VM_j$ virtual machine. It is also regarded that all the machines start at time zero when processing their assigned tasks. Every task assigned to $VM_j$ will need an execution time $Et_j$ to be executed by $VM_j$. The time the virtual machine $VM_j$ requires to execute all the assigned tasks until reaching the last assigned one is referred to as the completion time, denoted by $Et_j$. An allocation matrix has been used to specify the tasks assigned to a specific virtual machine (refer to Fig. 4), if the task $t_i$ is allocated to a specific virtual machine $VM_j$, the allocation matrix $A_{ij}$ is set to 1; otherwise, it is set to 0.

Ultimately, the makespan $MK$ is a significant metric, identified as the virtual machine with the highest total execution time among all the virtual machines, Eq. 4 represents the $MK$ calculation. This concept plays a crucial role in understanding the efficiency and performance of virtual machine processing.

$$MK = max Et_j, j = 1, ..., n \tag{4}$$

$$Et_j = \sum_{m}^{i=1} A_{ij} \times e_{ij} \tag{5}$$

From Eq. 5, $Et_j$is the total execution time of a virtual machine $VM_j$ meanwhile $e_{ij}$ refers to the execution time consumed by $i$th $VM$ on the $j$th task.

| | | $V_1$ | $V_2$ | ... | $V_j$ | ... | $V_m$ | |
|---|---|---|---|---|---|---|---|---|
| | $t_1$ | $A_{11}$ | $A_{12}$ | ... | $A_{1j}$ | ... | $A_{1m}$ | An A matrix with $n$ services and $m$ VM's |
| | $t_2$ | $A_{21}$ | $A_{22}$ | ... | $A_{2j}$ | ... | $A_{2m}$ | |
| $A =$ | . | ... | ... | ... | ... | ... | ... | |
| | $t_i$ | $A_{i1}$ | $A_{i2}$ | ... | $A_{ij}$ | ... | $A_{im}$ | |
| | . | ... | ... | ... | ... | ... | ... | |
| | $t_n$ | $A_{n1}$ | $A_{n2}$ | ... | $A_{nj}$ | ... | $A_{nm}$ | |

**Fig. 4** Expected time matrix ($A$)

### 4.4 Energy

In the context where each virtual machine has dual active or idle states, and the server's energy usage amounts to approximately 60% of its active state, the energy consumption attributed to $VM_j$ comprises the combined energy expenditure during its active and idle states. The idle time of $VM_j$ is determined by subtracting its execution time from the makespan. As for the calculation of the consumed energy $E(VM_j)$ for each virtual machine $VM_j$, it can be determined by the formula:

$$E(VM_j) = (Et_j \times b_j + (MK - Et_j) \times a_j)\times_j \tag{6}$$

$$b_j = 10^-8 \times s_j^2, a_j = 0.6 \times b_j \tag{7}$$

where $b_j$ is the consumed energy in terms of joules per M within the active state of each $VM$. The total energy ($\varphi$) consumed in the $FC$ system is computed by summing the energy consumed by each $VM$ as shown in Eq. 8.

$$\varphi = \sum_{n}^{j=1} E(VM_j) \tag{8}$$

### 4.5 Flow Time

The flow time($\eta$) is utilized to compute the total time taken by all tasks in their respective virtual machines (VMs) until completion. It is determined using the following formula:

$$\eta = \sum_{n}^{j=1} f_i \tag{9}$$

where $fi$ is the time needed within each $VM$ for task $i$. Due to the varied mix of energy sources utilized by each server for operation, the generation of carbon dioxide with different emission rates is computed as follows:

$$\gamma = \sum_{4}^{j=1} \varphi \times sh_k \times em_{rt} \times ratio \tag{10}$$

The power comprises a combination of coal, oil, natural gas, and non-fossil products, denoted by indices $k = 1, 2, 3, 4$, respectively. The participation of the $k^{th}$ energy source in the total energy consumption is represented by $sh_k$. Additionally, the emission factor of the $k_{th}$ energy source is denoted by $\alpha_k$, where the values are 0.5825 for oil, 0.7476 for coal, 0.4435 for natural gas, and 0 for non-fossil products

[14]. The ratio 44/12 signifies the conversion ratio from carbon to carbon dioxide, as per [15].

## 4.6 *Fitness Function*

As the makespan and energy consumption are identified as the two most influential factors affecting the carbon dioxide rate, along with time flow, a bi-objective function is introduced to assess the obtained solutions. The function is formulated as follows:

$$CostFunction = \lambda \times \varphi(1 - \lambda) \times MK \tag{11}$$

In this context, where λ is assigned the value of 0.8, and the weight allocated to makespan is set at 0.2, the emphasis is deliberately directed towards giving substantial importance to energy consumption. Algorithm 3 depicts the general pseudocode for solving task scheduling problems.

**Algorithm 3** General Procedure to solve Task scheduling problem

Initialize all parameters.
Define the cost function f(.)
Define upper bound and lower bound.
Set $itermax$. Set $iter = 1$.
Create the initial population according the upper and lower bounds of the problem.
Population size N.
Convert all continuous vectors to discrete vector
Evaluate the position of the population members or individuals.
**while** $iter : itermax$ **do**
    **for** $j = 1 : N$ **do**
        Apply the corresponding movement operators.
    **end for**
    Check the feasibility of new positions.
    Convert all continuous vectors to discrete vector
    Evaluate the fitness value of the new position of the members of the population.
**end while**
**End program**
Report the best solution value and its fitness value.

## 5 Experiments and Evaluations

Experiments were performed as simulations in MATLAB R2020a on a computer equipped with an AMD Ryzen 7 5800 with a 3.80 GHz processor, 64 GB of RAM, and running on Ubuntu 18.04 OS. The setup for the parameter configuration in all tested algorithms is presented in Table 1.

Our experiments considered two TSFC test scenarios (see Sect. 4). There, a pair of VMs are used to represent a heterogeneous system environment and a total of 35 repetitions of each scenario with each optimization algorithm. Scenario 1 is regarding a task of size 1200 the heterogeneous, in Table 2 are presented the fitness values for Scenario 1. The number of VMs is constant to only 30 VMs in total, whose VM IDs are $1, 2, \ldots, 30$. This scenario is considered for the evaluation of each algorithm's performance. In Scenario 1, the matrix A (expressing Expected Computational Time) regards random values with a complexity between 1000 and 6000. The velocity of processing is divided into two halves: the first with a velocity of 2500 and the second with 4000.

The tables related to Scenario 1 show the good performance of the SCA-NMS in comparison to the other algorithms, with the highlights being makespan, energy consumption, $CO_2$ emission rate and fitness. For Scenario 1, the number of VMs is 30, and the number of services or tasks is 1200. A comparative summary is shown from Tables 3, 4, and 5. The best results are remarked on bold text format. Moreover, the average fitness values are graphically presented in Fig. 5 and the overall fitness distribution is depicted in Fig. 6.

On the other hand, Scenario 2 relates to a task of size of 1500, with a random complexity between 1000 and 10000. The number of VMs is constant to only 30 VMs in total, whose VM id are 1,2,...,30. This scenario is used to evaluate the performance of the algorithms. In Scenario 2, the A matrix (Expected Computational Time ECT)

**Table 1** Algorithms parameters in the experimentation

| Algorithms | Parameters | Reference |
|---|---|---|
| CSA-NMS | $N_0 = 0.978$, $\tau = 0.618033$ $AP = 0.1$ and $fl = 2$ | Proposal |
| CSA | Crow Search Algorithm. $AP = 0.1$ and $fl = 2$ | Askarzadeh [12] |
| FPA | Flower Pollination Algorithm with the parameter $\beta = 0.75$ | Yang [16] |
| PSO | Particle Swarm Optimization. $c1 = 2.0$ and $c2 = 2.0$, $\omega = 0.9 - 0.2$ | Marini and Walczak [17] |
| GA | The mutation probability is set to $pm = 0.05$ and the crossover probability to $pc = 0.7$ | Michalewicz [18] |
| BAT | Initial sound intensity $A_i = 0.7$, and the emission rate $R_0 = 0.7$ $\alpha = 0.9$ | Yang and Gandomi [19] |

**Table 2** Fitness value of the scenario 1 with a task size of 1200

| Algorithms | Min | Avg | Max |
|---|---|---|---|
| CSA-NMS | **7.123E+05** | **6.816E+05** | **7.343E+05** |
| CSA | 7.601E+05 | 7.454E+05 | 7.768E+05 |
| GA | 8.235E+05 | 7.753E+05 | 8.822E+05 |
| PSO | 9.784E+05 | 9.055E+05 | 1.057E+06 |
| FPA | 9.253E+05 | 8.685E+05 | 9.726E+05 |
| BAT | 9.099E+05 | 8.681E+05 | 9.850E+05 |

**Table 3** $CO_2$ value of the scenario 1 with a task size of 1200

| Algorithms | Min | Avg | Max |
|---|---|---|---|
| CSA-NMS | **2.003E+06** | **2.096E+06** | **2.159E+06** |
| CSA | 2.191E+06 | 2.236E+06 | 2.284E+06 |
| GA | 2.722E+06 | 2.942E+06 | 3.262E+06 |
| PSO | 2.279E+06 | 2.422E+06 | 2.593E+06 |
| FPA | 2.553E+06 | 2.722E+06 | 2.859E+06 |
| BAT | 2.552E+06 | 2.677E+06 | 2.896E+06 |

**Table 4** Makespan value of the scenario 1 with a task size of 1200

| Algorithms | Min | Avg | Max |
|---|---|---|---|
| CSA NMS | **61.8** | **86.33** | **109.69** |
| CSA | 69.92 | 92.78 | 115.26 |
| GA | 79.25 | 104.66 | 135.79 |
| PSO | 92.68 | 122.09 | 147.92 |
| FPA | 98.07 | 131.21 | 163.19 |
| BAT | 102.79 | 128.79 | 153.77 |

**Table 5** Energy value in Joules of the scenario 1 with a task size of 1200

| Algorithms | Min | Avg | Max |
|---|---|---|---|
| CSA-NMS | **8.512E+05** | **8.909E+05** | **9.178E+05** |
| CSA | 9.310E+05 | 9.502E+05 | 9.706E+05 |
| GA | 9.684E+05 | 1.029E+06 | 1.102E+06 |
| PSO | 1.157E+06 | 1.250E+06 | 1.386E+06 |
| FPA | 1.085E+06 | 1.157E+06 | 1.215E+06 |
| BAT | 1.085E+06 | 1.138E+06 | 1.231E+06 |

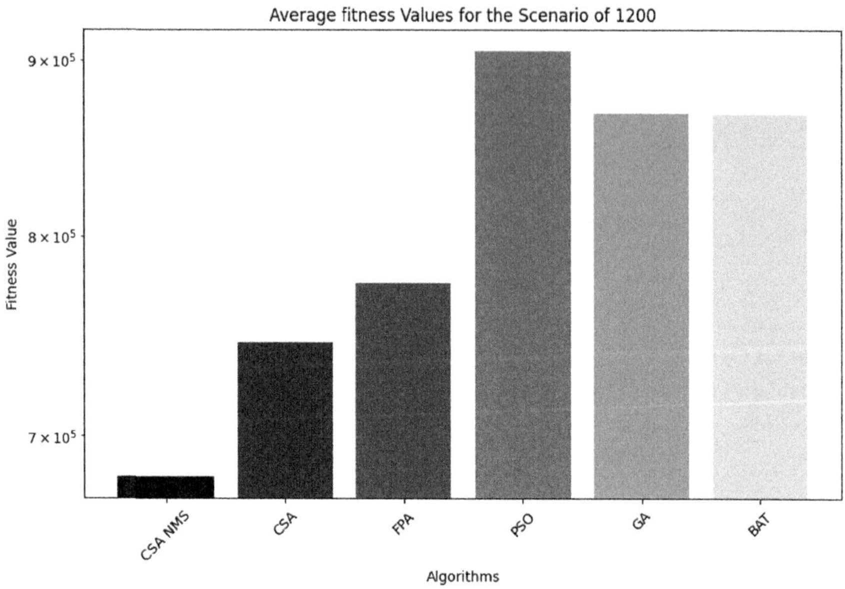


**Fig. 5** Average fitness value of the scenario 1 with a task size of 1200

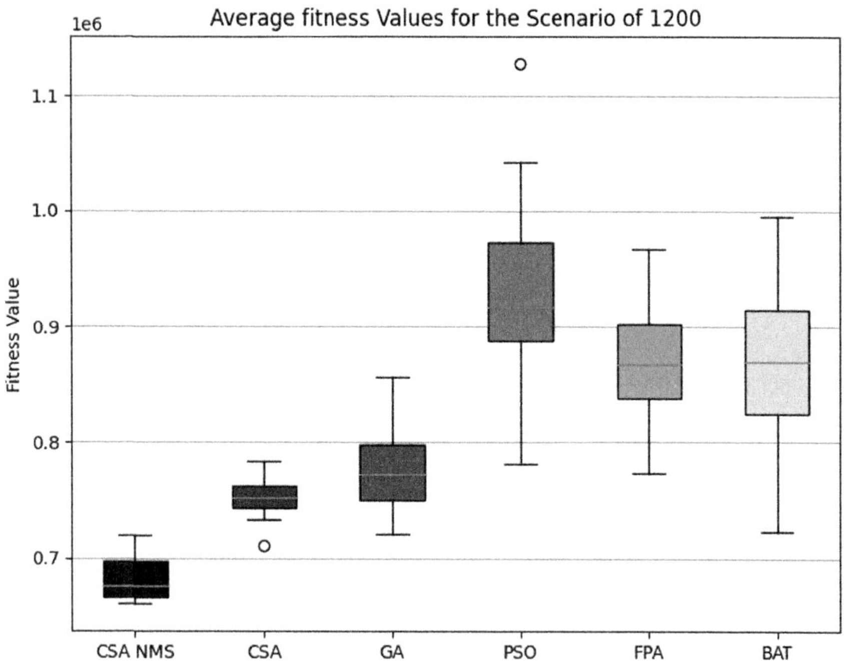


**Fig. 6** Fitness value distribution of the scenario 1 with a task size of 1200

**Table 6** Fitness value of the scenario 2 with a task size of 1500

| Algorithms | Min | Avg | Max |
|---|---|---|---|
| CSA-NMS | **8.839E+05** | **9.186E+05** | **9.463E+05** |
| CSA | 9.636E+05 | 9.837E+05 | 1.004E+06 |
| GA | 9.925E+05 | 1.049E+06 | 1.103E+06 |
| PSO | 1.096E+06 | 1.186E+06 | 1.237E+06 |
| FPA | 1.166E+06 | 1.246E+06 | 1.364E+06 |
| BAT | 1.107E+06 | 1.142E+06 | 1.188E+06 |

**Table 7** $CO_2$ value of the scenario 2 with a task size of 1500

| Algorithms | Min | Avg | Max |
|---|---|---|---|
| CSA-NMS | **2.599E+06** | **2.701E+06** | **2.781E+06** |
| CSA | 2.834E+06 | 2.893E+06 | 2.951E+06 |
| GA | 2.919E+06 | 3.085E+06 | 3.243E+06 |
| PSO | 3.222E+06 | 3.487E+06 | 3.636E+06 |
| FPA | 3.257E+06 | 3.359E+06 | 3.493E+06 |
| BAT | 3.429E+06 | 3.700E+06 | 4.107E+06 |

regards random values with a complexity between 1000 and 10,000. The velocity of processing is divided into two halves: the first with a velocity of 2500 and the second with 4000. The maximum number of iterations for all algorithms was set to 1200, with a population size of 30 in all cases. The algorithms calculated makespan and energy consumption using Eqs. 4–refeq:PowerEq. Table 6 presents the fitness value for a task size of 1500.

The reported results related to Scenario 2 also evidence the good performance of the SCA-NMS in contrast the other algorithms regarding to makespan, energy consumption, $CO_2$ emission rate and fitness. For this scenario, the number of VMs was 30 and the number of services or tasks is 1500. A comparative summary is shown from Tables 6, 7, 8, and 9. The best results are remarked on bold text format. In turn, Fig. 7 presents the average fitness values for a task size of 1500. The overall fitness distribution of Scenario 2 is shown in Fig. 8.

## *5.1 Wilcoxon Test Results*

Non-parametric statistics is a branch of statistical analysis that refrains from making specific assumptions about data distribution. Instead, it relies on methods applicable to any type of data, regardless of whether they follow a specific distribution.

The Wilcoxon method, also known as the Wilcoxon's signed-rank is a non-parametric test deviced to compare two related or paired samples [20]. It is employed

**Table 8** Makespan value of the scenario 2 with a task size of 1500

| Algorithms | Min | Avg | Max |
|---|---|---|---|
| CSA-NMS | **86.98** | **118.33** | **140.71** |
| CSA | 103.36 | 126.33 | 147.39 |
| GA | 108.69 | 140.42 | 162.29 |
| PSO | 131.32 | 166.07 | 199.54 |
| FPA | 121.95 | 157.12 | 180.06 |
| BAT | 129.84 | 161.61 | 178.25 |

**Table 9** Energy value in Joules of the scenario 2 with a task size of 1500

| Algorithms | Min | Avg | Max |
|---|---|---|---|
| CSA-NMS | **9.686E+05** | **1.029E+06** | **1.103E+06** |
| CSA | 1.157E+06 | 1.251E+06 | 1.386E+06 |
| GA | 1.085E+06 | 1.138E+06 | 1.231E+06 |
| PSO | 1.085E+06 | 1.157E+06 | 1.216E+06 |
| FPA | 1.042E+06 | 1.091E+06 | 1.120E+06 |
| BAT | 1.187E+06 | 1.142E+06 | 1.108E+06 |

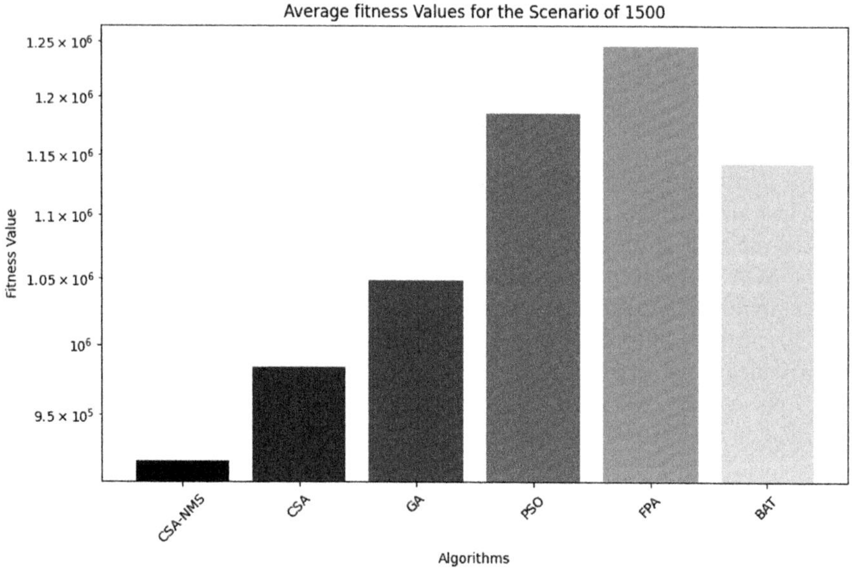


**Fig. 7** Average fitness value of the scenario 2 with a task size of 1500

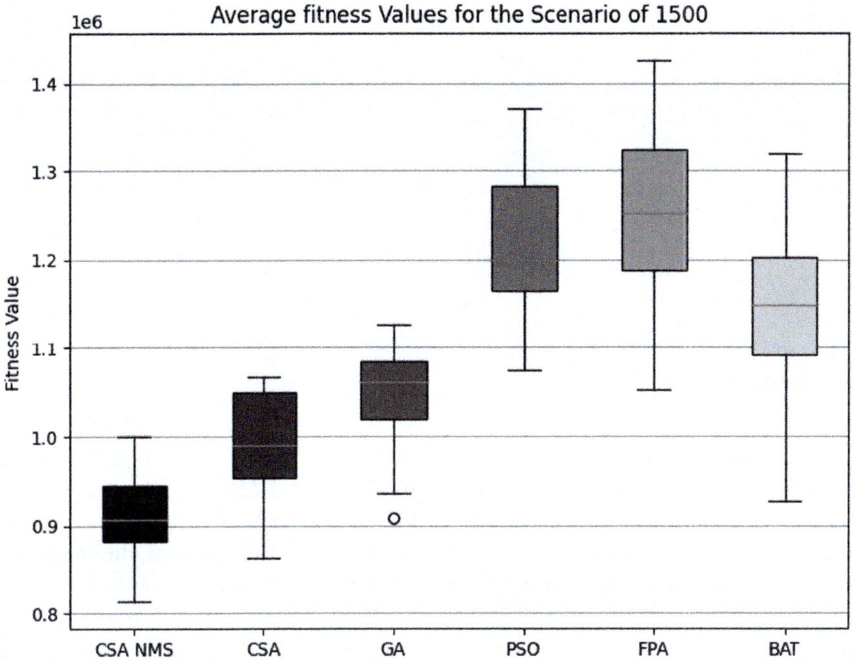


**Fig. 8** Fitness value distribution of the scenario 2 with a task size of 1500

when the data do not meet the normality assumptions required by parametric tests such as the Student's T-test.

In the context of validating experimentation in metaheuristic algorithms, the Wilcoxon method can be utilized to compare the performance of two metaheuristic algorithms across different datasets or multiple runs of the same dataset. It helps to determine the existence of statistically significant differences between the tested algorithms' results, thereby allowing for the evaluation and comparison of their effectiveness in solving optimization problems.

When applying the Wilcoxon method, a p-value is obtained, indicating the probability of obtaining the observed results if there is no real difference between the algorithms. If the p-value is less than a predefined threshold (e.g., 0.05), it can be concluded that there is a significant difference between the evaluated algorithms. Consequently, this method aids in supporting experimental validity and making informed decisions regarding the selection and improvement of metaheuristic algorithms. Table 10 displays the results of 35 independent repetitions of the 2 scenarios with their respective p-values from the Wilcoxon signed-rank test. The algorithm that won in each of these scenarios is highlighted in bold.

**Table 10** Wilcoxon test for the scenarios 1 and 2

| Algorithms | Scenario 1 | Scenario 2 |
|---|---|---|
| CSA-NMS | **8.839E+05** | **9.186E+05** |
| CSA | 9.636E+05 | 9.837E+05 |
| FPA | 9.925E+05 | 1.049E+06 |
| PSO | 1.096E+06 | 1.186E+06 |
| GA | 1.166E+06 | 1.246E+06 |
| BAT | 1.107E+06 | 1.142E+06 |

# 6 Conclusions

This research is dedicated to enhancing the efficiency of task scheduling for Internet of Things (IoT) systems. A cutting-edge model, CSA-NMS, has been introduced. This model, a fusion of CSA and NMS, is tailored to tackle TSFC issues. In a rigorous comparison with six other optimization algorithms, including the traditional CSA, CSA-NMS has emerged as the clear winner. It has outperformed all other algorithms in two highly challenging scenarios, particularly in terms of optimizing features such as makespan, energy consumption, and $CO_2$ emission rate. Looking ahead, CSA-NMS will be further developed to address task scheduling on fog servers in delayed and larger virtual machine (VM) scenarios.

# References

1. Ramson, S.J., Vishnu, S., Shanmugam, M.: Applications of Internet of Things (IoT)—an overview. In: 2020 5th International Conference on Devices, Circuits and Systems (ICDCS), pp. 92–95. IEEE (2020)
2. Kumar, R.P., Smys, S.: A novel report on architecture, protocols and applications in Internet of Things (IoT). In: 2018 2nd International Conference on Inventive Systems and Control (ICISC), pp. 1156–1161. IEEE (2018)
3. Singh, S.P., Nayyar, A., Kumar, R., et al.: Fog computing: from architecture to edge computing and big data processing. J. Supercomput. **75**, 2070–2105 (2019)
4. Singh, J., Singh, P., Gill, S.S.: Fog computing: a taxonomy, systematic review, current trends and research challenges. J. Parallel Distrib. Comput. **157**, 56–85 (2021)
5. Subbaraj, S., Thiyagarajan, R., Rengaraj, M.: A smart fog computing based real-time secure resource allocation and scheduling strategy using multi-objective crow search algorithm. J. Ambient Intell. Humanized Comput. **14**, 1003–1015 (2023)
6. Liu, W., Li, C., Zheng, A., Zheng, Z., Zhang, Z., Xiao, Y.: Fog computing resource-scheduling strategy in IoT based on artificial bee colony algorithm. Electronics **12**(7), 1511 (2023)
7. Saber, M., Abdelhamid, A.A., Ibrahim, A.: Metaheuristic optimization review: algorithms and applications. Full Length Article **3**(1), 21–30 (2023)
8. Arshed, J.U., Ahmed, M., Muhammad, T., Afzal, M., Arif, M., Bazezew, B.: Ga-irace: genetic algorithm-based improved resource aware cost-efficient scheduler for cloud fog computing environment. Wirel. Commun. Mobile Comput. (2022)

9. Ghafari, R., Mansouri, N.: Cost-aware and energy-efficient task scheduling based on grey wolf optimizer. J. Mahani Math. Res. **12**(1), 257–288 (2023)
10. Subramoney, D., Nyirenda, C.N.: Multi-swarm PSO algorithm for static workflow scheduling in cloud-fog environments. IEEE Access **10**, 117199–117214 (2022)
11. Ibrahim, R.A., Abualigah, L., Ewees, A.A., Al-Qaness, M.A., Yousri, D., Alshathri, S., Abd Elaziz, M.: An electric fish-based arithmetic optimization algorithm for feature selection. Entropy, **23**(9), 1189 (2021)
12. Askarzadeh, A.: A novel metaheuristic method for solving constrained engineering optimization problems: crow search algorithm. Comput. Struct. **169**, 1–12 (2016)
13. Abualigah, L., Al-qaness, M.A.A., Abd Elaziz, M., et al.: The non-monopolize search (no): a novel single-based local search optimization algorithm. Neural Comput. Appl. **36**, 5305–5332 (2024)
14. Abdel-Basset, M., El-Shahat, D., Elhoseny, M., Song, H.: Energy-aware metaheuristic algorithm for industrial-internet-of-things task scheduling problems in fog computing applications. IEEE Internet of Things J. **8**(16), 12638–12649 (2021)
15. Zhang, X., Myhrvold, N.P., Caldeira, K.: Key factors for assessing climate benefits of natural gas versus coal electricity generation. Environ. Res. Lett. **9**(11) (2014)
16. Yang, X.-S.: Flower pollination algorithm for global optimization. Unconventional Computation and Natural Computation, Lecture Notes in Computer Science, vol. 7445, pp. 240–249 (2012)
17. Marini, F., Walczak, B.: Particle swarm optimization (PSO). a tutorial. Chemometrics Intell. Laboratory Syst. **149**, 153–165 (2015)
18. Michalewicz, Z.: Genetic algorithms + data structures = evolution programs. Springer, New York (1992)
19. Yang, X.-S., Gandomi, A.: Bat algorithm: a novel approach for global engineering optimization. Eng. Comput. **29**(5), 464–483 (2012)
20. Wilcoxon, F.: Individual comparisons by ranking methods. Biometrics **1**, 80–83 (1945)

# Mean-Shift Clustering for Failure Detection in Quadcopter Unmanned Aerial Vehicles

**Itzel Aranguren, Arturo Valdivia, and Judith Navarro**

**Abstract** Since their inception, unmanned aerial vehicles (UAVs) or drones have demonstrate great potential for various applications due to their flexibility and versatility. Nonetheless, there are safety concerns about using UAVs in public areas, where a malfunction during flight can harm individuals, property, or the UAVs. Hence, it is important to develop safety standards and procedures to decrease the impact of such failures. The proposed study presents a new system for detecting failures in UAVs as the first stage of an autonomous emergency landing safety framework. The proposal explores the capabilities of the mean-shift clustering algorithm alongside the sensing of vibration data. The measurement of vibration signals was accomplished through multiple flight tests conducted under different faulty propeller scenarios using a customized hardware system. The collected data was analyzed to identify the optimal configuration of acceleration and gyroscope parameters for accurate failure detection in quadcopter propellers. The study revealed that the most accurate method to detect and ensure a safe emergency landing in case of a propeller fault or failure is to consider the gyroscope parameter in the vertical direction (gZ) and the accelerometer paramcter in the same direction (aZ). This approach enables quick detection and timely engagement of the safety framework. Scatter plots and confusion matrices were generated based on the parameter set (gZ-aZ), and the mean-shift clustering algorithm was applied to categorize the vibration data into three clusters representing different fitness states: normal, faulty, and failure. The proposed system was validated through real-time flight experiments, successfully identifying faults and failures using the established clusters.

I. Aranguren (✉) · A. Valdivia · J. Navarro
Depto. de Innovación Basada en la Información y el Conocimiento, Universidad de Guadalajara, CUCEI, Av. Revolución 1500, Guadalajara, Jalisco, Mexico
e-mail: itzel.aranguren@academicos.udg.mx

A. Valdivia
e-mail: arturo.valdivia@academicos.udg.mx

J. Navarro
e-mail: judith.navarro4332@alumnos.udg.mx



D. Oliva et al. (eds.), *Advances in Optimization Algorithms for Multidisciplinary Engineering Applications: From Classical Methods to AI-Enhanced Solutions*, Studies in Computational Intelligence 806, https://doi.org/10.1007/978-3-031-78440-8_35

# 1 Introduction

Unmanned aerial vehicles (UAVs), which are commonly known as drones, have emerged as one of the most promising technologies in recent years due to their low operating costs, small size, and ability to enter narrow indoor spaces [1]. Initially, UAVs were mainly used in military missions for surveillance, weather observations, photogrammetry, reconnaissance, remote sensing of crops, and other hazardous operations [2]. Due to advancements in micro and nano-technology, UAVs have evolved significantly and have revolutionized different industries, becoming vital tools for aerial photography, mapping, and enhancing professional job activities [3]. More improvements to UAVs have focused on hardware to enhance affordability, durability, and functionality, including integrating GPS and cameras. However, improving fault prediction methods remains critical to ensuring safety and preventing accidents.

The widespread use of UAVs in public spaces is still limited by safety circumstances, mainly related to the potential hazards associated with mid-flight failures [4]. There is a significant regulatory concern that UAVs could crash and cause severe damage to individuals and property, which makes it challenging to adopt them on a large scale. Additionally, failures can have an impact beyond the immediate surroundings, affecting the UAV itself and its payload. These failures can result from pre-existing faults or poor health of the UAV's main components. In order to enhance the safety of UAVs, they have been examined and tested in various conditions and applications, such as target tracking [5], fault-tolerant control [6], and formation flight [7].

In the context of fault-tolerant control (FTC), researchers have focused on developing more sophisticated and less error-prone equipment and novel algorithms to detect and predict potential software and hardware malfunctions. FTC methods aim to maintain the system's minimum required performance and continue its function in case of faults in hardware or software features [8]. Generally, FTC procedures are divided into two classes: passive and active. Passive FTC methods focus on improving the design and hardware of the controller to diminish faults and make the UAV robust enough to withstand critical situations. In turn, active FTC approaches center on developing algorithms that help to detect and identify the event of the fault and, hence, adjust the control scheme to maintain safe function. An initial phase of active FTC schemes uses a fault diagnosis mechanism to determine the faults' type, magnitude, and location. An accurate fault diagnosis (FD) system enables early detection of the quadcopter's anomalous state, which is essential for aircraft flight safety.

FD systems have been widely researched and can be divided into three methodologies: analytical model-based, signal processing-based, and knowledge-based [9]. The analytical model-based method relies on a precise mathematical model and observable inputs and outputs to create residual signals that detect discrepancies between the system's desired behavior and its actual operation mode. Two primary FD analytical model-based strategies are state estimation [10] and parameter estimation [11]. The state estimation-based FD techniques depend on filters and observers, which use the system's state estimation to define whether it operates within acceptable limits.

In contrast, FD analytical models based on parameter estimation analyze parameter changes in the scheme. In [12], multiple fault detection filters are proposed for improved detection performance when multiple faults occur in a quadrotor vehicle. The procedure considers missing data and nonlinearities for fault detection. Another approach is depicted in [13], where a robust controller for a quadrotor UAV uses sliding mode and backstepping techniques for trajectory tracking and fault estimation. In [14], a hybrid FD strategy was presented for three successive faults of UAV tilt inertial sensors to improve the system tolerance to multiple consecutive failures during flight. Study [15] proposes an adaptive three-phase Kalman filter that distinguishes aircraft vehicle faults and exterior interference with a low computational cost. Establishing accurate mathematical models in practical situations is often difficult, even though analytical model-based methods detect failures effectively.

The second FD methodology is signal processing-based approaches, which avoid the definition of intricate mathematical models of the system by extracting attributes from the measurement signals. Normally, processing approaches acquire the signals from device sensors and actuators and analyze them via transform functions. Authors in [16] proposed a tool for diagnosing faults in high-temperature solid oxide fuel cells using Wavelet transformation to decompose the battery's voltage signal. The work [17] employed the fast Fourier transform and artificial neural network architecture based on vibration signal data to classify defects in three-phase induction motors. In [18], a rational dilation Wavelet transform is applied to diagnose multiple faults in squirrel cage induction motors using a high-frequency signal from a microwave sensor. The study presented in [19] collects signals from UAV sensors, such as the air data system, the global positioning system, and the inertial measurement unit. It analyzes them with a short-time Fourier transform and a convolutional neural network to interpret system faults. Signal processing-based methods are useful for diagnosing specific UAV sensor faults; however, due to the narrow space of the quadcopter airframe, it is challenging to use additional sensors to pick up other important fault signals.

On their behalf, knowledge-based FD strategies simulate expert knowledge and reasoning processes to detect and isolate faults in systems, such as fault trees, fuzzy inference, and expert systems. In [20], a fault tree examination analyzes the reliability of communications among UAVs or between UAVs and ground station control. A fault tree approach for diagnosing incidents in cyber-physical systems is proposed in [21]. The diagnostic analysis is split into type causality, detection, and actual causality. Researchers in [22] present a real-time detection of faults in UAV navigation sensors onboard through a data-driven adaptive neuro-fuzzy inference system. The study [23] presents an FD comparative with three methods, fuzzy logic, neuro-fuzzy, and neural networks, to detect faults in the multirotor arms via the extraction of vibration data. The work introduced in [24] focuses on the diagnostic of the remote control and telemetry systems of the UAV, where an expert system based on a back propagation neural network is employed to detect the faults. In this study [25], researchers used noise measurements to develop an artificial neural network model for detecting unbalanced blades in UAV propellers. Although knowledge-based systems are efficient in detecting faults, most models tend to be complex, and fault detection accuracy highly relies on the level of expert knowledge.

As noticed above, different failure situations in UAVs are evaluated using various FD techniques, and every element of the UAV's structure can be analyzed independently. Moreover, as UAVs evolve in hardware and software, it is necessary to continue developing robust diagnostic systems, improving the efficiency and quality of timely fault detection of each sensor and actuator involved and increasing UAV safety and integrity. With this aim, this paper proposes a system for detecting failures in UAVs as the initial phase of an autonomous emergency landing safety framework. The proposed system explores the capabilities of the mean-shift clustering method [26], a non-parametric centroid-based algorithm for locating the highest density of data points within a given region. The mean-shift algorithm has been widely used for image processing and computer vision literature [27] due to its robust capacity to search regions of interest in an image. As far as we know, the mean-shift technique has not been utilized to detect faults in UAVs. Therefore, exploring its characteristics and advantages, such as its proficiency to separate clusters without prior knowledge and being a computationally inexpensive algorithm, can be useful for schemes related to fault detection.

For the experimental scheme, multiple flight tests were conducted using a custom-built hardware system to acquire vibration data under different faulty propeller scenarios. The collected data is then analyzed to identify the optimal configuration of acceleration and gyroscope parameters, enabling accurate propeller failure detection in quadcopter UAVs. The mean-shift algorithm categorizes the vibration data into three fitness states, namely normal, faulty, and failure, to depict the health condition of the UAV. Scatter plots and confusion matrices were generated based on the parameter set of the accelerometer and gyroscope sensors. According to the analysis, the most accurate configuration for detecting and ensuring the safe emergency landing of faulty UAVs is considering the gyroscope parameter in the vertical direction (gZ) and the accelerometer parameter in the same direction (aZ). The rest of the paper is arranged as follows:

Section 2 describes the mean-shift clustering algorithm.
Section 3 depicts the experimental scheme related to the proposed custom hardware and the propeller damages.
Section 4 introduces the proposed mean-shift-based UAV malfunction system.
Section 5 displays the overall results.
Section 6 closes the study with the conclusions and future work.

## 2 Mean-Shift Algorithm

In 1975, Hostetler and Fukunaga proposed the term Mean-Shift to develop the idea of clustering points and their density [26]. The algorithm yields a term called kernel density, which is used to estimate the density of a set of points, as described below:

$$f(\mathbf{x}) = \frac{1}{nh^d} \sum_{i=1}^{n} K\left(\frac{\mathbf{x} - \mathbf{x}_i}{h}\right) \tag{1}$$

where

$f(\mathbf{x})$ is the density at point $x$
$n$ represents the number of data points
$d$ is the dimension of the space
$\mathbf{x}_i$ are the observed data points
$h$ is the bandwidth
$K$ represents the kernel function.

Equation 1 performs a sum of the amount of data displayed around a point considering its distance. The term $\left(\frac{x-x_i}{h}\right)$ will calculate the distance between a point and each of the data adjusted to the bandwidth size $h$. Then, the kernel function $K$ will take the distance and convert it into a value representing each point considering the distance starting from point $x$. Finally, the data will be summed and adjusted according to the number of points and the bandwidth size, thus obtaining the density. The function of the Kernel density within mean-shift will then quantify the concentrations of points by prioritizing proximity. To ensure that the kernel estimation $K$ is accurate, it must meet the following criteria:

$$\int K(\mathbf{x})\, d\mathbf{x} = 1 \quad \text{where} \quad K(\mathbf{x}) = K(|\mathbf{x}|) \text{ for full values of } \mathbf{x} \tag{2}$$

The criteria represented in Eq. 2 serve to assure that the estimation function is normalized and to verify that there is spatial symmetry. In this work, the Gaussian kernel is employed to fulfill these criteria. The Gaussian kernel is widespread in mean-shift applications due to its smoothness and gradual reduction towards zero; these features provide more stable results and prevent any discontinuities in the computations. The kernel is determined by Eq. 3:

$$K(\mathbf{x}) = \frac{1}{(2\pi)^{\frac{d}{2}}} e^{-\frac{\mathbf{x}^2}{2}} \tag{3}$$

In Eq. 3, the $\mathbf{x}$ value depicts the distance between the search window and the central point. The kernel value decreases exponentially as $\mathbf{x}$ increases; this implies that those points that are distant from the center will have less impact on the mean displacement estimate. In turn, the Gaussian kernel function is symmetric about the central point and reaches its maximum when $\mathbf{x} = 0$.

## *2.1 Application of Mean-Shift in Clustering*

As previously mentioned, mean-shift is an algorithm designed to identify the points with the maximum density, which makes it an ideal approach for clustering tasks. The clustering approach is a non-supervised learning technique that divides a dataset into clusters based on the similarity between data points. The approach involves updating points to areas of higher density in a dataset. The clustering process through a mean-shift algorithm is performed as depicted in the next steps:

**Steps to perform mean-shift clustering**

1. **Initialization**: It consists of defining a starting point in any of the data sets used to establish the initial centroids.
2. **Selection of the bandwidth**: The algorithm defines the size of the area around the points where it will calculate how dense they are. It can be more easily understood if you think of this value as representing an approach; if the width is large, the viewing window will cover more space, so it might seem that the different groups of points are closer together than they actually are. On the other hand, if the width is small, it would imply that the points are more separated. Choosing a large value in $h$ can simplify the visualization when dealing with too much data. In contrast, a small value in $h$ could make the representation more complex by creating too many groups.
3. **Mean-shift vectors**: To calculate a vector, it is necessary to imagine that there is a circle, which we define as a value of $h$ in step 1, which is the bandwidth. Consider the other points within each point and make a weighted average of the positions. This average will be influenced by the other values close to the central point. The vectorization of the mean-shift algorithm will represent the direction those values take. The vector will indicate where it will be necessary to move the current point to bring it closer to the area of greater density, in this case of the Gaussian Kernel, which consists of measuring the importance of nearby points based on the current point. The process will be applied to all points and lead to clusters forming.
4. **Updating the position of a point**: In this step, the location will be adjusted to an area where more points are grouped. This is achieved after obtaining the average points in a specific area around a base point defined in the bandwidth. Subsequently, the position will be changed to where there is a higher concentration.
5. **Repetition**: Replicate steps 3 and 4 until the points converge; that is, the movement of the points is less significant. This concentration means that the points will no longer relocate in such a significant way.
6. **Cluster allocation**: Once the iterative process finishes, every point is assigned to the closer centroid, and the final clusters are formed.

Among the advantages of a mean-shift clustering algorithm is that it is unnecessary to specify the number of groups or clusters, and it can find groups with irregular shapes. Its foundation is excellent since it is based on probability density. In its application to images, mean-shift clustering has proven effective in segmenting images by clustering pixels by color or location in different regions [28]. It is adaptable to represent data densities and has ease of implementation. On the contrary, the disadvantages could be that they depend significantly on the bandwidth parameter; this parameter can influence the algorithm's sensitivity to data distribution and density mode detection.

## 3 Experimental Scheme

This section introduces the important issues for the experimental setup. Section 3.1 presents the design characteristics of the proposed hardware, while Sect. 3.2 defines the propeller fault cases.

### *3.1 Hardware System*

A customized hardware system has been developed to gauge vibration and detect possible faults in the quadcopter. The customized design consists of implementing an ESP32 system on a chip (SoC) in conjunction with a GY-521 gyroscope and accelerometer sensor, DC alimentation through a 9V battery, an RGB-LED with a 1K ohm resistor, and a piezoelectric buzzer. All the components are mounted on a mini breadboard and placed inside a prototype case. This case is then placed onto the top of a professional Parrot ANAFI quadcopter through square fasteners. The Parrot quadcopter is weightless, smoothly deployable, and foldable. Figure 1 illustrates where the custom hardware is mounted on the Parrot quadcopter.

The ESP32 is a microcontroller family developed by Espressif Systems. These devices are designed to be low-power, affordable, reduced size, and open-source. The microcontroller comes equipped with integrated Wi-Fi and dual-mode Bluetooth capabilities. The ESP32 series delivers different microprocessor possibilities, including the dual-core and single-core variations of the Tensilica Xtensa LX6, the dual-core Xtensa LX7, or a single-core RISC-V microprocessor. Furthermore, the ESP32 SoC includes important features, such as antenna switches, RF balun, power amplifiers, low-noise receive amplifiers, filters, and power-management modules. All these features make the ESP32 a competitive system for multi-application development, and it is even far superior in capabilities compared to other popular SoCs, such as the ESP8266 and the Arduino ONE. For this study, the ESP32 chip was programmed with the open-source software Thonny IDE. Thonny IDE operates the MicroPython language to develop microcontroller algorithms capable of reading sensor data through written code.

**Fig. 1** Location of the custom hardware on the quadcopter

For quadcopter data acquisition, a GY-521 sensor module is used in the custom hardware; this sensor consists of a 3-axis accelerometer, a 3-axis gyroscope, a digital motion processor, and a temperature sensor. Specifically, the GY-521 module collects the vibration signals from the quadcopter and measures the acceleration in the $y-$, and $z-$ directions. In the same context, the gyroscope information is gauged in the $y-$, and $z-$ directions. The sensor module is soldered to a pin header, which is then attached to the mini breadboard of the proposed hardware. The information acquired from the accelerometer is processed into the variables $aY$, and $aZ$, which represent the acceleration values obtained in the $y$, and $z$ directions, respectively. Likewise, $gY$, and $gZ$ represent the obtained gyroscope data in their respective direction. Thanks to the GY-521 module, precise and detailed data on the quadcopter's movements and vibrations can be obtained.

All vibration information collected is transferred to the computer via the Bluetooth module included in the ESP32. To detect the quadcopter's current condition, the mean-shift algorithm categorizes the information into three fitness states: normal, faulty, and failure. The proposed system incorporates an RGB LED and a piezoelectric buzzer to visually and audibly indicate the quadcopter's fitness state during flight (This procedure is explained in Sect. 4). A 9V battery powers the entire hardware system. Figure 2 depicts a diagram of how the hardware system is interconnected.

## 3.2 *Propeller Damage Scenarios*

Multiple test flights were performed to gather essential vibration data to detect failures, prevent unintended crashes, and develop the landing safety framework. The

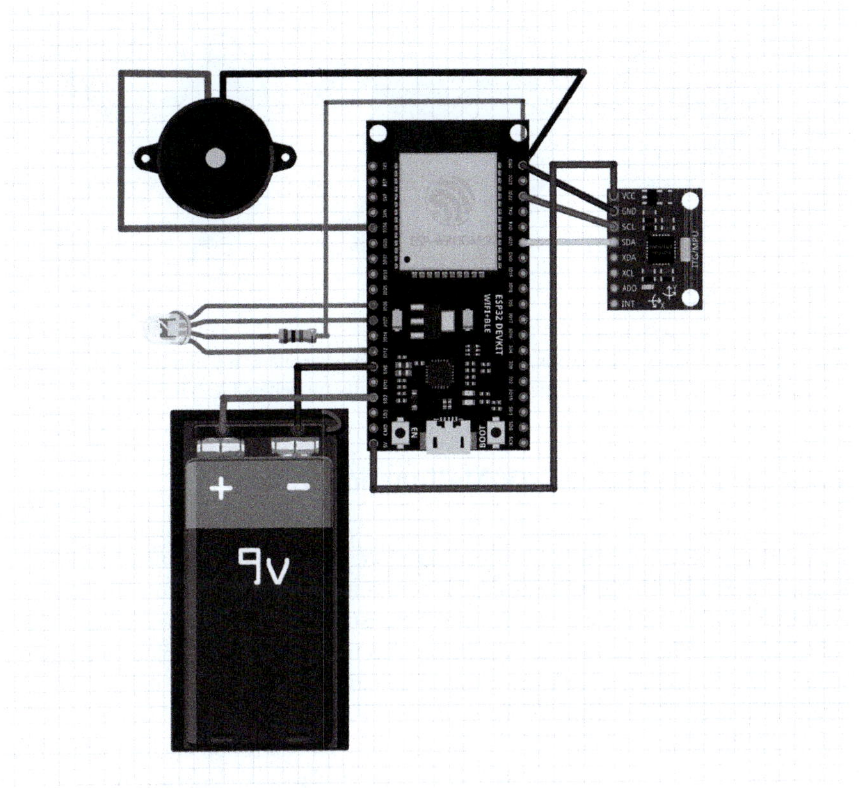


**Fig. 2** Custom hardware to measure the quadcopter vibration signals

main objective of the flight tests was to investigate how different propeller failure scenarios could affect the quadcopter's vibration levels. The propeller failure scenarios were simulated by replacing the normal (good) propellers with 6mm cut faulty propellers with different configurations between each flight. Three different configurations of the propeller faults were compared to study the effects of faulty propellers on the vibration levels of the UAV quadcopter, as shown in Fig. 3, which portrays each fault test. Figure 3a depicts the optimal configuration with non-faulty propellers (Normal state). The second scenario applied one set of faulty propellers (Fig. 3b). Finally, Fig. 3c presents the third test, which evaluates the vibrations with two faulty propeller sets. By analyzing and comparing the vibration data collected from these different scenarios, we aimed to identify early warning signs of potential failures and prevent unintended crashes, thus ensuring the safety of UAV operations.

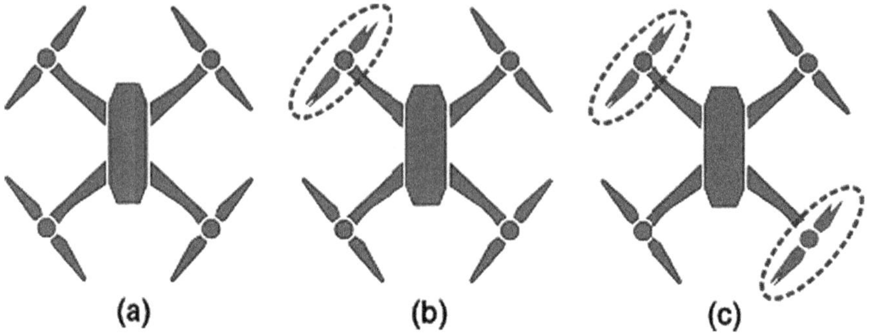


**Fig. 3** Propeller damage scenarios. **a** Normal condition propellers, **b** Condition with one set of faulty propellers, **c** Two faulty propellers condition. Faulty propellers has 6 mm cut

# 4 Mean-Shift-Based UAV Malfunction System

This study presents a system for detecting UAV propeller failures as part of a comprehensive safety framework to address community safety concerns. The principal objective is to minimize the occurrence of catastrophic UAV crashes in case of propeller damage. The proposed approach for effective failure detection consists of five steps: flight procedures, vibration measurement, data extraction, mean-shift clustering, and decision-making. Figure 4 summarizes the order of the steps in a block diagram. The flight procedures were performed in a controlled environment. There were fifteen test flights total and the three propeller damage scenarios (Refer to Sect. 3.2) had five flight tests overall, each lasting four minutes.

## *4.1 Vibration Measure System*

The quadcopter movement coordinates are depicted in Fig. 5, which is based on the Cartesian coordinate system, with the $x$ and $y$ axes moving aligned along the horizontal plane and the $z$-axis moving orthogonal to the horizontal plane. These arrangements define the position and orientation of the quadcopter in 3D space for maneuvering and control purposes. The $y$-axis represents forward or backward movement, the $x$-axis is left or right, and the $z$-axis is up or down. As for the hardware system, it plans to face according to the structure of the coordinate system in function.

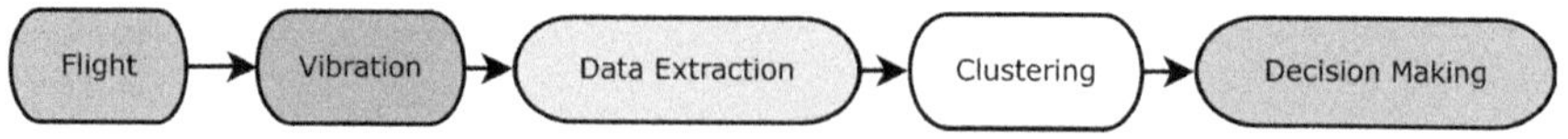


**Fig. 4** Block diagram of the proposed failure detection approach

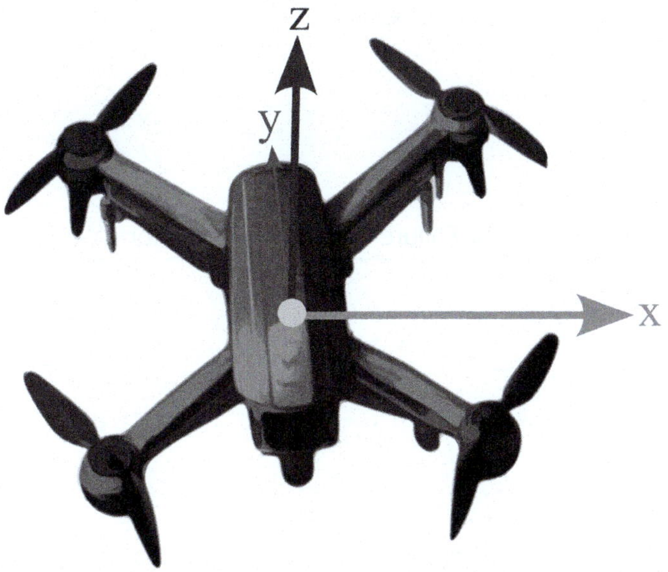


**Fig. 5** Coordinate system of the quadcopter

The sensor GY-521 from the hardware system measures acceleration and gyro in-flight vibration signals in the $y$ and $z$ directions; thus, four parameters are measured during each flight test: two for the acceleration ($aY$ and $aZ$) and two for the gyroscope sensor ($gY$ and $gZ$). The $x$ direction is not considered for comparisons since it has been demonstrated in previous works [CR] that its measurement does not provide meaningful information for vibration-based fault detection systems. The variable that stands independently for the system is the time step, measured in minutes. Meanwhile, the four measured parameters $aY$, $aZ$, $gY$, and $gZ$ are the dependent variables, denoted in the least significant bit (LSB). Subsequently, the ESP32 microcontroller transmits the vibration data to the computer through its built-in Bluetooth module (Refer to Fig. 2) to note how the hardware system is interconnected). Once the data are extracted from the ESP32, a customized Python script receives the vibration measurements, which are then processed and, depending on whether a fault exists or not, split into three distinct propeller fault output configurations: normal (NS), one faulty propeller (1FPS), and two faulty propellers (2FPS).

The mean-shift clustering algorithm previously described in the Sect. 2 is applied to process the obtained vibration data. The information on the four parameters ($aY$, $aZ$, $gY$, $gZ$) was combined to assess the best parameter configuration for failure detection in UAVs. From Eq. 1, the variable $n$ is set to 1024 and depicts the data points to clusterized through mean-shift, where the observed data points are denoted as ($x1$, ..., $xn$), and the maximum number of dimensions is $d = 2$. Table 1 presents the mixtures of parameter groups used in the research, where the 1-dimensional

**Table 1** Test parameter configurations and dimensions

| Dimension | Parameter set |
|---|---|
| 1 | aY |
| 1 | aZ |
| 2 | gY-aY |
| 2 | gZ-aZ |

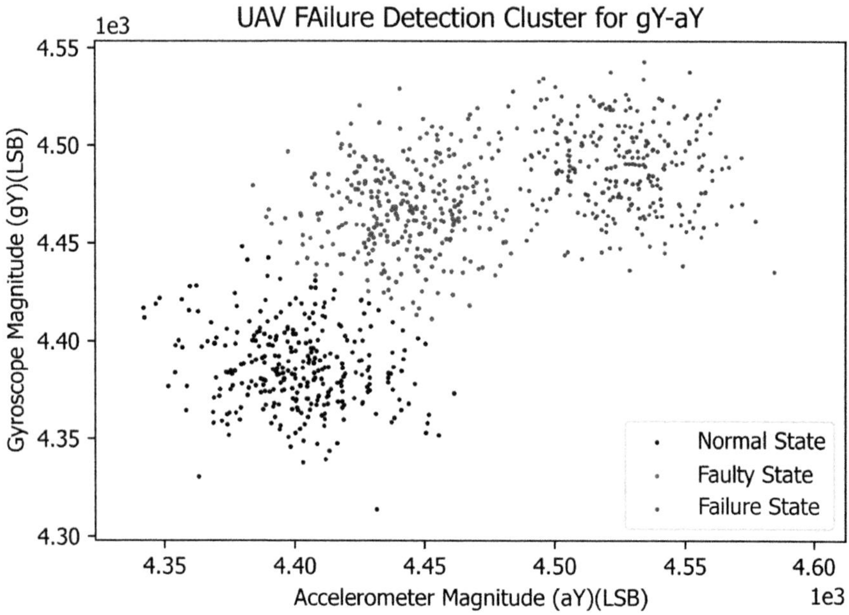


**Fig. 6** Example of clustering process denoting the three failure states: normal, faulty, and failure

problem considers only one sensor's information as a parameter. In turn, the 2-dimensional problem combines the accelerometer and gyroscope data to monitor their behavior in axes $Y$ and $Z$.

For each experiment, plots were generated to assess how the n-data points are clusterized visually. An example of this is presented in Fig. 6, where the information is exhibited in three distinct clusters, denoting the normal state in black color, the faulty state in red color, and the failure state in green color. The failure detection scheme employs $3 \times 3$ confusion matrices for each generated plot to contrast the health states as per the actual labels against those determined by clustering. Table 2 depicts the general confusion matrix, where the three health states were represented by the letters N, F, and A, which indicated "normal," "failure," and "faulty" states, respectively. Data points were grouped by mean-shift algorithm into different clusters and labeled "clustering." Both the actual label and the clustering label included the three health states. The actual label used the first letter in each cell, while the clustering label

**Table 2** Confusion matrix of the actual and clustering labels for the three failure states

| Actual label | Clustering label | | |
|---|---|---|---|
| | Failure state | Faulty state | Normal state |
| Failure state | FF | FA | FN |
| Faulty state | AF | AA | AN |
| Normal state | NF | NA | NN |

used the second letter. Each data point was used to compare the health state between the actual and the clustering labels for the confusion matrix. For example, "NA" indicated that the actual label was the normal state (N), while the clustering label was the faulty state (A).

## *4.2 Metrics to Assess the Effectiveness*

Equations (4)–(7) provide a set of quantitative measures to evaluate the clustering model's performance. These metrics include recall, accuracy, F-score, and precision and are calculated based on a $3 \times 3$ confusion matrix that summarizes the counts of true positives, true negatives, false positives, and false negatives for a given set of predictions. These metrics are essential in machine learning processes, as they identify a model's performance, provide better diagnosis of problems, improve code settings, validate model efficacy, and ultimately make correct decisions to ensure that an analysis is applicable in a real-world environment.

Among these metrics are those for binary classification, which often refer to a positive and negative class. These classes are usually imbalanced, as it would be unusual for such events to have a similar frequency in the data. The confusion matrix will help represent binary classification in a dataset to understand the effectiveness evaluation metrics. A confusion matrix is organized in a table where the predicted and actual values are placed. Where:

**Important terms**

**TP = True Positive**. Refers to a value that the algorithm classifies as positive, and it is positive.
**TN = True Negative**. These are values that the algorithm considers to be negative, and they are indeed negative in reality.
**FP = False Positive**. It refers to those values that the algorithm considers as positive, but in reality, they are negative.
**FN = False Negative**. The algorithm identifies it as negative, but in reality, it is positive.

The subsequent passage provides a detailed description of the specific metrics that have been utilized to assess the performance level of the failure scheme.

**Accuracy**: Corresponds to the ratio of correct predictions to the number of predictions made. It is not the best metric for unequal data; that is when one dataset is more frequent than another.

$$Accuracy = \frac{True\,Predictions}{All\,samples} \tag{4}$$

**Precision**: This metric takes into account false positives. It is defined as the proportion of true positives among all positive predictions.

$$Precision = \frac{True\,positive}{True\,Positive + False\,positive} \tag{5}$$

**Recall**: Takes into account the error in false negatives. Recall is the proportion of true positives to the number of true positives plus the number of false negatives.

$$Recall = \frac{True\,positive}{True\,Positive + False\,negative} \tag{6}$$

**F1 Score**: Represents a balance between precision and recall. It is an average that offers a balance between these two metrics. It is beneficial since it is considered more accurate than a scenario where its classes are unbalanced.

$$F = 2 \times \frac{Precision \times Recall}{Precision + Recall} \tag{7}$$

The choice of these metrics will depend on the specific context of the model, as all have advantages and some limitations in giving levels of importance to binary classes within a dataset.

## *4.3 Flight Failure Detection System*

To detect in-flight failures, RGB LED and piezoelectric buzzer subsystems are incorporated into the general custom hardware. These subsystems serve as a visual and audible representation of the UAV's different states and confirm the failure detection system's feasibility. The boundary conditions in these subsystems specify the lines separating the three health state regions from the quadcopter failure detection plot. The blue, yellow, and red LEDs are employed to identify the normal, faulty, and failure states, respectively. Moreover, an audible alarm is incorporated to depict

**Table 3** Confusion matrix of the accelerometer sensor in direction to the Y-axis (configuration $aY$)

| Actual label | Clustering label | | |
|---|---|---|---|
| | Failure state | Faulty state | Normal state |
| Failure state | 87 | 120 | 43 |
| Faulty state | 49 | 130 | 71 |
| Normal state | 3 | 56 | 191 |

the failure state and indicate that an emergency landing is required. The flowchart presented in Fig. 7 provides a detailed overview of the various processes involved in developing the proposed UAV propeller failure detection system. One of the most important aspects of this system is the EMERGENCY MODE, which is designed to protect the UAV, individuals, and property in the event of a critical failure. The EMERGENCY MODE will be activated automatically when a critical failure is detected, and it will initiate an emergency landing to avoid any mishap. During this mode, a buzzer alarm will alert individuals nearby, and a bright red LED will flash as a visual indicator. This will help ensure everyone is aware of the situation and can take appropriate action to stay safe.

## 5 Results

A new system for detecting failures in unmanned aerial vehicles has been proposed using custom hardware and the mean-shift clustering algorithm. Fifteen flights were conducted to evaluate the system's effectiveness, and the results were analyzed using confusion matrices and visual plots. An RGB LED and a buzzer were incorporated as a visual and audible subsystem, respectively, to indicate the current state of the UAV. The BLUE color indicates a normal state, YELLOW indicates a faulty state, and RED indicates a critical failure. The experimental results are presented below, starting with the confusion matrices and plots, continuing with the quantitative results of the metrics of recall, precision, accuracy, and F-score, and finalizing with the flight attempts using the visual and audible subsystem.

### *5.1 Confusion Matrix Results*

The confusion matrices were implemented for each parameter configuration presented in Table 1. The $3 \times 3$ confusion matrix for each parameter configuration is presented in Tables 3, 4, 5, and 6. The confusion matrices assess the health status of each data point (set n = 1024) by comparing the actual label with the clustering label in three cluster states: normal, fault, and failure. Tables 3 and 4 present the confusion

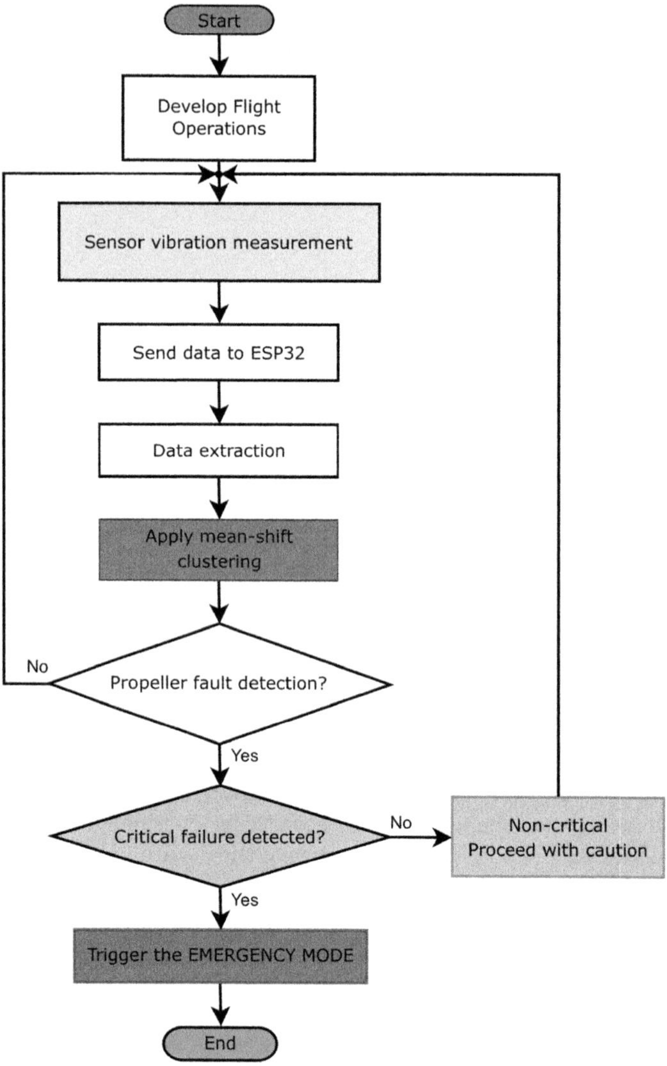


**Fig. 7** UAV malfunction system flow-diagram

matrix for the cases were the accelerometer parameter is only measured in directions Y-axis and Z-axis, respectively. In turn, Tables 5 and 6 present two different parameter configurations; the first in Table 5 combines the gyroscope and accelerometer sensors with a measure direction in the Y-axis. Table 6 combines again the gyroscope and acceleration sensors but assesses the performance on the Z-axis. From the presented confusion matrices, it can be observed that the most promising parameter configuration based on the actual and clustering labels is combining the gyroscope and acceleration sensors and evaluating its vibration values following the Z-axis;

**Table 4** Confusion matrix of the accelerometer sensor in direction to the Z-axis (configuration *aZ*)

| Actual label | Clustering label | | |
|---|---|---|---|
| | Failure state | Faulty state | Normal state |
| Failure state | 139 | 110 | 1 |
| Faulty state | 59 | 190 | 1 |
| Normal state | 1 | 13 | 236 |

**Table 5** Confusion matrix of the gyroscope sensor in direction to the Y-axis and the accelerometer sensor in the same direction (configuration *gY-aY*)

| Actual label | Clustering label | | |
|---|---|---|---|
| | Failure state | Faulty state | Normal state |
| Failure state | 165 | 80 | 5 |
| Faulty state | 45 | 169 | 36 |
| Normal state | 1 | 23 | 226 |

**Table 6** Confusion matrix of the gyroscope sensor in direction to the Z-axis and the accelerometer sensor in the same direction (configuration *gZ-aZ*)

| Actual label | Clustering label | | |
|---|---|---|---|
| | Failure state | Faulty state | Normal state |
| Failure state | 229 | 21 | 0 |
| Faulty state | 15 | 217 | 18 |
| Normal state | 0 | 6 | 244 |

this configuration is depicted as *gZ-aZ* and presented in Table 6. In Fig. 8 are presented the normal, faulty, and failure states clusters obtained through the mean-shift algorithm for the parameter configuration *gZ-aZ*.

## 5.2 Metrics Effectiveness Results

The percentage for each metric result of the mean-shift algorithm performance is presented in Table 7. The metrics involved in assessing the effectiveness are recall, precision, accuracy, and F-score; refer to Sect. 5.2 for a deeper understanding of the description of each metric. The setting with the highest percentage in the results was gZ-aZ, where the recall metric correctly identified 93.1% of all actual positive cases. The precision was 0.922; this means that when the algorithm predicts a case as positive, it is correct 92% of the time. The accuracy was 0.920, indicating that the algorithm correctly identified 92% of all cases (both positive and negative). The F

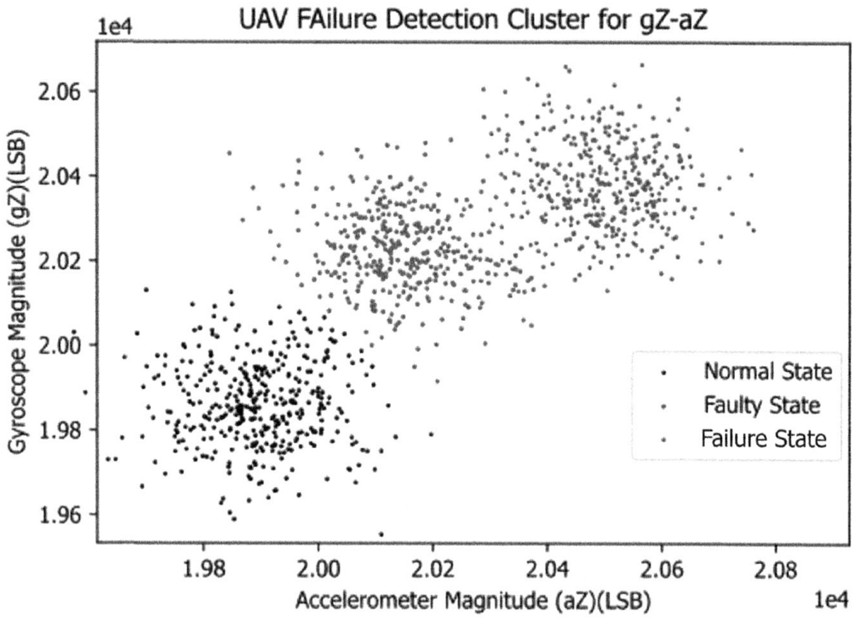


**Fig. 8** Normal, faulty, and failure states clusters for parameter configuration *gZ-aZ*

**Table 7** Metrics effectiveness evaluation for each parameter configuration

| n | Parameter set | Recall | Precision | Accuracy | F score |
|---|---|---|---|---|---|
| 1 | aY | 0.612 | 0.317 | 0.531 | 0.418 |
| 1 | aZ | 0.730 | 0.574 | 0.759 | 0.633 |
| 2 | gY-aY | 0.811 | 0.695 | 0.758 | 0.741 |
| 2 | gZ-aZ | 0.931 | 0.922 | 0.920 | 0.928 |

Score was 0.928. Considering their trade-off, an F Score of 92% indicates a balance between precision and recall.

Since gZ-aZ was the parameter set with the best results, this configuration will continue the fault detection procedure using LEDs. The margin is established with equations to segment and calculate the slopes that will define the zones of normal, defective, and failure states.

**Table 8** Tracking the RGB LED status from 6 flights

| | Attempt | Red color | Yellow color | Blue color |
|---|---|---|---|---|
| F1 | 1 | 0 | 6 | 94 |
| | 2 | 0 | 6 | 94 |
| | 3 | 0 | 4 | 96 |
| F2 | 1 | 90 | 8 | 0 |
| | 2 | 89 | 7 | 0 |
| | 3 | 91 | 11 | 0 |

### 5.3 *Visual Subsystem Results*

The plots for the $aY$, $aZ$, $gY$-$aY$, and $gZ$-$aZ$ parameters compared the quadcopter's vibration. Six flights were conducted where the drone could ascend according to the 2 propeller situations, with a total of 3 attempts each. Table X compiles the results of two flights. Flight F1 used healthy propellers, and flight F2 used 1 defective propeller. For each attempt, 100 samples were taken to track the number of times the LED lights were activated. The LED turn in a red color indicates a faulty state, which is critical and would trigger the emergency mode. The yellow color of the LED depicts a failure state that is non-critical and allows the flight to continue with caution. The LED turns a blue color when the state is optimal, which is the representation of the normal state where there are no faults or failures in the propellers of the UAV (Table 8).

The parameter set gZ-aZ was chosen to move forward with decision-making in the UAV failure detection system, as it showed the highest accuracy of 92% (see Table 7). To activate the visual subsystem, specific constraints were defined. Boundary conditions were created with equations for two linear lines that separate the three different cluster regions: normal-state region, faulty-state region, and failure-state region. The equations were determined in slope-intercept form by using two points on a line that separated two regions, and their slope was calculated. Figure 9 shows the plot for the parameter configuration $gZ$-$aZ$ with the cluster regions indicated along with their linear equations.

## 6 Conclusions

In this study, a new system for detecting failures in unmanned aerial vehicles (UAV) has been proposed. The system employs custom hardware that involves an ESP32 microcontroller, a gyroscope sensor, and an accelerometer sensor to measure the vibration of the Parrot ANAFI quadcopter during flight. By implementing the mean-shift clustering algorithm, the system can detect faults in the UAV and reduce the risks associated with failures. Fifteen flights, each lasting 4 min, were conducted to evaluate the level of vibration during propeller damage. The first five flights were carried out under normal conditions without any propeller damage. The following five flights were conducted with one propeller damaged, while the last five flights were evaluated with two damaged propellers. With the obtained data, the mean-shift algorithm was implemented to cluster to cluster the information in normal, faulty, and failure states. Confusion matrices were generated to compare the actual label of the data points and the cluster label designed by the mean-shift algorithm. Moreover, plots were generated to observe how the clusters are being designed visually.

At last, six flights of one minute each were performed to assess the failure detection system's feasibility through an incorporated visual and audible subsystem. This subsystem involves an RGB LED to visually represent the current state of the UAV and a buzzer to alarm when a failure occurs. The BLUE color of the LED indicates a normal state, and the YELLOW color denotes a faulty state in which the UAV system presents a fault but is not critical so it can continue the flight. The RED color indicates a critical failure in which a critical mode activates, and the buzzer starts sounding an audible alarm.

The experimentation found that the optimal configuration to assess the vibration is to consider the acceleration sensor in a direction to the Z-axis and the gyroscope

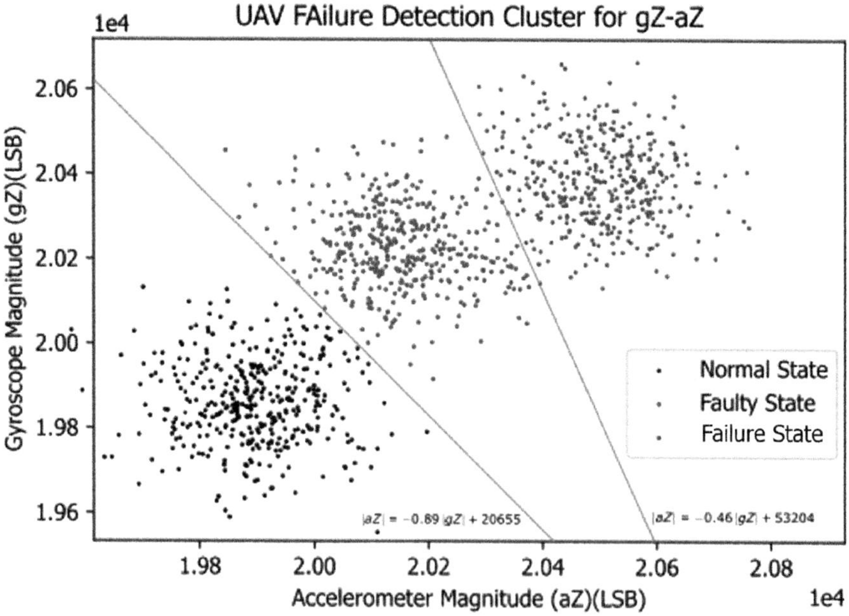


**Fig. 9** Boundary constraints for the visual subsystem

sensor in the same direction. Moreover, the visual and audible subsystem depicts the robustness of the proposed UAV malfunction system. Future work will focus on refining the model to add more stages of the emergency landing safety framework, aiming to improve further UAV failure detection in pursuit of adding reliability and ensuring that UAVs are safe in public spaces.

## References


1. Valavanis, K.P., Vachtsevanos, G.J.: Handbook of Unmanned Aerial Vehicles. Springer, Dordrecht, The Netherlands (2015)
2. Newcome, L.R.: Unmanned Aviation: A Brief History of Unmanned Aerial Vehicles. American Institute of Aeronautics and Astronautics, Reston, VA, USA (2004)
3. Rao, B., Gopi, A.G., Maione, R.: The societal impact of commercial drones. Technol. Soc. **45**, 83–90 (2016)
4. Kandaswamy, G., Balamuralidhar, P.: Health monitoring and failure detection of electronic and structural components in small unmanned aerial vehicles. Int. J. Mech. Mechatron. Eng. **11**, 1081–1089 (2017)
5. Hao, J., Zhou, Y. Zhang, G., Lv, Q., Wu, Q.: A review of target tracking algorithm based on UAV. In: 2018 IEEE International Conference on Cyborg and Bionic Systems (CBS), pp. 328–333. IEEE (2018)
6. Fourlas, G.K., Karras, G.C.: A survey on fault diagnosis and fault-tolerant control methods for unmanned aerial vehicles. Machines **9**(9), 197 (2021)
7. Zhang, Y., Mehrjerdi, H.: A survey on multiple unmanned vehicles formation control and coordination: Normal and fault situations. In: 2013 International conference on unmanned aircraft systems (ICUAS), pp. 1087–1096. IEEE (2013)
8. Qi, X., Theilliol, D., Qi, J., Zhang, Y., Han, J., Song, D. Wang, L., Xia, Y.: Fault diagnosis and fault tolerant control methods for manned and unmanned helicopters: a literature review. In: 2013 Conference on Control and Fault-Tolerant Systems (SysTol), pp. 132–139. IEEE (2013)
9. Aleem, S.A., Shahid, N., Naqvi, I.H.: Methodologies in power systems fault detection and diagnosis. Energy Syst. **6**, 85–108 (2015)
10. Kazemi, Hamed, Yazdizadeh, Alireza: Optimal state estimation and fault diagnosis for a class of nonlinear systems. IEEE/CAA J. Automatica Sinica **7**(2), 517–526 (2020)
11. Zhai, Shouchao, Wang, Wei, Ye, Hao: Fault diagnosis based on parameter estimation in closed-loop systems. IET Control Theory Appl. **9**(7), 1146–1153 (2015)
12. Liu, Fuqiang, Huang, Ji., Shi, Yang, Demin, Xu.: Fault detection for discrete-time systems with randomly occurring nonlinearity and data missing: a quadrotor vehicle example. J. Franklin Inst. **350**(9), 2474–2493 (2013)
13. Chen, Fuyang, Jiang, Rongqiang, Zhang, Kangkang, Jiang, Bin, Tao, Gang: Robust backstepping sliding-mode control and observer-based fault estimation for a quadrotor UAV. IEEE Trans. Industrial Electron. **63**(8), 5044–5056 (2016)
14. Yoon, Seungho, Kim, Seungkeun, Bae, Jonghee, Kim, Youdan, Kim, Eungtai: Experimental evaluation of fault diagnosis in a skew-configured UAV sensor system. Control Eng. Practice **19**(2), 158–173 (2011)
15. Zhong, Yujiang, Zhang, Youmin, Zhang, Wei, Zuo, Junyi, Zhan, Hao: Robust actuator fault detection and diagnosis for a quadrotor UAV with external disturbances. IEEE Access **6**, 48169–48180 (2018)
16. Pahon, E., Steiner, N.Y., Jemei, S., Hissel, D., Péra, M.C., Wang, K., Moçoteguy, P.: Solid oxide fuel cell fault diagnosis and ageing estimation based on wavelet transform approach. Int. J. Hydrogen Energy **41**(31), 13678–13687 (2016)

17. Glowacz, A., Glowacz, W., Kozik, J., Piech, K., Gutten, M., Caesarendra, W., Liu, H., Brumercik, F., Irfan, M., Khan, Z.F.: Detection of deterioration of three-phase induction motor using vibration signals. Measurement Sci. Rev. **19**(6), 241–249 (2019)
18. Madhusudhana Reddy Barusu and Meganathan Deivasigamani: Non-invasive vibration measurement for diagnosis of bearing faults in 3-phase squirrel cage induction motor using microwave sensor. IEEE Sens. J. **21**(2), 1026–1039 (2020)
19. Guo, Dingfei, Zhong, Maiying, Ji, Hongquan, Liu, Yang, Yang, Rui: A hybrid feature model and deep learning based fault diagnosis for unmanned aerial vehicle sensors. Neurocomputing **319**, 155–163 (2018)
20. Abdallah, R., Kouta, R., Sarraf, C., Gaber, J., Wack, M.: Fault tree analysis for the communication of a fleet formation flight of UAVs. In: 2017 2nd International Conference on System Reliability and Safety (ICSRS), pp. 202–206. IEEE (2017)
21. Zibaei, E., Banescu, S., Pretschner, A.: Diagnosis of safety incidents for cyber-physical systems: a UAV example. In: 2018 3rd International Conference on System Reliability and Safety (ICSRS), pp. 120–129. IEEE (2018)
22. Sun, R., Cheng, Q., Wang, G., Ochieng, W.Y.: A novel online data-driven algorithm for detecting UAV navigation sensor faults. Sensors **17**(10), 2243 (2017)
23. Ghazali, Mohamad Hazwan Mohd., Rahiman, Wan: Vibration-based fault detection in drone using artificial intelligence. IEEE Sens. J. **22**(9), 8439–8448 (2022)
24. Xiao, L., Sun, D., Liu, Y., Huang, Y.: A combined method based on expert system and bp neural network for UAV systems fault diagnosis. In: 2010 International Conference on Artificial Intelligence and Computational Intelligence, vol. 3, pp. 3–6. IEEE (2010)
25. Gino, I., Giuseppe, C., Amelia, T.: Fault diagnosis for uav blades using artificial neural network. Robotics **8**(3), 59 (2019)
26. Fukunaga, Keinosuke, Hostetler, Larry: The estimation of the gradient of a density function, with applications in pattern recognition. IEEE Trans. inf. theory **21**(1), 32–40 (1975)
27. Comaniciu, D., Meer, P.: Mean shift analysis and applications. In: Proceedings of the Seventh IEEE International Conference on Computer Vision, vol. 2, pp. 1197–1203. IEEE (1999)
28. Park, J.H., Lee, G.S., Park, S.Y.: Color image segmentation using adaptive mean shift and statistical model-based methods. Comput. Math. Appl. **57**(6), 970–980 (2009)

The manufacturer's authorised representative in the EU is Springer Nature Customer Service Centre GmbH, Europaplatz 3, 69115 Heidelberg, Germany. If you have any concerns regarding our products, please contact ProductSafety@springernature.com

Printed and bound by CPI Group (UK) Ltd, Croydon, CR0 4YY

01/10/2026

02225853-0014